Maxwell's Kinetic Theory of a Simple Monatomic Gas

This is a volume in
PURE AND APPLIED MATHEMATICS

A Series of Monographs and Textbooks

Editors: SAMUEL EILENBERG AND HYMAN BASS

A list of recent titles in this series appears at the end of this volume.

Fundamentals of Maxwell's Kinetic Theory of a Simple Monatomic Gas

TREATED AS A BRANCH
OF RATIONAL MECHANICS

C. TRUESDELL

&

R. G. MUNCASTER

1980

ACADEMIC PRESS
A Subsidiary of Harcourt Brace Jovanovich, Publishers
New York   London   Toronto   Sydney   San Francisco

COPYRIGHT © 1980, BY ACADEMIC PRESS, INC.
ALL RIGHTS RESERVED.
NO PART OF THIS PUBLICATION MAY BE REPRODUCED OR
TRANSMITTED IN ANY FORM OR BY ANY MEANS, ELECTRONIC
OR MECHANICAL, INCLUDING PHOTOCOPY, RECORDING, OR ANY
INFORMATION STORAGE AND RETRIEVAL SYSTEM, WITHOUT
PERMISSION IN WRITING FROM THE PUBLISHER.

ACADEMIC PRESS, INC.
111 Fifth Avenue, New York, New York 10003

United Kingdom Edition published by
ACADEMIC PRESS, INC. (LONDON) LTD.
24/28 Oval Road, London NW1 7DX

Library of Congress Cataloging in Publication Data

Truesdell, Clifford Ambrose, Date
    Fundamentals of Maxwell's kinetic theory of a
simple monatomic gas.

    (Pure and applied mathematics, a series of
monographs and textbooks ;   )
    Bibliography: p.
    Includes indexes.
    1. Gases, Kinetic theory of. I. Muncaster,
R. G., joint author. II. Title. III. Series.
QA3.P8 [QC175]   510'.8s [533'.7]   79–506
ISBN 0–12–701350–4

PRINTED IN THE UNITED STATES OF AMERICA

80 81 82 83   9 8 7 6 5 4 3 2 1

To
Charlotte Truesdell
and
Nancy Muncaster

Contents

PROLOGUE	xv
ACKNOWLEDGMENTS	xxi
NOTATION	xxiii
LIST OF SPECIAL SYMBOLS	xxv

Part A   Continuum Thermomechanics

Chapter I   Continuum Theories of Fluids

(i)	Basic concepts. Field equations	3
(ii)	Constitutive relations. The Navier–Stokes–Fourier theory of viscous fluids	4
(iii)	Thermodynamic quantities. The Maxwell number and the caloric	7
(iv)	More general constitutive assumptions and principles	8
(v)	Thermodynamics: The Clausius–Duhem inequality and its special cases and generalizations	12

Chapter II   The Stokes–Kirchhoff Gas, Some of Its Peculiarities, and Some of Its Flows

(i)	The Stokes–Kirchhoff and Euler–Hadamard theories of ideal gases	19
(ii)	Some parameters that control dynamical similarity	21
(iii)	The caloric of an ideal gas	22
(iv)	Equilibrium	22
(v)	Some particular homo-energetic flows: dilatation, affine flow, simple shearing, extension	23

Part B Basic Structures of the Kinetic Theory

Chapter III The Molecular Density, the Definitions of Gross Fields, and the Equation of Evolution

(i)	The molecular density and the number density	39
(ii)	Expectations. The thirteen basic fields	41
(iii)	The higher moments	45
(iv)	The retrogressors	46
(v)	The equation of evolution	49

Chapter IV Some Limits of Agreement between Kinetic Theories and Classical Fluid Mechanics

(i)	Failure of constitutive relations in the sense of continuum mechanics	51
(ii)	Limitations on the shear viscosity	52
(iii)	Vanishing of the bulk viscosity	57
(iv)	Disagreement between the kinetic theory and the Stokes–Kirchhoff theory for flows in which $T \geqq 1$	57

Chapter V The Differential Operators of the Kinetic Theory

(i)	The retrogressor and the retrogression reviewed	61
(ii)	Differentiation along a trajectory. Mild and strong derivatives	62
(iii)	Local forms of the equation of evolution	64
(iv)	Moments of the strong derivative of a function	66

Chapter VI The Dynamics of Molecular Encounters

(i)	Binary encounters	69
(ii)	Summational invariants and the Boltzmann–Gronwall theorem	71
(iii)	The encounter problem and its solutions	74
(iv)	The encounter operator and its properties	84
	Appendix A Proof of the lemma	88

Chapter VII The Maxwell Collisions Operator. Kinetic Constitutive Relations. The Total Collisions Operator and Bilinear Form

(i)	The collisions operator	91
(ii)	Kinetic constitutive quantities	93

CONTENTS

(iii)	Alternative forms of the collisions operator	95
(iv)	The bilinear form	98
(v)	Orthogonal invariance of the collisions operator and the bilinear form	101
(vi)	Inconsistency of Maxwell's kinetic theory with Newtonian mechanics	102

Chapter VIII Boltzmann's Monotonicity Theorem. The Maxwellian Density. Analogues of the Caloric and Its Flux

(i)	The Boltzmann monotonicity theorem	105
(ii)	Properties of the Maxwellian density	107
(iii)	Degree to which a Maxwellian expectation approximates a general one	112
(iv)	The caloric of a kinetic gas: Boltzmann's field h	116
(v)	Bounds for h and for its flux s	119
(vi)	Grossly determined functions, momentally determined functions	126

Part C The Maxwell–Boltzmann Equation and Its Elementary Consequences

Chapter IX The Maxwell–Boltzmann Equation. Maxwell's Consistency Theorem and Equation of Transfer   131

Chapter X Kinetic Equilibrium and Gross Equilibrium. Locally Maxwellian Solutions   137

Chapter XI Boltzmann's H-Theorem

(i)	The formal broad H-theorem and the formal narrow H-theorem	146
(ii)	Comparison and contrast of the formal H-theorem with the Clausius–Duhem inequality and the heat-bath inequality of thermomechanics	149
(iii)	The concept of a solid boundary in the kinetic theory	152
(iv)	The formal narrow H-theorem or the heat-bath inequality as a consequence of boundary conditions	161
(v)	Traditional interpretation of the formal narrow H-theorem. The ultra-narrow trend to equilibrium. Statement of corresponding rigorous propositions	166
(vi)	Difficulties faced in interpretation of the more general narrow H-theorem and the strict trend to equilibrium	168
(vii)	Lack of interpretation for the broad H-theorem	171

Part D Particular Molecular Models and Exact Solutions for Moments

Chapter XII The Collisions Operator for Some Special Kinetic Constitutive Relations, Especially Maxwellian Molecules — 175

Chapter XIII The Pressures and the Energy Flux in a Gas of Maxwellian Molecules. Maxwell's Relaxation Theorem and Evaluation of Viscosity and Thermal Conductivity

- (i) General equations for the pressures and energy flux — 187
- (ii) Maxwell's relaxation theorem — 189
- (iii) Implications of Maxwell's relaxation theorem on constitutive relations in the sense of continuum mechanics — 192
- (iv) Maxwell's evaluation of viscosity and thermal conductivity — 193

Chapter XIV Homo-energetic Simple Shearing of a Gas of Maxwellian Molecules

- (i) Homo-energetic simple shearing — 197
- (ii) The pressures as functions of time — 199
- (iii) The dominant pressures and their gross determination — 202
- (iv) Definition and rigorous evaluation of the viscosity of the kinetic gas — 203
- (v) Reduced viscometric functions of the Maxwellian gas — 204
- (vi) Comparison of the pressures as functions of time with their counterparts according to the Stokes–Kirchhoff theory — 205
- (vii) Asymptotic forms for fast shearing or rarefied gases — 207
- (viii) Solution for the energy flux. Instability — 208
- (ix) Entropy. Dissipation — 212
- (x) The principal solutions — 214

Chapter XV General Solution for the Pressures in Homo-energetic Affine Flows of a Gas of Maxwellian Molecules

- (i) Affine flows in general — 219
- (ii) Homo-energetic dilatation — 222
- (iii) Homo-energetic extension, I. The general solution for the pressures — 224
- (iv) Homo-energetic extension, II. The principal solutions — 227
- (v) Homo-energetic extension, III. Asymptotic status of the Stokes–Kirchhoff solution — 231
- (vi) Retrospect — 233

Part E The System of Equations for the Moments

Chapter XVI The General System of Equations for the Moments in a Gas of Maxwellian Molecules. Ikenberry's Theorem on the Structure of Collisions Integrals

(i)	Explicit collisions integrals for a gas of Maxwellian molecules	237
(ii)	Ikenberry's theorem: The structure of collisions integrals	244
(iii)	The general system of equations for the moments	249
	Appendix A Integration formulae and the proof of Ikenberry's theorem	251
	Appendix B Multi-indices	258

Chapter XVII Grad's Formal Evaluation of Collisions Integrals, and His Method of Approximating the Initial-value Problem

(i)	Grad's expansion and equations of transfer for the Hermite coefficients	261
(ii)	Contrast and comparison of Grad's formal expansion with Ikenberry's theorem	270
(iii)	Grad's method of truncation. His 13-moment system and his 20-moment system	272
(iv)	Comparison of solutions of Grad's systems with corresponding exact solutions for shearing	277
(v)	The relaxation theorem for Grad's 13-moment system. Grad's derivation of Enskog's first approximation to the viscosity and the Maxwell number	278
	Appendix A Conversion formulae	281
	Appendix B Exact solutions of the Maxwell–Boltzmann equation for a gas of Maxwellian molecules	284

Part F Existence, Uniqueness, and Qualitative Behavior

Chapter XVIII Existence Theory for the General Initial-value Problem. Part I: Molecules with Intermolecular Forces of Infinite Range

(i)	Prolegomena to existence theory	295
(ii)	Spatially homogeneous solutions for a gas of Maxwellian molecules: existence, uniqueness, and the trend to equilibrium	297
(iii)	Estimate of the rates of approach to equilibrium	299
(iv)	Retrospect	302

Chapter XIX Convergence Theorems and the Domain of the Collisions Operator

(i)	Preliminaries	306
(ii)	Restrictions on the growth of the integrand	307
(iii)	Convergence theorems	309
(iv)	Inverse K^{th}-power molecules	316

Chapter XX Existence Theory for the General Initial-value Problem. Part II: Place-dependent Solutions for Molecules with a Cut-off

(i)	Integral forms of the Maxwell–Boltzmann equation	319
(ii)	Survey of possibly place-dependent solutions	320
(iii)	A class of body forces	323
(iv)	Preliminary estimates	323
(v)	Glikson's theorem	327

Chapter XXI Existence Theory for the General Initial-value Problem. Part III: Spatially Homogeneous Solutions for Molecules with a Cut-off

(i)	Survey of spatially homogeneous solutions	335
(ii)	General results on existence and regularity	338
(iii)	A modified collisions operator and its properties	347
(iv)	An existence theorem for spatially homogeneous solutions	349
(v)	Proof of the ultra-narrow H-theorem	356
(vi)	Proof of the ultra-narrow trend to equilibrium	363
	Appendix A Estimation of fourth moments	368

Part G *Grossly and Momentally Determined Solutions and the Iterative Procedures of the Kinetic Theory*

Chapter XXII Hilbert's Formal Iterative Procedure for Calculating Gas-dynamic Solutions. The Assertion of Gross Causality. The Hilbert Mapping

(i)	Hilbert's formal iterative procedure	375
(ii)	Proof of effectiveness	380
(iii)	Hilbert's assertion of gross causality	384
(iv)	Properties of Hilbert's formal solutions. The Hilbert mapping	386
(v)	Locally Maxwellian solutions	390

(vi)	Proof that Hilbert's solutions are grossly determined	394
(vii)	Retrospect	395

Chapter XXIII   Grossly Determined Solutions. The Equations of Gross Determinism

(i)	Gas-dynamic solutions. The importance of grossly determined solutions	398
(ii)	Methods of determining gas flows	401
(iii)	The Maxwell–Boltzmann equation for grossly determined solutions	403
(iv)	The equations of gross determinism and properties of gross determiners	405
(v)	Principles of local action and the domain of the gross determiner	407
(vi)	A space of functions for the principal moment	411
(vii)	Gross determiners depending upon the body force. The generalized equations of gross determinism and the equation of transfer for gross determiners	413
(viii)	Gross determinism for affine flows	420
	Appendix A Calculus in Banach spaces	424

Chapter XXIV   The Method of Stretched Fields for Approximating Gross Determiners. Use of It to Obtain the Results of Enskog's Procedure

(i)	Enskog's procedure	430
(ii)	The method of stretched fields	434
(iii)	The basic expansion of gross determiners	437
(iv)	Approximate gross determiners	441
(v)	The expansion coefficients	445
(vi)	Derivation of the iterative system for the gross determiner when $\mathbf{b} = \mathbf{0}$	446
(vii)	Structure of the iterative system. Proof of effectiveness	452
(viii)	Properties of some of the expansion coefficients	455
(ix)	The formulae of Enskog, Burnett, Chapman & Cowling, and Boltzmann	461
(x)	Extension to take account of the body force	472
(xi)	Explicit results for Maxwellian molecules	479
(xii)	Explicit first approximations for general molecular models	481
(xiii)	Retrospect	488
	Appendix A Derivation of the iterative system	488
	Appendix B Computational formulae	494

Chapter XXV   The Maxwellian Iteration of Ikenberry & Truesdell

(i)	Exact results to which Maxwellian iteration is applied	508
(ii)	The scheme of Maxwellian iteration	512
(iii)	Illustration of the idea of Maxwellian iteration, applied to an ordinary differential equation	514

xiv   CONTENTS

   (iv) The first two stages of Maxwellian iteration: The Maxwell second
approximation to **P** and its companion for **q**   515
   (v) Comments on the results, origin, and nature of Maxwellian iteration   518
  (vi) The third stage of Maxwellian iteration   523
 (vii) Proof of effectiveness   524
(viii) Example: Homo-energetic simple shearing of a gas of Maxwellian molecules   526
 (ix) Atemporal Maxwellian iteration   528
  (x) Use of differential iteration to generate and improve Grad's method of truncation   537
 (xi) Retrospect upon formal methods of approximation   539

Chapter XXVI   Convergence and Divergence of Atemporal Maxwellian Iteration in Flows for Which an Exact Solution Is Known. Failure of the Higher Iterates to Improve the Asymptotic Approximation

   (i) Homo-energetic affine flows in general   542
  (ii) Homo-energetic dilatation   546
 (iii) Homo-energetic simple shearing   546
 (iv) Homo-energetic extension   548
  (v) Failure of the classical approach to approximate solution   549
 (vi) Retrospect   556

Epilogue   559

List of Works Cited   569

INDEX OF AUTHORS CITED   579
INDEX OF MATTERS TREATED   582

Prologue

The ancient Greek philosophers speculated whether matter were an assembly of tiny, invisible, and immutable particles, or a continuous expanse. As the quantitative, mathematical science of the West developed, the debate continued but became more and more definite and detailed. The great theorists proposed specific mathematical theories, restricted to certain specific kinds and circumstances of bodies, for example, to "aeriform fluids" subject to moderate pressures.

Until the first decades of this century it seemed possible that one or another theory would turn out to be the final one, the one that would explain everything about matter and thus be universally accepted as "correct", while all competitors would be defeated. Far from being borne out, this hope now seems childish. Our picture of nature has become less naïve. While in the nineteenth century more and more aspects of the sensible world were shown to be mere appearances, mere "applications" of a few fundamental "laws" of physics or biology, the recent enormous production of experimental data has undeceived us of our former simplisms. The line between the living and the inanimate has been blurred if not erased. Within the once indivisible atoms has been found an ever-growing host of mysterious "elementary particles" whose nature and function are scarcely clearer than those of dryads and familiar spirits.

Of course these discoveries have brought with them different attitudes toward theories of nature. Those who push forward the frontiers of experiment cannot wait for the thoughtful, critical, and hence cautious and slow analysis that mathematics has always demanded. Mathematicians, for their part, cannot afford to waste their time on physical theories of merely temporary interest.

These contrasting standpoints are reconciled by a keener appraisal of the role a theory is to play. A theory is not a gospel to be believed and sworn upon as an article of faith, nor must two different and seemingly contradictory theories battle each other to the death. A theory is a mathematical model for an aspect of nature. One good theory extracts and exaggerates some facets of the truth. Another good theory may idealize other facets. A theory cannot duplicate nature, for if it did so in all respects, it would be isomorphic to nature itself and hence useless, a mere repetition of all the complexity which nature presents to us, that very complexity we frame theories to penetrate and set aside.

If a theory were not simpler than the phenomena it was designed to model, it would serve no purpose. Like a portrait, it can represent only a part of the subject it pictures. This part it exaggerates, if only because it leaves out the rest. Its simplicity is its virtue, provided the aspect it portrays be that which we wish to study. If, on the other hand, our concern is an aspect of nature which a particular theory leaves out of account, then that theory is for us not wrong but simply irrelevant. For example, if we would analyse the stagnation of traffic in the streets, to take into account the behavior of the elementary particles that make up the engine, the body, the tires, and the driver of each automobile, however "fundamental" the physicists like to call those particles, would be useless even if it were not insuperably difficult. The quantum theory of individual particles is not wrong in studies of the deformation of large samples of air; it is simply a model for something else, something irrelevant to matter in gross.

With this sober and critical understanding of what a theory is, we need not see any philosophical conflict between two theories, one of which represents a gas as a plenum, the other as a numerous assembly of punctual masses. According to the physicists, a real gas such as air or hydrogen is neither of these, nothing so simple. Models of either kind represent aspects of real gases; if they represent those properly, they should entail many of the same conclusions, though of course not all.

A mathematical model for an aspect of nature is a mathematical structure, and as such it must be studied. A theory is not a duplicate of the real world but a diagram of what some simpler but in part similar world might be. Here lies the virtue of theory; for the real world, NEWTON taught us to expect, is simple if only we can learn to find its simplicity. We "understand" and "control" the real world in terms of pictures which make it seem simpler than it is.

Should we wish to trace the paths of a few hundred molecules, we might be able to do so by a theory that represents them individually. If, on the other hand, we design to describe the resistance offered to a solid body by a streaming gas made up of many billions of molecules, the fates of those individual molecules are by their mere number if for no other reason inaccessible to the mind of any one man, even were the most keenly merchandised and costly computer to endigit, ingest, and expectorate them in print. However discrete may be nature itself, the mathematics of a very numerous discrete system remains even today beyond anyone's capacity. To analyse the large, we replace it by the infinite, because the properties of the infinite are simpler and easier to manage. The mathematics of large systems is the infinitesimal calculus, the analysis of functions which are defined on infinite sets and whose values range over infinite sets. We need to differentiate and integrate functions. Otherwise we are hamstrung if we wish to deal effectively, precisely, with more than a few dozen objects able to interact with each other. Thus, somehow, we must introduce the continuum.

There are two ways the mathematics of the continuum may be brought to bear on large assemblies of discrete particles. One is to neglect their discreteness and finite cardinality outright, to represent them as being infinitely numerous and smoothly packed in every part, however small, of a certain region of space. This is the way of continuum mechanics. Aspects of matter such as mass, velocity, momentum, and energy are represented by smooth fields, introduced as primitive quantities and delimited by axioms which set forth their mathematical properties and thus allow us to operate with them[1]. It is this way's advantage to rule out, automatically and axiomatically, the "exceptional" or "unusual" cases of molecular behavior, without ever mentioning a molecule. Aside from this limitation, it is extremely general as well as superbly elegant if kept in the right hands. Its disadvantages are two: it excludes, *a fortiori*, any mathematical test of its own range of validity in application to discrete systems, however large, and, second, it affords at most heuristic means to make use of what we may know about the molecular constitution of bodies.

The second way to bring the mathematics of the continuum into the mechanics of the discrete is usually called "statistics". Instead of stating the positions and velocities of all the molecules, we allow the possibility that these may vary for some reason—be it because we lack precise information, be it because we wish only some average in time or in space, be it because we are content to represent the result of averaging over many repetitions. We thus allow the quantities which describe molecular actions to range over a continuum of values—again, something inherently beyond the range of any experiment by man. We can then assign a probability to each quantity and calculate the values expected according to that probability. Though the system itself is finite, by this broader view we subject it to the mathematics of

---

[1] Of course the smoothness to which I refer may be interrupted. Ever since EULER showed that natural problems in the propagation of plane waves do not generally admit solutions which are universally and permanently continuous, we have known that LEIBNIZ's "continuity" cannot be a "law". Discontinuities, which in today's climate of cultural suicide are grown popular under the lurid title "catastrophe", are sometimes inevitable companions or consequences of the field viewpoint. These "catastrophes" need not represent anything outlandish on the molecular scale. The model of a gas as an assembly of ideal spheres presumes that all molecular actions which give rise to dissipation are "catastrophic"; the model of a gas as an assembly of point molecules subject to mutual forces of infinite range presumes that no molecular actions, whether or not they give rise to dissipation, are "catastrophic". On the gross scale the results obtained from one model differ only in relatively minor details from those obtained from the other.

the continuum. The great advantage of this way is that it lets us take direct account of some aspects of the specification of individual molecules. Its disadvantage lies in the resulting mathematical problems, for they, if taken seriously and not merely nibbled or truncated or mused over or "solved" by declaration, are of enormous difficulty.

The models obtained in these two ways neglect much of nature. As the physicists say, they are "approximate". What one kind neglects, the other partly includes. They start from different assumptions, assumptions which can neither confirm nor contradict one another because they are set in different conceptual frames. However, their ranges of intended application to nature may be the same, or nearly so. Thus we can sometimes compare their conclusions, at least in part, by interpreting both within one and the same range of experience with natural bodies.

To compare conclusions, we must first have them. The only way to get a conclusion from a mathematical theory is by logic, by mathematical steps. Any conclusion gotten otherwise, as for example by "physical intuition", blind teamwork on huge machines, or other gilded guessing is really not a conclusion from the theory. At best it is itself some other theory, not a consequence of the one we study. At worst, it is wrong. Once we recognize that a theory is a mathematical model, we recognize also that only rigorous mathematical conclusions from a theory can be accepted in tests of the justice of that theory. If some conclusion of ours is not strict, and if we find it does not square with observations about nature, then we do not know whether to impute its failure to an inexactness of the theory or to our own incapacity as mathematicians. This latter, shared by us all in one degree or another, is to be regretted, but that bears no whit upon the theory itself. Even less is our mathematical incapacity just grounds for the measureless boasting that seems now to be the union label worn by those who regard themselves as gifted in applying mathematics. A proved theorem in a physical theory shows how some part of nature might behave. Mathematical botchery proves no more than the failure of the theorist.

Therefore, in physical theory mathematical rigor is of the essence. Being human beings and hence fallible, we may not always achieve this rigor, but we must attempt it. A result partly proved and honestly presented as such, like a tunnel drilled partly through a mountain, may be useful in giving the next man a better place from which to start, or, if fortune frowns, in showing him that to drive further in this direction is futile. A proof mathematically strict except for certain gaps made plain as the sites for future bridges is not at all the same as a "physical" or "intuitive" argument which claims to be a proof but is no more than a drug to whirl us over high mountains and across deep gorges by illusion, illusion which drops us when we awake at just the point where we started.

This book treats of only one theory: MAXWELL'S second kinetic theory of a moderately rarefied, simple, monatomic gas. This was the first molecular theory sufficiently definite and detailed to allow, at least in principle, prediction of the kind of effects and phenomena that the plenum theory of aeriform fluids had already described. MAXWELL's theory is now over 100 years old. Kinetic–molecular theories of much greater complication have been proposed and studied, but MAXWELL's theory remains today the only one that is both consistent with the mechanics of gross bodies and also simple enough in mathematical structure to yield the kind of conclusions we call "theorems"; conclusions proved strictly from the assumptions, by logical steps alone.

It is such conclusions, and such only, that this book designs to develop. We do not attempt to found MAXWELL's theory upon any other one; in particular, we do not "derive" it by imposing "approximations" upon a more refined or less specific statistical mechanics of molecular assemblies[2]. Rather, we set forth the primitive quantities, postulates, and definitions of MAXWELL's kinetic theory in the ordinary mathematical way, along with some motivation and some words about the physical circumstances the theory is intended to represent. Once the axioms of the theory shall have been stated, we will not alter or "approximate" them. In particular, we shall not mention linearized replacements such as are sometimes claimed valid when this or that quantity is "small", nor shall we consider "models" which replace the theory by another one which is similar to it but mathematically easier. These "approximations" and "models" now occupy so large a part of the literature on the kinetic theory that a beginner may easily gain an impression, altogether false, that the kinetic theory consists of nothing else—an instance of the daily more and more familiar principle that a pile of fake diamonds outshines a small gold nugget.

In fact, there is an exact, mathematical kinetic theory. Although MAXWELL framed the basic definitions and axioms over a century ago, and although in the past thirty years much fine work has been done in this field, the body of results obtained remains small. Thus we can present nearly all of it in this one book. Even so, many of the analyses contain important gaps, as we shall soon see.

Our first purpose is to uncover these gaps and to illuminate them as challenges to future research by mathematicians. To examine a gap, you must first have a sound, strong road to at least one side of it. To find such roads, we shall let the conceptual and logical structure of the kinetic theory speak for itself.

We have attempted to apply to MAXWELL's theory the standards of conceptual analysis, logical hygiene, and mathematical rigor to which the rational thermomechanics of continua developed in the past quarter-century

---

[2]Those not familiar with the physical concepts on which the theory rests may find sufficient background in the history written by TRUESDELL [1975], which develops the subject from its origins up to the year before MAXWELL formulated the definitive theory to which this book is devoted.

has accustomed us. The perspective gained from that discipline has been our guide at every turn.

Anybody who does a clean job nowadays with concepts, assertions, and proofs will see his work dismissed by some as mere "axiomatics". While criticism of this kind deserves no reply, we invite those inclined to it to cast their eyes upon the body of formal, explicit calculation we print here. In Chapters XVI and XVII they will see a larger list of exact collisions integrals than has ever been published before; if they study the text, they will learn how to calculate as many more of them as they have time and patience for. In Chapters XXIV and XXV they will see more explicit terms for "transport effects" than ever before published, along with formally exact evaluations of transport coefficients never before determined except through unassessable approximations; if they study the text, they will learn how to calculate as many more of these coefficients as their stomachs will bear. That our calculations are general and are exact at least formally; that our procedures are systematic, explicit, and demonstrably effective rather than "*ad hoc* juggling" of the first few terms in some series; that we specify what we desire to approximate before we start to calculate approximations—"axiomatics" these may seem to some, but they do not render the results themselves any less "physical" than if they had been obtained by the exhortatory and devotional manipulations heretofore accepted in this subject.

Acknowledgments

Much of this book is based upon a course of fifteen lectures I delivered in May and June, 1972, under the auspices of the Coordenação dos Programas de Pos-Graduação de Engenharia and the Instituto de Matemática at the Universidade Federal do Rio de Janeiro. I express my gratitude to Messrs. ALBERTO COIMBRA and GUILHERME DE LA PENHA for their generous hospitality, which gave me the leisure to write out consecutively a portion of the course on statistical mechanics I have taught every other year, more or less, since 1950. This book is in part a product of their dedicated effort, destined to be frustrated all too soon by the jealousy of inferior colleagues, to construct in Brazil a new American center of rational mechanics. My lectures as delivered in Rio de Janeiro were published in 1973 as a report of the Mathematical Institute there.

My tenure of a Lincean Professorship at the Scuola Normale Superiore, Pisa, in 1974 provided me a welcome opportunity to revise much of the text as I presented its contents then.

I am grateful to several of my students over the past quarter century for their criticism, suggestions, and assistance, and in particular to those of 1972–1973: R. C. BATRA and R. G. MUNCASTER. Subsequently Mr. MUNCASTER has supplied Chapters VI, XV–XVII, and XIX–XXIV, all of them consisting partly of his own researches and some entirely, to the point that much of this book is in first line his. In those chapters, credit for results that are here published for the first time is his alone; in particular, Chapters XXIII and XXIV grew from and supersede his doctoral dissertation, accepted by The Johns Hopkins University in May, 1975, and for these two chapters and for Chapters XIX–XXI he bears sole responsibility. He recalculated all previously known explicit collisions integrals of polynomials and adjoined to them some of higher degree; these results of his are presented as Equations (XVI.23) and (XVII.34). He supplied also Section (x) of Chapter XXV. The remaining text we have together hammered, straightened, cleaned, and polished, line by line and word by word; even Section (ix) of Chapter XXV, while its final form rests squarely on Mr. MUNCASTER's discoveries presented in Chapter XXIII and was written mainly by him, must be regarded as composed jointly because the results derived there have stood in all earlier draughts of the book, obtained by a method less clear and less efficient. Our scrutiny of the literature is likewise joint.

Like the rest of my work since 1961, the research presented here has been generously supported by the U.S. National Science Foundation through grants to The Johns Hopkins University. One of these supported also the doctoral and post-doctoral studies of Mr. MUNCASTER until 1976. Until recently he has been associated with the Department of Mathematics at Heriot-Watt University, Edinburgh, in which he held a research associateship through support from the Science Research Council of Great Britain.

We are indebted to Messrs. DAFERMOS and PITTERI for assistance in reading the proofsheets and for valuable criticism of some passages.

The cognoscenti of MAXWELL's second kinetic theory are few, too few even to make up a club. *A fortiori,* they are not yet professionalised, not yet incarcerated within and shielded by a carapace of union rules. Thus they lie in peril. I think this treatise is the first on the subject to deserve the qualification "mathematical". In these times, dark as they are for the mathematics of natural science, I fear it may be the last. Earnestly I hope, on the contrary, that it will open a new day in the kinetic theory, promote a new floraison, to the point that this fascinating branch of rational mechanics will at last surrender its mysteries.

"Il Palazzetto"   C. TRUESDELL
Baltimore
November 1, 1979

Notation

The advantages of direct notation for vectors and tensors are well known. In this book we use it wherever those advantages prevail. In some parts of the kinetic theory, unfortunately, they are outweighed by the burden we should bear, were we to introduce a different symbol for each symmetry operator, inner product, and trace of a tensor of order greater than 2. Therefore, we use Cartesian tensor notation in some chapters, and so for clarity we introduce that notation from the start in the definitions of quantities. A particular Cartesian co-ordinate system in an inertial framing is selected once and for all; in a few passages we discuss questions of invariance that may be interpreted as appropriate to changes of framing. The phrase "the vector v_k" shall mean "the vector $\mathbf{v}$ whose components in the Cartesian co-ordinate system are v_k"; likewise, a tensor $\mathbf{T}$ is sometimes denoted by its components T_{km}. Indices $k, m, r, \ldots$ are dummies which may be chosen as 1, 2, or 3. Repeated dummies $aa, bb, cc, \ldots$ are understood summed from 1 to 3. We shall usually write v for $|\mathbf{v}|$, the magnitude of $\mathbf{v}$; of course $v^2 = v_a v_a$. In the running text we shall often put $\mathbf{v}/A$ for $A^{-1}\mathbf{v}$.

Hollow letters denote quantities that refer directly to the molecules of which the gas is conceived to consist. Minuscules $\mathbb{m}$, $\mathbb{d}$, $\mathbb{g}$, $\mathbb{k}$, *etc.*, are molecular constants; majuscules $\mathbb{C}$, $\overline{\mathbb{C}}$, $\mathbb{E}$, *etc.*, are molecular functions or operators.

Fraktur letters $\mathfrak{d}$, $\mathfrak{D}$, $\mathfrak{r}$, $\mathfrak{R}$, $\mathfrak{L}$, $\mathfrak{P}$, $\mathfrak{q}$, *etc.*, denote either operators that refer to gross quantities only or mappings that are or resemble constitutive responses of materials.

Script capital letters $\mathscr{E}$, $\mathscr{S}$, $\mathscr{V}$, *etc.*, denote regions, sets, and function spaces.

Lightface Greek minuscules denote scalars having thermodynamic interpretations ($\gamma, \varepsilon, \rho, \theta$), material functions ($\eta, \lambda, \mu$), principal moments ($\rho_\alpha$), angles, and various other variables, mostly in usages confined to a single chapter. Indices $\alpha, \beta, \gamma, \ldots$ are dummies which may be chosen as 0, 1, 2, 3, 4. Repeated dummies of this type are understood summed from 0 to 4.

Boldface Greek minuscules $\boldsymbol{\rho}$, $\boldsymbol{\sigma}$, *etc.*, denote vectors formed from sets of the foregoing scalars.

Sans-serif upright letters M, T, Tr, Re, *etc.*, denote scaling parameters.

Roman type in formulae usually denotes a label (such as subscript r for "repeated") or a multi-index (s for $k_1 k_2 \ldots k_s$). The special symbol M denotes the Maxwell number.

List of Special Symbols

(Some symbols have two different definitions or names, one for the kinetic theory and the other for continuum thermomechanics. The latter definition or name, and its first place of occurrence, are given within square brackets to the right. Deviations from this list of special symbols sometimes occur within a single chapter.)

Symbol	Name	Equation or section of first occurrence
A_s, A_s	Hermite coefficients	(XVII.9)
$\mathfrak{a}, a_k$	unit vector-valued function	(VI.40)
$a_\perp$	normal component of $\mathfrak{a}$	(I.32)
$\mathbb{A}_{n\vert m}$		(XII.16)
$\mathbb{A}_{p\vert n\vert m}$		(XVII.31)
$\mathfrak{a}$	molecular constant	(XII.19)
$\mathbf{b}, b_k$	extrinsic force or body force [body force]	(III.21) [I_i]
$\mathbb{B}$		(XII.13)
$\mathscr{B}$	subset of Euclidean space	III$_i$ [I_v]
$\mathbf{c}, c_k$	random velocity	(III.7)
$\mathbb{C}$	collisions operator	III$_v$, VII$_i$
$\mathbb{C}(\,,\,)$	bilinear form	(VII.13)
$\bar{\mathbb{C}}_{G,H}, \bar{\mathbb{C}}_F, \bar{\mathbb{C}}$	total collisions operator	(VII.15), (VII.16), (IX.7)
$c_{2r\vert s}$	molecular constant	(XVI.33)
$\mathscr{C}(\mathscr{X})$	space of continuous functions	XX$_v$
$\mathfrak{d}$	molecular diameter or cut-off radius	IV$_{ii}$
$\mathfrak{d}$	mild derivative	(V.1)
$\mathfrak{D}$	strong derivative	(V.10)
E, E_{km}	distortion	(I.5)
E	expansion	(I.5)
$\mathbb{E}$	encounter operator	(VI.63)
$\mathscr{E}$	three-dimensional Euclidean space	III$_i$
F	molecular density	III$_i$
F_*, F', F'_*		VII$_i$
F_M, F_U	Maxwellian densities	(VIII.4), (X.17)
$\mathbb{F}$	frequency operator	(VII.4)
$\mathfrak{f}$	intermolecular force at unit distance	IV$_{ii}$

LIST OF SPECIAL SYMBOLS

Symbol	Name	Equation or section of first occurrence
$\mathbf{G}, G_{km}$	velocity gradient in affine flow	(II.41)
$\bar{g}$	average or expectation	(III.4)
g_*, g', g'_*		(VII.1)
$\mathbb{G}, \mathbb{Z}$	gross determiners	(VIII.69), (XXIII.51)
$\mathfrak{g}$	intermolecular force constant	(IV.4)
$\mathbf{H}_s, H_s$	Hermite polynomials	(XVII.1)
$h, h[F], [\eta]$	Boltzmann's function [caloric]	(VIII.46), (XI.60) [(I.18)]
h_M, h_U		(VIII.43), XI$_v$
$\mathbf{I}, I_\alpha$	principal invariants	(VI.20)
K	amount of shearing	(II.57)
$\mathbb{k}$		(IV.4)
$\mathscr{L}^1(\mathscr{X})$	space of Lebesgue-integrable functions	XIX$_i$
$\mathfrak{L}$	transport operator	(IX.13)
$\mathbf{M}, M_{km}$	pressure tensor	(III.9) [I$_i$]
$^n\mathbf{M}, M_{k_1\ldots k_n}$	relative moment of order n	(III.19)
$\mathbf{M}$	Maxwell number	(XIII.18) [(I.17)]
$\mathfrak{m}$	molecular mass	III$_i$
$\mathscr{M}(\mathscr{X})$	space of measurable functions	XXI$_{ii}$
M	local Mach number	(II.10)
n	number density	(III.2)
$\mathbb{O}$		(XVI.5)
$\mathbf{P}, P_{km}$	pressure deviator	(III.10) [I$_i$]
$P_{0|kmr}$		(XIII.6)
$\mathbf{P}_{2r|s}, P_{2r|s}$	spherical moment	(XVI.28)
$P^{(0)}_{2r|s}$	spherical moment of a Maxwellian density	(XVI.30)
P_s	Legendre polynomial	(XVI.33)
p	mean normal pressure	(III.13) [I$_i$]
$\mathbb{P}$	partial collisions operator	(VII.6)
$\mathscr{P}$	plane normal to velocity of approach	VI$_{iii}$
$\mathfrak{P}, \mathfrak{P}_{km}$	gross determiner for the pressure deviator	(VIII.71)
$\mathbf{Q}, Q_{km}$	orthogonal tensor	I$_{iv}$
$\mathbf{q}, q_k$	energy flux vector	(III.12) [I$_i$]
$\mathfrak{q}, \mathfrak{q}_k$	gross determiner for the energy flux vector	(VIII.71)
$\mathbf{r}, r_k$	displacement vector	(II.21)
$\mathbf{r}^e, r_k^e$	encounter parameter	(VI.55)
r	encounter parameter	(VI.55)
$\mathscr{R}$	the set of real numbers	VI$_{iii}$
R	local Reynolds number	(II.10)
$\mathfrak{R}_s, \mathfrak{r}_s$	retrogressors	(III.30), (III.25)
S	summational invariant	(VI.5)
$\mathbf{s}, s_k$	flux of h	(VIII.52)
$\mathbb{S}$	scattering factor	(VII.8)
$\mathscr{S}$	cross-section	VI$_{iii}$
$\mathfrak{S}$	shifter	(XXIII.11)
$\mathscr{T}$	one-dimensional Euclidean space	III$_i$
T	tension number	(II.60)
Tr	truncation number	(II.55)
U	body force potential	(II.15)

LIST OF SPECIAL SYMBOLS

Symbol	Name	Equation or section of first occurrence
$\mathbf{u}, u_k$	gross velocity [velocity]	(III.6) [I$_i$]
$\mathbf{v}, \mathbf{v}_*$	molecular velocities	III$_i$, VI$_i$
$\mathbf{v}', \mathbf{v}'_*$	asymptotic molecular velocities	VI$_i$
$\mathscr{V}$	three-dimensional inner-product space	III$_i$
$\mathbf{w}, w_k$	velocity of approach	(VI.32)
$\mathscr{W}$	translation space	XXII$_{vi}$
$\mathbf{Y}_{2r|s}, Y_{2r|s}$	spherical harmonics	(XVI.26)
γ	ratio of specific heats	(I.16)
ε	energetic	(III.14) [I$_i$]
ζ	encounter parameter	(VI.40)
θ	encounter parameter [temperature]	(VI.40) [I$_i$]
κ, κ_k	dimensionless random velocity	(XVII.6)
μ	viscosity [shear viscosity]	(XIII.14) [(I.7)]
ρ, ρ_α	principal moment	(VI.21)
ρ	mass density	(III.5) [I$_i$]
σ	gross condition	(VIII.68)
τ	time of relaxation	(XIII.3) [(II.52)]
ϕ	stretched field	(XXIV.5)
$\frown$	unit hemisphere	VI$_{iii}$

Part A

Continuum Thermomechanics

Chapter I

Continuum Theories of Fluids

The kinetic gas, though MAXWELL's genius framed it upon an astute combination of molecular and statistical ideas, is a continuous medium. The theory of the kinetic gas is a field theory, a theory in which each physical quantity is represented by a smooth function of time t and place $\mathbf{x}$.

(i) Basic concepts. Field equations

Not only is the kinetic theory a field theory, but also the kinetic gas is endowed with the very same field descriptors as continuum mechanics[1] introduces for its primitive concepts. Chief among these are the following fields:

Name	Symbol	Components	Dimensions
Mass density	ρ	ρ	mass/volume
Velocity	$\mathbf{u}$	u_k	length/time
Pressure tensor	$\mathbf{M}$	M_{km}	force/area
Pressure deviator	$\mathbf{P}$	P_{km}	force/area
Body force	$\mathbf{b}$	b_k	force/mass
Energy flux vector	$\mathbf{q}$	q_k	energy/(area)(time)
Mean normal pressure	p	p	force/area
Temperature[2]	θ	θ	temperature
Energetic (internal energy per unit mass)	ε	ε	energy/mass

---

[1] For an elementary introduction to the basic concepts and assumptions of continuum mechanics the reader may consult the book of C. TRUESDELL, *A First Course in Rational Continuum Mechanics*, Volume 1, New York *etc.*, Academic Press, 1977.

[2] This notation was introduced by FOURIER and adopted by MAXWELL [1867, Eq. (103)]; for some years now it has been in general use in works on continuum mechanics.

The components are taken with respect to a rectangular Cartesian co-ordinate system in the Euclidean 3-dimensional space of places assigned by some inertial framing.

As we shall see in Chapter IX, in the kinetic theory these fields are subject to exactly the same **Field Equations of Balance**, term by term, as those that the axioms of continuum mechanics make them satisfy:

Mass: $\qquad\qquad\dot\rho = -\rho E;$ (I.1)

Linear Momentum: $\qquad\rho\dot{\mathbf{u}} = -\operatorname{grad} p - \operatorname{div} \mathbf{P} + \rho\mathbf{b};$ (I.2)

Rotational Momentum: $\qquad\mathbf{P} = \mathbf{P}^T;$ (I.3)

Energy: $\qquad\qquad\rho\dot\varepsilon = -pE - \mathbf{P}\cdot\mathbf{E} - \operatorname{div}\mathbf{q}.$ (I.4)

Here and henceforth the comma indicates a partial derivative:

$$E \equiv \operatorname{div} \mathbf{u} = u_{a,a},$$
$$\mathbf{E} \equiv \tfrac{1}{2}(\operatorname{grad}\mathbf{u} + (\operatorname{grad}\mathbf{u})^T) - \tfrac{1}{3}E\mathbf{1},$$
$$E_{km} \equiv \tfrac{1}{2}(u_{k,m} + u_{m,k}) - \tfrac{1}{3}E\,\delta_{km}, \qquad (I.5)$$
$$p \equiv \tfrac{1}{3}\operatorname{tr}\mathbf{M} = \tfrac{1}{3}M_{aa},$$
$$\mathbf{P} \equiv \mathbf{M} - p\mathbf{1}, \qquad P_{km} \equiv M_{km} - p\,\delta_{km},$$

and the dot denotes the *material derivative*:

$$\dot f \equiv \partial_t f + \mathbf{u}\cdot\operatorname{grad} f = \partial_t f + u_a f_{,a}. \qquad (I.6)$$

Thus E is the *expansion*, $\mathbf{E}$ is the *distortion* (the deviator of the stretching), and $\dot{\mathbf{u}}$ is the *acceleration*. We shall use the letter u to denote the speed $\sqrt{u_a u_a}$. The *pressure* on a surface whose outer unit normal is $\mathbf{n}$ is the vector $\mathbf{Mn}$, the components of which are $M_{ka}n_a$.

(ii) Constitutive relations. The Navier–Stokes–Fourier theory of viscous fluids

In continuum mechanics we relate the field variables by specific axioms called *constitutive relations*. These relations define particular *ideal materials*. Typically, they determine the mean normal pressure p, the pressure deviator $\mathbf{P}$, the energy flux vector $\mathbf{q}$, the free energetic ψ, and the energetic ε as maps of quantities determined from the fields of density ρ, velocity $\mathbf{u}$, and temperature θ. For example, the constitutive relations of the thermomechanical theory of *linearly viscous fluids* are these:

Navier–Stokes Relations: $\qquad p = \varpi - (\lambda + \tfrac{2}{3}\mu)E,$

$$\mathbf{P} = -2\mu\mathbf{E}. \qquad (I.7)$$

Fourier Relation: $\mathbf{q} = -\kappa \operatorname{grad} \theta.$ (I.8)

Thermodynamic Relations:
$$\varpi = \rho^2 \, \partial_\rho \psi,$$
$$\varepsilon = \psi - \theta \, \partial_\theta \psi,$$ (I.9)

in which the symbols λ, μ, κ, and ψ denote assigned or "empirical" functions of the positive variables ρ and θ. These four functions are the *constitutive functions*, by choice of which one particular linearly viscous fluid is distinguished from another.

A relation that gives the *thermodynamic pressure* ϖ as a function of ρ and θ is called a *thermal equation of state*, while a relation that gives the *free energetic* ψ as a function of ρ and θ is called a *caloric equation of state*. As is plain from (I.9), a caloric equation of state determines the thermal equation of state, but the converse is false. Generally (I.9)$_2$ is assumed invertible for θ as a function of ε and ρ. In this case the coefficients ϖ, λ, μ, and κ may be expressed as functions of ε and ρ, so the Navier–Stokes–Fourier relations (I.7) and (I.8) may be stated in terms of mechanical units alone.

The right-hand sides of (I.7)–(I.9) are determined, once the scalars λ, μ, κ, and ψ be given, by the fields ρ, $\mathbf{u}$, and ε. Because at any fixed time the equations of balance (I.1)–(I.4) impose no restriction upon those fields, they appear in the constitutive relations (I.7)–(I.9) as *absolutely independent variables*.

Once the constitutive equations shall have been specified, we may solve particular problems. These will be set, usually, by assignment of **b** and of some specific conditions at one instant throughout a region of space and others at all instants upon the boundary of that region. Sometimes they are set instead by restriction to a particular class of motions: "steady flow", "plane waves", "simple shearing", *etc.*, which is thereafter shown to be compatible with the equations of balance and the constitutive relations if it is suitably specialized. For linearly viscous fluids the system of differential equations to be solved consists in (I.1) and the result of using (I.7)–(I.9) to eliminate p, **P**, and **q** from (I.2) and (I.4). Since **E** is symmetric, (I.3) is satisfied identically. The system of five scalar equations so obtained is often called "the Navier–Stokes–Fourier equations", but that term is misleading because the thermodynamic relations (I.9), which are essential, did not appear in fluid mechanics until late in the nineteenth century. The system, whatever we may choose to call it, imposes severe restrictions upon ρ, **u**, and ε; for many particular problems it determines them uniquely.

The word *flow* is used loosely in the literature of gas dynamics. Sometimes it means arbitrarily chosen fields ρ, **u**, and ε; other times it means a set of those fields that satisfies the Navier–Stokes–Fourier equations for a fluid defined by given constitutive functions λ, μ, κ, and ψ.

If $E = 0$, the volume of each portion of the fluid body remains constant in time. The flow is then said to be *isochoric*. A flow such that $\mathbf{E} = \mathbf{0}$ is *distorsionless*. A *dilatation* $\mathbf{u} = f(t)(\mathbf{x} - \mathbf{x}_0)$ provides an example of distorsionless flow.

From thermodynamic principles, as we shall show in Section (v) of this chapter, the *Stokes–Duhem inequalities* follow:

$$\mu \geq 0, \qquad \lambda + \tfrac{2}{3}\mu \geq 0, \qquad \kappa \geq 0. \tag{I.10}$$

We shall see in Section (iii) of this chapter that the first two of these inequalities follow also from a direct assumption: No action of viscosity can effect positive work. The third inequality follows at once from the direct assumption $\mathbf{q} \cdot \operatorname{grad} \theta \leq 0$, which asserts that heat does not flow from a colder part to a hotter. The names of the three constitutive functions that (I.10) asserts to be non-negative are the *shear viscosity*, the *bulk viscosity*, and the *thermal conductivity*, in that order. If $\lambda = \mu = 0$, the fluid is called *perfect*; otherwise, it is called *viscous*. A fluid for which $\kappa = 0$ is a *non-conductor* of heat; a fluid for which $\kappa \neq 0$ is a *conductor*.

The thermodynamic pressure ϖ is determined from a given caloric equation of state by $(I.9)_1$. It is a function of ρ and ε alone, independent of the velocity field. The mean normal pressure p is not generally such a function. Indeed, by substituting for E from (I.1) into the Navier–Stokes relation $(I.7)_1$ we see that for a given linearly viscous fluid $p - \varpi$ is the value of a specific, explicitly known function of ρ, ε, and $\dot{\rho}$:

$$p - \varpi = (\lambda + \tfrac{2}{3}\mu)\dot{\rho}/\rho. \tag{I.11}$$

If $\dot{\rho} = 0$ throughout an interval of time, the *flow* is isochoric, and then (I.11) reduces to $p = \varpi$, but otherwise in general $p \neq \varpi$. For those *special fluids* that have a null bulk viscosity:

$$\lambda + \tfrac{2}{3}\mu = 0, \tag{I.12}$$

it follows from (I.11) that

$$\varpi = p, \tag{I.13}$$

and only for such fluids does this equality subsist in all flows. Conversely, if p is a function of ρ and ε alone, the left-hand side of (I.11) becomes such a function; since the value of $\dot{\rho}$ at a given place and time is in no way restricted by the values of ρ and ε there and then, (I.11) requires that (I.12) shall hold. We can express in somewhat more specific terms the condition just proved: *In order that the mean normal pressure p shall be a function of ρ and ε alone, it is necessary and sufficient that the bulk viscosity vanish. In this case the thermodynamic pressure and the mean normal pressure coincide in all flows.* The literature of hydrodynamics calls (I.12) "the Stokes relation".

We note that *for a fluid of null bulk viscosity, a distorsionless flow at uniform temperature takes place according to the theory of perfect non-conductors.* This conclusion follows at once by setting $\lambda + \tfrac{2}{3}\mu = 0$, $\mathbf{E} = \mathbf{0}$, and $\operatorname{grad} \theta = \mathbf{0}$ in (I.7) and (I.8), so no matter what be the value of the expansion E, all terms in which κ, λ, or μ appear reduce to 0.

(iii) Thermodynamic quantities. The Maxwell number and the caloric

Further important constitutive functions in the theory of linearly viscous fluids are the *specific heats* δ_v and δ_ϖ at constant specific volume and constant pressure, respectively. For the present purpose the former of these may be defined as follows:

$$\delta_v \equiv -\theta \, \partial_\theta^2 \psi, \tag{I.14}$$

and the latter may be determined from the relation

$$\delta_\varpi - \delta_v = \frac{\theta(\partial_\theta \varpi)^2}{\rho^2 \, \partial_\rho \varpi}, \tag{I.15}$$

which is of thermostatic origin. Thus both δ_v and δ_ϖ are determined by ψ. It is customary to assume that $\partial_\rho \varpi > 0$. The *ratio of specific heats* is denoted by γ:

$$\gamma \equiv \frac{\delta_\varpi}{\delta_v}. \tag{I.16}$$

It is generally a function of ρ and ε; if $\partial_\rho \varpi > 0$ and $\delta_v > 0$, from (I.15) it follows that $\gamma \geq 1$ and that $\gamma = 1$ only for such values of ρ and ε as make $\partial_\theta \varpi = 0$. Values of γ obtained by experiment on gases seem not to exceed $\frac{5}{3}$, while for liquids values greater than 2 have been reported. Another important dimensionless ratio is the *Maxwell number*[3] **M**:

$$\mathbf{M} \equiv \frac{\kappa}{\mu \, \delta_v}, \tag{I.17}$$

on the understanding that $\mathbf{M} \equiv \infty$ if $\mu = 0$. Thus $0 \leq \mathbf{M} \leq \infty$. The magnitude of **M** serves as a rough guide to the relative importance of the two dissipative mechanisms represented by the theory of a viscous, heat-conducting fluid. Generally **M** is a function of ρ and ε, although in special cases it may reduce to a constant.

The *specific entropy* or *caloric* η may be defined for the purposes of this theory as follows:

$$\eta \equiv (\varepsilon - \psi)/\theta. \tag{I.18}$$

From the thermostatic relations (I.9) it is an easy matter to show that

$$\rho \theta \dot\eta = \rho \dot\varepsilon - \varpi \dot\rho / \rho. \tag{I.19}$$

---

[3] The literature of statistical mechanics refers to M as "the Eucken factor"; the literature of fluid mechanics calls M/γ "the Prandtl number" or "the Nüsselt number". M was introduced and evaluated in the kinetic theory by MAXWELL [1867, Eq. (149)] before any of the others after whom it is now commonly named were born.

Then by use of (I.1) and (I.4) we obtain a differential relation satisfied by η:

$$\rho\theta\dot{\eta} = (\varpi - p)E - \mathbf{P} \cdot \mathbf{E} - \text{div } \mathbf{q}. \tag{I.20}$$

To apply this result to the classical theory of fluids, we express $\varpi - p$ and $\mathbf{P}$ by the Navier–Stokes relations (I.7) and $\mathbf{q}$ by the Fourier relation (I.8), so obtaining

$$\rho\theta\dot{\eta} = (\lambda + \tfrac{2}{3}\mu)E^2 + 2\mu|\mathbf{E}|^2 + \text{div}(\kappa \text{ grad } \theta). \tag{I.21}$$

The value of the right-hand side is the *rate of dissipation of energy* per unit volume according to classical fluid mechanics. It may be used to motivate adopting the inequalities $(I.10)_{1,2}$ as postulates. The rate of dissipation vanishes if and only if $\dot\eta = 0$; that is, *a flow of a linearly viscous fluid is dissipationless if and only if it is isocaloric*. In general, as (I.21) shows, the dissipation due to viscosity is the sum of two non-negative parts: one associated with bulk viscosity, and one associated with shear viscosity. In a fluid such that the shear viscosity or the bulk viscosity vanishes, the corresponding rate of dissipation of energy is null in all flows. More generally, as E and $\mathbf{E}$ may be assigned independently, from (I.21) we see that $(I.10)_{1,2}$ are equivalent to the requirement that *the mechanical part of the rate of dissipation of energy be non-negative in every flow*.

(iv) More general constitutive assumptions and principles

The formal structure we have just presented and explained defines *classical fluid mechanics*. In the next chapter we shall develop the special case of this theory that lends itself to comparisons with the kinetic theory of gases. Classical fluid mechanics, however, is by no means the only continuum theory to which the kinetic theory provides analogies and contrasts. Out of the many other continuum theories that have been developed in recent years as well as in earlier times we now select the few to which we shall refer when we come to treat of the advanced parts of the kinetic theory.

In modern continuum theories a body is conceived as a manifold of bodypoints X which in the course of time are mapped onto places $\mathbf{x}$ and temperatures θ:

$$\mathbf{x} = \chi(t, X), \qquad \theta = \tau(t, X). \tag{I.22}$$

The pair of mappings (χ, τ) may be called the *thermomechanical motion*. Most theories of materials rest upon a **Principle of Determinism** which can be stated more easily in words than with symbols: *At the place* $\mathbf{x}$ *occupied by the bodypoint* X *at the time t, the values of the mean normal pressure p, the pressure deviator* $\mathbf{P}$, *the energy flux vector* $\mathbf{q}$, *the free energetic* ψ, *and the energetic* ε *are determined by the histories of the thermomechanical motion* (χ, τ) *evaluated at the*

(iv) GENERAL CONSTITUTIVE RELATIONS

body-points that the body comprises. Usually the body is described in terms of some fixed invertible mapping κ of body-points X onto places $\mathbf{X}$ in Euclidean space. That is, $\mathbf{X} = \kappa(X)$, $X = \kappa^{-1}(\mathbf{X})$. The motion (χ, τ) is then represented by mappings χ_κ and τ_κ defined as follows: $\chi_\kappa(t, \mathbf{X}) \equiv \chi(t, \kappa^{-1}(\mathbf{X}))$, $\tau_\kappa(t, \mathbf{X}) \equiv \tau(t, \kappa^{-1}(\mathbf{X}))$. This pair of mappings is the *thermomechanical transplacement*. In cases where no confusion is likely, we shall use θ to stand not only for the value of τ but also for the spatial field $\tau \circ \chi^{-1}$, which can be written also as $\tau_\kappa \circ \chi_\kappa^{-1}$.

The statements that p, $\mathbf{P}$, $\mathbf{q}$, ψ, and ε are particular maps of some set of histories of the thermomechanical transplacement are the *constitutive relations* of a particular material. The mappings themselves are called the *response* of the material. In the Navier–Stokes–Fourier relations (I.7), (I.8), and (I.9) we have seen a very simple example.

Many of the studies in continuum mechanics today refer to *simple materials*[4]. For these, the dependence of p, $\mathbf{P}$, $\mathbf{q}$, ψ, and ε upon the body-points comprised by the body reduces to dependence upon the first derivatives of χ_κ and τ_κ at $\mathbf{X}$ itself, as well as dependence upon θ there. For such a material the outcomes of experiments on affine transplacements suffice to determine in principle the outcomes of all experiments. Such bearing as the theory of simple materials is presently known to have upon the kinetic theory of gases—bearing limited but important—we shall discuss in Chapter XIV.

Modern continuum mechanics considers also materials that are *not simple*. An early example is that proposed by KORTEWEG to describe the structure of capillarity. Its constitutive relations generalize the Navier–Stokes relations (I.7) as follows:

$$p = \varpi - (\lambda + \tfrac{2}{3}\mu)E + \alpha \text{ grad } \rho \cdot \text{grad } \rho - \omega \text{ div grad } \rho,$$

$$\mathbf{P} = -2\mu\mathbf{E} + \beta(\text{grad } \rho \otimes \text{grad } \rho - \tfrac{1}{3} \text{ grad } \rho \cdot \text{grad } \rho \mathbf{1}) \quad \text{(I.23)}$$
$$- \delta(\text{grad grad } \rho - \tfrac{1}{3} \text{ div grad } \rho \mathbf{1}),$$

the coefficients α, β, ω, and δ being scalar functions of ρ and θ. Were the fluid defined by (I.23) simple, then, as we have stated above, the values of p and $\mathbf{P}$ in every motion would be determined by the totality of their values for affine transplacements. In every affine transplacement ρ is a function of t alone, so grad $\rho = \mathbf{0}$, and hence the portions of p and $\mathbf{P}$ proportional to the coefficients α, β, ω, and δ are null. Therefore, the class of affine motions does not suffice to determine α, β, ω, and δ through measurements of p and $\mathbf{P}$. We have shown in detail that *the fluid defined by* KORTEWEG'S *constitutive relations* (I.23) *is not*

---

[4] A general exposition of subject up to 1965 may be found in the treatise of C. TRUESDELL & W. NOLL, *The Non-linear Field Theories of Mechanics*, in Volume III/3 of FLÜGGE's *Encyclopedia of Physics*, Berlin etc., Springer-Verlag, 1965. Some of the important later work on the thermodynamics of deformation is presented in the book of W. A. DAY, *The Thermodynamics of Simple Materials with Fading Memory*, New York etc., Springer-Verlag, 1972.

simple. In Chapters XXIV and XXV we shall see that relations similar to (I.23) but far more complicated emerge from the kinetic theory.

More generally, a class of constitutive relations which we shall find to include a counterpart of results that follow from the kinetic theory may be defined as follows:

$$p(t, \mathbf{x}) = \underset{\mathbf{r},s}{\mathfrak{p}} \left(\rho(t - s, \mathbf{x} + \mathbf{r}), \mathbf{u}(t - s, \mathbf{x} + \mathbf{r}), \theta(t - s, \mathbf{x} + \mathbf{r}) \right),$$

$$\mathbf{P}(t, \mathbf{x}) = \underset{\mathbf{r},s}{\mathfrak{P}} \left(\rho(t - s, \mathbf{x} + \mathbf{r}), \mathbf{u}(t - s, \mathbf{x} + \mathbf{r}), \theta(t - s, \mathbf{x} + \mathbf{r}) \right),$$

$$\mathbf{q}(t, \mathbf{x}) = \underset{\mathbf{r},s}{\mathfrak{q}} \left(\rho(t - s, \mathbf{x} + \mathbf{r}), \mathbf{u}(t - s, \mathbf{x} + \mathbf{r}), \theta(t - s, \mathbf{x} + \mathbf{r}) \right), \qquad (I.24)$$

$$\psi(t, \mathbf{x}) = \underset{\mathbf{r},s}{\mathfrak{s}} \left(\rho(t - s, \mathbf{x} + \mathbf{r}), \mathbf{u}(t - s, \mathbf{x} + \mathbf{r}), \theta(t - s, \mathbf{x} + \mathbf{r}) \right),$$

$$\varepsilon(t, \mathbf{x}) = \underset{\mathbf{r},s}{\mathfrak{e}} \left(\rho(t - s, \mathbf{x} + \mathbf{r}), \mathbf{u}(t - s, \mathbf{x} + \mathbf{r}), \theta(t - s, \mathbf{x} + \mathbf{r}) \right).$$

The notation indicates that $\mathbf{r}$ runs over all vectors such as to render $\mathbf{x} + \mathbf{r}$ a place occupied by some point of the body at the time t and that s runs over all non-negative real numbers. If the response $\mathfrak{p}, \mathfrak{P}, \mathfrak{q}, \mathfrak{s}, \mathfrak{e}$ reduces to a quintuple of functions of a certain number of derivatives of ρ, $\mathbf{u}$, and θ at the time t and at the present place $\mathbf{x}$, the material is said to be of the *differential type*. We may express the constitutive equation for the pressure deviator of a material of the differential type symbolically in terms of a function $\mathfrak{p}$ and the operators ∂_t and grad:

$$\mathbf{P}(t, \mathbf{x}) = \mathfrak{p}(\partial_t, \text{grad}, \rho(t, \mathbf{x}), \mathbf{u}(t, \mathbf{x}), \theta(t, \mathbf{x})). \qquad (I.25)$$

The constitutive equations for p, $\mathbf{q}$, ψ, and ε are of the same kind.

Only a few simple materials are included in the subclass defined by (I.25), the Navier–Stokes–Fourier fluid being one of them. A more typical but still very special example of (I.25) is provided by KORTEWEG's relation (I.23), which pertains to a fluid of the differential type that is not simple.

In continuum thermomechanics two restrictions are imposed in addition to the principle of determinism, as follows. The first is the **Principle of Local Action**: *Two thermomechanical motions that always agree inside a neighborhood of the body-point X give rise to the same values of p, $\mathbf{P}$, $\mathbf{q}$, ψ, and ε at $\chi(t, X)$, $\tau(t, X)$.* As applied to (I.24) this principle asserts that for a fixed $\mathbf{x}$ the mappings $\mathfrak{p}, \mathfrak{P}, \ldots$, have the same values as do their restrictions to any neighborhood of $\mathbf{x}$. Any function of the gradients of ρ, $\mathbf{u}$, and θ satisfies this restriction. In particular, (I.25) does.

The second is the **Principle of Material Frame-indifference**. This principle asserts that the response of a material is the same for all observers. That is, if we subject the history of χ to an arbitrary time-dependent rigid motion, described by an orthogonal tensor-valued function $\mathbf{Q}$, the response of the material is such that the corresponding values of p, $\mathbf{P}$, $\mathbf{q}$, ψ, and ε are p, $\mathbf{QPQ}^T$, $\mathbf{Qq}$, ψ, and ε, respectively. For the class of materials defined by (I.24) this requirement may be expressed as a simple functional equation to be satisfied by the response. An example suffices to make the idea clear. Let us consider a very special case of (I.25):

$$\mathbf{P}(t, \mathbf{x}) = \mathfrak{l}(\rho(t, \mathbf{x}), \mathbf{u}(t, \mathbf{x}), E(t, \mathbf{x}), \mathbf{E}(t, \mathbf{x}), \mathbf{W}(t, \mathbf{x}), \theta(t, \mathbf{x})), \quad (\text{I.26})$$

in which $\mathbf{W}$ denotes the skew part of the velocity gradient, and $\mathfrak{l}$ is a linear function of E and $\mathbf{E}$. Then in order that $\mathfrak{l}$ satisfy the principle of material frame-indifference *it is necessary and sufficient*[5] *that* (I.26) *shall reduce to the Navier–Stokes relation* $(\text{I.7})_2$.

Other examples of frame-indifferent relations are easy to work out. One such is KORTEWEG's proposal (I.23).

We may reduce the general constitutive equation for materials of the differential type to an automatically frame-indifferent form, but that would serve little here. Let it suffice to remark that once a putative constitutive relation of the differential type be given, anyone can easily see whether it is frame-indifferent or not. For example, we show at once that under a change of frame

$$\rho^* = \rho, \qquad E^* = E,$$
$$\operatorname{grad}^* \rho^* = \mathbf{Q} \operatorname{grad} \rho, \qquad \operatorname{div}^* \operatorname{grad}^* \rho^* = \operatorname{div} \operatorname{grad} \rho, \quad (\text{I.27})$$
$$\mathbf{E}^* = \mathbf{Q} \mathbf{E} \mathbf{Q}^\mathrm{T},$$

the asterisk denoting quantities relative to the new framing. Using these facts, we see at a glance that KORTEWEG's equations (I.23) are indeed frame-indifferent.

The principle of determinism itself is sometimes broadened, or, from another point of view, weakened. Taking up and rendering more general a suggestion made by MAXWELL[6] so as to make his second kinetic theory seem more acceptable to the physicists of his day, who were skeptical about molecules, students of continuum mechanics have postulated and studied theories of the *rate type*[7]. In these a certain time-derivative of the pressure

---

[5] To verify this conclusion we note first that for (I.26) the formal expression of the principle at a fixed time t and a fixed place $\mathbf{x}$ is the functional equation

$$\mathfrak{l}(\rho, \mathbf{Q}\mathbf{u} + \dot{\mathbf{Q}}\mathbf{Q}^\mathrm{T}(\mathbf{x}^* - \mathbf{x}_0^*) + \dot{\mathbf{x}}_0^*, E, \mathbf{Q}\mathbf{E}\mathbf{Q}^\mathrm{T}, \mathbf{Q}\mathbf{W}\mathbf{Q}^\mathrm{T} + \dot{\mathbf{Q}}\mathbf{Q}^\mathrm{T}, \theta) = \mathbf{Q}\mathfrak{l}(\rho, \mathbf{u}, E, \mathbf{E}, \mathbf{W}, \theta)\mathbf{Q}^\mathrm{T};$$

the asterisk indicates the transformed position:

$$\mathbf{x}^* - \mathbf{x}_0^* = \mathbf{Q}(\mathbf{x} - \mathbf{x}_0).$$

If $\mathfrak{l}$ is linear in E and $\mathbf{E}$, the general solution of this functional equation leads to $(\text{I.7})_2$. A proof is given in §IV.4 of TRUESDELL's book cited in Footnote 1.

[6] MAXWELL [1867, pp. 30–31 of the reprint of his *Papers*].

[7] For a fairly general description of these, but without the argument θ or any consideration of $\mathbf{q}$, ψ, or ε, see §36 of the work by TRUESDELL & NOLL, cited above in Footnote 4.

tensor is assumed to be determined by the time-derivatives of lower order and the history of the thermomechanical motion. Also in this more general concept of material there is a principle of material frame-indifference. Without going into the details we may say roughly that it allows the time-derivatives of **P** to enter the constitutive relations only through certain frame-indifferent time-fluxes[8]. An example useful for comparisons with the kinetic theory is

$$\dot{\mathbf{P}} - \mathbf{WP} + \mathbf{PW} = \mathfrak{f}(\mathbf{P}, \mathbf{E}, \rho, \varepsilon, \operatorname{grad} \rho, \operatorname{grad} \varepsilon), \tag{I.28}$$

the response $\mathfrak{f}$ being *isotropic*:

$$\mathfrak{f}(\mathbf{QPQ}^T, \mathbf{QEQ}^T, \rho, \varepsilon, \mathbf{Q} \operatorname{grad} \rho, \mathbf{Q} \operatorname{grad} \varepsilon)$$
$$= \mathbf{Q}\mathfrak{f}(\mathbf{P}, \mathbf{E}, \rho, \varepsilon, \operatorname{grad} \rho, \operatorname{grad} \varepsilon)\mathbf{Q}^T \tag{I.29}$$

for all orthogonal tensors **Q**. The peculiar combination of tensors on the left-hand side of (I.28) is a frame-indifferent time-flux.

The reader of this book is expected to recall constantly the use for which constitutive relations of continuum mechanics are designed. This use we explained succinctly in Section (ii) of this chapter. There we referred only to the Navier–Stokes–Fourier constitutive equations (I.7) and (I.8), for that example makes the idea entirely clear. More general constitutive relations such as those we have just now presented lead to possibly different or additional initial conditions and possibly different boundary conditions as well as to analytical problems of much greater difficulty, but the approach is just the same.

(v) *Thermodynamics: The Clausius–Duhem inequality and its special cases and generalizations*

To complete this sketch of the aspects of modern continuum thermomechanics that will serve us in our study of the kinetic theory—for the understanding of which, today, the rudimentary continuum theories of the last century, despite their essential service to MAXWELL and BOLTZMANN, are far from sufficient—we remark that *continuum thermomechanics does not generally assume the existence of a caloric equation of state*. Indeed, as we have stated in the preceding section of this chapter, that equation is superseded by far more general relations such as (I.24)$_4$, subject indeed to certain interconnections, but nothing so special as (I.9). *A fortiori*, continuum thermomechanics allows the possibility that a thermal equation of state hold even if there be no caloric equation of state. For later comparisons with the kinetic theory we mention the possibility that (I.24)$_5$ might degenerate to $\varepsilon = \delta_v \theta$, (I.24)$_1$ might degenerate to

---

[8] Such fluxes are discussed in general and determined explicitly in §36 of the work of TRUESDELL & NOLL cited above in Footnote 4.

$p = r\rho\theta$, in which δ_v and r are constants, yet $(I.24)_4$ might assert a very complicated dependence of ψ upon the fields ρ, **u**, and θ in a neighborhood of **x** at the time t. Continuum mechanics offers no objection here. The student of continuum mechanics is not astonished by complicated functional dependences. What may astonish him about the kinetic theory is the fact that it reduces to extremely *special* forms some of the mappings he is accustomed to regarding as being possibly far more general. Herein lies the special charm, the ever-florescent novelty of the kinetic theory.

Continuum thermomechanics does indeed restrict the mappings that appear in the constitutive relations (I.24), but in this book we shall not need to develop or even summarise the results of this kind that have been obtained[9]. Certainly continuum thermomechanics does not generally make a fluid satisfy a relation like (I.21); hence, certainly, continuum thermomechanics does not require dissipationless flows to be isocaloric. As we shall see further on in this book, the same may be said of the kinetic theory of gases.

Continuum thermomechanics now provides both a general theory of constitutive relations and techniques for finding important special cases and for developing their properties. A central tool is an additional basic principle, assumed in addition to the generic axioms of balance of mass, linear momentum, rotational momentum, and energy, rendered effective through the constitutive principles of determinism, local action, and material frame-indifference. This is the *dissipation inequality*. Various general statements have been proposed for this axiom, statements which reduce in the case of classical fluid mechanics to one and the same form, namely, the inequalities (I.10), but differ in their implications for more general constitutive relations. The one most commonly studied is the **Clausius–Duhem Inequality**. To state it, we define the *caloric* or *specific entropy* η of the material in terms of ε, ψ, and θ by (I.18). Then from $(I.24)_{4,5}$ we see that η obeys a constitutive relation of the form

$$\eta(t, \mathbf{x}) = \underset{r,s}{\mathfrak{h}}\left(\rho(t-s, \mathbf{x}+\mathbf{r}), \mathbf{u}(t-s, \mathbf{x}+\mathbf{r}), \theta(t-s, \mathbf{x}+\mathbf{r})\right), \qquad (I.30)$$

and we may take this relation as a starting point if we please to. Next we define the present *calory* or *total entropy* of the body that presently occupies the region $\mathscr{B}$ as follows:

$$\mathfrak{H} \equiv \int_{\mathscr{B}} \rho\eta \, dV. \qquad (I.31)$$

For a given body of given material in a given flow, $\mathfrak{H}$ is a function of t alone. If **n** is the outer unit normal to the boundary $\partial\mathscr{B}$ of $\mathscr{B}$, we use the notation $a_\perp$ to

---

[9] The reader may consult the book of DAY, cited in Footnote 4, or the broader, discursive, and far less systematic earlier survey by C. TRUESDELL, *Rational Thermodynamics, a Course of Lectures on Selected Topics*, New York *etc.*, McGraw-Hill, 1969.

denote the signed length of the projection of **a** upon **n**, and the remainder of the vector **a** we denote by $\mathbf{a}_t$:

$$a_\perp \equiv \mathbf{n} \cdot \mathbf{a} = n_b a_b, \qquad \mathbf{a}_t \equiv \mathbf{a} - a_\perp \mathbf{n}. \tag{I.32}$$

In particular, the scalar quantity $q_\perp$ is the *flux of energy* out of the body. The *Clausius–Duhem inequality* asserts that for every motion of any body that presently occupies the arbitrary region $\mathscr{B}$

$$\dot{\mathfrak{H}} \geqq -\int_{\partial\mathscr{B}} \frac{q_\perp}{\theta}\, dA. \tag{I.33}$$

In stating it we have not included a term to represent radiation or other local sources of heat, for no such possibility is allowed in the kinetic theory of MAXWELL. For the same reason we have omitted such a term also in writing the equation of balance of energy (I.4). While opinions differ as to what the "Second Law of Thermodynamics" is, the Clausius–Duhem inequality is the most commonly accepted mathematical statement of that law in the rather scant literature that comes to grips with dissipative processes in deformable continua of some generality. A local equivalent to (I.33), for sufficiently smooth fields, is

$$\rho\theta\dot{\eta} + \operatorname{div} \mathbf{q} - \theta^{-1}\mathbf{q} \cdot \operatorname{grad} \theta \geqq 0. \tag{I.34}_1$$

Because of (I.20) we may write this relation in the alternative form

$$-(p - \varpi)E - \mathbf{P} \cdot \mathbf{E} - \theta^{-1}\mathbf{q} \cdot \operatorname{grad} \theta \geqq 0. \tag{I.34}_2$$

Thus the following conditions are sufficient for the Clausius–Duhem inequality to hold: Frictional pressure never does positive work, and heat never flows from a colder to a hotter part. "Frictional pressure" here means $\mathbf{P} + (p - \varpi)\mathbf{1}$, and ϖ is related to ψ and ε through (I.9).

If we adopt the Navier–Stokes–Fourier constitutive relations (I.7) and (I.8), $(1.34)_2$ assumes the form

$$(\lambda + \tfrac{2}{3}\mu)E^2 + 2\mu|\mathbf{E}|^2 + \kappa\theta^{-1}|\operatorname{grad}\theta|^2 \geqq 0. \tag{I.35}$$

Because we may select the fields E, $\mathbf{E}$, and $\operatorname{grad}\theta$ independently, (I.35) is equivalent to the Stokes–Duhem inequalities (I.10). Nothing so special can be proved on the basis of no more than the comprehensive constitutive relations (I.24).

Some authors contend that the "Second Law" does not refer to all processes or all regions, but only to special circumstances. A statement weaker than the Clausius–Duhem inequality can be obtained by restricting attention to some particular boundary $\partial\mathscr{B}$, with which is associated a particular field of *wall temperature* θ^W, not necessarily the same as the temperature field θ of the body at points on $\partial\mathcal{B}$. The assertion that for the particular boundary $\partial\mathcal{B}$

$$\dot{\mathfrak{H}} \geq -\int_{\partial\mathcal{B}} \frac{q_\perp}{\theta^W} dA, \tag{I.36}$$

we shall call the **Heat-bath Inequality**.

The two basic inequalities, (I.33) and (I.36), can be connected. As

$$-\frac{q_\perp}{\theta} = -\frac{q_\perp}{\theta^W} + \frac{-q_\perp(\theta^W - \theta)}{\theta^W \theta}, \tag{I.37}$$

the condition

$$q_\perp(\theta^W - \theta) \leq 0 \quad \text{on} \quad \partial\mathcal{B} \tag{I.38}$$

is sufficient that *the heat-bath inequality shall be a consequence of the Clausius–Duhem inequality*. The converse, of course, could not hold, since the heat-bath inequality refers only to some particular region, while the Clausius–Duhem inequality refers to all material regions. The inequality (I.38) itself, sometimes called MAXWELL's *heat-transfer condition*, asserts that at each point on $\partial\mathcal{B}$ where $\theta \neq \theta^W$, heat flows into $\partial\mathcal{B}$ or out of $\partial\mathcal{B}$ according as $\theta < \theta^W$ or $\theta > \theta^W$. The statement that *heat never flows from the colder body to the hotter* is sometimes regarded as implied by or being a part of the "Second Law". We have encountered it above in our first interpretation of $(I.10)_3$. In the same way the heat-transfer condition (I.38) has a status, whether we choose to regard it as directly founded on experience or as a part of some sort of protophysics.

No inconsistency between the heat-bath inequality and the Clausius–Duhem inequality has been found here. The heat-transfer condition (I.38) is compatible with both. If we choose to regard $\partial\mathcal{B}$ as a surface of discontinuity interior to a body larger than that which occupies $\mathcal{B}$, we can apply to that larger body the principle which for smooth fields is expressed by the Clausius–Duhem inequality. Then $\mathbf{q}$ and θ in (I.33) stand for interior limits; θ^W, the exterior limit corresponding to the latter. Denoting by $\mathbf{q}^W$ the corresponding exterior limit of $\mathbf{q}$, we obtain[10] the condition $q_\perp/\theta = q_\perp^W/\theta^W$. This statement shows that in the heat-transfer condition (I.38) we may replace $q_\perp$ by $q_\perp^W$, as the physical idea that that condition is intended to express suggests.

While we may sometimes find the heat-bath inequality easier to verify, that fact by no means implies that the Clausius–Duhem inequality fails to hold.

The inequality

$$\dot{\mathfrak{H}} \geq 0 \tag{I.39}$$

---

[10] We suppose that $\partial\mathcal{B}$ is a stationary surface devoid of sources of mass or caloric, and neither a shock nor a propagating vortex sheet. The assertion in the text then follows from Eq. (258.4) of the treatise by C. TRUESDELL & R. TOUPIN, *The Classical Field Theories*, in Volume III/1 of FLÜGGE's *Encyclopedia of Physics*, Berlin etc., Springer-Verlag, 1960. From Eqs. (205.5) and (241.5) we obtain corresponding jump conditions expressing the balance of momentum and energy, on the further assumption that $\partial\mathcal{B}$ is devoid of sources of momentum or energy:

$$\mathbf{Mn} = \mathbf{M}^W\mathbf{n}, \quad (\mathbf{u} - \mathbf{u}^W) \cdot \mathbf{Mn} = q_\perp^W - q_\perp.$$

—in words, the calory does not decrease—may be called the *Clausius Inequality*. It follows from the Clausius–Duhem inequality in either of the following two circumstances:

$$q_\perp \leq 0 \quad \text{on} \quad \partial \mathscr{B}, \quad \text{or}$$
$$\int_{\partial \mathscr{B}} q_\perp \, dA \leq 0 \quad \text{and} \quad \theta = \text{const.} \quad \text{on} \quad \partial \mathscr{B}. \tag{I.40}$$

Likewise, it follows from the heat-bath inequality if

$$q_\perp \leq 0 \quad \text{on} \quad \partial \mathscr{B}, \quad \text{or}$$
$$\int_{\partial \mathscr{B}} q_\perp \, dA \leq 0 \quad \text{and} \quad \theta^W = \text{const.} \quad \text{on} \quad \partial \mathscr{B}. \tag{I.41}$$

A deductive development of continuum thermomechanics would take $\mathfrak{H}$ rather than η as the starting point for its treatment of calory. Indeed, mass and force and energy rather than their densities enter the basic axioms of mechanics, from which the field equations (I.1)–(I.4) follow as consequences when the fields themselves are sufficiently smooth. As our purpose in this chapter is merely to collect some ideas and formulae useful for reference in the remainder of the book, we have dealt with the densities. In terms of them we may define as follows the *total energy* $\mathfrak{E}$ of the body that presently occupies the region $\mathscr{B}$:

$$\mathfrak{E} \equiv \int_{\mathscr{B}} \rho(\varepsilon + \tfrac{1}{2}u^2) \, dV. \tag{I.42}$$

Then from (I.1)–(I.5) we easily obtain an integrated statement of the balance of energy:

$$\dot{\mathfrak{E}} = -\int_{\partial \mathscr{B}} (\mathbf{u} \cdot \mathbf{Mn} + q_\perp) \, dA + \int_{\mathscr{B}} \rho \mathbf{u} \cdot \mathbf{b} \, dV. \tag{I.43}$$

Sometimes this statement is called the "First Law of Thermodynamics". In stating it we have not included a term to represent radiation or other local sources of heat, for, as we have mentioned already, no such possibility is allowed in the kinetic theory formulated by MAXWELL.

A body is *isolated* if such forces as act upon it do no work:

$$-\int_{\partial \mathscr{B}} \mathbf{u} \cdot \mathbf{Mn} \, dA + \int_{\mathscr{B}} \rho \mathbf{u} \cdot \mathbf{b} \, dV = 0. \tag{I.44}$$

We may define the normal pressure $p_\perp$ and the tangential pressure $\mathbf{p}_t$ upon $\partial \mathscr{B}$ by putting $\mathbf{Mn}$ for $\mathbf{a}$ in (I.32). Then the integrand in the surface integral of (I.44) is $u_\perp p_\perp + \mathbf{u}_t \cdot \mathbf{p}_t$. Consequently we have the following conditions together sufficient that the body in $\mathscr{B}$ be isolated:

(a) The body force vanishes: $\mathbf{b} = \mathbf{0}$ in $\mathscr{B}$.
(b) $\partial \mathscr{B}$ is a stationary boundary: $u_\perp = 0$ on $\partial \mathscr{B}$.
(c) $\partial \mathscr{B}$ experiences no shear pressure: $\mathbf{p}_t = \mathbf{0}$ on $\partial \mathscr{B}$.

(v) ISOLATED BODIES

In an isolated body the total energy changes only as a result of such flux of energy as there may be:

$$\dot{\mathfrak{E}} = -\int_{\partial \mathcal{B}} q_\perp \, dA. \tag{I.45}$$

This fact allows us to relate the growth of energy to the growth of calory. For example, if we adopt (I.38), we conclude that $q_\perp \leq 0$ on $\partial \mathcal{B}$ if $\theta^W > \theta$ on $\partial \mathcal{B}$. Then (I.36) implies the Clausius inequality (I.39), while (I.45) implies that $\dot{\mathfrak{E}} \geq 0$. We have demonstrated a major implication of the heat-bath inequality in conjunction with the heat-transfer condition: *If each point of the boundary of an isolated body is at a temperature lower than that of the environment there, neither the calory nor the total energy of that body can decrease.*

CLAUSIUS is the author of the sybillic utterance, "The energy of the universe is constant; the entropy of the universe tends to a maximum." The objectives of continuum thermomechanics stop far short of explaining the "universe", but within that theory we may derive easily an explicit statement in some ways reminiscent of CLAUSIUS' but referring only to a modest object: an isolated body of finite size. We may call the boundary $\partial \mathcal{B}$ of the region occupied by a body *adiabatic* if

$$\int_{\partial \mathcal{B}} q_\perp \, dA = 0; \tag{I.46}$$

insulated if

$$q_\perp = 0 \quad \text{on} \quad \partial \mathcal{B}. \tag{I.47}$$

The latter condition implies the former, but not conversely. Either of these conditions suffices that (I.45) shall reduce to $\dot{\mathfrak{E}} = 0$; the latter makes $\dot{\mathfrak{H}} \geq 0$ a consequence both of (I.33) and of (I.36); the former does likewise if we assume further that $\theta = $ const. and $\theta^W = $ const. at each time. We have thus established the following **Theorem.** *Let either the heat-bath inequality or the Clausius–Duhem inequality be adopted; then if the boundary of an isolated body is either*

(a) *insulated, or*
(b) *adiabatic and at uniform temperature,*

the total energy of the body remains constant, and its calory does not decrease. This theorem, which does not rest upon the heat-transfer condition and hence attributes to the exterior of the body in question nothing in case (a) and only uniform temperature in case (b), may give some substance as well as sense to CLAUSIUS' pronouncement about the universe.

The kinetic theory, as we shall see in following chapters of this book, delivers through mathematical proof formal analogues of most of the general principles of continuum thermomechanics, but it approaches them in an altogether different way, for its basic assumptions are couched in terms of molecular motions and stochastic expectations. Strict analogues of the field equations (I.1)–(I.4) are easy to obtain; we shall derive them in Chapter IX. Partial analogues of the basic inequalities (I.33) and (I.36) provide the subject of Chapter XI. Formulae deceivingly similar to constitutive relations but in fact quite different in nature are the most important products of the kinetic theory. They provide the central theme of this book; we shall first encounter them in Chapter XIII, and much of our text from there onward will concern them.

Chapter II

The Stokes–Kirchhoff Gas, Some of Its Peculiarities, and Some of Its Flows

For comparisons with the kinetic theory classical fluid mechanics serves only when the fluid it concerns is an ideal gas. We shall see now that *to formulate a theory of the flow of ideal viscous gases, we do not need thermodynamics and caloric equations of state.*

(i) *The Stokes–Kirchhoff and Euler–Hadamard theories of ideal gases*

An *ideal gas* is defined by the particular thermal equation of state

$$\varpi = r\rho\theta; \qquad (\text{II.1})$$

r is the constitutive constant of the gas. By (I.14), (I.15), and (I.9) we may show that this condition implies the following relations:

$$\varepsilon = \varepsilon(\theta), \qquad \delta_v(\gamma - 1) = r, \qquad \delta_v = \varepsilon'(\theta). \qquad (\text{II.2})$$

Thus δ_ϖ is constant if and only if δ_v is, and in that case (II.2) makes ε an affine function of θ. For reasons that are easy to motivate in the kinetic theory but can also be given a phenomenological basis we take the zero of temperature as being also the zero of energy. Then

$$\varepsilon = \delta_v \theta. \qquad (\text{II.3})$$

Therefore, *for an ideal gas with constant specific heats, temperature and energetic are interchangeable;* the one serves only as substitute for the other, and the specific heat δ_v may be regarded as *a factor for converting units of temperature to units of energetic*. It thus becomes easy to formulate the theory of such a gas in purely mechanical terms, with no reference to the temperature at all, and for later use we do so.

Of course, an ideal gas may be also a perfect fluid, but it need not be. There are many different kinds of internal friction which an ideal gas might have. The simplest kind is linear viscosity, represented by (I.7). Again for ease of reference when we come to the kinetic theory, in comparisons with classical fluid mechanics we shall limit attention to fluids whose bulk viscosity vanishes:

$$\lambda + \tfrac{2}{3}\mu = 0; \qquad (I.12)_r$$

equivalently, in all flows

$$\varpi = p. \qquad (I.13)_r$$

The resulting theory, that of *an ideal, linearly viscous gas with constant specific heats and null bulk viscosity*, we shall call the **Stokes–Kirchhoff Theory**.

By use of (II.3), (I.13)$_1$, and (II.2)$_2$ we may rewrite (II.1) in purely mechanical terms:

$$p = (\gamma - 1)\rho\varepsilon. \qquad (II.4)$$

Hence we can write the equation of balance of energy (I.4) for an ideal gas in the forms

$$\rho\dot\varepsilon - (\gamma - 1)\varepsilon\dot\rho = -\mathbf{P}\cdot\mathbf{E} - \operatorname{div}\mathbf{q},$$

$$\frac{\rho^\gamma}{\gamma - 1}(p\rho^{-\gamma})^{\cdot} = -\mathbf{P}\cdot\mathbf{E} - \operatorname{div}\mathbf{q}. \qquad (II.5)$$

We shall call *dissipationless* any flow of such a gas in which

$$\mathbf{P}\cdot\mathbf{E} + \operatorname{div}\mathbf{q} = 0. \qquad (II.6)$$

This usage is consistent with that in the discussion following (I.21) but independent of it, for here we have not introduced η or anything at all from thermodynamics. Comparing (II.5) with (II.6), we see that *a flow of an ideal gas with constant specific heats is dissipationless if and only if*

$$p\rho^{-\gamma} = \text{const.} \qquad (II.7)$$

for each fluid point. This statement is one form of POISSON's *law of adiabatic change.* As we may show by substituting (I.1) into (II.5)$_1$, an equivalent statement is

$$E = -\frac{1}{\gamma - 1}\frac{\dot\varepsilon}{\varepsilon}. \qquad (II.8)$$

It is easy to show that if (II.7) and (II.8) are both satisfied, so is (I.1).

Furthermore, the Fourier relation (I.8) is easy to write explicitly in terms of ε:

$$\mathbf{q} = -M\mu \operatorname{grad}\varepsilon, \qquad (II.9)$$

and the Maxwell number M, which is defined by (I.17), is a given function of ε and ρ.

The special case of the Stokes–Kirchhoff theory obtained by taking μ and $M\mu$ as 0 we shall call the *Euler–Hadamard Theory*. It represents the fluid as being an ideal perfect gas that does not conduct heat, so all its flows are dissipationless. In that theory we may apply Hugoniot's theorem and conclude that the intrinsic speed of propagation of weak waves is $\sqrt{\gamma(\gamma - 1)\varepsilon}$; the value of this scalar field at a place and time is the corresponding *speed of sound*. Kirchhoff showed that according to the linearized Stokes–Kirchhoff theory a sinusoidal forced plane disturbance propagates at nearly this same speed, provided its frequency be low enough.

(ii) Some parameters that control dynamical similarity

A theory of *scaling* or *dynamical similarity* can be constructed for the Stokes–Kirchhoff theory[1]. If the fields ρ, **u**, and ε provide a solution of the differential equations corresponding to given values of the constitutive functions μ and M and the constitutive constants δ_v and γ, constant multiples of those fields do not generally provide solutions for a fluid having the same or proportional constitutive parameters. A necessary condition is that μ be proportional to a product of powers of ρ and ε, the same powers for both fluids, and that γ be the same for both. Second, the Maxwell number M, defined by (I.17), must be constant and have the same value for both fluids. Finally, the *local Mach number* M and the *local Reynolds number* R for the two flows must be the same. These two dimensionless scalar fields are defined as follows in terms of the speed u and other quantities:

$$\mathsf{M} \equiv \frac{u}{\sqrt{\gamma(\gamma - 1)\varepsilon}}, \qquad \mathsf{R} \equiv \frac{\rho u^2}{\mu D}. \tag{II.10}$$

M is the ratio of the speed of flow to the local speed of sound. The scalar field D in (II.10)$_2$ is some positive scalar function of **E** and E that is homogeneous of degree 1, so its physical dimensions are those of a pure rate. Often some "characteristic length" l is suggested by a particular problem, and then D is taken as u/l. For comparisons with the kinetic theory it is preferable to do the opposite, namely, define an intrinsic length in terms of D. For example, the ratio of the speed of sound to D provides an intrinsic length: $l = \sqrt{\gamma(\gamma - 1)\varepsilon}/D$. The assumption that the fields ρ, **u**, ε, and μ for the two flows be proportional makes M and R for the two flows be the same fields if and only if at some one time they have the same values at some one pair of corresponding places. If we attach a subscript 0 to indicate evaluation at a fixed place and time, we can

---

[1] For details the reader may consult Chapter GII of the article by J. Serrin, *Mathematical principles of classical fluid mechanics*, in Volume VIII/1 of Flügge's *Encyclopedia of Physics* (co-editor C. Truesdell), Berlin *etc.*, Springer-Verlag, 1959.

regard u_0/D_0 as a fixed length l and so write R in the more familiar form $\mathsf{R} = \rho_0 u_0 l/\mu_0$.

Of greater importance for the kinetic theory is the fact that

$$\frac{\mathsf{M}^2}{\mathsf{R}} = \frac{1}{\gamma}\frac{\mu}{p}D, \tag{II.11}$$

as follows at once from (II.10) and (II.4).

(iii) The caloric of an ideal gas

From (I.18), (I.9), and (II.3) it is an easy matter to calculate the caloric of the Stokes–Kirchhoff gas, provided we be willing to assume that it has one. Two convenient forms of the resulting caloric equation of state are

$$\begin{aligned}\eta/\delta_v &= \log(p\rho^{-\gamma}) + \text{const.,} \\ \eta/r &= \log(\varepsilon^{1/(\gamma-1)}/\rho) + \text{const.,}\end{aligned} \tag{II.12}$$

both right-hand sides being simple expressions in terms of mechanical quantities alone. By use of them we could have obtained some of the results given above, for example, (II.7), but we have left calculation of η until the last because in fact we do not need it.

Although the theory of the Stokes–Kirchhoff gas is consistent with the principles of classical thermostatics, it may be formulated completely without ever using them, and without ever introducing the temperature θ, the free energetic ψ, or the caloric η. The reader may verify this fact easily by taking $\lambda + \frac{2}{3}\mu = 0$ first, so that $\varpi = p$, then assuming (II.4) as the thermal (or, more properly, energetic) equation of state for ϖ. The entire theory is defined by constitutive relations specified in terms of three non-negative scalar parameters: the constant $\gamma - 1$, the shear viscosity μ, and the Maxwell number M, these last two being assigned functions of the positive scalars ρ and ε. While the remaining variables of classical fluid mechanics such as θ, ψ, and η may be introduced for convenience or interpretation, they are not needed either to formulate the theory or to solve special problems.

Although conceptually meager, the purely formal approach to the Stokes–Kirchhoff theory presented at the beginning of this chapter has the advantage of offering a major point of contact with the kinetic theory. That we shall see in Chapter III.

(iv) Equilibrium

Equilibrium of a fluid body may be defined by the requirements that the velocity field vanish, that the energetic be uniform, and that there be no resultant accelerating force beyond that of the pressure and no change of energetic

(v) HOMO-ENERGETIC DILATATION   23

due to flux of energy. That is, by definition,

$$\mathbf{u} = \mathbf{0}, \quad \text{grad } \varepsilon = 0, \quad \text{div } \mathbf{P} = \mathbf{0}, \quad \text{div } \mathbf{q} = 0. \tag{II.13}$$

Putting (II.13)$_1$ into (I.1) shows that $\partial_t \rho = 0$. Putting (II.13)$_{1,4}$ into (I.4) shows that $\partial_t \varepsilon = 0$, so ε is altogether constant. If p satisfies a thermal equation of state $p = f(\rho, \varepsilon)$, it follows that $\partial_t p = 0$. Also, from (I.2) we see that

$$\rho \mathbf{b} = \text{grad } p = \frac{\partial f}{\partial \rho} \text{ grad } \rho, \quad \text{so} \quad \mathbf{b} = \text{grad}\left(\int \frac{1}{\rho} \frac{\partial f}{\partial \rho} d\rho\right). \tag{II.14}$$

Therefore, in order that equilibrium be possible, it is sufficient and essentially necessary that the body force $\mathbf{b}$ be a steady lamellar field with single-valued potential:

$$\mathbf{b} = -\text{grad } U, \quad U = U(\mathbf{x}). \tag{II.15}$$

If this condition is satisfied, (II.14) may be integrated for the density. For example, if the pressure p is that of an ideal gas with constant specific heats, (II.4) holds, so the general solution of (II.14)$_3$, regarded as being a differential equation for ρ, is

$$\frac{\rho(\mathbf{x})}{\rho_R} = \exp\left(\frac{-U(\mathbf{x})}{(\gamma - 1)\varepsilon}\right), \tag{II.16}$$

in which ρ_R is the value of ρ at a reference place $\mathbf{x}$ where $U(\mathbf{x}) = 0$. The relation (II.16) solves the problem of *equilibrium according to classical aerostatics*.

A body in equilibrium subject to null body force we shall describe as being in *free equilibrium*. From (II.14)$_{1,2}$ we see that *in free equilibrium, not only ε but also p is constant in space*; if p satisfies a thermal equation of state, then ε, p, and ρ are constant in space and time. The results obtained so far in this part of the chapter flow from (II.13) and, at the end, from (II.4).

In the classical theory of linearly viscous and linearly heat-conducting fluids, which are defined by the constitutive relations (I.7) and (I.8), the weaker conditions

$$\mathbf{u} = \mathbf{0}, \quad \text{grad } \varepsilon = 0 \tag{II.17}$$

suffice to define equilibrium. Indeed, from (I.7)$_2$ we see that (II.17)$_1$ implies (II.13)$_3$, while from (I.8) we see that (II.17)$_2$ implies (II.13)$_4$. Thus, in particular, to assume (II.17) reduces the local content of the Stokes–Kirchhoff theory to (II.16).

(v) *Some particular homo-energetic flows: dilatation, affine flow, simple shearing, extension*

Even with a theory so simple as that of the Stokes–Kirchhoff gas, few specific problems have been solved. We shall close our discussion of this theory by determining three particular families of flows possible for such a gas. All three of them are *homo-energetic*: ε is taken from the outset as being a function of t only. Equivalently, such a flow is *homothermal*. Moreover, in all of them

$$\text{div } \mathbf{P} = \mathbf{0}, \qquad \text{div } \mathbf{q} = 0. \tag{II.18}$$

Thus these flows satisfy three of the four conditions (II.13) that define equilibrium, and for them the equations of balance of linear momentum and energy, (I.2) and (II.5), become, respectively,

$$\dot{\mathbf{u}} = -(\gamma - 1)\varepsilon \text{ grad log } \rho + \mathbf{b},$$

$$\rho \dot{\varepsilon} + pE = \frac{\rho^\gamma}{\gamma - 1}(p\rho^{-\gamma})^\cdot = -\mathbf{P} \cdot \mathbf{E}. \tag{II.19}$$

We must satisfy also (I.1), which expresses the conservation of mass. While we shall treat an ideal gas with constant specific heats, so (II.4) holds, beyond that we shall push our analysis as far as we can *on the basis of the conditions* (II.18) *alone*. In each class of flows, most of the results found will be *independent of constitutive relations* except (II.4). We shall make good use of this fact in Chapters X, XIV, and XV when we come to determine what the kinetic theory predicts if we start from the same assumptions but use its apparatus instead of the Stokes–Kirchhoff constitutive relations. Here, of course, we shall at the end of each analysis reduce the results to the explicit forms valid when the Stokes–Kirchhoff constitutive equations are applied, for these forms will provide useful comparisons with those we shall later obtain for the kinetic gas.

Example 1. *Homo-energetic dilatation*. The kinematical differential equation $\mathbf{E} = \mathbf{0}$, or, equivalently,

$$\text{grad } \mathbf{u} + (\text{grad } \mathbf{u})^T = \tfrac{2}{3}E\mathbf{1}, \tag{II.20}$$

was integrated by CAUCHY. He showed first that E was necessarily a linear function of the position vector $\mathbf{r}$ with respect to some particular place $\mathbf{x}_0$:

$$\mathbf{r} \equiv \mathbf{x} - \mathbf{x}_0; \tag{II.21}$$

his general solution of (II.20), which may easily be verified, is

$$\mathbf{u} = (\mathbf{a} \cdot \mathbf{r} + b)\mathbf{r} - \tfrac{1}{2}r^2\mathbf{a} + \mathbf{g} + \mathbf{W}\mathbf{r}, \tag{II.22}$$

in which the scalar b, the vectors $\mathbf{a}$ and $\mathbf{g}$, and the skew tensor $\mathbf{W}$ are functions of t alone. For this velocity field $E = 3(\mathbf{a} \cdot \mathbf{r} + b)$. The place $\mathbf{x}_0$ will remain arbitrary.

In a dissipationless flow of an ideal gas for which $\gamma = \text{const.}$, (II.19)$_3$ reduces to POISSON's law of adiabatic change, which we may write in the form (II.8). If the flow is homo-energetic, (II.8) shows that E is a function of t alone. Combining these two results with the foregoing, we conclude that *in a homo-energetic dissipationless dilatation of an ideal gas with constant specific heats*

$$\mathbf{u} = \mathbf{g} + \tfrac{1}{3}E\mathbf{r} + \mathbf{W}\mathbf{r}, \qquad \mathbf{W} = -\mathbf{W}^T; \tag{II.23}$$

g, **W**, and ε are functions of t only, and E, of course, is determined by ε through (II.8). In general ρ remains a function of **r** as well as of t.

These results apply to the Stokes–Kirchhoff gas. We see from (I.7)$_2$ that **P** = **0**; from (II.9), that **q** = **0**. Thus every *homo-energetic dilatation of a Stokes–Kirchhoff gas is dissipationless*, and so for this gas (II.8) and (II.23) hold in every dilatation. Since the bulk viscosity of a Stokes–Kirchhoff gas is zero, we certainly should have expected to find, as we just have, that in such a gas *a dilatation, provided it be homo-energetic, proceeds unhindered by the action of viscosity;* that is, *for such a motion the Stokes–Kirchhoff theory and the Euler–Hadamard theory agree*. However, to obtain (II.23) for dilatations it is merely sufficient, *not necessary*, to use the constitutive relations of the Stokes–Kirchhoff theory. The essential conditions are (II.18), and any theory of ideal gases that delivers (II.18) for homo-energetic dilatation will yield (II.23) also. We shall make good use of this fact in Chapters X and XV.

The problem before us then is to determine all functions ρ, ε, **g**, E, and **W** for which the equations of balance, now reduced to (I.1), (II.8), and (II.19)$_1$, are satisfied, the velocity **u** in these being given by (II.23). To solve this problem is not a simple matter. Several people[2] have studied it in the context of the kinetic theory, but their analysis is not general. We present now the complete solution. That is, we exhibit all possible homo-energetic dissipationless dilatations of an ideal gas which satisfy the requirements (II.18). Since we have demonstrated already that a Stokes–Kirchhoff gas in homo-energetic dilatation does satisfy (II.18), our results will apply in particular to such a gas. They will apply also to the kinetic theory, for a condition formally the same as (II.18) appears there, too, as we shall see in Chapter X.

We may interpret (II.8) as a formula for determining E once ε is known. Thus we may eliminate (II.8) from the problem by using it to eliminate E from the other equations. The velocity field (II.23) becomes

$$\mathbf{u} = \mathbf{g} - \frac{1}{2}\lambda \frac{\dot{\varepsilon}}{\varepsilon}\mathbf{r} + \mathbf{W}\mathbf{r}, \tag{II.24}$$

in which

$$\lambda \equiv \frac{2}{3(\gamma - 1)}. \tag{II.25}$$

We have shown already that the balance of mass and the balance of energy for dissipationless flows are together equivalent to (II.7) and (II.8). Use of (II.4) and (II.25) enables us to write (II.7) in the form

$$(\rho \varepsilon^{-\frac{3}{2}\lambda})^{\cdot} = 0. \tag{II.26}$$

---

[2] BOLTZMANN [1876, §III] assumed from the outset that **b** = − grad U, U being a function of **x** only. Later authors in the kinetic theory, so far as we have seen, assume that **b** = **0** or that the velocity field is steady. BOLTZMANN's concluding words are "It would be difficult to treat the case in which the forces acting on the gas from the outside are functions of time also."

Using (I.6) to calculate the acceleration $\dot{\mathbf{u}}$ corresponding to (II.24), we find that

$$\dot{\mathbf{u}} = -\operatorname{grad}\{-[\varepsilon^{\lambda}(\varepsilon^{-\lambda}\mathbf{g})^{\cdot} - \tfrac{1}{2}\varepsilon^{\lambda}(\varepsilon^{-\lambda})^{\cdot}\mathbf{g} + \mathbf{W}\mathbf{g}]\cdot\mathbf{r} \\ - \tfrac{1}{4}[\varepsilon^{\lambda}(\varepsilon^{-\lambda})^{\cdot\cdot} - \tfrac{1}{2}(\varepsilon^{\lambda}(\varepsilon^{-\lambda})^{\cdot})^{2}]r^{2} + \tfrac{1}{2}|\mathbf{W}\mathbf{r}|^{2}\} + \varepsilon^{\lambda}(\varepsilon^{-\lambda}\mathbf{W})^{\cdot}\mathbf{r}. \quad \text{(II.27)}$$

Comparison of this expression for $\dot{\mathbf{u}}$ with the one provided by (II.19)$_1$ delivers the following restriction on the body force $\mathbf{b}$: *In order that* (II.19)$_1$ *shall be satisfied for some homo-energetic dissipationless dilatation, it is necessary that there be a scalar field* $U(t, \mathbf{x})$ *and a skew tensor-valued function* $\mathbf{B}(t)$ *such that*

$$\mathbf{b} = -\operatorname{grad} U + \mathbf{B}\mathbf{r}. \quad \text{(II.28)}$$

If the body force is not of this form, our problem has no solution[3]. If, however, the body force is of this kind, then we see from (II.27) and (II.28) that the balance of linear momentum (II.19)$_1$ is equivalent to the conditions

$$U = -(\gamma - 1)\varepsilon \log \rho - [\varepsilon^{\lambda}(\varepsilon^{-\lambda}\mathbf{g})^{\cdot} - \tfrac{1}{2}\varepsilon^{\lambda}(\varepsilon^{-\lambda})^{\cdot}\mathbf{g} + \mathbf{W}\mathbf{g}]\cdot\mathbf{r} \\ - \tfrac{1}{4}[\varepsilon^{\lambda}(\varepsilon^{-\lambda})^{\cdot\cdot} - \tfrac{1}{2}(\varepsilon^{\lambda}(\varepsilon^{-\lambda})^{\cdot})^{2}]r^{2} + \tfrac{1}{2}|\mathbf{W}\mathbf{r}|^{2} + \phi, \\ \mathbf{B} = \varepsilon^{\lambda}(\varepsilon^{-\lambda}\mathbf{W})^{\cdot}, \quad \text{(II.29)}$$

ϕ being an arbitrary function of t alone. Before we analyse (II.29), we turn to the balance of mass.

It is easy to interpret (II.26): The quantity $\rho\varepsilon^{-\frac{3}{2}\lambda}$ is constant along the trajectory of each fluid-point. Thus, if the fluid-point that is at $\mathbf{x}$ at time t occupies the place $\mathbf{X}$ at time 0, the general solution of (II.26) is

$$\rho(t, \mathbf{x}) = \left(\frac{\varepsilon(t)}{\varepsilon(0)}\right)^{\frac{3}{2}\lambda} \rho_0(\mathbf{X}). \quad \text{(II.30)}$$

We are free to specify the initial value ρ_0 appearing in this formula. The problem now is to express $\mathbf{X}$ in terms of the spatial variables t and $\mathbf{x}$. To do so, we write $\mathbf{x} = \boldsymbol{\chi}(t, \mathbf{X})$ for the motion that gives rise to the velocity field (II.24). This means that $\mathbf{u} = \partial_t \boldsymbol{\chi}$, and so we see from (II.24) and (II.21) that $\boldsymbol{\chi}$ satisfies the equations

$$\partial_t \boldsymbol{\chi} = \mathbf{g} - \frac{1}{2}\lambda\frac{\dot{\varepsilon}}{\varepsilon}(\boldsymbol{\chi} - \mathbf{x}_0) + \mathbf{W}(\boldsymbol{\chi} - \mathbf{x}_0), \\ \boldsymbol{\chi}(0, \mathbf{X}) = \mathbf{X}, \quad \text{(II.31)}$$

in which we regard $\mathbf{g}$, ε, and $\mathbf{W}$ as known functions of t. If we let $\mathbf{R}$ be the orthogonal tensor-valued function of time that solves the differential equation and initial condition

$$\dot{\mathbf{R}} = -\mathbf{R}\mathbf{W}, \quad \mathbf{R}(0) = \mathbf{1}, \quad \text{(II.32)}$$

---

[3] This result was obtained by W.-L. YIN in research done at the Johns Hopkins University in 1969–1970. His unpublished manuscript does not contain our solution (II.34) but was helpful to us in our study of the problem.

(v) HOMO-ENERGETIC DILATATION

then it is a simple matter to verify that the solution of (II.31) is

$$\chi(t, \mathbf{X}) = \mathbf{x}_0 + \left(\frac{\varepsilon(t)}{\varepsilon(0)}\right)^{-\frac{1}{2}\lambda} \mathbf{R}(t)^{\mathrm{T}}(\mathbf{X} - \mathbf{x}_0) + \int_0^t \left(\frac{\varepsilon(t)}{\varepsilon(s)}\right)^{-\frac{1}{2}\lambda} \mathbf{R}(t)^{\mathrm{T}} \mathbf{R}(s) \mathbf{g}(s)\, ds. \quad \text{(II.33)}$$

As $\mathbf{R}$ is orthogonal, we can easily invert χ to obtain $\mathbf{X} = \chi^{-1}(t, \mathbf{x})$, and we find that (II.30) becomes

$$\rho(t, \mathbf{x}) = \left(\frac{\varepsilon(t)}{\varepsilon(0)}\right)^{\frac{3}{2}\lambda} \rho_0 \left(\mathbf{x}_0 + \left(\frac{\varepsilon(t)}{\varepsilon(0)}\right)^{\frac{1}{2}\lambda} \mathbf{R}(t) \mathbf{r} - \int_0^t \left(\frac{\varepsilon(s)}{\varepsilon(0)}\right)^{\frac{1}{2}\lambda} \mathbf{R}(s) \mathbf{g}(s)\, ds\right). \quad \text{(II.34)}$$

We may interpret the foregoing results as follows. If we prescribe the functions ε, $\mathbf{g}$, $\mathbf{W}$, and ρ_0, then the velocity field is determined by (II.24). Since $\mathbf{W}$ is given, we can use (II.32) to find the orthogonal tensor $\mathbf{R}$, and then the field of density is given by (II.34). Thus the problem of finding all possible homo-energetic dilatations of a Stokes–Kirchhoff gas is completely solved. The condition (II.29), which expresses the balance of linear momentum, simply provides us with formulae for determining the potential U and the skew tensor $\mathbf{B}$ such that the body force $\mathbf{b}$ given by (II.28) will produce the homo-energetic dilatation already specified.

If, on the other hand, the body force is prescribed first, (II.29) provides certain restrictions on the functions ε, $\mathbf{g}$, $\mathbf{W}$, ϕ, and ρ_0, and so we may not prescribe these arbitrarily. To illustrate the form of these restrictions, we consider the case in which the body force vanishes: $U = 0$, $\mathbf{B} = \mathbf{0}$. We shall work out the details only when $\lambda = 1$, since 1 will be found later to be the only value of λ that the kinetic theory allows. The results we give now are all due to BOLTZMANN. Since $\mathbf{B} = \mathbf{0}$, we see from (II.29)$_2$ that

$$\mathbf{W} = \varepsilon(t)\mathbf{S}, \quad \text{(II.35)}$$

the constant $\mathbf{S}$ being an arbitrary skew tensor. Since $U = 0$, we may satisfy (II.29)$_1$ by interpreting it as giving ρ as a function of ε, $\mathbf{g}$, $\mathbf{W}$, and ϕ. By substituting this formula for ρ into (II.26) we obtain after some calculation

$$(\varepsilon^{-1})^{\cdots} r^2 + 4(\varepsilon^{-1}\mathbf{g})^{\cdot\cdot} \cdot \mathbf{r} - 4(\varepsilon^{-1}\phi - \tfrac{1}{2}\varepsilon^{-1}g^2 - \log \varepsilon)^{\cdot} = 0. \quad \text{(II.36)}$$

This equation requires that $(\varepsilon^{-1})^{\cdots} = 0$ and $(\varepsilon^{-1}\mathbf{g})^{\cdot\cdot} = \mathbf{0}$; with ε and $\mathbf{g}$ so selected, (II.36) becomes a differential equation for ϕ. We may write the general solutions of these three differential equations in the forms

$$\varepsilon^{-1} = \alpha^{-1}\left[\left(1 + \frac{t}{T}\right)^2 + \beta\frac{t}{T}\right],$$

$$\varepsilon^{-1}\mathbf{g} = \left(1 + \frac{t}{T}\right)\boldsymbol{\mu} + \mathbf{v}, \quad \text{(II.37)}$$

$$\varepsilon^{-1}\phi = \tfrac{1}{2}\varepsilon^{-1}g^2 + \log\frac{\varepsilon}{\alpha} + \delta,$$

α, β, T, δ, μ, and $\mathbf{v}$ being constants. As $\varepsilon > 0$, it is necessary that $\alpha > 0$. Moreover, the coefficient of t in ε^{-1} is $(\beta + 2)/T\alpha$, and so by allowing T to be either positive or negative we may assume with no loss of generality that $\beta + 2 > 0$. The fields ρ and $\mathbf{u}$ can be calculated now directly from (II.24), (II.29)$_1$, (II.35), and (II.37):

$$\mathbf{u} = \frac{1}{(t+T)^2 + \beta tT}[(t+T)(\mathbf{r} + \alpha T\boldsymbol{\mu}) + \tfrac{1}{2}\beta T\mathbf{r} + \alpha T^2\mathbf{v} + \alpha T^2 \mathbf{Sr}],$$

$$\rho = \left[\left(1 + \frac{t}{T}\right)^2 + \beta\frac{t}{T}\right]^{-\tfrac{3}{2}} \exp\left\{\frac{3}{2}\frac{1}{(t+T)^2 + \beta tT}\left[(t+T)^2(\tfrac{1}{2}\alpha\mu^2 + \delta)\right.\right.$$
$$+ (t+T)(\alpha\boldsymbol{\mu}\cdot\mathbf{v} + \beta\,\delta T - \tfrac{1}{2}\beta\boldsymbol{\mu}\cdot\mathbf{r} + \mathbf{v}\cdot\mathbf{r} - \alpha T\mathbf{S}\boldsymbol{\mu}\cdot\mathbf{r}) + \tfrac{1}{2}\alpha v^2 - \beta\,\delta T^2$$
$$\left.\left.+ \beta T\boldsymbol{\mu}\cdot\mathbf{r} + \tfrac{1}{2}\beta T\mathbf{v}\cdot\mathbf{r} - \alpha T^2\mathbf{Sv}\cdot\mathbf{r} + \frac{(4+\beta)\beta}{8\alpha}r^2 + \tfrac{1}{2}\alpha T^2|\mathbf{Sr}|^2\right]\right\}.$$
(II.38)

When ρ is assumed to be a function of t only, these results simplify greatly, and as so simplified they will be of special importance to us in Chapter XV. Since $\beta + 2 > 0$, we see from (II.38)$_2$ that ρ is independent of $\mathbf{r}$ if and only if $\beta = 0$, $\mathbf{v} = \mathbf{0}$, and $\mathbf{S} = \mathbf{0}$. Hence the only homochoric flows included in (II.38) are of the form

$$\rho = \rho(0)\left(1 + \frac{t}{T}\right)^{-3},$$

$$\mathbf{u} = (t+T)^{-1}\mathbf{r}, \qquad (\text{II.39})$$

$$\varepsilon = \varepsilon(0)\left(1 + \frac{t}{T}\right)^{-2}.$$

Since $\gamma = \tfrac{5}{3}$ when $\lambda = 1$, the field of pressure defined by (II.4) becomes

$$p = p(0)\left(1 + \frac{t}{T}\right)^{-5}. \qquad (\text{II.40})$$

Even this special solution shows that the behavior of a Stokes–Kirchhoff gas in homo-energetic dilatation may be various. If $T > 0$, then u, ε, p, and ρ all tend to 0 at $\mathbf{x}$ as $t \to \infty$. The gas flows radially outward from $\mathbf{x}_0$, ever more slowly at any given place except $\mathbf{x}_0$. Since $\rho \to 0$, we may picture the gas as flowing out of every bounded part of space and accumulating at ∞ as $t \to \infty$. On the other hand, if $T < 0$, the solution exists only in $[0, -T)$; the gas flows inward toward $\mathbf{x}_0$, and ever more rapidly at any given place except $\mathbf{x}_0$; and as $t \to -T$ the quantities u, ε, p, and ρ all tend to ∞. We may picture all of the gas as accumulating at $\mathbf{x}_0$ after the lapse of time $-T$.

Example 2. *Homo-energetic affine flows in general.* If at each time the velocity field is assumed to be an affine function of position, the conditions of

(v) HOMO-ENERGETIC AFFINE FLOWS

homogeneity (II.18) enable us to solve the equations of balance explicitly, irrespective of constitutive relations. The following formal conditions define a *homo-energetic affine flow*[4]:

$$\mathbf{u} = \mathbf{G}(t)\mathbf{r} + \mathbf{g}(t), \qquad u_k = G_{ka}(t)r_a + g_k(t); \qquad (\text{II.41})$$

$$\text{div } \mathbf{q} = 0; \qquad (\text{II.18})_{2\,r}$$

$$\rho, \varepsilon, \text{ and } \mathbf{P} \text{ are functions of } t \text{ alone}; \qquad (\text{II.42})$$

$$\mathbf{b} = \text{const.} \qquad (\text{II.43})$$

Many of the few explicit solutions known in the dynamics of viscous gases pertain to this class of flows; in Chapters XIV and XV we shall see that also all of the explicit solutions presently known in the kinetic theory that are not dissipationless pertain to this same class. It subsumes also the subclass of homo-energetic dilatations treated at the very end of the preceding example, those in which we assumed that $\mathbf{b}$ was null and ρ was a function of t only, so (II.39) followed.

Here we provide for this whole class a groundwork that applies to the Stokes-Kirchhoff theory and the kinetic theory alike.

We expect the reader to bear in mind that (II.41) makes $\mathbf{E}$ and E functions of t alone.

Balance of mass. Because ρ has been assumed to be a function of t alone, we may integrate (I.1) as follows:

$$\rho = \rho(0) \exp\left(-\int_0^t E \, ds\right). \qquad (\text{II.44})$$

Balance of linear momentum. Since ρ and ε are functions of t alone, by use of (I.6) and (II.43) we reduce (II.19)$_1$ to the following ordinary differential equations for determining $\mathbf{G}$ and $\mathbf{g}$:

$$\begin{aligned} \dot{\mathbf{G}} + \mathbf{G}^2 &= \mathbf{0}, \\ \dot{\mathbf{g}} + \mathbf{G}\mathbf{g} &= \mathbf{b}. \end{aligned} \qquad (\text{II.45})$$

The general solution of this system is

$$\begin{aligned} \mathbf{G} &= [\mathbf{1} + t\mathbf{G}(0)]^{-1}\mathbf{G}(0), \\ \mathbf{g} &= [\mathbf{1} + t\mathbf{G}(0)]^{-1}[\tfrac{1}{2}t^2\mathbf{G}(0)\mathbf{b} + t\mathbf{b} + \mathbf{g}(0)]. \end{aligned} \qquad (\text{II.46})$$

The solution exists for such t as render $\mathbf{1} + t\mathbf{G}(0)$ invertible. If $\mathbf{G}(0)$ has positive proper numbers, let the largest of these be denoted by $-1/t_-$; if it has none, we

---

[4] The analysis here extends somewhat the treatment of the purely mechanical aspects by C. TRUESDELL, "The simplest rate theory of pure elasticity", *Communications on Pure and Applied Mathematics* **8**, 123-132 (1955). *Cf.* also GALKIN [1958] and NIKOL'SKII [1965, **2**, §2].

write $t_- = -\infty$; if $\mathbf{G}(0)$ has negative proper numbers, let the smallest of these be denoted by $-1/t_+$; if it has none, we write $t_+ = +\infty$. Then the interval of time in which the solution (II.46) exists is (t_-, t_+). At any finite endpoint of that interval, $\mathbf{G}$ ceases to exist, and at least one of its components tends to $\pm\infty$. If $\mathbf{G}(0)$ is nilpotent, its only proper number is 0, so *the interval in which* (II.46) *exists is* $(-\infty, +\infty)$; in fact

$$\mathbf{G} = [\mathbf{1} - t\mathbf{G}(0)]\mathbf{G}(0), \tag{II.47}$$

and

$$\mathbf{G}^2 = \mathbf{G}(0)^2. \tag{II.48}$$

All the principal invariants of a nilpotent tensor are null. In particular, $E = 0$, so *the flow is isochoric*.

Directly from (II.45) we see that *for the flow to be steady, it is necessary and sufficient that*

$$\mathbf{G}^2 = \mathbf{0}. \tag{II.49}$$

This subclass of flows satisfies both (II.48) and (II.47) trivially.

Balance of energy. By use of (II.4) we may reduce (II.19)$_2$ to the form

$$\dot\varepsilon/\varepsilon = -(\gamma - 1)(E + \mathbf{P} \cdot \mathbf{E}/p). \tag{II.50}$$

Because ε, E, $\mathbf{E}$, p, and $\mathbf{P}$ are functions of t alone, we see from (II.50) that

$$\varepsilon = \varepsilon(0) \exp\left[-(\gamma - 1)\int_0^t (E + \mathbf{P} \cdot \mathbf{E}/p)\, ds\right]. \tag{II.51}$$

This formula exhibits the opposing effects of expansion and of work done against internal friction. We generally expect that friction shall use up work. Then $\mathbf{P} \cdot \mathbf{E} < 0$. If so, then $\varepsilon(t) \to \infty$ as $t \to \infty$ unless E becomes large enough to make $E + \mathbf{P} \cdot \mathbf{E}/p$ a positive quantity. In a flow of this kind that does not entail very rapid positive expansion, the temperature will tend to ∞ as time goes on. This phenomenon is easy to understand, for both of the ordinary mechanisms for drawing off heat are absent: the temperature gradient is null and there are no sinks of heat: grad $\varepsilon = \mathbf{0}$, and div $\mathbf{q} = 0$.

While E and $\mathbf{E}$ are determined from (II.46), we have as yet no means of determining $\mathbf{P}$. Thus (II.51) does not generally deliver ε explicitly.

The reader will easily verify that in a dilatation for which ρ is a function of time alone the solution (II.46)$_1$ is equivalent to $E = E(0)/[1 + \frac{1}{3}E(0)t]$, by use of which we may obtain generalizations of (II.39) to arbitrary values of γ.

The foregoing results make no use of constitutive relations beyond (II.4).

Before we specialize the results to the Stokes-Kirchhoff theory, we note some simple parameters that we can define in terms of an arbitrary function μ of ρ and ε. We assign to μ the physical dimensions of mass/(length)(time), and later we shall identify it with the shear viscosity. For the time being, however, we continue to avoid all constitutive relations other than (II.4).

The first parameter is τ, a time defined as follows:

$$\tau \equiv \frac{\mu}{p} = \frac{\mu}{(\gamma - 1)\rho\varepsilon}. \tag{II.52}$$

It will be of great use in the kinetic theory; we shall encounter it in that context in Chapter XIII and in several chapters thereafter. We regard it as a given function of ρ and ε. Since ρ and ε are functions of t alone in homo-energetic affine flows, so also is τ. If μ happens to be proportional to the temperature, say

$$\mu = C\varepsilon, \tag{II.53}$$

then we can use (II.44) to determine the time-dependence of τ explicitly:

$$\tau = \frac{C}{(\gamma - 1)\rho} = \tau(0) \exp\left(\int_0^t E \, ds\right). \tag{II.54}$$

The second important parameter is the *truncation number* Tr, defined as follows[5]:

$$\text{Tr} \equiv \sqrt{2}\,\tau\,|\mathbf{E}|. \tag{II.55}$$

Tr has appeared in effect in (I.21); it serves as a dimensionless measure of the rate of dissipation of energy through distorsion. We shall encounter it again in Chapter IV. In a homo-energetic affine flow the factor $\sqrt{2}\,|\mathbf{E}|$ is a known function of t, determined by (II.46)$_1$. If (II.53) holds, then Tr is also a known function of t, determined by $\rho(0)$ and $\mathbf{G}(0)$.

At last we are ready to bring the constitutive relations of the Stokes-Kirchhoff theory to bear. If we substitute (I.7)$_2$ into (II.51) and then use (II.52) and (II.55), we obtain

$$\varepsilon = \varepsilon(0) \exp\left[-(\gamma - 1)\int_0^t (E - \text{Tr}^2/\tau) \, ds\right]. \tag{II.56}$$

By good fortune, the special case when (II.53) holds is the very one that is most useful for comparison with results in the kinetic theory. In that special case, as we have seen, the integrand in (II.56) becomes a known function of t. Therefore (II.56) becomes an explicit determination of ε. We have established the following **Theorem.** *For a Stokes-Kirchhoff gas whose viscosity is proportional to the*

---

[5] Several numbers of this kind, with different numerical factors and sometimes based upon $\mathbf{E} + \frac{1}{3}E\mathbf{1}$ rather than upon $\mathbf{E}$, have been introduced by TRUESDELL, the earliest being given by Eq. (13) of "On the differential equations of slip flow", *Proceedings of the National Academy of Sciences* (U.S.A.) **34**, 342–347 (1948). The symbol $\mathfrak{Tr}$ defined by TRUESDELL [1969, *2*] is not the same as the Tr defined here. The name "truncation number" is motivated by the fact that if this number is small, so is every component of $\mathbf{E}$, and thus a series in powers of those components may be truncated, with good approximation for small values of $|\mathbf{E}|$, after the first term that does not vanish.

temperature, algebraic operations and two quadratures suffice to exhibit every homo-energetic affine flow. The initial values $\rho(0)$, $\mathbf{G}(0)$, $\mathbf{g}(0)$, and $\varepsilon(0)$ determine ρ, $\mathbf{u}$, and ε uniquely. The solution exists in a non-empty interval of time which contains $t = 0$ and is determined by $\mathbf{G}(0)$. The solution is set forth in (II.44), (II.46), and (II.56).

For the Stokes-Kirchhoff theory the assumptions we have made are pleonastic: From the defining relations (I.7), (II.9), and (I.12) we see that it is unnecessary to assume the conditions (II.18)$_2$ and (II.42)$_3$, for they are consequences of (II.41) and (II.42)$_{1,2}$.

We proceed now to work out the details for two simple and illuminating special classes of homo-energetic affine flows. For purposes of comparison with the kinetic theory we maintain a divided analysis, presenting in each case first the consequences that follow from no more than the equations of balance and the defining relation (II.4) of an ideal gas with constant specific heats, and only in a consequent and more detailed study descending to use of the constitutive relations (I.7), (I.12), and (II.9).

Example 3. *Homo-energetic simple shearing.* We consider the velocity field given by

$$u_1 = Kx_2, \qquad u_2 = 0, \qquad u_3 = 0, \qquad K = \text{const.} \tag{II.57}$$

(Figure II.1). This is a steady isochoric flow:

$$\|G_{km}\| = \begin{Vmatrix} 0 & K & 0 \\ 0 & 0 & 0 \\ 0 & 0 & 0 \end{Vmatrix}, \qquad \|E_{km}\| = \begin{Vmatrix} 0 & \tfrac{1}{2}K & 0 \\ \tfrac{1}{2}K & 0 & 0 \\ 0 & 0 & 0 \end{Vmatrix}, \tag{II.58}$$

so $\mathbf{G}^2 = \mathbf{0}$ and $E = 0$. The constant K, which without loss of generality can be taken as positive, is called the *amount of shearing*. We suppose that $\mathbf{b} = \mathbf{0}$ and that the essential conditions (II.18) and (II.42) are satisfied. Then we can apply the solution of the equations of balance that (II.44), (II.47), and (II.51) express.

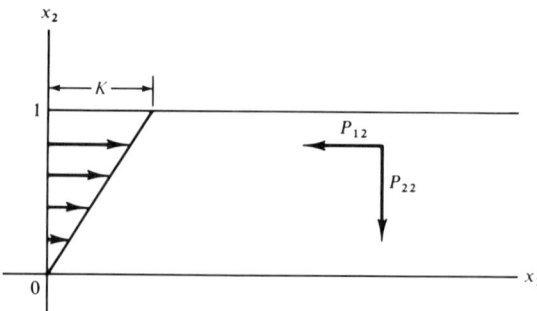

Figure II.1

(v) HOMO-ENERGETIC SIMPLE SHEARING

The first of these shows that ρ is constant altogether; the second is automatically satisfied; and the third becomes

$$\rho\dot{\varepsilon} = -KP_{12}. \tag{II.59}$$

Thus far we have not used any constitutive relation except (II.4).

Before we specialize the results to the Stokes–Kirchhoff theory, we note some simple parameters defined in terms of μ. The first of these is τ as defined generally by (II.52). Another is the dimensionless number T defined as follows for this flow:

$$\mathsf{T} \equiv \tau K. \tag{II.60}$$

It turns out that $\mathsf{T} = \mathsf{Tr}$ as the latter is defined generally by (II.55). In terms of T we may write (II.59) as follows:

$$\tau\dot{\varepsilon}/\varepsilon = \tau\dot{p}/p = -(\gamma - 1)\mathsf{T}P_{12}/p. \tag{II.61}$$

This form of the differential equation of energy does not call upon the Stokes–Kirchhoff theory; it is merely a preparatory rearrangement of (II.59).

We now make the specializing assumption (II.53). Since $E = 0$, from (II.54) we see that $\tau = $ const., and from (II.55) or (II.60) it follows that $\mathsf{T} = $ const.

Everything we have done so far concerning this example follows from the equations of balance, from (II.4), from definitions in terms of a function μ, and from the assumption that $\mu \propto \varepsilon$—nothing more.

At last we are ready to invoke the constitutive equations of the Stokes–Kirchhoff theory through their consequence (II.56). Thence we at once obtain the *explicit solution for homo-energetic shearing of a Stokes–Kirchhoff gas whose viscosity is proportional to the temperature*:

$$p/p(0) = \varepsilon/\varepsilon(0) = e^{Rt/\tau}, \qquad R \equiv (\gamma - 1)\mathsf{T}^2. \tag{II.62}$$

Because of the constitutive relation $(I.7)_2$, for this flow

$$P_{12}/p = P_{21}/p = -\mathsf{T}, \tag{II.63}$$

and of course the other components of **P** vanish.

Example 4. Homo-energetic extension. We consider now the velocity field given by

$$u_1 = \xi(t)(x_1 - x_0), \qquad u_2 = 0, \qquad u_3 = 0, \tag{II.64}$$

x_0 being a constant. In such a flow the fluid-points move along lines parallel to the x_1-axis. The points on each plane $x_1 = $ const. move with the same velocity, either toward or away from the plane $x_1 = x_0$. For this flow

$$\|G_{km}\| = \xi \begin{Vmatrix} 1 & 0 & 0 \\ 0 & 0 & 0 \\ 0 & 0 & 0 \end{Vmatrix}, \qquad \|E_{km}\| = \xi \begin{Vmatrix} \tfrac{2}{3} & 0 & 0 \\ 0 & -\tfrac{1}{3} & 0 \\ 0 & 0 & -\tfrac{1}{3} \end{Vmatrix}, \qquad E = \xi. \tag{II.65}$$

In the preceding example **G** was steady and nilpotent; in this example it is neither. We suppose that **b** = **0** and that the essential conditions (II.18) and (II.42) are satisfied. Then we can apply the solution of the equations of balance that (II.44), (II.46), and (II.51) express. According to the second of these,

$$E = \xi(t) = \frac{1}{t+T}, \qquad (\text{II.66})$$

T being a constant which (II.46) forbids from being 0. Placing this result into (II.64), we see that if $T > 0$, the gas flows away from the plane $x_1 = x_0$, but ever more slowly, and $u \to 0$ as $t \to \infty$, while if $T < 0$, the gas flows toward the plane $x_1 = x_0$, and ever more swiftly, and $u \to \infty$ as $t \to -T$. The former case corresponds to a slow expansion of the gas into all of space; the latter, to condensation of all the gas. These conclusions are illuminated by the behavior of ρ, which follows at once from (II.44):

$$\rho = \rho(0)\left(1 + \frac{t}{T}\right)^{-1}. \qquad (\text{II.67})$$

In the expansive flows the gas grows steadily rarer: At every fixed place, $\rho \to 0$ as $t \to \infty$, so we may say that ultimately no gas remains. In the condensing flows, on the other hand, $\rho \to \infty$ at every place as $t \to -T$. By comparing (II.66) and (II.67) we see that

$$\rho/\xi = T\rho(0). \qquad (\text{II.68})$$

This fact reflects a more general property of affine flows. Directly from (I.1) and (II.45)$_1$ we see that

$$\rho(E/\rho)^{\cdot} = (\operatorname{tr} \mathbf{G})^2 - \operatorname{tr} \mathbf{G}^2, \qquad (\text{II.69})$$

so $E/\rho = E(0)/\rho(0)$ if and only if $(\operatorname{tr} \mathbf{G})^2 = \operatorname{tr} \mathbf{G}^2$.

The only remaining condition imposed by the equations of balance is (II.50), which for this flow reduces to

$$\frac{\dot{\varepsilon}}{\varepsilon} = -(\gamma - 1)\left(1 + \frac{P_{11}}{p}\right)\xi = -(\gamma - 1)\frac{1 + P_{11}/p}{t + T}. \qquad (\text{II.70})$$

Thus far we have not used any constitutive relation except (II.4).

Before we specialize the results to the Stokes-Kirchhoff gas, we note some simple parameters defined in terms of μ. The first of these is τ as defined generally by (II.52). Because of (II.67) it follows that

$$\tau\xi = \frac{1}{(\gamma - 1)T\rho(0)}\frac{\mu}{\varepsilon}. \qquad (\text{II.71})$$

Another convenient parameter is T, defined as follows:

$$\mathsf{T} \equiv \tfrac{4}{3}\tau\xi, \qquad (\text{II.72})$$

(v) HOMO-ENERGETIC EXTENSION

a dimensionless quantity. We may express in terms of it the parameter Tr, which is defined generally by (II.55):

$$\mathsf{T}r^2/\tau = \mathsf{T}\xi. \tag{II.73}$$

Thus

$$E - \mathsf{T}r^2/\tau = (1 - \mathsf{T})\xi,$$
$$= \frac{1-\mathsf{T}}{t+T}. \tag{II.74}$$

The formulae (II.71)–(II.74) do not call upon the Stokes–Kirchhoff theory but express merely preparatory rearrangements.

If we now make the specializing assumption (II.53), we conclude from (II.66) and (II.54) that

$$\tau = \tau(0)\left(1 + \frac{t}{T}\right). \tag{II.75}$$

From (II.72) we conclude that $\mathsf{T} = $ const.; indeed

$$\mathsf{T} = \frac{4}{3}\frac{\tau(0)}{T}. \tag{II.76}$$

At last we are ready to invoke the Stokes–Kirchhoff constitutive equations through their consequence (II.56). Thence we obtain the *explicit solution for homo-energetic extension of a Stokes–Kirchhoff gas whose viscosity is proportional to the temperature*:

$$\varepsilon = \varepsilon(0)\left(1 + \frac{t}{T}\right)^{R+1}, \tag{II.77}$$

in which

$$R + 1 \equiv -(\gamma - 1)(1 - \mathsf{T}). \tag{II.78}$$

Also

$$p = p(0)\left(1 + \frac{t}{T}\right)^{R}. \tag{II.79}$$

Directly from $(I.7)_2$, $(II.65)_2$, and (II.72) we see that

$$P_{11} = -\mathsf{T}p, \qquad P_{22} = P_{33} = \tfrac{1}{2}\mathsf{T}p; \tag{II.80}$$

all other components of **P** vanish.

In this example the results are of two kinds, according to the signs of T and $1 - \mathsf{T}$: As t approaches the upper limit of the interval of existence

$$\varepsilon(t) \to \begin{cases} 0 & \text{if } T(1-\mathsf{T}) > 0, \\ \infty & \text{if } T(1-\mathsf{T}) < 0. \end{cases} \tag{II.81}$$

In the former case expansion predominates over internal friction: The density of the gas decreases so rapidly that the quantity of gas available for internal friction to work upon is insufficient to receive an increase of temperature thereby. In the latter case just the reverse occurs, but the approach to ∞ is then infinitely slower than it is in simple shearing. If $\mathsf{T} = 1$, the effects of expansion and internal friction are just balanced, and $\varepsilon = $ const.

The three examples we have derived and analysed illustrate different aspects of classical gas dynamics. The dilatations, being dissipationless, are the very flows of a Stokes–Kirchhoff gas that are unaffected by viscosity. For them, the Stokes–Kirchhoff theory falls into agreement with the Euler–Hadamard theory of perfect fluids. Simple shearings, on the contrary, are indeterminate within the theory of perfect fluids; it is only a mechanism of internal friction that can provide a non-zero shear stress and thereby lead to a specific theory of the sliding of one part of a fluid body along another. In simple shearing we see the operation of shear viscosity, unhindered and unaided by the effects of expansion. The result is an exponential increase of temperature as $t \to \infty$. In an extension we see both expansion and shear viscosity at work. The initial conditions determine whether one or the other shall predominate or the two shall balance one another. Accordingly, the temperature tends to 0 or to ∞ or remains constant. The rates are far slower than exponential.

Although homo-energetic affine flows are so simple as to be almost degenerate, our treatment of them illustrates the usual way in which specific problems of continuum mechanics are set up and solved. First, the field equations expressing the balance of mass, momentum, and energy are laid down. These lead to the simple and specific conclusions (II.44), (II.46), and (II.51). Second, constitutive equations defining a particular material are invoked. These when substituted into the equations of balance lead to *equations of motion*. For homo-energetic affine flows of a Stokes–Kirchhoff gas these reduce to (II.56) alone. The third step is to *integrate the equations of motion* for arbitrary initial conditions. For a Stokes–Kirchhoff gas in which (II.53) holds, the relation (II.56) reduces to a quadrature of known functions.

Although the kinetic theory deals with fields having just the same interpretations as ρ, **u**, ε, p, **P**, and **q**, it approaches specific problems in an entirely different way. We shall see what the kinetic theory predicts about homo-energetic simple shearing in Chapter XIV and what it predicts about homo-energetic extension in Chapter XV. Homo-energetic dilatation according to the kinetic theory will be studied in Chapters X and XV. The results we have presented in this chapter are so arranged as to provide the groundwork for the corresponding calculations in the kinetic theory as well as to allow immediate contrast of the kinetic results with those that the Stokes–Kirchhoff theory has been shown here to deliver.

Part B

Basic Structures of the Kinetic Theory

Chapter III

The Molecular Density, the Definitions of Gross Fields, and the Equation of Evolution

We begin now our study of the kinetic theory of MAXWELL. This kinetic theory is the only one that we shall examine, and all the rest of this book will concern it and nothing else. The contents of the two preceding chapters will serve us for motivation and reference only. We shall often use again such symbols as **u** and **P** and **q**, symbols which have been used in the chapters on continuum thermomechanics, but they will be defined afresh in terms of kinetic concepts alone[1]. The duplicating notation will serve to motivate definitions and to promote correspondence between the two theories, no more. In strict logic continuum thermomechanics and the kinetic theory have only three primitive concepts in common: time t, place **x**, and body force **b** per unit mass.

The development will always be self-sufficient. In this part of the book we present the basic definitions and axioms that make up the kinetic theory. Following modern mathematical practice, once the primitive quantities and definitions have been set down we develop the properties of the kinetic theory that flow from them alone, disregarding alternative definitions and more general concepts. This will allow us to see at a glance the roots from which stem the properties of the kinetic gas that we shall deduce in subsequent parts of the theory.

(i) *The molecular density and the number density*

The kinetic theory of gases is a mathematical model in which a gas is envisaged as a collection of molecules, each tracing out in the course of time a trajectory in space determined by such forces as the molecule may experience.

---

[1] MAXWELL [1867, *passim*; *e.g.* Eqs. (63) and (71)]. Some of the definitions and concepts derive from earlier work by MAXWELL himself and by various predecessors. *Cf.* TRUESDELL [1975].

The basic variables that we use in discussing molecular motions are a *time t*, a *place* **x**, and a *velocity* **v**:

$t \in \mathcal{T}$, a 1-dimensional Euclidean space,

$\mathbf{x} \in \mathcal{E}$, a 3-dimensional Euclidean space,

$\mathbf{v} \in \mathcal{V}$, a 3-dimensional inner-product space.

The time t, the co-ordinates x_m of the place **x**, and the components v_k of the velocity **v** are assigned physical dimensions of time, length, and length/time, respectively. So as to represent a gas whose molecules are alike, sometimes called a *simple gas*, we introduce a constant $\mathfrak{m}$ bearing the units of mass. We call it the *molecular mass*. In this book we treat only a simple gas.

We will not follow the motions of individual molecules; rather, we shall suppose that their positions and velocities are distributed randomly, according to a specific rule. This rule is described by means of a *molecular density F*, a real-valued function defined and measurable over $\mathcal{T} \times \mathcal{E} \times \mathcal{V}$, integrable over $\mathcal{V}$.

If $\mathcal{B} \subset \mathcal{E}$ and $\mathcal{U} \subset \mathcal{V}$, then

$$\int_{\mathcal{U}} \int_{\mathcal{B}} F(t, \mathbf{x}, \mathbf{v}) \, d\mathbf{x} \, d\mathbf{v}$$

is interpreted as the *expected number* or *average number* of molecules in $\mathcal{B}$ with velocities in $\mathcal{U}$ at the time t. Here $d\mathbf{x}$ and $d\mathbf{v}$ denote volume measures in $\mathcal{E}$ and $\mathcal{V}$, respectively. Following this interpretation, we assume that

$$F \geq 0, \tag{III.1}$$

and assign to F units of $(\text{time})^3/(\text{length})^6$. In this book we shall need to use only averages over all possible molecular velocities. That is, $\mathcal{U} = \mathcal{V}$. Henceforth we shall write $\int$ for $\int_{\mathcal{V}} \cdots d\mathbf{v}$.

The *number density* $n(t, \mathbf{x})$ is the value of the scalar field defined as follows:

$$n \equiv \int F. \tag{III.2}$$

Of course (III.1) implies that $n \geq 0$. Let $\mathcal{D} \subset \mathcal{T} \times \mathcal{E}$ be the set of places and times for which n exists and is positive:

$$n > 0. \tag{III.3}$$

We call $\mathcal{D} \times \mathcal{V}$ the *essential domain* of F. Then n is a positive scalar field over $\mathcal{D}$. For each pair $(t, \mathbf{x}) \in \mathcal{D}$ the function $\mathbf{v} \mapsto F(t, \mathbf{x}, \mathbf{v})/n(t, \mathbf{x})$ is a probability density over $\mathcal{V}$.

Hereinafter we assume $\mathcal{D}$ is non-empty and open, and we consider only pairs $(t, \mathbf{x})$ in $\mathcal{D}$. We could, if we so wished, take F as defined and measurable

(ii) Expectations. The thirteen basic fields

only on what we have just called its essential domain. Given such an F, we may easily extend its domain to all of $\mathscr{T} \times \mathscr{E} \times \mathscr{V}$ by assigning it the value 0 at all points not contained in $\mathscr{D} \times \mathscr{V}$.

If g is a real-valued function defined on $\mathscr{D} \times \mathscr{V}$, and if $\mathbf{v} \mapsto F(t, \mathbf{x}, \mathbf{v})g(t, \mathbf{x}, \mathbf{v})$ is integrable over $\mathscr{V}$, then the *average* or *expectation* $\bar{g}$ of the function g at the place $\mathbf{x}$ and at the time t is defined as follows:

$$\bar{g} \equiv \frac{1}{n}\int Fg. \tag{III.4}$$

The expectation $\bar{g}$ is a field over $\mathscr{D}$, and the field $\bar{g}$ is to be interpreted in the same way as are the fields introduced directly in continuum thermomechanics. Our concern is with just a few expectations, certain fields that have immediate and useful counterparts in continuum thermomechanics. Generally we assume that these particular fields are smooth, though we shall see later that such an assumption cannot justly be imposed until more specific knowledge about the kinetic theory than is now available shall have been established.

The first expectation of direct interest is the *mass density* ρ, namely, the expected total mass of the molecules per unit volume at $\mathbf{x}$ and t, whatever be their velocities:

$$\rho \equiv \mathfrak{m}n. \tag{III.5}$$

The second important expectation is the *velocity field* $\mathbf{u}$ of the gas; the value of this field is the *gross velocity*, which is merely the average of the speeds of all molecules at the place $\mathbf{x}$ and the time t:

$$\mathbf{u} \equiv \bar{\mathbf{v}}, \qquad u_k \equiv \bar{v_k}. \tag{III.6}$$

The *random velocity*[2] $\mathbf{c}$ of a molecule is its velocity relative to that of the gas:

$$\mathbf{c} \equiv \mathbf{v} - \mathbf{u}, \qquad c_k \equiv v_k - u_k; \tag{III.7}$$

hence

$$\bar{\mathbf{c}} = \mathbf{0}. \tag{III.8}$$

In the same sense we may speak of the *random momentum* $\mathfrak{m}\mathbf{c}$ and the *random kinetic energy* $\frac{1}{2}\mathfrak{m}c^2$ of a molecule whose velocity is $\mathbf{v}$. The next important expectations are the *pressure tensor field* $\mathbf{M}$ and the *pressure deviator field* $\mathbf{P}$:

$$\mathbf{M} \equiv \rho\overline{\mathbf{c} \otimes \mathbf{c}} = \mathfrak{m}\int F\mathbf{c} \otimes \mathbf{c}, \qquad M_{km} \equiv \rho\overline{c_k c_m} = \mathfrak{m}\int Fc_k c_m, \tag{III.9}$$

$$\mathbf{P} \equiv \mathbf{M} - \tfrac{1}{3}(\operatorname{tr}\mathbf{M})\mathbf{1}, \qquad P_{km} \equiv M_{km} - \tfrac{1}{3}M_{aa}\delta_{km}. \tag{III.10}$$

---

[2] MAXWELL's term is "the motion of agitation".

Thus the field of pressure vectors **Mn** on a surface whose outward unit normal field is **n** is the surface density of expected transport of random momentum per unit time by the molecules that cross the surface, which is conceived as being in motion with the gross velocity **u** of the gas; the components of **Mn** are $M_{ka}n_a$. In particular, the definition (III.9) makes the *pressure tensor symmetric*:

$$\mathbf{M}^T = \mathbf{M}. \tag{III.11}$$

In the same vein, the *energy flux vector* **q** is defined in terms of the expected transport of random kinetic energy per unit time:

$$\mathbf{q} \equiv \tfrac{1}{2}\rho\overline{c^2\mathbf{c}} = \tfrac{1}{2}\mathfrak{m}\int Fc^2\mathbf{c}, \qquad q_k \equiv \tfrac{1}{2}\rho\overline{c^2 c_k} = \tfrac{1}{2}\mathfrak{m}\int Fc^2 c_k. \tag{III.12}$$

The field of *energy flux* on the surface just mentioned is $\mathbf{q} \cdot \mathbf{n}$.

The entire purpose of the kinetic theory is to relate the 13 *scalar fields* ρ, u_k, M_{pm}, *and* q_r *to various circumstances of the kinetic gas.* All these fields are determined by the molecular density F, so if we know F, we know in principle all that we wish to know. Conversely, however, it is plain from the definitions that these few fields cannot determine F. To any one particular set ρ, **u**, **M**, and **q** correspond infinitely many F. Thus F contains and is capable of expressing vastly more information than we shall ever wish to have about the motion of the kinetic gas. In order to answer a particular question, we may use any one of the infinitely many appropriate F; we need not find any of the others. The particular beauty of the kinetic theory lies in this superabundance of solutions and the consequent problem of selecting some one F that is good enough for a given task. In many cases F itself is never found, but only some useful relations among some of the quantities it determines. Viewed in the light of its desired applications, the kinetic theory can never be reduced to problems of boundary values or initial values of any classical type, though the study of such problems may provide handy tools for probing its deeper reaches. We shall come back to this basic conceptual problem many times in the course of this book. Here we have first touched upon it, but we are not yet in any position to describe it clearly, let alone define it.

The velocity field **u** is often called a *flow* of the kinetic gas. We may construct from it the motions of "particles" of the gas in the sense of classical hydrodynamics[3] and so visualize the gross motion as being generated by those particles. To avoid verbal confusion with the molecules of the gas, we shall call these by the name that now in continuum mechanics commonly denotes "particles", namely *fluid-points*. Using in (I.6) the gross velocity **u** of the kinetic gas, we define a *material derivative* just as in continuum mechanics, and we interpret $\dot{f}$ as the time derivative of f apparent to an observer following the motion of the fluid-point presently at **x**.

---

[3] *E.g.* §13 of H. LAMB, *Hydrodynamics*, 2nd ed., Cambridge University Press, 1895, or any later edition.

(ii) ENERGETIC

Some scalars derived from the 13 basic fields are important for their interpretation. One of these[4] is the *mean normal pressure* p:

$$p \equiv \tfrac{1}{3} \operatorname{tr} \mathbf{M} = \tfrac{1}{3} M_{aa} = \tfrac{1}{3}\rho \overline{c^2}. \tag{III.13}$$

Now the expected random kinetic energy per unit mass is given by

$$\varepsilon \equiv \tfrac{1}{2}\overline{c^2}. \tag{III.14}$$

Thus

$$p = \tfrac{2}{3}\rho\varepsilon. \tag{III.15}$$

If we interpret ε as being the *energetic* and assume that the mean pressure p may be confused with the thermodynamic pressure ϖ, this relation agrees with (II.4) from the Stokes-Kirchhoff theory of ideal gases, specialized by giving γ the value $\tfrac{5}{3}$. In the following chapter we shall provide a reason for our having assumed here that $p = \varpi$.

As seems reasonable according to the kinetic model of gas pressure, $p > 0$ and $\varepsilon > 0$ unless almost all molecules are moving at the gross velocity $\mathbf{u}$.

By adopting the definition (III.14) and calling ε the energetic in the kinetic theory, we agree to regard the entire internal energy of the kinetic gas as being neither more nor less than the expectation of the kinetic energy of the random translational motion of the molecules. Since the molecules are conceived as punctual masses, without bulk or internal structure, the definition seems appropriate, provided there be no intermolecular potential energy. We are thus led to perceive that the theory framed in this way cannot represent all the static effects of intermolecular forces. If we look back at the definitions of $\mathbf{M}$ and $\mathbf{q}$, namely (III.9) and (III.12), respectively, we see that we may make the same observation in more general form in connection with those fields, too, since they take into account only the transfer of translational momentum and translational kinetic energy. The theory is thus set up as a model for a *moderately rarefied point gas*. The adjective "moderately" merely reflects the obvious fact that if the gas is so rarefied that molecules never interact, the intended interpretations fail or become uninteresting; indeed, in such a gas the molecules are free bodies, or nearly so, and thus their motions are easily determined by analytical dynamics. It is scarcely fair to use the term "monatomic gas", since the atom of the physicists is not a point. Should we use even so simple a model for an atom as a solid sphere which can exchange spin with other atoms, we should have to take rotational momentum and rotational energy into account also.

These remarks make it plain that the very definitions are so framed as to yield a mathematical model, not an exact physical description of any real gas. The properties of this model may have, should have, and in fact do have interesting interpretations, and in some regards they furnish a good picture of the behavior of some real gases. The model itself is a mathematical one, and its theory forms a major chapter in rational mechanics.

---

[4] This famous and central formula, which here appears as a mere definition, was first obtained by WATERSTON in a great memoir rejected in 1845 by the Royal Society of London but finally published by it in 1893. By then WATERSTON had been dead for ten years. *Cf.* TRUESDELL [1975]. WATERSTON's derivation of his formula for the mean pressure from concepts of momentum transfer through the random motion of molecules if slightly generalized provides motivation for taking the tensor $\mathbf{M}$, as defined by (III.9), to be the kinetic counterpart, for a gas visualized as a numerous assembly of mass-points, of the pressure tensor that continuum mechanics introduces as in essence a primitive quantity.

A real gas is hot or cold, so a model of a gas must be capable of representing the temperature. We can introduce temperature and specific heats in the kinetic theory by the formal process outlined in connection with the Stokes–Kirchhoff theory in Chapter II. This process is now motivated by the mechanical view of heat, according to which temperature is a mere manifestation of the energy of intestine motion, and the "absolute" temperature is 0 when that motion ceases. The only such motion represented in MAXWELL's kinetic theory is random translation, so we must suppose that the temperature θ is proportional to ε:

$$\theta \equiv \varepsilon/\delta_v, \qquad \delta_v > 0. \tag{III.16}$$

Comparison with (II.3) allows us to interpret δ_v as the specific heat at constant specific volume. Since θ is measured in units independent of those of mass, length, and time, these being the only ones the kinetic theory employs, the factor δ_v must cancel those units out. For reasons outside the kinetic theory we assume that δ_v is inversely proportional to $\mathfrak{m}$ and write

$$\delta_v = \tfrac{3}{2} k/\mathfrak{m}; \tag{III.17}$$

k, which is independent of all molecular properties, is called BOLTZMANN's *constant*. Then from (III.15) we have

$$p = \mathfrak{r}\rho\theta, \qquad \mathfrak{r} \equiv k/\mathfrak{m}. \tag{III.18}$$

Therefore, *the mean normal pressure is related to the temperature and the density according to the thermal equation of state* (II.1) *for an ideal gas;* moreover, *the gas constant $\mathfrak{r}$ is inversely proportional to the molecular mass*. Of course, if the pressure tensor is hydrostatic, then $\mathbf{M} = p\mathbf{1}$, so the mean normal pressure is the whole pressure on any surface.

We shall see in Chapter X that certain molecular densities F in the kinetic theory do indeed have pressure tensors that are hydrostatic. In this sense the kinetic theory is consistent with thermostatics.

By writing down an equation of the same form as (II.2)$_2$ we may define γ in terms of δ_v and $\mathfrak{r}$. By use of (III.17) and (III.18) we conclude again that $\gamma = \tfrac{5}{3}$.

We see that all the quantities which figure in the theory of ideal gases may be introduced by appropriate definitions in the kinetic theory. Those definitions imply that *for every molecular density F, the kinetic theory represents an ideal gas with constant specific heats, the ratio of which is* $\tfrac{5}{3}$.

In the kinetic theory everything that concerns temperature is a superfluous concept. It may be brought in by the definition (III.16), but we could equally well omit that definition and all mention of the temperature. Although the temperature is important for visualizing various results and for comparing them with counterparts in continuum mechanics, it is of no use in obtaining them. The kinetic theory is purely mechanical. It concerns velocity, mass, momentum, and energy; everything else about the gas and its condition may be interpreted in terms of properties of the basic mechanical fields.

In this aspect the kinetic theory is like the Stokes–Kirchhoff theory of ideal gases, with a special case of which, as we have seen, it agrees in part. Other points of similarity with the Stokes–Kirchhoff theory will appear in later chapters.

(iii) The higher moments

Looking back at the definitions of the fields of interest, we see that all of them are proportional to certain moments of F, regarded as a function of $\mathbf{v}$. Indeed $\rho/\mathfrak{m}$ is the moment of order 0, and $\mathbf{u}$ is the mean or moment of order 1. The further moments of interest are calculated with respect to the mean. For later use we introduce the *relative moment* $^n\mathbf{M}$ *of order n*:

$$^n\mathbf{M} \equiv \overline{\rho \mathbf{c} \otimes \cdots \otimes \mathbf{c}} = \mathfrak{m} \int F \mathbf{c} \otimes \cdots \otimes \mathbf{c},$$

$$M_{k_1 \ldots k_n} \equiv \overline{\rho c_{k_1} \cdots c_{k_n}} = \mathfrak{m} \int F c_{k_1} \cdots c_{k_n}.$$

(III.19)

It is customary in the kinetic theory to assume that every molecular density F possesses relative moments $^n\mathbf{M}$ of all orders although, as we shall see, such need not be the case.

In the notation (III.19) the components of $^2\mathbf{M}$ are the quantities M_{km} as defined above by (III.9)$_{3,4}$, and according to (III.8) and (III.12)

$$0 = M_k,$$
$$q_k = \tfrac{1}{2} M_{kaa}.$$

(III.20)

The relative moments beyond these first few will be called the *higher moments*. Their interpretation in mechanics is obscure, and it might seem superfluous to mention them. On the contrary, we shall learn that except in a few elementary considerations it is not possible to avoid them. They play a major role in the solution of problems in the kinetic theory, and the great mathematical problem of that theory, seen from one aspect, is to overcome the difficulties to which their presence in the basic equations gives rise.

Here we make only one further remark about the moments, a remark which is obvious but nevertheless of great importance throughout the kinetic theory. Namely, while F determines all its moments uniquely, no finite set of moments determines a unique F. That is, *any given finite set of moments corresponds to infinitely many different molecular densities*.

Nothing presented in this chapter has been proved from the kinetic theory. The axiom on which the theory rests is yet to be formulated. We have seen nothing more than definitions, and our discussion serves only to interpret them and to show what they imply by themselves, no matter what be the molecular density F. Maxwell's kinetic theory of gases takes these definitions as basic; it does not extend them, and it cannot avoid them. They are built into it from the start. If we prefer, we may interpret them as expressions of the physical ideas with which that theory is designed from the outset to be consistent.

(iv) The retrogressors

We have defined the quantities we wish to determine from the molecular density F, but how do we find F? The earliest kinetic theorists put forward special assumptions about the probability that a gas take on a certain condition. Thus they derived or otherwise inferred particular densities F. Different assumptions and conjectures led to different answers, of course. MAXWELL's *first* kinetic theory (1860) was of this kind. To replace such guesswork by a secure basis for accurate and detailed calculation, it was his magnificent idea to lay down a single basic assumption, partly stochastic and partly mechanical in origin, regarding the evolution of F in time. This axiom, which defines MAXWELL's *second*[5] kinetic theory and provides the subject of this book, is the **Maxwell-Boltzmann Integro-differential Equation**. Now we shall take the first step toward specifying that equation.

To do so, we presuppose an underlying picture of the kinetic gas as a collection of molecules moving in the point space $\mathscr{E}$ under the influence of a system of forces. We view this motion as a set of curves in the space $\mathscr{T} \times \mathscr{E} \times \mathscr{V}$; one of these curves passes through each point $(t, \mathbf{x}, \mathbf{v})$. According to the analytical dynamics of systems of mass-points this curve if assumed sufficiently smooth is determined by any particular point $(t, \mathbf{x}, \mathbf{v})$ through which it passes.

We consider the total force acting upon a typical molecule of the gas to have two components. The first is the value of a field $\mathbf{b}$ of *extrinsic* force per unit mass. We picture this force as one that acts upon the gas as a whole; thus we expect to be able to identify it with the body force $\mathbf{b}$ that appears in the equation of balance of momentum of a continuum, and in Chapter IX we shall prove that we may. The second component force is the intermolecular, *mutual* force, the force exerted upon the molecule by its fellows. Analytical dynamics would require us to add these two and so determine the motion of each and every molecule individually by appeal to the differential equations of motion. In the kinetic theory, however, we treat the two forces differently. We take the former just as Newtonian dynamics would require, but we approximate the latter by a schema of stochastic kind.

In this chapter we shall consider only the effect of the extrinsic force $\mathbf{b}$. We shall take it as the value of a field *that depends upon position alone*:

$$\mathbf{b} = \mathbf{b}(\mathbf{x}). \tag{III.21}$$

---

[5] ENSKOG was accurate in always referring to this theory as MAXWELL's *second* kinetic theory. In conversation among physicists and chemists today the term "kinetic theory" usually refers to the first theory and hence has no direct bearing on the subject of this book. TRUESDELL [1975] has described MAXWELL's first theory and the earlier work which it partly subsumed and partly superseded.

(iv) THE RETROGRESSORS   47

If all of $\mathscr{E}$ were empty except for a single molecule subject to the extrinsic force **b**, then, should the molecule at time t occupy the place **x** and have the velocity **v**, its position at another time s would be given by a function $s \mapsto \chi[t, \mathbf{x}, \mathbf{v}](s)$, and χ would be determined as the solution of the single Newtonian equation

$$\chi'' = \mathbf{b}(\chi) \tag{III.22}$$

such as to satisfy the initial conditions

$$\chi[t, \mathbf{x}, \mathbf{v}](t) = \mathbf{x}, \qquad \chi'[t, \mathbf{x}, \mathbf{v}](t) = \mathbf{v}. \tag{III.23}$$

The resulting *extrinsic trajectory* in $\mathscr{T} \times \mathscr{E} \times \mathscr{V}$ would be the curve defined parametrically as follows: $s \mapsto (s, \chi[t, \mathbf{x}, \mathbf{v}](s), \chi'[t, \mathbf{x}, \mathbf{v}](s))$. We assume **b** smooth enough to let us call upon the theorems on differential equations that deliver the mapping χ and make it unique in such intervals of time as we shall consider in this book. For example, if **b** is a constant field and if the region in which the molecules may move is all of $\mathscr{E}$, then

$$\chi[t, \mathbf{x}, \mathbf{v}](s) = \mathbf{x} - (t - s)\mathbf{v} + \tfrac{1}{2}(t - s)^2 \mathbf{b}. \tag{III.24}$$

This case suffices for most purposes of the kinetic theory; very often the still more special case in which $\mathbf{b} = \mathbf{0}$ is considered.

Returning to a general **b**, we shall henceforth regard the extrinsic trajectories as known. With any such trajectory we may associate a transformation which maps the point on it that corresponds to the time t onto the point on this same trajectory that corresponds to the time $t - s$. Specifically, for each real number s we define an operator $\mathbf{r}_s$ on $\mathscr{T} \times \mathscr{E} \times \mathscr{V}$ as follows:

$$\mathbf{r}_s(t, \mathbf{x}, \mathbf{v}) \equiv (t - s, \chi[t, \mathbf{x}, \mathbf{v}](t - s), \chi'[t, \mathbf{x}, \mathbf{v}](t - s)). \tag{III.25}$$

If $s > 0$, $\mathbf{r}_s$ goes back to the point in $\mathscr{T} \times \mathscr{E} \times \mathscr{V}$ that the molecule would have occupied s units of time earlier, had it been subject to the extrinsic force **b** alone. For example, to consider again the case in which **b** is a constant field, we substitute (III.24) into (III.25) and so obtain

$$\mathbf{r}_s(t, \mathbf{x}, \mathbf{v}) = (t - s, \mathbf{x} - s\mathbf{v} + \tfrac{1}{2}s^2 \mathbf{b}, \mathbf{v} - s\mathbf{b}). \tag{III.26}$$

We may call $\mathbf{r}_s$ in general a *retrogressor* induced by **b**. For a sufficiently smooth **b** the uniqueness of χ implies that

$$\mathbf{r}_{s_1} \mathbf{r}_{s_2} = \mathbf{r}_{s_1 + s_2}. \tag{III.27}$$

Moreover, the initial conditions (III.23) show that

$$\mathbf{r}_0 = \text{identity}, \tag{III.28}$$

so (III.27) implies that

$$\mathbf{r}_s \mathbf{r}_{-s} = \text{identity}. \tag{III.29}$$

These conclusions show that the retrogressors $\mathbf{r}_s$ corresponding to **b** form a group acting on $\mathscr{T} \times \mathscr{E} \times \mathscr{V}$.

III. MOLECULAR DENSITY. GROSS FIELDS

If we use (III.22) and (III.23) as the only basis for defining r_s, we presume tacitly that all of the point space $\mathscr{E}$ is accessible to the molecule whose motion is χ. Should we wish to describe a gas confined within a vessel, which we may represent as a given subset $\mathscr{B}$ of $\mathscr{E}$ or of $\mathscr{T} \times \mathscr{E}$, we should have to adjoin to (III.22) and (III.23), considered now as equations valid in the interior of $\mathscr{B}$, certain conditions which would specify the interaction of a molecule with the walls of that vessel. Simple conditions of this kind are easy to formulate, but in the kinetic theory they involve us in a new difficulty, namely, they must be accompanied by corresponding conditions that restrict the values F may assume upon the walls. In Section (iii) of Chapter XI we shall consider such conditions, but otherwise we shall not in this book enter into the complications that walls may introduce.

Using the group r_s, we can define a second group, one that acts upon all functions defined on some subset of $\mathscr{T} \times \mathscr{E} \times \mathscr{V}$. Namely, if G is such a function, we define as follows the transform $\mathfrak{R}_s G$ of G effected by the composition of G with r_s:

$$\mathfrak{R}_s G \equiv G \circ r_s. \tag{III.30}$$

The value $\mathfrak{R}_s G(t, \mathbf{x}, \mathbf{v})$ is found by going back along the extrinsic trajectory through $(t, \mathbf{x}, \mathbf{v})$ to the point corresponding to the time $t - s$ and then evaluating G there. Since r_s is induced by $\mathbf{b}$, we may use (III.25) to obtain the following explicit expression for $\mathfrak{R}_s G$:

$$\mathfrak{R}_s G(t, \mathbf{x}, \mathbf{v}) \equiv G(r_s(t, \mathbf{x}, \mathbf{v})),$$
$$= G(t - s, \chi[t, \mathbf{x}, \mathbf{v}](t - s), \chi'[t, \mathbf{x}, \mathbf{v}](t - s)). \tag{III.31}$$

When $\mathbf{b}$ is constant, so r_s is given by (III.26), then (III.31) reduces to

$$\mathfrak{R}_s G(t, \mathbf{x}, \mathbf{v}) = G(t - s, \mathbf{x} - s\mathbf{v} + \tfrac{1}{2}s^2\mathbf{b}, \mathbf{v} - s\mathbf{b}). \tag{III.32}$$

Of course the domain of $\mathfrak{R}_s G$ can be found by applying r_{-s} to all points in the domain of G.

After these illustrations we return to the general idea and call $\mathfrak{R}_s G$ the *retrogression* of G that corresponds to the given extrinsic trajectories. Without fear of confusion we may call $\mathfrak{R}_s$ by the same name as r_s, the *retrogressor*[6].

The following properties of $\mathfrak{R}_s$ are easily verified:

$$\mathfrak{R}_s(\alpha G + \beta H) = \alpha \mathfrak{R}_s G + \beta \mathfrak{R}_s H, \qquad \mathfrak{R}_s(GH) = (\mathfrak{R}_s G)(\mathfrak{R}_s H),$$
$$\mathfrak{R}_s G \leq \mathfrak{R}_s H \quad \text{if} \quad G \leq H; \tag{III.33}$$

here α and β are real numbers, and G and H are real-valued functions defined on a common subset of $\mathscr{T} \times \mathscr{E} \times \mathscr{V}$. Another property which is easy to establish is

$$\mathfrak{R}_{t-t_0} G(t, \mathbf{x}, \mathbf{v}) = \mathfrak{R}_{t-t_0} G_0(\mathbf{x}, \mathbf{v}) \tag{III.34}$$

---

[6] Special cases of $\mathfrak{R}_s$, in increasing degrees of generality, were introduced by ENSKOG [1928, Eq. (69)], WILD [1951], GRAD [1958, §19], POVZNER [1962, §1], and GLIKSON [1972, *1*, §2].

for each t, $\mathbf{x}$, and $\mathbf{v}$, G_0 being the restriction of G to the time t_0. Indeed, (III.31)$_2$ shows that the left-hand side is simply $G(t_0, \chi[t, \mathbf{x}, \mathbf{v}](t_0), \chi'[t, \mathbf{x}, \mathbf{v}](t_0))$ while the right-hand side is $G_0(\chi[t, \mathbf{x}, \mathbf{v}](t_0), \chi'[t, \mathbf{x}, \mathbf{v}](t_0))$, and by our choice of G_0 these are equal. It is slightly less obvious that *for any G and any real number s such that $(t, \mathbf{x}, \mathbf{v})$ lies in the domain of $\mathfrak{R}_{t-s}G$ the function $(t, \mathbf{x}, \mathbf{v}) \mapsto \mathfrak{R}_{t-s}G(t, \mathbf{x}, \mathbf{v})$ is constant along each extrinsic trajectory.* To prove this fact, let us choose a point $(t_0, \mathbf{x}_0, \mathbf{v}_0)$ and fix it. Then by (III.23) and (III.25) a general point $(t, \mathbf{x}, \mathbf{v})$ on the extrinsic trajectory through this fixed point will be given in terms of $\mathfrak{r}_s$ by

$$(t, \mathbf{x}, \mathbf{v}) = \mathfrak{r}_{t_0 - t}(t_0, \mathbf{x}_0, \mathbf{v}_0). \tag{III.35}$$

This equation defines $\mathbf{x}$ and $\mathbf{v}$ as functions of t such as to make $(t, \mathbf{x}, \mathbf{v})$ lie upon the trajectory through $(t_0, \mathbf{x}_0, \mathbf{v}_0)$. Denoting these functions by $t \mapsto \mathbf{x}(t)$ and $t \mapsto \mathbf{v}(t)$, respectively, we find by use of (III.31)$_1$ and (III.27) that

$$\begin{aligned}
\mathfrak{R}_{t-s}G(t, \mathbf{x}(t), \mathbf{v}(t)) &= G(\mathfrak{r}_{t-s}(t, \mathbf{x}(t), \mathbf{v}(t))), \\
&= G(\mathfrak{r}_{t-s}\mathfrak{r}_{t_0 - t}(t_0, \mathbf{x}_0, \mathbf{v}_0)), \\
&= G(\mathfrak{r}_{t_0 - s}(t_0, \mathbf{x}_0, \mathbf{v}_0)), \\
&= \mathfrak{R}_{t_0 - s}G(t_0, \mathbf{x}_0, \mathbf{v}_0).
\end{aligned} \tag{III.36}$$

Because the last right-hand side does not depend upon t, the proposition stated above in italics is established.

(v) *The equation of evolution*

We are now ready to consider the evolution of the molecular density function F. If a set of molecules is subject to no force but that of the extrinsic field $\mathbf{b}$, that is, if the molecules do not interact with each other, their extrinsic trajectories are their actual trajectories. Then we wish to regard all expectations associated with that set as unchanged in the course of time, provided only we follow those molecules in their motion. That is, we demand that the molecular density be conserved on each actual trajectory. If expressed in terms of the retrogressor $\mathfrak{R}_s$, this assumption is neither more nor less than

$$F(t, \mathbf{x}, \mathbf{v}) = \mathfrak{R}_{t - t_0} F(t, \mathbf{x}, \mathbf{v}) \tag{III.37}$$

for some fixed time t_0 and for every t, $\mathbf{x}$, and $\mathbf{v}$. However, the kinetic theory is designed to represent the effects of molecular interactions, whether through mutual forces acting at a distance or through impacts of the molecules. A treatment in accord with analytical dynamics would require that mutual forces, if they were present, should be added to the right-hand side of (III.22) for each molecule, and that such impacts as might occur should follow assignable rules. Impacts, in general, would destroy the smoothness of the molecular motion.

The kinetic theory rests on a different idea. It represents the interaction of molecules, of whatever kind they be, as having a significant effect only in intervals of time so short as to make molecules seem to appear or disappear instantly at $(t, \mathbf{x}, \mathbf{v})$ at a rate that depends in some specified way on F. This rate is assumed determined by an operator $\mathbb{C}$, called the *collisions operator*, which maps a given F onto another function of t, $\mathbf{x}$, and $\mathbf{v}$, a function we shall denote by $\mathbb{C}F$. Of course the effect of collisions is cumulative as time goes on. The kinetic theory represents the result of this accumulation by an integral along the extrinsic trajectory, the trajectory a molecule would follow, did it not interact with other molecules. Thus, the *basic axiom* of the kinetic theory is the **Equation of Evolution**:

$$F(t, \mathbf{x}, \mathbf{v}) = \mathfrak{R}_{t-t_0} F(t, \mathbf{x}, \mathbf{v}) + \int_{t_0}^{t} \mathfrak{R}_{t-s}(\mathbb{C}F)(t, \mathbf{x}, \mathbf{v}) \, ds. \qquad (\text{III.38})$$

In arriving at this equation we contemplated a specific point $(t, \mathbf{x}, \mathbf{v})$ and worked on the extrinsic trajectory through it. However, this point was arbitrary, and therefore we shall require that (III.38) be satisfied for all points $(t, \mathbf{x}, \mathbf{v})$, $t > t_0$. We shall regard the restriction F_0 of F to the time t_0 as an initial value which can be assigned arbitrarily in some class. Once F_0 has been assigned, the equation of evolution becomes a functional-integral equation to be satisfied by the molecular density F.

To complete the formulation of the kinetic theory, we shall need to define $\mathbb{C}$ and to develop its properties. At least in some schematic way, $\mathbb{C}$ should reflect the laws of the mechanics of mass-points. How it does so, we shall discuss in Chapter VII, and only there shall we arrive at the defining axiom of MAXWELL's kinetic theory.

Before that we shall see what conclusions may be drawn from the structures set down so far. Mathematically these structures are meager, but conceptually they are not. As we shall see in the following chapter, it is possible to prove some simple theorems which assess the nature of results any theory based upon them may deliver. These theorems will rest upon two general assumptions, first, that the equation of evolution (III.38) has solutions for a broad class of initial values F_0, and, second, that the collisions operator $\mathbb{C}$ is defined in terms of the outcomes of molecular encounters consistent at least in general terms with the laws of classical dynamics. We shall substantiate these assumptions further on in the book. A particular operator $\mathbb{C}$ will be specified formally in Chapter VII; its existence will be demonstrated in Chapter XIX; and existence theorems for the equation of evolution will be proved in Chapters XX and XXI.

Chapter IV

Some Limits of Agreement between Kinetic Theories and Classical Fluid Mechanics

One of the traditional objectives of the kinetic theory has been to calculate the viscosities μ and λ and the Maxwell number M of a viscous fluid on the basis of particular molecular models. Of course, a necessary step toward this end is to prove that the kinetic theory itself delivers quantities formally analogous to μ, λ, and M. We recall that in the kinetic theory the mean normal pressure p, the mass density ρ, the velocity field $\mathbf{u}$, and the energetic ε are defined directly in terms of the molecular density F by means of (III.13), (III.5), (III.6), and (III.14), respectively.

(i) *Failure of constitutive relations in the sense of continuum mechanics*

In order to establish in the kinetic theory the truth of relations having the same form as the classical constitutive equations:

$$p = \varpi - (\lambda + \tfrac{2}{3}\mu)E,$$
$$\mathbf{P} = -2\mu\mathbf{E}; \qquad (\text{I.7})_r$$
$$\mathbf{q} = -M\mu \operatorname{grad} \varepsilon, \qquad (\text{II.9})_r$$

in which, of course, E and $\mathbf{E}$ are calculated from $\mathbf{u}$ by the definitions $(\text{I.5})_{1,3}$, we should have to show first that the equation of evolution, namely,

$$F(t, \mathbf{x}, \mathbf{v}) = \mathfrak{R}_{t-t_0} F(t, \mathbf{x}, \mathbf{v}) + \int_{t_0}^{t} \mathfrak{R}_{t-s}(\mathbb{C}F)(t, \mathbf{x}, \mathbf{v})\, ds, \qquad (\text{III.38})_r$$

forces $\mathbf{P}$ and $\mathbf{q}$ to be related to ρ, $\mathbf{u}$, and ε in these special ways. No such thing can be true for all solutions. Indeed, if a law of molecular interaction is prescribed, we expect that $\mathbb{C}F$, the rate of increase of F due to collisions, should be uniquely determined. Such is the case for the particular collisions operator that we introduce later, the one on which MAXWELL's second kinetic theory is based. This being so, the equation of evolution (III.38) determines $F(t)$ when $F(t_0)$ is known, t_0 being any given fixed time. Here $F(t)$ denotes the restriction of F to the time t. The function $F(t_0)$, interpreted as the initial value of $F(t)$, is essentially arbitrary. *Thus the equation of evolution imposes no condition at all upon the value $F(t_0)$ for any one given t_0.* Since the fields ρ, **u**, ε, **P**, and **q** at the time t_0 are determined by $F(t_0)$, and since it is a trivial matter to construct an initial value having any particular moments of orders 0, 1, 2, and 3, MAXWELL's *second kinetic theory cannot* in general *imply relations having the same form as the constitutive relations of classical fluid mechanics*. That is, since it is possible to construct values of F which do not satisfy (I.7) and (II.9) at some one instant, those relations cannot generally follow from the kinetic theory.

<small>In making these statements we have assumed that the equation of evolution has solutions in some interval $[t_0, T)$ for each element of a suitable set of initial values. This set must be small enough that all of its elements possess moments of orders 0, 1, 2, and 3; at the same time it must be sufficiently large to contain at least one element corresponding to each choice of the basic 13 moments; it must also be small enough to ensure that the solutions to which it gives rise also possess moments of orders 0, 1, 2, and 3. Otherwise the set is unrestricted. In Chapter XX, when we present existence theorems for (III.38), we shall prove that such a set of initial values does exist, at least for a certain class of operators $\mathbb{C}$.</small>

We can reach a much more general conclusion in this regard. The foregoing argument makes no use of any special feature of (I.7) and (II.9) but rests upon the fact that they deliver one moment in terms of others. Thus the argument applies equally well to the principle of determinism as described in Section (iv) of Chapter I. According to that principle, now rephrased as it would have to be were we to try to establish it in the kinetic theory, the histories of some moments of F determine other moments at the arbitrary time t. Since the moments, according to the kinetic theory, may be assigned arbitrarily at any single instant desired, *constitutive relations such as those assumed in continuum thermomechanics cannot hold for all solutions in the kinetic theory.*

That such constitutive relations cannot follow exactly for all solutions of the equation of evolution (III.38), does not make it impossible that they do follow exactly for *a special class of solutions, or even for a broad class of solutions in some sense of limit or approximation*. In this chapter we shall find some restrictions upon any such agreement.

(ii) *Limitations on the shear viscosity*

First of all[1], the collisions operator $\mathbb{C}$ is to be defined somehow in terms of the outcomes of collisions. Therefore the dimension-bearing constants that

---

[1] The arguments here and in the next section of the chapter clarify, consolidate, and generalize those given by TRUESDELL [1952].

enter the specification of $\mathbb{C}$ can be no more than those that enter the specification of the outcome of an encounter. Encounters are described by classical mechanics, in which science occur but three independent physical dimensions: mass, length, and time. Thus at most three independent dimension-bearing constants can enter the specification of $\mathbb{C}$. No molecular constant enters the specification of $\mathfrak{R}_s$. Accordingly, a solution F of the equation of evolution (III.38) can depend upon at most three independent dimension-bearing constants. In the kinetic theory it is convenient to choose these as follows:

$\mathfrak{m}$, the mass of a molecule;

$\mathfrak{d}$, the diameter or some other length associated with a molecule;

$\mathfrak{f}$, the intermolecular force at unit distance.

In special cases the second or the third of these may want altogether, or the two may enter only in a particular combination.

Let us suppose now that somehow we can derive from the kinetic theory a relation of the form $(I.7)_2$, say as a limit, the first term in a series expansion, or otherwise. The left-hand side, **P**, is defined from a solution F (whether particular or general) of (III.38); so is **E** on the right-hand side. Consequently, the scalar coefficient μ is determined by F. Therefore μ is the value of a functional of F and hence is a function of the dimension-bearing constants upon which F depends. In other words, if there is a μ such that $(I.7)_2$ holds, then

$$\mu = f(\rho, \varepsilon; \mathfrak{m}, \mathfrak{d}, \mathfrak{f}). \qquad (IV.1)$$

We shall reduce the scalar relation (IV.1) among scalar fields and constants by use of the Pi-theorem of dimensional analysis[2]. In general, the dimensional matrix of the 6 quantities entering (IV.1) is

	M	L	T
μ	1	-1	-1
ρ	1	-3	0
ε	0	2	-2
$\mathfrak{m}$	1	0	0
$\mathfrak{d}$	0	1	0
$\mathfrak{f}$	1	1	-2

Of course we assume here that the dimension-bearing constants $\mathfrak{m}$, $\mathfrak{d}$, and $\mathfrak{f}$ are neither zero nor infinite, for otherwise the ensuing analysis would be subject to trivial and tedious exceptions. The rank of the dimensional matrix is 3. Since

---

[2] D. E. CARLSON, "On some new results in dimensional analysis", *Archive for Rational Mechanics and Analysis* **68**, 191–210 (1978).

the number of quantities entering the matrix is 6, and since $6 - 3 = 3$, the Pi-theorem shows us that (IV.1) is equivalent to the vanishing of a function of 3 dimensionless ratios formed from μ, ρ, ε, $\mathfrak{m}$, $\mathfrak{d}$, and $\mathfrak{f}$, and that any three linearly independent ratios will do. Since $\varepsilon > 0$ and $\mathfrak{m} > 0$, the following ratios always exist:

$$\frac{\mu \mathfrak{d}^2}{\mathfrak{m}\sqrt{\varepsilon}}, \quad \frac{\rho \mathfrak{d}^3}{\mathfrak{m}}, \quad \frac{\mathfrak{f}\mathfrak{d}}{\mathfrak{m}\varepsilon},$$

and they are obviously independent. Thus (IV.1) is equivalent to a relation expressed as follows in terms of an arbitrary dimensionless function g:

$$\mu = \frac{\mathfrak{m}\sqrt{\varepsilon}}{\mathfrak{d}^2} g\left(\frac{\rho \mathfrak{d}^3}{\mathfrak{m}}, \frac{\mathfrak{f}\mathfrak{d}}{\mathfrak{m}\varepsilon}\right). \tag{IV.2}$$

This formula enables us to survey qualitatively the effects of varying the dimension-bearing molecular parameters, other things being equal. For example, (III.5) shows that the first argument of g is $n\mathfrak{d}^3$, which is independent of the mass $\mathfrak{m}$ and is the ratio of the volume of the molecules themselves, if pictured as cubes of edge-length $\mathfrak{d}$, to the volume of the region in which they move. Likewise, since ε is proportional to the temperature, the way in which the temperature-dependence of μ is connected with the nature of the intermolecular force can be surveyed also.

The most interesting cases are those in which the maximum number of independent constants is not achieved. For example, if the molecules exert no finite forces on each other, there will be no such constant as $\mathfrak{f}$. Accordingly, we delete the row of the dimensional matrix corresponding to $\mathfrak{f}$. Since the rank remains 3 but the number of quantities is now 5, the relation (IV.1) is equivalent to one connecting 2 dimensionless ratios. Thus we conclude that

$$\mu = \frac{\mathfrak{m}\sqrt{\varepsilon}}{\mathfrak{d}^2} g\left(\frac{\rho \mathfrak{d}^3}{\mathfrak{m}}\right). \tag{IV.3}$$

That is, *in a gas whose molecules do not exert finite forces upon one another, the viscosity is proportional to the square root of the energetic.* Such is the case, for example, with molecules conceived as spheres of diameter $\mathfrak{d}$ which move like free bodies except when they actually collide with each other. For the purposes of the present argument, the laws of collision need not be specified. When they are the same as those of frictionless, perfectly hard[3] balls, the molecules may be called *ideal spheres*.

---

[3] MAXWELL in his first work (1860) called such spheres "small, hard, and perfectly elastic". Later [1879, above Eq. (13)] he used the more apt term "rigid-elastic". Physicists continue to use the word "elastic" in this connection. In the mechanics of real materials, of course, "elastic" means something else.

(ii) INVERSE $\mathbb{k}^{th}$-POWER MOLECULES

Another case of interest is that of molecules which repel each other with a central force whose magnitude f per unit mass is inversely proportional to the $\mathbb{k}^{th}$ power of the distance d between them:

$$\mathbb{m}f = \frac{\mathbb{g}}{d^{\mathbb{k}}}, \qquad \mathbb{k} > 1, \quad \mathbb{g} > 0. \tag{IV.4}$$

Molecules of this kind are called *inverse $\mathbb{k}^{th}$-power molecules*. The dimensions of $\mathbb{g}$ are $ML^{\mathbb{k}+1}T^{-2}$, and the units of force and distance enter (IV.1) only in this one combination. Thus the dimensional matrix of the relation (IV.1) becomes

	M	L	T
μ	1	-1	-1
ρ	1	-3	0
ε	0	2	-2
$\mathbb{m}$	1	0	0
$\mathbb{g}$	1	$\mathbb{k}+1$	-2

The rank of this matrix is 3, so again (IV.1) is equivalent to a dimensionless relation connecting 2 dimensionless ratios. Since $\mathbb{m} > 0$, $\varepsilon > 0$, and $\mathbb{g} > 0$, the following ratios always exist:

$$\frac{\mu}{\mathbb{m}\sqrt{\varepsilon}\left(\frac{\mathbb{m}\varepsilon}{\mathbb{g}}\right)^{2/(\mathbb{k}-1)}}, \qquad \frac{\mathbb{m}\varepsilon}{\mathbb{g}}\left(\frac{\rho}{\mathbb{m}}\right)^{(1-\mathbb{k})/3},$$

and they are obviously independent. Thus for inverse $\mathbb{k}^{th}$-power molecules (IV.1) is equivalent to[4]

$$\mu = \mathbb{m}\sqrt{\varepsilon}\left(\frac{\mathbb{m}\varepsilon}{\mathbb{g}}\right)^{2/(\mathbb{k}-1)} g\left(\frac{\mathbb{m}\varepsilon}{\mathbb{g}}\left(\frac{\rho}{\mathbb{m}}\right)^{(1-\mathbb{k})/3}\right). \tag{IV.5}$$

---

[4] For the case when g reduces to a constant, the result (IV.5) was obtained by RAYLEIGH [1900], but he can scarcely be said to have proved it, since he simply assumed μ to be proportional to some power of ε and then adjusted that power so as to balance dimensions, a trivial thing to do. The Pi-theorem proves that the power law is the only one possible, provided the viscosity be independent of the density.

If the dimension-bearing constants are $\mathbb{m}$ and $\mathfrak{f}$ only, without any constant having the dimension of length, then a fresh dimensional analysis shows that

$$\mu = \mathbb{m}^{1/3}\rho^{2/3}\sqrt{\varepsilon g}\left(\frac{\mathfrak{f}^2}{\varepsilon^2 \mathbb{m}^{4/3}\rho^{2/3}}\right),$$

but this observation seems to serve nothing since no molecular model studied in the kinetic theory is of this kind.

The conclusions we have reached so far are quite general in that they derive from the mere definitions used in the kinetic theory, not requiring the equation of evolution to be rendered specific. In MAXWELL's kinetic theory it turns out that μ is always independent of ρ—at least, such is the case if we can believe the purely formal manipulations that are all we have on this subject even today. The details are presented in Chapter XXIV; the fact just mentioned is a consequence of (XXIV.91). If we are willing to grant that ρ drops out[5], then the results (IV.3) and (IV.5) assume particularly simple forms[6]:

$$\mu \propto \frac{\mathfrak{m}\sqrt{\varepsilon}}{\mathfrak{d}^2} \qquad \text{for ideal spheres,}$$

(IV.6)

$$\mu \propto \mathfrak{m}\sqrt{\varepsilon}\left(\frac{\mathfrak{m}\varepsilon}{\mathfrak{g}}\right)^{2/(\mathfrak{k}-1)} \qquad \text{for inverse } \mathfrak{k}^{\text{th}}\text{-power molecules.}$$

Those experienced in the kinetic theory expect results for ideal spheres to follow from their counterparts for inverse $\mathfrak{k}^{\text{th}}$-power molecules in the limit as $\mathfrak{k} \to \infty$, which is so for (IV.6).

Many of the comparisons between the kinetic theory and experiment refer to functional dependences exemplified by (IV.6). The temperature and the viscosity of a gas may be measured grossly, without knowledge of molecular constants such as $\mathfrak{m}$ and $\mathfrak{g}$ and $\mathfrak{k}$. If the dependence found experimentally is $\mu \propto \theta^s$, then the experimenter, after looking at (IV.6)$_2$, may be tempted to assert that he has confirmed the kinetic theory and has proved that $\frac{1}{2} + 2/(\mathfrak{k} - 1) = s$. However, insofar as such comparisons rest only upon proportions and do not call in question the constants of proportionality and the values of molecular constants, they afford no test of MAXWELL's kinetic theory at all, since the results themselves do not rest upon any specific property of that theory, which we have not yet even formulated.

The remarks just made are nearly obvious yet not without importance. They illustrate the fact, which we first encounter here and shall reconfirm many times in the sequel, that *it is only the specific, numerical predictions of the kinetic theory that are not mere illustrations of continuum mechanics.* Once we have decided upon a molecular model such as ideal spheres or inverse $\mathfrak{k}^{\text{th}}$-power molecules, the functional dependence of μ is determined by dimensional reasoning alone. The kinetic theory should provide more, namely, the *numerical coefficients* in relations such as (IV.6). We shall see in Chapters XIV, XXIV, and XXV that it does.

---

[5] This is not an idle remark. Up to this point, all our results hold for various generalizations of MAXWELL's theory, *e.g.* for ENSKOG's theory of dense gases, in which μ does depend upon ρ.

[6] By applying SERRIN's theorem stated just before (II.10) we conclude from (IV.6) that *the Stokes-Kirchhoff approximation for a gas of ideal spheres or inverse $\mathfrak{k}^{\text{th}}$-power molecules has a theory of scaling.*

(iii) Vanishing of the bulk viscosity

Of course we could apply the same reasoning to the second viscosity λ, since its dimensions are those of μ, but we can do better and evaluate λ explicitly. Indeed, from (III.15) we see that *the mean normal pressure p is by its very definition a function of ρ and ε*. If the Navier–Stokes constitutive equations (I.7) hold, we may apply the theorem proved just after (I.13) and so conclude that

$$\varpi = p \quad \text{and} \quad \lambda + \tfrac{2}{3}\mu = 0. \qquad (\text{I.13})_r, (\text{I.12})_r$$

Here, of course, ϖ and λ stand for some quantities to be defined somehow in terms of the molecular density F. The theorem asserts that if in the kinetic theory there are any quantities ϖ and λ such as to satisfy (I.7), they can be only p and $-\tfrac{2}{3}\mu$, respectively. *The kinetic gas, insofar as it is a linearly viscous fluid, has bulk viscosity zero*[7].

In view of the matters set forth in Section (i) of Chapter II, we may express the foregoing results more succinctly: *Insofar as the kinetic gas conforms with classical gas dynamics, it is a Stokes–Kirchhoff gas for which $\gamma = \tfrac{5}{3}$.*

(iv) Disagreement between the kinetic theory and the Stokes–Kirchhoff theory for flows in which $\mathsf{T} \geq 1$

We have derived two conditions necessary for MAXWELL's kinetic theory and the Stokes–Kirchhoff continuum theory to agree. We shall now derive a third one[8]. From MAXWELL's definition (III.9)$_1$ of the pressure tensor **M** we see that at any point $(t, \mathbf{x}) \in \mathscr{D}$, $\mathscr{D} \times \mathscr{V}$ being the essential domain of F, the normal pressure $p_\perp$ acting upon a surface normal to the unit vector **n** satisfies the relation

$$p_\perp \equiv \mathbf{n} \cdot \mathbf{M}\mathbf{n} = \overline{\rho(\mathbf{c} \cdot \mathbf{n})^2} > 0, \qquad (\text{IV.7})$$

and (IV.7) asserts that this pressure is always positive: *The kinetic gas can never support tension, at any place or time in any flow whatever.* No such conclusion follows from the Stokes–Kirchhoff theory in general. Indeed, from (I.5)$_7$ and (I.7)$_2$ we obtain

$$p_\perp = p - 2\mu \mathbf{n} \cdot \mathbf{E}\mathbf{n}. \qquad (\text{IV.8})$$

Hence (IV.7) requires that

$$p > 2\mu \mathbf{n} \cdot \mathbf{E}\mathbf{n} \qquad (\text{IV.9})$$

---

[7] TRUESDELL [1952].
[8] TRUESDELL [1969, 2].

for all unit vectors **n**. To render the quantity on the right-hand side of this inequality a maximum at any given place and time, we take for **n** the proper vector of **E** that corresponds to the greatest proper number, which is one third of the sum of the two greatest principal shearings[9]. Let the principal stretchings d_k be so ordered that $d_1 \geq d_2 \geq d_3$, and let us define the *field of tension numbers* T as follows:

$$\mathsf{T} \equiv \frac{2}{3}\frac{\mu}{p}[(d_1 - d_2) + (d_1 - d_3)], \tag{IV.10}$$

both summands on the right-hand side being non-negative. Since $\mu > 0$ and $p > 0$, in order that $p_\perp > 0$ for all **n** it is necessary and sufficient that

$$\mathsf{T} < 1. \tag{IV.11}$$

Thus we have proved that *the kinetic theory of gases cannot possibly square with the Stokes–Kirchhoff theory in any flow such that* $\mathsf{T} \geq 1$.

We have seen already that no constitutive equation of continuum mechanics can follow both exactly and in general from the kinetic theory. The result just established gives a necessary condition for approximate or limiting consistency with the Stokes–Kirchhoff theory: *Only for flows in which the value of the tension number field* T *is small could the kinetic theory confirm classical fluid mechanics.*

We remark in passing that

$$\mathsf{T} = 0 \quad \text{if and only if} \quad d_1 = d_2 = d_3. \tag{IV.12}$$

Thus the tension number vanishes in a region if and only if the gross flow there is a dilatation. As we saw in Chapter I, a dilatation for a fluid with null bulk viscosity is dissipationless and indistinguishable from the corresponding flow of an inviscid fluid, and in Chapter II we outlined the determination of the possible fields of velocity and energetic and density for a Stokes–Kirchhoff gas undergoing a dilatation. In Chapter X we shall see that there is a class of dilatations compatible with MAXWELL's kinetic theory and that for it the predictions of the kinetic theory agree exactly both with the Stokes–Kirchhoff theory and with the Euler–Hadamard theory. It may be conjectured that these dilatations are the only flows in which the kinetic theory provides solutions that agree with those of the Stokes–Kirchhoff theory, those solutions being of course special ones in the class of all solutions compatible with given gross fields ρ, **u**, and ε.

---

[9] The *principal stretchings* at a place are the extremal rates of increase of length per unit length of material line segments. These are the proper numbers of $\frac{1}{2}(\text{grad } \mathbf{u} + (\text{grad } \mathbf{u})^\mathsf{T})$. The *principal shearings* at a place are the extremal rates of increase of angle between material line segments that are presently orthogonal. If the principal stretchings are d_1, d_2, d_3, the principal shearings are $\pm(d_2 - d_3)$, $\pm(d_3 - d_1)$, $\pm(d_1 - d_2)$. For details see §83 of the work by TRUESDELL & TOUPIN cited in Footnote 10 to Chapter 1 and §46 of J. L. ERICKSEN's Appendix to that work.

(iv) TRUNCATION NUMBER

A convenient expression for T is

$$\mathsf{T} = \frac{2\mu}{p}(d_1 - \tfrac{1}{3}E). \tag{IV.13}$$

We have already encountered T in two special cases. In a simple shearing, defined by (II.57), we see easily that $d_1 = \tfrac{1}{2}K > 0$, $d_2 = 0$, $d_3 = -\tfrac{1}{2}K$, so the tension number field as defined by (IV.10) reduces to $(\mu/p)K$. Thus the quantity T defined by (II.60) is in fact the tension number of the simple shearing. Likewise, in an extension, as defined by (II.64), the tension number field defined by (IV.10) reduces to $\tfrac{4}{3}(\mu/p)\xi$ if $\xi > 0$. Thus the quantity T defined by (II.72) is in fact the tension number of the flow in question, provided that $\xi > 0$. However, if $\xi < 0$, the tension number of the flow is $-\tfrac{2}{3}(\mu/p)\xi$ and thus is $-\tfrac{1}{2}$ times the quantity defined by (II.72).

The tension number has appeared of itself in the course of our solution of a simple but central problem; that solution has suggested the definition (IV.10). We may interpret T in terms more familiar in the literature of rarefied gases[10]. To define the Reynolds number (II.10)$_2$, let us select the same scalar rate as is used to define T:

$$D \equiv \tfrac{2}{3}(2d_1 - d_2 - d_3). \tag{IV.14}$$

Then comparison of (IV.10) with (II.11) yields

$$\mathsf{T} = \frac{5}{3}\frac{\mathsf{M}^2}{\mathsf{R}}. \tag{IV.15}$$

Small values of M^2/R are said to correspond to "the regime of continuum flow", while large values of M^2/R correspond to "free molecular flow", and values of M^2/R near to 1 correspond to "the transition regime". We see that these "regimes" correspond, respectively, to small values of T, large values of T, and values of T near to 1. As we have remarked above, in simple shearing $D = K$; if we like, we can write $K = V/L$, V being a typical relative velocity and L the corresponding distance; then the simple statements made above reduce exactly to their counterparts in the heuristic literature.

A truncation number Tr has been defined by (II.55). In view of (II.52) we may write its definition as follows:

$$\mathsf{Tr} \equiv \sqrt{2}\frac{\mu}{p}|\mathbf{E}|. \tag{IV.16}$$

Clearly

$$\mathsf{Tr} = 0 \quad \text{if and only if} \quad d_1 = d_2 = d_3: \tag{IV.17}$$

---

[10] These considerations generalize those of ŁUNC [1963], which refer only to approximations.

The truncation number vanishes in a region if and only if the gross flow there is a dilatation. Since T and Tr vanish together and are continuous functions of the same variables, to make one sufficiently small is to make the other sufficiently small. Indeed, as

$$2\mathsf{Tr}^2 = \mathsf{T}^2 + \left(\frac{2\mu}{p}\right)^2 [(d_2 - \tfrac{1}{3}E)^2 + (d_3 - \tfrac{1}{3}E)^2], \qquad (\text{IV}.18)$$

it is clear that

$$\mathsf{T} < \sqrt{2}\,\mathsf{Tr} \qquad (\text{IV}.19)$$

unless $d_1 = d_2 = d_3$, in which case T = Tr = 0. An explicit upper bound for the ratio Tr/T is not so easy to obtain. In a simple shearing T = Tr.

The advantage in using Tr rather than T is that to obtain it we do not have first to calculate the proper numbers of **E**. We shall see in Footnote 7 to Chapter VIII and more convincingly in (XXIV. 113) that Tr appears as a natural measure of departure from equilibrium in the kinetic theory. The results we have just obtained regarding the meaning of T will then take on added value because they apply *grosso modo* to Tr as well.

We have gone as far as we can in general terms, using only the basic ideas of the kinetic theory. It is time to enter into the details. In order to do so, we must develop the properties of the retrogressor $\mathfrak{R}_s$ and specify the collisions operator $\mathbb{C}$, in terms of which the equation of evolution (III.38) was stated. This we do in the following three chapters.

Chapter V

The Differential Operators of the Kinetic Theory

(i) The retrogressor and the retrogression reviewed

We have explained already that the *equation of evolution* (III.38) describes a balance between two changes of F in the course of time, one due to the extrinsic force, usually called the "body force", the other due to intermolecular forces. We represent the effect of the body force $\mathbf{b}$ per unit mass on the evolution of F by means of a functional operator $\Re_s$. For each positive s the value of the *retrogression* $\Re_s F$ at a given point $(t, \mathbf{x}, \mathbf{v})$ is found by transforming this point into the one occurring s units of time earlier on the same extrinsic trajectory, and then evaluating F at this earlier point. In order to describe this transformation backward along trajectories, we considered the Newtonian differential equation of motion

$$\chi'' = \mathbf{b}(\chi) \qquad (\text{III.22})_r$$

subject to the initial conditions

$$\chi[t, \mathbf{x}, \mathbf{v}](t) = \mathbf{x}, \qquad \chi'[t, \mathbf{x}, \mathbf{v}](t) = \mathbf{v}; \qquad (\text{III.23})_r$$

we assumed that a unique solution existed in the interval $[t - s, t]$, and in terms of it we defined as follows the *retrogressor* $\mathfrak{r}_s$:

$$\mathfrak{r}_s(t, \mathbf{x}, \mathbf{v}) \equiv (t - s, \chi[t, \mathbf{x}, \mathbf{v}](t - s), \chi'[t, \mathbf{x}, \mathbf{v}](t - s)). \qquad (\text{III.25})_r$$

The transformation along trajectories is given, then, explicitly by $(t, \mathbf{x}, \mathbf{v}) \mapsto \mathfrak{r}_s(t, \mathbf{x}, \mathbf{v})$, and so

$$\Re_s G(t, \mathbf{x}, \mathbf{v}) \equiv G(\mathfrak{r}_s(t, \mathbf{x}, \mathbf{v})),$$

$$= G(t - s, \chi[t, \mathbf{x}, \mathbf{v}](t - s), \chi'[t, \mathbf{x}, \mathbf{v}](t - s)). \qquad (\text{III.31})_r$$

V. DIFFERENTIAL OPERATORS

We noted that the retrogressor r_s enjoyed the following properties:

$$r_{s_1} r_{s_2} = r_{s_1 + s_2}; \tag{III.27}_r$$

$$r_0 = \text{identity}; \tag{III.28}_r$$

$$r_s r_{-s} = \text{identity}. \tag{III.29}_r$$

Of course, in view of $(III.31)_1$ also the functional *retrogressor* $\Re_s$ has these properties.

(ii) Differentiation along a trajectory. Mild and strong derivatives

Let H be a real-valued function defined on a subset of $\mathcal{T} \times \mathcal{E} \times \mathcal{V}$. We say that H is *differentiable along the extrinsic trajectories* if for fixed t, $\mathbf{x}$, and $\mathbf{v}$ the function $s \mapsto \Re_s H(t, \mathbf{x}, \mathbf{v})$ is differentiable. On the space of such functions we define an operator $\eth$, called the *mild derivative*, as follows:

$$\eth H \equiv -\frac{d}{ds} \Re_s H \bigg|_{s=0}. \tag{V.1}$$

Using $(III.31)_2$, we see that the values of the new function $\eth H$ are given explicitly by

$$\eth H(t, \mathbf{x}, \mathbf{v}) = \frac{d}{ds} H(s, \chi[t, \mathbf{x}, \mathbf{v}](s), \chi'[t, \mathbf{x}, \mathbf{v}](s)) \bigg|_{s=t}. \tag{V.2}$$

Thus the function $\eth H$ equals the derivative, evaluated at $(t, \mathbf{x}, \mathbf{v})$, of H along the extrinsic trajectory through $(t, \mathbf{x}, \mathbf{v})$.

In terms of the mild derivative we may express the property of the retrogressor stated just before (III.35) in the following concise form:

$$\eth \Re_{t-s} G(t, \mathbf{x}, \mathbf{v}) = 0 \tag{V.3}$$

for all functions G, all real numbers s, and all points $(t, \mathbf{x}, \mathbf{v})$ in the domain of $\Re_{t-s} G$. Two further properties of the mild derivative and the retrogressor will be useful later. The first is

$$\Re_{t-s} \eth H(t, \mathbf{x}, \mathbf{v}) = \frac{d}{ds} \Re_{t-s} H(t, \mathbf{x}, \mathbf{v}). \tag{V.4}$$

(ii) MILD AND STRONG DERIVATIVES

We may prove it directly from the definitions:

$$\begin{aligned}
\mathfrak{R}_{t-s}\mathfrak{d}H(t, \mathbf{x}, \mathbf{v}) &= -\mathfrak{R}_{t-s}\frac{d}{dr}\mathfrak{R}_r H(t, \mathbf{x}, \mathbf{v})\bigg|_{r=0}, \\
&= -\frac{d}{dr}H(\mathfrak{r}_r\mathfrak{r}_{t-s}(t, \mathbf{x}, \mathbf{v}))\bigg|_{r=0}, \\
&= -\frac{d}{dr}H(\mathfrak{r}_{t-(s-r)}(t, \mathbf{x}, \mathbf{v}))\bigg|_{r=0}, \\
&= \frac{d}{dr}\mathfrak{R}_{t-r}H(t, \mathbf{x}, \mathbf{v})\bigg|_{r=s}.
\end{aligned} \quad (V.5)$$

The second may be stated as follows: *Let H_s be a one-parameter family of functions, each differentiable along the extrinsic trajectories, such that for fixed t, $\mathbf{x}$, and $\mathbf{v}$ the function $s \mapsto H_s(t, \mathbf{x}, \mathbf{v})$ is continuous. Then*

$$\mathfrak{d}\int_{t_0}^{t} H_s(t, \mathbf{x}, \mathbf{v})\, ds = H_t(t, \mathbf{x}, \mathbf{v}) + \int_{t_0}^{t}\mathfrak{d}H_s(t, \mathbf{x}, \mathbf{v})\, ds. \quad (V.6)$$

This is a direct consequence of (V.2), (III.31), (III.28), and the simple identity

$$\begin{aligned}
\frac{d}{dr}\int_{t_0}^{r} &H_s(r, \chi[t, \mathbf{x}, \mathbf{v}](r), \chi'[t, \mathbf{x}, \mathbf{v}](r))\, ds\bigg|_{r=t} \\
&= H_r(r, \chi[t, \mathbf{x}, \mathbf{v}](r), \chi'[t, \mathbf{x}, \mathbf{v}](r))\bigg|_{r=t} \\
&\quad + \int_{t_0}^{r}\frac{d}{dr}H_s(r, \chi[t, \mathbf{x}, \mathbf{v}](r), \chi'[t, \mathbf{x}, \mathbf{v}](r))\, ds\bigg|_{r=t}. \quad (V.7)
\end{aligned}$$

Specializing this result, we take some function G and define H_s as follows: $H_s(t, \mathbf{x}, \mathbf{v}) \equiv \mathfrak{R}_{t-s}G(t, \mathbf{x}, \mathbf{v})$. From (V.3), (V.6), and (III.28) we see that *if G be such as to render $s \mapsto \mathfrak{R}_{t-s}G(t, \mathbf{x}, \mathbf{v})$ continuous, then*

$$\mathfrak{d}\int_{t_0}^{t}\mathfrak{R}_{t-s}G(t, \mathbf{x}, \mathbf{v})\, ds = G(t, \mathbf{x}, \mathbf{v}): \quad (V.8)$$

Considered as an operator, the mild derivative is a left-inverse of the integral along an extrinsic trajectory.

If H is a real-valued function which is defined on an open subset of $\mathcal{T} \times \mathcal{E} \times \mathcal{V}$ and which is differentiable with respect to each of its arguments, we may calculate the right-hand side of (V.2) with the aid of the chain rule and so obtain

$$\begin{aligned}
\mathfrak{d}H(t, \mathbf{x}, \mathbf{v}) = \partial_t H(t, \mathbf{x}, \mathbf{v}) &+ \mathbf{v} \cdot \partial_{\mathbf{x}} H(t, \mathbf{x}, \mathbf{v}) \\
&+ \chi''[t, \mathbf{x}, \mathbf{v}](t) \cdot \partial_{\mathbf{v}} H(t, \mathbf{x}, \mathbf{v}). \quad (V.9)
\end{aligned}$$

Here ∂_x and ∂_v denote derivatives with respect to x and v, respectively, and we have made use of (III.28), or, what is the same thing, (III.23). Since χ is the unique solution of (III.22) and (III.23),

$$\mathfrak{d}H = \mathfrak{D}H, \tag{V.10}$$

$\mathfrak{D}$ being defined by

$$\mathfrak{D}H \equiv \partial_t H + \mathbf{v} \cdot \partial_x H + \mathbf{b} \cdot \partial_v H. \tag{V.11}$$

This differential operator is the one we shall use most often in the kinetic theory. Since it acts only on functions that are smooth, we shall call it the *strong derivative*.

(iii) *Local forms of the equation of evolution*

In Section (v) of Chapter III we laid down the basic axiom of the kinetic theory, the *equation of evolution*:

$$F(t, \mathbf{x}, \mathbf{v}) = \mathfrak{R}_{t-t_0} F(t, \mathbf{x}, \mathbf{v}) + \int_{t_0}^t \mathfrak{R}_{t-s}(\mathbb{C}F)(t, \mathbf{x}, \mathbf{v})\, ds. \tag{III.38}_r$$

Properties of the retrogressor imply a number of formal equivalents of this axiom, equations to which it reduces for such solutions as possess certain degrees of regularity. Because use of some of these equivalent forms is at present unavoidable in some calculations in the kinetic theory, we calculate them now and classify the solutions they govern.

Let us denote the restriction of F to the fixed time t_0 by F_0. We call F_0 the *initial value of F*, t_0 the *initial time*. Because of (III.34) we may now write (III.38), with the argument $(t, \mathbf{x}, \mathbf{v})$ suppressed, in the form

$$F = \mathfrak{R}_{t-t_0} F_0 + \int_{t_0}^t \mathfrak{R}_{t-s} \mathbb{C} F\, ds. \tag{V.12}$$

If the function $s \mapsto \mathfrak{R}_{t-s} \mathbb{C} F$ is integrable, an assumption that we shall always make implicitly, we may evaluate both sides of this equation at $t = t_0$ and then by use of (III.28) obtain

$$F\bigg|_{t=t_0} = F_0, \tag{V.13}$$

as we should expect. This conclusion reaffirms our statement in Chapter III that the equation of evolution imposes no restriction on the value of F at some one fixed time t_0. The equation (V.12) is a functional-integral equation for the molecular density F. As we shall see once the specific collisions operator $\mathbb{C}$ defining MAXWELL's kinetic theory has been set down, the solutions of this

(iii) EQUATION OF EVOLUTION, LOCAL FORMS

equation need not possess any special regularity. For this reason we call them *weak solutions* according to the kinetic theory. In later chapters we shall encounter other functional-integral equations, ones which are formally equivalent to (V.12) and which also do not impose any special regularity on F. The term "weak" will be applied to their solutions as well.

A molecular density F that is differentiable along trajectories and satisfies the functional-differential equation

$$\flat F = \mathbb{C} F \qquad (\text{V}.14)$$

is called a *mild solution*. The equation (V.14) is a local form of the equation of evolution (III.38). To see why, let us assume that F satisfies (III.38) and that for fixed t, $\mathbf{x}$, and $\mathbf{v}$ the function $s \mapsto \mathfrak{R}_{t-s} \mathbb{C} F(t, \mathbf{x}, \mathbf{v})$ is continuous. Then by (V.8) the second term on the right-hand side of (III.38) is differentiable along trajectories, and its mild derivative is simply $\mathbb{C} F$. Likewise, (V.3) assures us that the first term on the right-hand side of (III.38) is differentiable along the extrinsic trajectories, and its mild derivative is null. We conclude that F is differentiable along the extrinsic trajectories and that its mild derivative is given by (V.14). Indeed, we have established the fact that *a weak solution F such as to render $s \mapsto \mathfrak{R}_{t-s} \mathbb{C} F(t, \mathbf{x}, \mathbf{v})$ continuous is a mild solution*. Conversely, if we apply $\mathfrak{R}_{t-s}$ to each side of (V.14) and then evaluate the result at $(t, \mathbf{x}, \mathbf{v})$, in view of (V.4) we obtain

$$\frac{d}{ds} \mathfrak{R}_{t-s} F(t, \mathbf{x}, \mathbf{v}) = \mathfrak{R}_{t-s} \mathbb{C} F(t, \mathbf{x}, \mathbf{v}). \qquad (\text{V}.15)$$

Integrating with respect to s now from t_0 to t and using (III.28), we obtain (III.38). Thus, not surprisingly, *a mild solution is also a weak solution*. We have shown that *on an extrinsic trajectory such as to make the mapping $s \mapsto \mathfrak{R}_{t-s} \mathbb{C} F(t, \mathbf{x}, \mathbf{v})$ continuous, the equation of evolution* (III.38) *is equivalent to the functional-differential equation* (V.14).

A molecular density F that is smooth and satisfies the functional-differential equation

$$\mathfrak{D} F = \mathbb{C} F \qquad (\text{V}.16)$$

is called a *strong solution*. Here $\mathfrak{D}$ is the strong derivative defined by (V.11). In view of (V.10) *a strong solution is a mild solution*, but in general we do not know whether, or under what additional restrictions, a mild solution is also a strong one. The local form (V.16) of the equation of evolution is the standard one in the kinetic theory, and for various purposes the use of it seems at present unavoidable. Whenever we use it we shall assume implicitly that F is smooth, even though to assume outright that it is reflects no physical principle and is in no way natural in a theory employing ideas associated with probability.

(iv) Moments of the strong derivative of a function

We pause now to verify and record formulae for some of the moments of $\mathfrak{D}F$, and then we derive major conditions from them. In all the following calculations we take F to be a differentiable function of t, $\mathbf{x}$, and $\mathbf{v}$ such that it and its derivatives $\partial_t F$ and $\partial_{\mathbf{x}} F$ possess moments of orders 0, 1, 2, and 3.

First we remark that

$$\int \mathfrak{D}F = \int (\partial_t F + \mathbf{v} \cdot \partial_{\mathbf{x}} F + \mathbf{b} \cdot \partial_{\mathbf{v}} F),$$

$$= \partial_t \int F + \operatorname{div} \int F\mathbf{v} + \mathbf{b} \cdot \int \partial_{\mathbf{v}} F, \qquad (V.17)$$

$$= \partial_t n + \operatorname{div}(n\mathbf{u}).$$

In these manipulations we have used (V.11), (III.2), and (III.6), along with the assumptions

$$\mathbf{b} = \mathbf{b}(\mathbf{x}); \quad \text{and either} \qquad (III.21)_r$$

$$\mathbf{b} = \mathbf{0} \quad \text{or} \quad F = o(v^{-2}) \quad \text{as} \quad v \to \infty. \qquad (V.18)$$

The former of these assumptions, which we have made in connection with $\mathfrak{R}_s$, restricts attention to steady extrinsic body forces, independent of the velocities of the molecules. We shall retain this assumption henceforth, though it is not necessary for many of the results we shall obtain. About the assumption (V.18), which requires that unless the body force vanishes, F shall vanish at a certain rate as $v \to \infty$, we shall have more to say later.

Next, by use of (III.7) and (III.8) we note that

$$\int (\mathfrak{D}F)\mathbf{v} = \int (\partial_t F + \mathbf{v} \cdot \partial_{\mathbf{x}} F + \mathbf{b} \cdot \partial_{\mathbf{v}} F)\mathbf{v},$$

$$= \partial_t \int F\mathbf{v} + \operatorname{div} \int F\mathbf{v} \otimes \mathbf{v} + \left(\int \mathbf{v} \otimes \partial_{\mathbf{v}} F \right) \mathbf{b},$$

$$= \partial_t \int F\mathbf{v} + \operatorname{div} \left[\left(\int F \right) \mathbf{u} \otimes \mathbf{u} + \int F\mathbf{c} \otimes \mathbf{c} \right] + \left[\int \partial_{\mathbf{v}}(F\mathbf{v}) - \left(\int F \right) \mathbf{1} \right] \mathbf{b}, \qquad (V.19)$$

$$= \partial_t(n\mathbf{u}) + \operatorname{div}(n\mathbf{u} \otimes \mathbf{u}) + \frac{1}{m} \operatorname{div} \mathbf{M} - n\mathbf{b}.$$

In these manipulations we have used the definitions (III.2), (III.6), and (III.9), and we have replaced the assumption (V.18) by a stronger one:

$$\mathbf{b} = \mathbf{0} \quad \text{or} \quad F = o(v^{-3}) \quad \text{as} \quad v \to \infty. \qquad (V.20)$$

(iv) MOMENTS OF THE DERIVATIVE

Similarly we note that

$$\tfrac{1}{2}\int (\mathfrak{D}F)v^2 = \tfrac{1}{2}\int (\partial_t F + \mathbf{v}\cdot\partial_x F + \mathbf{b}\cdot\partial_v F)v^2,$$

$$= \tfrac{1}{2}\partial_t \int Fv^2 + \tfrac{1}{2}\operatorname{div}\int Fv^2\mathbf{v} + \tfrac{1}{2}\mathbf{b}\cdot\int v^2\,\partial_v F,$$

$$= \tfrac{1}{2}\partial_t\left[\int F(u^2 + c^2)\right] + \tfrac{1}{2}\operatorname{div}\left[\left(\int F\right)u^2\mathbf{u} + \left(\int Fc^2\right)\mathbf{u}\right.$$
$$\left. + 2\left(\int F\mathbf{c}\otimes\mathbf{c}\right)\mathbf{u} + \int Fc^2\mathbf{c}\right] + \tfrac{1}{2}\mathbf{b}\cdot\int[\partial_v(Fv^2) - 2F\mathbf{v}], \quad (\text{V.21})$$

$$= \partial_t[n(\varepsilon + \tfrac{1}{2}u^2)] + \operatorname{div}[n(\varepsilon + \tfrac{1}{2}u^2)\mathbf{u}]$$
$$+ \frac{1}{\mathfrak{m}}\operatorname{div}(\mathbf{M}\mathbf{u}) + \frac{1}{\mathfrak{m}}\operatorname{div}\mathbf{q} - n\mathbf{b}\cdot\mathbf{u}.$$

In these manipulations we have used the definitions (III.2), (III.6), (III.9), (III.12), and (III.14), and we have replaced the assumption (V.18) by a still stronger one:

$$\mathbf{b} = \mathbf{0} \quad \text{or} \quad F = o(v^{-4}) \quad \text{as} \quad v \to \infty. \quad (\text{V.22})$$

We may collect these results and write them more compactly by use of the notations (I.5) and (I.6):

$$\mathfrak{m}\int \mathfrak{D}F = \dot\rho + \rho E,$$

$$\mathfrak{m}\int (\mathfrak{D}F)\mathbf{v} = (\dot\rho + \rho E)\mathbf{u} + \rho\dot{\mathbf{u}} + \operatorname{grad} p + \operatorname{div}\mathbf{P} - \rho\mathbf{b}, \quad (\text{V.23})$$

$$\tfrac{1}{2}\mathfrak{m}\int (\mathfrak{D}F)v^2 = (\dot\rho + \rho E)(\varepsilon + \tfrac{1}{2}u^2) + \mathbf{u}\cdot(\rho\dot{\mathbf{u}} + \operatorname{grad} p + \operatorname{div}\mathbf{P} - \rho\mathbf{b})$$
$$+ \rho\dot\varepsilon + pE + \mathbf{P}\cdot\mathbf{E} + \operatorname{div}\mathbf{q}.$$

Here and henceforth the superimposed dot is defined by (I.6) as before, except that in the kinetic theory the velocity $\mathbf{u}$ is of course the gross velocity, defined by (III.6). If we compare (V.23) with (I.1), (I.2), and (I.4), we read off the following major result[1]: *The conditions*

$$\int \mathfrak{D}F = 0, \quad \int (\mathfrak{D}F)\mathbf{v} = \mathbf{0}, \quad \int (\mathfrak{D}F)v^2 = 0, \quad (\text{V.24})$$

---

[1] In principle, though not in detail and by no means in statement, this result is due to MAXWELL [1867, Eqs. (74), (76), (94)]. HILBERT [1912, Eq. (27)], stating it somewhat obscurely, attributed it to him.

are necessary and sufficient that the fields ρ, $\mathbf{u}$, ε, p, $\mathbf{P}$, and $\mathbf{q}$ defined from F satisfy the field equations of continuum mechanics, namely, (I.1), (I.2), and (I.4). The reader will recall that to obtain this result we have taken the body force $\mathbf{b}$ to be a given function of place only, and if that function is not the zero function we have assumed further that $F = o(v^{-4})$ as $v \to \infty$. In either case F has been assumed differentiable, and it and its derivatives $\partial_t F$ and $\partial_x F$ have been assumed to possess moments of orders 0, 1, 2, and 3.

Chapter VI

The Dynamics of Molecular Encounters

The collisions operator $\mathbb{C}$ is to be defined in terms of the details of molecular encounters. Before we can set down its definition, we must look at the dynamics of molecular interactions, delimit those types of encounters that the kinetic theory is designed to consider, and develop their main features.

(i) *Binary encounters*

MAXWELL's kinetic theory represents a rarefied gas. The term "rarefied" means, essentially, that if a molecule interacts with another one at a certain time, then it interacts with no other. Encounters subject to this restriction are called *binary*. Sometimes the encounters are visualized as actual collisions, like the impact of two balls. We shall prefer to employ consistently the model of a gas as an assembly of mass-points obeying the laws of analytical mechanics. The mutual forces are then actions at a distance. These actions may extend to ∞, as they do for mass-points which repel each other with central and pairwise equilibrated forces whose magnitude is some function of the distance between the places occupied by the points. Another possibility is a mutual force which vanishes outside spheres of diameter d centered upon the positions of the two mass-points. Such molecules are said to have a *cut-off*. A special case is provided by forces which are infinite upon those spheres, so that one molecule can never penetrate the sphere of action of another. If these infinite forces are such as to make the centers of the spheres move as though the spheres were themselves rigid and frictionless balls that bounced off each other according to the rules of "elastic" impact, the molecules are said to be *ideal spheres*. We have met them already in Chapter IV. Here we shall examine certain features of binary encounters which are common to molecules of all these kinds.

Let all space be empty except for two mass-points, supposed to repel one another according to some conservative law of force, which, along with its

VI. MOLECULAR ENCOUNTERS

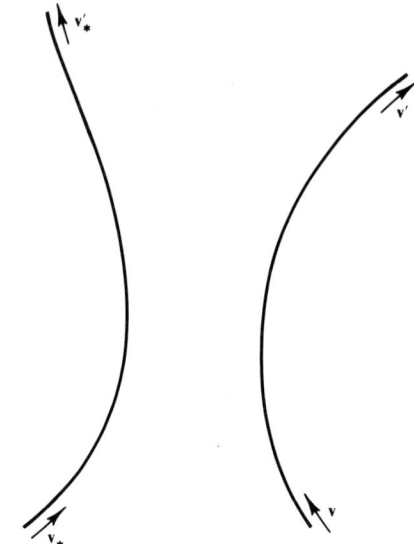

Figure VI.1 Encounter of two molecules

potential, is assumed to vanish as the distance between the mass-points becomes infinite. Let the law of force be such as to make the velocity vectors of the two mass-points approach limits $\mathbf{v}$ and $\mathbf{v}_*$ as $t \to -\infty$ and limits $\mathbf{v}'$ and $\mathbf{v}'_*$ as $t \to +\infty$. Figure VI.1 represents the general nature of an encounter. Figure VI.2 is a diagram of an encounter between two mass-points whose mutual forces have a cut-off. If $\mathbf{b} = \mathbf{0}$, two such mass-points when outside each other's spheres of action travel in straight lines at uniform speeds, so the values $\mathbf{v}, \mathbf{v}_*, \mathbf{v}'$,

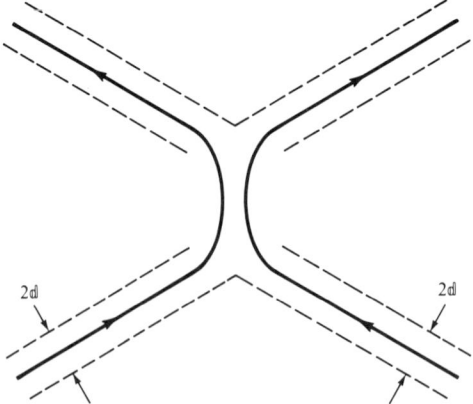

Figure VI.2 Encounter between molecules with a cut-off. The two molecules travel with uniform velocity until the tubes of diameter 2d centered upon their trajectories intersect

and $\mathbf{v}'_*$ are not just limits but are actually assumed and maintained most of the time.

Because the two masses are equal and the potential energy of the mutual forces vanishes at ∞, the laws of conservation of momentum and energy imply that

$$\mathbf{v} + \mathbf{v}_* = \mathbf{v}' + \mathbf{v}'_*; \tag{VI.1}$$

$$v^2 + v_*^2 = v'^2 + v_*'^2. \tag{VI.2}$$

These two requirements on the asymptotic velocities are fundamental to the kinetic theory. For a given choice of $\mathbf{v}$ and $\mathbf{v}_*$ they provide four scalar equations which the six components of $\mathbf{v}'$ and $\mathbf{v}'_*$ must satisfy. Thus they tell us a great deal, though not everything, about the outcome of any single encounter. Their main importance lies in their being, by their very statement, *independent of the law of intermolecular force*. Therefore any result that we might derive on the basis of them alone will be a *universal* property in the kinetic theory, one which is valid for all gases that the kinetic theory envisions. We shall encounter a number of such properties in this chapter and in the two following.

There are two things that we may do in a detailed study of encounters. On the one hand we may attempt to characterize all possible encounters, that is, all possible pairs $(\mathbf{v}, \mathbf{v}_*)$ and $(\mathbf{v}', \mathbf{v}'_*)$ that satisfy (VI.1) and (VI.2). This is called the *encounter problem*, and we shall confront it in Section (iii) of this chapter. We should expect one characterization to come out of the full solution of the two-body problem that initially led us to (VI.1) and (VI.2). On the other hand we may look at functions of $\mathbf{v}$ and ask how their values before and after an encounter are related. Of particular interest in this regard are functions which in some sense are unaffected by encounters. It is to this problem that we turn first.

(ii) Summational invariants and the Boltzmann–Gronwall theorem

A real-valued function C on $\mathscr{V} \times \mathscr{V}$ that satisfies the condition

$$C(\mathbf{v}, \mathbf{v}_*) = C(\mathbf{v}', \mathbf{v}'_*) \tag{VI.3}$$

for all $(\mathbf{v}, \mathbf{v}_*)$ and $(\mathbf{v}', \mathbf{v}'_*)$ related through the conservation laws (VI.1) and (VI.2) is called a *collisional invariant*. A simple characterization of such functions is the following: *C is a collisional invariant if and only if it is of the form*

$$C(\mathbf{v}, \mathbf{v}_*) = f(\mathbf{v} + \mathbf{v}_*, v^2 + v_*^2). \tag{VI.4}$$

The kinetic theory concerns functions defined on $\mathscr{V}$ rather than on $\mathscr{V} \times \mathscr{V}$, and for these we introduce a corresponding concept: A real-valued function S on $\mathscr{V}$ such that

$$S(\mathbf{v}) + S(\mathbf{v}_*) = S(\mathbf{v}') + S(\mathbf{v}'_*) \tag{VI.5}$$

for all $(\mathbf{v}, \mathbf{v}_*)$ and $(\mathbf{v}', \mathbf{v}'_*)$ related through (VI.1) and (VI.2) is a *summational invariant*. For such an S, of course, $S(\mathbf{v}) + S(\mathbf{v}_*)$ is a collisional invariant. The summational invariants form a subspace of any linear space of real-valued functions of $\mathbf{v}$ that is large enough to include them. We may call this subspace the *invariant subspace*. We shall see now that its dimension is 5. Indeed, a basis of 5 elements for the invariant subspace is provided by the **Boltzmann–Gronwall Theorem on Summational Invariants**. *A measurable function S over $\mathscr{V}$ is a summational invariant if and only if it is an affine combination of momentum and energy*:

$$S(\mathbf{v}) = \alpha v^2 + \boldsymbol{\mu} \cdot \mathbf{v} + \beta, \qquad (\text{VI.6})$$

α and β being scalars and $\boldsymbol{\mu}$ a vector.

It is obvious that (VI.6) is sufficient for S to be a summational invariant. Our proof of necessity makes use of certain properties of additive functions. It is well known[1] that a function which is measurable and additive on the real line is linear. This result can be extended to functions of several real variables and to functions defined on the non-negative real line. A further extension, appropriate for our purposes, is stated in the following **Lemma**. *Let g be a real-valued measurable function on $\mathscr{V}$ which is odd*:

$$g(-\mathbf{v}) = -g(\mathbf{v}), \qquad (\text{VI.7})$$

and which is additive on orthogonal pairs:

$$g(\mathbf{v} + \mathbf{v}_*) = g(\mathbf{v}) + g(\mathbf{v}_*) \qquad \text{if} \quad \mathbf{v} \cdot \mathbf{v}_* = 0. \qquad (\text{VI.8})$$

Then g is linear on $\mathscr{V}$. In the appendix to this chapter we supply a proof of this lemma.

Proof of the Boltzmann–Gronwall Theorem[2]. In view of (VI.4) we are to find all measurable functions S such that

$$S(\mathbf{v}) + S(\mathbf{v}_*) = f(\mathbf{v} + \mathbf{v}_*, v^2 + v_*^2) \qquad (\text{VI.9})$$

for all $\mathbf{v}$ and $\mathbf{v}_*$. If S is such a function, so also is $S - S(\mathbf{0})$, so with no loss in generality we may impose the conditions

$$S(\mathbf{0}) = 0, \qquad f(\mathbf{0}, 0) = 0. \qquad (\text{VI.10})$$

---

[1] E.g. §19 of W. MAAK, *Fastperiodische Funktionen*, Berlin etc., Springer, 1950. Indeed KESTELMAN has proved that an additive function which is bounded above on a set of positive measure is necessarily linear. Cf. his paper, "On the functional equation $f(x + y) = f(x) + f(y)$", *Fundamenta Mathematica* **34**, 144–147 (1953).

[2] The proof of BOLTZMANN [1875, §I] [1876, §II] presumes S differentiable. GRONWALL [1915] [1916] was the first to reduce the problem to requirement that certain functions be additive. Although he applied that requirement only to S bounded near $\mathbf{0}$, his proof suffices to establish the theorem in the generality natural to the kinetic theory, namely, for measurable S. A proof due to CARLESON & FROSTMAN is included in the posthumously published booklet of CARLEMAN [1957, §6]. We give here a proof which is essentially new.

(ii) BOLTZMANN–GRONWALL THEOREM

Putting $\mathbf{v}_* = \mathbf{0}$ in (VI.9) then yields

$$S(\mathbf{v}) = f(\mathbf{v}, v^2), \tag{VI.11}$$

and so the problem is reduced to determination of all measurable functions f such that

$$f(\mathbf{v}, v^2) + f(\mathbf{v}_*, v_*^2) = f(\mathbf{v} + \mathbf{v}_*, v^2 + v_*^2) \tag{VI.12}$$

for all $\mathbf{v}$ and $\mathbf{v}_*$.

If we set $\mathbf{v}_* = -\mathbf{v}$ in (VI.12), we find that

$$f(\mathbf{0}, 2v^2) = f(\mathbf{v}, v^2) + f(-\mathbf{v}, v^2). \tag{VI.13}$$

If $\mathbf{v} \cdot \mathbf{v}_* = 0$, then (VI.12) and (VI.13) imply that

$$\begin{aligned} f(\mathbf{0}, 2v^2 + 2v_*^2) &= f(\mathbf{v} + \mathbf{v}_*, v^2 + v_*^2) + f(-\mathbf{v} - \mathbf{v}_*, v^2 + v_*^2), \\ &= f(\mathbf{v}, v^2) + f(\mathbf{v}_*, v_*^2) + f(-\mathbf{v}, v^2) + f(-\mathbf{v}_*, v_*^2), \quad \text{(VI.14)} \\ &= f(\mathbf{0}, 2v^2) + f(\mathbf{0}, 2v_*^2). \end{aligned}$$

Thus for all non-negative numbers λ and λ_*

$$f(\mathbf{0}, \lambda + \lambda_*) = f(\mathbf{0}, \lambda) + f(\mathbf{0}, \lambda_*). \tag{VI.15}$$

A known theorem on additive functions states that $\lambda \mapsto f(\mathbf{0}, \lambda)$ is linear on the set of non-negative real numbers. Hence there is a constant α such that

$$f(\mathbf{0}, v') = \alpha v'. \tag{VI.16}$$

Now we define as follows a function g:

$$g(\mathbf{v}) \equiv f(\mathbf{v}, v^2) - f(\mathbf{0}, v^2). \tag{VI.17}$$

It follows directly from (VI.13) and (VI.16) that g is measurable and odd. Moreover it follows directly from (VI.12) and (VI.14)$_3$ that g is additive on orthogonal pairs. We conclude from the lemma that g is linear:

$$g(\mathbf{v}) = \boldsymbol{\mu} \cdot \mathbf{v}. \tag{VI.18}$$

From (VI.11), (VI.16), (VI.17), and (VI.18) we obtain

$$S(\mathbf{v}) = \alpha v^2 + \boldsymbol{\mu} \cdot \mathbf{v}. \quad \triangle \tag{VI.19}$$

The Boltzmann–Gronwall theorem is used as a tool again and again in the kinetic theory. It can be interpreted as a statement that 1, v_k, and v^2 form a basis in the linear space of summational invariants. We shall call these five quantities henceforth the *principal invariants* and denote them by the symbol I_α:

$$I_\alpha \equiv \begin{cases} 1 & \text{if } \alpha = 0, \\ v_\alpha & \text{if } \alpha = 1, 2, 3, \\ v^2 & \text{if } \alpha = 4. \end{cases} \tag{VI.20}$$

Sometimes we shall use the symbol **I** to denote all of these invariants collectively. We denote by ρ_α the product of ρ by the expectation of I_α:

$$\rho_\alpha \equiv \rho \bar{I}_\alpha = \mathfrak{m} \int F I_\alpha, \qquad \alpha = 0, 1, \ldots, 4, \tag{VI.21}$$

or, collectively,

$$\boldsymbol{\rho} = \rho \bar{\mathbf{I}}; \tag{VI.22}$$

we call $\boldsymbol{\rho}$ the *principal moment* of F. It follows easily from (VI.21) and the basic definitions (III.5), (III.6), and (III.14) that

$$\rho_\alpha = \begin{cases} \rho & \text{if } \alpha = 0, \\ \rho u_\alpha & \text{if } \alpha = 1, 2, 3, \\ \rho(2\varepsilon + u^2) & \text{if } \alpha = 4. \end{cases} \tag{VI.23}$$

The five fields ρ, u_k, and ε determine the principal moment $\boldsymbol{\rho}$ uniquely; conversely, $\boldsymbol{\rho}$ determines them.

Classical gas dynamics is described in terms of the fields ρ, u_k, and ε. We think of these fields as determining everything of interest and as being themselves determined by their own initial values, once boundary conditions have been set. Such determinism presumes that also the body force **b** has been specified. Against the need to discuss similar matters in the kinetic theory we shall call the ordered pair $(\boldsymbol{\rho}, \mathbf{b})$ the *gross condition* of the kinetic gas and denote it by $\boldsymbol{\sigma}$. We shall use this concept many times in the succeeding chapters.

(iii) The encounter problem and its solutions

We turn now to the *encounter problem*: to determine all pairs $(\mathbf{v}, \mathbf{v}_*)$ and $(\mathbf{v}', \mathbf{v}'_*)$ that satisfy the laws of conservation (VI.1) and (VI.2). The first thing we must do is decide what we shall mean by a solution of this problem.

We have noted already that for fixed $\mathbf{v}$ and $\mathbf{v}_*$, the laws of conservation provide four scalar equations to be satisfied by the six components of $\mathbf{v}'$ and $\mathbf{v}'_*$. Therefore we expect to obtain a two-parameter family of solutions $(\mathbf{v}', \mathbf{v}'_*)$ corresponding to each choice of $(\mathbf{v}, \mathbf{v}_*)$. Any solution, then, of the encounter problem should include a parameter set $\mathscr{K}$ and a function $\boldsymbol{\phi} \colon \mathscr{V} \times \mathscr{V} \times \mathscr{K} \to \mathscr{V} \times \mathscr{V}$ with the property that $(\mathbf{v}, \mathbf{v}_*)$ and $(\mathbf{v}', \mathbf{v}'_*)$ satisfy (VI.1) and (VI.2) if and only if

$$(\mathbf{v}', \mathbf{v}'_*) = \boldsymbol{\phi}(\mathbf{v}, \mathbf{v}_*, \mathbf{k}) \tag{VI.24}$$

for some $\mathbf{k} \in \mathscr{K}$. This, however, is only part of what we require.

Let us note that the laws of conservation are unaltered when we interchange the pair $(\mathbf{v}, \mathbf{v}_*)$ with the pair $(\mathbf{v}', \mathbf{v}'_*)$, and also when we interchange $(\mathbf{v}, \mathbf{v}')$ with

$(\mathbf{v}_*, \mathbf{v}'_*)$. These symmetries are extremely important in the kinetic theory, and it is useful to state them in terms of properties of the function ϕ. To be specific, if $(\mathbf{v}', \mathbf{v}'_*)$ is related to $(\mathbf{v}, \mathbf{v}_*)$ by the rule (VI.24), then, since the laws of conservation are invariant with respect to an interchange of these pairs, we see that $(\mathbf{v}, \mathbf{v}_*)$ will be related to $(\mathbf{v}', \mathbf{v}'_*)$ by the same rule, though perhaps with a different value, say $\mathbf{k}'$, of the parameter:

$$(\mathbf{v}, \mathbf{v}_*) = \phi(\mathbf{v}', \mathbf{v}'_*, \mathbf{k}'). \tag{VI.25}$$

Of course $\mathbf{k}'$ is determined by all four asymptotic velocities, and since these are already related by (VI.24), there is a function κ' such that

$$\mathbf{k}' = \kappa'(\mathbf{v}, \mathbf{v}_*, \mathbf{k}). \tag{VI.26}$$

Similarly, by interchanging the roles of $(\mathbf{v}, \mathbf{v}')$ and $(\mathbf{v}_*, \mathbf{v}'_*)$ in (VI.1) and (VI.2) we see that there is another value $\mathbf{k}_*$ of the parameter for which

$$(\mathbf{v}'_*, \mathbf{v}') = \phi(\mathbf{v}_*, \mathbf{v}, \mathbf{k}_*), \tag{VI.27}$$

and this parameter is the value of another function κ_*:

$$\mathbf{k}_* = \kappa_*(\mathbf{v}, \mathbf{v}_*, \mathbf{k}). \tag{VI.28}$$

The functions κ' and κ_* characterize the symmetries of (VI.1) and (VI.2) by expressing them in the forms (VI.25) and (VI.27). The existence of these two functions follows from that of ϕ. It will be important later to know explicitly what they are, and so by a *solution* of the encounter problem we shall mean a parameter set $\mathscr{K}$ and functions ϕ, κ', and κ_* with the properties described above.

There are a number of solutions of the encounter problem, each with its own special uses. We shall present here three of them. The first one[3] is fundamental in that for it the functions ϕ, κ', and κ_* are simple and explicit, and the other solutions are obtained from this one by making certain invertible changes of the parameter $\mathbf{k}$. For applications in the kinetic theory the second and third solutions are the most important ones, and in the next chapter we shall see what roles they play in the definition and properties of the collisions operator $\mathbb{C}$.

Solution 1. Suppose first that $\mathbf{v}' = \mathbf{v}$. Then (VI.1) yields $\mathbf{v}_* = \mathbf{v}'_*$, indicating that no encounter has occurred. Whenever there is an encounter, then, we may define a unit vector $\boldsymbol{\alpha}$ in the direction of the change of momentum experienced by the first mass-point:

$$\boldsymbol{\alpha} \equiv \frac{1}{A}(\mathbf{v}' - \mathbf{v}), \quad A \equiv |\mathbf{v}' - \mathbf{v}|. \tag{VI.29}$$

---

[3] BOLTZMANN [1896, §3], GRAD [1949, Appendix I].

These definitions and (VI.1) yield

$$\mathbf{v}' = \mathbf{v} + A\boldsymbol{\alpha}, \qquad \mathbf{v}'_* = \mathbf{v}_* - A\boldsymbol{\alpha}. \tag{VI.30}$$

Hence

$$v'^2 + v'^2_* = v^2 + v^2_* - 2\boldsymbol{\alpha} \cdot (\mathbf{v}_* - \mathbf{v})A + 2A^2, \tag{VI.31}$$

and so by (VI.2) we see that

$$A = \boldsymbol{\alpha} \cdot (\mathbf{v}_* - \mathbf{v}) = \boldsymbol{\alpha} \cdot \mathbf{w}, \tag{VI.32}$$

in which $\mathbf{w} \equiv \mathbf{v}_* - \mathbf{v}$. As $\mathbf{w}$ is the velocity of the second mass-point relative to the first before the encounter, we shall call it the *velocity of approach* of the *oncoming* molecule.

Now (VI.30) becomes

$$\mathbf{v}' = \mathbf{v} + (\mathbf{w} \cdot \boldsymbol{\alpha})\boldsymbol{\alpha}, \qquad \mathbf{v}'_* = \mathbf{v}_* - (\mathbf{w} \cdot \boldsymbol{\alpha})\boldsymbol{\alpha}. \tag{VI.33}$$

For a given velocity of approach $\mathbf{w}$, all possible outcomes of encounters are obtained by letting the terminus of the unit vector $\boldsymbol{\alpha}$ vary over the unit sphere. Thus, if we take $\boldsymbol{\alpha}$ to be the parameter $\mathbf{k}$ and the unit sphere to be the parameter set $\mathscr{K}$, (VI.33) may be interpreted as an explicit form of (VI.24) and so defines the function $\boldsymbol{\phi}$ for this solution.

We may obtain a graphical interpretation of (VI.33) by noting that the vector $(\mathbf{w} \cdot \boldsymbol{\alpha})\boldsymbol{\alpha}$ is parallel to $\boldsymbol{\alpha}$ and that its length equals that of the projection of $\mathbf{w}$ onto $\boldsymbol{\alpha}$. Hence if we draw a sphere such that the termini of $\mathbf{v}$ and $\mathbf{v}_*$ are antipodal points (Figure VI.3), any other pair of antipodal points on this same

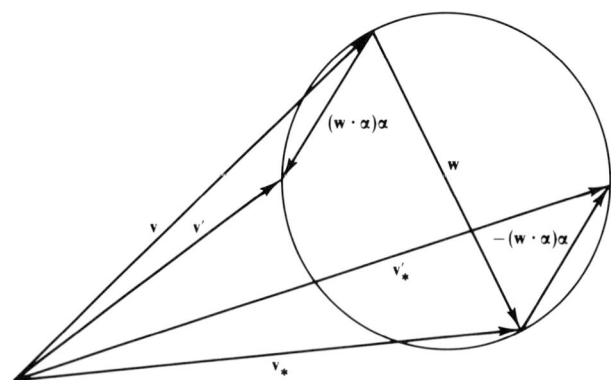

Figure VI.3 Construction for the possible outcomes of a binary encounter

sphere are termini of vectors $\mathbf{v}'$ and $\mathbf{v}'_*$ representing possible outcomes of the encounter.

Because the pairs of vectors $(\mathbf{v}, \mathbf{v}_*)$ and $(\mathbf{v}', \mathbf{v}'_*)$ appear symmetrically in (VI.1) and (VI.2), we may also write

$$\mathbf{v} = \mathbf{v}' + (\mathbf{w}' \cdot \boldsymbol{\alpha}')\boldsymbol{\alpha}', \qquad \mathbf{v}_* = \mathbf{v}'_* - (\mathbf{w}' \cdot \boldsymbol{\alpha}')\boldsymbol{\alpha}', \tag{VI.34}$$

in which $\mathbf{w}' \equiv \mathbf{v}'_* - \mathbf{v}'$ and $\boldsymbol{\alpha}'$ is a unit vector in the direction of $\mathbf{v} - \mathbf{v}'$. Because $\boldsymbol{\alpha}$ has the same direction as $\mathbf{v}' - \mathbf{v}$,

$$\boldsymbol{\alpha}' = -\boldsymbol{\alpha}. \tag{VI.35}$$

Since (VI.33) is the explicit form of (VI.24), we may interpret (VI.34) as the explicit form of (VI.25) provided we choose the parameter $\mathbf{k}'$ to be $\boldsymbol{\alpha}'$. Therefore (VI.35) defines the function κ' appearing in (VI.26).

Because the pairs $(\mathbf{v}, \mathbf{v}')$ and $(\mathbf{v}_*, \mathbf{v}'_*)$ also appear symmetrically in (VI.1) and (VI.2), we may write

$$\mathbf{v}'_* = \mathbf{v}_* + (\mathbf{w}_* \cdot \boldsymbol{\alpha}_*)\boldsymbol{\alpha}_*, \qquad \mathbf{v}' = \mathbf{v} - (\mathbf{w}_* \cdot \boldsymbol{\alpha}_*)\boldsymbol{\alpha}_*, \tag{VI.36}$$

in which $\mathbf{w}_* \equiv \mathbf{v} - \mathbf{v}_* = -\mathbf{w}$ and $\boldsymbol{\alpha}_*$ is a unit vector in the direction of $\mathbf{v}'_* - \mathbf{v}_*$. From (VI.1) we see that $\boldsymbol{\alpha}$ is a unit vector in the direction of $\mathbf{v}_* - \mathbf{v}'_*$, and therefore

$$\boldsymbol{\alpha}_* = -\boldsymbol{\alpha}. \tag{VI.37}$$

Identifying (VI.36) with (VI.27), and hence $\mathbf{k}_*$ with $\boldsymbol{\alpha}_*$, we see that (VI.37) defines the function κ_* appearing in (VI.28). Thus (VI.33), (VI.35), and (VI.37) express one solution of the encounter problem.

Solution 2. The preceding solution has the disadvantage of giving the outcomes of collisions in terms of three variables which are not independent: α_1, α_2, α_3. This disadvantage can be remedied[4] by describing $\boldsymbol{\alpha}$ in terms of two angles. We note first that if $\mathbf{v}$ and $\mathbf{v}_*$ are given, then (VI.33) shows that $\boldsymbol{\alpha}$ and $-\boldsymbol{\alpha}$ define the same asymptotic velocities $\mathbf{v}'$ and $\mathbf{v}'_*$, and therefore we may restrict the terminus of $\boldsymbol{\alpha}$ to lie on some hemisphere. As we shall see later, one choice of this hemisphere is particularly useful. Let $\mathscr{P}$ denote the plane which is perpendicular to $\mathbf{w}$ and which passes through some fixed point $\mathbf{x}$, and let $\boldsymbol{\alpha}$ describe points on the unit hemisphere centered at $\mathbf{x}$ and lying upon the side of $\mathscr{P}$ away from which $\mathbf{w}$ points. In order to express $\boldsymbol{\alpha}$ in terms of angular variables, for each $\mathbf{w}$ let $\boldsymbol{\xi}(\mathbf{w})$ and $\boldsymbol{\eta}(\mathbf{w})$ be an orthonormal pair of vectors lying in the plane $\mathscr{P}$:

$$\mathbf{w} \cdot \boldsymbol{\xi}(\mathbf{w}) = \mathbf{w} \cdot \boldsymbol{\eta}(\mathbf{w}) = \boldsymbol{\xi}(\mathbf{w}) \cdot \boldsymbol{\eta}(\mathbf{w}) = 0,$$
$$\boldsymbol{\xi}(\mathbf{w}) \cdot \boldsymbol{\xi}(\mathbf{w}) = \boldsymbol{\eta}(\mathbf{w}) \cdot \boldsymbol{\eta}(\mathbf{w}) = 1. \tag{VI.38}$$

One possible choice of these functions is

$$\xi_1 = \frac{\omega}{w}, \qquad \xi_2 = -\frac{w_1 w_2}{\omega w}, \qquad \xi_3 = -\frac{w_1 w_3}{\omega w},$$
$$\eta_1 = 0, \qquad \eta_2 = \frac{w_3}{\omega}, \qquad \eta_3 = -\frac{w_2}{\omega}, \tag{VI.39}$$
$$\omega \equiv \sqrt{w_2^2 + w_3^2}.$$

---

[4] Although the rest of this chapter presents ideas which certainly are of great age, we cannot cite any source for the details of our development of them.

We introduce angles θ and ζ such that

$$\boldsymbol{\alpha} = \mathbf{a}(\theta, \zeta, \mathbf{w}) \equiv \cos\theta \frac{1}{w}\mathbf{w} - \sin\theta\cos\zeta\,\boldsymbol{\xi}(\mathbf{w}) - \sin\theta\sin\zeta\,\boldsymbol{\eta}(\mathbf{w}). \quad \text{(VI.40)}$$

This definition shows that θ is the angle between $\boldsymbol{\alpha}$ and $\mathbf{w}$ while ζ is the angular variable in a system of polar co-ordinates centered at $\mathbf{x}$ in the plane $\mathscr{P}$. Since $\boldsymbol{\alpha}$ must lie on a hemisphere, the following ranges of the angles suffice: $0 \leq \zeta < 2\pi$, $0 \leq \theta \leq \frac{1}{2}\pi$. The points of this hemisphere constitute the parameter set $\mathscr{K}$ for this solution; we shall denote this $\mathscr{K}$ by the symbol $\triangle$. If we place (VI.40) into (VI.33), we obtain

$$\begin{aligned}\mathbf{v}' &= \mathbf{v} + [\mathbf{w}\cdot\mathbf{a}(\theta, \zeta, \mathbf{w})]\mathbf{a}(\theta, \zeta, \mathbf{w}),\\ \mathbf{v}'_* &= \mathbf{v}_* - [\mathbf{w}\cdot\mathbf{a}(\theta, \zeta, \mathbf{w})]\mathbf{a}(\theta, \zeta, \mathbf{w}),\end{aligned} \quad \text{(VI.41)}$$

and these define the corresponding function $\boldsymbol{\phi}$ for the solution, the parameter $\mathbf{k}$ being the ordered pair (θ, ζ).

We may introduce in the same way angles θ' and ζ' for which

$$\boldsymbol{\alpha}' = \mathbf{a}(\theta', \zeta', \mathbf{w}'), \qquad \mathbf{w}' \equiv \mathbf{v}'_* - \mathbf{v}'. \quad \text{(VI.42)}$$

By (VI.35), θ' and ζ' are related to θ and ζ through the equation

$$\mathbf{a}(\theta', \zeta', \mathbf{w}') = -\mathbf{a}(\theta, \zeta, \mathbf{w}). \quad \text{(VI.43)}$$

By solving (VI.41) and (VI.43) we obtain θ' and ζ' as certain functions of θ, ζ, and $\mathbf{w}$:

$$\theta' = \Theta'(\theta, \zeta, \mathbf{w}), \qquad \zeta' = Z'(\theta, \zeta, \mathbf{w}). \quad \text{(VI.44)}$$

These define the function κ' appearing in (VI.26).

It is not difficult to derive the functions Θ' and Z' explicitly, and for future reference we do so now. Using (VI.41), we note the following useful identities:

$$\begin{aligned}\mathbf{w}' &= \mathbf{w} - 2[\mathbf{w}\cdot\mathbf{a}(\theta, \zeta, \mathbf{w})]\mathbf{a}(\theta, \zeta, \mathbf{w}),\\ w' &= w,\\ \mathbf{w}'\cdot\mathbf{a}(\theta', \zeta', \mathbf{w}') &= \mathbf{w}\cdot\mathbf{a}(\theta, \zeta, \mathbf{w}),\\ \mathbf{w}'\wedge\mathbf{a}(\theta', \zeta', \mathbf{w}') &= -\mathbf{w}\wedge\mathbf{a}(\theta, \zeta, \mathbf{w}),\end{aligned} \quad \text{(VI.45)}$$

the third and fourth following from the first two by use of (VI.43).

By (VI.38) and (VI.40), $\mathbf{w}\cdot\mathbf{a}(\theta, \zeta, \mathbf{w}) = w\cos\theta$, so it follows from (VI.45)$_{2,3}$ that $\cos\theta' = \cos\theta$. Because θ and θ' both lie between 0 and $\frac{1}{2}\pi$, it follows that

$$\theta' = \theta. \quad \text{(VI.46)}$$

This is the explicit form of (VI.44)$_1$. Next we note from (VI.39) that $\mathbf{w}\wedge\boldsymbol{\xi}(\mathbf{w})/w = \boldsymbol{\eta}(\mathbf{w})$ and $\mathbf{w}\wedge\boldsymbol{\eta}(\mathbf{w})/w = -\boldsymbol{\xi}(\mathbf{w})$. Thus, by use of the definition of $\mathbf{a}$

(iii) ENCOUNTER PROBLEM, SOLUTION 2

in both sides of (VI.45)$_4$, we obtain

$$-[\sin \zeta' \, \xi(\mathbf{w}') - \cos \zeta' \, \eta(\mathbf{w}')] = \sin \zeta \, \xi(\mathbf{w}) - \cos \zeta \, \eta(\mathbf{w}). \qquad \text{(VI.47)}$$

Using (VI.39), we can write out explicitly the first and second components of this equation. By eliminating $\mathbf{w}'$ through use of (VI.45)$_1$ we obtain the following formulae:

$$\sin \zeta' = -\frac{\omega}{\omega'} \sin \zeta,$$

$$\cos \zeta' = \frac{1}{\omega'} (w_1 \sin 2\theta + \omega \cos 2\theta \cos \zeta), \qquad \text{(VI.48)}$$

$$\omega' \equiv \sqrt{\omega^2 \sin^2 \zeta + (w_1 \sin 2\theta + \omega \cos 2\theta \cos \zeta)^2}.$$

These give us the explicit form of the function Z' in (VI.44)$_2$.

The transformation from unprimed to primed variables as given by (VI.41) and (VI.44) has one more property that will be useful later. Namely, it is locally volume-preserving:

$$\left| \frac{\partial(\mathbf{v}', \mathbf{v}'_*, \theta', \zeta')}{\partial(\mathbf{v}, \mathbf{v}_*, \theta, \zeta)} \right| = 1. \qquad \text{(VI.49)}$$

That such is indeed the case, may be verified by direct if lengthy calculation. We have collected the basic formulae that are needed, those being (VI.41), (VI.46), and (VI.48). We leave the details to the reader[5].

Now let us introduce two new angles θ_* and ζ_* such that

$$\alpha_* = \mathbf{a}(\theta_*, \zeta_*, \mathbf{w}_*), \qquad \mathbf{w}_* \equiv \mathbf{v} - \mathbf{v}_* = -\mathbf{w}. \qquad \text{(VI.50)}$$

By (VI.37), θ_* and ζ_* are related to θ and ζ through the equation

$$\mathbf{a}(\theta_*, \zeta_*, -\mathbf{w}) = -\mathbf{a}(\theta, \zeta, \mathbf{w}). \qquad \text{(VI.51)}$$

By solving this for θ_* and ζ_* we obtain a result of the form

$$\begin{aligned} \theta_* &= \Theta_*(\theta, \zeta, \mathbf{w}), \\ \zeta_* &= Z_*(\theta, \zeta, \mathbf{w}), \end{aligned} \qquad \text{(VI.52)}$$

---

[5] SCHNUTE [1975, §3] has remarked that it is not uncommon in the kinetic theory to infer (VI.49) immediately as a consequence of the fact that the transformation $(\mathbf{v}, \mathbf{v}_*, \theta, \zeta) \mapsto (\mathbf{v}', \mathbf{v}'_*, \theta', \zeta')$ is its own inverse. That this argument is false, can be seen from the simple example

$$(x', y') = f(x, y) \equiv \left(x, \frac{1}{y}\right), \qquad -\infty < x < \infty, \quad 0 < \frac{1}{a} \leq y \leq a < \infty,$$

for which $f \circ f =$ identity, but $|\partial(x', y')/\partial(x, y)| = 1/y^2 \neq 1$. SCHNUTE [1975, §§3–5] presents another approach to characterizing encounters between mass-points, and in the case in which his "collision law" is specialized to satisfy (VI.1) and (VI.2) he has supplied a geometric proof of (VI.49).

for certain functions Θ_* and Z_*. These define the function κ_* that appears in (VI.28). It is not difficult to determine Θ_* and Z_* explicitly. We note from (VI.39) that $\xi(-\mathbf{w}) = \xi(\mathbf{w})$ and $\eta(-\mathbf{w}) = -\eta(\mathbf{w})$. Thus, using (VI.40), we find that (VI.51) is equivalent to the conditions

$$\cos \theta_* = \cos \theta, \qquad \cos \zeta_* = -\cos \zeta, \qquad \sin \zeta_* = \sin \zeta. \qquad \text{(VI.53)}$$

The unique solution of this system for which $0 \leq \theta_* \leq \tfrac{1}{2}\pi$ and $0 \leq \zeta_* < 2\pi$ whenever $0 \leq \theta \leq \tfrac{1}{2}\pi$ and $0 \leq \zeta < 2\pi$ is

$$\theta_* = \theta,$$

$$\zeta_* = \begin{cases} \pi - \zeta & \text{if } 0 \leq \zeta \leq \pi, \\ 3\pi - \zeta & \text{if } \pi < \zeta < 2\pi. \end{cases} \qquad \text{(VI.54)}$$

This is the explicit form of (VI.52). The equations (VI.41), (VI.46), (VI.48), and (VI.54) constitute the second solution of the encounter problem.

For future reference we note from (VI.54) that the mapping $\zeta \mapsto \zeta_*$ is piecewise smooth, and at its points of continuity the Jacobian of the mapping $(\mathbf{v}, \mathbf{v}_*, \theta, \zeta) \mapsto (\mathbf{v}_*, \mathbf{v}, \theta_*, \zeta_*)$ has absolute value 1. The function Z' defined implicitly by (VI.48) and the requirement that $0 \leq \zeta' < 2\pi$ whenever $0 \leq \zeta < 2\pi$ is also only piecewise smooth, so the local condition (VI.49) is valid only at points where Z' is differentiable.

Solution 3. In the two solutions we have presented, (VI.1) and (VI.2) were viewed simply as a pair of algebraic equations for $\mathbf{v}'$ and $\mathbf{v}'_*$, independent of their physical interpretation. A third solution, one reflecting more closely the picture of two molecules colliding, comes out of the solution of a two-body problem of analytical dynamics.

Considering two molecules, one with asymptotic velocity $\mathbf{v}$ and the other with asymptotic velocity $\mathbf{v}_*$, we ask first if they may experience an encounter. If their mutual forces are of infinite range, an encounter does occur. Indeed, according to the laws of analytical dynamics, the two molecules influence each other's motion always, so the encounter lasts forever. If the molecules have a cut-off, only those whose motions are such as to make them enter each other's sphere of action experience encounters, and the duration of encounters may be finite. For example, two ideal spheres may pass by each other without interacting at all. If the positions of these two spheres at some time are $\mathbf{x}$ and $\mathbf{x}_*$, if their diameters are d, and if we describe a right cylinder of radius d and axis through $\mathbf{x}_*$ parallel to $\mathbf{v}_* - \mathbf{v}$, then an encounter occurs according as that cylinder does or does not contain $\mathbf{x}$ (Figure VI.4).

A similar statement can be made for general encounters. As for two ideal spheres, it is easiest to picture an encounter by choosing a framing in which one molecule is always at rest at some fixed place $\mathbf{x}$. The trajectory of a second molecule in this framing is its trajectory relative to that at $\mathbf{x}$. This trajectory is assumed to have an asymptote at $t = -\infty$, and that asymptote is of course

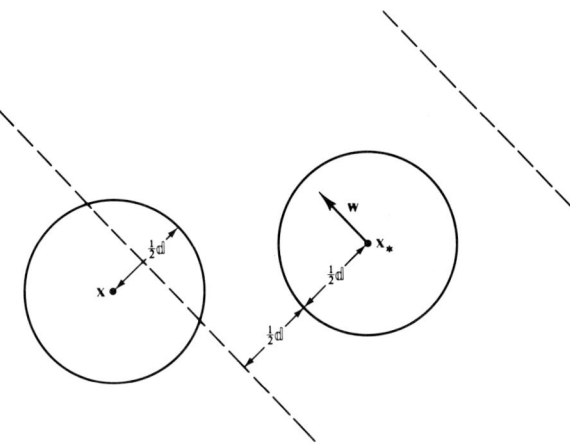

Figure VI.4

parallel to the velocity of approach **w**. If as before we denote by $\mathscr{P}$ the plane through **x** normal to **w**, the asymptote intersects $\mathscr{P}$ at some point $\mathbf{x}^e$ as is shown in Figure VI.5. Also $\mathbf{r}^e \equiv \mathbf{x}^e - \mathbf{x}$, and $r \equiv |\mathbf{r}^e|$. In terms of the point $\mathbf{x}^e$ we may associate a sphere of action with a molecule. Knowledge of the specific details of an encounter or of the law of intermolecular force should enable us to determine a set $\mathscr{S}$ lying in the plane $\mathscr{P}$ such that if $\mathbf{x}^e$ lies within $\mathscr{S}$ a collision occurs, while if $\mathbf{x}^e$ lies outside of $\mathscr{S}$ a collision does not occur. We should expect that if the intermolecular force extended to ∞, then $\mathscr{S}$ would be the whole plane $\mathscr{P}$. On the other hand, for molecules with a cut-off a collision does not take place if $\mathbf{x}^e$ lies outside some circle of finite radius d, and so $\mathscr{S}$ will be the interior of this circle. We call $\mathscr{S}$ the *cross-section* for encounters. In the solution of the encounter problem that we are constructing here we choose as parameters the two components on the plane $\mathscr{P}$ of the displacement vector $\mathbf{r}^e$, or

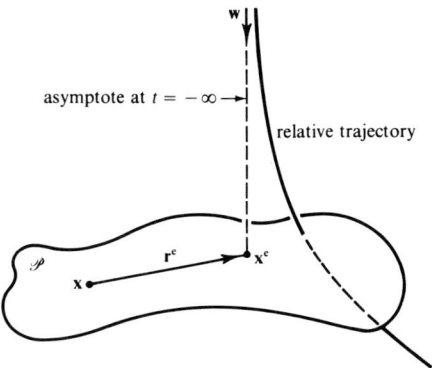

Figure VI.5

equivalently, the polar co-ordinates (r, ζ) of $\mathbf{r}^e$ with respect to $\boldsymbol{\xi}(\mathbf{w})$ and $\boldsymbol{\eta}(\mathbf{w})$:

$$\mathbf{r}^e = r(\cos \zeta \, \boldsymbol{\xi}(\mathbf{w}) + \sin \zeta \, \boldsymbol{\eta}(\mathbf{w})). \tag{VI.55}$$

This means that the parameter set $\mathscr{K}$ will be a subset of $\mathscr{R}^2$ containing the origin, $\mathscr{R}$ being the set of real numbers. For analytical convenience we shall simply denote this parameter set by the symbol $\mathscr{S}$. When we wish to emphasize the physical picture of an encounter, we shall revert to the equivalent interpretation of $\mathscr{S}$ set down originally, namely, as a subset of the plane $\mathscr{P}$ containing $\mathbf{x}$.

In this book we shall consider only encounters which are *planar*: The relative trajectory of a molecule lies in the plane containing $\mathbf{w}$, $\mathbf{x}$, and $\mathbf{x}^e$. This will be the case, for example, when the intermolecular force is central. We assume also that the relative trajectory has an asymptote at $t = +\infty$. Such an asymptote is parallel, of course, to $\mathbf{w}'$. This means that $\mathbf{w}'$ also lies in the plane containing $\mathbf{w}$, $\mathbf{x}$, and $\mathbf{x}^e$. The relations (VI.40), (VI.45), and (VI.55) justify Figure VI.6 as a representation of all parameters used in describing encounters.

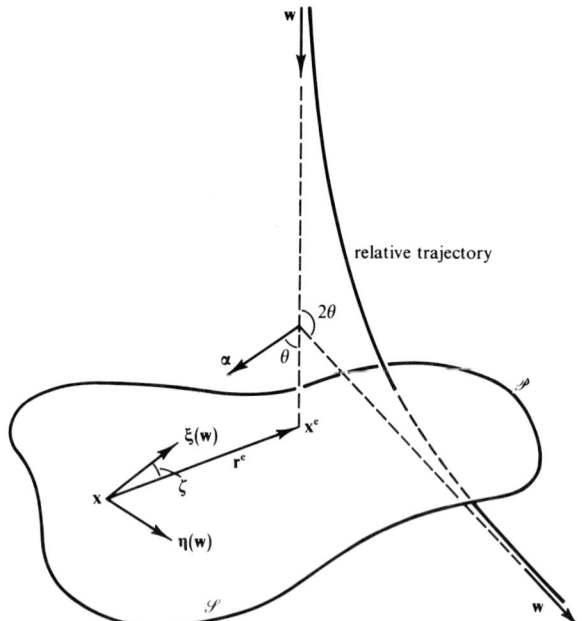

Figure VI.6 Variables used to describe an encounter

Once $\mathbf{w}$ and $\mathbf{r}^e$ shall have been prescribed, the outcome of an encounter will be known when the angle 2θ between $\mathbf{w}$ and $\mathbf{w}'$ is known. Of course θ will be different for different relative trajectories. We expect that a change in the relative trajectory may occur in response to a change in the velocity of approach $\mathbf{w}$, or to a different value of r, or indeed to a change in the plane in which the encounter takes place. Thus, we expect θ to be a function of the parameters that specify these effects:

$$\theta = \Theta(r, \zeta, \mathbf{w}). \tag{VI.56}$$

Even for planar encounters the class of molecules described by (VI.56) is quite general, indeed too general to be incorporated without restriction in the kinetic theory. Accordingly we shall restrict our attention henceforth to a subclass, that corresponding to *spherically symmetric* molecules[6]. For these the general relation (VI.56) reduces to

$$\theta = \Theta(r, w), \tag{VI.57}$$

and we shall take this as their definition. Analytically we require more, namely that Θ be smooth and that for each positive w we may invert (VI.57) and so express r in terms of θ and w. For convenience we take these restrictions to be part of the definition of spherically symmetric molecules. According to (VI.57) the relative trajectories in all the planes $\zeta = $ const. look the same. We may obtain one from another by a rotation about an axis that is parallel to $\mathbf{w}$ and passes through $\mathbf{x}$. As a consequence the cross-section $\mathscr{S}$ is symmetric with respect to such rotations. Hence it is either a disk or an entire plane.

To say more about an encounter than (VI.57) asserts, we must have specific knowledge of the relative trajectory; to obtain it, we must make use of analytical dynamics. Through the solution of a two-body problem, either by using the specific details of an encounter as for ideal spheres, or by solving an initial-value problem as when the intermolecular force is smooth and extends to ∞, we expect to be able to determine the relative trajectory and thence to determine the function Θ. In Chapter XII we shall exhibit some specimens of this calculation. However, to erect the general structure of the kinetic theory, we need not descend beyond (VI.57).

To that end we need a third solution of the encounter problem. It follows almost immediately from (VI.57), which allows us to express the unit vector $\boldsymbol{\alpha}$ in terms of the parameters r and ζ rather than θ and ζ. By reference to (VI.40) and (VI.57) we define a function $\mathbf{e}(r, \zeta, \mathbf{w})$ as follows:

$$\boldsymbol{\alpha} = \mathbf{e}(r, \zeta, \mathbf{w}) \equiv \mathbf{a}(\Theta(r, w), \zeta, \mathbf{w}). \tag{VI.58}$$

---

[6] Although molecules which are not spherically symmetric have been considered by several authors, as far as we have seen they have been used only in studies in equilibrium statistical mechanics, never in the kinetic theory. Indeed, the transformations (VII.7) and (VII.11), which are invoked in nearly every study of the kinetic theory, are not generally valid for molecules with a cut-off that are not spherically symmetric. This is due to the fact that the cross-section $\mathscr{S}$ is no longer a fixed set but instead varies with the orientations of the two molecules involved in an encounter.

Using **e** to determine **α** in (VI.33), we obtain
$$\begin{aligned}\mathbf{v}' &= \mathbf{v} + [\mathbf{w} \cdot \mathbf{e}(r, \zeta, \mathbf{w})]\mathbf{e}(r, \zeta, \mathbf{w}), \\ \mathbf{v}'_* &= \mathbf{v}_* - [\mathbf{w} \cdot \mathbf{e}(r, \zeta, \mathbf{w})]\mathbf{e}(r, \zeta, \mathbf{w}),\end{aligned} \qquad (\text{VI}.59)$$
and these define the function ϕ in (VI.24) for this third solution.

Next we take (r', ζ') as the parameter $\mathbf{k}'$, and we note from (VI.44), (VI.46), and (VI.57) that r' and ζ' are related to r and ζ by
$$\begin{aligned}\Theta(r', w') &= \Theta'(\Theta(r, w), \zeta, \mathbf{w}) = \Theta(r, w), \\ \zeta' &= Z'(\Theta(r, w), \zeta, \mathbf{w}).\end{aligned} \qquad (\text{VI}.60)$$
We know already by (VI.45)$_2$ that $w' = w$, so from the invertibility of Θ for fixed w we find that (VI.60)$_2$ reduces to
$$r' = r. \qquad (\text{VI}.61)$$
Moreover, the function Z' is already known, being given implicitly by (VI.48), so (VI.60)$_3$ is an explicit formula for ζ' as a function of r, ζ, and $\mathbf{w}$. Therefore (VI.60)$_3$ and (VI.61) define the function κ' appearing in (VI.26).

Finally we take (r_*, ζ_*) as the parameter $\mathbf{k}_*$, and we note that r_* and ζ_* will be related to r and ζ by (VI.54) after we substitute $\theta_* = \Theta(r_*, w_*)$ and $\theta = \Theta(r, w)$. Because $\mathbf{w}_* = -\mathbf{w}$, and hence $w_* = w$, we find again from the invertibility of Θ for fixed w that
$$r_* = r, \qquad (\text{VI}.62)$$
and ζ_* is given as a function of r, ζ, and $\mathbf{w}$ by (VI.54)$_{2,3}$. Hence (VI.54)$_{2,3}$ and (VI.62) define the function κ_* appearing in (VI.28). This completes our third solution of the encounter problem.

(iv) The encounter operator and its properties

Of the three solutions of the encounter problem that we have just seen, the third one is by far the most important. This is so not only because it is the only one in which the physical process of an encounter plays a role, but also because it will be used explicitly in the definition of the collisions operator $\mathbb{C}$. Its special status impels us to summarize it in terms of an operator $\mathbb{E}$ which we call the **Encounter Operator**[7]:
$$\begin{aligned}\mathbb{E}(\mathbf{v}, \mathbf{v}_*, r, \zeta) \equiv (&\mathbf{v} + [\mathbf{w} \cdot \mathbf{e}(r, \zeta, \mathbf{w})]\mathbf{e}(r, \zeta, \mathbf{w}), \\ &\mathbf{v}_* - [\mathbf{w} \cdot \mathbf{e}(r, \zeta, \mathbf{w})]\mathbf{e}(r, \zeta, \mathbf{w}), \\ &r, Z'(\Theta(r, w), \zeta, \mathbf{w})).\end{aligned} \qquad (\text{VI}.63)$$

---

[7] $\mathbb{E}$ was introduced by TRUESDELL [1973, Chapter VI], generalizing the operator J used by ARKERYD [1972, *1*, §0] and many earlier, informal descriptions of encounters. ARKERYD's J is smooth, while $\mathbb{E}$ need be only piecewise smooth, as our discussion makes plain.

(iv) THE ENCOUNTER OPERATOR

$\mathbb{E}$ is a piecewise smooth mapping of $\mathscr{V} \times \mathscr{V} \times \mathscr{P}$ into itself. The value of $\mathbb{E}$ at the argument $(\mathbf{v}, \mathbf{v}_*, r, \zeta)$ we denote by attaching primes:

$$(\mathbf{v}', \mathbf{v}'_*, r', \zeta') \equiv \mathbb{E}(\mathbf{v}, \mathbf{v}_*, r, \zeta). \tag{VI.64}$$

Then (VI.59), (VI.61), (VI.60)$_3$, and (VI.63) show us that $\mathbf{v}'$, $\mathbf{v}'_*$, r', and ζ' have the interpretations we have given them in preceding passages of this chapter. These primed variables have been defined in such a way that if we apply $\mathbb{E}$ to them, we recover $\mathbf{v}$, $\mathbf{v}_*$, r, and ζ. More concisely, we have **Property 1**. $\mathbb{E}$ *is its own inverse*:

$$\mathbb{E} \circ \mathbb{E} = \text{identity}. \tag{VI.65}$$

The fact that $\mathbf{v}$, $\mathbf{v}_*$, $\mathbf{v}'$, and $\mathbf{v}'_*$ related through (VI.64) satisfy the laws of conservation is summarized compactly in **Property 2**:

$$\mathbb{M} \circ \mathbb{E} = \mathbb{M}, \qquad \mathbb{K} \circ \mathbb{E} = \mathbb{K}, \tag{VI.66}$$

the operators $\mathbb{M}$ and $\mathbb{K}$ being defined by[8]

$$\begin{aligned} \mathbb{M}(\mathbf{v}, \mathbf{v}_*, r, \zeta) &\equiv \mathbf{v} + \mathbf{v}_*, \\ \mathbb{K}(\mathbf{v}, \mathbf{v}_*, r, \zeta) &\equiv v^2 + v_*^2. \end{aligned} \tag{VI.67}$$

Of course the letters $\mathbb{M}$ and $\mathbb{K}$ recall "momentum" and "kinetic energy". The parameters r_* and ζ_* have been defined in such a way that if we change from $(\mathbf{v}, \mathbf{v}_*, r, \zeta)$ to $(\mathbf{v}_*, \mathbf{v}, r_*, \zeta_*)$ in the argument of $\mathbb{E}$, a similar change occurs in its value. That is, $\mathbb{E}$ has **Property 3**:

$$\mathbb{X} \circ \mathbb{E} = \mathbb{E} \circ \mathbb{X}, \tag{VI.68}$$

the operator $\mathbb{X}$ being defined by

$$\mathbb{X}(\mathbf{v}, \mathbf{v}_*, r, \zeta) \equiv (\mathbf{v}_*, \mathbf{v}, r_*, \zeta_*), \tag{VI.69}$$

and r_* and ζ_* being given as functions of r and ζ by (VI.62) and (VI.54)$_{2,3}$. The letter $\mathbb{X}$ should recall "exchange".

Two additional properties of $\mathbb{E}$ that will be important later we pause now to record and prove. The first of these is **Property 4**: $\mathbb{E}$ *is locally volume-preserving*. Since area measure in the plane $\mathscr{P}$ is $r \, dr \, d\zeta$, we may state this property as follows:

$$\frac{r'}{r} \left| \frac{\partial(\mathbf{v}', \mathbf{v}'_*, r', \zeta')}{\partial(\mathbf{v}, \mathbf{v}_*, r, \zeta)} \right| = 1 \tag{VI.70}$$

at points where $\mathbb{E}$ is differentiable.

---

[8] ARKERYD [1972, *1* §0].

Proof. Let

$$r = R(\theta, w) \tag{VI.71}$$

denote the inverse of (VI.57). This implies immediately that

$$\partial_\theta R(\theta, w) \, \partial_r \Theta(r, w) = 1 \tag{VI.72}$$

for all r, θ, and w satisfying (VI.71). Using (VI.49), we find that

$$\frac{r'}{r} \left| \frac{\partial(\mathbf{v}', \mathbf{v}'_*, r', \zeta')}{\partial(\mathbf{v}, \mathbf{v}_*, r, \zeta)} \right| = \frac{r'}{r} \left| \frac{\partial(\mathbf{v}', \mathbf{v}'_*, r', \zeta')}{\partial(\mathbf{v}', \mathbf{v}'_*, \theta', \zeta')} \right|$$

$$\times \left| \frac{\partial(\mathbf{v}', \mathbf{v}'_*, \theta', \zeta')}{\partial(\mathbf{v}, \mathbf{v}_*, \theta, \zeta)} \right| \times \left| \frac{\partial(\mathbf{v}, \mathbf{v}_*, \theta, \zeta)}{\partial(\mathbf{v}, \mathbf{v}_*, r, \zeta)} \right|, \tag{VI.73}$$

$$= \frac{r'}{r} \partial_\theta R(\theta', w') \, \partial_r \Theta(r, w).$$

But we know by (VI.61), (VI.46), and (VI.45)$_2$ that $r' = r$, $\theta' = \theta$, and $w' = w$. Therefore (VI.70) follows from (VI.72) and (VI.73). △

The next property of $\mathbb{E}$ is its orthogonal invariance. It is well known that the equations of analytical dynamics are invariant under constant orthogonal transformations. For the purposes of our study of encounters we may take this to mean that by applying an orthogonal transformation $\mathbf{Q}$ to the asymptotes of the trajectory in Figure VI.6, we obtain a new relative trajectory for some encounter. If the first encounter is determined by $\mathbf{v}$, $\mathbf{v}_*$, and $\mathbf{r}^e$, we expect that the second will be determined by $\mathbf{Qv}$, $\mathbf{Qv}_*$, and $\mathbf{Qr}^e$. In order to state this conjecture in terms of $\mathbb{E}$, we let $\mathbf{Q}$ denote not only an orthogonal tensor over $\mathscr{V}$ but also the action of that tensor on points of $\mathscr{V} \times \mathscr{V} \times \mathscr{S}$, defined as follows:

$$\mathbf{Q}(\mathbf{v}, \mathbf{v}_*, \mathbf{r}^e) \equiv (\mathbf{Qv}, \mathbf{Qv}_*, \mathbf{Qr}^e), \tag{VI.74}$$

and we view $\mathbb{E}$ now as a function of $\mathbf{v}$, $\mathbf{v}_*$, and $\mathbf{r}^e$ rather than of $\mathbf{v}$, $\mathbf{v}_*$, r, and ζ. This is possible because of the simple relation (VI.55). Our conjecture may now be stated as **Property 5.** $\mathbb{E}$ *is orthogonally invariant:*

$$\mathbf{Q} \circ \mathbb{E} = \mathbb{E} \circ \mathbf{Q} \quad \text{for all orthogonal } \mathbf{Q}. \tag{VI.75}$$

Proof. First we note that there is an angle $\Phi(\mathbf{Q}, \mathbf{w})$ such that

$$\mathbf{Q}\boldsymbol{\xi}(\mathbf{w}) = \cos \Phi(\mathbf{Q}, \mathbf{w}) \, \boldsymbol{\xi}(\mathbf{Qw}) + \sin \Phi(\mathbf{Q}, \mathbf{w}) \, \boldsymbol{\eta}(\mathbf{Qw}),$$
$$\mathbf{Q}\boldsymbol{\eta}(\mathbf{w}) = -\sin \Phi(\mathbf{Q}, \mathbf{w}) \, \boldsymbol{\xi}(\mathbf{Qw}) + \cos \Phi(\mathbf{Q}, \mathbf{w}) \, \boldsymbol{\eta}(\mathbf{Qw}), \tag{VI.76}$$

for all vectors $\mathbf{w}$ and for all orthogonal tensors $\mathbf{Q}$. This is a consequence of two facts: $\mathbf{Q}\boldsymbol{\xi}(\mathbf{w})$ and $\mathbf{Q}\boldsymbol{\eta}(\mathbf{w})$ are orthonormal and lie in the plane orthogonal to $\mathbf{Qw}$, and, by definition of $\boldsymbol{\xi}$ and $\boldsymbol{\eta}$, $\boldsymbol{\xi}(\mathbf{Qw})$ and $\boldsymbol{\eta}(\mathbf{Qw})$ are also an orthonormal pair lying in this same plane. The angle $\Phi(\mathbf{Q}, \mathbf{w})$ defines a rotation that takes one

(iv) THE ENCOUNTER OPERATOR

pair into the other. If we apply $\mathbf{Q}$ to (VI.55) and use (VI.76), we find that

$$\mathbf{Qr}^e = r\{\cos[\zeta + \Phi(\mathbf{Q}, \mathbf{w})]\,\xi(\mathbf{Qw}) + \sin[\zeta + \Phi(\mathbf{Q}, \mathbf{w})]\,\eta(\mathbf{Qw})\}. \quad \text{(VI.77)}$$

Since the polar co-ordinates of $\mathbf{r}^{e'}$ are r' and ζ' as given by (VI.61) and (VI.60)$_3$, by similar reasoning we find that

$$\mathbf{Qr}^{e'} = r\{\cos[\zeta' + \Phi(\mathbf{Q}, \mathbf{w}')]\,\xi(\mathbf{Qw}') + \sin[\zeta' + \Phi(\mathbf{Q}, \mathbf{w}')]\,\eta(\mathbf{Qw}')\}. \quad \text{(VI.78)}$$

If we apply $\mathbf{Q}$ to both sides of (VI.59), we see from (VI.58) and (VI.40) that the result will be precisely (VI.75) provided we can prove that $(\mathbf{Qr}^e)' = \mathbf{Qr}^{e'}$. Equivalently, we must show that $(\mathbf{Qr}^e)'$ and $\mathbf{Qr}^{e'}$ have the same polar co-ordinates. From (VI.77) the polar co-ordinates of $\mathbf{Qr}^e$ are r and $\zeta + \Phi(\mathbf{Q}, \mathbf{w})$; hence, by (VI.61) and (VI.60)$_3$, those of $(\mathbf{Qr}^e)'$ are r and $Z'(\Theta(r, \mathbf{w}), \zeta + \Phi(\mathbf{Q}, \mathbf{w}), \mathbf{Qw})$. Comparing these with the polar co-ordinates of $\mathbf{Qr}^{e'}$ given by (VI.78), we reduce our problem to proof that

$$Z'(\Theta(r, \mathbf{w}), \zeta, \mathbf{w}) + \Phi(\mathbf{Q}, \mathbf{w}') = Z'(\Theta(r, \mathbf{w}), \zeta + \Phi(\mathbf{Q}, \mathbf{w}), \mathbf{Qw}). \quad \text{(VI.79)}$$

Recalling now that we obtained the function Z' by solving (VI.47) for ζ' as a function of θ, ζ, $\mathbf{w}$, we apply $\mathbf{Q}$ to this equation. Using (VI.76), we find that

$$\begin{aligned}-\{\sin[\zeta' + \Phi(\mathbf{Q}, \mathbf{w}')]\,\xi(\mathbf{Qw}') - \cos[\zeta' + \Phi(\mathbf{Q}, \mathbf{w}')]\,\eta(\mathbf{Qw}')\} \\ = \sin[\zeta + \Phi(\mathbf{Q}, \mathbf{w})]\,\xi(\mathbf{Qw}) - \cos[\zeta + \Phi(\mathbf{Q}, \mathbf{w})]\,\eta(\mathbf{Qw}).\end{aligned} \quad \text{(VI.80)}$$

But this is equivalent, by (VI.47) and its solution (VI.44)$_2$, to

$$\zeta' + \Phi(\mathbf{Q}, \mathbf{w}') = Z'(\Theta(r, \mathbf{w}), \zeta + \Phi(\mathbf{Q}, \mathbf{w}), \mathbf{Qw}). \quad \text{(VI.81)}$$

Since ζ' is given by (VI.44)$_2$ also, we arrive at (VI.79). This completes the proof of (VI.75). △

We have emphasized the third solution of the encounter problem because it lies close to the two-body problem of analytical dynamics. This solution is actually the general member of an infinite family of solutions of the encounter problem, one for each choice of Θ in the defining relation (VI.57). For each one there is a corresponding encounter operator $\mathbb{E}$. The first and second solutions, though less important, nevertheless do play a role. In some calculations we shall find it useful to be able to move back and forth between one solution and another. Locally we may do this by a simple change of parameter, for this is precisely how each was initially derived. Globally, however, we must be careful that the parameter sets $\mathscr{K}$ for the two solutions agree under this change. To be specific, let us look at what happens when we change from r and ζ as parameters to θ and ζ. In both cases ζ may take on any value in the interval

$$0 \leqq \zeta < 2\pi. \quad \text{(VI.82)}$$

Corresponding intervals for r and θ, however, will depend upon the properties of the function Θ. We consider two cases:

Case 1. Molecules with intermolecular forces of infinite range. In this case $\mathscr{S}$ is an entire plane, and so r takes on values in the interval $0 \leq r < \infty$. For such molecules we assume always that

$$\begin{aligned} \Theta(r, w) &\to \tfrac{1}{2}\pi \quad \text{as} \quad r \to \infty, \\ \Theta(r, w) &\to 0 \quad \text{as} \quad r \to 0. \end{aligned} \tag{VI.83}$$

This being so, θ may take on any value in the interval

$$0 \leq \theta \leq \tfrac{1}{2}\pi. \tag{VI.84}$$

Case 2. Molecules with a cut-off. Here $\mathscr{S}$ is a disk of finite radius $\mathbb{d}$, and so r lies in the interval $0 \leq r \leq \mathbb{d}$. We retain (VI.83)$_2$, and in place of (VI.83)$_1$ we introduce a cut-off angle by means of the function $\mathbb{A}$ defined by

$$\mathbb{A}(w) \equiv \Theta(\mathbb{d}, w). \tag{VI.85}$$

Then θ may take on values in the interval

$$0 \leq \theta \leq \mathbb{A}(w). \tag{VI.86}$$

Corresponding values of the unit vector $\boldsymbol{\alpha}$ are readily deduced from (VI.82), (VI.84), (VI.86), and (VI.40).

Appendix A

Proof of the Lemma

Lemma. *Let g be a real-valued measurable function on $\mathscr{V}$ which is odd:*

$$g(-\mathbf{v}) = -g(\mathbf{v}), \tag{VI.7}_r$$

and which is additive on orthogonal pairs:

$$g(\mathbf{v} + \mathbf{v}_*) = g(\mathbf{v}) + g(\mathbf{v}_*) \quad \text{if} \quad \mathbf{v} \cdot \mathbf{v}_* = 0. \tag{VI.8}_r$$

Then g is linear on $\mathscr{V}$.

Proof. It suffices to show that g is additive on $\mathscr{V}$. First we show that g is additive on parallel pairs.

Let $\mathbf{v}$ and $\mathbf{v}_*$ be any two parallel vectors. If $\mathbf{v} = \mathbf{0}$ or $\mathbf{v} = -\mathbf{v}_*$, (VI.7) and (VI.8) make the assertion trivial. Thus there is no loss in generality if we suppose that $\mathbf{v} \neq \mathbf{0}$ and that $\mathbf{v}_*$ is longer than $\mathbf{v}$ unless it is $\mathbf{v}$. We let $\mathbf{n}$ be the unit vector such that

$$\mathbf{v} = \lambda \mathbf{n}, \quad \lambda > 0, \quad \mathbf{v}_* = \lambda_* \mathbf{n}. \tag{VIA.1}$$

APPENDIX A: PROOF OF LEMMA

Then either $\lambda_* \geq \lambda$ or $\lambda_* < -\lambda$.

Case 1. $\lambda_* > 0$. Then for any unit vector **m** orthogonal to **n**

$$\lambda \mathbf{n} \cdot \sqrt{\lambda \lambda_*}\, \mathbf{m} = 0, \qquad \lambda_* \mathbf{n} \cdot (-\sqrt{\lambda \lambda_*}\, \mathbf{m}) = 0,$$
$$(\lambda \mathbf{n} + \sqrt{\lambda \lambda_*}\, \mathbf{m}) \cdot (\lambda_* \mathbf{n} - \sqrt{\lambda \lambda_*}\, \mathbf{m}) = 0, \qquad \text{(VIA.2)}$$

and so from (VI.7) and (VI.8) we conclude that

$$\begin{aligned} g(\mathbf{v} + \mathbf{v}_*) &= g([\lambda \mathbf{n} + \sqrt{\lambda \lambda_*}\, \mathbf{m}] + [\lambda_* \mathbf{n} - \sqrt{\lambda \lambda_*}\, \mathbf{m}]), \\ &= g(\lambda \mathbf{n} + \sqrt{\lambda \lambda_*}\, \mathbf{m}) + g(\lambda_* \mathbf{n} - \sqrt{\lambda \lambda_*}\, \mathbf{m}), \\ &= g(\lambda \mathbf{n}) + g(\sqrt{\lambda \lambda_*}\, \mathbf{m}) + g(\lambda_* \mathbf{n}) + g(-\sqrt{\lambda \lambda_*}\, \mathbf{m}), \\ &= g(\mathbf{v}) + g(\mathbf{v}_*). \end{aligned} \qquad \text{(VIA.3)}$$

Case 2. $\lambda_* < -\lambda$. Then

$$\lambda \mathbf{n} \cdot [-(\lambda + \lambda_*)\mathbf{n}] = -\lambda(\lambda + \lambda_*) > 0, \qquad \text{(VIA.4)}$$

and so by (VI.7) and Case 1 we find that

$$\begin{aligned} g(\mathbf{v}_*) &= -g(-\lambda_* \mathbf{n}), \\ &= -g(-[\lambda + \lambda_*]\mathbf{n} + \lambda \mathbf{n}), \\ &= -g(-[\lambda + \lambda_*]\mathbf{n}) - g(\lambda \mathbf{n}), \\ &= g(\mathbf{v} + \mathbf{v}_*) - g(\mathbf{v}). \end{aligned} \qquad \text{(VIA.5)}$$

Having shown that g is additive on parallel pairs as well as orthogonal pairs, we now take any two vectors **v** and $\mathbf{v}_*$ and resolve the latter into components parallel and perpendicular to the former:

$$\mathbf{v}_* = \xi \mathbf{v} + \mathbf{m}, \qquad \mathbf{m} \cdot \mathbf{v} = 0. \qquad \text{(VIA.6)}$$

Then

$$\begin{aligned} g(\mathbf{v} + \mathbf{v}_*) &= g(\mathbf{v} + \xi \mathbf{v} + \mathbf{m}), \\ &= g(\mathbf{v} + \xi \mathbf{v}) + g(\mathbf{m}), \\ &= g(\mathbf{v}) + g(\xi \mathbf{v}) + g(\mathbf{m}), \\ &= g(\mathbf{v}) + g(\xi \mathbf{v} + \mathbf{m}), \\ &= g(\mathbf{v}) + g(\mathbf{v}_*). \quad \triangle \end{aligned} \qquad \text{(VIA.7)}$$

This proof makes no use of the dimension of $\mathscr{V}$, so the lemma is valid for any finite-dimensional inner-product space $\mathscr{V}$.

Chapter VII

The Maxwell Collisions Operator. Kinetic Constitutive Relations. The Total Collisions Operator and Bilinear Form

(i) The collisions operator

In Section (v) of Chapter III we introduced the symbol $\mathbb{C}$ to denote an operator whose action upon a molecular density F is designed to deliver the rate $\mathbb{C}F$ at which F is increasing through effects of collisions. MAXWELL presented[1] an astute and ingenious mixture of dynamic and stochastic ideas to give the *collisions operator* $\mathbb{C}$ a definite structure in terms of the encounter operator $\mathbb{E}$ and the plane $\mathscr{P}$. In this book we shall take the particular form adopted for $\mathbb{C}$ as *an axiom on which to base the mathematical theory*. In order to make that axiom clear enough to be useful, we must offer a few words of motivation.

MAXWELL pictured the gas as being sufficiently rarefied that all encounters would be binary, and that the outcome of each encounter between two molecules would always be just the same as if no other molecules existed. Furthermore, he regarded the asymptotic velocities v, v_*, v', and v'_* as indistinguishable from the actual velocities of the molecules. Thus, at a place x and time t, an encounter between molecules of velocities v and v_* imparts velocities v' and v'_* to them at once. Every pair of molecules at x and x_* with the velocities v and v_*, respectively, is thus instantly converted into a pair with velocities v' and v'_*. Although classical mechanics delivers this transformation, in general, only as the result of the interaction *for all time* of two bodies alone in

---

[1] MAXWELL [1867, Eqs. (1), (2), (3)] introduced what we call the total collisions operator $\bar{\mathbb{C}}_F$ in the form obtained by putting $F = G = H$ in (VII.20)$_4$. BOLTZMANN [1872, Eqs. (16), (44)] [1875, §1] perceived the importance of the kernel of $\bar{\mathbb{C}}_F$, which we here denote by $\mathbb{C}$. *Cf.* also MAXWELL [1879, Eq. (8)]. The standard "derivation" is that of BOLTZMANN [1896, §§15–16]. Another, perhaps more sophisticated, is given and supplemented with a critique by GRAD [1958, §§7–14].

infinite space—an interaction which typically yields smooth trajectories—
MAXWELL applied it as an *instantaneous* conversion in a space full of arbitrarily
many bodies, some of which could be arbitrarily close to each other.

Moreover, when it comes to calculating the part $\mathscr{S}$ of the plane $\mathscr{P}$ that
corresponds to possible encounters, MAXWELL's picture allows us to include or
exclude a point $\mathbf{x}^e$ on $\mathscr{P}$ according as it does or does not lie within the sphere of
action of the molecule at $\mathbf{x}$. For this purpose, that is, the relative trajectory is
replaced by its asymptote at $t = -\infty$.

MAXWELL supposed also that any two molecules would be sufficiently near
to one another during an encounter that the difference between their positions
$\mathbf{x}$ and $\mathbf{x}_*$ could be neglected in the arguments of F. Thus the expectation of a set
of velocities is assumed not to change much from place to place in the gas, at
any one time.

With these draconic simplifications, MAXWELL replaced the laws of analytical mechanics by a schema of *instantaneous and discontinuous conversions of velocities*. The conversions themselves he took as being those of the asymptotic velocities that result from solution of the problem of two bodies in infinite space according to Newtonian mechanics, some particular law of mutual force being laid down.

Finally, MAXWELL assumed that the probability density for a pair of
molecules with velocities $\mathbf{v}_1$ and $\mathbf{v}_2$ at $(t, \mathbf{x})$ was proportional to
$F(t, \mathbf{x}, \mathbf{v}_1) F(t, \mathbf{x}, \mathbf{v}_2)$. That is, the pairs are stochastically independent. The
physicists use the term "molecular chaos" to refer to this assumption[2]. It
represents a sort of fading memory, since it wipes out the past histories of
the two molecules as far as encounters are concerned.

To express the result of applying these ideas, we introduce a condensed
notation for the arguments of a function g of t, $\mathbf{x}$, and $\mathbf{v}$:

$$
\begin{aligned}
g &\quad \text{stands for} \quad g(t, \mathbf{x}, \mathbf{v}), \\
g_* &\quad \text{stands for} \quad g(t, \mathbf{x}, \mathbf{v}_*), \\
g' &\quad \text{stands for} \quad g(t, \mathbf{x}, \mathbf{v}'), \\
g'_* &\quad \text{stands for} \quad g(t, \mathbf{x}, \mathbf{v}'_*).
\end{aligned}
\tag{VII.1}
$$

According to MAXWELL's picture of a gas, the density of the rate of decrease of
F at $(t, \mathbf{x})$ due to collisions with molecules of velocity $\mathbf{v}_*$ is proportional to FF_*
evaluated at $(t, \mathbf{x})$. On the other hand, molecules of velocity $\mathbf{v}$ are being
produced as the result of collisions. Since the equations of analytical dynamics
are invariant under time reversal, the pair of velocities $(\mathbf{v}, \mathbf{v}_*)$ results as the
outcome of a collision between molecules of velocities $(\mathbf{v}', \mathbf{v}'_*)$ as determined by
the encounter operator $\mathbb{E}$. Thus the density of the rate of increase of F at $(t, \mathbf{x})$
due to collisions is proportional to $F'F'_*$, evaluated at $(t, \mathbf{x})$. The net rate of

---

[2] *Cf.* the discussion by GRAD [1958, §11].

increase in the density of molecules with velocity **v** by collisions is proportional to the difference $F'F'_* - FF_*$, weighted by an appropriate factor to convert it into a molecular density, and then integrated over all velocities $\mathbf{v}_*$ and all possible relative positions. The appropriate factor is $w\, dS$, in which dS is the element of area in the plane $\mathscr{P}$. This is so because in the time dt the oncoming molecule travels a distance $w\, dt$, so the volume swept out in that time by all the molecules above $\mathscr{S}$ on $\mathscr{P}$ is $(\int_{\mathscr{S}} w\, dS)\, dt$, the cross-section $\mathscr{S}$ being the part of $\mathscr{P}$ that corresponds to the possible encounters. Thus, finally the **Maxwell Collisions Operator** is defined as follows:

$$\mathbb{C}F \equiv \int_{\mathscr{V}} \int_{\mathscr{S}} w(F'F'_* - FF_*)\, dS\, d\mathbf{v}_*. \qquad (\text{VII.2})$$

We shall henceforth drop from the notation dS, the element of area on $\mathscr{P}$, and also $d\mathbf{v}_*$ and the space $\mathscr{V}$, thereby writing $\mathbb{C}F$ in the abbreviated form

$$\mathbb{C}F = \int \int_{\mathscr{S}} w(F'F'_* - FF_*). \qquad (\text{VII.3})$$

The reader will not confuse this convention with that of Section (i) of Chapter III for integrals of functions of **v** only, for here we refer to functions of **v**, $\mathbf{v}_*$, and position on $\mathscr{P}$, and we hold **v** fixed while integrating over the other variables.

The foregoing description of the Maxwell collisions operator is traditional. However, recognizing the many questionable aspects in MAXWELL's schematic picture of a gas, in this book we prefer a detached, mathematical approach. *The definition* (VII.2) *of* $\mathbb{C}$, *while it is motivated by certain somewhat plausible inferences, is from the standpoint of rational mechanics an axiom.* Since every term in it is unequivocally defined once a particular encounter operator $\mathbb{E}$ and cross-section $\mathscr{S}$ shall have been specified, there is no difficulty in principle in calculating $\mathbb{C}F$ for a given F, provided the integrals converge. Thus $\mathbb{C}$ is a well-defined functional operator, and as such we shall treat it.

(ii) Kinetic constitutive quantities

It is also traditional to refer $\mathbb{E}$ and $\mathscr{S}$ to the specification and solution of a two-body problem of analytical dynamics, *e.g.* the problem of encounter of two ideal spheres of diameter $\mathbb{d}$, or the problem of calculating the asymptotic relative velocities of a mass-point repelled by another with a force inversely proportional to the $\mathbb{k}^{\text{th}}$ power of the intervening distance. The solution of this problem then determines the basic formula (VI.57) and hence by use of (VI.58) determines also $\mathbb{E}$ as specified by (VI.63). We shall later record some important special cases obtained in this way. However, in contrast with this traditional point of view, we choose to regard as fundamental not the two-body problem but rather $\mathbb{E}$ and $\mathscr{S}$ themselves. Accordingly we shall redefine these now as primitive quantities: A *Cross-section* $\mathscr{S}$ is a non-empty open sphere in $\mathscr{R}^2$ centered at the origin, $\mathscr{R}$ denoting the real line. If $\mathscr{S}$ is bounded, we shall refer to it simply as a disk. An *Encounter Operator* $\mathbb{E}$ is any piecewise continuously differentiable function

$$\mathbb{E}: \mathscr{V} \times \mathscr{V} \times \mathscr{S} \to \mathscr{V} \times \mathscr{V} \times \mathscr{S}$$

which satisfies

$$\mathbb{E} \circ \mathbb{E} = \text{identity}; \tag{VI.65}_r$$

$$\mathbb{M} \circ \mathbb{E} = \mathbb{M},$$

$$\mathbb{K} \circ \mathbb{E} = \mathbb{K}; \tag{VI.66}_r$$

$$\mathbb{X} \circ \mathbb{E} = \mathbb{E} \circ \mathbb{X}; \tag{VI.68}_r$$

$$\mathbf{Q} \circ \mathbb{E} = \mathbb{E} \circ \mathbf{Q} \quad \text{for all orthogonal } \mathbf{Q}; \tag{VI.75}_r$$

$$\frac{r'}{r} \left| \frac{\partial(\mathbf{v}', \mathbf{v}'_*, r', \zeta')}{\partial(\mathbf{v}, \mathbf{v}_*, r, \zeta)} \right| = 1; \tag{VI.70}_r$$

the operators $\mathbb{M}$, $\mathbb{K}$, and $\mathbb{X}$ are defined as follows:

$$\mathbb{M}(\mathbf{v}, \mathbf{v}_*, r, \zeta) \equiv \mathbf{v} + \mathbf{v}_*,$$
$$\mathbb{K}(\mathbf{v}, \mathbf{v}_*, r, \zeta) \equiv v^2 + v_*^2; \tag{VI.67}_r$$

$$\mathbb{X}(\mathbf{v}, \mathbf{v}_*, r, \zeta) \equiv (\mathbf{v}_*, \mathbf{v}, r_*, \zeta_*); \tag{VI.69}_r$$

and the action of an orthogonal tensor $\mathbf{Q}$ on points of $\mathscr{V} \times \mathscr{V} \times \mathscr{S}$ is denoted also by $\mathbf{Q}$:

$$\mathbf{Q}(\mathbf{v}, \mathbf{v}_*, \mathbf{r}^e) \equiv (\mathbf{Q}\mathbf{v}, \mathbf{Q}\mathbf{v}_*, \mathbf{Q}\mathbf{r}^e). \tag{VI.74}_r$$

The quantities $\mathbf{v}$, $\mathbf{v}_*$, r, ζ, r_*, ζ_*, and $\mathbf{r}^e$ are to be interpreted as in Chapter VI, and of course

$$(\mathbf{v}', \mathbf{v}'_*, r', \zeta') \equiv \mathbb{E}(\mathbf{v}, \mathbf{v}_*, r, \zeta). \tag{VI.64}_r$$

Although formally $\mathbf{v}$, $\mathbf{v}_*$, $\mathbf{v}'$, and $\mathbf{v}'_*$ are just vectors, to promote the intended interpretation we shall continue to call them "asymptotic velocities of an encounter". According to (VI.66) and (VI.67), an encounter operator respects conservation of the linear momentum and the energy of a pair of mass-points of equal mass. By (VI.65), $\mathbb{E}$ is its own inverse, and by (VI.70) it is locally volume-preserving. In view of (VI.62) and (VI.54)$_{2,3}$ the transformation $\mathbb{X}$ is its own inverse, and it is also locally volume-preserving. Moreover (VI.68) expresses the fact that when $\mathbf{v}$ and $\mathbf{v}_*$ are interchanged, $\mathbf{v}'$ and $\mathbf{v}'_*$ are also interchanged. Finally $\mathbb{E}$ commutes with any orthogonal tensor $\mathbf{Q}$. Since the operator $\mathbb{E}$ defined by (VI.63) has all these properties, we see that this general definition includes as a special case those operators $\mathbb{E}$ which arise from the solution of some two-body problem of analytical mechanics, the bodies being spherically symmetric in the sense defined by (VI.57).

For the purposes of the rational kinetic theory, we may state our position somewhat more abstractly by listing the **Constitutive Quantities of the Kinetic Theory**:

(a) The *molecular mass* $\mathbb{m}$. This assignable constant is used in the definitions of the mass density, the pressure tensor, and the energy flux, namely, (III.5), (III.9), and (III.12).

(b) The *encounter operator* $\mathbb{E}$. This operator determines $\mathbf{v}'$ and $\mathbf{v}'_*$ when $\mathbf{v}$, $\mathbf{v}_*$, r, and ζ are assigned arbitrarily. $\mathbb{E}$ may itself depend upon an intermolecular force-constant $\mathbb{g}$ and a molecular diameter or other characteristic length $\mathbb{d}$. Calculation of $\mathbb{E}$ from a particular problem of analytical dynamics may require that $\mathbb{m}$ be specified. On the other hand, $\mathbb{E}$ may be introduced *a priori* as an operator satisfying (VI.65), (VI.66), (VI.68), (VI.70), and (VI.75).

(c) The *cross-section* $\mathscr{S}$. This set may be either a disk or all of $\mathscr{R}^2$. Its specification may depend upon a molecular diameter $\mathbb{d}$.

The field of body force **b**, which defines the retrogressor, is regarded as external rather than constitutive—that is, as specifying the circumstances of the gas rather than its constitution. We always presume it to be a prescribed function of place alone.

(iii) *Alternative forms of the collisions operator*

If $\mathscr{S}$ is a disk, the intermolecular forces are said to have a *cut-off*. The classical example of molecules whose intermolecular forces have a cut-off is provided by ideal spheres[3], but it is possible to imagine intermolecular forces of any sort whatever within a sphere of diameter $\mathbb{d}$ about each molecule, and then specify that they fall straight off to zero as that sphere is crossed. If the intermolecular forces have a cut-off, it is reasonable, following the conventions of notation in (VII.3), to set

$$\mathbb{F}F \equiv \int\!\!\int_{\mathscr{S}} wF_*, \qquad \text{(VII.4)}$$

---

[3] Rough development of a kinetic theory which presumes the molecules to be spherical or at least takes such molecules as prototype goes back to the work of HERAPATH, WATERSTON, and CLAUSIUS. The idea of a more general kind of molecule with a cut-off was suggested by WATERSTON, who used the term "surface of powerful molecular repulsion". *Cf.* TRUESDELL [1975, §3].

VII. THE COLLISIONS OPERATOR

and to expect that the integrals on the right-hand side converge[4]. The value of $\mathbb{F}F$ for a given F is a scalar field. If $v \equiv \mathbb{F}F$, we easily interpret $v(t, \mathbf{x}, \mathbf{v})$ as the expected number of collisions per unit time that a molecule located at the place $\mathbf{x}$ and having the velocity $\mathbf{v}$ at the time t will experience; that is, the field v is the *collision frequency*[5] corresponding to F. Thus $\mathbb{F}$ may be called the *frequency operator*. As we have said, for molecules with a cut-off we expect $\mathbb{F}$ to be defined on a broad domain of functions F, but if the intermolecular forces extend to ∞, $\mathbb{F}$ generally fails to exist for any function F other than 0.

For molecules with a cut-off, then, we can expect $\mathbb{C}F$ to be the difference of two convergent integrals. If so, we shall be able to write (VII.2) as follows in terms of $\mathbb{F}$:

$$\mathbb{C}F = -(\mathbb{F}F)F + \mathbb{P}F. \qquad \text{(VII.5)}$$

$\mathbb{P}$, which is given by the definition

$$\mathbb{P}F \equiv \int_{\mathscr{S}} \int wF'F'_* , \qquad \text{(VII.6)}$$

is the *partial collisions operator*. The value of $\mathbb{P}F$ is the rate of increase of F due to molecules caused to appear with velocity $\mathbf{v}$ at $(t, \mathbf{x})$ by encounters. *If $\mathscr{S}$ is not bounded, the integrals that define $\mathbb{F}$ and $\mathbb{P}$ generally diverge, and the decomposition* (VII.5) *is then not possible*.

When the encounter operator $\mathbb{E}$ is derived from a two-body problem of analytical dynamics, the collisions operator $\mathbb{C}$ can be expressed in an alternative form by use of the angles θ and ζ rather than r and ζ to describe collisions.

---

[4] BOLTZMANN always began his discussions leading to the collisions operator with statements about $\int_{\mathscr{V}} \int_{\mathscr{S}} wFF_* \, dS \, d\mathbf{v}_*$ and $\int_{\mathscr{V}} \int_{\mathscr{S}} wF'F'_* \, dS \, d\mathbf{v}_*$, thus seeming to presume that both of these integrals converged. However, in assessing work by physicists of the last century we must take care not to let ourselves be misled by the loose language then current. It can scarcely have escaped BOLTZMANN that for a gas of inverse k^{th}-power molecules (*cf.* Chapter IV), which he commonly considered, these integrals diverge (*cf.* the next footnote), so he must have intended his remarks about them to be merely heuristic steps toward (VII.2).

[5] We may, if we like, interpret $c/v(t, \mathbf{x}, \mathbf{v})$ as the *mean free path* for a molecule of random velocity $\mathbf{c}$ at $(t, \mathbf{x})$. A mean free path independent of $\mathbf{c}$ may be defined in various ways; for example, $\overline{c/v}$ is such a quantity. In Footnote 3 to Chapter XII we shall calculate this expectation under particular assumptions. The mean free path played a great role in the early, informal kinetic theories, and the term is still common in the conversation of physicists and aerospatial engineers.

The laws of intermolecular force studied in the mathematical kinetic theory, which is the main subject of this book, are often of infinite range, *e.g.* for a gas of inverse k^{th}-power molecules. For them, as MAXWELL [1879, first footnote] remarked in effect, the integral defining v generally diverges; we may say, if we like, that the collision frequency is infinite and the mean free path is 0, but, as MAXWELL also remarked, other characteristic lengths associated with a gas flow may be derived. It must be one of these to which physicists refer when they use the term "mean free path". We shall pursue the matter further in Footnote 3 to Chapter VIII.

(iii) ALTERNATIVE FORMS OF $\mathbb{C}$

First consider molecules subject to forces that extend to ∞. Then $\mathscr{S} = \mathscr{R}^2$. With R denoting the inverse of Θ as in (VI.71), from (VI.82) and (VI.84) we see that

$$\int_{\mathscr{S}} \cdots w \, dS = \int_0^{2\pi} d\zeta \int_0^{\infty} dr \cdots rw = \int_0^{2\pi} d\zeta \int_0^{\frac{1}{2}\pi} d\theta \sin \theta \cdots \mathbb{S}(\theta, w), \quad \text{(VII.7)}$$

the *Scattering Factor* $\mathbb{S}$ being defined through the relation

$$\sin \theta \, \mathbb{S}(\theta, w) \equiv wR(\theta, w)|\partial_\theta R(\theta, w)|. \quad \text{(VII.8)}$$

Thus

$$\mathbb{S}(\theta, w) > 0 \quad \text{if} \quad 0 < \theta \leq \tfrac{1}{2}\pi, \quad w \neq 0. \quad \text{(VII.9)}$$

When the foregoing calculations and inversions are permissible, the integral with respect to surface area on $\mathscr{S}$ may be expressed as an integral with respect to solid angle over the unit hemisphere $\varominus$:

$$\mathbb{C}F = \int\int_{\varominus} \mathbb{S}(F'F'_* - FF_*). \quad \text{(VII.10)}$$

Of course $\int_{\varominus}$ stands for $\int_{\varominus} \cdots \sin \theta \, d\theta \, d\zeta$.

For molecules with a cut-off $\mathscr{S}$ is a disk. We lose no generality by denoting the radius of that disk by $\mathbb{d}$ in general. For molecules that are ideal spheres $\mathbb{d}$ will be their common diameter. For any molecules with a cut-off, $\mathbb{d}$ may serve as the "characteristic length" we have mentioned above. Then (VI.86) applies, and (VII.7) is replaced by

$$\int_{\mathscr{S}} \cdots w \, dS = \int_0^{2\pi} d\zeta \int_0^{\mathbb{d}} dr \cdots rw = \int_0^{2\pi} d\zeta \int_0^{\mathbb{A}(w)} d\theta \sin \theta \cdots \mathbb{S}(\theta, w), \quad \text{(VII.11)}$$

$\mathbb{S}$ still being given by (VII.8). The operators $\mathbb{F}$ and $\mathbb{P}$ now have the forms

$$\mathbb{F}F = \int\int_0^{2\pi} \int_0^{\mathbb{A}(w)} F_* \mathbb{S} \sin \theta \, d\theta \, d\zeta,$$

$$\mathbb{P}F = \int\int_0^{2\pi} \int_0^{\mathbb{A}(w)} F'F'_* \mathbb{S} \sin \theta \, d\theta \, d\zeta. \quad \text{(VII.12)}$$

Thus we may say roughly that for both kinds of molecules, *an alternative set of constitutive quantities* in the kinetic theory consists in $\mathbb{m}$, $\mathbb{S}$, and $\mathbb{A}$, the last of these being given by (VI.85). In Chapter XII we shall consider the determination of $\mathbb{S}$ and $\mathbb{A}$ from certain simple laws of intermolecular force.

(iv) *The bilinear form*

Generalizing a little the collisions operator $\mathbb{C}$ defined by (VII.2), we denote by $\mathbb{C}(\,,\,)$ the bilinear operator defined as follows:

$$\mathbb{C}(G, H) \equiv \tfrac{1}{2} \int\!\!\int_{\mathscr{S}} w(G'H'_* + G'_* H' - GH_* - G_* H), \tag{VII.13}$$

the functions G and H being any such as to render the integral convergent. In Chapter XIX we shall prove that there are many such functions. Referring back to (VII.2), we see that

$$\mathbb{C}F = \mathbb{C}(F, F). \tag{VII.14}$$

The molecular density F is of interest to us only through various expectations it enables us to calculate. Thus the main importance of the collisions operator $\mathbb{C}$ flows from its effect upon various functions we may wish to average. So as to discuss this effect, we define the *total collisions operators*[6] $\bar{\mathbb{C}}_{G,H}$ and $\bar{\mathbb{C}}_F$ as follows:

$$\bar{\mathbb{C}}_{G,H} g \equiv \int g \mathbb{C}(G, H),$$

$$= \tfrac{1}{2} \int\!\!\int\!\!\int_{\mathscr{S}} wg(G'H'_* + G'_* H' - GH_* - G_* H), \tag{VII.15}$$

$$\bar{\mathbb{C}}_F g \equiv \bar{\mathbb{C}}_{F,F} g = \int g \mathbb{C}F,$$

$$= \int\!\!\int\!\!\int_{\mathscr{S}} wg(F'F'_* - FF_*). \tag{VII.16}$$

We assume that the functions g, G, H, and F are such as to render the several integrals convergent.

The values of $\bar{\mathbb{C}}_{G,H} g$ and $\bar{\mathbb{C}}_F g$ are functions of t and $\mathbf{x}$, that is, fields in the ordinary sense of continuum mechanics. As we have remarked before, the kinetic theory has no other purpose than to deliver fields that can be interpreted in the context of gas flows, so we may expect the properties of $\bar{\mathbb{C}}_F g$ for various functions g and appropriate molecular densities F to be of central importance.

---

[6] The operator $\bar{\mathbb{C}}_{G,H}$ arises naturally in connection with the kinetic theory of gas mixtures and thus appears implicitly in MAXWELL's great work of discovery [1867, Eqs. (2) and (3)], though only in the form (VII.20)$_4$. The definition (VII.15), much preferable, was used by HILBERT [1912, just before Eq. (13)].

(iv) THE BILINEAR FORM

We now derive a number of other forms[7] of (VII.15) and hence also of (VII.16). First, if we change from $(\mathbf{v}, \mathbf{v}_*, r, \zeta)$ to $(\mathbf{v}_*, \mathbf{v}, r_*, \zeta_*)$, then the integrand in (VII.15)$_2$ is unaltered except that wg is replaced by $w_* g_*$. Since this transformation is locally volume-preserving, and since $w_* = w$ as a consequence of (VI.50)$_3$, we see that

$$\bar{\mathbb{C}}_{G,H} g = \tfrac{1}{2} \iint \int_{\mathscr{S}} w g_* (G'H'_* + G'_* H' - GH_* - G_* H). \qquad \text{(VII.17)}$$

If we interchange $(\mathbf{v}, \mathbf{v}_*, r, \zeta)$ with $(\mathbf{v}', \mathbf{v}'_*, r', \zeta')$, the integrand in (VII.15)$_2$ is again unaltered except that now wg is replaced by $-w'g'$. In virtue of (VI.70) and (VI.45)$_2$ we find that

$$\bar{\mathbb{C}}_{G,H} g = -\tfrac{1}{2} \iint \int_{\mathscr{S}} w g' (G'H'_* + G'_* H' - GH_* - G_* H). \qquad \text{(VII.18)}$$

By use of (VII.17) we now obtain

$$\bar{\mathbb{C}}_{G,H} g = -\tfrac{1}{2} \iint \int_{\mathscr{S}} w g'_* (G'H'_* + G'_* H' - GH_* - G_* H). \qquad \text{(VII.19)}$$

Simple combinations of these results yield

$$\bar{\mathbb{C}}_{G,H} g = \tfrac{1}{4} \iint \int_{\mathscr{S}} w(g + g_*)(G'H'_* + G'_* H' - GH_* - G_* H),$$

$$= \tfrac{1}{8} \iint \int_{\mathscr{S}} w(g + g_* - g' - g'_*)(G'H'_* + G'_* H' - GH_* - G_* H),$$

$$\text{(VII.20)}$$

$$= \tfrac{1}{4} \iint \int_{\mathscr{S}} w(g' + g'_* - g - g_*)(GH_* + G_* H),$$

$$= \tfrac{1}{2} \iint \int_{\mathscr{S}} w(g' - g)(GH_* + G_* H).$$

In (VII.17) to (VII.20)$_2$ the convergence of the integrals which appear is implied by the convergence of the defining relation (VII.15)$_2$. We may pass from (VII.20)$_2$ to (VII.20)$_3$ and (VII.20)$_4$ if the integral in (VII.20)$_2$ is the difference of two convergent integrals, one with integrand $w(g + g_* - g' - g'_*)(G'H'_* + G'_* H')$ and the other with integrand $w(g + g_* - g' - g'_*) \times$

---

[7] HILBERT [1912, Eq. (13) and the unnumbered equations preceding it] published (VII.17), (VII.18), (VII.19), and (VII.20)$_2$. He did not include (VII.20)$_{3,4}$, concerning which, as we remark in the text above, there are difficulties of convergence. BOLTZMANN [1896, §17], following MAXWELL (cf. Footnote 1 to this chapter), started from (VII.20)$_4$ as the definition. Earlier BOLTZMANN [1872, Eqs. (18)–(24)] [1875, §1] had obtained these transformations for the special case on which his H-theorem is based (cf. Chapter XI).

$(GH_* + G_*H)$. However, the convergence of these two integrals is not generally implied by that of the defining integral (VII.15)$_2$. For the time being, we shall simply assume that all the formulae (VII.20) are valid for a suitable class of functions g, G, and H and for such kinetic constitutive relations as will be used. In Chapter XIX we shall study the convergence of $\mathbb{C}$ and related integral operators, and there we will prove that for a gas composed either of molecules with a cut-off or of inverse k^{th}-power molecules for which $\mathsf{k} > 3$, both (VII.20)$_3$ and (VII.20)$_4$ are valid if g, G, and H belong to a certain broad class.

From (VII.20)$_2$ and (VI.5) we see at once that when g is a summational invariant S, then $\bar{\mathbb{C}}_{G,H} S = 0$, no matter what be the densities G and H. To prove a converse, we choose both G and H in (VII.20)$_2$ to be Ce^{kg}, C and k constants, so

$$\bar{\mathbb{C}}_F g = \tfrac{1}{4} C^2 \iint \int_{\mathscr{S}} w(g + g_* - g' - g'_*)(e^{k(g' + g_{*}')} - e^{k(g + g_*)}), \quad \text{(VII.21)}$$

F being taken as Ce^{kg}. If $k < 0$, the integrand is never negative. If g is not a summational invariant, the integrand is positive on a set of positive measure in $\mathscr{V} \times \mathscr{V} \times \mathscr{S}$, so the integral, if it exists, cannot have the value zero. A parallel argument holds if $k > 0$, the integrand then being never positive. We can state as follows the results just proved: *For any summational invariant S,*

$$\bar{\mathbb{C}}_{G,H} S = 0 \quad \text{for all} \quad G, H. \quad \text{(VII.22)}$$

In particular,

$$\bar{\mathbb{C}}_F S = 0 \quad \text{for all} \quad F. \quad \text{(VII.23)}$$

If g is a function such as to render (VII.21) *convergent*[8] *for some k, then $\bar{\mathbb{C}}_F g = 0$ only if g is a summational invariant.* Roughly speaking, we can say that both (VII.22) and (VII.23) are *necessary and sufficient conditions that S be a summational invariant.*

We may interpret the simple identity (VII.23) as follows: No matter what be the kinetic constitutive relations and the molecular density, the total effect of collisions upon a function invariant under collision is null. Stated in this way, the result seems self-evident, but the proof shows it is not. Rather, that the desired conclusion (VII.23) does follow strictly, serves as a check on the reasonableness of MAXWELL's choice (VII.2) for the collisions operator.

---

[8] A different sufficient condition follows if we take $G = e^{-kv^2}$ and $H = ge^{-kv^2}$, namely, that

$$\iint \int_{\mathscr{S}} we^{-k(v^2 + v_*^2)} (g + g_* - g' - g'_*)^2$$

be convergent. HILBERT [1912, after Eq. (26)] called the corresponding proof beautiful and attributed it to HECKE.

(v) *Orthogonal invariance of the collisions operator and the bilinear form*

We now develop the property of $\mathbb{E}$ which we have expressed abstractly as (VI.75) and (VI.74): *If* $\mathbf{v}$, $\mathbf{v}_*$, $\mathbf{v}'$, *and* $\mathbf{v}'_*$ *are asymptotic velocities for some encounter, then for every orthogonal tensor* $\mathbf{Q}$ *the vectors* $\mathbf{Qv}$, $\mathbf{Qv}_*$, $\mathbf{Qv}'$, *and* $\mathbf{Qv}'_*$ *are also asymptotic velocities for an encounter*. We show now that this property of $\mathbb{E}$ determines the invariance[9] of the collisions operator $\mathbb{C}$ under rigid changes of framing. In order to do so, we write out explicitly the variables of integration appearing in $\mathbb{C}$. First we define operators $\mathbb{E}_1$, $\mathbb{E}_2$, and $\mathbb{E}_3$ such as to pick out the parts of the value of $\mathbb{E}$ at a particular argument:

$$\mathbb{E}_1(\mathbf{v}, \mathbf{v}_*, \mathbf{r}^e) \equiv \mathbf{v}', \qquad \mathbb{E}_2(\mathbf{v}, \mathbf{v}_*, \mathbf{r}^e) \equiv \mathbf{v}'_*, \qquad \mathbb{E}_3(\mathbf{v}, \mathbf{v}_*, \mathbf{r}^e) \equiv \mathbf{r}^{e'}. \qquad \text{(VII.24)}$$

We also write $\mathscr{S}_\rho(\mathbf{w})$ for the disk of radius ρ in the plane with normal $\mathbf{w}$. The value of ρ may be ∞. Then $\mathbb{C}$ as defined by (VII.2) has the explicit form

$$\mathbb{C}F(\mathbf{v}) = \int_{\mathscr{V}} \int_{\mathscr{S}_\rho(\mathbf{v}_* - \mathbf{v})} |\mathbf{v}_* - \mathbf{v}| [F(\mathbb{E}_1(\mathbf{v}, \mathbf{v}_*, \mathbf{r}^e))F(\mathbb{E}_2(\mathbf{v}, \mathbf{v}_*, \mathbf{r}^e))$$
$$- F(\mathbf{v})F(\mathbf{v}_*)]\, dS\, d\mathbf{v}_*. \qquad \text{(VII.25)}$$

Given an orthogonal tensor $\mathbf{Q}$, we denote by a caret superimposed upon a vector the result of operating with $\mathbf{Q}$ on that vector: $\hat{\mathbf{a}} \equiv \mathbf{Qa}$. Then

$$\mathbb{C}(F \circ \mathbf{Q})(\mathbf{v}) = \int_{\mathscr{V}} \int_{\mathscr{S}_\rho(\mathbf{Q}^T[\hat{\mathbf{v}}_* - \hat{\mathbf{v}}])} |\hat{\mathbf{v}}_* - \hat{\mathbf{v}}| [F(\mathbf{Q}\mathbb{E}_1(\mathbf{v}, \mathbf{v}_*, \mathbf{r}^e))$$
$$\times F(\mathbf{Q}\mathbb{E}_2(\mathbf{v}, \mathbf{v}_*, \mathbf{r}^e)) - F(\mathbf{Qv})F(\mathbf{Qv}_*)]\, dS\, d\mathbf{v}_*,$$

$$= \int_{\mathscr{V}} \int_{\mathscr{S}_\rho(\mathbf{Q}^T[\hat{\mathbf{v}}_* - \hat{\mathbf{v}}])} |\hat{\mathbf{v}}_* - \hat{\mathbf{v}}| [F(\mathbf{Q}\mathbb{E}_1(\mathbf{Q}^T\hat{\mathbf{v}}, \mathbf{Q}^T\hat{\mathbf{v}}_*, \mathbf{Q}^T\hat{\mathbf{r}}^e))$$
$$\times F(\mathbf{Q}\mathbb{E}_2(\mathbf{Q}^T\hat{\mathbf{v}}, \mathbf{Q}^T\hat{\mathbf{v}}_*, \mathbf{Q}^T\hat{\mathbf{r}}^e)) - F(\hat{\mathbf{v}})F(\hat{\mathbf{v}}_*)]\, dS\, d\mathbf{v}_*, \qquad \text{(VII.26)}$$

$$= \int_{\mathscr{V}} \int_{\mathscr{S}_\rho(\mathbf{Q}^T[\hat{\mathbf{v}}_* - \hat{\mathbf{v}}])} |\hat{\mathbf{v}}_* - \hat{\mathbf{v}}| [F(\mathbb{E}_1(\hat{\mathbf{v}}, \hat{\mathbf{v}}_*, \hat{\mathbf{r}}^e))F(\mathbb{E}_2(\hat{\mathbf{v}}, \hat{\mathbf{v}}_*, \hat{\mathbf{r}}^e))$$
$$- F(\hat{\mathbf{v}})F(\hat{\mathbf{v}}_*)]\, dS\, d\mathbf{v}_*;$$

the last step follows by use of (VI.75). We change variables now from $\mathbf{r}^e$ to $\hat{\mathbf{r}}^e = \mathbf{Qr}^e$. If the polar co-ordinates of $\mathbf{r}^e$ are r and ζ, we see from (VI.77) that those of $\hat{\mathbf{r}}^e$ are r and $\zeta + \Phi(\mathbf{Q}, \mathbf{w})$. Thus the area measure dS is unchanged by this transformation. Also, if $\mathbf{r}^e \in \mathscr{S}_\rho(\mathbf{Q}^T[\hat{\mathbf{v}}_* - \hat{\mathbf{v}}])$, then $\hat{\mathbf{r}}^e \in \mathscr{S}_\rho(\hat{\mathbf{v}}_* - \hat{\mathbf{v}})$. We may then change variables from $\mathbf{v}_*$ to $\hat{\mathbf{v}}_* = \mathbf{Qv}_*$, this transformation being volume-

---

[9] WANG [1976, §2].

preserving in $\mathscr{V}$. Therefore, we may drop all carets except those on $\mathbf{v}$ and so obtain

$$\mathbb{C}(F \circ \mathbf{Q})(\mathbf{v}) = \int_{\mathscr{V}} \int_{\mathscr{S}_\rho(\mathbf{v}_* - \hat{\mathbf{v}})} |\mathbf{v}_* - \hat{\mathbf{v}}| [F(\mathbb{E}_1(\hat{\mathbf{v}}, \mathbf{v}_*, \mathbf{r}^e))F(\mathbb{E}_2(\hat{\mathbf{v}}, \mathbf{v}_*, \mathbf{r}^e))$$
$$- F(\hat{\mathbf{v}})F(\mathbf{v}_*)] \, dS \, d\mathbf{v}_*, \qquad (VII.27)$$
$$= \mathbb{C}F(\hat{\mathbf{v}}).$$

That is, *the collisions operator commutes with orthogonal tensors*:

$$\mathbb{C}(F \circ \mathbf{Q}) = (\mathbb{C}F) \circ \mathbf{Q}. \qquad (VII.28)$$

More generally, *for all F and G, and for any orthogonal tensor* $\mathbf{Q}$,

$$\mathbb{C}(F \circ \mathbf{Q}, G \circ \mathbf{Q}) = \mathbb{C}(F, G) \circ \mathbf{Q}. \qquad (VII.29)$$

All these operations refer to a fixed time. Accordingly, the orthogonal tensor $\mathbf{Q}$ may be taken as any function of time we please. Thus we may interpret the results as showing that *the collisions operator is frame-indifferent*.

(vi) Inconsistency of Maxwell's kinetic theory with Newtonian mechanics

In concluding this chapter we reconsider the basis of the axiom (VII.2). Clearly the molecular schema employed, however plausible it may seem, is not consistent with the principles of Newtonian mechanics. If the intermolecular forces extend to infinity, then no encounter can be binary, and the motions of any two molecules are influenced, though possibly indeed not much, by the motions of all the rest. If the intermolecular forces have a cut-off, then binary encounters indeed become possible, but nevertheless we are not justified in assuming that all encounters are of this kind. The motions of a set of mass-points subject to specified mutual forces are determined uniquely by the initial conditions and the dynamical equations. Thus we are not at liberty to assume anything about those motions. Whether or not all encounters of an assembly of mass-points be binary, is a matter for mathematical analysis and proof, not for guesswork. Indeed, if the molecules have finite spheres of action, analytical dynamics leads us to expect that ultimately three such spheres will intersect, not merely two, unless we start the system of molecules in some exceptional way. In general, therefore, MAXWELL's *assumptions regarding the molecular motion contradict the laws of analytical dynamics.*

If MAXWELL's assumptions do not generally follow from the principles of mechanics, there is no reason to hope to draw them from the theory of probability, either. There is no general theory of stochastic mechanics, and if there were one, we could not expect it to yield the kind of specific, determinate outcome of an encounter we are here assuming as a prerequisite for even stating the form of $\mathbb{C}F$.

Rather, the operator $\mathbb{C}$ as we have specified it must be regarded as a *mathematical model* for the likely effect of the outcome of many collisions, not as the result of calculating that effect or of calculating any mathematical expectation of that effect. MAXWELL's kinetic theory is a consequence neither of classical mechanics nor of the axioms of probability theory. Though it is motivated by a masterly and suggestive combination of mechanical and stochastic ideas, it is an independent model of a gas. As such it is to be respected and studied mathematically. The proof of the model lies in its product. The two models of a dissipative fluid that have proved their value again and again are the Navier-Stokes theory of linearly viscous fluids and MAXWELL's kinetic theory. Both are far better in product than any argument ever used to motivate or infer them might suggest. Each involves a special kind of non-linearity that seems somehow to reflect much, though by no means all, of the phenomena seen in natural fluids. These two non-linearities offer perfect challenges to the student of rational mechanics: They are genuine yet concrete, mathematically so difficult as to afford anyone, no matter how expert he may be, opportunity for a lifetime of study, yet not so difficult as to blank all rational inquiry. They enrich most of all the understanding of him who can weigh and value them both. In their predictions they partly agree but mostly disagree widely. They model partly similar and partly different aspects of natural gases. One of the purposes of this book is to reveal and delimit their range of agreement and to display their disagreement.

The integral operator $\mathbb{C}$ defined by (VII.2) is characteristic of MAXWELL's kinetic theory. Any other operator would refer to a different theory, not the one we are studying in this book. Our only freedom is to choose the *constitutive quantities of the kinetic theory*, namely, to specify $\mathbb{m}$, $\mathbb{E}$, and $\mathscr{S}$. Adopting common usage, we shall often refer to the kinetic constitutive assumptions collectively as describing a *molecular model*. Once a molecular model has been selected, everything else follows by mathematical process alone, or it does not follow at all.

Before we consider the effects of different choices of molecular model, we turn to an analysis of the most striking property of MAXWELL's operator $\mathbb{C}$, the *irreversibility* somehow incorporated in it. We will make no attempt to trace the source of this irreversibility in more general theories or physico-philosophical speculations. Rather, in the spirit of rational mechanics, we shall attempt to determine its specific and rigorous mathematical nature and consequences.

Chapter VIII

Boltzmann's Monotonicity Theorem. The Maxwellian Density. Analogues of the Caloric and Its Flux

(i) *The Boltzmann monotonicity theorem*

The Maxwell collisions operator $\mathbb{C}$ has been defined:

$$\mathbb{C}F = \iint_{\mathscr{S}} w(F'F'_* - FF_*), \qquad \text{(VII.3)}_r$$

and in terms of it the total collisions operator $\bar{\mathbb{C}}$:

$$\bar{\mathbb{C}}_F g \equiv \bar{\mathbb{C}}_{F,F}\, g = \int g\mathbb{C}F,$$

$$= \iint\int_{\mathscr{S}} wg(F'F'_* - FF_*). \qquad \text{(VII.16)}_r$$

Other expressions for this operator may be obtained by putting F for both G and H in (VII.17)–(VII.20).

The most important example of $\bar{\mathbb{C}}_F g$ is provided by the choice $g = \log F$ in the essential domain of F. For it, use of (VII.20)$_2$ yields

$$\bar{\mathbb{C}}_F \log F = \tfrac{1}{4}\iint\int_{\mathscr{S}} w(F'F'_* - FF_*)\log\frac{FF_*}{F'F'_*}. \qquad \text{(VIII.1)}$$

If $F'F'_* \geqq FF_*$, the logarithm in the integrand is not positive; if $F'F'_* \leqq FF_*$, that logarithm is not negative; hence the integrand is never positive. Consequently the integral vanishes if and only if the integrand vanishes almost everywhere; that is, if and only if

$$\log F + \log F_* = \log F' + \log F'_*. \qquad \text{(VIII.2)}$$

This equation is of the same form as (VI.5), so we have proved that $\bar{\mathbb{C}}_F \log F < 0$ unless $\log F$ is a summational invariant. In virtue of the Boltzmann–Gronwall theorem, stated and proved in Section (ii) of Chapter VI, we conclude that $\bar{\mathbb{C}}_F \log F < 0$ for every positive measurable function F except one whose logarithm is an affine combination of momentum and energy:

$$\log F = \alpha v^2 + \boldsymbol{\mu} \cdot \mathbf{v} + \beta. \tag{VIII.3}$$

We may write such an F as follows in terms of fields a, b, and $\mathbf{u}$:

$$F = F_M \equiv ae^{-bc^2} = ae^{-b|\mathbf{v}-\mathbf{u}|^2}; \tag{VIII.4}$$

as a molecular density must be positive on some essential domain and integrable over $\mathscr{V}$, it is necessary that $a > 0$ and $b > 0$, and these conditions are also sufficient that F_M shall be a positive, integrable function. The function F_M as given by (VIII.4)$_2$ is the *Maxwellian molecular density* corresponding to a, b, and $\mathbf{u}$. The conclusion of the foregoing argument is summarized in the **Boltzmann Monotonicity Theorem**[1]:

$$\bar{\mathbb{C}}_F \log F \leq 0 \quad \text{for all} \quad F. \tag{VIII.5}$$

Moreover,

$$\bar{\mathbb{C}}_F \log F = 0 \quad \Leftrightarrow \quad F = F_M. \tag{VIII.6}$$

The physical dimensions of $\bar{\mathbb{C}}_F \log F$ are those of the time-rate of change of the number density n. Thus, roughly, for any given molecular density F *the nature of* MAXWELL'S *collisions operator is such as to prevent a certain number density from increasing in time.*

This theorem concerns all functions F such as to make $\bar{\mathbb{C}}_F \log F$ exist, not merely those F that are appropriate to problems of the kinetic theory of gases. It expresses the most important property of the total collisions operator $\bar{\mathbb{C}}_F$. Because the form of that operator is introduced into the kinetic theory as an axiom, the monotonicity theorem is itself a corollary of that axiom alone, independent of the others.

As we shall see further on, the Boltzmann monotonicity theorem serves as basis from which to derive various irreversibilities in time implied by the kinetic theory. The reversibility theorem of analytical dynamics shows that if a certain motion of a conservative dynamical system is possible, so is the reversed motion, at least locally. Thus no kind of irreversibility can follow by correct mathematics from the analytical dynamics of a single conservative system— any such system. In order for any kinetic theory to predict some irreversibility, some assumption of that theory must introduce that irreversibility and must thus contradict analytical dynamics. As the Boltzmann monoticity theorem

---

[1] BOLTZMANN [1875, §I] [1876, §II]. A more general inequality was obtained by CHAPMAN & COWLING [1939, §4.41].

shows, in MAXWELL's kinetic theory *it is the assumed form* (VII.2) *for the collisions operator* $\mathbb{C}$ *that introduces irreversibility*.

The development and interpretation of the particular irreversibilities implied by MAXWELL's basic assumptions is the most fascinating problem the kinetic theory offers. Much of the following contents of this book will concern it.

(ii) *Properties of the Maxwellian density*

The Boltzmann monotonicity theorem has led us naturally to the Maxwellian density F_M, defined by (VIII.4)$_2$. In Chapter X we shall see that F_M is connected with the concept of equilibrium in the kinetic theory. More generally, the Maxwellian density has always played a central part in the mathematical developments, both for comparisons with the results of gas dynamics and as a basis for approximations, be they formal schemes or rigorous estimates. Accordingly, we now develop some of its properties.

A glance at (VIII.4), (VII.2), (VI.1), and (VI.2) shows that $\mathbb{C}F_M = 0$. Conversely, suppose that $\mathbb{C}F = 0$. Then $\bar{\mathbb{C}}_F g = 0$ for all g. In particular, we may put log F for g in the Boltzmann monotonicity theorem (VIII.6) and so conclude that $F = F_M$. Thus we have proved *Maxwell's Assertion*[2]:

$$\mathbb{C}F = 0 \quad \Leftrightarrow \quad F = F_M: \tag{VIII.7}$$

A molecular density is unaffected by collisions if and only if it is Maxwellian. Thus the condition of the gas as represented by F_M is in a sense permanent. MAXWELL therefore regarded F_M as being appropriate to equilibrium. We shall substantiate this view of his in Chapter X, but only after restricting it severely. Indeed, as we shall see now, a Maxwellian density (VIII.4) by no means implies a gas in uniform motion but may be fitted to any gross condition whatsoever.

The parameters a, b, and $\mathbf{u}$ in a Maxwellian density (VIII.4) are easy to interpret. First, as the notation suggests, $\mathbf{u}$ is in fact the gross velocity field, defined generally by (III.6). Second, it is a routine matter to find from the definitions (III.2) and (III.14) of n and ε, respectively, that

$$a = \frac{n}{(\frac{4}{3}\pi\varepsilon)^{\frac{3}{2}}}, \qquad b = \frac{3}{4\varepsilon}. \tag{VIII.8}$$

In the terms introduced in Section (ii) of Chapter VI, for a gas of known molecular mass $\mathfrak{m}$ *the principal moment of a Maxwellian density may be assigned at will; it determines that density uniquely.*

---

[2] MAXWELL [1867, Eq. (26)ff.]. As BOLTZMANN [1872, §I] remarked, MAXWELL's argument to show that F_M is the only density unaffected by collisions is insufficient. BOLTZMANN's own proof in the same paper rests on a primitive form of his *H*-theorem (Chapter XI) and also is insufficient.

VIII. BOLTZMANN'S MONOTONICITY THEOREM

Such being the case, every moment of F_M that exists is determined by the principal moment ρ. For example, in consequence of the definitions of the pressure tensor **M** and the energy flux vector **q**, namely,

$$\mathbf{M} \equiv \overline{\rho \mathbf{c} \otimes \mathbf{c}}, \qquad (\text{III.9})_{1r}$$

$$\mathbf{q} \equiv \tfrac{1}{2}\overline{\rho c^2 \mathbf{c}}, \qquad (\text{III.12})_{1r}$$

a glance at (VIII.4) shows that if $F = F_M$, then

$$\mathbf{M} = p\mathbf{1} = \tfrac{2}{3}\rho\varepsilon\mathbf{1},$$
$$\mathbf{P} = \mathbf{0}, \quad \mathbf{q} = \mathbf{0}: \qquad (\text{VIII.9})$$

A Maxwellian density makes the pressure tensor hydrostatic and the energy flux null. Thus *the Maxwellian density represents the kinetic gas as obeying the constitutive relations of the Euler–Hadamard theory.*

It is easy to show that all moments of a Maxwellian density exist and to exhibit the way they are determined by ρ. It is convenient to do so in terms of the relative moments ${}^n\mathbf{M}$, defined by (III.19). Those of odd order vanish because F_M is an even function of **c**:

$$^{2s+1}\mathbf{M} = \mathbf{0}, \quad s = 0, 1, \ldots. \qquad (\text{VIII.10})$$

Those of even order are easily calculated from the recursion formula for their components:

$$M_{k_1 k_2 \ldots k_s} = \tfrac{2}{3}\varepsilon \sum_{m=1}^{s-1} \delta_{k_m k_s} M_{k_1 k_2 \ldots k_{m-1} k_{m+1} \ldots k_{s-1}}. \qquad (\text{VIII.11})$$

For example, the components of ${}^4\mathbf{M}$ and ${}^6\mathbf{M}$ are, respectively,

$$M_{kmrs} = \rho(\tfrac{2}{3}\varepsilon)^2 (\delta_{ks}\delta_{mr} + \delta_{ms}\delta_{kr} + \delta_{rs}\delta_{km}),$$

$$M_{kmrstu} = \rho(\tfrac{2}{3}\varepsilon)^3 [\delta_{ku}(\delta_{mt}\delta_{rs} + \delta_{rt}\delta_{ms} + \delta_{st}\delta_{mr}) \qquad (\text{VIII.12})$$
$$+ \delta_{mu}(\delta_{kt}\delta_{rs} + \delta_{rt}\delta_{ks} + \delta_{st}\delta_{kr})$$
$$+ \delta_{ru}(\delta_{kt}\delta_{ms} + \delta_{mt}\delta_{ks} + \delta_{st}\delta_{km})$$
$$+ \delta_{su}(\delta_{kt}\delta_{mr} + \delta_{mt}\delta_{kr} + \delta_{rt}\delta_{km})$$
$$+ \delta_{tu}(\delta_{ks}\delta_{mr} + \delta_{ms}\delta_{kr} + \delta_{rs}\delta_{km})].$$

It is easy to calculate the mean random speed $\bar{c}$ corresponding to F_M:

$$\bar{c} = \frac{2}{\sqrt{\pi b}} = \frac{4}{\sqrt{3\pi}}\sqrt{\varepsilon}. \qquad (\text{VIII.13})$$

According to the Euler–Hadamard theory of a gas for which $\gamma = \tfrac{5}{3}$, the speed of sound is $(\sqrt{10/3})\sqrt{\varepsilon}$, very nearly $\tfrac{4}{5}$ of the above value of $\bar{c}$ for those Maxwellian densities that correspond to ε, of which there are, of course, infinitely many.

(ii) THE KNUDSEN NUMBER

This comparison and the formula (VIII.13), from which it follows, may serve as a basis for introducing the "Knudsen number", which is useful for rough interpretation of the results of experiments on rarefied gases[3].

[3] WANG CHANG & UHLENBECK [1948, §II] defined the *Knudsen number* K as the ratio of the "mean free path" L to a "characteristic length" l:

$$K \equiv L/l. \tag{A}$$

The experimentist may have his own reasons for choosing these lengths in various ways, which may or may not be compatible with one or another theory. In the formal kinetic theory, as we remarked in Footnote 5 to Chapter VII, a mean free path does not generally exist. However, the informal, heuristic, "elementary" theory presumes its existence and calculates other things in terms of it. For example, as presented by CHAPMAN & COWLING [1939, Eq. (6.2.1)] the "elementary" theory delivers the following expression for the shear viscosity of a kinetic gas:

$$\mu = \tfrac{1}{2} k \rho \bar{c} L, \quad k \approx .998. \tag{B}$$

We can take this formula as the *definition* of a length L; we can regard μ as a function determined by experiment, or we can use the value delivered by the kinetic theory itself for an appropriate molecular model (Chapters XIII, XXIV, XXV). Then

$$L \equiv \frac{2}{k} \frac{\mu}{\rho \bar{c}}, \tag{C}$$

and hence

$$K = \frac{2}{k} \frac{\mu}{\rho \bar{c} l}. \tag{D}$$

WANG CHANG & UHLENBECK expressed K in terms of a "molecular Mach number" M_m and a quantity R_m that we could call the "molecular Reynolds number":

$$M_m \equiv \frac{u}{\bar{c}}, \quad R_m \equiv \frac{\rho u^2 l}{\mu \bar{c}}. \tag{E}$$

These numbers are defined in just such a way to make both l and $\bar{c}$ cancel neatly out of (D) and yield

$$K = \frac{2}{k} \frac{M_m^2}{R_m} \approx 2.004 \frac{M_m^2}{R_m}. \tag{F}$$

To use this formula, we need not specify either l or the molecular density used to define $\bar{c}$.

While M_m and R_m arise more or less naturally from the concepts of the kinetic theory, it is possible to express K in terms of the usual M and R defined by (II.10). As remarked by TRUESDELL [1969, 2, Appendix], we may appeal to the fact that the squared speed of sound according to classical aerodynamics is $(10/9)\varepsilon$ and so write (D) in the form

$$K = \frac{2}{k} \frac{\tfrac{10}{9} \varepsilon}{\bar{c} l D} \frac{M^2}{R}. \tag{G}$$

We may choose for l the intrinsic length suggested in the discussion following (II.10):

$$lD = \sqrt{10\varepsilon}/3. \tag{H}$$

(*continued*)

VIII. BOLTZMANN'S MONOTONICITY THEOREM

Now let us consider a body of gas grossly at rest, and let a stationary plane divide it in imagination into two parts, one of which is governed by the Maxwellian density $a_1 e^{-b_1 c^2}$ and the other by the Maxwellian density $a_2 e^{-b_2 c^2}$. Formally, we write F_M^1 and F_M^2 for these two Maxwellian densities; we let **n** be the unit normal to the plane that points from the region governed by F_M^1 toward the region governed by F_M^2; and $c_\perp \equiv \mathbf{c} \cdot \mathbf{n}$. Then if $\mathbf{x}_0$ is some fixed place on the plane,

$$F = \begin{cases} F_M^1 & \text{if } \begin{cases} \mathbf{n} \cdot (\mathbf{x} - \mathbf{x}_0) < 0 & \text{or} \\ \mathbf{n} \cdot (\mathbf{x} - \mathbf{x}_0) = 0 & \text{and} \quad c_\perp > 0; \end{cases} \\ F_M^2 & \text{if } \begin{cases} \mathbf{n} \cdot (\mathbf{x} - \mathbf{x}_0) > 0 & \text{or} \\ \mathbf{n} \cdot (\mathbf{x} - \mathbf{x}_0) = 0 & \text{and} \quad c_\perp < 0. \end{cases} \end{cases} \qquad \text{(VIII.14)}$$

We let n_1 and ε_1 be determined from a_1 and b_1 according to (VIII.8), and we define n_2 and ε_2 analogously. Both ε and n generally suffer discontinuities at the dividing plane. At points on that plane the statement (VIII.14) regards the gas as composed of two kinds, one for which $c_\perp < 0$ and the other for which $c_\perp > 0$.

If we also adopt for $\bar{c}$ the value (VIII.13) appropriate to a Maxwellian density, from (G) we obtain

$$K = \frac{1}{k} \sqrt{\frac{5\pi}{6}} \frac{M^2}{R},$$
$$\approx 1.62 \frac{M^2}{R}. \qquad \text{(I)}$$

Other choices of l lead to different numerical factors in place of 1.62. Often the factor is taken as 1. GRAD [1972, p. 3] refers to the result of this rather arbitrary choice as "a basic and mysterious formula of kinetic theory". One way to obtain it would be to use (D) in (II.11) so as to obtain

$$\frac{M^2}{R} = \frac{9}{20} \frac{k \bar{c} l D}{\varepsilon} K, \qquad \text{(J)}$$

leaving $\bar{c}$ and l arbitrary, then choose D in such a way as to make the coefficient of K reduce to 1.

Arbitrariness of this kind is unnecessary. Referring back to Section (iv) of Chapter IV, we recall that with the intrinsic choice (IV.14) for D both in the definition of R and in the definition (IV.10) of the tension number T, we obtain the not inelegant formula

$$T = \frac{5}{3} \frac{M^2}{R}; \qquad \text{(IV.15)}_r$$

if we take K as given by (I), then

$$K \approx 0.97 T. \qquad \text{(K)}$$

Thus those who prefer to think in terms of a Knudsen number may regard the tension number T as being very nearly that. Since T presents itself naturally in the solutions of various problems of the kinetic theory and does not require recourse to the unsound concept of the mean free path, some nebulous replacement for it, the notorious quicksands of the "elementary" kinetic theory, or the special choice of F needed to get (VIII.13), we stay with T.

(ii) MAXWELLIAN DENSITIES AT A WALL

The number densities for the two kinds are $\frac{1}{2}n_1$ and $\frac{1}{2}n_2$, respectively, and (VIII.14) shows that at points on the dividing plane

$$n = \tfrac{1}{2}n_1 + \tfrac{1}{2}n_2, \qquad p = \tfrac{1}{2}p_1 + \tfrac{1}{2}p_2, \qquad n\varepsilon = \tfrac{1}{2}n_1\varepsilon_1 + \tfrac{1}{2}n_2\varepsilon_2. \qquad \text{(VIII.15)}$$

These formulae illustrate a general rule for evaluating the expectation of any even function of **c** at a point upon the dividing plane: $n\bar{g} = \tfrac{1}{2}n_1\bar{g}_1 + \tfrac{1}{2}n_2\bar{g}_2$. The expectations of odd functions have more interesting properties. For example, $\mathbf{q} = \mathbf{0}$ at points not on the plane, but we easily show that $\mathbf{q} \neq \mathbf{0}$ on it. To do so, we first apply the condition $\bar{c}_\perp = 0$ at points on the plane. Using (VIII.14), we write this condition, which we may interpret as a statement that the dividing plane does not create or destroy molecules, in the form

$$0 = a_1 \int_{c_\perp > 0} c_\perp e^{-b_1 c^2}\, d\mathbf{c} + a_2 \int_{c_\perp < 0} c_\perp e^{-b_2 c^2}\, d\mathbf{c}. \qquad \text{(VIII.16)}$$

The value of the first integral is $\tfrac{1}{2}\pi a_1/b_1^2$; that of the second, $-\tfrac{1}{2}\pi a_2/b_2^2$. Thus

$$a_1/b_1^2 = a_2/b_2^2; \qquad \text{(VIII.17)}$$

because of (VIII.8) this condition may be written as

$$n_1\sqrt{\varepsilon_1} = n_2\sqrt{\varepsilon_2}. \qquad \text{(VIII.18)}$$

Hence we may reduce (VIII.15)$_3$ to the form

$$\varepsilon = (n_1/n_2)\varepsilon_1 = (n_2/n_1)\varepsilon_2. \qquad \text{(VIII.19)}$$

We may also calculate the resultant pressure acting upon the side of the plane that borders the region labelled with a 1:

$$\begin{aligned}
-(p - p_1) = p - p_2 &= \tfrac{1}{2}(p_1 - p_2), \\
&= \tfrac{1}{3}\mathfrak{m}(n_1\varepsilon_1 - n_2\varepsilon_2), \\
&= \tfrac{1}{3}\rho_1\sqrt{\varepsilon_1}(\sqrt{\varepsilon_1} - \sqrt{\varepsilon_2}).
\end{aligned} \qquad \text{(VIII.20)}$$

Thus the dividing plane suffers a pressure that pushes it from the hotter side toward the colder.

As the discontinuity of ε suggests, heat should flow across the dividing plane, and we proceed to calculate the value there of $\mathbf{q} \cdot \mathbf{n}$, which we denote by $q_\perp$. From (III.12) and (VIII.14) we find that

$$\frac{2q_\perp}{\mathfrak{m}} = a_1 \int_{c_\perp > 0} c_\perp c^2 e^{-b_1 c^2}\, d\mathbf{c} + a_2 \int_{c_\perp < 0} c_\perp c^2 e^{-b_2 c^2}\, d\mathbf{c}. \qquad \text{(VIII.21)}$$

The value of the first integral is $\pi a_1/b_1^3$. Thus

$$\begin{aligned}
\frac{2q_\perp}{\mathfrak{m}} &= \pi\left(\frac{a_1}{b_1^3} - \frac{a_2}{b_2^3}\right), \\
&= \pi\frac{a_1}{b_1^2}\left(\frac{1}{b_1} - \frac{1}{b_2}\right),
\end{aligned} \qquad \text{(VIII.22)}$$

the second step being a consequence of (VIII.17). Use of (VIII.8) yields

$$q_\perp = \sqrt{2/\pi}(\tfrac{2}{3})^{\frac{3}{2}} \rho_1 \sqrt{\varepsilon_1}(\varepsilon_1 - \varepsilon_2). \tag{VIII.23}$$

The reader will recognize this statement as being an instance of what the literature, with scant historical justice, calls "NEWTON's Law of Cooling". We have proved it only for Maxwellian molecular densities: *In a body of gas at rest, governed by Maxwellian densities on the two sides of a dividing plane, energy flows out of the hotter side of the plane into the cooler side at a rate proportional to the difference of temperatures.* As a corollary, the heat-transfer condition (I.38) is satisfied. "NEWTON's Law" was inferred from experiments on solids radiating heat into a steady stream of cool air. Here we find a parallel statement for two adjacent, still bodies of gas. The coefficient, which FOURIER called "the superficial conductivity", in units such that $k/\mathfrak{m} = 1$ has the explicit value $\sqrt{2/\pi}\rho_1\sqrt{\theta_1}$, as we see from (III.16) and (III.17). Contrary to the impression an uninformed person might gather from our summary in words, the striking formula (VIII.23) does not reflect any law of evolution of the gas. It follows from *nothing more than the definitions of* $\mathbf{q}$ and F_M and the requirement that $\bar{c}_\perp = 0$; we have not called even upon the definition of $\mathbb{C}$. Like the Boltzmann monotonicity theorem, (VIII.23) suggests that a Maxwellian density might serve to represent equilibrium of the kinetic gas.

> The reader will observe also that the result continues to hold for a moving surface material with respect to the gross flow; that is, the velocity of the surface normal to itself is $u_\perp \mathbf{n}$, the component of the velocity in that direction of the gas upon it. The tangential velocities of the gas on the two sides of the wall need not be zero or equal to each other. We shall encounter surfaces of this kind in Chapter XI.
>
> Results such as those we have just obtained are not inconsistent with three-dimensional continuum thermomechanics. However, in that theory they cannot be calculated but rather must be imposed as boundary conditions. Specifically, if ε experiences a jump at a boundary, three-dimensional continuum thermomechanics suggests nothing useful in assigning to it a value at points on that boundary; if we are to regard that boundary as the seat of sources of energy flux, we cannot calculate those sources; if we require them, we may assign them directly. The kinetic theory, on the contrary, regards ε and $\mathbf{q}$ as quantities to be calculated from F, and the simple but striking results we have just obtained illustrate not only the specific quality of the kinetic theory but also the possibility it affords that the gross fields it delivers may suffer strong discontinuities.

(iii) Degree to which a Maxwellian expectation approximates a general one

The importance of the Maxwellian density, especially in relation to the concept of equilibrium, makes it desirable to estimate the deviations of various general averages from their counterparts for a Maxwellian density. A method of doing so has been noticed by ŁUNC[4]. Let F be a function of t, $\mathbf{x}$, and $\mathbf{v}$, not

---

[4] ŁUNC [1963]. Earlier ŁUNC [1962] indicated a more primitive approach.

(iii) MAXWELLIAN AND GENERAL EXPECTATIONS

necessarily a molecular density, and suppose $F\mathbf{I}$ is integrable over $\mathscr{V}$. Then there is one and only one Maxwellian density F_M that has the same principal moment ρ as F. We may call that F_M the Maxwellian density that *corresponds* to F. We can then define a function G such that[5]

$$F = F_M(1 + G). \qquad \text{(VIII.24)}$$

Clearly $F_M G$ is a function whose principal moment is $\mathbf{0}$. Equivalently

$$\int GF_M S = 0 \qquad \text{(VIII.25)}$$

for every summational invariant S. If we use crotchets to denote expectations calculated from the F_M that corresponds to F:

$$\langle g \rangle \equiv \frac{1}{n} \int F_M g, \qquad \text{(VIII.26)}$$

then, because n is the same for both F and F_M,

$$\bar{g} - \langle g \rangle = \frac{1}{n} \int F_M G g. \qquad \text{(VIII.27)}$$

In virtue of the Schwarz inequality

$$\left(\frac{1}{n}\int F_M G g\right)^2 = \left(\frac{1}{n}\int \sqrt{F_M}g\sqrt{F_M}G\right)^2,$$
$$\leq \left(\frac{1}{n}\int F_M g^2\right)\left(\frac{1}{n}\int F_M G^2\right), \qquad \text{(VIII.28)}$$
$$= \langle g^2\rangle\langle G^2\rangle.$$

Comparison of (VIII.27) with (VIII.28) shows that

$$|\bar{g} - \langle g \rangle| \leq \sqrt{\langle G^2 \rangle}\sqrt{\langle g^2 \rangle}. \qquad \text{(VIII.29)}$$

The dimensionless factor $\sqrt{\langle G^2 \rangle}$ depends upon F alone and is the same for all functions g, the deviations of whose averages $\bar{g}$ from their corresponding Maxwellian averages $\langle g \rangle$ are to be estimated. A molecular density F such that $\langle G^2 \rangle$ is small might be called "nearly Maxwellian".

An example will show the ease with which the estimate (VIII.29) may be rendered explicit. In order to give it, we construct a function G that would seem

---

[5] BOLTZMANN [1872, §III], MAXWELL [1879, Eq. (12)] ("with Boltzmann"), and BOLTZMANN [1880, Eq. (18) in §III] introduced formulae like (VIII.24) with various polynomials for G. They did not state that F and F_M should have the same principal moment; rather, their complicated calculations seem, at least in some cases, to produce this agreement as an outcome of supposing F to be in one or another purely formal sense an approximate solution of the Maxwell–Boltzmann equation (IX.1).

potentially useful, namely,

$$G = \mathbf{c} \cdot \mathbf{Kc}, \qquad \mathbf{K} = \mathbf{K}^T. \tag{VIII.30}$$

Then by use of the definitions (III.9) and (VIII.26), assuming always that F and F_M have the same principal moment, we see that the pressure tensor corresponding to F is given by

$$\begin{aligned}\mathbf{M} &= \rho \langle \mathbf{c} \otimes \mathbf{c} \rangle + \rho \langle (\mathbf{c} \cdot \mathbf{Kc})\mathbf{c} \otimes \mathbf{c} \rangle, \\ &= p\mathbf{1} + \rho(\tfrac{2}{3}\varepsilon)^2[(\mathrm{tr}\,\mathbf{K})\mathbf{1} + 2\mathbf{K}];\end{aligned} \tag{VIII.31}$$

the second step is a consequence of $(\text{VIII.12})_1$. Taking the trace of (VIII.31) and appealing to (III.13), we conclude that $\mathrm{tr}\,\mathbf{K} = 0$ and thence by use of (III.10) that

$$\mathbf{P} = \tfrac{4}{3}p\varepsilon\mathbf{K}. \tag{VIII.32}$$

We have shown that for a given principal moment and a given traceless symmetric tensor field $\mathbf{P}$, the function

$$F_M\left(1 + \frac{3}{4p\varepsilon}\mathbf{c} \cdot \mathbf{Pc}\right) \tag{VIII.33}$$

has $\mathbf{P}$ as its pressure deviator, and the reader will easily show that it is the only multiple of F_M by a quadratic function of $\mathbf{c}$ to do so, subject to the condition (VIII.25). However, if this function is to be a molecular density, it must be non-negative. Thus $\mathbf{c} \cdot \mathbf{Pc} \geq -4p\varepsilon/3$ for all $\mathbf{c}$; hence $\mathbf{c} \cdot \mathbf{Pc} \geq 0$. The only such traceless tensor $\mathbf{P}$ is $\mathbf{0}$. Thus (VIII.33) with a traceless tensor $\mathbf{P}$ *is a molecular density if and only if it reduces to F_M*. Once and for all we remark that it is no good saying (VIII.33) is "nearly" Maxwellian when $|\mathbf{c}|$ is "small". The only use in a molecular density is to integrate over all values of $\mathbf{c}$, and it makes no sense to restrict the range of $\mathbf{c}$ when talking of (VIII.33). The function defined by (VIII.33) is not a molecular density, and that is that. It may, indeed, approximate a molecular density, and therefore to discuss the values of averages calculated by means of it may make sense.

Before doing that, we generalize the approach so as to allow arbitrary values of the flux of energy $\mathbf{q}$, as well as of $\mathbf{P}$. If we start with

$$G = (1 + Lc^2)\mathbf{k} \cdot \mathbf{c}, \tag{VIII.34}$$

we may calculate $\mathbf{q}$ and find that

$$2\mathbf{q} = \rho\langle c^2\mathbf{c}\rangle + \rho\langle c^2\mathbf{c} \otimes \mathbf{c}\rangle\mathbf{k} + \rho L\langle c^4\mathbf{c} \otimes \mathbf{c}\rangle\mathbf{k}. \tag{VIII.35}$$

By use of $(\text{VIII.12})_{1,2}$ and (VIII.25) we easily calculate $\mathbf{k}$ and L. Combining the result with (VIII.32), we obtain the function G_G defined as follows:

$$G_G \equiv \frac{3}{4p\varepsilon}\mathbf{c} \cdot \mathbf{Pc} - \frac{3}{2p\varepsilon}\mathbf{q} \cdot \mathbf{c}\left(1 - \frac{3}{10\varepsilon}c^2\right); \tag{VIII.36}$$

(iii) MAXWELLIAN AND GENERAL EXPECTATIONS

the function F_G obtained by substituting G_G as given by (VIII.36) for G in (VIII.24) *has* ρ *as its principal moment,* **P** *as its pressure deviator, and* **q** *as its flux of energy*. The function F_G was noticed by GRAD[6], and we shall encounter it again in Chapter XVII. Although the words just used would suggest the contrary, F_G *is not a molecular density* unless it reduces to F_M. Indeed, we have already proved this fact for the case in which $\mathbf{q} = \mathbf{0}$. If $\mathbf{q} \neq \mathbf{0}$, then G_G as given by (VIII.36) is a cubic polynomial; since no cubic polynomial is bounded below, the condition $G_G > -1$ for all **c** cannot be satisfied.

We are now ready to calculate $\langle G_G^2 \rangle$. By use of (VIII.11) it is a simple but tedious matter to do so. The result is

$$\langle G_G^2 \rangle = \frac{1}{2} \frac{|\mathbf{P}|^2}{p^2} + \frac{3}{5} \frac{q^2}{p^2 \varepsilon}. \tag{VIII.37}$$

If put back into (VIII.29), this estimate delivers the far from surprising conclusion that when both $|\mathbf{P}|^2/p^2$ and $q^2/p^2\varepsilon$ are small, the fractional error that results from using F_M instead of F_G to calculate expectations is uniformly small[7].

More generally, if G is a polynomial of odd degree in **c**, then $F_M(1 + G)$ is obviously not a molecular density. However there are certain polynomials G of even degree greater than 2 such that $F_M(1 + G) > 0$ and (VIII.25) is satisfied.

Estimates of $\langle G^2 \rangle$ for two classes of functions G are easy to obtain.

Class 1. If there is a positive constant α and a non-negative integer m such that

$$|G| \leq \alpha c^m, \tag{VIII.38}$$

---

[6] GRAD [1949, Eq. (5.8)]. While the properties of this function may be suggested in retrospect by some calculations of ENSKOG [1917, Eq. (65) and the following page], they seem not to have been noticed in general terms by authors prior to GRAD.

[7] ŁUNC [1963] obtained the first term in (VIII.37) and chose to interpret it by applying the Navier-Stokes-Fourier constitutive relations (I.7) and (II.9). Then, indeed,

$$\frac{1}{2}\frac{|\mathbf{P}|^2}{p^2} + \frac{3}{5}\frac{q^2}{p^2\varepsilon} = 2\frac{\mu^2}{p^2}|\mathbf{E}|^2 + \frac{3}{5}\mathsf{M}^2\frac{\mu^2}{p^2}\frac{|\text{grad }\varepsilon|^2}{\varepsilon},$$

$$= \mathsf{Tr}^2 + \tfrac{3}{5}\mathsf{H}^2,$$

if the truncation number Tr is defined by (IV.16), while the *heat-transfer number* H is defined as follows:

$$\mathsf{H} \equiv \frac{\mathsf{M}\mu}{p}\frac{|\text{grad }\varepsilon|}{\sqrt{\varepsilon}}.$$

This kind of argument, however, is circular, since it is the smallness of certain parameters that should make the kinetic theory deliver the Stokes-Kirchhoff theory approximately.

then

$$\langle G^2 \rangle = (\tfrac{4}{3}\pi\varepsilon)^{-\frac{3}{2}} \int e^{-bc^2} G^2,$$

$$\leq (\tfrac{4}{3}\pi\varepsilon)^{-\frac{3}{2}} \alpha^2 \int c^{2m} e^{-bc^2},$$

$$= (\tfrac{4}{3}\pi\varepsilon)^{-\frac{3}{2}} 4\pi\alpha^2 \int_0^\infty c^{2m+2} e^{-bc^2} \, dc, \qquad \text{(VIII.39)}$$

$$= (2m+1)!! \, \alpha^2 (\tfrac{2}{3}\varepsilon)^m.$$

Class 2. If there are constants α and β such that

$$|G| \leq \alpha e^{\beta c^2}, \qquad \beta < \tfrac{1}{2} b, \quad \alpha > 0, \qquad \text{(VIII.40)}$$

then

$$\langle G^2 \rangle \leq (\tfrac{4}{3}\pi\varepsilon)^{-\frac{3}{2}} \alpha^2 \left(\frac{\pi}{b - 2\beta} \right)^{\frac{3}{2}},$$

$$= \alpha^2 \left(1 + 2 \frac{\varepsilon}{\varepsilon_\beta} \right)^{-\frac{3}{2}}, \qquad \text{(VIII.41)}$$

ε_β being the energetic that corresponds to $e^{\beta c^2}$ considered as a molecular density: $\beta = -\tfrac{3}{4}/\varepsilon_\beta$. If $\beta < 0$, the corresponding F is "nearly Maxwellian" in an obvious sense.

In rough summary of these results, we may say that *Maxwellian expectations are good approximations to general ones of two kinds*:

(a) When $|F/F_M - 1|$ is bounded by a power of c, and when the temperature of the gas is sufficiently low.

(b) When $|F/F_M - 1|$ is bounded by a Maxwellian density.

In proof of these results the condition $G > -1$ has not been used. Thus they compare Maxwellian averages with averages weighted by functions F having the same principal moment, whether or not those F be molecular densities. This observation is useful because, as we shall see in Chapter XXIV, some processes of approximation to solutions F represent them as limits of functions, not all of which are positive.

(iv) *The caloric of a kinetic gas: Boltzmann's field h*

The definitions of pressure and temperature force the kinetic gas, no matter what be its molecular density F, to obey the thermal equation of state

$$p = \tfrac{2}{3}\rho\varepsilon. \qquad \text{(III.15)}_r$$

In this sense the kinetic gas is always an ideal gas. The classical thermostatics of fluids rests upon the assumption that there is a caloric equation of state relating the *caloric* or *specific entropy* η to the mass density ρ and the temperature θ. This equation is "fundamental" in the sense that from it we may derive (III.15) and all other relations between the variables describing the gas at rest, provided we have the general structure of thermostatics at our disposal. In the kinetic theory we do not have that structure. The caloric, if we are to get it, must be defined from the molecular density, and the starting point for proof of such relations between various quantities as are true can be nothing else than the definitions of those quantities as expectations calculated from a molecular density.

We first consider equilibrium subject to no body force. Although the apparatus so far assembled does not suffice to provide a good definition of kinetic equilibrium, we can say this much: Any idea of free equilibrium must at least include the possibility that a body of gas may remain endowed with constant, uniform density, gross velocity, and energetic, unaffected by collisions. Inspection of (VIII.7) and (VIII.8) shows that a Maxwellian molecular density with uniform constant parameters a, b, and $\mathbf{u}$ satisfies this requirement, and that no non-Maxwellian density does so. Thus in considering kinetic equilibrium it *suffices* to consider only Maxwellian densities. In fact, the argument we shall give now does not require the parameters a, b, and $\mathbf{u}$ to be constant.

We wish to find a function g such that if $\eta \equiv \bar{g}$, then when $F = F_M$ the resulting η shall be the caloric of an ideal perfect gas for which $\gamma = \frac{5}{3}$. For this caloric η the more general expression (II.12) yields $\eta/r = \log(\varepsilon^{\frac{3}{2}}/\rho) + \text{const.}$ The considerations in Chapter III lead us to expect that r should be replaced by $\mathfrak{r}$ as defined by (III.18)$_2$. Therefore it is η defined thus:

$$\eta/\mathfrak{r} \equiv \log(\varepsilon^{\frac{3}{2}}/\rho) + \text{const.,} \tag{VIII.42}$$

that we seek to express as an expectation with respect to F_M. By (VIII.26)

$$\langle \log F_M \rangle = \langle \log a \rangle - \langle bc^2 \rangle,$$

$$= \log a - \tfrac{3}{2}, \tag{VIII.43}$$

$$= \log \frac{n}{(\tfrac{4}{3}\pi\varepsilon)^{\frac{3}{2}}} - \tfrac{3}{2} \equiv h_M,$$

say. Comparison with (VIII.42) shows that if we are to define in the kinetic theory a quantity η that corresponds to the caloric of continuum thermostatics, then to within an additive constant we must have

$$\eta \equiv -\mathfrak{r}h, \quad h \equiv \overline{\log F} \quad \text{when} \quad F = F_M. \tag{VIII.44}$$

118   VIII. BOLTZMANN'S MONOTONICITY THEOREM

In terms of the decomposition (VIII.24), this requirement is equivalent to

$$h \equiv \overline{\log F} + \overline{\mathfrak{h}(G; t, \mathbf{x}, \mathbf{c})}, \qquad \text{(VIII.45)}$$

$\mathfrak{h}$ being such that the second term reduces to a constant if $G = 0$. An example of such an $\mathfrak{h}$ is $f(G(t, \mathbf{x}, \mathbf{c}) - G(t, \mathbf{x}, -\mathbf{c}))$ if $f(0) = 0$. As we have remarked in Chapter III, the kinetic theory does not allow us to introduce quantities that bear thermal or caloric units except by the addition of a dimensional factor which serves to cancel these units out again. Such a factor is $\mathfrak{r}$, and in $(\text{VIII}.44)_1$ it does exactly that. Just as nothing is gained in the kinetic theory by introducing θ, its purpose being served neatly by the purely mechanical quantity ε, so also we have no need of η, for its purpose is served by the purely mechanical quantity h. When we wish to compare theorems of the kinetic theory with phenomenological statements about entropy, we must compare $-\mathfrak{r}h$ with η.

The tradition of the kinetic theory, following BOLTZMANN[8], chooses the simplest possibility:

$$h \equiv \overline{\log F}. \qquad \text{(VIII.46)}$$

We shall follow BOLTZMANN further in giving the field $\overline{\log F}$ no name but that of the letter that denotes it.

Since F is not dimensionless, the value of h will depend upon the choice of units. This fact occasions no difficulty, for the difference of the values of h in two different systems of units is a constant, and an additive constant makes no difference in any use that h may serve.

The analogy established between classical thermodynamics and the kinetic theory refers to a Maxwellian density. The considerations presented in Section (i) of Chapter IV apply here and show that for a general F the field h as defined by (VIII.46) will not be a function of the fields ρ and ε. In other words, MAXWELL's *kinetic theory does not in general deliver caloric equations of state* such as (I.9). The reader will recall that the Stokes–Kirchhoff theory as formulated in Chapter II likewise evades classical thermodynamics. The kinetic theory leaves open the possibility that h may be determined by ρ, ε, and $\mathbf{u}$ in a region, at least for special solutions. We shall return to this possibility in Chapter XXIV.

Whatever may be the physical interpretation of the choice we have made in adopting (VIII.46), it lends itself to mathematical analysis based on the convexity of various functions defined in terms of the logarithm, as we shall see now.

---

[8] While we have not searched every one of BOLTZMANN's works, in our study of his major papers we have not found any claim that (VIII.46) provided anything more than an analogy to $-\eta/\mathfrak{r}$. Later students have been less cautious, and the literature of the kinetic theory calls $-\mathfrak{r}h$ with h defined by (VIII.46) "the" entropy according to the kinetic theory.

(v) Bounds for h and for its flux s

Anyone who has studied statistical mechanics will remember a celebrated theorem of GIBBS[9]: "*If an ensemble of systems is canonically distributed in phase, the average index of probability is less than in any other distribution of the ensemble having the same average energy.*" This statement, a corollary of no more than the definitions of the terms used, was intended to represent in some way or perhaps replace the pronouncement of CLAUSIUS about the trend of entropy that we have quoted in Section (v) of Chapter I. In the simpler situation envisaged in the kinetic theory, we can prove a similar theorem: *Among all molecular densities having the same principal moment, the Maxwellian density gives h its smallest value.* We shall prove this fact as a corollary of simple properties of the logarithm[10] and of the definitions of h and F_M.

Beginning from (VIII.24), we find that

$$F \log F - F_M \log F_M = F_M G \log F_M + F_M[(1 + G) \log(1 + G) - G] + F_M G.$$
(VIII.47)

Now it is easy to show that if $-1 < x < \infty$,

$$0 \leq (1 + x) \log(1 + x) - x \leq x^2;$$
(VIII.48)

equality holds below or above if and only if $x = 0$, but also it is approached above in the limit as $x \to -1$. Because F is a molecular density, we know that $G > -1$, so we can apply (VIII.48) and thus obtain

$$F_M G + F_M G \log F_M \leq F \log F - F_M \log F_M \leq F_M G \log F_M + F_M G^2 + F_M G.$$
(VIII.49)

This relation is satisfied by *every molecular density F*; equality holds *if and only if $F = F_M$*. If we assume that $F \log F$ is integrable over $\mathscr{V}$, we may first inte-

---

[9] Theorem II in Chapter XI of *Elementary Principles in Statistical Mechanics developed with special reference to the Rational Foundation of Thermodynamics*, New Haven, 1901, reprinted in Volume I of *The Collected Works of J. WILLARD GIBBS*. In attaching phenomenological names to statistical quantities GIBBS was even more cautious than BOLTZMANN.

GIBBS' proof rests upon the fact that $ye^y + 1 - e^y > 0$ if $y \neq 0$. The proof of the somewhat similar theorem in the text above uses also the further fact that $(e^y - 1)^2 > ye^y + 1 - e^y$ if $y \neq 0$.

[10] The analysis we present here has not been published before, so far as we know. We were led to it in an attempt to substantiate some remarks of GRAD [1965, 3, Eq. (1.5)].

Our theorem, like GIBBS's, is specific in statement, rigorous in proof based upon explicit inequalities. It is not to be confused with the variational hocus-pocus sometimes adduced to support imprecise claims of a vaguely similar ring, *e.g.*, by TOLMAN [1938, §50].

grate (VIII.49) and (VIII.47) and then by appeal to (VIII.25) and (VIII.3) conclude that

$$0 \leq \int F \log F - \int F_M \log F_M = \int F_M[(1 + G) \log(1 + G) - G], \tag{VIII.50}$$

$$\leq \int F_M G^2;$$

these estimates, in view of (VIII.46) and (VIII.43), we may write in the form

$$0 \leq h - h_M \leq \langle G^2 \rangle. \tag{VIII.51}$$

Equality holds if and only if $G = 0$ for almost all $\mathbf{v}$. The left-hand inequality asserts that F_M does indeed give h the smallest value it can attain for molecular densities corresponding to the same gross condition. The right-hand side provides an upper bound for the deviation of h from its value for the Maxwellian density that corresponds to F. It is interesting to note that this bound is the square of that which (VIII.29) gives for $|\bar{g} - \langle g \rangle|/\sqrt{\langle g^2 \rangle}$, the function g being arbitrary. Thus for "nearly Maxwellian" molecular densities the Euler-Hadamard theory gives much better approximations to the caloric as obtained from (VIII.46) than it has been shown to provide for expectations in general. Estimates of $\langle G^2 \rangle$ for two classes of functions G have been recorded above as (VIII.39) and (VIII.41). They must be used with caution[11] here, since the argument leading to (VIII.51) rests essentially upon the assumption that $G > -1$.

[11] We are not justified in using for G here the function G_G as defined by (VIII.36), because $G_G \ngtr -1$. Nevertheless it is interesting to notice that GRAD [1949, Eq. (7.13)] arrived at an approximate formula for an h seeming to correspond to $F_M(1 + G_G)$ which amounts to asserting equality upon the right-hand side of (VIII.51) and using (VIII.37) for $\langle G^2 \rangle$:

$$h - h_M \approx \frac{1}{2} \frac{|\mathbf{P}|^2}{p^2} + \frac{3}{5} \frac{q^2}{p^2 \varepsilon}.$$

This approximation, however, is not correct. GRAD obtained it by replacing the function $\log(1 + x)$ by the first term in its Taylor expansion $\log(1 + x) = x + \cdots$, and so he arrived at

$$(1 + x) \log(1 + x) - x = (1 + x)(x + \cdots) - x,$$
$$= x^2 + \cdots,$$

which, apart from the terms neglected, conforms with the inequality (VIII.48). However, some of the terms neglected have an important effect. Indeed, if instead we use $\log(1 + x) = x - \frac{1}{2}x^2 + \cdots$, we find that

$$(1 + x) \log(1 + x) - x = (1 + x)(x - \frac{1}{2}x^2 + \cdots) - x,$$
$$= \frac{1}{2}x^2 + \cdots,$$

the dots denoting terms of order $O(x^3)$. Thus GRAD's approximate formula for $h - h_M$ is twice as great as what he should have gotten from his analysis.

(v) BOUNDS FOR s

We define a vector field s as follows:

$$\mathbf{s} \equiv \rho \overline{(\log F)\mathbf{c}}, \qquad \text{(VIII.52)}$$

and we call it the *flux of h*. In Chapter XI the reader will see why we choose this name, and why s should be compared with $-\mathbf{q}/(\tfrac{2}{3}\varepsilon)$. Of course the interest in this latter quantity lies in the fact that its product by $\mathfrak{r}$ is $-\mathbf{q}/\theta$, the very field that appears in the flux integral of the Clausius–Duhem inequality (I.33). We shall show now that for nearly Maxwellian densities the two quantities to be compared are nearly equal. Indeed,

$$F(\log F)\mathbf{c} + \frac{\tfrac{1}{2}c^2 F}{\tfrac{2}{3}\varepsilon}\mathbf{c} = F\left(\log F + \frac{3}{4}\frac{c^2}{\varepsilon}\right)\mathbf{c},$$

$$= F\left[\log(1 + G) + \log a + \left(\frac{3}{4\varepsilon} - b\right)c^2\right]\mathbf{c}. \qquad \text{(VIII.53)}$$

As both F and F_M give rise to the same field ε, from (VIII.8)$_2$ we see that (VIII.53) reduces to

$$F(\log F)\mathbf{c} + \frac{\tfrac{1}{2}c^2 F}{\tfrac{2}{3}\varepsilon}\mathbf{c} = [F_M(1 + G)\log(1 + G) + F\log a]\mathbf{c}. \qquad \text{(VIII.54)}$$

By integrating this relation and appealing once more to (VIII.25) we obtain

$$\mathbf{s} + \frac{1}{\tfrac{2}{3}\varepsilon}\mathbf{q} = \rho\langle[(1 + G)\log(1 + G) - G]\mathbf{c}\rangle. \qquad \text{(VIII.55)}$$

Again using (VIII.48), we see that[12]

$$\left|s_k + \frac{q_k}{\tfrac{2}{3}\varepsilon}\right| \leqq \rho\langle|c_k|G^2\rangle \leqq \rho\langle cG^2\rangle, \qquad k = 1, 2, 3. \qquad \text{(VIII.56)}$$

It is tempting to assert that GRAD's approximation, when suitably corrected, provides an exact upper bound corresponding to his approximate molecular density defined by (VIII.36), but no such thing has been proved here, because the condition $G > -1$, which is needed in order to apply (VIII.48) and so prove (VIII.51), is violated by G_G. Indeed, $F_M(1 + G_G)$ does not lead to any h at all, since $\log(1 + G_G)$ does not exist on all of $\mathscr{V}$.

Nevertheless, we shall see in Footnote 27 to Chapter XXIV that GRAD's result here, after correction, agrees precisely with the outcome of a systematic procedure.

[12] For GRAD's function G_G, given by (VIII.36), we may use (VIII.11) and so conclude that

$$\langle G_G^2 \mathbf{c}\rangle = \tfrac{4}{5}\mathbf{Pq}/p^2.$$

GRAD [1949, Eq. (7.14)] obtained an approximate formula for s which amounts to replacing $(1 + G)\log(1 + G) - G$ in (VIII.55) by its upper bound G^2 and then using the above expression for $\langle G_G^2 \mathbf{c}\rangle$. In view of the remarks in Footnote 11 this approximation is twice what he should have obtained from his analysis. The caution expressed in Footnote 11 should also be repeated here. So should the last sentence of that footnote.

Again referring to functions G such as to satisfy (VIII.38) or (VIII.40), we easily show that, respectively,

$$\langle cG^2 \rangle \leqq \begin{cases} \dfrac{2}{\sqrt{\pi}} (m+1)! \, \alpha^2 (\tfrac{4}{3}\varepsilon)^{m+\frac{1}{2}}, \\ \dfrac{4\alpha^2 \sqrt{\varepsilon}}{\sqrt{3\pi}} (1 - \tfrac{8}{3}\beta\varepsilon)^{-2}. \end{cases} \quad \text{(VIII.57)}$$

Summing up all this, we compare the estimates (VIII.57) with those provided by (VIII.39) and (VIII.41) and so see that *for certain clearly specified "nearly Maxwellian" functions F the field h, if it exists, is very nearly the field appropriate to equilibrium at the present value of the principal moment of F, and the field* **s**, *if it exists, is very nearly the field* $-\mathbf{q}/(\tfrac{2}{3}\varepsilon)$. Of course for a Maxwellian density $\mathbf{s} = \mathbf{0}$ and $\mathbf{q} = \mathbf{0}$.

<small>The inequalities given above do not suffice to show that h and **s** exist, because the integrands of the integrals defining these fields need not be non-negative. For example, in order to prove that h exists it is necessary to prove that both the positive and the negative parts of $F \log F$ are integrable.</small>

At the end of Section (ii) of this chapter we considered a body of gas grossly at rest and governed by two generally different Maxwellian densities, $a_1 e^{-b_1 c^2}$ and $a_2 e^{-b_2 c^2}$, to the one side and the other of a plane. We expressed through (VIII.17) the statement that the wall neither creates nor captures molecules, and in (VIII.15) and (VIII.19) we exhibited the discontinuity experienced by ε at points on the plane. We easily do the same for h. By use of (VIII.43)$_2$, (VIII.17), and (VIII.8)$_2$ we see that

$$\begin{aligned} h_1 - h_2 &= \log \frac{a_1}{a_2}, \\ &= 2 \log \frac{b_1}{b_2}, \quad \text{(VIII.58)} \\ &= -2 \log \frac{\varepsilon_1}{\varepsilon_2}. \end{aligned}$$

As $-\mathfrak{r}h$ is commonly thought to represent the caloric, we may interpret this result as stating that *the part of the gas having the greater temperature has also the greater caloric*. An easy calculation based on the fact that $\log F_M$ is even enables us to show that the value of h at points on the wall obeys a rule just like (VIII.15)$_3$, from which by the use of (VIII.18) and (VIII.58) we find that

$$h = h_1 + \frac{2\sqrt{\varepsilon_1}}{\sqrt{\varepsilon_1} + \sqrt{\varepsilon_2}} \log \frac{\varepsilon_1}{\varepsilon_2} = h_2 + \frac{2\sqrt{\varepsilon_2}}{\sqrt{\varepsilon_1} + \sqrt{\varepsilon_2}} \log \frac{\varepsilon_2}{\varepsilon_1}, \quad \text{(VIII.59)}$$

a value intermediate between h_1 and h_2. In Section (ii) of this chapter we obtained the energy flux (VIII.23), which shows that heat flows across the plane, from the hotter side to the colder. We can evaluate also the normal flux of h, namely, $s_\perp$, as follows[13]:

$$\frac{s_\perp}{m} = a_1 \int_{c_\perp > 0} (\log a_1 - b_1 c^2) c_\perp e^{-b_1 c^2}\, d\mathbf{c}$$

$$+ a_2 \int_{c_\perp < 0} (\log a_2 - b_2 c^2) c_\perp e^{-b_2 c^2}\, d\mathbf{c}. \qquad \text{(VIII.60)}$$

Using again the evaluations of quadratures stated just after (VIII.16) and (VIII.21), we find that the value of the first quadrature here is $(\frac{1}{2}\pi a_1 \log a_1)/b_1^2 - \pi a_1/b_1^2$, so by use of (VIII.17) we obtain

$$\frac{s_\perp}{m} = \frac{1}{2}\pi \frac{a_1}{b_1^2} \log \frac{a_1}{a_2},$$

$$= \pi \frac{a_1}{b_1^2} \log \frac{b_1}{b_2}; \qquad \text{(VIII.61)}$$

the second step follows from the first by a second application of (VIII.17). Now referring back to (VIII.22), we see that if $b_1 = b_2$, then $q_\perp = 0$ and $s_\perp = 0$. Assuming that $b_1 \neq b_2$, that is, $\varepsilon_1 \neq \varepsilon_2$, we find that

$$s_\perp = -2q_\perp \frac{\log \frac{1}{b_1} - \log \frac{1}{b_2}}{\frac{1}{b_1} - \frac{1}{b_2}}, \qquad \text{(VIII.62)}$$

$$= -\frac{q_\perp}{\frac{2}{3}} \frac{\log \varepsilon_2 - \log \varepsilon_1}{\varepsilon_2 - \varepsilon_1}.$$

It follows that *if a body of gas is governed by Maxwellian densities on the two sides of a plane, the vectors* **s** *and* **q** *on that plane point in opposite directions.* Because it is $-r h$ that is traditionally interpreted as the caloric, we may think of $-r s$ as proportional to the flux of caloric. On either side of the wall, the energetic and the caloric are uniform, and the fluxes of both vanish. The wall itself, however, if the temperatures on the two sides of it are different, is the seat of sources of heat and caloric: $q_\perp \neq 0$, $s_\perp \neq 0$. In the spirit of the statement we made just after (VIII.23), we may say that in this example *entropy, like heat, flows from the hotter side toward the colder side*; equivalently, from the *side having greater caloric to the side having lesser*. This statement expresses a kind of stability: The flow of energy is such as to render both temperature and caloric more nearly uniform.

---

[13] We are indebted to Mr. C.-S. MAN for part of the work here.

We can use (VIII.19) to relate the values of $s_\perp$ and ε at one and the same point on the wall:

$$s_\perp = -\frac{q_\perp}{\frac{2}{3}\varepsilon} \frac{\log n_2 - \log n_1}{n(n_2 - n_1)/(n_1 n_2)}. \tag{VIII.63}$$

If we consider the limit as $\varepsilon_1 \to \varepsilon_2$, of course $s_\perp \to 0$ and $q_\perp \to 0$. A sharper statement follows from (VIII.62)$_2$:

$$\lim_{\varepsilon_1 \to \varepsilon_2} \frac{s_\perp}{-q_\perp/(\frac{2}{3}\varepsilon)} = 1. \tag{VIII.64}$$

The tradition of classical thermodynamics takes the quotient of any source of heat by the temperature at which it operates as being also the strength of a corresponding source of entropy or at least a lower bound for that strength. The former possibility is confirmed by (VIII.64) in the limit as the temperature becomes nearly continuous—confirmed, that is, if we adopt the traditional interpretations for h and $\mathbf{s}$.

We may pursue the latter possibility also if we are willing to consider the temperature in question as being that appropriate to a boundary, not necessarily the same temperature as that of the gas in contact with that boundary. We have explained this idea from the standpoint of continuum thermomechanics in Section (v) of Chapter I. The counterpart here would regard the dividing plane as a wall and the unit normal $\mathbf{n}$ as pointing outward. The energetic of the wall would then be ε_2, while of course ε, the energetic of the gas at points on the wall, is given by (VIII.19). Turning to the details, we go on to estimate the magnitude of $s_\perp$, as follows. It is easy to show that if $x > y > 0$, then

$$\frac{1}{x} < \frac{\log x - \log y}{x - y} < \frac{1}{y}. \tag{VIII.65}$$

Hence

$$\left.\begin{array}{c}\dfrac{1}{\varepsilon_1}\\[6pt]\dfrac{1}{\varepsilon_2}\end{array}\right\} < \frac{\log \varepsilon_1 - \log \varepsilon_2}{\varepsilon_1 - \varepsilon_2} < \left\{\begin{array}{ll}\dfrac{1}{\varepsilon_2} & \text{if } \varepsilon_1 > \varepsilon_2,\\[6pt]\dfrac{1}{\varepsilon_1} & \text{if } \varepsilon_1 < \varepsilon_2.\end{array}\right. \tag{VIII.66}$$

A glance at (VIII.23) shows that $q_\perp > 0$ if the upper alternative holds, $q_\perp < 0$ if the lower does. Multiplying (VIII.66) by $-q_\perp/\frac{2}{3}$ and then using (VIII.62)$_2$, for both alternatives we obtain the estimate

$$\frac{-q_\perp}{\frac{2}{3}\varepsilon_1} > s_\perp > \frac{-q_\perp}{\frac{2}{3}\varepsilon_2}. \tag{VIII.67}$$

Thus this particular example conforms with the ideas of classical thermodynamics: *At a point on the wall the inward flux of energy divided by the temperature of the wall provides a lower bound for the inward flux of caloric.*

All these results flow from nothing more than the definitions of **s**, **q**, and F_M and from use of the particular F defined by (VIII.14). In Section (iv) of Chapter XI we shall see first how to interpret (VIII.23) and (VIII.67)$_2$ as representing the effect of a rough boundary upon a confined body of gas in equilibrium, and at the same time we shall see that (VIII.67)$_2$ expresses only a special case of a far more general fact. For the time being we should note that despite the arbitrary way we chose to consider a molecular density patched together from two Maxwellian ones, that possibility is compatible with any kinetic theory based upon the equation of evolution (III.38). That equation restricts the initial value F_0 in no way. We may take our Maxwellian densities upon the two sides of a stationary plane as representing an initial molecular density. The equation of evolution (III.38) may be expected to determine future values of F. However it may do so, *the result will correspond to an initial flux of energy compatible with* "NEWTON's *Law of Cooling*" *and with a flux of h that has the opposite sign and satisfies* (VIII.67)$_2$. The definitions of ρ, **u**, ε, **q**, and **s** are so framed as to ensure that such shall be the case.

It is clear from (VIII.51) that when F is not Maxwellian, h as defined by (VIII.46) will not generally make $-h$ equal the right-hand side of (VIII.42). Some authors of the physical kind have based upon this fact a claim that the concepts of thermodynamics and continuum mechanics fail to be applicable except for conditions near to equilibrium. The conclusion may be correct, but the reasoning, if such there be, is not. It is conceivable that h might turn out to be determined from the principal moment in a time-space neighborhood of the time and place in question, not merely at the time and place themselves; such determination, expressed by (I.30), is common in continuum thermodynamics today. Second, the apparent restriction to equilibrium is certainly a result of the definition (VIII.46). That definition was motivated, indeed, by the requirement that in equilibrium it square with the classical thermodynamics of ideal gases, but that requirement is expressed by (VIII.45), of which (VIII.46) provided only the simplest case. Moreover if we choose for $\mathfrak{h}$ in (VIII.45) a function whose values are nonpositive, we shall obtain an h which for a given F is not greater than that given by (VIII.46), so again we shall conclude that for a given principal moment the Maxwellian density delivers a smaller h than does any other. With some choice of $\mathfrak{h}$ other than 0 we might obtain, not relinquishing any of the classical requirements of an entropy, an h such as to make $-\mathfrak{r}h$ just one of the entropies considered in modern rational thermomechanics.

(vi) *Grossly determined functions, momentally determined functions*

To complete this chapter, we look once more at the Maxwellian density and recall that it is determined uniquely at $(t, \mathbf{x})$ by the value $\rho(t, \mathbf{x})$ of its own principal moment. We cannot expect to find other densities of this kind, but the broader possibility that F be determined by the field $\rho(t, \mathbf{x} + \cdot)$ will be considered often in later chapters of this book. It will be useful to allow such an F to depend upon the body force $\mathbf{b}$ as well as upon ρ. We may call the pair $(\rho, \mathbf{b})$ the *gross condition* σ of the gas whose molecular density is F:

$$\sigma \equiv (\rho, \mathbf{b}). \tag{VIII.68}$$

For the purposes of this book, we recall, $\mathbf{b}$ is a known function of $\mathbf{x}$ alone. A molecular density that is determined for each $\mathbf{v}$ by the gross condition at t will be called *grossly determined*[14]:

$$F(t, \mathbf{x}, \mathbf{v}) = \mathbb{G}(\mathbf{v}; \sigma(t, \mathbf{x} + \cdot)). \tag{VIII.69}$$

$\mathbb{G}$, which for each fixed $\mathbf{v}$ maps $\sigma(t, \mathbf{x} + \cdot)$ onto $F(t, \mathbf{x}, \mathbf{v})$, is the *gross determiner* of F.

We shall say that an ordinary field, that is, a function of t and $\mathbf{x}$, is *grossly determined*[15] if the gross condition determines it at each fixed t:

$$K(t, \mathbf{x}) = \mathfrak{K}(\sigma(t, \mathbf{x} + \cdot)). \tag{VIII.70}$$

It is an obvious and very important property of a grossly determined molecular density that its pressure deviator and energy flux are grossly determined:

$$\begin{aligned} \mathbf{P}(t, \mathbf{x}) &= \mathfrak{P}(\sigma(t, \mathbf{x} + \cdot)), \\ \mathbf{q}(t, \mathbf{x}) &= \mathfrak{q}(\sigma(t, \mathbf{x} + \cdot)), \end{aligned} \tag{VIII.71}$$

$\mathfrak{P}$ and $\mathfrak{q}$ being defined as follows in terms of $\mathbb{G}$:

$$\begin{aligned} \mathfrak{P}(\sigma(t, \mathbf{x} + \cdot)) &\equiv \mathfrak{m} \int \mathbb{G}(\mathbf{v}; \sigma(t, \mathbf{x} + \cdot))(\mathbf{c} \otimes \mathbf{c} - \tfrac{1}{3}c^2 \mathbf{1}), \\ \mathfrak{q}(\sigma(t, \mathbf{x} + \cdot)) &\equiv \tfrac{1}{2}\mathfrak{m} \int \mathbb{G}(\mathbf{v}; \sigma(t, \mathbf{x} + \cdot))c^2 \mathbf{c}. \end{aligned} \tag{VIII.72}$$

As we shall see in Chapters XIII, XIV, XV, and XXIII, the kinetic theory gives a special role to grossly determined molecular densities F, and to calculate the corresponding *gross determiners* $\mathfrak{P}$ and $\mathfrak{q}$ has been since the very beginning of the kinetic theory its major problem.

---

[14] MUNCASTER [1975, Chapter IV] introduced the term in the special case when $\mathbf{b} = 0$.

[15] We prefer this term to "functions of the static gross condition", which IKENBERRY & TRUESDELL [1956, §15] used in introducing the concept, specialized to functions of the derivatives of ρ and $\mathbf{b}$ of all orders.

(vi) GROSS DETERMINISM, MOMENTAL DETERMINISM

Another special class of solutions is suggested in Chapter XXV, one that leads to fields **P** and **q** whose values at $(t, \mathbf{x})$ are determined by the principal moment ρ in a space-time neighborhood of $(t, \mathbf{x})$. These we shall call *momentally determined*[16]. The formal properties of a momentally determined molecular density F and a momentally determined field B are

$$F(t, \mathbf{x}, \mathbf{v}) = \mathbb{M}(\mathbf{v}; \rho(t + \cdot, \mathbf{x} + \cdot)), \qquad B(t, \mathbf{x}) = \mathfrak{B}(\rho(t + \cdot, \mathbf{x} + \cdot)). \qquad \text{(VIII.73)}$$

$\mathbb{M}$ and $\mathfrak{B}$ are the *momental determiners* of F and B, respectively.

The difference between these two special kinds of determination is made clear by two examples, examples far from irrelevant to our later studies.

Example of a grossly determined **P**:

$$\mathbf{P} = \mathbf{F}(\rho, \text{grad } \rho, \text{grad } \mathbf{b}). \qquad \text{(VIII.74)}$$

Example of a momentally determined **P**:

$$\mathbf{P} = \mathbf{G}(\rho, \dot{\rho}, \text{grad } \rho). \qquad \text{(VIII.75)}$$

The symbols **F** and **G** here denote functions of the arguments indicated.

If we look back now at the sketch of continuum mechanics given in Chapter I, we see that (VIII.75) is a special instance of (I.25), which is the starting point for obtaining constitutive relations that define materials of the differential type. For (VIII.74) the occurrence of grad **b** makes comparisons with continuum mechanics doubtful, for in continuum mechanics the theorist conceives contact forces and body forces as being different in kind[17].

The simplest continuum theories and the simplest molecular density fall into both special classes. The Maxwellian density F_M is an explicit function of ρ, and for it $\mathbf{P} = 0$ and $\mathbf{q} = 0$. Thus F_M is both grossly and momentally determined. Likewise, the constitutive relations (I.7), (I.8), and (I.9) of the theory of linearly viscous fluids presume that **P** and **q** are both grossly and momentally determined. In both these examples ρ is an arbitrary smooth field.

With this chapter we complete our elementary study of the concepts, definitions, and operators that enter MAXWELL's kinetic theory of gases. We are ready to state the defining equation of that theory and to develop its consequences.

---

[16] In introducing the concept, specialized to functions of the derivatives of all orders, IKENBERRY & TRUESDELL [1956, §2] used the term "functions of the gross condition".

[17] The distinction is brought out clearly by the researches of GURTIN, NOLL, and WILLIAMS as reported in §§III.1 and III.3 of TRUESDELL's book cited in Footnote 1 to Chapter I.

Part C

The Maxwell–Boltzmann Equation and Its Elementary Consequences

Chapter IX

The Maxwell–Boltzmann Equation. Maxwell's Consistency Theorem and Equation of Transfer

By substituting (V.11) and (VII.3) into (V.16) we obtain the *Maxwell–Boltzmann Equation*[1], the integro-differential equation that governs the evolution of the molecular density F:

$$\mathfrak{D}F = \partial_t F + \mathbf{v} \cdot \partial_\mathbf{x} F + \mathbf{b} \cdot \partial_\mathbf{v} F = \int\!\!\int_{\mathscr{S}} w(F'F'_* - FF_*) = \mathbb{C}F. \quad \text{(IX.1)}$$

We have not yet called upon this equation. All our discussion so far has been preliminary in the sense that a function F from a class specified or implied only by conditions of smoothness and integrability, an F which is not necessarily a solution of the Maxwell–Boltzmann equation and hence not necessarily a true molecular density for the kinetic theory, has been used. We have determined some properties of the operators $\mathfrak{D}$ and $\mathbb{C}$, but we have not yet restricted attention to densities F such that $\mathfrak{D}F = \mathbb{C}F$ over an interval of time. Hitherto, that is, we have analysed definitions and assumptions but have not put them together. We now begin our study of solutions according to the kinetic theory.

For the time being we shall let (IX.1) stand as the basis of the kinetic theory, although we shall note for later use that the underlying assumptions of that theory are better expressed by the integral-functional equation that results from substituting (VII.3) into (III.38).

---

[1] BOLTZMANN [1872, Eq. (12)]. MAXWELL [1867, Eqs. (3) and (73)] had inferred the equation of transfer (IX.2) directly from considerations regarding molecular encounters. More specifically, MAXWELL used the form of $\bar{\mathbb{C}}_F g$ given by (VII.20)$_4$ with $G = H = F$. From his working immediately with this form of the total collisions operator it might appear that he avoided the questions of convergence mentioned following (VII.20), but such is not the case. Indeed, in order to proceed backward from (VII.20)$_2$ to (VII.20)$_1$ we must assume that $w(g + g_*) \times (G'H'_* + G'_*H' - GH_* - G_*H)$ and $w(g' + g'_*)(G'H'_* + G'_*H' - GH_* - G_*H)$ are integrable, and we confront then essentially the same problem that occurs in proceeding from (VII.20)$_2$ to (VII.20)$_3$ (*cf.* our remarks in Chapter XIX).

IX. THE MAXWELL–BOLTZMANN EQUATION

If we multiply (IX.1) by a function g and integrate over $\mathscr{V}$, we obtain the **Basic Equation of Transfer** for g:

$$\int (\mathfrak{D}F)g = \bar{\mathbb{C}}_F\, g, \tag{IX.2}$$

in which $\bar{\mathbb{C}}_F$ denotes the total collisions operator (VII.16). Before developing the equation of transfer we read off its simplest but nevertheless central consequences.

In the first place let us take g to be a summational invariant S. By (VII.23), for such a g the right-hand side of (IX.2) vanishes. Therefore, F must be such that for these same g the left-hand side vanishes likewise. The function F here is simply any molecular density that satisfies (IX.2) for all g; in particular, every F that satisfies the Maxwell–Boltzmann equation (IX.1) is included. Looking back at the result following from (V.24), we conclude that the fields ρ, $\mathbf{u}$, ε, p, $\mathbf{P}$, $\mathbf{q}$ corresponding to any solution of the Maxwell–Boltzmann equation satisfy the field equations (I.1), (I.2), (I.4) of continuum mechanics. We write these equations again here, and with new reference numbers, since they have now been derived from the kinetic theory:

$$\dot{\rho} = -\rho E; \tag{IX.3}$$

$$\rho \dot{\mathbf{u}} = -\operatorname{grad} p - \operatorname{div} \mathbf{P} + \rho \mathbf{b}; \tag{IX.4}$$

$$\rho \dot{\varepsilon} + pE = \tfrac{3}{2}\rho^{5/3}(p\rho^{-5/3})^{\cdot} = -\mathbf{P}\cdot\mathbf{E} - \operatorname{div}\mathbf{q}. \tag{IX.5}$$

As we remarked in Section (ii) of Chapter III, the definitions of the kinetic theory force it to represent an ideal gas with constant specific heats whose ratio is $\tfrac{5}{3}$. Of course (IX.5) is a form of the equation of energy appropriate to such a gas. These results express **Maxwell's Consistency Theorem**[2]: *Every smooth solution of the Maxwell–Boltzmann equation represents the kinetic gas as a continuous medium, indeed an ideal gas with constant specific heats whose ratio is $\tfrac{5}{3}$.* The term "ideal gas" is used exactly as defined in Section (i) of Chapter II, with no reference to thermodynamics, to a caloric equation of state, or to constitutive relations for $\mathbf{P}$ and $\mathbf{q}$.

We can scarcely hope to obtain any other universal identities in this way. Certainly the result stated after (VII.23) shows that to within questions of convergence of the integral in (VII.21), the summational invariants provide all functions g that cause the right-hand side of (IX.2) to vanish for all molecular densities F.

It is often claimed that a sufficiently rarefied gas does not behave like a continuum, so that for it the equations of continuum mechanics may fail to hold. Such may be the case in physics, but MAXWELL's consistency theorem shows that if so, the kinetic theory also cannot apply to these gas flows. *The*

---

[2] MAXWELL [1867, Eqs. (74), (76), (94)], specialized here to the case of a simple gas. Textbooks by physicists make little of this theorem. *Cf.* CHAPMAN & COWLING [1939, §3.21].

kinetic theory is a continuum theory. It represents the gas, no matter how rarefied it be, as *a continuum which obeys the ordinary field equations of continuum mechanics* as well as the special relations $p = \frac{2}{3}\rho\varepsilon$ and $\gamma = \frac{5}{3}$.

MAXWELL's consistency theorem says nothing about constitutive relations for the pressure tensor and the energy flux. They are another matter entirely.

First of all we repeat some of the observations of Chapter IV about kinetic theories in general, since the specific Maxwell-Boltzmann equation (IX.1) renders those observations explicit for MAXWELL's theory. The equation of evolution (III.38) suggests an *Initial-value Problem*, which becomes definite with the particular choice of $\mathbb{C}$ that defines MAXWELL's theory: *If we prescribe* $F(t_0)$, *does* (IX.1) *determine a unique* $F(t)$ *when* $t > t_0$? Recall that $F(t)$ denotes the restriction of F to the time t. The heuristic background of the kinetic theory suggests that this question should have an affirmative answer, at least for a short period of time. No general existence theorem of this kind has been proved as yet. In Chapters XX and XXI we shall develop most of the positive results presently known, but these refer only to a gas of molecules with a cut-off. A related result for a very special kind of intermolecular force of infinite range will be derived in Chapter XVIII.

In Section (iv) of Chapter IV we have seen that insofar as the kinetic gas can be a linearly viscous fluid, it must be a Stokes-Kirchhoff gas, and the gas flow must be such that $\mathsf{T} < 1$. In Section (i) of Chapter IV we have shown also that constitutive relations of continuum mechanics cannot follow both exactly and universally from any theory based on (III.38). A fortiori, *no constitutive relation of continuum mechanics can be both an exact and a general consequence of* MAXWELL's *kinetic theory*. It remains possible that such relations can be appropriate to special circumstances, special classes of solutions, limits of solutions, or other approximations to solutions. It is in this sense that the mathematician must seek and delimit ranges of agreement and disagreement between the Stokes-Kirchhoff theory of ideal gases and MAXWELL's kinetic theory. Such is the meaning that must be given, in particular, to a determination of the viscosity μ and the Maxwell number M in terms of a kinetic-molecular model. Such is the meaning, more generally, of kinetic "corrections" to the Navier-Stokes-Fourier constitutive relations.

Even results valid for special classes of solutions must be interpreted with caution. MAXWELL's consistency theorem has a corollary that students of the kinetic theory often overlook. Because *every* solution of the Maxwell-Boltzmann equation gives rise to fields ρ, **u**, ε, p, **P**, and **q** such as to satisfy (IX.3)-(IX.5) when **b** is given, *the kinetic theory can prove* nothing *about gross fields that are* not *so related*. For example, if $\mathbf{b} = \mathbf{0}$, we can *never* in any relation obtained from the kinetic theory justify by use of that theory any inquiry about the result of varying the fields ρ, **u**, ε, p, **P**, and **q** independently during an interval of time. While it is conceivable that relations similar in form to the constitutive relations (I.24) of a class of continuum theories of fluids may be demonstrated in the kinetic theory—and in Chapters XXII–XXV we shall see that indeed they can—they cannot be interpreted in the same way, for the variables on the right-hand sides *cannot* be varied independently. We cannot even ask if such relations satisfy the principle of material frame-indifference, stated and illustrated in Section (iv) of Chapter I, for that principle refers to arguments independently variable—independently variable not only at one time but *over an interval of time*. The kinetic theory *cannot* deliver constitutive relations in the sense of continuum mechanics although it may—and does— deliver formulae that look like constitutive relations but in fact are proved to hold only for severely restricted domains of their independent variables. Obviously the restriction of a function to a proper subdomain of its domain generally does not determine the function itself. Infinitely many different functions may have one and the same restriction to a proper subdomain of their original domains. Obvious as these facts are[3], writers on the kinetic theory have sometimes neglected them.

The equation of transfer (IX.2) is one of the tools through which the Maxwell–Boltzmann equation is exploited. Indeed, it was this equation, rather than the Maxwell–Boltzmann equation, that MAXWELL himself inferred, for a function of $\mathbf{v}$ alone. The requirement that F shall satisfy the equation of transfer for a class of functions g may be weaker mathematically than the condition that it satisfy the Maxwell–Boltzmann equation. MAXWELL's own attempts to determine properties of F were based upon examination of an explicit form for the left-hand side, $\int (\mathfrak{D}F)g$, a form which exhibited the expectations of g and some of its derivatives.

We shall obtain now a relation more general than MAXWELL's in that g need not be a function of $\mathbf{v}$ alone. By (V.11), (III.7), and (I.5) we easily show that

$$(\mathfrak{D}F)g = \partial_t(Fg) + \mathbf{u} \cdot \partial_\mathbf{x}(Fg) + EFg$$
$$+ \operatorname{tr} \partial_\mathbf{x}(Fg\mathbf{c}) - F\,\partial_t g - F\mathbf{v} \cdot \partial_\mathbf{x} g + (\mathbf{b} \cdot \partial_\mathbf{v} F)g. \qquad (\text{IX}.6)$$

Integration of this identity over $\mathscr{V}$, followed by use of the definition (III.4) of $\bar{g}$, enables us to write (IX.2) in the form

$$(\rho\bar{g})^{\cdot} + \rho E\bar{g} + \operatorname{div}(\rho\overline{g\mathbf{c}}) = \rho\,\overline{\partial_t g} + \rho\mathbf{v} \cdot \overline{\partial_\mathbf{x} g} - \mathfrak{m}\mathbf{b} \cdot \int (\partial_\mathbf{v} F)g + \mathfrak{m}\bar{\mathbb{C}}g, \qquad (\text{IX}.7)$$

which is **Enskog's Equation of Transfer**[4]. In writing it we have dropped the subscript F from $\bar{\mathbb{C}}_F$, because a given F must be understood in order that the overbars on the left-hand side, which denote expectations, shall mean anything.

The term proportional to the body force $\mathbf{b}$ may be simplified for certain g

---

[3] TRUESDELL [1977].
[4] ENSKOG [1911, 2].

THE EQUATION OF TRANSFER

and F. If $\mathcal{U}$ is a ball in $\mathcal{V}$, then, since $\mathbf{b}$ does not depend upon $\mathbf{v}$

$$\mathbf{b} \cdot \int_{\mathcal{U}} (\partial_v F) g = \int_{\mathcal{U}} [\partial_v \cdot (Fg\mathbf{b}) - F\mathbf{b} \cdot \partial_v g],$$

$$= \int_{\partial\mathcal{U}} Fg\mathbf{b} \cdot \mathbf{n} - \int_{\mathcal{U}} F\mathbf{b} \cdot \partial_v g, \tag{IX.8}$$

in which $\mathbf{n}$ is the unit outer normal to the boundary $\partial\mathcal{U}$. Conditions sufficient to make the surface integral vanish in the limit as the radius of $\mathcal{U}$ tends to ∞ are easy to formulate. Using them, we conclude that *if*

$$\text{either} \quad \mathbf{b} = 0 \quad \text{or} \quad Fg = o(v^{-2}) \quad \text{as} \quad v \to \infty, \tag{IX.9}$$

then (IX.7) *reduces to*

$$(\rho\bar{g})^{\cdot} + \rho E \bar{g} + \operatorname{div}(\rho\overline{g\mathbf{c}}) = \rho \overline{\partial_t g} + \rho \overline{\mathbf{v} \cdot \partial_x g} + \rho \mathbf{b} \cdot \overline{\partial_v g} + \mathfrak{m}\bar{\mathbb{C}}g. \tag{IX.10}$$

In this equation every term but the last one is an explicit expectation.
Because of (IX.3),

$$(\rho\bar{g})^{\cdot} + \rho E \bar{g} = \rho\dot{\bar{g}}, \tag{IX.11}$$

so we can write (IX.7) and (IX.10), respectively, in the forms

$$\rho\dot{\bar{g}} + \operatorname{div}(\rho\overline{g\mathbf{c}}) = \rho(\overline{\partial_t g} + \overline{\mathbf{v} \cdot \partial_x g}) + \begin{Bmatrix} -\mathfrak{m}\mathbf{b} \cdot \int (\partial_v F) g \\ + \rho\mathbf{b} \cdot \overline{\partial_v g} \end{Bmatrix} + \mathfrak{m}\bar{\mathbb{C}}g. \tag{IX.12}$$

When g is a function of the components of $\mathbf{c}$ alone, then $\partial_v g = \partial_\mathbf{c} g$ and also $\partial_x g = -(\operatorname{grad} \mathbf{u})^T \partial_\mathbf{c} g$, so (IX.12) may be simplified. With the aid of (IX.4) it is a simple matter to derive the following variant[5]

$$\mathfrak{L}g = \mathfrak{m}\bar{\mathbb{C}}g, \tag{IX.13}$$

in which $\mathfrak{L}$ is defined as follows:

$$\mathfrak{L}g \equiv (\rho\bar{g})^{\cdot} + \rho E\bar{g} + \overline{\rho\mathbf{c} \otimes \operatorname{grad} g} \cdot (\operatorname{grad} \mathbf{u})^T$$
$$- \overline{\operatorname{grad} g} \cdot \operatorname{div} \mathbf{M} + \operatorname{div}(\rho\overline{g\mathbf{c}}). \tag{IX.14}$$

Since g is a function of $\mathbf{c}$ only, there is no ambiguity in writing grad g here for $\partial_\mathbf{c} g$.

This equation of transfer is not equivalent, even formally, to the Maxwell-Boltzmann equation. In order for a function g that satisfies (IX.13) to satisfy (IX.12) also, it is necessary and sufficient that the body force $\mathbf{b}$ be determined from (IX.4). Therefore, a formal statement of the governing principles of the kinetic theory may be gotten by adjoining to (IX.13) the equation expressing balance of linear momentum.

[5] IKENBERRY & TRUESDELL [1956, Eq. (4.2)].

It is possible that the notorious frame-dependence of the kinetic theory might be set aside by considering (IX.13) as it stands, extended to fields that need not satisfy (IX.4), for (IX.13) so extended makes sense, and from it we do not see any frame-dependent consequences right off. Moreover, the fact that conclusions drawn from (IX.13) alone will not refer to **b** in any way makes it conceivable that contact actions in the kinetic theory, like those in continuum mechanics, could be determined by relations in which **b** does not appear.

It was these possibilities that caused IKENBERRY & TRUESDELL to base their study of momentally determined relations in the kinetic theory upon (IX.13). Their work is discussed in Chapter XXV.

In order to render the Maxwell-Boltzmann equation definite, we must specify a molecular model. Once we have done so, then, provided only the model be such as to render the various expectations based upon use of that model sufficiently smooth for the particular solution F we choose to consider, we may proceed through the various transformations from which we have drawn the preceding conclusions, MAXWELL's consistency theorem in particular. In this sense the conclusions deduced in this chapter are independent of the molecular model.

The equation of transfer is a relation among fields. It expresses the way the field $\bar{g}$ changes in time. In this respect it recalls the field equations (IX.3), (IX.4), and (IX.5), which, as we have seen at the beginning of this chapter, it contains as special cases obtained by choosing g as appropriate summational invariants. Since it is relations among fields that the kinetic theory is designed to deliver, the equation of transfer seems a likely tool, but it offers two characteristic difficulties, very different in kind from each other.

The first of these difficulties is afforded by the term $\text{div}(\rho \overline{g\mathbf{c}})$; if $\rho \bar{g}$ is a tensor of order n, then $\rho \overline{g\mathbf{c}}$ is a tensor of order $n + 1$. For example, if $g = \mathbf{c} \otimes \mathbf{c}$, then $\rho \bar{g} = \mathbf{M}$, so the equation of transfer gives an explicit expression for the time derivative $\dot{\mathbf{M}}$. However, in this case, $g\mathbf{c} = \mathbf{c} \otimes \mathbf{c} \otimes \mathbf{c}$, so $\rho \overline{g\mathbf{c}} = {}^3\mathbf{M}$. Thus the equation of transfer expresses the time derivatives of the second moment $\mathbf{M}$ in terms of the divergence of the third moment ${}^3\mathbf{M}$. Any attempt to study the moments of F successively runs straight onto this difficulty, sometimes called the "forward coupling of the equations of moments". More generally, whether or not g be a polynomial, the equation of transfer necessarily connects $\dot{\bar{g}}$ to the divergence of $\overline{g\mathbf{c}}$, an expectation of a more complicated kind. The term $\text{div}(\rho \overline{g\mathbf{c}})$ is the first of the two hero-villains of the kinetic theory. Its form is explicit and independent of the kinetic constitutive relations.

The second hero-villain, of course, is the total collisions operator $\bar{\mathbb{C}}$. Every attempt to get specific answers out of the kinetic theory must face the problem of evaluating or estimating $\bar{\mathbb{C}}g$ for certain functions g. As we shall see in subsequent chapters, the results depend very much on the molecular model.

Before considering further the equation of transfer in general, we shall use some of the results of this chapter to develop the concept of equilibrium in the kinetic theory and to relate it to equilibrium according to aerostatics.

Chapter X

Kinetic Equilibrium and Gross Equilibrium. Locally Maxwellian Solutions

The Maxwellian kinetic gas is a continuum, but its behavior is determined by properties of molecules. Statements which refer explicitly to F, $\mathfrak{m}$, and other molecular properties we shall henceforth denote as *kinetic*, while statements phrased solely in terms of $\mathbf{b}$, ρ, $\mathbf{u}$, ε, p, $\mathbf{P}$, and $\mathbf{q}$ and other fields over $\mathcal{T} \times \mathcal{E}$ we shall denote as *gross*. The central problem of the kinetic theory is to demonstrate from kinetic specification of a gas the gross properties of the flows which a body of that gas may suffer. We begin by considering the difference between kinetic and gross concepts of equilibrium.

The kinetic gas is said to be *in gross equilibrium* if its gross fields satisfy the formal analogues of the conditions that define equilibrium for a continuous fluid body: In a region of space,

$$\mathbf{u} = \mathbf{0}, \qquad \text{grad } \varepsilon = \mathbf{0}, \qquad \text{div } \mathbf{P} = \mathbf{0}, \qquad \text{div } \mathbf{q} = 0. \qquad (\text{II}.13)_r$$

These conditions reduce (IX.4) and (IX.5) to forms precisely the same as those to which (I.2) and (I.4) reduce when $\mathbf{u} = \mathbf{0}$, $\mathbf{P} = \mathbf{0}$, and $\mathbf{q} = \mathbf{0}$. Therefore, *in gross equilibrium the kinetic theory and the Euler–Hadamard theory deliver exactly the same fields ρ, $\mathbf{u}$, and ε*. In Chapter XIII we shall see that the same cannot be said of the fields $\mathbf{P}$ and $\mathbf{q}$, for the conditions div $\mathbf{P} = \mathbf{0}$ and div $\mathbf{q} = 0$ do not imply that $\mathbf{P} = \mathbf{0}$ and $\mathbf{q} = \mathbf{0}$ according to the kinetic theory.

Specializing the foregoing statements, we may call *gross free equilibrium* the special case of gross equilibrium in which $\mathbf{b} = \mathbf{0}$. Just as in the Euler–Hadamard theory, we invoke (IX.3), (IX.4), (IX.5), and (III.15) so as to conclude that in *gross free equilibrium of the kinetic gas, the fields ρ, p, and ε are constant both in space and in time*.

All the classical theories of fluids, including of course the Stokes–Kirchhoff theory, rest upon constitutive equations which *reduce in a state of rest to those of Eulerian aerostatics*; that is, if (II.13)$_1$ holds in a region, then $\mathbf{P} = \mathbf{0}$ there, and hence, trivially, (II.13)$_3$ holds also. As we have seen in Section (iv) of Chapter II, for equilibrium to be possible in a body of ideal gas it is necessary and sufficient that **b** be a steady lamellar field with single-valued potential. The general solution of the field equations is then given by (II.16). Therefore we may apply that result[1] to the kinetic theory, giving γ the value $\frac{5}{3}$ of course:

$$\frac{\rho(\mathbf{x})}{\rho_R} = \exp\left(-\frac{U(\mathbf{x})}{\frac{2}{3}\varepsilon}\right); \tag{X.1}$$

U is the potential of the body force, the energetic ε is an assigned constant, and ρ_R is the value of the density ρ at a reference place $\mathbf{x}$ where $U(\mathbf{x}) = 0$. To conclude that (X.1) does hold in the kinetic theory, then, all that remains is to show that for the kinetic gas in equilibrium the defining conditions (II.13) are indeed satisfied. It is clearly sufficient to show that the velocity vanishes, that the pressure tensor is hydrostatic, that the energy flux vanishes, and that the energetic is uniform in space. *A fortiori*, in order to prove anything of the sort, we must first define what we mean by equilibrium in the kinetic theory.

MAXWELL seemed to regard permanence of F under the action of collisions as the defining idea of kinetic equilibrium. He claimed to prove that *the only molecular densities unaffected by collisions are the Maxwellian ones*. In Chapter VIII we have stated this assertion as (VIII.7) and have proved it.

As we have seen in Section (ii) of Chapter VIII, a Maxwellian density represents the kinetic gas as obeying the constitutive equations of the Euler-Hadamard theory, restricted of course to arguments that satisfy the field equations (IX.3), (IX.4), and (IX.5). That fact implies (II.13)$_{3,4}$ but allows **u** and ε to be arbitrary fields and thus does not ensure that the other two classical requirements, namely, (II.13)$_{1,2}$, be satisfied. If we like, we may simply adjoin them and regard the result as being a definition of kinetic equilibrium. If we do so, we may assert that in equilibrium the Maxwellian kinetic gas is grossly an Euler-Hadamard gas for which $\gamma = \frac{5}{3}$. Such a definition of equilibrium in the kinetic theory is unsatisfying because it mixes gross and kinetic ideas: F is Maxwellian, **u** = **0**, and grad $\varepsilon = \mathbf{0}$. Not only that, characterization of kinetic equilibrium in this way does not even appeal to the basic principle of the kinetic theory, the Maxwell-Boltzmann equation. Indeed, since it is now easy to verify that that equation is satisfied, some assumption in the reasoning leading to an F_M with **u** = **0** and **b** = const. must have been redundant, for the Maxwell-Boltzmann equation, which governs the whole kinetic theory, ought to be used to find solutions to problems, not merely checked as an after-thought. Definitions in the kinetic theory should rest on kinetic ideas alone. To achieve a satisfactory concept of kinetic equilibrium, we now retreat to a more general vantage point.

First of all, we ask if MAXWELL's requirement $F = F_M$, combined with the requirement that F_M satisfy the Maxwell-Boltzmann equation, can provide a

---

[1] MAXWELL [1867, Addition]. In the paper as sent to press MAXWELL had slipped here, and also objections arising from physical considerations were raised later against MAXWELL's conclusion that ε was constant in the equilibrium of a heavy gas.

suitable definition of kinetic equilibrium. Since $\mathbb{C}F_M = 0$, the Maxwell-Boltzmann equation (IX.1) becomes

$$\partial_t F_M + \mathbf{v} \cdot \partial_x F_M + \mathbf{b} \cdot \partial_v F_M = 0. \qquad (X.2)$$

A solution F_M of this kind is called *locally Maxwellian*. The fields a, b, and $\mathbf{u}$ in (VIII.4) are no longer arbitrary but are required to be such as to make (X.2) be satisfied when (VIII.4) is substituted into it. In view of (VIII.8) and MAXWELL's consistency theorem in Chapter IX we see that *for any locally Maxwellian solution, the gross behavior of the kinetic gas is exactly that of an Euler-Hadamard gas with constant specific heats whose ratio is* 5/3. In particular, by $(IX.5)_2$ we see that these solutions obey POISSON's law of adiabatic change: *For each fluid-point*

$$p\rho^{-5/3} = \text{const.}, \qquad \rho\varepsilon^{-3/2} = \text{const.} \qquad (X.3)$$

From $(VIII.8)_1$ it follows that

$$a = \text{const.} \quad \text{for each fluid-point.} \qquad (X.4)$$

The locally Maxwellian solutions are not the only ones for which the kinetic theory agrees precisely with aerodynamics according to the Euler-Hadamard theory. In Appendix B to Chapter XVII we shall present a kinetic solution corresponding to the dilatation (II.39) which is distinct from the locally Maxwellian solution but also agrees precisely with the results of classical gas dynamics.

We have considered so far only the restrictions delivered by MAXWELL's consistency theorem, but these are merely necessary, not sufficient for (X.2) to hold. It is not difficult to determine all the locally Maxwellian solutions, but for our present purpose they are too numerous. The velocity fields $\mathbf{u}$ that belong to locally Maxwellian solutions include not only rigid rotations with steady axis and steady angular speed but also certain dilatations. Thus the class of solutions F such that $\mathbb{C}F = 0$ is too large to correspond to anything like gross equilibrium.

BOLTZMANN[2] considered locally Maxwellian solutions but obtained only those that can correspond to steady lamellar body force. We show now how to get a complete solution of the problem.

We put (VIII.4) into (X.2); since $\mathbf{b}$ does not depend upon $\mathbf{v}$, we obtain a cubic polynomial in the components of $\mathbf{v}$ which must vanish for all $\mathbf{v}$. Hence each of its four terms must vanish identically. Two of the conditions so obtained merely specialize MAXWELL's consistency theorem to the particular problem at hand. The cubic term vanishes if and only if b is a function of t alone. Therefore, by $(VIII.8)_2$, *a locally Maxwellian solution is necessarily homo-energetic*. Taking account of this fact, we find that the quadratic term vanishes if and only if the motion is a dilatation, the expansion of which is $\tfrac{3}{2}\dot{b}/b$. Thus, first, if $\mathbf{r} \equiv \mathbf{x} - \mathbf{x}_0$, $\mathbf{x}_0$ being an arbitrary place, then

$$\mathbf{u} = \mathbf{g} + \tfrac{1}{3}E\mathbf{r} + \mathbf{W}\mathbf{r}, \qquad \mathbf{W} = -\mathbf{W}^T; \qquad (II.23)_r$$

the translational velocity $\mathbf{g}$, the expansion E, and the spin $\mathbf{W}$ are functions of t only. At this point the kinetic problem reduces to a special case of a problem of continuum gas dynamics that we have solved in Section (v) of Chapter II. In the results given there we have only to take γ as $\tfrac{5}{3}$;

---

[2] BOLTZMANN [1876, §III].

equivalently, $\lambda = 1$. From (II.8) we see that

$$\frac{1}{3} E = -\frac{\dot{\varepsilon}}{2\varepsilon}. \tag{X.5}$$

If we prescribe the functions ε, $\mathbf{g}$, and $\mathbf{W}$, then (II.23)$_1$ and (X.5) determine the velocity $\mathbf{u}$, and (II.34) becomes

$$\rho(t, \mathbf{x}) = \left(\frac{\varepsilon(t)}{\varepsilon(0)}\right)^{\frac{3}{2}} \rho_0\left(\mathbf{x}_0 + \sqrt{\frac{\varepsilon(t)}{\varepsilon(0)}} \mathbf{R}(t)\mathbf{r} - \int_0^t \sqrt{\frac{\varepsilon(s)}{\varepsilon(0)}} \mathbf{R}(s)\mathbf{g}(s) \, ds\right), \tag{X.6}$$

the initial value ρ_0 being any positive function, and $\mathbf{R}$ being the unique orthogonal tensor-valued function of t that satisfies the conditions

$$\dot{\mathbf{R}} = -\mathbf{RW}, \qquad \mathbf{R}(0) = \mathbf{1}. \tag{II.32}_r$$

Also, the body force $\mathbf{b}$ required to produce this solution can be calculated explicitly in terms of ε, $\mathbf{g}$, $\mathbf{W}$, and ρ_0 by means of (II.28) and (II.29), of course in the special case when $\lambda = 1$. Conversely, any assignment of ε, $\mathbf{g}$, $\mathbf{W}$, and ρ_0 yields a locally Maxwellian solution corresponding to the $\mathbf{b}$ they determine.

The free locally Maxwellian solutions, those that exist when $\mathbf{b} = \mathbf{0}$, were calculated by BOLTZMANN. They are given in terms of their fields of density, gross velocity, and energetic by (II.38) and (II.37)$_1$. The special case in which ρ is a function of time only supplies some particularly valuable information. The corresponding fields for it are given by (II.39). Since ε, too, is a function of t only, POISSON's law (X.3) makes $p\rho^{-5/3}$ constant in both space and time. This fact we may interpret as an illustration of the kinetic theory's failure to provide a positive bulk viscosity: A homo-energetic dilatation proceeds unhindered by dissipation. Likewise, by (VIII.8)$_1$, a reduces to a constant. Moreover (VIII.8)$_2$ and (II.39) yield

$$bc^2 = \frac{3}{4\varepsilon(0)T^2} |(t + T)\mathbf{v} - \mathbf{r}|^2, \tag{X.7}$$

T being an arbitrary non-zero constant. There are two possibilities:

(a) $T > 0$. Then for each $\mathbf{x}$ and $\mathbf{v}$

$$F_M \to 0 \quad \text{as} \quad t \to \infty. \tag{X.8}$$

(b) $T < 0$. Then for each $\mathbf{x}$ and $\mathbf{v}$

$$F_M \to a \exp\left(-\frac{3}{4\varepsilon(0)T^2} r^2\right) \quad \text{as} \quad t \to -T. \tag{X.9}$$

The second case shows that *a solution of the Maxwell–Boltzmann equation cannot be expected to exist for all time.* Because 0 does not have a positive integral over $\mathscr{V}$, while a non-zero constant is not integrable over $\mathscr{V}$, in both cases F_M approaches pointwise a well-defined limit, but a limit that is not only not Maxwellian, *it is not even a molecular density! A fortiori,* we cannot expect in general that solutions of the Maxwell–Boltzmann equation tend with time to approach Maxwellian densities, though such is known to be the case for some solutions (Appendix B to Chapter XVII, Chapter XXI).

The locally Maxwellian solutions are the least interesting ones the kinetic theory affords, since for them collisions have no overall effect, and so the very mechanism the kinetic theory is designed to model does not come into play. In this book we shall not study the locally Maxwellian solutions further; rather, we now turn our attention back to the concept of equilibrium. A glance at $(II.23)_1$, (X.5), and (X.6) suffices to convince us that if kinetic equilibrium is to stand in any simple relation to gross equilibrium, it cannot reduce to the bare requirement that F be a locally Maxwellian solution.

MAXWELL and BOLTZMANN both saw that a proper definition of kinetic equilibrium would have to lead to some special locally Maxwellian solution; they imposed more or less *ad hoc* further restrictions so as to obtain a result which squared properly with aerostatics. C.-C. WANG defined **Kinetic Equilibrium** as follows:

$$F = F(\mathbf{x}, \mathbf{v}) = F(\mathbf{x}, -\mathbf{v}). \qquad (X.10)$$

According to this concept, *equilibrium corresponds to solutions of the Maxwell–Boltzmann equation that are steady, even functions of the molecular velocity*. Of course the definition implies outright that $\mathbf{u} = \mathbf{0}$, but from it alone we cannot conclude that ρ and ε are constant.

We can easily determine all densities that can correspond to kinetic equilibrium. Multiplying the Maxwell–Boltzmann equation (IX.1) by $\log F$ and integrating the result over $\mathscr{V}$ yields

$$\int (\log F)\mathbf{v} \cdot \partial_{\mathbf{x}} F + \int (\log F)\mathbf{b} \cdot \partial_{\mathbf{v}} F = \bar{\mathbb{C}} \log F. \qquad (X.11)$$

If F is an even function of $\mathbf{v}$, so is $(\log F)\, \partial_{\mathbf{x}} F$, and thus the value of the first integral on the left-hand side is 0. Now since $\mathbf{b}$ is independent of $\mathbf{v}$,

$$\int (\log F)\mathbf{b} \cdot \partial_{\mathbf{v}} F = \mathbf{b} \cdot \int \partial_{\mathbf{v}}(F \log F - F), \qquad (X.12)$$

and since the integrand is odd, this integral vanishes also. Thus $\bar{\mathbb{C}} \log F = 0$. By the Boltzmann monotonicity theorem in Chapter VIII, F is Maxwellian. Since $\mathbf{u} = \mathbf{0}$,

$$F = ae^{-bv^2}, \qquad (X.13)$$

and the steady fields a and b must satisfy (X.2), which here reduces to

$$\mathbf{v} \cdot \partial_{\mathbf{x}}(ae^{-bv^2}) + \mathbf{b} \cdot \partial_{\mathbf{v}}(ae^{-bv^2}) = 0. \qquad (X.14)$$

(Note that b here does not denote $|\mathbf{b}|$.)

Dividing the analysis into two parts, we first consider the case of *free kinetic equilibrium*: $\mathbf{b} = \mathbf{0}$. Then (X.14) requires that F_M be independent of $\mathbf{x}$. Thus the parameters a and b are constant not only in time but also in space. Such an F_M is called a *uniform Maxwellian density* and is denoted by F_U. For it, of course, (VIII.8) shows that ρ and ε are both constant. *The kinetic gas in free kinetic equilibrium is governed by a uniform Maxwellian density. The corresponding fields of energetic, density, and pressure are everywhere and always constant; the gross velocity vanishes; the pressure tensor is hydrostatic, and the energy flux vanishes.* A fortiori, *the gas is in free gross equilibrium.*

Let us now turn to the case in which $\mathbf{b} \neq \mathbf{0}$. We have remarked already that if F is Maxwellian, the pressure tensor is hydrostatic, so our analysis in Chapter II applies and shows that $\mathbf{b}$ must be lamellar[3] in order that $\mathbf{u} = \mathbf{0}$. Thus we must set $\mathbf{b} = -\operatorname{grad} U$ in (X.14). Doing so, we obtain

$$\mathbf{v} \cdot [\operatorname{grad} \log a - v^2 \operatorname{grad} b] + 2b\mathbf{v} \cdot \operatorname{grad} U = 0. \tag{X.15}$$

Hence

$b = \text{const.},$ and hence $\varepsilon = \text{const.},$ $\log a = -2bU + \text{const.},$ (X.16)

so

$$F = F_U e^{-2bU}. \tag{X.17}$$

Since $b = \text{const.}$, it follows from (VIII.8)$_2$ that $F/F_U = \rho/\rho_R$ so again we obtain (X.1). Thus kinetic equilibrium is impossible unless $\mathbf{b}$ is steady and lamellar; *the kinetic gas in kinetic equilibrium is also in gross equilibrium*. These results constitute the **Maxwell–Boltzmann–Wang Theorem on Kinetic Equilibrium**[4].

We have shown that kinetic equilibrium implies gross equilibrium. In Chapters XIII and XVIII we shall see that the converse is false: Corresponding to any constant fields ρ and ε as well as the null velocity field $\mathbf{u} = \mathbf{0}$, there are infinitely many sets of moments $^n\mathbf{M}$, $n = 2, 3, \ldots$, that differ from those of a

---

[3] The reader should note that this conclusion refers only to Maxwellian densities for which $\mathbf{u} = \mathbf{0}$. The F_M that correspond to the velocity field (II.23) are compatible with the more general fields of body force determined by (II.28) and (II.29), λ being taken as 1. If $\mathbf{B} \neq \mathbf{0}$ and if U is not the sum of a function of $\mathbf{x}$ alone and a function of t alone, such a body force is incompatible with gross equilibrium.

[4] MAXWELL [1867, Eq. (22)] in effect defined the "permanent condition" of a gas, subject to any body force, by the condition $\mathfrak{C}F = 0$, not trying to determine whether any such solution were compatible with an assigned non-zero field of body force. BOLTZMANN [1868, §§2–5], with his usual awkward verbosity, observed that (X.17) was unaffected by collisions; subsequently [1871, §2] he convinced himself that since the H-theorem (cf. Chapter XI) remained valid when a conservative body force acted, it implied that likewise Maxwellian densities were the only densities appropriate to equilibrium, and he used the condition $\mathfrak{D}F = 0$ to infer (X.17). Later MAXWELL [1873] showed that (X.17) satisfied the equation $\mathfrak{D}F = 0$, but he did not state conditions sufficient to make it the most general such solution. The tentative quality of all this early work is reflected by the way physicists often describe how they obtain (X.17): "by inserting a Boltzmann factor".

In 1965, in connection with the Seminar in Rational Mechanics at the Johns Hopkins University, C.-C. WANG formulated his definition of kinetic equilibrium and proved the theorem we present above. His statements were published by TRUESDELL [1969, *1*, Lecture 8]; his simple proof, by TRUESDELL [1973, Chapter X].

Maxwellian density and hence do not correspond to kinetic equilibrium. Before facing the calculations necessary to establish this fact, we shall consider one further general aspect of the kinetic theory, namely, the interpretation of irreversibility in terms of a quantity like the negative of the caloric of thermomechanics. This is the field h, which we have introduced and estimated in Chapter VIII.

Chapter XI

Boltzmann's *H*-Theorem

In Chapter VIII we have motivated and laid down BOLTZMANN's definition of h, namely,

$$h \equiv \overline{\log F}, \qquad \text{(VIII.46)}_r$$

as the simplest expectation such as to reduce when F is Maxwellian to $-\eta/\mathfrak{r}$, η as given by (VIII.42) being the classical caloric of an ideal gas for which $\gamma = \frac{5}{3}$. Although we remarked there that classical thermostatics truly motivates only the far more general definition (VIII.45), we fixed our attention upon h, partly for its own interest and partly out of respect for the tradition of our subject, and here, for the same reasons, we continue to do so. In Section (v) of Chapter VIII we defined also the field **s**,

$$\mathbf{s} \equiv \rho \overline{(\log F)\mathbf{c}}, \qquad \text{(VIII.52)}_r$$

and we called it "the flux of h", though we gave no reason for that name.

The considerations in Chapter VIII rested upon the bare definitions of h and **s**. Those considerations are relevant to the kinetic theory in view of what we saw in Chapter IV. Because any non-negative integrable function having a positive integral may be taken as the initial value of the molecular density F, both h and **s** may be assigned at will at the initial instant—that is, at any time, subject to the *caveat* stated in small print toward the end of Section (i) of Chapter IV.

Thus, in general, neither h nor **s** can be determined by other expectations such as ρ, **u**, ε, **P**, or **q** at the same time. It remains possible that the evolution of h and **s** in time can be restricted or even determined by other expectations in a time-space neighborhood of some given time and place. Accordingly, we now derive some properties of h and **s** that follow from the fact that F must be a solution of the Maxwell–Boltzmann equation.

(i) *The formal broad H-theorem and the formal narrow H-theorem*

To obtain the rate of change of h for a given fluid-point, we appeal to ENSKOG's general equation of transfer, namely

$$\rho \dot{\bar{g}} + \text{div}(\rho \overline{g\mathbf{c}}) = \rho(\overline{\partial_t g} + \overline{\mathbf{v} \cdot \partial_x g}) - \mathfrak{m}\mathbf{b} \cdot \int (\partial_v F)g + \mathfrak{m}\bar{C}g, \quad \text{(IX.12)}_{1r}$$

and in it we substitute log F for g. For this choice of g the term $\rho(\overline{\partial_t g} + \overline{\mathbf{v} \cdot \partial_x g})$ reduces to $\partial_t \rho + \text{div}(\rho \mathbf{u})$, a quantity which vanishes because of (IX.3). The term proportional to $\mathbf{b}$ vanishes, of course, if $\mathbf{b} = \mathbf{0}$; it vanishes also if $F \log F - F = o(v^{-2})$ as $v \to \infty$. Supposing one or the other of these sufficient conditions to hold, by making use of the definitions (VIII.46) and (VIII.52) we obtain

$$\rho \dot{h} + \text{div } \mathbf{s} = \Sigma, \qquad \text{(XI.1)}$$

in which

$$\Sigma = \mathfrak{m}\bar{C} \log F. \qquad \text{(XI.2)}$$

In analogy to the total entropy in thermodynamics we define as follows a quantity H associated with the kinetic gas occupying a region $\mathscr{B}$ of $\mathscr{E}$:

$$H \equiv \int_{\mathscr{B}} \rho h \, dV. \qquad \text{(XI.3)}$$

Supposing now that $\mathscr{B}$ be the shape at time t of some particular material region, that is, a generally time-dependent region of space swept out by fluid-points that move always with the gross velocity $\mathbf{u}$ of the gas at the places they occupy, we make H a function of time alone. To calculate its rate of change $\dot{H}$, we assume that $\mathscr{B}$ is a smooth, bounded region, integrate (XI.1) over it, and so obtain

$$\dot{H} + \int_{\partial \mathscr{B}} s_\perp \, dA = \int_{\mathscr{B}} \Sigma \, dV. \qquad \text{(XI.4)}$$

As usual, we here write $a_\perp$ for $\mathbf{a} \cdot \mathbf{n}$, denoting by $\mathbf{n}$ the outer unit normal to the bounding surface $\partial \mathscr{B}$. The relation (XI.4) is a consequence of the equation of transfer, subject to the hypotheses of smoothness we have mentioned, to which must be added hypotheses sufficient to justify use of the divergence theorem. It motivates our having called the field $\mathbf{s}$ "the flux of h".

To obtain (XI.4), we have used (IX.12), and (IX.12) presumes much smoothness in F. It might be possible to establish (XI.4) directly, without assuming that derivatives of expectations exist. Then discontinuities in h and $\mathbf{s}$ might arise.

In most parts of mathematical physics assumptions of smoothness are so freely allowed as to be accepted without even being stated. In the kinetic theory more caution is called for. There is no physical reason to assume that F be continuous. A discontinuous F can easily give rise to discontinuous $\mathbf{s}$, $\mathbf{q}$, and ε. For an example we need only refer to Chapter VIII, where we considered a wall

(i) FORMAL NARROW H-THEOREM

separating two regions, each governed by a uniform Maxwellian molecular density. If those two densities are different, the corresponding F is discontinuous at the wall. For that case we showed that **s** and **q** vanished on either side of the wall but not on it, and we calculated explicitly the discontinuities of ε and h. Since **s** is discontinuous at the wall, div **s** does not exist there. Because any non-negative integrable function on $\mathscr{E} \times \mathscr{V}$ having a positive integral may serve as the initial value of F in the equation of evolution, we cannot reject the example.

The Boltzmann monotonicity theorem in Chapter VIII may be restated as follows: $\Sigma < 0$ unless $F = F_M$, in which case $\Sigma = 0$. We recall also that if $F = F_M$, then $\mathbf{s} = \mathbf{0}$. Consequently the right-hand side of (XI.4) is negative unless $F = F_M$ almost everywhere. Thus we have established the *Formal Broad H-Theorem*, usually attributed to BOLTZMANN[1]:

$$\dot{H} < -\int_{\partial \mathscr{B}} s_\perp \, dA \quad \text{if} \quad F \neq F_M, \quad \text{(XI.5)}$$

$$\dot{H} = 0 \quad \text{if} \quad F = F_M.$$

If circumstances are such that for the boundary $\partial \mathscr{B}$ of the region $\mathscr{B}$

$$-\int_{\partial \mathscr{B}} s_\perp \, dA \leqq 0 \quad \text{(XI.6)}$$

at some instant, then (XI.5) implies that at that instant

$$\dot{H} < 0 \quad \text{if} \quad F \neq F_M,$$

$$\dot{H} - 0 \quad \text{if} \quad F = F_M. \quad \text{(XI.7)}$$

That is, for the given material region $\mathscr{B}$, at an instant when (XI.6) holds H *must decrease unless F is Maxwellian*. This statement is the *Formal Narrow H-Theorem*. If (XI.6) holds over an interval of time, so does (XI.7), and the formal narrow H-theorem then asserts that *so long as F is not Maxwellian, H decreases*. This statement is a kinetic analogue of the Clausius inequality (I.39).

As the proof shows, this *trend in time* reflects the Boltzmann monotonicity theorem. That theorem in turn follows directly from the specific nature of the collisions operator. The irreversibility expressed by the narrow H-theorem is not to be expected on the basis of the molecular model, which is formulated in terms of a dynamical system whose every motion is at least locally reversible. Those who have attempted to found an explanation of irreversibility without denying the

---

[1] BOLTZMANN came to his results on this subject slowly, with what now appear to be painful detours. First considering the density of molecular energy rather than F, through a long chain of difficult transformations he obtained a special case of what we call below the formal narrow H-theorem (BOLTZMANN [1872, §1]). The theorem as we give it generalizes that of BOLTZMANN [1875, §1] [1896, §18] for a gas in a stationary, smooth, vessel; under BOLTZMANN's assumptions, as we shall demonstrate below in Section (iii), the formal narrow H-theorem follows. We cannot ascertain the earliest source for the explicit surface integral; it may be found incidentally in the discussion of CHAPMAN & COWLING [1939, §4.13] for a stationary vessel with smooth walls. The treatment in the text above, taken from TRUESDELL [1973, Chapter X], is equivalent to that of GRAD [1958, §18] except for his unnecessary assumption that $\mathbf{b} = \mathbf{0}$. The fact that $\mathbf{b}$ need not vanish is of great importance for comparison of the H-theorem with other assertions of irreversibility.

laws of analytical dynamics have had recourse to various expedients. One of these, perhaps the most successful, has been to envision a sequence of similar dynamical systems like in kind but increasing in their number of degrees of freedom and to calculate limits at a fixed time as that number tends to ∞. BOLTZMANN repeatedly stressed a statistical interpretation. He claimed that the Maxwell–Boltzmann equation described, not the actual motion of a system of mass-points, but rather the "most probable" motion of such a system, regarded as infinitely numerous. Just what kind of probability should be introduced and what kind of limit as the number of mass-points becomes infinite should be taken in order to make BOLTZMANN's contention precise and hence subject to mathematical proof or disproof, remains to this day a matter of dispute and doubt. Considerations of this kind do not fall within the scope of this book.

We see immediately two simple conditions sufficient that the formal narrow H-theorem shall follow from the formal broad H-theorem.

Case 1. *F is independent of place.* If F does not depend upon $\mathbf{x}$, then for every g that is independent of $\mathbf{x}$ the expectation $\bar{g}$ is a function of t only, so $\int_{\partial\mathcal{B}} \bar{g}\mathbf{n}\, dA = \bar{g}\int_{\partial\mathcal{B}} \mathbf{n}\, dA = 0$. By taking $(\log F)\mathbf{c}$ for g we conclude that (XI.6) holds, so *the formal narrow H-theorem follows.*

Moreover, the definitions (III.2), (III.5), (III.6), (III.9), (III.14), and (III.12) make the fields ρ, $\mathbf{u}$, $\mathbf{M}$, ε, and $\mathbf{q}$ independent of place. Therefore, the field equations (IX.3), (IX.4), and (IX.5) reduce to

$$\partial_t \rho = 0, \qquad \partial_t \mathbf{u} = \mathbf{b}, \qquad \partial_t \varepsilon = 0, \qquad (\text{XI.8})$$

so both ρ and ε are constant fields. Because of (III.15), so also is p. If $\mathbf{b} = \mathbf{0}$, we lose no generality if we take $\mathbf{u}$ as $\mathbf{0}$. The condition of the body of gas is then one of gross free equilibrium as defined at the beginning of Chapter X. As $\mathbf{q} = \mathbf{0}$, (I.47) is satisfied. It follows, then, from the kinetic counterpart of (I.43) that $\mathfrak{E} = 0$: *The total energy of the gas in any bounded region remains constant.* Thus any bounded region $\mathcal{B}$ in which $\mathbf{b}$ vanishes and F is independent of place exhibits the behavior CLAUSIUS attributed to the "universe".

Indeed, some physicists have adduced this particular formal narrow H-theorem as evidence in favor of CLAUSIUS' pronouncement, though nobody, so far as we can learn, has yet provided experimental data sufficient to justify our taking as a model for the universe a bounded region occupied by a moderately rarefied monatomic gas which conforms to a spatially homogeneous molecular density.

Anyway, all that has been shown is that if $F \neq F_M$ and if (XI.6) holds, then $-\mathfrak{r}H$ increases as t increases, just as continuum thermomechanics requires $\mathfrak{H}$ to increase if $\int_{\partial\mathcal{B}} q_\perp \, dA \leq 0$. There is no reason to conclude that the two increasing quantities are the same functions of t for a given body of gas. Of course, if $F = F_M$, then $-\mathfrak{r}H = \mathfrak{H} = $ const., on the assumption that ρ and ε as defined in the kinetic theory are identified with the quantities denoted by the same letters in gas dynamics. The definition of h was motivated in Chapter VIII just so as to make this statement hold. That fact tells us nothing at all about h when it is obtained from an F that is not Maxwellian.

(ii) FORMAL H-THEOREMS AND THERMOMECHANICS   149

Case 2. *The molecular density F exists in all of $\mathscr{E}$ and vanishes at ∞ sufficiently rapidly.* By the latter condition we mean that $s_\perp = o(r^{-2})$, r being the radius of a sphere centered at a fixed point in $\mathscr{E}$. Then we can take $\mathscr{B}$ as all of $\mathscr{E}$ and conclude that the infinite integral defining H exists, that the integral on the left-hand side of (XI.6) is null, that $\dot{H} < 0$ unless $F = F_M$. Such is the case, for example, if F vanishes outside a certain fixed sphere. Solutions of this kind, if they exist, may be interpreted as representing a body of gas that occupies only a finite part of space.

(ii) Comparison and contrast of the formal H-theorem with the Clausius–Duhem inequality and the heat-bath inequality of thermomechanics

It is tempting to compare the formal H-theorem with the Clausius–Duhem inequality (I.33) in continuum thermomechanics, of course after the latter has been specialized to an ideal gas for which $\gamma = \tfrac{5}{3}$:

$$-\frac{\dot{\mathfrak{H}}}{r} \leq \int_{\partial\mathscr{B}} \frac{q_\perp}{\tfrac{2}{3}\varepsilon}\, dA. \tag{XI.9}$$

If, as usual, we identify $\mathbf{q}$ and ε in the kinetic theory with the fields denoted by the same symbols in continuum mechanics, and if we identify $-rH$ in the kinetic theory with $\mathfrak{H}$ in continuum thermomechanics, then the kinetic analogue of (XI.9) is

$$\dot{H} \leq \int_{\partial\mathscr{B}} \frac{q_\perp}{\tfrac{2}{3}\varepsilon}\, dA. \tag{XI.10}$$

Henceforth by "the Clausius–Duhem inequality" we shall mean (XI.10), as it is the only special case of the inequality that is known by that name in continuum thermomechanics to which the kinetic theory could conceivably be expected to provide a strict analogue.

For a perfect analogy we should have to establish (XI.10) for all material regions $\mathscr{B}$. Instead, we could look for a weaker analogue, an inequality of just the same form but restricted to a particular material region $\mathscr{B}$, a region bounded by a particular wall $\partial\mathscr{B}$ on which the symbol ε would stand for the interior limit of the field generally denoted by that symbol. In the same spirit the inequality

$$\dot{H} \leq \int_{\partial\mathscr{B}} \frac{q_\perp}{\tfrac{2}{3}\varepsilon^W}\, dA \tag{XI.11}$$

would be the kinetic analogue of the heat-bath inequality (I.36). Then ε^W would be the energetic prescribed upon the confining wall $\partial\mathscr{B}$, and there would be no reason to expect that $\varepsilon^W = \varepsilon$ on that wall. Relations between inequalities of the forms (XI.10) and (XI.11) have been established in Chapter I. One of them is important enough to restate in the present notation: *If the heat-bath inequality* (XI.11) *holds, then either of the following two conditions suffices that also the formal narrow H-theorem shall hold:*

(a)   *At almost all points of $\partial \mathscr{B}$ no heat flows outward; that is,* $q_\perp \leqq 0$.
(b)   *The temperature of the wall is uniform at each time, and no heat flows out of $\mathscr{B}$; that is,* $\varepsilon^W = f(t)$ *and* $\int_{\partial \mathscr{B}} q_\perp \, dA \leqq 0$.

We note that these same conditions would make the formal narrow H-theorem a consequence of the Clausius–Duhem inequality (XI.10), should the latter have been proved to hold for the region $\mathscr{B}$. We recall also that if the heat-transfer condition (I.38) holds—and in the terms we use for the kinetic theory that inequality reads

$$-q_\perp(\varepsilon - \varepsilon^W) \leqq 0 \quad \text{on} \quad \partial \mathscr{B} \tag{XI.12}$$

—then the Clausius–Duhem inequality would imply the heat-bath inequality. That is, *if the heat-bath inequality holds for a region $\mathscr{B}$, then the consequences we may derive from it would follow also from the Clausius–Duhem inequality.*

Analogues to (XI.10) and (XI.11) we shall seek in the formal broad H-theorem (XI.5) or the slightly weaker statement

$$\dot{H} \leqq -\int_{\partial \mathscr{B}} s_\perp \, dA, \tag{XI.13}$$

first of all because it, too, expresses *a trend in time* that is limited by what is occurring upon the boundary of the body occupying the region $\mathscr{B}$, and secondly because of the formal analogy we have already shown to hold when F is Maxwellian or independent of place. When $\partial \mathscr{B}$ is a particular prescribed boundary, $s_\perp$ may be related to ε^W, and if so (XI.5) like (XI.11) will express *the response of a body of gas to a specified environment.*

In Chapter VIII and in the preceding section of this chapter we have found circumstances sufficient that the corresponding members of (XI.5) and (XI.10) should agree exactly or approximately. In most of these the molecular density was either Maxwellian or nearly so, and thus h was either h_M or close to it. We now abandon any attempt to relate H to the thermostatic entropy. Letting H be what it may, we ask for circumstances in which *the kinetic theory confirms the Clausius–Duhem inequality or the heat-bath inequality.* We take the formal broad H-theorem (XI.5) as the basis of all our arguments.

As we have remarked, the kinetic theory cannot generally require that $s_\perp + q_\perp / \tfrac{2}{3}\varepsilon = 0$. The same reasoning shows that for an arbitrary region $\mathscr{B}$ the kinetic theory cannot imply for all solutions F the inequality

$$-\int_{\partial \mathscr{B}} s_\perp \, dA \leqq \int_{\partial \mathscr{B}} \frac{q_\perp}{\tfrac{2}{3}\varepsilon} \, dA. \tag{XI.14}$$

Thus (XI.13) does not generally imply (XI.10): *The formal broad H-theorem does not generally support the Clausius–Duhem inequality. Neither does it, in general, support the heat-bath inequality* (XI.11). All this follows *a fortiori* from the fact that s for a solution of the Maxwell–Boltzmann equation may be at the initial instant *any vector field we please*.

It remains possible that the kinetic theory may imply relations between $s_\perp$ and $q_\perp/\tfrac{2}{3}\varepsilon$ for *special solutions*. These solutions may be conceived as corresponding to particular conditions appropriate to particular kinds of boundaries $\partial\mathscr{B}$. Only two such conditions have been shown to deliver concrete results.

Condition 1. *At almost all points on $\partial\mathscr{B}$*

$$s_\perp = 0, \qquad q_\perp = 0. \tag{XI.15}$$

The first statement reduces the formal broad H-theorem (XI.5) to the formal narrow H-theorem (XI.7), neither more nor less. The second statement reduces both the heat-bath inequality (XI.11) and the Clausius–Duhem inequality (XI.10) likewise to the formal narrow H-theorem, neither more nor less. That is, *the H-theorem and the Clausius–Duhem inequality become indistinguishable*, always on the assumption that $-rH$ and $\mathfrak{H}$ should be regarded as representing one and the same thing in physics.

Condition 2. *At almost all points on $\partial\mathscr{B}$*

$$s_\perp \geqq -\frac{q_\perp}{\tfrac{2}{3}\varepsilon^{\mathrm{w}}}, \tag{XI.16}$$

in which ε^{w} denotes the energetic assigned to $\partial\mathscr{B}$. To prescribe ε^{w} on the wall $\partial\mathscr{B}$ amounts to prescribing the temperature of that wall. The reader will easily understand that the kinetic theory should allow us to represent a gas in contact with a wall that is maintained at prescribed temperature, and in the following section we shall give examples of boundary conditions that may be interpreted as doing just that. As in continuum thermomechanics, it is possible that $\varepsilon \neq \varepsilon^{\mathrm{w}}$. Subject to this interpretation, the condition (XI.16) is *sufficient that the formal broad H-theorem* (XI.5) *shall imply the heat-bath inequality* (XI.11). Condition 2 has an important corollary, a local counterpart of the first of the conditions stated in italics a few lines after (XI.11): If $q_\perp \leqq 0$, then $s_\perp \geqq 0$. That is, *if Condition 2 is satisfied and if there is no flux of energy outward through any point of the boundary, the formal narrow H-theorem follows*. In other words, a certain condition which in continuum thermomechanics would imply that $\dot{\mathfrak{H}} \geqq 0$ suffices to conclude in the kinetic theory that $-r\dot{H} \geqq 0$. Like Condition 1, the far weaker Condition 2 provides *a domain of agreement between the kinetic theory and continuum thermomechanics*, always on the assumption that it is $-rh$ we should regard as the kinetic analogue of η.

Neither Condition 1 nor Condition 2 necessarily makes H reduce to its value for a Maxwellian density having the same principal moment as F; indeed, neither of them even suggests that h should be a grossly determined field.

Both Condition 1 and Condition 2 are local. Obvious conditions referring to integrals over $\partial\mathscr{B}$ would suffice for the conclusions we have drawn from them, and such conditions are what the problem naturally suggests. There may well be circumstances under which the integral conditions hold even though the corresponding local conditions do not hold at every point. Future research may establish broader agreement between the kinetic theory and continuum thermomechanics than is now secure, as all presently established connections refer to circumstances in which one or other of the local conditions (XI.15) or (XI.16) is valid.

In Section (i) of this chapter we have noticed that solutions independent of place satisfy Condition 1, no matter what region $\mathscr{B}$ we consider. This case, while of interest, is not typical. The H-theorem provides in general an upper bound for $\dot{H}$ which depends upon the nature of the boundary $\partial\mathscr{B}$ of the region for which H is calculated. Accordingly, we address ourselves now to statement of boundary conditions for the kinetic theory.

(iii)   The concept of a solid boundary in the kinetic theory

Boundary conditions, like field equations, are proposed by theorists who dare to represent nature by mathematical hypotheses. "Draw from the model and imitate the antique," said RUBENS. The tradition shows us that only after a theory has been formulated can existence theorems be proved. In framing boundary conditions, just as in framing field equations, the theorist outlines Nature as best he can from what little of herself she lets him see through the fogs with which she modestly covers her sincerity. To do so, he follows the forms and practices that his masters, the great theorists of old, have taught him by example. He demonstrates the properties that solutions must have in order to satisfy his conditions. Like his great forebeers he runs the risk that solutions of the kind he analyses may not exist: that all his labor may be spent on describing a few of the countless attributes of the null set. The tradition gives him hope as well as example. In that spirit we now approach boundary conditions in the kinetic theory.

MAXWELL was the first to do so. His discussion[2], which makes excellent reading, still dominates the subject.

[2] MAXWELL [1879, Appendix] added this discussion at the request of the referee, some of whose suggestions it adopts and improves. From the inimitable style in which the referee expressed himself, MAXWELL, who was then mortally ill, must have recognized him as being his friend THOMSON, later Lord KELVIN. MAXWELL did not discuss the H-theorem, for his interest lay in the phenomenon of the radiometer. He acknowledged that a paper by REYNOLDS had drawn his

(iii) IDEALLY SMOOTH BOUNDARY

If we were really allowed to make statements about the molecular motions, we could prescribe the way the molecules rebound from a wall. For example, we could suppose that upon striking it a molecule would spring back with unchanged tangential velocity and reversed normal velocity. However, such "specular reflection" would correspond to a condition imposed in analytical dynamics, the very discipline that the kinetic theory designs to circumvent. The program of the kinetic theory requires that we express all gross statements in terms of expectations. Consequently the effects of the wall must be modelled by an alteration of the molecular density F.

To this end we consider a fixed place upon the wall at a fixed time, and we drop t and $\mathbf{x}$ from the notation. Referring to the decomposition (I.32), we let $a_\perp \mathbf{n}$ and $\mathbf{a}_t$ denote the normal and tangential parts of a vector $\mathbf{a}$ at a point on the wall. We may suppose that point to move with a velocity $\mathbf{u}^W$ whose normal component is $u_\perp^W$. For example, at a stationary wall $u_\perp^W = 0$. Molecules such that $v_\perp > u_\perp^W$ are *incident* upon the wall; molecules such that $v_\perp < u_\perp^W$ are *emitted* by the wall. We shall denote by F^e and F^i, respectively, the molecular densities for the emitted and the incident molecules. That is, at points upon the wall $\partial \mathscr{B}$

$$F(t, \mathbf{x}, \mathbf{v}) = \begin{cases} F^i(t, \mathbf{x}, \mathbf{v}) & \text{if } v_\perp > u_\perp^W, \\ F^e(t, \mathbf{x}, \mathbf{v}) & \text{if } v_\perp < u_\perp^W. \end{cases} \qquad (\text{XI}.17)$$

Since F is used only to calculate expectations, and since the set of velocities such that $v_\perp = u_\perp^W$ is of measure 0, there is no need to define F for those arguments. If we look back now at (VIII.14), we recognize it as including specification of a condition at a boundary: F^i and F^e are both Maxwellian densities.

The simplest kinetic boundary condition represents specular reflection from a stationary wall[3]: $u_\perp^W = 0$, and "the properties of the incident and reflected gas are symmetrical with respect to the tangent plane of the surface." We express

---

attention to the importance of boundary conditions in explaining the effect REYNOLDS called "thermal transpiration". MAXWELL had been a referee of REYNOLDS' paper, which the Royal Society had returned to its author for revision. Thus MAXWELL's appendix criticises and improves parts of a paper by REYNOLDS that had not yet been published. It is small wonder that REYNOLDS was angered, but MAXWELL's time was short.

The matter is described in detail by S. G. BRUSH & C. W. F. EVERITT, "Maxwell, Osborne Reynolds, and the radiometer," *Historical Studies in the Physical Sciences* **1**, 105–125 (1969) = §5.5 of S. G. BRUSH, *The Kind of Motion We Call Heat*, 2 vols., Amsterdam etc., North-Holland Publishing Co., 1976. The reader should be warned, however, that their description of MAXWELL's work is incorrect at one point: "a constant temperature gradient... would not by itself produce an inequality of pressure." We do not find that statement in MAXWELL's paper, nor is it true, as the reader may see at once from (XXV.14), below, which restores the terms proportional to the gradient of μ, terms MAXWELL stated that he chose to neglect.

[3] MAXWELL [1879, Appendix].

this idea as follows:

$$F^e(v_\perp \mathbf{n} + \mathbf{v_t}) = F^i(-v_\perp \mathbf{n} + \mathbf{v_t}), \qquad v_\perp < 0. \tag{XI.18}$$

Such a wall is *ideally smooth*. For any given normal speed and any given tangential velocity, the expected number of molecules emitted per unit volume in $\mathscr{E} \times \mathscr{V}$ shall equal the expected number incident. Phrased in terms of F alone, the condition (XI.18) is not a constraint in the sense of analytical dynamics: It does not require that the wall be impermeable, only that in accepting and emitting molecules it follow in mean the quality of a perfect mirror.

We may write (XI.18) more conveniently as a functional equation for F over all of $\mathscr{V}$:

$$F(v_\perp \mathbf{n} + \mathbf{v_t}) = F(-v_\perp \mathbf{n} + \mathbf{v_t}). \tag{XI.19}$$

In calculating expectations we can integrate first over $v_\perp$, holding $\mathbf{v_t}$ fixed. Using superscript i and e to refer to the incident and the reflected portions of the gas, and denoting by $\bar{g}^i$ and $\bar{g}^e$ averages with respect to the one and the other portion, we see from (XI.19) that

$$n^i = n^e,$$

$$\bar{g} = \begin{Bmatrix} \bar{g}^i = \bar{g}^e \\ 0 \end{Bmatrix} \quad \text{if} \quad g \text{ is an } \begin{Bmatrix} \text{even} \\ \text{odd} \end{Bmatrix} \text{function of } v_\perp. \tag{XI.20}$$

Hence

$$u_\perp \equiv \bar{v}_\perp = 0, \qquad \mathbf{u_t} \equiv \bar{\mathbf{v}}_\mathbf{t} = \bar{\mathbf{v}}_\mathbf{t}^i = \bar{\mathbf{v}}_\mathbf{t}^e. \tag{XI.21}$$

The first group of these conditions shows that the ideally smooth surface is indeed a boundary of the gross motion of the gas; the second group shows that the incident and the emitted parts of the body of gas slip along the wall at the same velocity. We can calculate the pressure vector at a point on the wall:

$$\mathbf{p} \equiv \mathbf{Mn} = \mathfrak{m} \int F c_\perp \mathbf{c} = \mathfrak{m} \int F(c_\perp^2 \mathbf{n} + c_\perp \mathbf{c_t}) = \mathfrak{m} \int F c_\perp^2 \mathbf{n}. \tag{XI.22}$$

Consequently

$$\mathbf{p_t} = \mathbf{0}: \tag{XI.23}$$

An ideally smooth wall suffers no shear stress from the gas in contact with it. For this reason MAXWELL rejected the boundary condition (XI.19): "Since gases can actually exert oblique stress against real surfaces, such surfaces cannot be represented as perfectly reflecting surfaces."

There are deeper objections to the ideally smooth wall. A thorough devotion to the kinetic-molecular view of matter would insist that not only the gas but also any solid body with which the gas might come in contact should be conceived as an assembly of molecules in motion. Molecules of the gas and molecules of the solid would interdiffuse. Only a statistical definition could even prescribe the location of a boundary of this kind. Such complexity is insuperable except, perhaps, in some very special cases. After rejecting the ideally smooth wall, MAXWELL[4] chose to conceive a solid boundary as a layer of ideal and immovable spheres, against which the molecules of the gas, for this purpose conceived likewise as ideal spheres, would strike, and from which they would rebound. Even this simple model is too complicated to allow precise treatment. MAXWELL used it only to motivate specific conditions to be applied upon a boundary conceived as a mathematical surface. Such a boundary lends itself to formulation of boundary conditions for solutions of the Maxwell-Boltzmann equation and thus may lead to a theory of such solutions, considered in themselves, without reference to some other molecular theory in adjacent regions.

MAXWELL's specific model suggested to him that F^e should be Maxwellian:

$$F^e = a^W e^{-b^W |\mathbf{v} - \mathbf{u}^W|^2}. \qquad (XI.24)$$

Such a wall is *ideally rough*. We may express a^W and b^W in terms of n^W and ε^W by use of (VIII.8), but we must recall that F^e is to be used for only those velocities that point inward from the wall. Thus $n^e = \frac{1}{2} n^W$, while of course $\varepsilon^e = \varepsilon^W$. We shall expect that in general $\varepsilon^i \neq \varepsilon^e$ and $\mathbf{u}^i \neq \mathbf{u}^e \neq \mathbf{u}^W$. In the case envisioned by MAXWELL the wall was stationary: $\mathbf{u}^W = \mathbf{0}$.

In this book we have already encountered the ideally rough wall without naming it. In Sections (ii) and (v) of Chapter VIII we have analysed a condition of the gas in which a plane divides a region governed by the Maxwellian density F_M^1 from one governed by the Maxwellian density F_M^2. Either portion of the body of gas may be considered as a body in kinetic equilibrium, bounded by an ideally rough wall. If F_M^2 is interpreted as appropriate to the wall, then its

---

[4] MAXWELL [1879, Appendix]:

If a molecule, whose velocity is given in direction and magnitude, but whose line of motion is not given in position, strikes a fixed elastic sphere, its velocity after rebound may with equal probability be in any direction.

Consider, therefore, a stratum in which fixed elastic spheres are placed so far apart from one another that any one sphere is not to any sensible extent protected by any other sphere from the impact of molecules, and let the stratum be so deep that no molecule can pass through it without striking one or more of the spheres, and let this stratum of fixed spheres be spread over the surface of the solid we have been considering, then every molecule which comes from the gas towards the surface must strike one or more of the spheres, after which all directions of its velocity become equally probable.

When, at last, it leaves the stratum of spheres and returns into the gas, its velocity must of course be *from* the surface, but the probability of any particular magnitude and direction of the velocity will be the same as in a gas at rest with respect to the surface.

The distribution of velocity among the molecules which are leaving the surface will therefore be the same as if, instead of the solid, there were a portion of gas at rest, having the temperature of the solid, and a density such that the number of molecules which pass from it through the surface in a given time is equal to the number of molecules of the real gas outside which strike the surface.

energetic ε_2 is the prescribed energetic of the wall. The result (VIII.18) illustrates our remark in Section (ii) of this chapter that the assigned energetic of the wall at a point need not be the energetic of the gas at that same point. It shows also that the ideally rough wall may be the seat of discontinuities and sources of flux of energy and flux of h. The example, two Maxwellian equilibrium densities governing regions contiguous through an ideally rough wall, is not claimed to be a solution of the Maxwell–Boltzmann equation. Rather, it is admissible as an initial value for a solution. The fact that $\mathbf{q}$ and $\mathbf{s}$ vanish except upon the wall, and that ε and h suffer discontinuities there, suggests that a solution corresponding to this initial value will represent a gas coming gradually to kinetic equilibrium at a value of ε intermediate between ε_1 and ε_2; that is, that F for arguments on either side of the wall should approach in time a common uniform Maxwellian density different from F_M^1 and F_M^2; but that is conjecture.

Still using his model of stationary ideal spheres, MAXWELL[5] suggested that a condition closer to reality might be obtained by supposing that a fraction f of the incident molecules would be emitted as by an ideally rough wall, while the remaining fraction $1-f$ would be emitted as by an ideally smooth wall. That is, for a stationary wall,

$$F^e = f a^W e^{-b^W v^2} + (1-f) F^i(-v_\perp \mathbf{n} + \mathbf{v}_t). \tag{XI.25}$$

---

[5] MAXWELL [1879, Appendix]:
If the spheres are so near together that a considerable part of the surface of each sphere of the outer layer is shielded from the direct impact of the incident molecules by the spheres which lie next to it, then if we call that point of each sphere which lies furthest from the solid the *pole* of the sphere, a greater proportion of molecules will strike any one of the outer layer of spheres near its pole than near its equator, and the greater the obliquity of incidence of the molecule, the greater will be the probability that it will strike a sphere near its pole.

The direction of the rebounding molecule will no longer be with equal probability in all directions, but there will be a greater probability of the tangential part of its velocity being in the direction of the motion before impact, and of its normal part being opposite to the normal part before impact.

The condition of the molecules which leave the surface will therefore be intermediate between that of evaporated gas and that of reflected gas, approaching most nearly to evaporated gas at normal incidence and most nearly to reflected gas at grazing incidence.

If the spheres, instead of being hard elastic bodies, are supposed to act on the molecules at finite, though small distances, and if they are so close together that their spheres of action intersect, then the gas which leaves the surface will be still more like reflected gas, and less like evaporated gas.

We might also consider a surface on which there are a great number of minute asperities of any given form, but since in this case there is considerable difficulty in calculating the effect when the direction of rebound from the first impact is such as to lead to a second or third impact, I have preferred to treat the surface as something intermediate between a perfectly reflecting and a perfectly absorbing surface, and, in particular, to suppose that of every unit of area a portion f absorbs all the incident molecules, and afterwards allows them to evaporate with velocities corresponding to those in still gas at the temperature of the solid, while a portion $1-f$ perfectly reflects all the molecules incident upon it.

(iii) GENERAL BOUNDARY CONDITION

The adjustable *accommodation coefficient* f is intended to model a degree of asperity between the ideally smooth, for which $f = 0$, and the ideally rough, for which $f = 1$. This condition has been used again and again to interpret the results of experiment and to make crude calculations[6]. It is not presently regarded as good for much else.

We can phrase MAXWELL's approach in its naked essentials, divested of the three specializing assumptions he considered in turn. In analytical dynamics the way a particle rebounds from a wall is determined by the velocity with which it strikes the wall and by the nature of that wall. MAXWELL modelled this fact in the kinetic theory by supposing that *the molecular density for the emitted portion of the gas would be determined by the molecular density for the incident portion* at the same place and time. Thus he replaced the solid wall by a body of gas of a special kind, a gas described by a molecular density about which we have some knowledge. Formally, the **General Boundary Condition** of the kinetic theory is

$$F^e(t, \mathbf{x}, \mathbf{v}) = \mathbb{W}(t, \mathbf{x}, \mathbf{v}; F^i) \quad \text{if} \quad v_\perp < u_\perp^W, \tag{XI.26}$$

for all $\mathbf{x}$ on the boundary at the time t; the mapping $\mathbb{W}$ is the *wall operator*. The solid boundary is modelled in terms of the concepts of the kinetic theory, so the theorist may confront boundary-value problems of that theory by itself.

Many kinds of interaction between the portions of gas on the two sides of such a boundary are easy to imagine. A physical wall may absorb more gas than it emits, may emit more than it absorbs, or may just balance emission and incidence. It may be hotter or colder than the gas it confines. The gas may adhere to it or may slip along it. It may yield to the forces exerted by the gas upon it, or it may be conceived so strong as to stand firm against any pressure and torque, however great.

Only one thing is certain. If we are to apply the H-theorem as we have developed it in Section (i) of this chapter, the surface $\partial \mathcal{B}$ must be a *material surface*[7] for the gross motion. Thus at a given point upon $\partial \mathcal{B}$ the speed of displacement $u_\perp^W$ of $\partial \mathcal{B}$ must equal the normal component of the gross velocity there:

$$u_\perp = u_\perp^W. \tag{XI.27}$$

---

[6] MAXWELL [1879, Appendix] took for F^i the approximation that had led him to an expression for the interior stress in a rarefied gas that we find below to be included in (XXV.14). He recognized his method as being imperfect, its results "liable to important corrections". Be that as it may, his work on this subject stands today as a tower in the field, and the heuristic process by which he reached his results is of a kind reminiscent of some brilliant passages in the works of HUYGENS and NEWTON.

We do not follow the matter in this book because our only reason for considering boundary conditions at all is to apply them to the formal broad H-theorem.

A good discussion of boundary conditions in general has been written by CERCIGNANI [1975, Chapter III].

[7] *Cf.*, *e.g.*, §II.6 of TRUESDELL's book cited in Footnote 1 to Chapter 1.

In mean, the wall neither captures molecules of the gas nor supplies molecules to it.

We may apply (XI.17) and (XI.26) to the general definitions given in Chapter III and so express expectations at points on $\partial \mathscr{B}$ as sums of two parts, one of which arises from the actions of the incident molecules and the other from the emitted ones. Using superscripts to distinguish these, we first obtain the obvious relations[8]

$$n = n^i + n^e, \quad n^i \equiv \int_{v_\perp > u_\perp^w} F^i \, d\mathbf{v}, \quad n^e \equiv \int_{v_\perp < u_\perp^w} F^e \, d\mathbf{v}. \quad (XI.28)$$

Next,

$$\bar{g} = \frac{n^i}{n} \bar{g}^i + \frac{n^e}{n} \bar{g}^e, \quad \bar{g}^i \equiv \frac{1}{n^i} \int_{v_\perp > u_\perp^w} F^i g \, d\mathbf{v}, \quad \bar{g}^e \equiv \frac{1}{n^e} \int_{v_\perp < u_\perp^w} F^e g \, d\mathbf{v}. \quad (XI.29)$$

In particular,

$$\mathbf{u} = \frac{n^i}{n} \mathbf{u}^i + \frac{n^e}{n} \mathbf{u}^e, \quad (XI.30)$$

so the condition (XI.27) may be written in the form

$$\frac{n^i}{n} u_\perp^i + \frac{n^e}{n} u_\perp^e = u_\perp^w. \quad (XI.31)$$

We remark in passing that $\bar{g}$ lies between $\bar{g}^e$ and $\bar{g}^i$, that

$$\bar{g}^e \geqq \bar{g} \quad \text{if and only if} \quad \bar{g}^e \geqq \bar{g}^i, \quad (XI.32)$$

and that the signs of equality and inequality are associated.

The condition (XI.31) may be written in the form

$$n u_\perp^w = \int_{v_\perp > u_\perp^w} v_\perp F^i \, d\mathbf{v} + \int_{v_\perp < u_\perp^w} v_\perp F^e \, d\mathbf{v}. \quad (XI.33)$$

Another, more convenient way to write it results if we begin from the relation $\bar{\mathbf{c}} = \mathbf{0}$, which is a trivial consequence of the definition of $\mathbf{c}$. Equally trivially, if $\mathbf{m}$ is any unit vector whatever, we may split the integral defining $\bar{\mathbf{c}}$ into two and so obtain

$$\mathbf{0} = \int_{\mathbf{c} \cdot \mathbf{m} > 0} F \mathbf{c} \, d\mathbf{c} + \int_{\mathbf{c} \cdot \mathbf{m} < 0} F \mathbf{c} \, d\mathbf{c}. \quad (XI.34)$$

---

[8] The reader familiar with the theory of gas mixtures will recognize these relations. Here the two constituents of the mixture differ from one another not in mass but in the signs of their normal velocities at the boundary.

(iii) GENERAL BOUNDARY CONDITION

On $\partial \mathcal{B}$ we may apply the condition (XI.27) and so conclude that $v_\perp - u_\perp = v_\perp - u_\perp^W$. If we take for **m** the outer unit normal **n** to $\partial \mathcal{B}$, then the restrictions of F that appear in the two integrals are F^i and F^e, respectively, so (XI.34) becomes

$$0 = \int_{c_\perp > 0} F^i \mathbf{c}\, d\mathbf{c} + \int_{c_\perp < 0} F^e \mathbf{c}\, d\mathbf{c}. \tag{XI.35}$$

The scalar product of this relation and **n** is (XI.33), written as follows:

$$0 = \int_{c_\perp > 0} c_\perp F^i \, d\mathbf{c} + \int_{c_\perp < 0} c_\perp F^e \, d\mathbf{c}. \tag{XI.36}$$

From (XI.30) we see that $\mathbf{u}_t \neq \mathbf{u}_t^W$ in general, so the gas slips past the wall. In the same way we may calculate the pressure vector **p** and the normal flux of energy $q_\perp$:

$$\begin{aligned} \frac{1}{m}\mathbf{p} &= \int_{c_\perp > 0} F^i c_\perp \mathbf{c}\, d\mathbf{c} + \int_{c_\perp < 0} F^e c_\perp \mathbf{c}\, d\mathbf{c}, \\ \frac{2}{m} q_\perp &= \int_{c_\perp > 0} F^i c^2 c_\perp \, d\mathbf{c} + \int_{c_\perp < 0} F^e c^2 c_\perp \, d\mathbf{c}. \end{aligned} \tag{XI.37}$$

DARROZES & GUIRAUD[9] proposed for consideration the special case of (XI.26) that results when the wall operator $\mathbb{W}$ is linear. It suffices to express this idea in terms of a representation for linear mappings generated by a non-negative Lebesgue–Stieltjes measure $d_{\mathbf{v}*}\mathbb{R}(\mathbf{v})$ that models the roughness of the wall: If $c_\perp < 0$, then

$$-c_\perp F^e(\mathbf{v}) = \int_{c_\perp* > 0} c_\perp^* F^i(\mathbf{v}^*) \, d_{\mathbf{v}*}\mathbb{R}(\mathbf{v}). \tag{XI.38}$$

The notation indicates that the measure $d\mathbb{R}$ is defined over the half of $\mathscr{V}$ for which $c_\perp^* > 0$ and depends upon **v** as a parameter. The condition (XI.36) that the wall be a material surface imposes the requirement

$$\begin{aligned} \int_{c_\perp* > 0} c_\perp^* F^i(\mathbf{v}^*)\, d\mathbf{v}^* &= -\int_{c_\perp < 0} c_\perp F^e(\mathbf{v})\, d\mathbf{v}, \\ &= \int_{c_\perp < 0} \left[\int_{c_\perp* > 0} c_\perp^* F^i(\mathbf{v}^*)\, d_{\mathbf{v}*}\mathbb{R}(\mathbf{v})\right] d\mathbf{v}, \\ &= \int_{c_\perp* > 0} c_\perp^* F^i(\mathbf{v}^*) \int_{c_\perp < 0} d_{\mathbf{v}*}\mathbb{R}(\mathbf{v})\, d\mathbf{v}. \end{aligned} \tag{XI.39}$$

---

[9] DARROZES & GUIRAUD [1966]. The proposal had been made earlier by GRAD [1958, Eq. (19.5)], but he did not develop it, obtain the condition (XI.40), or impose the requirement (XI.41).

For this relation to hold for all F^i, it is necessary and sufficient that the measures $d\mathbf{v}^*$ and $d_{\mathbf{v}*}\mathbb{R}(\mathbf{v})$ be related as follows:

$$\int_{c_\perp < 0} d_{\mathbf{v}*} \mathbb{R}(\mathbf{v}) \, d\mathbf{v} = d\mathbf{v}^*; \tag{XI.40}$$

that is, that $d_{\mathbf{v}*}\mathbb{R}(\mathbf{v})$ be a probability density. Thus (XI.38) may be interpreted as assigning a probability to the emission at velocity $\mathbf{v}$ of the molecules incident at velocity $\mathbf{v}^*$. (The ratio $-c_\perp^*/c_\perp$ could have been absorbed into the measure $d_{\mathbf{v}*}\mathbb{R}(\mathbf{v})$, but in the interest of the simple statement (XI.40) it was not.) Finally, DARROZES & GUIRAUD associate a Maxwellian density F^W with the wall by imposing the following requirement upon $d\mathbb{R}$: If $c_\perp < 0$, then

$$\int_{c_\perp * > 0} c_\perp^* F^W(\mathbf{v}^*) \, d_{\mathbf{v}*} \mathbb{R}(\mathbf{v}) = -c_\perp F^W(\mathbf{v}). \tag{XI.41}$$

That is, if the incident gas is governed by F^W, so is the emitted gas. We may regard the wall itself as a gas in equilibrium with the Maxwellian molecular density F^W; in general the influence of the wall upon the gas it confines causes the emitted gas to conform with a molecular density F^e which is different from F^W, but in the special event that $F^i = F^W$, then also $F^e = F^W$. In this regard the ideally rough wall is not typical, because for it $F^e = F^W$ no matter what F^i may be, and also the emitted gas has the same temperature as the wall: $\varepsilon^e = \varepsilon^W$. The more general condition (XI.40) allows us to define energetics ε^W as being those of the Maxwellian densities F^W that satisfy (XI.41). Even if all of these densities have a common ε^W, generally $\varepsilon^e \neq \varepsilon^W$.

DARROZES & GUIRAUD require that the solution F^W of (XI.41) be unique to within its number density. Thus the roughness measure $d\mathbb{R}$ determines b^W and $\mathbf{u}^W$ but not a^W. Because of (VIII.8)$_2$, all these F^W give rise then to one and the same ε^W. Such uniqueness follows the intended interpretation of $d\mathbb{R}$ as providing a model for a wall which has an assigned temperature and velocity and is able to serve as a boundary for a body of gas having whatever number density it will. Formally, n^W is adjustable so as to make n^e satisfy the relation (XI.28)$_1$, namely, $n^e = n - n^i$, whatever be n^i. However reasonable such uniqueness may be, in the use of DARROZES & GUIRAUD's condition we shall presently make we do not need to presume it.

We shall say that the conditions (XI.38), (XI.40), and (XI.41) describe a *linear wall*.

The reader will easily verify that the linear wall includes as special cases the ideally smooth wall, the ideally rough wall, and the wall defined by (XI.25) in terms of MAXWELL's accommodation coefficient. For the first and third kinds the measure $d_{\mathbf{v}*}\mathbb{R}(\mathbf{v})$ has an atom on the hyperplane $v_\perp^* = -v_\perp$, $\mathbf{v}_t^* = \mathbf{v}_t$. For the second and third kinds the absolutely continuous part of $d_{\mathbf{v}*}\mathbb{R}(\mathbf{v})$ is $-Ac_\perp F^W(\mathbf{v}) \, d\mathbf{v}^*$, in which the constant A is so chosen as to make (XI.40) hold.

(iv) The formal narrow H-theorem or the heat-bath inequality as a consequence of boundary conditions

We shall consider always a fixed time, and in much that follows in this section we shall take up also a fixed place on a boundary wall. Whenever we do so, we shall drop t and $\mathbf{x}$ from the notation and write $F(\mathbf{v})$ for $F(t, \mathbf{x}, \mathbf{v})$. We shall consider in turn the three boundary conditions set forth in the preceding section of this chapter: the ideally smooth wall, the ideally rough wall, and the linear wall, and we shall show that a body of kinetic gas confined in a vessel whose bounding surface is of any one of these kinds obeys the heat-bath inequality. For the ideally smooth wall and for the ideally rough wall we shall demonstrate more than that.

Case 1. *An ideally smooth vessel*[10]. By use of (XI.20)$_2$ we see at a glance that $q_\perp$ and $s_\perp$ both vanish on $\partial\mathcal{B}$. Hence Condition 1 in Section (ii) of this chapter is satisfied on $\partial\mathcal{B}$, and so it follows that *for a body of gas confined in an ideally smooth vessel the formal narrow H-theorem holds.* Also, if $\mathbf{b} = \mathbf{0}$, we see from (XI.21)$_2$, (XI.23), and the result following (I.44) that a body of gas within an ideally smooth vessel is isolated as well as insulated. Therefore, by appeal to MAXWELL's consistency theorem, we may use (I.45) and so conclude that if $\mathbf{b} = \mathbf{0}$, *the total internal energy of a body of gas within an ideally smooth vessel remains constant.* Thus the kinetic gas in an ideally smooth vessel exhibits the behavior CLAUSIUS attributed to the universe.

Case 2. *An ideally rough vessel*[11]. Now F^e is given by (XI.24), which we shall denote here by F_M.

Because $F_M \mathbf{c}$ and $F_M c^2 \mathbf{c}$ are odd functions of $\mathbf{c}$, we may write the resulting forms of (XI.36) and (XI.37)$_2$ as follows: At a point on $\partial\mathcal{B}$

$$0 = \int_{c_\perp > 0} (F^i - F_M) c_\perp \, d\mathbf{c}, \qquad \frac{2q_\perp}{m} = \int_{c_\perp > 0} (F^i - F_M) c_\perp c^2 \, d\mathbf{c}. \qquad \text{(XI.42)}$$

Now if $x > 0$ and $y > 0$,

$$x \log x \geq x \log y + x - y, \qquad \text{(XI.43)}$$

and equality holds if and only if $x = y$. Taking F^i for x and F_M for y, we show that at a point on $\partial\mathcal{B}$

---

[10] BOLTZMANN [1875, §1] [1896, §18], CHAPMAN & COWLING [1939, §4.13].

[11] We generalize results due to CARLESON, which we shall cite in Footnote 12. For a part of the argument we give here we are indebted to Professor IKENBERRY.

$$\frac{s_\perp}{m} = \int F c_\perp \log F,$$

$$= \int_{c_\perp > 0} F^i c_\perp \log F^i \, d\mathbf{c} + \int_{c_\perp < 0} F_M c_\perp \log F_M \, d\mathbf{c},$$

$$= \int_{c_\perp > 0} c_\perp (F^i \log F^i - F_M \log F_M) \, d\mathbf{c},$$

$$\geqq \int_{c_\perp > 0} c_\perp (F^i \log F_M + F^i - F_M - F_M \log F_M) \, d\mathbf{c}, \qquad (XI.44)$$

$$= \int_{c_\perp > 0} c_\perp (F^i - F_M) \log F_M \, d\mathbf{c},$$

$$= \int_{c_\perp > 0} c_\perp (F^i - F_M)(\log a^w - b^w c^2) \, d\mathbf{c},$$

$$= -b^w \int_{c_\perp > 0} c_\perp c^2 (F^i - F_M) \, d\mathbf{c},$$

$$= -2 b^w q_\perp / m.$$

Making use of (VIII.8)$_2$, we see that

$$s_\perp \geqq -\frac{q_\perp}{\tfrac{2}{3}\varepsilon^w}. \qquad (XI.16)_r$$

Thus Condition 2 of Section (ii) of this chapter is satisfied. Consequently *the heat-bath inequality* (XI.11) *holds for a body of gas confined by ideally rough walls*. It is not hard to see that equality holds in (XI.44) if and only if $F = F_M$. Then, of course, $\mathbf{s} = \mathbf{0}$ and $\mathbf{q} = \mathbf{0}$. F_M here stands for the particular Maxwellian density assumed to govern the emitted molecules. If F for the incident molecules is a different Maxwellian density, equality will not hold in (XI.44).

> To verify this last statement, the reader may reread the discussion at the end of Section (ii) of Chapter VIII and interpret it as referring to a body of gas by itself in kinetic equilibrium but in contact with an ideally rough wall. From (VIII.23) he will see that if $\varepsilon^i \neq \varepsilon^w$, then $q_\perp \neq 0$. He may refer to the explicit formula (VIII.62)$_2$ for $s_\perp$, which shows that inequality holds in (XI.16). Of course he must not regard these results as describing the behavior of the gas over an interval of time. Rather, as we have remarked above, a condition of a body of gas that is in equilibrium yet is confined by an ideally rough wall at a different energetic should be regarded as an initial condition for a solution of the Maxwell–Boltzmann equation.

In Section (ii) of this chapter we have noticed two interesting applications of (XI.16). According to the first of them, the inequality $q_\perp \leqq 0$ suffices for the truth of the formal narrow H-theorem. This is the result established by

CARLESON[12], who was the first to demonstrate mathematically anything about ideally rough vessels.

CARLESON suggested that *if an ideally rough wall were hot enough, then $q_\perp \leq 0$ upon it.* We may prove this fact as follows. Writing $F_M = ae^{-bc^2}$ for brevity and using (XI.29)$_2$ and the evaluations of quadratures listed following (VIII.16) and (VIII.21), we express (XI.42) thus:

$$n^i \bar{c}^i_\perp = \tfrac{1}{2}\pi a/b^2, \qquad 2q_\perp = 2q^i_\perp - \pi \mathfrak{m} a/b^3. \tag{XI.45}$$

Eliminating a between these two relations shows that

$$q_\perp = q^i_\perp - \tfrac{4}{3}\mathfrak{m} n^i \bar{c}^i_\perp \varepsilon^W, \tag{XI.46}$$

in which we have restored W in the notation. Hence the condition $q_\perp \leq 0$ is equivalent to

$$\varepsilon^W \geq q^i_\perp / (\tfrac{4}{3}\mathfrak{m} n^i \bar{c}^i_\perp). \tag{XI.47}$$

The lower bound would be difficult to relate to ε^i, but it is determined by the molecular density F^i for the incident molecules alone; thus if F^i is given, the inequality (XI.47) is satisfied by infinitely many scalar fields ε^W over $\partial\mathscr{B}$. As $\partial\mathscr{B}$ is a closed and bounded set, the lower bound in (XI.47) if continuous attains a maximum upon it, and any constant field ε^W not less than that maximum will suffice to ensure that $q_\perp \leq 0$ and hence to deliver the formal narrow H-theorem. However, the second alternative in the corollary stated just after (XI.11) shows that the same conclusion follows from weaker assumptions: ε^W may have any constant value over $\partial\mathscr{B}$ at each t, and there is no net flow of heat out of $\mathscr{B}$.

Case 3. A vessel having a linear wall. The main results concerning the ideally smooth and ideally rough vessels are special cases of a more general result obtained recently by several authors[13]:

Theorem. *Let F^e, F^i, and F^W satisfy the conditions* (XI.38), (XI.40), *and* (XI.41) *that define a linear wall. If both F^e and F^i are measurable and if F^i is bounded by some Maxwellian density for which $b \geq b^W$, then at each point on the wall*

$$s_\perp \geq -\frac{q_\perp}{\tfrac{2}{3}\varepsilon^W}. \tag{XI.16}_r$$

Here ε^W is the energetic of the Maxwellian density F^W. Condition 2 of Section (ii) is satisfied, so, subject to the conditions of the theorem, *a body of gas confined by linear walls obeys the heat-bath inequality.* As we have stated in our presentation of the conditions that define a linear wall, we expect that all solutions of (XI.41) will have the same energetic ε^W, but the proof of (XI.16) we shall give presently does not require such uniqueness.

---

[12] CARLESON included his work in the posthumously published booklet of CARLEMAN [1957, §§7-8].

[13] CERCIGNANI [1972, Appendix] [1973] [1975, Chapter III, §4] attributes the theorem to DARROZES & GUIRAUD [1966], whose presentation many readers have found far from easy to follow. Related work appears in numerous publications: GUIRAUD [1965] [1970] [1972] [1973] [1975], CERCIGNANI [1969, §5 of Chapter II], CERCIGNANI & LAMPIS [1971], and other physical literature.

Proof[14]. Jensen's inequality[15] implies that if the non-negative measure dP defines a probability upon $\mathscr{V}$:

$$\int_{\mathscr{V}} dP = 1, \qquad (XI.48)$$

and if C is a convex function on $[0, \infty)$, then

$$C\left(\int_{\mathscr{V}} g\, dP\right) \leq \int_{\mathscr{V}} C(g)\, dP \qquad (XI.49)$$

for every non-negative measurable function g which is everywhere finite and which is integrable over $\mathscr{V}$ with respect to dP. To apply it, we define a non-negative measure $d\mathbb{P}$ in terms of the roughness measure $d\mathbb{R}$ and the Maxwellian density F^W that governs the wall: If $c_\perp < 0$,

$$d_{\mathbf{v}*}\mathbb{P}(\mathbf{v}) \equiv \begin{cases} -\dfrac{c_\perp^* F^W(\mathbf{v}^*)}{c_\perp F^W(\mathbf{v})}\, d_{\mathbf{v}*}\mathbb{R}(\mathbf{v}) & \text{if } c_\perp^* > 0, \\ 0 & \text{if } c_\perp^* < 0. \end{cases} \qquad (XI.50)$$

Integrating (XI.50) with respect to $\mathbf{v}^*$ over $\mathscr{V}$, we see that (XI.48) follows from (XI.41), so $d_{\mathbf{v}*}\mathbb{P}(\mathbf{v})$ for each fixed $\mathbf{v}$ defines a probability over $\mathscr{V}$. Our assumptions regarding F^e, F^i, and F^W allow us to define as follows a function g:

$$g(\mathbf{v}) \equiv \begin{cases} F^i(\mathbf{v})/F^W(\mathbf{v}) & \text{if } c_\perp > 0, \\ F^e(\mathbf{v})/F^W(\mathbf{v}) & \text{if } c_\perp < 0, \end{cases} \qquad (XI.51)$$

and to prove that g is non-negative, everywhere finite, and integrable with respect to $d\mathbb{P}$ over $\mathscr{V}$. Then (XI.38) tells us that if $c_\perp < 0$,

$$g(\mathbf{v}) = \int_{c_\perp * > 0} g(\mathbf{v}^*)\, d_{\mathbf{v}*}\mathbb{P}(\mathbf{v}) = \int_{\mathscr{V}} g(\mathbf{v}^*)\, d_{\mathbf{v}*}\mathbb{P}(\mathbf{v}). \qquad (XI.52)$$

For this particular function g Jensen's inequality (XI.49) yields

$$C(g(\mathbf{v})) \leq \int_{\mathscr{V}} C(g(\mathbf{v}^*))\, d_{\mathbf{v}*}\mathbb{P}(\mathbf{v}) = \int_{c_\perp * > 0} C(g(\mathbf{v}^*))\, d_{\mathbf{v}*}\mathbb{P}(\mathbf{v}). \qquad (XI.53)$$

The relation (XI.40) takes the following form in terms of $d\mathbb{P}$: If $c_\perp^* > 0$,

$$\int_{c_\perp < 0} c_\perp F^W(\mathbf{v})\, d_{\mathbf{v}*}\mathbb{P}(\mathbf{v})\, d\mathbf{v} = -c_\perp^* F^W(\mathbf{v}^*)\, d\mathbf{v}^*. \qquad (XI.54)$$

---

[14] CERCIGNANI [1975, Chapter III, §4].

[15] *E.g.*, Theorem 3.3 of W. RUDIN's *Real and Complex Analysis*, New York etc., McGraw-Hill, 1966. That theorem refers only to a function which is defined and convex on an *open* interval (possibly infinite). Its proof applies also to any convex function C on $[0, \infty)$ provided that $g \not\equiv 0$ on a set of positive measure. However, if $g = 0$ almost everywhere, (XI.49) is satisfied trivially since (XI.48) reduces its right-hand side to $C(0)$, which is the value of its left-hand side.

CERCIGNANI refers to "a strictly convex continuous function" but does not specify any restriction upon the domain of such a function.

(iv) H-THEOREM FOR LINEAR WALLS

We now multiply (XI.53)$_2$ by $c_\perp F^W(\mathbf{v})$ and integrate the result over the half-space $c_\perp < 0$; by use of (XI.54) we conclude that

$$\int_{c_\perp < 0} c_\perp F^W(\mathbf{v}) C(g(\mathbf{v})) \, d\mathbf{v} \geqq \int_{c_\perp < 0} c_\perp F^W(\mathbf{v}) \int_{c_{\perp *} > 0} C(g(\mathbf{v}^*)) \, d_{\mathbf{v}*} \mathbb{P}(\mathbf{v}) \, d\mathbf{v},$$

$$= -\int_{c_{\perp *} > 0} c_\perp^* F^W(\mathbf{v}^*) C(g(\mathbf{v}^*)) \, d\mathbf{v}^*. \tag{XI.55}$$

Rearrangement yields the elegant inequality of CERCIGNANI[16]: *Let C be a convex function on $[0, \infty)$. Then at a point on a linear wall*

$$\int c_\perp F^W C(F/F^W) \geqq 0. \tag{XI.56}$$

For C we now take the particular convex function on $[0, \infty)$ defined as follows:

$$C(\lambda) = \begin{cases} \lambda \log \lambda & \text{if } \lambda > 0, \\ 0 & \text{if } \lambda = 0. \end{cases} \tag{XI.57}$$

Then

$$F^W C(F/F^W) = F(\log F - \log F^W),$$
$$= F(\log F - \log a^W + b^W c^2). \tag{XI.58}$$

The definitions of $\mathbf{s}$ and $\mathbf{q}$, namely (VIII.52) and (III.12), followed by use of (XI.36) show that

$$m \int c_\perp F^W C(F/F^W) = s_\perp + 2 b^W q_\perp. \tag{XI.59}$$

Comparing (XI.56) with (XI.59), we see that we have derived (XI.16) from the hypotheses stated. △

The foregoing proof shows that the Maxwell-Boltzmann equation has nothing to do with the result just obtained. It is a functional inequality, referring only to properties of F_M and the roughness measure $d\mathbb{R}$. It could have been presented in Chapter VIII.

For complete consistency with the ideas of classical thermodynamics regarding a body in its environment we should need to establish not only the heat-bath inequality, which the theorem just proved indeed does, but also MAXWELL's heat-transfer condition:

$$q_\perp(\theta^W - \theta) \leqq 0 \qquad \text{on } \partial\mathcal{B}. \tag{I.38}_r$$

---
[16] CERCIGNANI [1972, Appendix] [1975, Chapter III, Eq. (4.1)].

A counterexample provided by Dr. MARIO PITTERI shows that such is not always the case for an ideally rough wall. It follows *a fortiori* that *a gas in contact with a linear wall does not necessarily satisfy the heat-transfer inequality.*

(v) *Traditional interpretation of the formal narrow H-theorem. The ultra-narrow trend to equilibrium. Statement of corresponding rigorous propositions*

Traditionally the formal narrow H-theorem is thought to imply a trend to equilibrium. The argument, which descends from BOLTZMANN[17] himself, is plausible only in the most degenerate circumstances; namely, F is assumed from the outset to be independent of place, and $\mathbf{b} = \mathbf{0}$. As we have shown in Case 1 of Section (i) of this chapter, the formal narrow H-theorem holds for such solutions. The theorem itself for them is most naturally expressed in terms of h, which is a function of t alone: *If in some interval of time there is a solution F that is not Maxwellian, then h decreases until such time—be it finite or infinite—as it shall correspond to a Maxwellian F.* But, since the gas is in gross free equilibrium, one and only one Maxwellian solution is compatible with the circumstances, and it is a uniform one, say F_U. Hence, denoting by $h[F]$ the function h defined from F by (VIII.46), we conclude that

$$h[F](t) \downarrow h[F_U] \quad \text{as} \quad t \to \infty. \tag{XI.60}$$

Consequently

$$F \to F_U \quad \text{as} \quad t \to \infty. \tag{XI.61}$$

This last statement asserts the *strict trend to equilibrium* for the particular, extremely narrow circumstances considered.

In presenting this traditional argument CHAPMAN & COWLING[18] seem to have noticed at least one of its obvious gaps, for they have tried to close it. They assert that h is bounded below because it could be infinite only if $\int F \log F$ diverged. But $\int F c^2$ certainly converges, for it is proportional to ε. Hence if $\int F \log F$ were to diverge, $-\log F$ would have to tend to ∞ more rapidly than c^2 as $c \to \infty$. But this would imply that F tends to 0 more rapidly than e^{-c^2} as $c \to \infty$ in which case $\int F \log F$ would certainly converge. They conclude, " It therefore appears that when the density, mean velocity and temperature of a uniform gas are assigned, there is only one possible permanent mode of distribution of the molecular velocities, and that the actual mode, if different, will tend to approach this mode." We are unable to follow "proofs" of this kind; we

---

[17] BOLTZMANN [1875, 1]. It appears in a few lines on p. 14 of the reprint in his collected works.
[18] CHAPMAN & COWLING [1939, §4.1, especially the footnote] [1970, *ibid.*].

repeat this one, our one and only specimen of the logical standards heretofore accepted in the kinetic theory, for its quaintness. Dubious as are the steps of inference, the result is plausible. It asserts simply that *a gas in gross free equilibrium tends with time to kinetic equilibrium.*

In the kinetic theory there are different levels of rigor. First, we may assume that a solution of a given type exists and then rigorously deduce its properties. Such rigor is sometimes regarded as merely formal in that all functions are silently supposed smooth enough to admit the manipulations of the infinitesimal calculus. It is rigor of this kind that CARLESON and DARROZES & GUIRAUD claimed in deriving their H-theorems, and they were justified in doing so. On the other hand, a stricter rigor would require that the manipulations of calculus be proved correct at each stage. The critical mathematician must first prove that a solution of the type he wishes exists, and thereafter he must prove that it is sufficiently smooth. For the kinetic theory this higher logical precision is of supreme difficulty.

A treatment of the problem rigorous in the strict sense, even in the extremely special circumstances appropriate to the narrow H-theorem, would entail precise formulation and proof of theorems of three kinds. Each presumes an admissible class of functions, which would have to be specified in the strict statement of each case considered. F is assumed to be independent of x, and $F(t)$ denotes the restriction of F to the time t.

I. Existence and Uniqueness of Solutions of the Initial-value Problem. Given F_0 in an admissible class, if $t > t_0$ there is a solution $F(t)$ of the Maxwell-Boltzmann equation such that $F(t_0) = F_0$. Two solutions that agree at one time agree at all times for which both exist.

II. Rigorous Ultra-narrow H-Theorem. Let F be the solution delivered by Statement I. If $h[F_0]$ exists, so does $h[F](t)$, and

$$h[F](t_2) \leq h[F](t_1) \quad \text{when} \quad t_2 \geq t_1. \tag{XI.62}$$

III. Rigorous Ultra-narrow Trend to Equilibrium.

$$F \to F_U \quad \text{as} \quad t \to \infty. \tag{XI.61}_r$$

The reader will note that these statements refer to existence for infinite time and to a limit as $t \to \infty$. This is a reflection of experience in the proof of theorems regarding solutions independent of x. The heuristic background does not exclude the possibility that such limits might be approached in a finite time, at which the solution might cease to exist.

Traditionally the result asserted by Statement II is used as an element in the proof of Statement III. There is no logical reason for this, however. In Chapter XVIII we shall prove a version of Statement III by a method which gives no information about Statement II.

Only one step in the program set forth above is presently easy to make.

Namely, we first assume that $\mathbf{b} = \mathbf{0}$, and then from (XI.8) we see at once that a solution F which is independent of place determines uniquely a uniform Maxwellian density F_U having the same principal moment as F. Moreover, if $F \neq F_U$ and $h[F]$ exists, from (VIII.51)$_1$ we see that

$$h[F] > h[F_U] = \text{const.} \tag{XI.63}$$

Therefore, $h[F]$, being a monotone-decreasing function, bounded below in each interval of time through which it exists, must approach a limit from above:

$$h[F] \downarrow \text{ a limit} \geq h[F_U]. \tag{XI.64}$$

Without further information we cannot conclude that that limit is in fact $h[F_U]$, even if the interval of time is infinite.

(vi) *Difficulties faced in interpretation of the more general narrow H-theorem and the strict trend to equilibrium*

We have included "ultra-narrow" in the names given to Statements II and III to remind the reader that the problem as formulated in the preceding section of this chapter is narrower than even the narrow H-theorem would suggest, and Statement III asserts merely that in a condition assumed from the start to be gross free equilibrium, kinetic equilibrium is approached. Such a trend is far too special to model common experience. We see a body of real fluid, once it has been disturbed and then left to itself in a stationary vessel, settle into gross equilibrium. The Navier–Stokes theory of incompressible fluids provides a simple, elegant, and rigorous theorem which corresponds precisely to this fact. A theorem of the same sort ought to emerge from the kinetic theory as well.

In Section (iv) of this chapter we have derived formal narrow H-theorems for bodies of fluids in several circumstances, particularly when confined by walls of certain kinds. Existence and smoothness of the solutions used were presumed in the analysis there. A rigorous proof would have to justify the formal steps or replace them by some other kind of argument. The initial value F_0 of the molecular density would not be spatially uniform, and neither would the solution. The strict trend to equilibrium would then be a statement of this kind: *A body of gas for which the narrow H-theorem holds tends to kinetic equilibrium.* Such a theorem, which *would assert a simultaneous approach to gross equilibrium and kinetic equilibrium*, would reflect the fact of nature which the theory is intended to model.

We have seen that among free solutions independent of place there is only one Maxwellian solution for a given, constant principal moment. Also other circumstances may limit the possible Maxwellian solutions to one. Such seems to be the case for a gas within boundaries which are specified in terms of a particular Maxwellian density F^W, but the details are not clear. Certainly ideally smooth walls present a problem of a different kind. Any locally Maxwellian density is an even function of the components of **c**. Thus any locally Maxwellian solution such as to render the gross velocity tangent to $\partial \mathscr{B}$ will satisfy (XI.19). For example, let $\partial \mathscr{B}$ be a part of a circular cylinder contained between two cross-sections, and let the body of gas within it be in free rigid rotation about the axis of the cylinder. Then (XI.18) is satisfied on $\partial \mathscr{B}$. As we have seen in Chapter X, a free locally Maxwellian solution must be a dilatation and hence satisfy (II.37) and (II.38). Both the magnitude of the spin and the energetic remain arbitrary, even if we require them to be steady. Thus there is a two-parameter family of locally Maxwellian solutions for this problem. Each of these locally Maxwellian solutions exists for all time, and none of them tends toward another one. In particular, the rotation does not die out as time goes on. Thus there are infinitely many locally Maxwellian solutions compatible with the boundary conditions. For all of these, of course, $\dot{h} = 0$, and the narrow H-theorem holds. *Thus the narrow H-theorem is not a sufficient condition for the strict trend to equilibrium.*

It remains possible that the locally Maxwellian solutions for the problem of gas flow within a cylinder may attract all the other solutions. That is, any given solution may tend to one or another locally Maxwellian one. As GRAD[19] wrote,

> ... the H-theorem gives no indication that there actually will be an approach to absolute equilibrium since it gives no clue to the transition from local to absolute Maxwellian. That this is no empty worry can be easily shown. If the distribution should become and remain locally Maxwellian, the H-function would then reduce to the negative of the thermodynamic entropy, integrated over the physical domain as in fluid mechanics. The local Maxwellian implies vanishing stresses and heat flow. This in turn implies that the flow is governed by the classical nondissipative Euler equations of fluid dynamics. The thermodynamic entropy of the system does not change (this is consistent with $dH/dt = 0$), and there is no approach to equilibrium. Of course, the distribution will not be exactly a local Maxwellian at any finite time; the question is whether the deviation from a local Maxwellian, which is fed by molecular streaming in the presence of spatial inhomogeneity, is sufficiently strong to ultimately wipe out the inhomogeneity. We see that a valid proof of the approach to equilibrium in a spatially varying problem requires just the opposite of the procedure that is followed in a proof of the H-theorem, *viz.*, to show that the distribution function does not approach too closely to a local Maxwellian.
>
> A rigorous mathematical proof in this generality is extremely difficult. We need only mention that it would involve at least as much effort as a similar theorem for the nonlinear Navier–Stokes equations, and this is not yet accomplished.

The problem is complicated by the fact that solutions of the Maxwell–Boltzmann equation may cease to exist after a finite time has elapsed. Again the locally Maxwellian solutions, which we studied in the preceding chapter, serve as examples. For all these solutions $\dot{h} = 0$, so the narrow H-theorem holds. Far

---

[19] GRAD [1965, *3*, §1].

from implying that F_M shall tend with time to a uniform Maxwellian density, here the narrow H-theorem does not prevent F_M from ceasing to exist! In the special case governed by (X.7) we need only glance at (X.8) and (X.9) to see that F_M approaches with sufficient time—finite if $T < 0$, infinite if $T > 0$—a limit that is not a molecular density at all.

The variety of locally Maxwellian solutions suggests another possibility not inconsistent with the narrow H-theorem. A solution F might equal some one F_M at some particular time, or over an interval of time, but not before then or afterward. Then $\dot H$ for a material region would vanish at that time or during that interval, but afterward would become negative again, as indicated in Figure XI.1. A uniqueness theorem would exclude this behavior: Two solutions that coincide at one instant continue to do so as long as both exist. Such a theorem has been proved for a broad but far from exhaustive class of kinetic constitutive relations, as we shall see in Chapter XX.

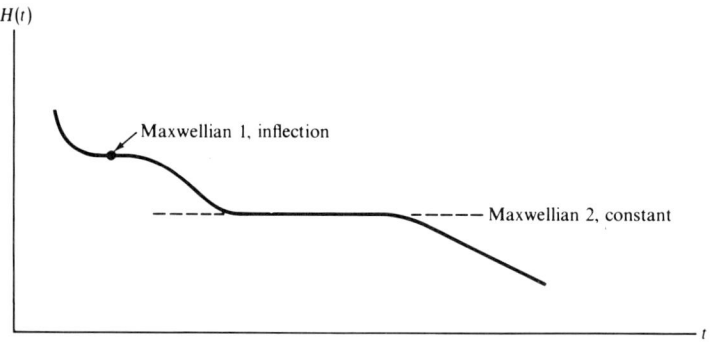

Figure XI.1 Sketch to illustrate behavior compatible with the narrow H-theorem

It remains an open problem *to find conditions upon a place-dependent solution sufficient to establish a strict trend to equilibrium*. Far from solving this problem, the narrow H-theorem seems to cast no light on it at all. For example, we can look back at the locally Maxwellian solution for free homo-energetic and homochoric dilatation, presented near the end of Chapter X, and write as follows the conclusion stated just before (X.7): There is a uniform Maxwellian density F_U such that

$$h[F_M] = h[F_U] \tag{XI.65}$$

everywhere and always. Nevertheless, $F_M \nrightarrow F_U$, no matter what be the value of the arbitrary constant T. Therefore, *the formal narrow H-theorem is not a sufficient condition for the strict trend to equilibrium*. Moreover, a unique uniform Maxwellian density is suggested by the problem, namely, that which corresponds to the principal moment $\rho(0)$.

The "proof" by CHAPMAN & COWLING which we have outlined just after (XI.61) does not seem to require that ρ, **u**, and ε be constant, merely that ε shall exist. If this interpretation is correct, the example given just above contradicts their conclusion and hence shows that their "proof" is false.

(vii) *Lack of interpretation for the broad H-theorem*

The H-theorem is commonly regarded as reflecting the irreversibility of phenomena observed in real gases. The physical concept of irreversibility is far more general than the mere trend of a confined and undisturbed body of fluid to settle to rest. A typical problem of gas dynamics concerns something much more interesting than that. It usually requires analysis of a steady flow produced by the continued action of forces, which the gaseous body transmits and by its consequent motion converts into other forces. The dissipative quality of internal friction makes certain of these flows stable, while others are not. The kinetic theory has not yet received mathematical cultivation sufficient to discover what precise information it can give about typical gas flows.

Deformation of a body should use work, not give it out. The Navier-Stokes theory of fluids incorporates this feature. In Section (ix) of Chapter XIV we shall see that the kinetic theory, on the contrary, does not always do so. The broad H-theorem (XI.5) expresses an irreversibility of some kind. What that irreversibility is, outside of circumstances in which (XI.5) may be approximated sufficiently by the Clausius-Duhem inequality, remains a mystery. We do not know of a single example which casts any light upon it, so the following elementary remarks may offer some interest.

In Section (v) of this chapter we have proved that if h_M corresponds to F_M in the decomposition (VIII.24), then

$$h - h_M \geq 0. \qquad \text{(VIII.51)}_{1r}$$

It is natural to ask if the difference $h - h_M$ at each fluid-point tends to dwindle or grow as time goes on. The change of h_M is governed by the equation of balance of energy (IX.5), which we may write in the form

$$-\tfrac{2}{3}\rho\varepsilon \dot{h}_M = -\mathbf{P}\cdot\mathbf{E} - \operatorname{div}\mathbf{q}. \qquad \text{(XI.66)}$$

In regard to the balance of energy, therefore, h_M plays the same role for the kinetic theory as $-\eta/r$ does in the dynamics of any gas for which the thermodynamic relations (I.9) hold and $\varpi = p$. The local statement of the formal broad H-theorem is

$$\rho \dot{h} \leq -\operatorname{div}\mathbf{s}, \qquad \text{(XI.67)}$$

as follows from (XI.1) and the fact that $\Sigma < 0$ except if $F = F_M$, in which case $\dot{h} = 0$. Using (XI.66) and (XI.67), we conclude that

$$\rho(\dot{h} - \dot{h}_M) \leq \frac{1}{\frac{2}{3}\varepsilon}\left(-\mathbf{P}\cdot\mathbf{E} - \frac{1}{\varepsilon}\mathbf{q}\cdot\text{grad }\varepsilon\right) - \text{div}\left(\mathbf{s} + \frac{1}{\frac{2}{3}\varepsilon}\mathbf{q}\right), \quad \text{(XI.68)}$$

and equality holds if and only if $F = F_M$, in which case both sides reduce to 0. The material derivative, denoted by a superimposed dot, means the same thing in both instances because F and F_M give rise to the same velocity field. The general considerations in Section (i) of Chapter IV show that *the right-hand side of* (XI.68) *may be positive, negative, or null* at any given instant. *Thus no general trend for the difference $h - h_M$ can be established* on the basis of (XI.68). It is possible, nevertheless, that for certain special solutions some definite conclusion may be drawn. We shall give an example below, just after (XXIV.113). Here we remark only that the classical dissipation inequalities or the Clausius–Duhem inequality would make the first summand on the right-hand side of (XI.68) non-negative. In the kinetic theory, however, it may be negative, as we shall see by an explicit example below in Section (ix) of Chapter XIV.

Part D

Particular Molecular Models and Exact Solutions for Moments

Chapter XII

The Collisions Operator for Some Special Kinetic Constitutive Relations, Especially Maxwellian Molecules

In Chapter VII we defined and explained MAXWELL's collisions operator in general terms. To make further progress, we must descend to special cases of that operator. If we follow the tradition of the kinetic theory, to calculate these forms we should prescribe a particular law of intermolecular force and then solve the two-body problem of analytical dynamics to which it leads. While physicists and chemists seem to take pleasure in the horrendous formal algebra, which for all but the simplest cases leads to the joys of numerical calculation on costly machines, students of the foundations often prefer to avoid it and to regard the encounter operator $\mathbb{E}$ itself, along with the cross-section $\mathscr{S}$ and of course the molecular mass $\mathfrak{m}$, as the constitutive quantities to be prescribed outright.

In this book we shall take a middle course. The reader must have some idea what is involved, or he will not estimate at its true ferocity the mathematical problem he faces. On the other hand, he must not let these nasty quadratures, which are called *collisions integrals*, bury the conceptual structure and the basic problems of the kinetic theory, as some books by physicists and chemists might lead him to do. Thus we will present the calculations, all of which are due to MAXWELL[1] himself, for the simplest cases: ideal spheres and inverse $\mathbb{k}^{th}$-power molecules. Only by inspecting the results for general $\mathbb{k}$ can the reader see what makes the value 5 for $\mathbb{k}$ simpler than any other. Molecules of this kind are called *Maxwellian*. Some of the most beautiful results now known are limited to them. Their prime importance to promote comprehension of the theory—an importance of which authors of the "physical" kind seem to be insensible—provides another reason that we present explicit forms of certain Maxwellian collisions integrals. Physicists and chemists will object that these laws of

---

[1] MAXWELL [1867, Eqs. (8)–(16) and the following text].

intermolecular force are "unrealistic"; they are right, but there seems to be little agreement, and little evidence for deciding, what the "true" intermolecular forces are[2], and anyhow we do not offer this book as a contribution to physics, to chemistry, or to computing.

The next following chapters do not refer to the detailed calculations which we shall present in this one, only to certain specific formulae obtained. Therefore the reader may, if he so wishes, pass over the proofs and simply look at and accept the results (XII.24), (XII.25)$_2$, and (XII.26) for Maxwellian molecules, because these three formulae are the only ones in this chapter that will be put to specific use immediately. In order to understand Chapter XVI and some later chapters, however, he will have to know other details of this one.

In Chapter VII we have given two forms, roughly equivalent, for MAXWELL's collisions operator:

$$\mathbb{C}F \equiv \int\int_{\mathscr{S}} w(F'F'_* - FF_*); \qquad \text{(VII.3)}_r$$

$$= \int\int_{\triangle} \mathbb{S}(F'F'_* - FF_*). \qquad \text{(VII.10)}_r$$

In these formulae F_*, F', and F'_* stand for F with its argument $\mathbf{v}$ replaced by $\mathbf{v}_*$, $\mathbf{v}'$, and $\mathbf{v}'_*$; $\mathbf{w} = \mathbf{v}_* - \mathbf{v}$; and $\int$ stands for integration with respect to $\mathbf{v}_*$ over

---

[2] *Cf.* S. G. BRUSH, "Interatomic forces and gas theory from Newton to Lennard-Jones", *Archive for Rational Mechanics and Analysis* **39**, 1–29 (1970). In evaluating the successes and failures of specific researches from the beginning of the kinetic theory down to the present time, BRUSH concludes that "if you are primarily interested in understanding the properties of matter or in predicting new properties, forget about using a 'realistic' force law and use the *simplest* model that still retains the crude qualitative physical properties that might be important . . . [T]he major effort should be put into ensuring that the predicted properties of matter are rigorous logical consequences of the model, whatever it may be . . . ."

While CHAPMAN & COWLING [1939, §1] might not today seem to have erected a pinnacle of strict mathematics, such may have been their aim, for they begin their book with the words

The purpose of this book is to elucidate some of the observed properties of the natural objects called gases. The method used is a mathematical one. . . .

The joint labours of experimental and theoretical physicists have suggested certain hypotheses regarding the structure and interaction of molecules: the details are, however, known for very few kinds of molecule. The mathematician has therefore to consider ideal systems, chosen as illustrating the particular features of actual gas-molecules that are to be studied, and to work out their properties as accurately as possible. The difficulty of this undertaking imposes limitations on the ideal systems which can be used.

CHAPMAN & COWLING [1939, §7] also relieve us of total surrender to the demands of real physics:

Our aim is to explain things that are seen and directly measurable by means of imagined things that are not seen and not directly measurable. . . .

It should be emphasized that only approximate agreement is to be expected between the kinetic-theory "interpreted" macroscopic relations and the observed macroscopic relations, because some divergence between the two sets of relations may reasonably be attributed to the imperfect representation of the actual molecules by the "model" molecules with which the mathematician works.

$\mathscr{V}$, while of course **v** is kept constant. In the first formula $\int_{\mathscr{S}}$ denotes integration with respect to surface area over a subset $\mathscr{S}$ of the plane $\mathscr{P}$ through **x** and normal to **w**, the cross-section $\mathscr{S}$ being either a disk or the entire plane $\mathscr{P}$. The vectors **v'** and $\mathbf{v'_*}$ in this first case are determined from **v**, $\mathbf{v_*}$, and a point $\mathbf{x}^e$ on the plane $\mathscr{P}$ by the encounter operator $\mathbb{E}$. Properties of $\mathbb{E}$ were set forth in Chapters VI and VII. Roughly speaking, $\mathbb{E}$ maps pairs of vectors $(\mathbf{v}, \mathbf{v_*})$ and the displacement vector $\mathbf{r}^e$ in the plane $\mathscr{P}$ into pairs of vectors $(\mathbf{v'}, \mathbf{v'_*})$ such that as $\mathbf{r}^e$ varies over the cross-section $\mathscr{S}$, **v'** and $\mathbf{v'_*}$ represent all possible velocities which respect conservation of momentum and energy in an encounter involving velocities **v** and $\mathbf{v_*}$:

$$\mathbf{v} + \mathbf{v_*} = \mathbf{v'} + \mathbf{v'_*}; \qquad (\text{VI.1})_r$$

$$v^2 + v_*^2 = v'^2 + v'^2_*. \qquad (\text{VI.2})_r$$

In (VII.10) $\int_{\Omega}$ stands for integration with respect to solid angle over the unit hemisphere with base on $\mathscr{P}$, center at **x**, and curved surface lying upon the side of $\mathscr{P}$ away from which **w** points. If θ and ζ denote colatitude and azimuthal angles, respectively, on this hemisphere, **v'** and $\mathbf{v'_*}$ are determined as follows by **v**, $\mathbf{v_*}$, θ, and ζ:

$$\mathbf{v'} = \mathbf{v} + [\mathbf{w} \cdot \mathbf{a}(\theta, \zeta, \mathbf{w})]\mathbf{a}(\theta, \zeta, \mathbf{w}),$$

$$\mathbf{v'_*} = \mathbf{v_*} - [\mathbf{w} \cdot \mathbf{a}(\theta, \zeta, \mathbf{w})]\mathbf{a}(\theta, \zeta, \mathbf{w}), \qquad (\text{VI.41})_r$$

a being given by

$$\mathbf{a}(\theta, \zeta, \mathbf{w}) \equiv \cos\theta \frac{1}{w}\mathbf{w} - \sin\theta \cos\zeta\, \boldsymbol{\xi}(\mathbf{w}) - \sin\theta \sin\zeta\, \boldsymbol{\eta}(\mathbf{w}), \quad (\text{VI.40})_{2r}$$

and $\mathbf{w}/w$, $\boldsymbol{\xi}(\mathbf{w})$, and $\boldsymbol{\eta}(\mathbf{w})$ being any orthonormal set of vectors. One choice of $\boldsymbol{\xi}$ and $\boldsymbol{\eta}$ is that expressed by (VI.39). The scattering factor $\mathbb{S}$ has been defined thus:

$$\sin\theta\, \mathbb{S}(\theta, w) \equiv wR(\theta, w)|\partial_\theta R(\theta, w)|, \qquad (\text{VII.8})_r$$

the magnitude r of $\mathbf{r}^e$ being presumed specified by (VI.71), obtained through the solution of some two-body problem in analytical dynamics.

The simplest kind of molecule with a cut-off is an ideal sphere. We have mentioned it already in Chapters IV, VI, and VII. For ideal spheres the calculation of $\mathbb{S}$ is straightforward. Figure XII.1 represents two such spheres of diameter d at the moment of collision; they are shown in section normal to $\mathscr{P}$ in the plane of the relative trajectory. If the path of the oncoming molecule intersects $\mathscr{P}$ at a point $\mathbf{x}^e$ within a disk of radius d about **x**, an encounter occurs.

XII. SPECIAL KINETIC CONSTITUTIVE RELATIONS

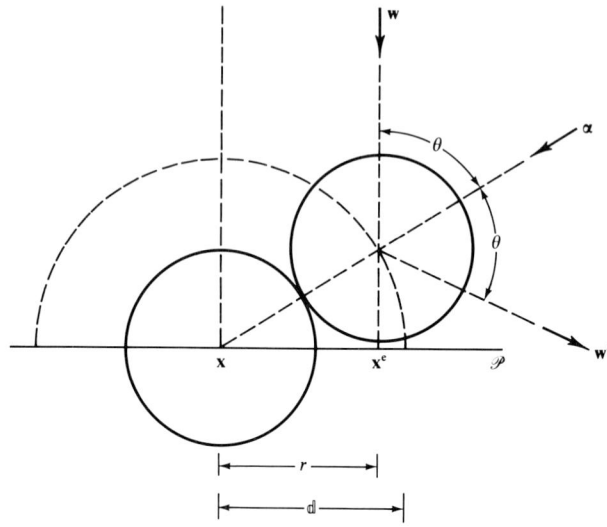

Figure XII.1 Encounter between two ideal spheres

Thus in this particular case (VI.71) assumes the form $r = d \sin \theta$, so (VII.8) and (VI.59) become[3]

$$\mathbb{S}(\theta, \mathbf{w}) = d^2 w \cos \theta,$$

$$v'_k = v_k + \left[1 - \left(\frac{r}{d}\right)^2\right] w_k - w\frac{r}{d}\sqrt{1 - \left(\frac{r}{d}\right)^2} \left[\cos \zeta \, \xi_k(\mathbf{w}) + \sin \zeta \, \eta_k(\mathbf{w})\right],$$

$$v'_{*k} = v_{*k} - \left[1 - \left(\frac{r}{d}\right)^2\right] w_k + w\frac{r}{d}\sqrt{1 - \left(\frac{r}{d}\right)^2} \left[\cos \zeta \, \underline{\xi}_k(\mathbf{w}) + \sin \zeta \, \eta_k(\mathbf{w})\right].$$

(XII.1)

Thus the forms (VII.3) and (VII.10) of $\mathbb{C}$ are rendered explicit.

---

[3] In Chapter VII we have remarked that for molecules with a cut-off we can define as follows the collision frequency v:

$$v \equiv \mathbb{F}F,$$

$\mathbb{F}$ being the frequency operator, which is defined by (VII.4). In terms of the scattering factor $\mathbb{S}$,

$$v = \int_{\mathscr{V}} \int_{\Omega} \mathbb{S} F_* \, d\mathbf{v}_*.$$

A classical example is provided by a gas of ideal spheres. For it we may use $(\text{XII}.1)_1$ and so obtain

$$v = \pi d^2 \int_{\mathscr{V}} w F_* \, d\mathbf{v}_*.$$

We shall now calculate $\mathfrak{S}$ for inverse $\mathfrak{k}^{\text{th}}$-power molecules, which are defined by the intermolecular force

$$\mathfrak{m}f = \frac{\mathfrak{g}}{d^{\mathfrak{k}}}, \qquad \mathfrak{k} > 1, \qquad \mathfrak{g} > 0, \tag{IV.4}_r$$

d being the distance between the two molecules. To describe the relative trajectory, we take polar co-ordinates (d, ϕ) in its plane, measuring the azimuth ϕ from the normal to the plane $\mathscr{P}$. These co-ordinates are shown in Figure XII.2, which is an expanded version of Figure VI.6. The integrals of rotational momentum and energy are

$$d^2\dot{\phi} = rw,$$

$$\tfrac{1}{2}(\dot{d}^2 + d^2\dot{\phi}^2) + \frac{2\mathfrak{g}}{\mathfrak{m}(\mathfrak{k} - 1)d^{\mathfrak{k}-1}} = \tfrac{1}{2}w^2. \tag{XII.2}$$

If F is Maxwellian, we may evaluate the quadrature easily:

$$v = \sqrt{\frac{\pi}{b}}\, n\mathrm{d}^2 g(\sqrt{2b}\, c), \qquad g(y) \equiv e^{-\frac{1}{2}y^2} + \left(y + \frac{1}{y}\right)\int_0^y e^{-\frac{1}{2}q^2}\, dq.$$

This result is due to O.-E. MEYER (1866).

The *mean free path* L corresponding to the random velocity c may be defined as follows:

$$L \equiv c/v.$$

If

$$L_\infty \equiv \frac{1}{\pi\mathrm{d}^2 n},$$

we can show from (VIII.13) that L/L_∞ is a function of $c/\bar{c}$ alone and that $L \to L_\infty$ as $c/\bar{c} \to \infty$. Thus L is inversely proportional to $n\mathrm{d}^2$, as WATERSTON had indicated in 1843. L depends upon ε through the ratio $c/\bar{c}$, and as a function of that argument increases monotonically from the value 0 when $c = 0$ to the value 1 when $c = \infty$. That $L < L_\infty$, was indicated in 1850 by CLAUSIUS, to whom is due the first calculation of L_∞, which he defined, in effect, as $L(\infty)$. Much argument resulted. TAIT (1887) proposed in effect to take the mean free path as the average of L over all values of c. MEYER's formula for v leads easily to the value of $\bar{L}$:

$$\frac{\bar{L}}{L_\infty} = 4\int_0^\infty \frac{x^3 e^{-x^2}\, dx}{e^{-x^2} + \left(2x + \dfrac{1}{x}\right)\int_0^x e^{-y^2}\, dy},$$

$$= 0.677\ldots.$$

We include these considerations here only to show how special and severely conditioned are the concept and the calculation of the mean free path. For details and related matters the reader may consult the account of TRUESDELL [1975]; however, Footnote 13 to his §8 needs correction. It was not $\bar{L}$ but a corresponding formula for μ that BOLTZMANN had published earlier. As for $\bar{L}$, BOLTZMANN declared in 1887 and more definitely in 1895 that he had found TAIT's result twenty years earlier but "out of annoyance" left it unpublished because he had gone on to demonstrate the insecurity of all statements based on use of the mean free path. TAIT's and BOLTZMANN's numerical calculations agree in the three digits printed above.

XII. SPECIAL KINETIC CONSTITUTIVE RELATIONS

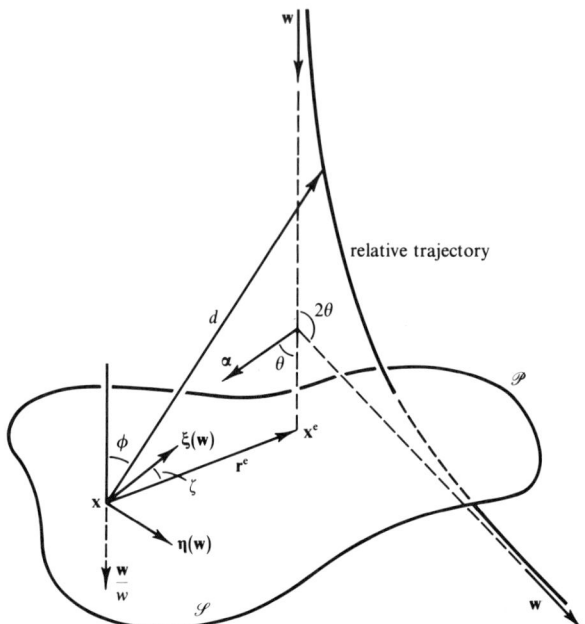

Figure XII.2 Variables used to describe an encounter

The factor 2 multiplying the potential energy arises because the motion of the oncoming molecule relative to the molecule at **x** is the same as that of a point repelled from a fixed center with doubled force per unit mass. If

$$\delta(t) \equiv \frac{r}{d(t)}, \quad z \equiv r\left(\frac{mw^2}{2g}\right)^{1/(k-1)}, \tag{XII.3}$$

from (XII.2) we obtain the orbital equation

$$\phi = \int_0^\delta \left[1 - u^2 - \frac{2}{k-1}\left(\frac{u}{z}\right)^{k-1}\right]^{-\frac{1}{2}} du. \tag{XII.4}$$

The angle ϕ at which d is a minimum, so δ is a maximum, necessarily bisects the angle between the asymptotes and hence equals θ. The values of δ such as to render $d\delta/d\phi = 0$ therefore make the integrand in (XII.4) infinite, so they are roots of the equation

$$1 - \delta^2 - \frac{2}{k-1}\left(\frac{\delta}{z}\right)^{k-1} = 0. \tag{XII.5}$$

Because $k > 1$, the function on the left-hand side of this equation decreases steadily from 1 to $-\infty$ as δ increases from 0 to ∞, so for a fixed value of z the equation itself has exactly one positive root $\delta = D(z)$. By (XII.4), then,

$$\theta = \int_0^{D(z)} \left[1 - u^2 - \frac{2}{\Bbbk - 1}\left(\frac{u}{z}\right)^{\Bbbk - 1}\right]^{-\frac{1}{2}} du. \tag{XII.6}$$

Replacing z in this formula by its definition (XII.3)$_2$, we obtain θ as some function Θ of r and w, just as we were led to expect in our analysis of encounters in general. The present result provides a specific, explicit example of (VI.57).

The integral in (XII.6) is of hyperelliptic type and may be inverted to give

$$z = Z(\theta). \tag{XII.7}$$

By (XII.3)$_2$, then,

$$r = \left(\frac{\mathfrak{m}w^2}{2\mathfrak{g}}\right)^{-1/(\Bbbk - 1)} Z(\theta). \tag{XII.8}$$

The function on the right-hand side is a specific example of the function R that appears in (VI.71). Using (VII.8), we find that

$$\sin \theta \, \mathfrak{S}(\theta, w) \, d\theta = wR(\theta, w)|\partial_\theta R(\theta, w)| \, d\theta,$$

$$= w\left(\frac{\mathfrak{m}w^2}{2\mathfrak{g}}\right)^{-2/(\Bbbk - 1)} Z(\theta)|Z'(\theta)| \, d\theta, \tag{XII.9}$$

$$= \left(\frac{2\mathfrak{g}}{\mathfrak{m}}\right)^{2/(\Bbbk - 1)} w^{(\Bbbk - 5)/(\Bbbk - 1)} z \, dz.$$

These results were obtained by MAXWELL. He saw at once that *the scattering factor $\mathfrak{S}$ is independent of w if and only if $\Bbbk = 5$*. Molecules such that $\Bbbk = 5$ are called *Maxwellian*. These are the only molecules with infinite cross-section $\mathscr{S}$ which have proved amenable, so far, to precise calculations.

Continuing with the consideration of results contained in principle in the work of MAXWELL, we shall show how to evaluate collisions integrals, at first considering a general molecular model. Because of (VI.39) and (VI.40) the components of **a** are

$$a_1 = \frac{w_1}{w} \cos \theta - \frac{\sqrt{w_2^2 + w_3^2}}{w} \sin \theta \cos \zeta,$$

$$a_2 = \frac{w_2}{w} \cos \theta + \frac{w_1 w_2}{w\sqrt{w_2^2 + w_3^2}} \sin \theta \cos \zeta - \frac{w_3}{\sqrt{w_2^2 + w_3^2}} \sin \theta \sin \zeta,$$

$$a_3 = \frac{w_3}{w} \cos \theta + \frac{w_1 w_3}{w\sqrt{w_2^2 + w_3^2}} \sin \theta \cos \zeta + \frac{w_2}{\sqrt{w_2^2 + w_3^2}} \sin \theta \sin \zeta.$$

$$\tag{XII.10}$$

Since $\boldsymbol{\alpha} \cdot \mathbf{w} = w \cos \theta$, by subtracting $\mathbf{u}$ from each side of (VI.41)$_1$ and (VI.41)$_2$ we get

$$c_1' = c_1 + w_1 \cos^2 \theta - \tfrac{1}{2}\sqrt{w_2^2 + w_3^2} \sin 2\theta \cos \zeta,$$

$$c_2' = c_2 + w_2 \cos^2 \theta + \frac{1}{2}\frac{w_1 w_2}{\sqrt{w_2^2 + w_3^2}} \sin 2\theta \cos \zeta$$

$$- \frac{1}{2}\frac{w w_3}{\sqrt{w_2^2 + w_3^2}} \sin 2\theta \sin \zeta,$$

$$c_3' = c_3 + w_3 \cos^2 \theta + \frac{1}{2}\frac{w_1 w_3}{\sqrt{w_2^2 + w_3^2}} \sin 2\theta \cos \zeta$$

$$+ \frac{1}{2}\frac{w w_2}{\sqrt{w_2^2 + w_3^2}} \sin 2\theta \sin \zeta, \qquad \text{(XII.11)}$$

$$c_{*1}' = c_{*1} - w_1 \cos^2 \theta + \tfrac{1}{2}\sqrt{w_2^2 + w_3^2} \sin 2\theta \cos \zeta,$$

$$c_{*2}' = c_{*2} - w_2 \cos^2 \theta - \frac{1}{2}\frac{w_1 w_2}{\sqrt{w_2^2 + w_3^2}} \sin 2\theta \cos \zeta$$

$$+ \frac{1}{2}\frac{w w_3}{\sqrt{w_2^2 + w_3^2}} \sin 2\theta \sin \zeta,$$

$$c_{*3}' = c_{*3} - w_3 \cos^2 \theta - \frac{1}{2}\frac{w_1 w_3}{\sqrt{w_2^2 + w_3^2}} \sin 2\theta \cos \zeta$$

$$- \frac{1}{2}\frac{w w_2}{\sqrt{w_2^2 + w_3^2}} \sin 2\theta \sin \zeta.$$

By taking $G = H = F$ in (VII.20)$_3$ we obtain the following expression for the total collisions operator:

$$\bar{\mathbb{C}}g = \tfrac{1}{2} \iint \int_{\mathscr{S}} (g' + g_*' - g - g_*)wFF_*, \qquad \text{(XII.12)}$$

providing the integrals converge. Again we have dropped the subscript F from $\bar{\mathbb{C}}_F$, since some given F is presumed here. We may rewrite (XII.12) as follows:

$$\bar{\mathbb{C}}g = \tfrac{1}{2} \int F(\mathbf{v}) \int F(\mathbf{v}_*) \mathbb{B}g,$$

$$\mathbb{B}g \equiv w \int_0^{2\pi} \int_0^\infty (g' + g_*' - g - g_*)r\, dr\, d\zeta. \qquad \text{(XII.13)}$$

If g is a polynomial in the components of $\mathbf{c}$, by (XII.11) it is plain that the integration with respect to ζ may be carried out explicitly. For example,

$$\mathbb{B}c_1^2 = w \int_0^\infty \{2\pi[2w_1(c_1 - c_{*1})\cos^2\theta + 2w_1^2 \cos^4\theta]$$

$$+ \tfrac{1}{2}\pi(w_2^2 + w_3^2)\sin^2 2\theta\} r \, dr, \qquad \text{(XII.14)}$$

$$= \tfrac{1}{2}\pi w(w_2^2 + w_3^2 - 2w_1^2) \int_0^\infty \sin^2 2\theta \, r \, dr.$$

This is one component of the general formula

$$\mathbb{B}c_k c_m = -\tfrac{3}{2}\pi w(w_k w_m - \tfrac{1}{3} w^2 \, \delta_{km}) \int_0^\infty \sin^2 2\theta \, r \, dr, \qquad \text{(XII.15)}$$

which may be verified in the same way.

For a general polynomial g we see from (XII.13)$_2$ and (XII.14)$_2$ that $\mathbb{B}g$ is a linear combination of the quantities $\mathbb{A}_{n|m}$ defined as follows:

$$\mathbb{A}_{n|m}(w) \equiv 2\pi w \int_0^\infty \cos^n\theta \, \sin^m\theta \, r \, dr,$$

$$= 2\pi \int_0^{\frac{1}{2}\pi} \cos^n\theta \, \sin^{m+1}\theta \, \mathbb{S}(\theta, w) \, d\theta, \qquad \text{(XII.16)}$$

the coefficients of that sum being functions of the components of $\mathbf{c}$ and $\mathbf{c}_*$. In view of (XII.11), $\mathbf{c}' - \mathbf{c}$ and $\mathbf{c}'_* - \mathbf{c}_*$ are proportional to $\cos\theta$, and so only those $\mathbb{A}_{n|m}$ for which $n > 0$ can enter[4] the expression for $\mathbb{B}g$. In (XII.15) only the function $\mathbb{A}_{2|2}$ appears. For any given molecular model $\mathbb{A}_{n|m}$ is in principle a known function. For example, for ideal spheres it follows from (XII.16)$_2$ and (XII.1)$_1$ that

$$\mathbb{A}_{2|2}(w) = \tfrac{1}{6}\pi d^2 w. \qquad \text{(XII.17)}$$

---

[4] It is important to be sure that if $n > 0$ the integrals in (XII.16) converge for the molecular models of interest. Since for any molecule with a cut-off the range of integration in (XII.16)$_1$ is finite, the integrals converge for such models. To examine the problem for molecules whose intermolecular force has infinite range, let us look ahead to Chapter XIX, in which we take up the convergence of integrals in general. We prove (*cf.* Theorem 6) that for inverse k^{th}-power molecules

$$\int_0^{\frac{1}{2}\pi} (\tfrac{1}{2}\pi - \theta) \sin\theta \, \mathbb{S}(\theta, w) \, d\theta < \infty \qquad \text{if } k > 3.$$

For these molecular models the convergence of (XII.16) if $n > 0$ follows directly from this fact and the inequality

$$\sin^m\theta \, \cos^n\theta \leq \cos\theta \leq \tfrac{1}{2}\pi - \theta \qquad \text{if } n > 0, \; 0 \leq \theta \leq \tfrac{1}{2}\pi.$$

For inverse $\Bbbk^{th}$-power molecules, by (XII.8) and (XII.9) we obtain

$$\mathbb{A}_{2|2}(w) = \tfrac{1}{2}\mathfrak{a}\left(\frac{2\mathfrak{g}}{\mathfrak{m}}\right)^{2/(\Bbbk-1)} w^{(\Bbbk-5)/(\Bbbk-1)}, \tag{XII.18}$$

$\mathfrak{a}$ being the following constant, which is in general different for each $\Bbbk$:

$$\mathfrak{a} \equiv \pi \int_0^\infty \sin^2 2\theta \, z \, dz; \tag{XII.19}$$

θ is given as a function of z according to (XII.6). For Maxwellian molecules numerical quadrature yields[5]

$$\mathfrak{a} = 1.3703\ldots. \tag{XII.20}$$

In view of (XII.16) we may regard (XII.15) as giving $\mathbb{B}c_k c_m$ explicitly as a function of $\mathbf{w}$. For example, putting (XII.18) into (XII.15) yields

$$\mathbb{B}c_k c_m = -\tfrac{3}{2}\mathfrak{a}\left(\frac{2\mathfrak{g}}{\mathfrak{m}}\right)^{2/(\Bbbk-1)} w^{(\Bbbk-5)/(\Bbbk-1)} (w_k w_m - \tfrac{1}{3}w^2 \delta_{km}). \tag{XII.21}$$

A similar but more complicated calculation yields

$$\begin{aligned}\mathbb{B}c_k c_m c_r &= -\tfrac{3}{2}\mathbb{A}_{2|2}(w)[(c_k + c_{*k})(w_m w_r - \tfrac{1}{3}w^2 \delta_{mr}) \\ &\quad + (c_m + c_{*m})(w_r w_k - \tfrac{1}{3}w^2 \delta_{rk}) + (c_r + c_{*r})(w_k w_m - \tfrac{1}{3}w^2 \delta_{km})], \\ &= -\tfrac{3}{4}\mathfrak{a}\left(\frac{2\mathfrak{g}}{\mathfrak{m}}\right)^{2/(\Bbbk-1)} w^{(\Bbbk-5)/(\Bbbk-1)}[(c_k + c_{*k})(w_m w_r - \tfrac{1}{3}w^2 \delta_{mr}) \\ &\quad + (c_m + c_{*m})(w_r w_k - \tfrac{1}{3}w^2 \delta_{rk}) + (c_r + c_{*r})(w_k w_m - \tfrac{1}{3}w^2 \delta_{km})],\end{aligned} \tag{XII.22}$$

the former expression being general, the latter, appropriate to inverse $\Bbbk^{th}$-power molecules.

Since $\Bbbk > 1$, no value of $\Bbbk$ other than 5 makes $(\Bbbk - 5)/(\Bbbk - 1)$ an even integer. Thus 5 is the only value of $\Bbbk$ for which the right-hand sides of (XII.21) and (XII.22)$_2$ are polynomials in the components of $\mathbf{c}$ and $\mathbf{c}_*$. For the case in

---

[5] Done for IKENBERRY & TRUESDELL [1956, Eq. (7.22)$_2$]. This value differs somewhat from the result of MAXWELL [1867, Eq. (16)] and all other published values of this number. In terms of Legendre polynomials P_s

$$3\mathfrak{a} = 2\pi \int_0^\infty [1 - P_2(\cos 2\theta)] z \, dz.$$

Other numbers calculated for IKENBERRY & TRUESDELL are as follows:

$$2\pi \int_0^\infty [1 - P_1(\cos 2\theta)] z \, dz = 2.6511\ldots, \quad 2\pi \int_0^\infty [1 - P_4(\cos 2\theta)] z \, dz = 4.9087\ldots.$$

The former differs slightly from the result of MAXWELL [1867, Eq. (15)].

which $g = c_k c_m$, if we take $\mathbb{k} = 5$ and substitute (XII.21) into (XII.13)$_1$, then by use of (III.9) and (III.13) we obtain

$$\overline{\mathbb{C}c_k c_m} = -\tfrac{3}{2}n^2 \mathbb{a} \sqrt{\frac{\mathfrak{g}}{2\mathfrak{m}}} [2\overline{c_k c_m} - 2\overline{c_k c_m} - \tfrac{1}{3} \delta_{km}(2\overline{c^2} - 2\overline{c_a c_a})],$$

$$= -3n^2 \mathbb{a} \sqrt{\frac{\mathfrak{g}}{2\mathfrak{m}}} (\overline{c_k c_m} - \tfrac{1}{3}\overline{c^2}\, \delta_{km}), \qquad (\text{XII.23})$$

$$= -3\rho \mathbb{a} \sqrt{\frac{\mathfrak{g}}{2\mathfrak{m}^5}} (M_{km} - p\, \delta_{km}).$$

Using (III.10) and (III.13), we arrive at[6]

$$\mathfrak{m}\overline{\mathbb{C}c_k c_m} = -3\rho \mathbb{a} \sqrt{\frac{\mathfrak{g}}{2\mathfrak{m}^3}} P_{km}. \qquad (\text{XII.24})$$

A similar formula for $\overline{\mathbb{C}c_k c_m c_r}$ follows easily by taking $\mathbb{k} = 5$ and substituting (XII.22)$_2$ into (XII.13)$_1$:

$$\mathfrak{m}\overline{\mathbb{C}c_k c_m c_r} = -\tfrac{1}{2}n^2 \mathfrak{m}\mathbb{a} \sqrt{\frac{\mathfrak{g}}{2\mathfrak{m}}} [9\overline{c_k c_m c_r} - (\overline{c_k c^2}\, \delta_{mr}$$
$$+ \overline{c_m c^2}\, \delta_{rk} + \overline{c_r c^2}\, \delta_{km})], \qquad (\text{XII.25})$$

$$= -\tfrac{1}{2}\rho \mathbb{a} \sqrt{\frac{\mathfrak{g}}{2\mathfrak{m}^3}} [9M_{kmr} - 2(q_k\, \delta_{mr} + q_m\, \delta_{rk} + q_r\, \delta_{km})].$$

Contraction on m and r yields[7]

$$\mathfrak{m}\overline{\mathbb{C}c^2 c_k} = -4\rho \mathbb{a} \sqrt{\frac{\mathfrak{g}}{2\mathfrak{m}^3}} q_k. \qquad (\text{XII.26})$$

The calculations needed to obtain these results are elaborate; for polynomials of degree greater than 3, they become essentially more difficult. In the next chapter we shall present MAXWELL's major conclusions from his formulae (XII.24) and (XII.26).

To conclude this chapter, we return to general kinetic constitutive relations so as to describe a special class of collisions operators whose properties are essentially simpler than all others. This class is suggested by the simplicity we have seen to result when the molecules are of one of the following two kinds:

(a) $\mathscr{S}$ is a disk (molecules with a cut-off).
(b) $\mathbb{S}$ is a function of θ alone, and the limits of integration over the variables θ and ζ in (VII.11)$_2$ are constants (Maxwellian molecules).

---

[6] MAXWELL [1867, Eqs. (41) and (42)].
[7] MAXWELL [1867, Eq. (43)]. In fact MAXWELL made a slip in passing from his Eq. (39) to his Eq. (43), which should have been (XII.26). His error was noticed by BOLTZMANN [1872, §III] and corrected by POINCARÉ [1893].

For molecules with a cut-off, we see from (VI.86) that in general the range of θ is not consistent with (b). For inverse $\Bbbk^{\text{th}}$-power molecules other than the Maxwellian, (XII.9) shows that $\mathbb{S}$ is not a function of θ alone. While it is traditional to specify both $\mathbb{S}$ and $\mathbb{A}$ by using velocities that result asymptotically from the solution of the two-body problem in analytical dynamics when some particular mutual force is selected, the theory itself does not dictate this limitation. We may, if we like, recalling the definition of "encounter operator" in Chapter VII, choose any such $\mathbb{E}$ we please and also any cut-off radius $\mathbb{d}$. If we allow ourselves this much freedom, we may set aside all direct reference to laws of intermolecular force and seek the scattering factor $\mathbb{S}$ and cut-off angle $\mathbb{A}$ most advantageous for mathematical arguments. Thus arise the following conditions:

(a) $\mathbb{A}(w) = \frac{1}{2}\pi$;
(b) $\mathbb{S}$ is a function of θ alone;
(c) $\int_{\triangle} \mathbb{S} < \infty$ (molecules with a cut-off).

It is unlikely there be any law of mutual force such as to make analytical dynamics compatible with these three properties. Even for a Maxwellian intermolecular force with a cut-off, namely

$$\mathbb{m}f = \begin{cases} \mathbb{g}/d^5 & \text{if } d \leq \mathbb{d} = \text{const.}, \\ 0 & \text{if } d > \mathbb{d}, \end{cases} \tag{XII.27}$$

while of course (b) and (c) are satisfied trivially, it is by no means clear that (a) is satisfied also. Nevertheless, the assumptions (a), (b), and (c) have been used frequently in analytical studies. We may say that they define *Maxwellian pseudomolecules*. For them

$$\mathbb{F}F = \left(\int_{\triangle} \mathbb{S}\right) \int F = An; \tag{XII.28}$$

here n is the number density, and A is the constant $\int_{\triangle} \mathbb{S}$.

The analytical simplicity that results from use of Maxwellian pseudomolecules does not require the assumptions (a), (b), and (c). Rather, (XII.28) is the key to it. Accordingly we may introduce the class of *pseudomolecules*, using (XII.28) as their defining property. Maxwellian pseudomolecules, often misleadingly[8] called "pseudo-Maxwellian molecules", make a special case.

As WILD[9] observed, if for molecules with a cut-off we select $\mathbb{E}$ arbitrarily and let $\mathbb{d}$ depend upon the relative speed w in the following particular way:

$$\mathbb{d}(w) = \alpha w^{-\frac{1}{2}}, \qquad \alpha = \text{const.} > 0, \tag{XII.29}$$

then $(\text{VII.11})_1$ gives us

$$\int_{\mathscr{S}} w \, dS = \pi\alpha^2 = \text{const.}, \tag{XII.30}$$

so (VII.4) reduces to (XII.28). This example shows that we may choose the constitutive quantities $\mathbb{m}$ and $\mathbb{E}$ in any way we like and then use (XII.29) to adjust the remaining constitutive quantity $\mathscr{S}$ so as to obtain pseudomolecules. Artificial as is the class of pseudomolecules, it is nevertheless a large one.

The earliest analytical studies of the kinetic theory show that for pseudomolecules it is rather easy to prove theorems of existence and uniqueness. By now, however, all such results have been extended to a broad class of molecules with a cut-off, so we shall not consider pseudomolecules further in this book. For Maxwellian molecules, on the other hand, such theorems seem to be as difficult, with one exception (Chapter XVIII), as for any other model, but the fact that collisions integrals may be evaluated exactly and explicitly makes it possible to consider in detail, and in some cases to solve, important problems presently inexpugnable when any other kinetic constitutive relation is laid down.

---

[8] KANIEL [1977, §1] uses the still more misleading name "generalized Maxwellian molecules".
[9] WILD [1951].

Chapter XIII

The Pressures and the Energy Flux in a Gas of Maxwellian Molecules. Maxwell's Relaxation Theorem and Evaluation of Viscosity and Thermal Conductivity

In this chapter and the two succeeding ones we shall develop some important theorems about a gas of Maxwellian molecules, theorems for which at the present time no rigorous extension to other kinetic constitutive relations is known. It is these theorems that make the Maxwellian molecular model important. While, as we shall see in later chapters, for other molecular models theorems of other kinds have been proved, the theorems about the gas of Maxwellian molecules are even today the only ones that cast much light on the gross motions of a kinetic gas.

(i) General equations for the pressures and energy flux

We begin from the equation of transfer in a special form appropriate to a function of **c** alone, a special form which is obtained from (IX.13) and (IX.14), namely,

$$(\rho \bar{g})^{\cdot} + \rho \bar{g} E + \rho \overline{c_a g}_{,b} u_{b,a} - \overline{g_{,a}} M_{ab,b} + (\rho \overline{c_a g})_{,a} = \mathfrak{m} \bar{C} g. \qquad \text{(XIII.1)}$$

To within questions of existence and smoothness, this equation is valid for any kinetic constitutive relation, but in this chapter we apply it only to a gas of Maxwellian molecules and at that only to two special choices of g, choices for which we evaluated $\bar{C}g$ in the preceding chapter. First we take $g = \mathbf{c} \otimes \mathbf{c}$. We have shown that for Maxwellian molecules

$$\mathfrak{m} \bar{C} c_k c_m = -3 \rho \mathfrak{a} \sqrt{\frac{g}{2\mathfrak{m}^3}} P_{km}; \qquad \mathfrak{a} = 1.3703 \ldots .$$

$$\text{(XII.24)}_r; \text{(XII.20)}_r$$

Therefore, in view of (III.8) and the definitions (III.9)$_2$ and (III.19), setting $g = \mathbf{c} \otimes \mathbf{c}$ on the left-hand side of (XIII.1) yields

$$\dot{M}_{km} + M_{km}E + M_{ka}u_{m,a} + M_{ma}u_{k,a} + M_{kma,a} = -\tau^{-1}P_{km}, \quad \text{(XIII.2)}$$

τ being defined as follows:

$$\tau \equiv \frac{1}{3n\mathbb{A}_{2|2}} = \frac{1}{3\mathfrak{a}}\sqrt{\frac{2\mathfrak{m}^3}{\mathfrak{g}}}\frac{1}{\rho}. \quad \text{(XIII.3)}$$

This equation relates the pressure tensor $\mathbf{M}$, its material derivative $\dot{\mathbf{M}}$, the velocity gradient, and the divergence of the third moments $^3\mathbf{M}$. To derive it, we have taken the molecules as Maxwellian and have assumed that the moments of orders 0, 1, 2, and 3 exist and are differentiable. If these presumptions be granted, it expresses an exact property of the kinetic gas.

For later use we record an alternative form, obtained by eliminating $\dot{p}$ by use of the equation of balance of energy in the form (IX.5)$_2$:

$$\dot{P}_{km} + P_{km}E - \tfrac{2}{3}P_{ab}E_{ab}\,\delta_{km} + P_{ak}u_{m,a}$$
$$+ P_{am}u_{k,a} + (M_{kma} - \tfrac{2}{3}q_a\,\delta_{km})_{,a} + 2pE_{km}$$
$$= -\tau^{-1}P_{km}. \quad \text{(XIII.4)}$$

This relation seems more complicated than (XIII.2) but is easier to use because in it the role of the pressure deviator $\mathbf{P}$ is separated from that of the mean normal pressure p.

In just the same way, taking g as $\mathbf{c} \otimes \mathbf{c} \otimes \mathbf{c}$ in (XIII.1) and using (XII.25)$_2$, after some reduction we may obtain a relation connecting the third moments $^3\mathbf{M}$, the velocity gradient, the pressure tensor and its gradient, and the divergence of the fourth moments $^4\mathbf{M}$:

$$\dot{M}_{kmr} + M_{kmr}E + M_{kam}u_{r,a} + M_{mar}u_{k,a} + M_{kar}u_{m,a} + M_{kmra,a}$$
$$-\frac{1}{\rho}M_{km}M_{ra,a} - \frac{1}{\rho}M_{kr}M_{ma,a} - \frac{1}{\rho}M_{rm}M_{ka,a} = -\tfrac{1}{2}\tau^{-1}[3M_{kmr}$$
$$-\tfrac{2}{3}(q_k\,\delta_{rm} + q_m\,\delta_{kr} + q_r\,\delta_{km})]. \quad \text{(XIII.5)}$$

For later use it is convenient to write this equation as two, one for the energy flux $\mathbf{q}$ and the other for the quantities $P_{0|kmr}$ defined as follows:

$$P_{0|kmr} \equiv M_{kmr} - \tfrac{1}{5}(M_{aak}\,\delta_{mr} + M_{aar}\,\delta_{km} + M_{aam}\,\delta_{kr}). \quad \text{(XIII.6)}$$

Any component of $^3\mathbf{M}$ is a linear combination of the components of $\mathbf{q}$ and the quantities $P_{0|kmr}$. The corresponding decomposition of (XIII.5) is

$$\dot{q}_k + \tfrac{5}{3}q_k E + P_{0|kab} E_{ab} + q_a u_{k,a} + \tfrac{4}{5}q_a E_{ka} - \frac{5p}{2\rho} p_{,k} - \frac{5p}{2\rho} P_{ka,a}$$

$$- \frac{1}{\rho} P_{ka} p_{,a} - \frac{1}{\rho} P_{ka} P_{ab,b} + \tfrac{1}{2} M_{kaab,b} = -\tfrac{2}{3}\tau^{-1} q_k, \qquad \text{(XIII.7)}$$

$$\dot{P}_{0|kmr} + P_{0|kmr} E + 3 P_{0|a(km} u_{r),a} - \tfrac{6}{5} P_{0|ab(k} \, \delta_{mr)} E_{ab} + \tfrac{12}{5} q_{(k} E_{mr)}$$

$$- \tfrac{24}{25} q_a E_{a(k} \, \delta_{mr)} - \frac{3}{\rho} P_{(km} p_{,r)} - \frac{3}{\rho} P_{(km} P_{r)a,a} + \frac{6}{5\rho} P_{a(k} \, \delta_{mr)} (p_{,a}$$

$$+ P_{ab,b}) + M_{kmra,a} - \tfrac{3}{5} \delta_{(km} M_{r)aab,b} = -\tfrac{3}{2}\tau^{-1} P_{0|kmr}.$$

To shorten the writing of the second equation we have denoted by parentheses enclosing three indices the sum over the 3! permutations of the indices, divided by 3!

From the remarkable relations (XIII.4) and (XIII.7), due in principle though not in entirely correct detail to MAXWELL[1] himself, follow many important consequences.

(ii) Maxwell's relaxation theorem

To begin with, these relations enable us to analyse the implications of gross equilibrium for the kinetic theory. In Chapter X we defined gross free equilibrium of the kinetic gas by the conditions $\mathbf{b} = \mathbf{0}$ and

$$\mathbf{u} = \mathbf{0}, \qquad \text{grad } \varepsilon = \mathbf{0}, \qquad \text{div } \mathbf{P} = \mathbf{0}, \qquad \text{div } \mathbf{q} = 0. \qquad \text{(II.13)}_r$$

The relations (XIII.4) and (XIII.7) show that in order to conclude anything at all about **P** and **q**, we shall have to know something about the third and fourth moments, ³**M** and ⁴**M**. We shall accordingly both extend and restrict the definition of gross free equilibrium as follows: *A condition of the gas when* $\mathbf{b} = \mathbf{0}$ *will be called grossly homogeneous of order n if the moments of F of orders* $0, 1, \ldots, n$ *exist and are constant in space.* A condition grossly homogeneous of order 3 or greater is necessarily one of gross free equilibrium. A molecular density F that is independent of place gives rise to a grossly homogeneous condition of every order n, provided the moments of orders $0, \ldots, n$ exist.

---

[1] MAXWELL [1867] did not write out all the terms in (XIII.2), but the remarks he put between his Eqs. (121) and (122) show that he had obtained them. For (XIII.7) we may consult his Eqs. (122) and (143), noting his discussion of the terms neglected. Later MAXWELL [1879, §§10–14] published more of the terms. GRAD [1949, Eqs. (5.13), (5.14), (5.15)], who used a different approach, seems to have been the first to publish (XIII.4) and (XIII.7) in full and correct form. IKENBERRY & TRUESDELL [1956, Eqs. (2.4), (2.5)] obtained them by restoring terms MAXWELL had neglected and correcting his slips in calculation.

Next we shall begin to determine for a kinetic gas of Maxwellian molecules a centrally important property of a grossly homogeneous condition of order n. First, if $n = 3$, all but the first of the terms on the left-hand side of (XIII.4) vanish, and so that relation becomes

$$\tau \dot{P}_{km} = -P_{km}. \tag{XIII.8}$$

Likewise, if $n = 4$, (XIII.7)$_1$ becomes

$$\tfrac{3}{2}\tau \dot{q}_k = -q_k. \tag{XIII.9}$$

Since **P** and **q** are functions of time only, and since τ is constant in a grossly homogeneous condition, the general solutions of these equations are, respectively,

$$P_{km} = P_{km}(0)e^{-t/\tau},$$

$$q_k = q_k(0)e^{-t/\frac{3}{2}\tau}. \tag{XIII.10}$$

Similarly it is easy to show from (XIII.7)$_2$ that

$$P_{0|kmr} = P_{0|kmr}(0)e^{-t/\frac{2}{3}\tau}. \tag{XIII.11}$$

The results (XIII.10) and (XIII.11) express **Maxwell's Relaxation Theorem**[2]. They show that *gross homogeneity does not imply kinetic equilibrium*, for in kinetic equilibrium, as we have seen, F is Maxwellian, so $\mathbf{P} = \mathbf{0}$ and $^3\mathbf{M} = \mathbf{0}$. In a grossly homogeneous condition of order 4 these moments may have at any instant any values we please. However, as $t \to \infty$, *their values all decrease exponentially to the values appropriate to kinetic equilibrium*, namely **0**. The rates at which they do so are governed by the parameter τ, which is called the *time of relaxation*[3] for the pressure deviator **P**. The logarithmic decrements are τ, $\tfrac{3}{2}\tau$, and $\tfrac{2}{3}\tau$, respectively. The time τ, as (XIII.3) shows, is determined by the molecular constants m and g and by the assigned, constant density ρ. Later in this chapter we shall show how to estimate τ from experimental data. In comparing two Maxwellian gases we may say that both greater mass and lesser intermolecular force per unit mass increase the time required for kinetic equilibrium to be sufficiently approximated. In comparing the behaviors of the same gas at different densities, we may say that the rarer is the gas, the slower is the approach to kinetic equilibrium, and that it may be made as slow as we like by simply rarefying the gas. That is, insofar as the phenomenon of relaxation is concerned, *the departure of the kinetic theory from the Stokes–Kirchhoff theory becomes more and more significant, the rarer is the gas.*

MAXWELL introduced the concept of relaxation of stress in the very way we have just indicated. Since his time, it has spread throughout much of physics

---

[2] MAXWELL [1867, Eq. (130)] [1879, §5].
[3] MAXWELL's term was "modulus of the time of relaxation".

and mechanics. However, until very recently MAXWELL's relaxation theorem was the only result concerning relaxation of stress and energy flux proved in the exact kinetic theory. Usually relaxation times are introduced by constitutive relations of continuum mechanics, by conjecture about the results of unperformed analysis or numerical work, by inspiration, or by all three combined[4]. Examples we shall give in Chapter XV show that when there is a trend to kinetic equilibrium, it need not be described by exponential decay. In Chapter XX we shall mention some very recent work that delivers exponential decay for a gas of molecules with a cut-off if the initial value of F is sufficiently near to a uniform Maxwellian density F_U. Toward the end of Chapter XVII we shall present a relaxation theorem that follows for a general molecular model when the exact equations of transfer are truncated in a way that will be specified there.

Even for Maxwellian molecules, it is only the moments of F of orders 2 and 3 that we have proved to approach their equilibrium values. The *assertion of the trend to equilibrium*, mentioned in connection with the *H*-theorem in Chapter XI, claims that in a grossly homogeneous condition, kinetic equilibrium is approached: $F \to F_U$ as $t \to \infty$, but we have not proved anything so strong as that. However, MAXWELL's relaxation theorem does support the assertion, since it shows that the quantities of greatest mechanical interest do decrease very rapidly to their values appropriate to kinetic equilibrium. In a much more concrete way than BOLTZMANN's *H*-theorem, they illustrate the *irreversibility* of the behavior of the kinetic gas. This irreversibility is particularly striking if we attempt to trace the origin of a grossly homogeneous condition by considering past times instead of future ones. Indeed *the magnitude of each component of* **P** *and* ^{3}**M** *that is not 0 at* $t = 0$ *tends to* ∞ *as* $t \to -\infty$. Thus in a grossly homogeneous condition any present departure from kinetic equilibrium must be the outcome of still greater departure in the past.

Conclusions drawn from MAXWELL's relaxation theorem are tarnished by a logical gap in the theorem itself. We have considered only the equations of second moments and third moments. The conditions derived are proved necessary for a solution of the kind sought, but it remains possible that the Maxwell-Boltzmann equation might impose restrictions upon them. The general considerations in Section (v) of Chapter III and in Section (i) of Chapter IV show that the initial values $\mathbf{P}(0)$ and $\mathbf{q}(0)$ are indeed arbitrary, but is there a solution F that corresponds to grossly homogeneous conditions for a Maxwellian gas, and does it deliver the results (XIII.10) and (XIII.11) with no restriction upon t and the molecular parameters? While this question is presently open, in Chapter XVIII we shall go a long way toward answering it by showing that the whole infinite system of moments for a Maxwellian gas in a grossly homogeneous condition does have a solution for all time. The results (XIII.10) and (XIII.11) will emerge as portions of that solution.

---

[4] For example, from the viscosity and pressure measured in experiments a time may be found by calculating their ratio (*cf.* (II.52)$_1$). This time is often called "the time of relaxation" and interpreted as if it could be put for τ in (XIII.10), though no such interpretation has ever been justified by the kinetic theory except for Maxwellian molecules or through some purely formal approximation such as that we give in Chapter XVII.

(iii) *Implications of Maxwell's relaxation theorem on constitutive relations in the sense of continuum mechanics*

A constitutive relation in continuum mechanics determines **P** and **q** from the histories of various kinematic and thermodynamic fields, but MAXWELL's relaxation theorem shows that no such determination is possible in the kinetic theory. *No constitutive relation in the sense of continuum mechanics can hold generally according to the kinetic theory*, as we have seen from other reasoning in Section (i) of Chapter IV. On the other hand, the theorem shows that as $t \to \infty$, *the values of* **P** *and* **q** *do tend, without exception, to the values they would have according to ordinary aerostatics*, namely, **0**. Thus, at least in this special case, the kinetic theory does allow gross relations agreeing with those used in continuum mechanics *to emerge as limits in time, for a given gas in a gross condition*: m, g, ρ, and **b** are fixed. Moreover, they emerge also at a fixed time as *asymptotic limits when a molecular parameter or the gross condition of the gas is varied*: If t is fixed at a positive value, then

$$P_{km} \to 0, \qquad q_k \to 0 \quad \text{as} \quad \tau \to 0. \tag{XIII.12}$$

If, however, t is fixed at a negative value, then all the components of **P** and **q** that are not 0 at that time approach $\pm \infty$ as $\tau \to 0$. Looking back at (XIII.3), we see several ways to interpret (XIII.12), of which three are worth emphasis. First, we may hold fixed m and g as well as t and take the limit as $\rho \to \infty$ or $n \to \infty$. Then we consider *a given gas at a given time and conclude that the denser it is, the better the Stokes–Kirchhoff theory applies to it*. Next we may fix ρ and m as well as t and take the limit as g $\to 0$. Then for *Maxwellian gases of the same molecular mass* (XIII.12) asserts that *the less is the intermolecular force per unit mass, the better the Stokes–Kirchhoff theory applies at a given time for a given gross condition*. Third, we may fix ρ or n and g as well as t and take the limit as m $\to 0$. Then we conclude that *for Maxwellian gases of a given intermolecular force per unit mass, the Stokes–Kirchhoff theory applies better at a given time and for a given gross condition, the lighter are the molecules*. All these statements refer to MAXWELL's relaxation theorem. The last, which concerns the limit as m $\to 0$, shows the greatest promise for extension to other circumstances because it does not require that the solution in question exist for all time or that the details of the mechanism of encounter be known. It makes sense for all gases in all conditions.

Finally, gross determinations which look like constitutive relations of continuum mechanics may emerge as properties of special solutions F. In the case of gross equilibrium, any continuum theory of fluids would yield **P** = **0**, **q** = **0**. Though, presumably, infinitely many solutions F are compatible with a grossly homogeneous condition, and MAXWELL's theorem suggests that these yield infinitely many different possible values of **P** and **q**, *one of these F, namely, the*

F_M that is determined uniquely by ρ and ε, does yield the results of the continuum theories of fluids. MAXWELL's relaxation theorem shows that *this solution F_U is the most important of all solutions that correspond to gross free equilibrium*, since it yields right off and for all time the result which follows from any other of them in the limit as $t/\tau \to \infty$.

MAXWELL's relaxation theorem—simple, explicit, elegant—suggests the central problem of the kinetic theory and illustrates the type of results we may hope to formulate and prove. It is older and more telling than BOLTZMANN's H-theorem, but it is restricted to Maxwellian molecules.

(iv) *Maxwell's evaluation of viscosity and thermal conductivity*

MAXWELL's objective in obtaining the specific equations of transfer (XIII.4) and (XIII.7) was to confirm the Navier–Stokes–Fourier constitutive relations (I.7) and (II.9) of nineteenth-century continuum mechanics and to evaluate the viscosities λ and μ and the Maxwell number M as specific rather than general functions of ρ and ε, functions whose nature reflected the molecular constitution of the gas. Indeed, (XIII.4) and (XIII.7)$_1$ seem to deliver **P** and **q**, yet *they do so in terms of their own time rates and of moments of higher order*, ^{3}**M** and ^{4}**M**, respectively.

If in (XIII.4) we throw away all but the last term on the left-hand side, that equation reduces to

$$2pE_{km} = -\tau^{-1}P_{km}. \qquad (XIII.13)$$

This agrees with STOKES's constitutive equation (I.7) for a linearly viscous fluid, provided[5]

$$\varpi = p; \qquad \lambda + \tfrac{2}{3}\mu = 0; \qquad (I.13)_r; (I.12)_r$$

$$\mu = p\tau = \frac{2}{9}\frac{\mathfrak{m}\varepsilon}{\mathbb{A}_{2|2}} = \frac{1}{3\mathfrak{a}}\sqrt{\frac{2\mathfrak{m}^3}{\mathfrak{g}}\frac{2}{3}\varepsilon}. \qquad (XIII.14)$$

These remarkable formulae express MAXWELL's *evaluations of the viscosities of a Maxwellian gas in terms of its molecular constants* $\mathfrak{m}$, $\mathfrak{g}$, and $\mathfrak{a}$. Conversely, if we accept (XIII.14) as providing a correct representation of nature, then, as MAXWELL noticed, we may use a measured value of μ to evaluate τ and so estimate the rates of relaxation according to (XIII.10) and (XIII.11). For air at ordinary atmospheric conditions MAXWELL concluded that $\tau \approx 2 \times 10^{-10}$ seconds; with modern data his calculation would yield approximately half that value.

---

[5] Clearly the conditions (I.12), (I.13), and (XIII.14) are sufficient for agreement. Although the argument here does not prove the relations (I.12) and (I.13) to be also necessary, in Section (iii) of Chapter IV we have shown that they are.

Looking back at the relaxation theorem, with the numerical value just obtained we see that according to MAXWELL's theorem the pressures and energy fluxes in a gas subject to grossly homogeneous conditions at atmospheric pressure will become indistinguishable from those required by classical aerostatics after about 10^{-8} sec at most. We can write (XIII.3)$_2$ in terms of any simultaneous values τ_0, ρ_0 of τ and ρ as follows: $\tau/\tau_0 = \rho_0/\rho$. A vacuum such that $\rho_0/\rho = 10^6$ is easy to produce in laboratories; it corresponds to the density of air about 100 miles above the earth. Under these conditions τ is about 10^{-4} sec if we accept 10^{-10} sec as the value of τ_0. A vacuum such that $\rho_0/\rho = 10^9$ is not uncommon in laboratory conditions. Then τ is roughly 10^{-1} sec, still a short time on the scale of human perception.

If we wish to get a linear relation from a non-linear one, we can always name circumstances in which the non-linear terms are "small" and claim that in those circumstances the non-linear terms can be "neglected". MAXWELL[6] gave arguments of this kind, but we shall not repeat them here, since his celebrated result (XIII.14) may now be derived in two better ways. These two ways we shall present in Chapters XIV and XXIV, respectively. Here we merely note that in order to disregard the terms MAXWELL disregarded, it suffices to replace them by their values in kinetic equilibrium since then $\mathbf{P} = \mathbf{0}$ and $\mathbf{q} = \mathbf{0}$.

The results themselves are in part just what we saw in Chapter IV they would have to be if delivered in any way by the kinetic theory: The thermodynamic and mean normal pressures equal one another, and the bulk viscosity vanishes. In addition, *the shear viscosity is a function of the energetic alone, independent of the density.* This fact, had we known it in advance, would have enabled us to conclude from a dimensional analysis parallel to that which led to (IV.5) but presuming the variable ρ absent from the start, that $\mu \propto \varepsilon$, but of course the specific numerical coefficient in (XIII.14)$_3$ could not be obtained in this way.

To get from (XIII.7)$_1$ an appropriate gross determination for the energy flux, MAXWELL again replaced all terms not proportional to τ^{-1} by their expressions in terms of ρ and ε when $F = F_M$. This time, however, not all terms vanish. Since $p = \tfrac{2}{3}\rho\varepsilon$, we see that

$$-\frac{5p}{2\rho}p_{,k} = -\tfrac{10}{9}\varepsilon(\rho\varepsilon)_{,k},$$

$$= -\tfrac{10}{9}\varepsilon(\varepsilon\rho_{,k} + \rho\varepsilon_{,k}). \qquad \text{(XIII.15)}$$

---

[6] MAXWELL [1867, remarks after Eqs. (122) and (143)]. The process introduced by BOLTZMANN [1872, §III] delivers MAXWELL's results by substituting into the Maxwell–Boltzmann equation the "Ansatz" (VIII.24) with G a quadratic or cubic polynomial, then determining the coefficients of that polynomial so as to make F a solution if terms of degree higher than some specific order are cast away. While it suggests the beginning of a systematic process of approximation, BOLTZMANN's argument is just as unconvincing as MAXWELL's, more complicated, and less perspicuous. In Chapter XXV we shall see that MAXWELL's evaluation of μ and M, the latter given by (XIII.18), can also be viewed as the first step in a systematic procedure.

Also[7], because of (VIII.12)$_1$,

$$\tfrac{1}{2}M_{kaab,b} = \tfrac{10}{9}\varepsilon(\varepsilon\rho_{,k} + 2\rho\varepsilon_{,k}). \qquad \text{(XIII.16)}$$

In obtaining this equation we have assumed that $F = F_M$; for other molecular densities, it does not hold. Adding (XIII.15) to (XIII.16) cancels the density gradients, and all other terms on the left-hand side of (XIII.7)$_1$ vanish when $F = F_M$, so (XIII.7)$_1$ reduces to

$$\tfrac{10}{9}\rho\varepsilon\varepsilon_{,k} = -\tfrac{2}{3}\tau^{-1}q_k. \qquad \text{(XIII.17)}$$

Because of (XIII.14) this result agrees with FOURIER's constitutive relation (II.9) for heat conduction and asserts that

$$\mathbf{M} = \tfrac{5}{2}. \qquad \text{(XIII.18)}$$

Thus the Maxwell number is determined.

Of all results derived from the kinetic theory, this one is the most celebrated. (In fact, MAXWELL[8] obtained $\tfrac{5}{3}$ instead of $\tfrac{5}{2}$ because of a slip in calculation, but the difference is unimportant for matters of principle.) Unlike the formula for μ, this one is free of the adjustable molecular constants $\mathfrak{m}$ and $\mathfrak{g}$. It is free also of the fields ρ and ε that were used in calculating it. Thus it is a *universal relation*, valid for all kinds of Maxwellian molecules and all densities and temperatures. Moreover, $\tfrac{5}{2}$ is close to the magnitudes of the Maxwell number measured in experiments on real monatomic gases.

Surveying MAXWELL's work nearly twenty years after his death, BOLTZMANN[9] wrote

> Many terms of the equations must be neglected in order to obtain the hydrodynamical equations in their usual form. Even if this course in most cases is justifiable, it cannot be rigorously proved that such is the case, and the mathematician is not satisfied. The following question arises, is this a defect of the theory of gases, or is it rather one of hydrodynamics? Are these terms required by the theory of gases not an essential correction of the equations of hydrodynamics?

Today we can phrase these questions more precisely:

(a) By what limit process or method of selection in the kinetic theory do gross determinations of **P** and **q** emerge?

(b) When do these gross determinations reduce to those of the Stokes-Kirchhoff theory, and what are the shear viscosities and Maxwell numbers that

---

[7] MAXWELL [1867, Eqs. (144)–(146)].
[8] MAXWELL [1867, Eq. (149)]. The error was detected by BOLTZMANN [1872, §III], and the details were explained by POINCARÉ [1893].
[9] BOLTZMANN [1895].

thus correspond to kinetic constitutive relations other than those of Maxwellian molecules?

(c) When the Stokes–Kirchhoff theory fails, what replaces it?

To these problems, and to the underlying problem of establishing that solutions of the Maxwell–Boltzmann equation exist, the remaining chapters of this book are devoted.

Chapter XIV

Homo-energetic Simple Shearing of a Gas of Maxwellian Molecules

One way to evaluate the shear viscosity of a fluid is to consider a particular flow in which the phenomenon of shearing is especially simple. As its name suggests, simple shearing is such a flow, and traditionally an analysis of simple shearing has formed a part of every study of the effects of the internal friction of fluids.

(i) Homo-energetic simple shearing

In Chapter II we have considered simple shearing briefly and have calculated the solution corresponding to it in a Stokes-Kirchhoff gas at uniform, time-dependent energetic. The body force was assumed to vanish, and the velocity field was assumed to be

$$u_1 = Kx_2, \quad u_2 = 0, \quad u_3 = 0, \quad K = \text{const.}, \qquad (\text{II}.57)_r$$

the *amount of shearing* K being taken as positive. We then considered the field equations expressing the balance of mass, momentum, and energy. We showed that for the flow in question, under the assumptions

$$\text{div } \mathbf{P} = \mathbf{0}, \quad \text{div } \mathbf{q} = 0, \qquad (\text{II}.18)_r$$

those equations reduced to the one simple equation (II.59). Introducing the numbers τ and T defined as follows:

$$\tau \equiv \frac{\mu}{p} = \frac{\mu}{(\gamma - 1)\rho\varepsilon}; \qquad (\text{II}.52)_r$$

$$\mathsf{T} \equiv \tau K, \qquad (\text{II}.60)_r$$

we expressed the governing equation (II.59) in the form

$$\tau\dot{\varepsilon}/\varepsilon = \tau\dot{p}/p = -(\gamma - 1)\mathsf{T}P_{12}/p. \qquad (\text{II}.61)_r$$

We noted also that if μ, which in this general part of the analysis was an arbitrary function of ρ and ε, should happen to be a linear function of ε alone, then both τ and T would reduce to constants. All this we did without use of any constitutive relation beyond the thermal equation of state of an ideal gas.

To this basic corpus we then applied the constitutive equations of the Stokes-Kirchhoff theory. They rendered the assumptions (II.18) satisfied; with the function μ taken as the shear viscosity of the gas and then assumed to be indeed a linear function of ε alone, they led to the explicit solutions (II.62) and (II.63), which determine ε and p and **P** in terms of the initial values $\varepsilon(0)$ and $p(0)$.

We shall now solve a corresponding problem for the kinetic gas of Maxwellian molecules. The same basic corpus, being a consequence only of the equations of balance, remains at our disposal. For the function μ, of course, we shall take that which MAXWELL's heuristic analysis has suggested, indeed a linear function of ε:

$$\mu = p\tau = \frac{2}{9}\frac{\mathfrak{m}\varepsilon}{\mathbb{A}_{2|2}} = \frac{1}{3\mathfrak{a}}\sqrt{\frac{2\mathfrak{m}^3}{\mathfrak{g}}}\frac{2}{3}\varepsilon, \qquad (XIII.14)_r$$

and for τ the corresponding special case of (II.52):

$$\tau \equiv \frac{1}{3n\mathbb{A}_{2|2}} = \frac{1}{3\mathfrak{a}}\sqrt{\frac{2\mathfrak{m}^3}{\mathfrak{g}}}\frac{1}{\rho}. \qquad (XIII.3)_r$$

Thus we shall have at our disposal the special case of (II.61) in which $\gamma = \frac{5}{3}$, and both T and τ will be constants.

For the kinetic gas, nevertheless, even one composed of Maxwellian molecules, the problem cannot be so easily specified, for our only information toward determining **P** and **q** is contained in (XIII.4) and (XIII.7)$_1$. In the former appears $M_{kma,a}$, and so something about the third moments ^{3}**M** must be known before **P** can be determined. Similarly, in the latter equation appears $M_{kaab,b}$, so something about the fourth moments ^{4}**M** must be known before (XIII.7)$_1$ can be used to find **q**. In special cases these troublesome terms may vanish. One such case is provided by grossly homogeneous conditions, which we began to study in the preceding chapter. It is not necessary to limit attention to this case, however, for the first appearance of the higher moments is in the equation of second moments; they do not appear in the preceding ones. Accordingly we shall suppose[1] that *the relative moments of orders 2, 3, and 4 are*

---

[1] Except as specifically noted to the contrary, all the analysis in this chapter is taken from the paper of TRUESDELL [1956, Chapter V], but only a part of the results obtained there are reported here. TRUESDELL [1956, §32] remarked that his approach would lead to exact solutions corresponding to all the steady homo-energetic affine flows (that is, **G**(*t*) = **G**(0) and **G**(0)^2 = **0** in (II.41)), but he did not work out the details. At about the same time GALKIN [1956] obtained this class of solutions, but he did not develop the properties of the solution for shearing or note its implications. In Chapter XV we shall present a general analysis that subsumes GALKIN's and TRUESDELL's.

functions of time only. Then $M_{kma,a} = 0$, and $M_{kaab,b} = 0$. If there are any solutions of this kind, for them (XIII.4) and (XIII.7) reduce to a system of differential equations for the moments **P** and ³**M** alone. In particular, the essential conditions (II.18) are satisfied, and so (II.61) with constant τ and T is again at our disposal.

At the corresponding point in Chapter II we used the constitutive relations of the Stokes-Kirchhoff gas. Now we have neither them nor any other constitutive relations of continuum mechanics. We must turn instead to MAXWELL's differential system (XIII.4) for the second moments, simplified by our present assumptions. That system, preceded by the appropriate special case of (II.61), we express neatly in terms of τ and T:

$$\tau\dot{p} + \tfrac{2}{3}TP_{12} = 0,$$
$$\tau\dot{P}_{12} + P_{12} + T(p + P_{22}) = 0,$$
$$\tau\dot{P}_{22} + P_{22} - \tfrac{2}{3}TP_{12} = 0, \qquad (XIV.1)$$
$$\tau\dot{P}_{11} + P_{11} + \tfrac{4}{3}TP_{12} = 0,$$
$$\tau\dot{P}_{13} + P_{13} + TP_{23} = 0,$$
$$\tau\dot{P}_{23} + P_{23} = 0.$$

This system of six ordinary differential equations with constant coefficients determines the six components of the pressure tensor $\mathbf{M}(t)$ that corresponds to the arbitrary initial value $\mathbf{M}(0)$.

(ii) The pressures as functions of time

The last two members of the system (XIV.1) can be integrated at sight[2]:

$$P_{23} = P_{23}(0)e^{-t/\tau},$$
$$P_{13} = [P_{13}(0) - (t/\tau)TP_{23}(0)]e^{-t/\tau}. \qquad (XIV.2)$$

We may eliminate P_{12} between (XIV.1)$_3$ and (XIV.1)$_4$, thus obtaining a differential equation which may be integrated at once to yield

$$P_{11} + 2P_{22} = [P_{11}(0) + 2P_{22}(0)]e^{-t/\tau},$$
$$P_{22} - P_{33} = [P_{22}(0) - P_{33}(0)]e^{-t/\tau}; \qquad (XIV.3)$$

the second formula follows from the first and the fact that $P_{aa} = 0$. These are the uninteresting parts of the solution, the parts independent of T or little

---

[2] Eqs. (XIV.1)$_5$ and (XIV.2)$_2$ correct their counterparts in the paper of TRUESDELL [1956, Eqs. (33.3)$_5$ and (33.4)$_1$]. The error there has no effect on any of the other conclusions drawn by TRUESDELL in that paper.

influenced by its value, since all the combinations of components of **P** thus obtained relax to zero much as they do for a grossly homogeneous condition.

The interesting parts of the solution are obtained from the first three members of (XIV.1), which form a system to determine p, P_{22}, and P_{12}. We see easily that there are solutions proportional to $\exp(\chi t/\tau)$ if the growth modulus χ satisfies the cubic equation

$$\chi(\chi + 1)^2 = \tfrac{2}{3}\mathsf{T}^2. \qquad (\text{XIV}.4)$$

So as to exclude the case in which $K = 0$, which we treated in the preceding chapter, *we shall assume henceforth, without restatement, that* $\mathsf{T} > 0$. Then the cubic equation (XIV.4) has one real root R; as a function of T, that root R increases monotonically and is unbounded. Of course it is easy, by use of Cardano's formulae, to exhibit R explicitly as an algebraic function of T, but we do not need the result here. Directly from (XIV.4) we see that R has an expansion in powers of T, *an expansion which converges if* $\mathsf{T} < \sqrt{2/3}$ *and diverges if* $\mathsf{T} > \sqrt{2/3}$. In particular,

$$R = \tfrac{2}{3}\mathsf{T}^2(1 - \tfrac{4}{3}\mathsf{T}^2) + O(\mathsf{T}^6) \qquad \text{as} \quad \mathsf{T} \to 0, \qquad (\text{XIV}.5)$$

so the result $(\text{II}.62)_3$ of the Stokes–Kirchhoff theory is recovered as an approximation valid for small values of T. The other two roots of (XIV.4) are $Q \pm iJ\mathsf{T}$, in which $Q = -(1 + \tfrac{1}{2}R)$ and $J\mathsf{T} = \sqrt{R(1 + \tfrac{3}{4}R)}$. Figure XIV.1 presents numerical values of R and J for a large range of values of T.

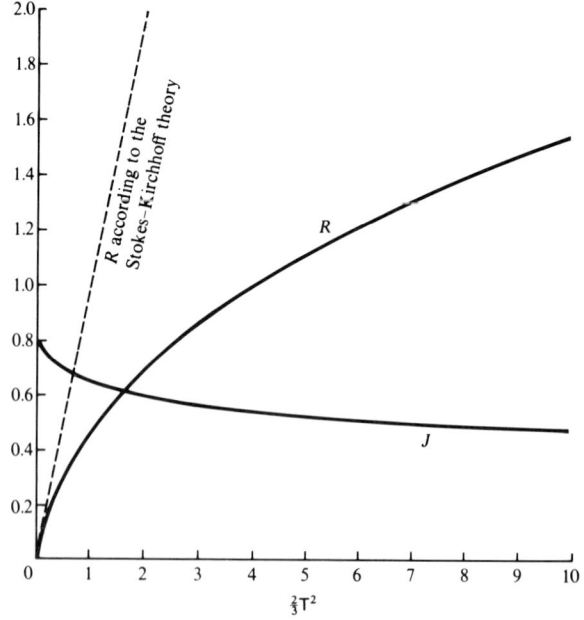

Figure XIV.1 The amplification factor R and the frequency factor J as functions of $\tfrac{2}{3}\mathsf{T}^2$

(ii) PRESSURES AS FUNCTIONS OF TIME

Routine calculation shows that the general solution of the system for p, P_{12}, and P_{22} is

$$\frac{p}{p(0)} = Ae^{Rt/\tau} + e^{-(1+\frac{1}{2}R)t/\tau}[B \cos JKt + C \sin JKt],$$

$$\frac{P_{12}}{p(0)} = \frac{1}{\frac{2}{3}T}\{-ARe^{Rt/\tau} + e^{-(1+\frac{1}{2}R)t/\tau}[(\{1 + \tfrac{1}{2}R\}B - JTC) \cos JKt$$

$$+ (\{1 + \tfrac{1}{2}R\}C + JTB) \sin JKt]\}, \tag{XIV.6}$$

$$\frac{P_{22}}{p(0)} = -\frac{1}{R+1}\bigg\{ARe^{Rt/\tau} + e^{-(1+\frac{1}{2}R)t/\tau}$$

$$\times \bigg[\bigg(\{\tfrac{3}{2} + R\}B + \frac{JT}{R}C\bigg) \cos JKt$$

$$+ \bigg(\{\tfrac{3}{2} + R\}C - \frac{JT}{R}B\bigg) \sin JKt\bigg]\bigg\},$$

in which the constants A, B, and C satisfy the following system of linear equations:

$$1 = A + B,$$

$$\frac{\tfrac{2}{3}TP_{12}(0)}{p(0)} = -AR + (1 + \tfrac{1}{2}R)B - JTC, \tag{XIV.7}$$

$$-\frac{(1+R)P_{22}(0)}{p(0)} = AR + (\tfrac{3}{2} + R)B + \frac{JT}{R}C.$$

Since the determinant of this system is $JT(3 + 1/R)$, when $T > 0$ unique values of A, B, and C are determined by the quantities $P_{12}(0)/p(0)$, $P_{22}(0)/p(0)$, and T. Directly from (XIV.6) it follows that $A > 0$, for otherwise the condition $p(t) > 0$ is violated as $t \to \infty$. The condition $A > 0$ is sufficient to satisfy also the requirement that $p + P_{22} > 0$, at least as $t \to \infty$.

From (XIV.6) we see that the behavior exemplified by MAXWELL's relaxation theorem need not hold in circumstances that are not grossly homogeneous. Since $A > 0$, the pressures $p(t)$, $P_{12}(t)$, and $P_{22}(t)$ do not approach any limits as $t \to \infty$. In this flow there is no trend to equilibrium, nor should there be, for work must be done continually on the gas in order to maintain the flow. A part of the solution, however, does approach zero: the *damped part*, namely, the summands proportional to $\exp[-(1 + \tfrac{1}{2}R)t/\tau]$. The remainder of the solution, since it is proportional to $\exp(Rt/\tau)$, increases exponentially as $t \to \infty$. We may call this part the *dominant part*, because for all choices of B and C it is the asymptotic form of the solution as $t/\tau \to \infty$.

(iii) The dominant pressures and their gross determination

MAXWELL's relaxation theorem shows us that if we are to find any agreement between the kinetic theory and the constitutive equations of continuum mechanics, we must employ either limit forms of solutions or special solutions. Accordingly, for such comparisons we shall limit our attention to the dominant parts, which we shall denote by attaching the superscript D. From (XIV.2), (XIV.3), and (XIV.6) we may write these down:

$$\frac{p^D}{p(0)} = A e^{Rt/\tau},$$

$$\frac{P^D_{12}}{p(0)} = \frac{-AR}{\frac{2}{3}T} e^{Rt/\tau},$$

$$\frac{P^D_{22}}{p(0)} = \frac{-AR}{R+1} e^{Rt/\tau}, \qquad \text{(XIV.8)}$$

$$P^D_{33} = P^D_{22},$$

$$P^D_{11} = -2P^D_{22},$$

$$P^D_{13} = P^D_{23} = 0.$$

All the dominant parts are time-dependent, and all depend upon the initial conditions, but all do so only through the common factor $p(0)Ae^{Rt/\tau}$. The constants B and C in (XIV.6), which are determined from the initial value $\mathbf{P}(0)/p(0)$ through (XIV.7), do not affect the determination of $\mathbf{P}^D(t)/p^D(t)$. The condition $A > 0$ makes all dominant normal pressures positive always.

In continuum mechanics the problem of viscometric flow has been studied intensively. For the simple fluid of NOLL, the most general possible system of pressures in a simple shearing of any kind is as follows[3]:

$$P_{12} = -\sigma_0(K; \rho, \varepsilon),$$

$$P_{22} - P_{33} = -\sigma_1(K; \rho, \varepsilon), \qquad \text{(XIV.9)}$$

$$P_{11} - P_{33} = -\sigma_2(K; \rho, \varepsilon);$$

σ_0, σ_1, and σ_2 are the *viscometric functions* of the fluid. As a function of K, the shear stress function σ_0 is odd, while the *normal-stress difference functions* σ_1 and σ_2 are even.

It is easy to see that (XIV.8) may be interpreted as a special case of (XIV.9).

---

[3] E.g. §111 of the book of TRUESDELL & NOLL cited in Footnote 4 to Chapter I. Dependence upon θ or ε as a parameter is not considered explicitly there but is common in later works on continuum thermodynamics.

If we use the symbol $o(1)$ to denote a quantity that approaches 0 as $t \to \infty$, then by use of (III.15) in (XIV.6)$_{1,2}$ we see that

$$P_{12} = -\rho\varepsilon \frac{R}{T} + o(1). \qquad \text{(XIV.10)}$$

We may express this result and its counterparts for the differences $P_{22} - P_{33}$ and $P_{11} - P_{33}$ as follows[4]: *Asymptotically as $t \to \infty$, the pressures in the kinetic gas of Maxwellian molecules in homo-energetic simple shearing are those of a simple fluid whose viscometric functions are*

$$\sigma_0 = \rho\varepsilon \frac{R}{T},$$

$$\sigma_1 = 0, \qquad \text{(XIV.11)}$$

$$\sigma_2 = -2\rho\varepsilon \frac{R}{1+R}.$$

Because R is a function of T alone, the right-hand sides of these expressions are functions of ρ, ε, and T. Because $T = \tau K$ and because, as we see from (XIII.3) and (II.57), $\tau \propto 1/\rho$ and $K = u_{1,2}$, those right-hand sides equal functions of ρ, grad $\mathbf{u}$, and ε. The formulae (XIV.11) show, then, that *the dominant part of the pressure tensor of the Maxwellian gas in homo-energetic simple shearing is grossly determined at $\mathbf{b} = \mathbf{0}$*. The particular functions appearing on the right-hand side of (XIV.11) are the *gross determiners* of the asymptotic system of pressures as $t \to \infty$, always supposing that $\mathbf{b} = \mathbf{0}$.

If $\mathbf{b} \neq \mathbf{0}$, we see from (IX.4) that grad p + div $\mathbf{P} \neq \mathbf{0}$. Thus the method used here gives us no information. We do not know whether simple shearing be possible subject to a general $\mathbf{b}$ or not. Thus we cannot determine generally whether any grossly determined functions be associated with simple shearing. For the same reason we are ignorant of momentally determined functions.

(iv) Definition and rigorous evaluation of the viscosity of the kinetic gas

One method of defining the shear viscosity μ_S in continuum mechanics is as the ratio of the shear pressure P_{12} to the negative of the shearing K in a simple shearing, for small values of K:

$$\mu_S \equiv -\lim_{K \to 0} \frac{P_{12}}{K}. \qquad \text{(XIV.12)}$$

---

[4] TRUESDELL [1969, *1*, Eq. (10.25)].

From (XIV.11) we see that such a limit should exist, provided we replace P_{12} by its dominant part P_{12}^D. Indeed, by use of (XIV.5) we see from (XIV.11) that for fixed τ, that is, for a given Maxwellian gas at a given density,

$$\sigma_0 = pT + O(T^3) = p\tau K + O(T^3),$$
$$\sigma_2 = -2pT^2 + O(T^4) \qquad \text{(XIV.13)}$$

as $T \to 0$. Comparison of (XIV.13)$_1$ with (XIV.12) yields $\mu_S = \mu$, which is MAXWELL's result (XIII.14) for μ. Thus in homo-energetic simple shearing *the gas of Maxwellian molecules never obeys the Stokes–Kirchhoff constitutive relation* (I.7)$_2$ *exactly, but it does so as an approximation when* T *is small*, τ being kept fixed. This result expresses just what the theorem on the tension number in Section (iv) of Chapter IV would lead us to expect. More than that, it gives *a precise, viscometric status*[5] to the quantity μ introduced formally through (XIII.13) and (XIII.14).

The measured values of μ for gases at ordinary temperatures are about 2×10^{-4} dyne·sec·cm^{-2}, very nearly independent of density. Thus a shearing of about 50 sec^{-1} corresponds roughly to the value $T = 10^{-2}$.

(v) Reduced viscometric functions of the Maxwellian gas

Of greater interest than (XIII.14) are the exact forms (XIV.11) for σ_0, σ_1, and σ_2. We note that while in the limit as $T \to 0$ the dependence of the shear stress on the density disappears, for general values of T the functions σ_0 and σ_2 are not constant and hence do depend upon ρ, because, as we see from (II.60) and (XIII.3), T is proportional to K/ρ. In particular, *for fixed p and small* T *the normal-stress difference function* σ_2 *is ultimately proportional to* K^2/ρ^2. Thus, other things being equal, appreciable normal-stress differences are obtained by rarefying the gas. The exact dependence upon ε is explicit: *All three viscometric functions of a Maxwellian gas are proportional to the energetic.* Since this dependence, which generalizes MAXWELL's conclusion that $\mu \propto \varepsilon$, is common, it can be divided out. We therefore introduce as follows the *dimensionless reduced viscometric functions*[6]:

$$V(T) \equiv \frac{\sigma_0}{p} = \frac{3}{2}\frac{R}{T},$$
$$N_1(T) \equiv \frac{\sigma_1}{p} = 0, \qquad \text{(XIV.14)}$$
$$N_2(T) \equiv \frac{\sigma_2}{p} = -\frac{3R}{1+R}.$$

---
[5] TRUESDELL [1973, Chapter XIII].
[6] TRUESDELL [1969, *1*, Eq. (10.28)].

(vi) COMPARISON WITH STOKES–KIRCHHOFF THEORY

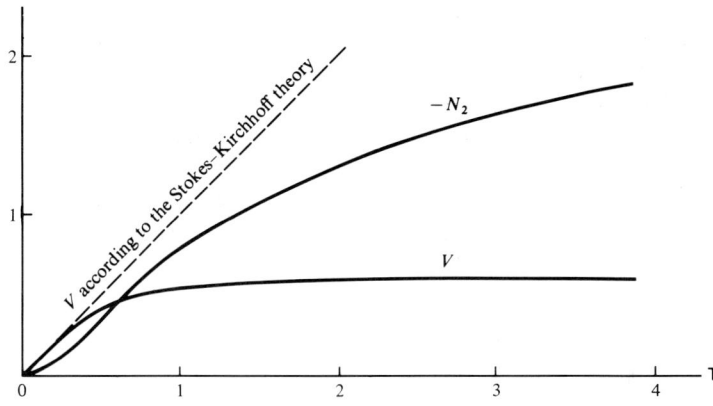

Figure XIV.2 Reduced viscometric functions of a Maxwellian gas

These may be presented on a diagram (Figure XIV.2). The shear-stress function V falls rapidly below its value for the Stokes–Kirchhoff theory, which is of course T. It experiences a maximum value $\sqrt{3/8}$ at $T = \sqrt{6}$, after which it tends to 0 as $T \to \infty$. The normal-stress difference function N_2, which is identically 0 in the Stokes–Kirchhoff theory, decreases fairly rapidly to the value -3. In Chapter IV we have seen that the kinetic theory cannot agree with the Stokes–Kirchhoff theory if $T \geq 1$. For the problem in question here, it never agrees exactly, but the viscometric functions of the two theories approximate each other fairly well if $T < 0.1$, though even in that range the dominant normal pressures M_{11}^D and M_{33}^D begin to differ noticeably from each other. In one particular regard, however, the agreement between the kinetic theory and the Stokes–Kirchhoff theory is exact: Since $\sigma_1 = 0$, the *normal pressures on all planes parallel to the velocity are equal*.

(vi) *Comparison of the pressures as functions of time with their counterparts according to the Stokes–Kirchhoff theory*

We have seen that for small values of T the Stokes–Kirchhoff constitutive relation is borne out fairly well by the kinetic theory. This does not at all imply that other aspects of the two theories be in good agreement. To understand that there can be a difference, we need only recall how the Stokes–Kirchhoff theory is used. The constitutive relation for the pressure deviator **P** is substituted into (I.2), which expresses the balance of linear momentum; in this way result the differential equations of motion, commonly called "the Navier–Stokes equations". These equations are then solved to yield, among other things, the pressures as functions of x and t. However, as a glance at (XIV.10) shows, to obtain viscometric functions we began with (XIV.6), which expresses the pressures as functions of time according to the kinetic theory, and we let t approach ∞. Thereafter, to get a range of agreement with the Stokes–Kirchhoff theory, we let T approach 0. There is no reason to expect that if we first let T approach 0 so as to obtain the constitutive relation, then calculate the corresponding solution as a function of t, agreement should again ensue as $t \to \infty$.

Indeed, it does not. In Chapter II we have calculated the solution for homo-energetic shearing according to the Stokes–Kirchhoff theory of a gas whose viscosity is proportional to ε. Taking γ as $\frac{5}{3}$, we may write the solutions (II.62) and (II.63) in the forms

$$p^{S-K} = p(0) e^{\frac{2}{3} T^2 t/\tau},$$

$$P_{12}^{S-K} = -p(0) T e^{\frac{2}{3} T^2 t/\tau}. \qquad (XIV.15)$$

Comparison with (XIV.8)$_{1,2}$ yields

$$\frac{p^{S-K}}{p^D} = \frac{1}{A} e^{(\frac{2}{3}T^2 - R)t/\tau},$$

$$\frac{P_{12}^{S-K}}{P_{12}^D} = \frac{\frac{2}{3}T^2}{AR} e^{(\frac{2}{3}T^2 - R)t/\tau}. \qquad (XIV.16)$$

Now from (XIV.4) it is obvious that $R < \frac{2}{3}T^2$. Therefore, if T is fixed,

$$\frac{P_{12}^{S-K}(t)}{P_{12}^D(t)} \to \infty \qquad \text{as} \quad t \to \infty. \qquad (XIV.17)$$

The same result holds for all the other non-vanishing components of **M**. *No matter how small is* T, *the Stokes–Kirchhoff solutions for the pressures become infinitely larger than the corresponding kinetic pressures as* $t \to \infty$. Thus the Stokes–Kirchhoff solution of the initial-value problem can give good results *at best for a short time*[7]. These results correspond to a fact well known from

---

[7] For a numerical estimate, let us suppose the initial conditions so adjusted that $P_{12}^{S-K}(0)/P_{12}^D(0) \approx 1$; if T is small, we see from (XIV.16) and (XIV.5) that this agreement obtains very nearly if $A = 1$. Then, again for small T, from (XIV.16) and (XIV.5) we find that

$$\frac{t}{\tau} \approx \frac{9}{8} T^{-4} \left(\frac{P_{12}^{S-K}(t)}{P_{12}^D(t)} - 1 \right).$$

For example, if we use the rough data adduced at the end of Section (iv) of this chapter, we conclude that a shearing of 50 sec^{-1} gives T the very small value 10^{-2}, whatever be the density of the gas. Then after about 10^8 times of relaxation $P_{12}^{S-K}(t)$ becomes double the correct value $P_{12}^D(t)$. Thus if for τ we adopt the value 10^{-10} sec, which we saw in Section (iv) of Chapter XIII to be appropriate to normal conditions, the lapse of time needed to produce an error of 100% is 10^{-2} sec. In a vacuum such that $\rho_0/\rho = 10^6$, no great strain upon laboratory conditions, the same error occurs after 10^{-8} sec, or, if we prefer, after 10^{-2} sec when the shearing has the very small value 5×10^{-5} sec^{-1}.

(vii) FAST SHEARING, RAREFIED GASES

continuum mechanics: An excellent approximation to the constitutive equation of a material may provide a poor approximation to the solution of the corresponding initial-value problem as t increases.

There remains, however, a possible sense of agreement between the solutions of the two initial-value problems. Namely, we may hold t fixed and take limits as $\mathsf{T} \to 0$. If in addition we hold fixed the two ratios $P_{12}(0)/p(0)$ and $P_{22}(0)/p(0)$, by use of (XIV.5) and the expression given below it for $J\mathsf{T}$ we see from (XIV.7) that

$$A \to 1, \qquad B \to 0, \qquad C \to 0. \qquad (XIV.18)$$

Using these results and (XIV.15) in (XIV.6), we see that for a fixed value of t/τ

$$\frac{p^{S-K}}{p} \to 1, \qquad \frac{P_{12}^{S-K}}{P_{12}} \to 1 \quad \text{as} \quad \mathsf{T} \to 0. \qquad (XIV.19)$$

Therefore if t/τ is held fixed, *not only does the Stokes–Kirchhoff constitutive relation emerge asymptotically as* $\mathsf{T} \to 0$, *but so also does the Stokes–Kirchhoff solution.*

(vii) Asymptotic forms for fast shearing or rarefied gases

Our study of limiting forms of the results (XIV.6) has referred only to small values of T, they being the only ones for which any agreement, however limited, with classical gas dynamics could be expected. One way to obtain results appropriate to a very rarefied gas, or to very fast shearings, is to consider limits as $\mathsf{T} \to \infty$.

First of all, from (XIV.4) it is not difficult to show that

$$R = \sqrt[3]{\tfrac{2}{3}}\mathsf{T}^{2/3} - \tfrac{2}{3} + O(\mathsf{T}^{-2/3}) \qquad \text{as} \quad \mathsf{T} \to \infty, \qquad (XIV.20)$$

and from (XIV.7), if $M_{22}(0)/p(0)$ and $P_{12}(0)/p(0)$ are held fixed, that

$$\frac{AR}{\tfrac{2}{3}\mathsf{T}} = \frac{1}{\sqrt{6}} (\tfrac{2}{3}\mathsf{T}^2)^{1/6} \frac{M_{22}(0)}{p(0)} - \frac{1}{3}\frac{P_{12}(0)}{p(0)} + O(\mathsf{T}^{-1/3}). \qquad (XIV.21)$$

The exact expressions (XIV.6) for the components of **M** are not functions of T alone but of T, t/τ, and Kt. From (II.60) we see that $\mathsf{T} = (\tau/t)(Kt)$; therefore, in order that $\mathsf{T} \to \infty$ it is necessary that either $\tau/t \to \infty$ or $Kt \to \infty$. The results that hold when τ/t is held fixed and $Kt \to \infty$ would seem appropriate to very rapid shearing; those for which Kt is held fixed and $\tau/t \to \infty$, to a very rarefied gas.

In the former case we easily see from (XIV.6) that the damped parts of $\mathbf{M}(t)$ are asymptotically null and that the dominant parts have the following asymptotic behavior:

$$p \sim \tfrac{1}{3}\sqrt[3]{\tfrac{2}{3}}\mathsf{T}^{2/3} M_{22}(0) \exp[(\sqrt[3]{\tfrac{2}{3}}\mathsf{T}^{2/3} - \tfrac{2}{3})t/\tau],$$
$$P_{12} \sim -(\tfrac{3}{2})^{2/3}\mathsf{T}^{-1/3} p(t),$$
$$M_{22} \sim \sqrt[3]{\tfrac{3}{2}}\mathsf{T}^{-2/3} p(t), \qquad (XIV.22)$$
$$M_{11} \sim 3p(t).$$

In this limiting case the normal pressure M_{11} on the planes normal to the flow becomes enormously greater than all other components of **M**. The magnitudes of all components of **M** are proportional to the initial normal pressure $M_{22}(0)$ on the planes being sheared. In interpreting these results we must recall that the whole system of initial pressures **M**(0) has been held fixed in the calculation of the asymptotic forms.

In the latter case, that appropriate to a very rarefied gas, the damped parts of (XIV.6) do not vanish in the limit, so a more delicate calculation is needed[8]. The result is

$$p \to p(0) - \tfrac{2}{3}P_{12}(0)Kt + \tfrac{1}{3}M_{22}(0)K^2t^2,$$

$$P_{12} \to P_{12}(0) - M_{22}(0)Kt,$$

$$M_{22} \to M_{22}(0), \qquad (XIV.23)$$

$$M_{11} \to M_{11}(0) - 2P_{12}(0)Kt + M_{22}(0)K^2t^2.$$

These limit values are just what a direct calculation delivers[9] for *free molecular flow*, that is, a flow governed by (III.37) instead of (III.38). When $\mathbf{b} = \mathbf{0}$, we may use (III.32) to reduce (III.37) to

$$F(t, \mathbf{x}, \mathbf{v}) = F_0(\mathbf{x} - t\mathbf{v}, \mathbf{v}). \qquad (XIV.24)$$

The discussion below in Section (viii) of Chapter XXIII suggests that for a homo-energetic shearing F_0 should be a function of **c** alone. Then there is a function G such that

$$F_0(\mathbf{x}, \mathbf{v}) = G(\mathbf{v} - \mathbf{u}(\mathbf{x})),$$

$$= G(v_1 - Kx_2, v_2, v_3), \qquad (XIV.25)$$

so (XIV.24) yields

$$F(t, \mathbf{x}, \mathbf{v}) = G(v_1 - K[x_2 - tv_2], v_2, v_3),$$

$$= G(c_1 + Ktc_2, c_2, c_3), \qquad (XIV.26)$$

also a function of **c** alone. We easily show that the functions p, P_{12}, M_{22}, and M_{11} corresponding to this F are precisely the right-hand sides of (XIV.23).

(viii) Solution for the energy flux. Instability

So far, we have discussed only the pressure tensor **M**. Since grad $\varepsilon = 0$ in the flow we are considering, in continuum thermomechanics we expect that the energy flux of such a flow shall vanish. Indeed, this is so not only according to the Fourier constitutive relation (II.9), but also, as a consequence of the Clausius–Duhem inequality, for a very broad class of continuum theories. We have assumed that $^3\mathbf{M}$ is independent of **x**, but it will not generally be independent also of t. From MAXWELL's relaxation theorem we know that in the kinetic theory the best we can expect is that $\mathbf{q}(t) \to \mathbf{0}$ as $t/\tau \to \infty$. In order to determine **q** in homo-energetic shearing, we need the equation of transfer (XIII.5) for all

---

[8] Some intermediate steps are presented by TRUESDELL [1956, §46], but some of his numerical coefficients are wrong, and so is his summary of the results.
[9] NIKOL'SKII [1965, 1].

(viii) ENERGY FLUX

the third moments. In terms of the notation

$$P_{0|kmr} \equiv M_{kmr} - \tfrac{1}{5}(M_{aak}\delta_{mr} + M_{aar}\delta_{km} + M_{aam}\delta_{kr}), \qquad \text{(XIII.6)}_r$$

it reduces to the following differential system:

$$\tau\dot{q}_1 + \tfrac{2}{3}q_1 + \tfrac{7}{5}Tq_2 + TP_{0|112} = 0,$$
$$\tau\dot{q}_2 + \tfrac{2}{3}q_2 + \tfrac{2}{5}Tq_1 + TP_{0|221} = 0,$$
$$\tau\dot{q}_3 + \tfrac{2}{3}q_3 + TP_{0|123} = 0,$$
$$\tau\dot{P}_{0|123} + \tfrac{3}{2}P_{0|123} + TP_{0|223} + \tfrac{2}{5}Tq_3 = 0,$$
$$\tau\dot{P}_{0|223} + \tfrac{3}{2}P_{0|223} - \tfrac{2}{5}TP_{0|123} = 0,$$
$$\tau\dot{P}_{0|113} + \tfrac{3}{2}P_{0|113} + \tfrac{8}{5}TP_{0|123} = 0,$$
$$\tau\dot{P}_{0|333} + \tfrac{3}{2}P_{0|333} - \tfrac{6}{5}TP_{0|123} = 0, \qquad \text{(XIV.27)}$$
$$\tau\dot{P}_{0|332} + \tfrac{3}{2}P_{0|332} - \tfrac{2}{5}TP_{0|221} - \tfrac{4}{25}Tq_1 = 0,$$
$$\tau\dot{P}_{0|112} + \tfrac{3}{2}P_{0|112} + \tfrac{8}{5}TP_{0|221} + \tfrac{16}{25}Tq_1 = 0,$$
$$\tau\dot{P}_{0|222} + \tfrac{3}{2}P_{0|222} - \tfrac{6}{5}TP_{0|221} - \tfrac{12}{25}Tq_1 = 0,$$
$$\tau\dot{P}_{0|221} + \tfrac{3}{2}P_{0|221} + TP_{0|222} - \tfrac{2}{5}TP_{0|112} + \tfrac{16}{25}Tq_2 = 0,$$
$$\tau\dot{P}_{0|331} + \tfrac{3}{2}P_{0|331} + TP_{0|332} - \tfrac{2}{5}TP_{0|112} - \tfrac{4}{25}Tq_2 = 0,$$
$$\tau\dot{P}_{0|111} + \tfrac{3}{2}P_{0|111} + \tfrac{9}{5}TP_{0|112} - \tfrac{12}{25}Tq_2 = 0.$$

Since $P_{0|kmr}$ is symmetric with respect to k, m, and r, these equations govern all the components of the third moment. Because of the identities

$$P_{0|111} + P_{0|122} + P_{0|133} = 0,$$
$$P_{0|211} + P_{0|222} + P_{0|233} = 0, \qquad \text{(XIV.28)}$$
$$P_{0|311} + P_{0|322} + P_{0|333} = 0,$$

the system (XIV.27) is of 10th order. It may be split into smaller systems that can be solved separately.

As they stand, (XIV.27)$_{3,4,5}$ form a system for q_3, $P_{0|123}$, and $P_{0|223}$. The characteristic equation of this system is

$$(\chi + \tfrac{3}{2})^2(\chi + \tfrac{2}{3}) = \tfrac{1}{3}T^2. \qquad \text{(XIV.29)}$$

To solve the remaining equations[10], we first eliminate $P_{0|123}$ from (XIV.27)$_{5,6}$ to obtain an equation for $P_{0|223} + \tfrac{1}{4}P_{0|113}$, and solve it; then eliminate $P_{0|122}$ and q_1 from (XIV.27)$_{8,9}$ to obtain an equation for $P_{0|233} + \tfrac{1}{4}P_{0|112}$, and solve

[10] MUNCASTER [1974].

that. Then we apply (XIV.28)$_{2,3}$ so as to determine $P_{0|333}$ and $P_{0|222}$. Thereafter we may eliminate $P_{0|222}$ between these results and (XIV.27)$_{11}$. We obtain the equation

$$\tau \dot{P}_{0|122} + \tfrac{3}{2}P_{0|122} - \tfrac{23}{30}TP_{0|112} + \tfrac{16}{25}Tq_2 = NTe^{-3t/\tau}, \qquad \text{(XIV.30)}$$

N being a constant of integration. This equation and (XIV.27)$_{1,2,9}$ constitute an inhomogeneous system of fourth order for $q_1, q_2, P_{0|112}$, and $P_{0|122}$. There is a particular integral proportional to $\exp(-\tfrac{3}{2}t/\tau)$; the corresponding homogeneous system has the characteristic equation

$$(\chi + \tfrac{3}{2})^2(\chi + \tfrac{2}{3})^2 - 2T^2(\chi + \tfrac{31}{36}) = 0. \qquad \text{(XIV.31)}$$

All the characteristic roots[11] of the system of moments of third order besides $-\tfrac{3}{2}$ are given by the roots of (XIV.29) and (XIV.31). The former equation has one and only one real root R_1. This root is an increasing function of T for which

$$\begin{aligned} R_1(0) &= -\tfrac{2}{3}, & R_1(T) &< 0 & \text{if } 0 \le T < 3/\sqrt{2}, \\ R_1(3/\sqrt{2}) &= 0, & R_1(T) &> 0 & \text{if } 3/\sqrt{2} < T. \end{aligned} \qquad \text{(XIV.32)}$$

The two complex roots of (XIV.29) have a negative real part which decreases monotonically with increasing T. A similar result follows from (XIV.31). Namely, there are two real roots, one of which is always negative and one, R_2 say, which is an increasing function of T such that

$$\begin{aligned} R_2(0) &= -\tfrac{2}{3}, & R_2(T) &< 0 & \text{if } 0 \le T < \sqrt{\tfrac{18}{31}}, \\ R_2(\sqrt{\tfrac{18}{31}}) &= 0, & R_2(T) &> 0 & \text{if } \sqrt{\tfrac{18}{31}} < T. \end{aligned} \qquad \text{(XIV.33)}$$

The two complex roots of (XIV.31) have a negative real part which decreases monotonically with increasing T. Denoting by $o(1)$ a quantity which approaches 0 as $t \to \infty$ no matter what be the value of T, we write the solution for the third moments in homo-energetic simple shearing as follows[12]:

$$q_1 = De^{R_2 t/\tau} + o(1),$$

$$q_2 = -\frac{1}{3}\frac{(R_2 + \tfrac{3}{2})(R_2 + \tfrac{2}{3})}{T(R_2 + \tfrac{19}{18})} De^{R_2 t/\tau} + o(1),$$

$$q_3 = Ge^{R_1 t/\tau} + o(1),$$

$$P_{0|123} = -\frac{R_1 + \tfrac{2}{3}}{T} Ge^{R_1 t/\tau} + o(1),$$

[11] TRUESDELL [1956, §51] obtained and analysed (XIV.29); his remarks on the remaining exponents are partly incorrect. MUNCASTER [1974] perceived the reduction indicated in the text above, obtained (XIV.31), determined the nature of its roots, and wrote out and analysed the entire explicit solution of the initial-value problem for the third moments.

[12] Eqs. (XIV.34)$_{8,11,12}$ correct their counterparts in the paper of MUNCASTER [1974, Eqs. (14)$_5$, (13)$_3$, (14)$_2$].

$$P_{0|223} = -\frac{2}{5}\frac{R_1 + \frac{2}{3}}{R_1 + \frac{3}{2}}Ge^{R_1 t/\tau} + o(1),$$

$$P_{0|113} = \frac{8}{5}\frac{R_1 + \frac{2}{3}}{R_1 + \frac{3}{2}}Ge^{R_1 t/\tau} + o(1),$$

$$P_{0|333} = -\frac{6}{5}\frac{R_1 + \frac{2}{3}}{R_1 + \frac{3}{2}}Ge^{R_1 t/\tau} + o(1), \qquad \text{(XIV.34)}$$

$$P_{0|233} = \frac{2}{15}\frac{(R_2 + \frac{2}{3})^2}{T(R_2 + \frac{19}{18})}De^{R_2 t/\tau} + o(1),$$

$$P_{0|112} = -\frac{8}{15}\frac{(R_2 + \frac{2}{3})^2}{T(R_2 + \frac{19}{18})}De^{R_2 t/\tau} + o(1),$$

$$P_{0|222} = \frac{2}{5}\frac{(R_2 + \frac{2}{3})^2}{T(R_2 + \frac{19}{18})}De^{R_2 t/\tau} + o(1),$$

$$P_{0|122} = -\frac{2}{5}\frac{(R_2 + \frac{2}{3})(R_2 + \frac{2}{9})}{(R_2 + \frac{3}{2})(R_2 + \frac{19}{18})}De^{R_2 t/\tau} + o(1),$$

$$P_{0|133} = -\frac{2}{5}\frac{(R_2 + \frac{2}{3})(R_2 + \frac{7}{9})}{(R_2 + \frac{3}{2})(R_2 + \frac{19}{18})}De^{R_2 t/\tau} + o(1),$$

$$P_{0|111} = \frac{4}{5}\frac{(R_2 + \frac{2}{3})(R_2 + \frac{1}{2})}{(R_2 + \frac{3}{2})(R_2 + \frac{19}{18})}De^{R_2 t/\tau} + o(1);$$

D and G are arbitrary constants.

From (XIV.32) and (XIV.34) we see that *if* $T > 3/\sqrt{2}$, *the solutions* q_3, $P_{0|123}$, $P_{0|223}$, $P_{0|113}$, *and* $P_{0|333}$ *grow exponentially in time*. These solutions all have 3 as one index. In particular, *if* $T > 3/\sqrt{2}$, *any initial flux of energy transversely across the sheared planes grows exponentially as the flow proceeds*. A similar result follows from (XIV.33) and (XIV.34), but the critical value of T is smaller. Namely, *if* $T > \sqrt{\frac{18}{31}}$, *any flux of energy in the direction of the flow or normal to the sheared planes grows exponentially as t increases*.

We can describe these facts loosely by saying that if $T > \sqrt{\frac{18}{31}}$, *the flux of energy is unstable*. We note that $\sqrt{\frac{18}{31}}$ is approximately $\frac{3}{4}$ and hence less than the value of T shown in Chapter IV to be critical for other reasons.

We can state in still another way the results just obtained, always in reference to behavior as $t \to \infty$.

Case 1. $T < \sqrt{\frac{18}{31}}$. The solution for the third moments has the grossly determined dominant part

$$M^D_{kmr} = 0. \qquad \text{(XIV.35)}$$

Case 2. $\sqrt{\frac{18}{31}} \leq T < 3/\sqrt{2}$. The components q_3, $P_{0|123}$, $P_{0|223}$, $P_{0|113}$, and $P_{0|333}$ have the grossly determined dominant part 0. The remaining third moments have dominant parts proportional to $D \exp(R_2 t/\tau)$. The ratio of any one of these to the dominant part of q_1 is grossly determined.

Case 3. $3/\sqrt{2} < T$. The components q_3, $P_{0|123}$, $P_{0|223}$, $P_{0|113}$, and $P_{0|333}$ have dominant parts proportional to $G \exp(R_1 t/\tau)$. The ratio of any one of these to the dominant part of q_3 is grossly determined. The behavior of the remaining third moments is as in Case 2.

We note that in all cases, *provided* T *be small enough, dominant parts exist, and certain ratios of them are grossly determined.*

To those who claim that in sufficiently rarefied conditions a real gas cannot be regarded as a continuous medium, the instability of the third moments when T is large may afford some comfort. Indeed, as MAXWELL showed, the kinetic theory represents the gross fields associated with the gas as being always those of a continuum in the ordinary sense, but this mathematical fact does not mean that the results need be borne out by observation. On the one hand, the instability of the energy flux for values of T above $\sqrt{\frac{18}{31}}$ may indicate simply that the kinetic gas as embodied in MAXWELL's theory does not provide a good model for rarefied gases. On the other, a hydrodynamicist of traditional cast might take a different view. To him, the instability of the particular velocity field considered might suggest only that when T is large enough, a different velocity field might model the motion of a real gas better in the circumstances that for small T give rise to homo-energetic simple shearing. All these statements are questions put in the conditional mood.

(ix) *Entropy. Dissipation*

We might think to find some relation between the results about shearing flow and the predictions of the thermomechanics of continuous media, but no way to do so suggests itself, for we have not determined F and hence have no evident way to calculate $\overline{\log F}$. In any case, we have seen from direct reasoning in Chapter XI that if we take $-rh$, h being defined by (VIII.46), as the kinetic analogue of the caloric, then that analogue cannot generally satisfy a relation having the same form as the Clausius–Duhem inequality (XI.10). Using the solution for homo-energetic shearing, we can confirm the general conclusion by exhibiting a particular disagreement between the two theories. Namely, since in this problem both **q** and ε are functions of time only, a thermodynamics that implies the thermodynamic relations (I.9) would require (I.35) to hold: The rate of dissipation of energy must be non-negative. Before the rise of continuum thermomechanics, students of gas dynamics were wont to impose this requirement directly, either as being a consequence of the general idea of friction or as a sort of condition of stability. In simple shearing the rate of dissipation is $-KP_{12}$, as we have seen in Chapter II, so to violate the condition all we need do is exhibit a solution in which $P_{12} > 0$: that is, *a solution in which the shearing stress assists rather than opposes the shearing motion.* Repugnant as such an idea may be to "common sense", it is perfectly compatible with the kinetic theory. After all, if, as MAXWELL's relaxation theorem shows to be the case, the kinetic gas can support shear stress when it is grossly at rest, why cannot it when sheared support a shear pressure that points "the wrong way"?

Indeed, it can. From $(XIV.7)_{1,2}$ we see that $P_{12}(0) > 0$ if and only if

$$JTC < 1 - A + \tfrac{1}{2}R(1 - 3A), \qquad (XIV.36)$$

on the presumption that $p(0) > 0$. The initial conditions must always be such as to satisfy (IV.7), so $M_{22}(0) > 0$. That is, $P_{22}(0)/p(0) > -1$, and by $(XIV.7)_{1,3}$ this requirement is

$$JTC < \tfrac{1}{2}R(3A - 1). \qquad (XIV.37)$$

In order to see that for a given value of T there are infinitely many positive values[13] of A and C such as to satisfy both (XIV.36) and (XIV.37), we may consider the positive quadrant of points (A, JTC). We seek points lying below both of the lines $JTC = -(1 + \tfrac{3}{2}R)A + (1 + \tfrac{1}{2}R)$ and $JTC = \tfrac{3}{2}RA - \tfrac{1}{2}R$. The former line has a negative slope, and its A-intercept is $(1 + \tfrac{1}{2}R)/(1 + \tfrac{3}{2}R)$, which quantity lies between $\tfrac{1}{3}$ and 1. The latter of the two lines has a positive slope and the A-intercept $\tfrac{1}{3}$. Thus the desired points fill the interior of a triangle, the length of whose base is $\tfrac{2}{3}/(1 + \tfrac{3}{2}R)$ and whose altitude is $R/(1 + 3R)$. All the initial conditions so selected make $P_{12}(0)$ positive. The corresponding solutions (XIV.6) will have negative mechanical dissipation of energy at the first instant and hence, by continuity, during an interval of time thereafter.

The total mechanical energy accumulated by a finite portion of the gas in a finite interval of time is finite. As time goes on, the dominant part P_{12}^D of the shear pressure according to $(XIV.6)_2$ becomes infinitely larger than the damped part, and $P_{12}^D < 0$ if and only if $A > 0$. Also, according to $(XIV.6)_3$, $P_{22}^D/p^D > -1$ if $A > 0$. Thus all natural requirements are satisfied by the solution (XIV.6) as time goes on. Whatever be the sign of $P_{12}(0)$, when t is sufficiently large $P_{12}(t)$ becomes and remains negative, so the shear pressure opposes the shearing, as we should expect, and as $t \to \infty$ an arbitrarily large amount of mechanical energy is dissipated by any finite amount of the kinetic gas.

---

[13] The argument here replaces a more elaborate one given by TRUESDELL [1956, §50]. TRUESDELL's inequality (50.10) asserts the opposite of (XIV.37) and thus is false; consequently, so is the statement just following it; but both were offered merely as comments in passing and were put to no use in his proof of the assertion we here demonstrate more simply.

Thus the predictions of the kinetic theory for homo-energetic simple shearing have the same quality as those for relaxation in a grossly homogeneous condition: Arbitrarily large departures from what continuum mechanics teaches us to expect are allowed at any one instant, but in time they are damped out. There are three important differences, however. First, the damped parts of the pressures are oscillatory, not monotone. Second, the dominant pressures agree not with the Navier–Stokes theory but with NOLL's theory of the simple fluid, including the at first astonishing but now familiar normal-stress effects characteristic of non-linear fluids. (Of course, the theory of the simple fluid confirms the Navier–Stokes theory when T is small enough.) Third, the energy flux and the other third moments have grossly determined dominant parts only if T is sufficiently small.

For still another way of casting the sum of these results, we could say that however various the effects allowed by the general solution of the initial-value problem, *the dominant parts of the solution conform with all the ordinary requirements of continuum mechanics*, so long as $T < \sqrt{\frac{18}{31}}$. If $T \geq \sqrt{\frac{18}{31}}$, however, the dominant parts of the energy flux are not generally 0, which is not at all what any continuum theory of heat conduction would lead us to expect for a uniform field of temperature.

(x) *The principal solutions*

To conclude this chapter, we consider again the qualities of the general solution we have obtained for the pressure tensor and the flux of energy in homo-energetic shearing of a gas of Maxwellian molecules. First, in Section (iii) we have shown that if T is given any fixed value, then as $t \to \infty$ the pressure deviator $\mathbf{P}(t)$ has an asymptotic form which is determined by ρ, K, and ε alone, independently of the initial conditions. In Section (viii) we have seen that the energy flux $\mathbf{q}$ satisfies, trivially, a relation of the same kind, but only if T is sufficiently small. In terms of the principal moment ρ, defined by (VI.21), and its gradient grad ρ, we can express as follows the foregoing facts in a generality that is concise if seemingly excessive.

First Main Theorem on homo-energetic simple shearing. *If* $T < \sqrt{\frac{18}{31}}$, *there are gross determiners* $\mathbf{P}^G$ *and* $\mathbf{q}^G$ *such that as* $t \to \infty$ *every solution* $\mathbf{P}$ *and* $\mathbf{q}$ *satisfies the asymptotic relation*

$$P_{km}(t, \mathbf{x}) = P_{km}^G(\rho(t, \mathbf{x}), \text{grad } \rho(t, \mathbf{x}))[1 + o(1)],$$
$$q_k(t, \mathbf{x}) = q_k^G(\rho(t, \mathbf{x}), \text{grad } \rho(t, \mathbf{x}))[1 + o(1)]; \quad \text{(XIV.38)}$$

if $T \geq \sqrt{\frac{18}{31}}$, *the former relation holds, but generally the latter does not*. We can read off from (XIV.8) and (XIV.35) expressions for these gross determiners:

$$P_{11}^G = \frac{2R}{R+1} p, \qquad P_{22}^G = -\frac{R}{R+1} p, \qquad P_{33}^G = -\frac{R}{R+1} p,$$

(x) ASYMPTOTIC GROSS DETERMINISM 215

$$P^G_{12} = -\frac{R}{\frac{2}{3}T}p, \qquad P^G_{13} = 0, \qquad P^G_{23} = 0, \qquad \text{(XIV.39)}$$

$$q^G_k = 0, \qquad k = 1, 2, 3.$$

An alternative form of the gross determiner $\mathbf{P}^G$ is given by (XIV.11) in terms of viscometric functions.

The asymptotic forms on the right-hand sides of (XIV.38) are not merely estimates but are in fact equal to solutions corresponding to particular initial conditions. One way to verify this conclusion for $\mathbf{P}$ is to take the explicit dominant pressures as given by (XIV.8) and to see that they satisfy (XIV.1) for a certain choice of initial conditions. They do, provided p be proportional to $e^{Rt/\tau}$. An easier way to reach the same conclusion is to notice that a suitable choice of initial conditions makes the damped terms in (XIV.6) vanish identically. Taking account of (XIV.7), we see that this choice of initial conditions must imply

$$A = 1, \qquad B = C = 0. \qquad \text{(XIV.40)}$$

Equivalently,

$$\frac{P_{12}(0)}{p(0)} = -\frac{R}{\frac{2}{3}T}, \qquad \frac{P_{22}(0)}{p(0)} = -\frac{R}{1+R}. \qquad \text{(XIV.41)}$$

Then we can choose the initial values of the remaining pressures such that

$$P_{11}(0) + 2P_{22}(0) = 0, \qquad P_{22}(0) - P_{33}(0) = 0, \qquad P_{13}(0) = P_{23}(0) = 0. \qquad \text{(XIV.42)}$$

With this very particular, uniquely determined choice of the initial value $\mathbf{P}(0)$, for each initial value of p there is one and only one solution of (XIV.1). We can read off this solution from (XIV.2), (XIV.3), and (XIV.6). We denote this special class of solutions, the *principal solutions*, by a superscript P to recall that they are given *exactly* and not merely asymptotically by the gross determiners:

$$\varepsilon^P = \varepsilon^P(0)e^{Rt/\tau},$$

$$P^P_{11} = \frac{4}{3}\frac{R}{1+R}\rho\varepsilon^P(0)e^{Rt/\tau},$$

$$P^P_{22} = -\frac{2}{3}\frac{R}{1+R}\rho\varepsilon^P(0)e^{Rt/\tau}, \qquad \text{(XIV.43)}$$

$$P^P_{33} = -\frac{2}{3}\frac{R}{1+R}\rho\varepsilon^P(0)e^{Rt/\tau},$$

$$P^P_{12} = -\frac{R}{T}\rho\varepsilon^P(0)e^{Rt/\tau},$$

$$P^P_{13} = 0, \qquad P^P_{23} = 0.$$

For the energy flux the corresponding results are trivial, since identically vanishing third moments satisfy (XIV.27) for null initial conditions. Thus for **q** there is one and only one principal solution, namely,

$$q_k^P = 0. \tag{XIV.44}$$

Again in a generality that may seem excessive, we may summarize these results as the

Second Main Theorem on homo-energetic simple shearing. *The asymptotic forms provided by* (XIV.38) *are the values of a principal solution. That is, whatever be the solution* **M**, **q**, *provided only that* $\mathsf{T} < \sqrt{\frac{18}{31}}$, *there is a principal solution* $\mathbf{M}^P$, $\mathbf{q}^P$ *such that as* $t \to \infty$

$$M_{km}(t, \mathbf{x}) = M_{km}^P(t, \mathbf{x})[1 + o(1)],$$
$$q_k(t, \mathbf{x}) = q_k^P(t, \mathbf{x})[1 + o(1)]; \tag{XIV.45}$$

if $\mathsf{T} \geq \sqrt{\frac{18}{31}}$, *the former result still holds, but there is no principal solution such as to satisfy the latter.*

The principal solutions are special in that for them **M** and **q** are grossly determined, this fact being a time-independent restriction upon them. As we see from (XIV.39), the choice of initial values specified by (XIV.41) and (XIV.42) merely expresses that restriction at the initial instant. The *gross determiners once laid down, a principal solution is determined uniquely by the initial value* ρ_0 *of its own principal moment* ρ. Let it not be thought, however, that the two solutions set into correspondence through (XIV.45) have their principal moments in common. Indeed, comparison of (XIV.43) with (XIV.6)$_1$ or (XIV.8)$_1$ shows that in order for (XIV.45)$_1$ to hold it is necessary and sufficient that

$$\varepsilon^P(0) = A\varepsilon(0), \tag{XIV.46}$$

$\varepsilon(0)$ being the initial value of the energetic that belongs to the general solution for **M**. Solving (XIV.7) for A, we obtain

$$A = \frac{(1 + R)\left(1 + R\dfrac{M_{22}(0)}{p(0)}\right) - \dfrac{2}{3}\mathsf{T}\dfrac{P_{12}(0)}{p(0)}}{1 + 3R}. \tag{XIV.47}$$

Therefore, in order to determine the particular principal solution that approximates a given solution, while we do not need to know the entire set of initial values $P_{km}(0)/p(0)$ that determine the given solution, we must know two of them. Generally, *the effect of the initial values* $\mathbf{P}(0)/p(0)$ *upon the solution* **P** *that they determine is not annulled in the course of time.* This conclusion contrasts strongly with MAXWELL's relaxation theorem for gross free equilibrium. For that problem there is one and only one principal solution, namely, $\mathbf{P}^P = \mathbf{0}$ and $\mathbf{q}^P = \mathbf{0}$, and it approximates *every* solution of the problem as $t \to \infty$.

However, just as we have given in Section (vi) of this chapter an asymptotic status to the Stokes–Kirchhoff solution, here we can give a corresponding asymptotic status to the use of a principal solution having the same initial principal moment as does a given solution to approximate that solution. From (XIV.47) and (XIV.5) we see that as $T \to 0$,

$$A = 1 - \frac{2}{3}\frac{P_{12}(0)}{p(0)}T + O(T^2). \tag{XIV.48}$$

Comparison with (XIV.46) shows that with error $O(T)$,

$$\varepsilon^P(0) = \varepsilon(0), \tag{XIV.49}$$

but the agreement does not persist to any higher order. Hence we conclude the

Third Main Theorem on homo-energetic simple shearing. *The principal solution that approximates a given solution as $t \to \infty$ is not determined by the principal moment of the given solution except as an approximation for very small values of T when the initial values are held constant.*

If we were to use a principal solution so as to solve the initial-value problem, we should expect to be able to use the standard data of gas dynamics, namely, the initial value ρ_0. The foregoing theorem shows such an expectation to be false. The initial datum ρ_0 suffices to find the asymptotic form of $\mathbf{P}(t)$ as $t \to \infty$ only as an approximation for small T; that is, roughly, to the same extent that use of the Stokes–Kirchhoff theory is justified. This fact confirms the status of the Stokes–Kirchhoff approximation but casts grave doubt on the use of principal solutions for any other purpose—in particular, upon use of them to improve or "correct" the results of solving the initial-value problem of classical gas dynamics.

The terms "normal solution" and "deterministic" and "causal" occur frequently in literature on the kinetic theory but have never been given precise definitions[14]. Our three main theorems on homo-energetic shearing refer to three senses in which the words sometimes are used. We do not employ the

---

[14] CHAPMAN & COWLING [1939, §7.2] introduced "normal solutions" as follows:
Suppose that initially a given mass of gas possesses an arbitrary velocity-distribution function. Let the gas be left to itself for a while; . . . f [in the notation of this book, F] will approach one of a series of "normal" values, each depending on certain parameters in a standard way. After a normal distribution of velocities has been reached, f will continue to take only normal values. On physical grounds, . . . [t]he normal solutions of Boltzmann's equation . . . depend only on n, $\mathbf{c}_0$, and T [in the notation of this book, the fields n, $\mathbf{u}$, and $\frac{2}{3}\varepsilon/r$], and vary with the time only through their dependence on n, $\mathbf{c}_0$, and T.
So far as we can find, this passage is the only one in which CHAPMAN & COWLING [1939] [1952] explain what they mean by "normal", which does not occur in their index.

(*continued*)

word "normal" at all, preferring the precisely defined terms "principal solution" and "grossly determined"; properties of these are set forth in the first two theorems. Our third theorem shows that solutions "normal" in a third sense of the word generally fail to exist[15].

In the following chapter we shall present two other explicit solutions, which in some measure complement and in some measure reinforce the conclusions drawn in this chapter.

GRAD [1958, §23] defined "normal" as follows: "What seems to be true is that the class of solutions of the Boltzmann equation which can be represented by power series in [some parameter] ε is very restricted, and in this class there is a one-to-one correspondence between solutions $f(\xi, \mathbf{x}, t)$ [$F(t, \mathbf{x}, \mathbf{v})$ in the notation of this book] and initial values $\rho_y(\mathbf{x}, 0)$. We shall refer to this as the 'Hilbert class' of solutions or the class of 'normal solutions'." In this book we present HILBERT's series in Chapter XXII.

Three decades after their first pronouncement CHAPMAN & COWLING [1970, §7.2] retreated to a vaguer position:

[I]n general f at each point in the gas begins by varying rapidly on account of molecular encounters, with a time-scale of variation comparable with the mean collision-interval. However, by analogy with the results for a gas in a uniform state, f can be expected quickly to approach a limiting form in which molecular encounters no longer produce such rapid time-variations. The limiting value of f is called a 'normal solution' of Boltzmann's equation, and the corresponding state of the gas is a 'normal state' . . .

If the gas is uniform, the normal solution is the Maxwellian distribution function. In a slightly non-uniform gas the normal solution is a function of $\mathbf{r}$ and t, though at any time it approximates locally to the Maxwellian form . . .

Normal solutions . . . are distinguished from non-normal solutions by having no "irregularities" which can be rapidly smoothed out by molecular encounters. In consequence, they depend on fewer parameters than the non-normal solutions. . . . we can expect that n, $\mathbf{c}_0$, and T . . . are the only local properties on which a normal solution can depend. Actually, since the state at the point P is appreciably affected only by the gas immediately adjacent (i.e. within a distance of a few mean free paths) a normal solution at P can be taken to depend on the values at P of n, $\mathbf{c}_0$, and T and their space derivatives; it may also depend on the force $\mathbf{F}$ acting on the molecules.

These quotations may serve as specimens of the flights of romantic imagination which the kinetic theory has inspired in otherwise sober academic circles.

[15] Our wording throughout differs from that used by TRUESDELL [1956, §§34 and 37], but the results we give are precisely his. In explaining the one expressed here by our third theorem he wrote, "For shearing flow of a Maxwellian gas, no 'normal' solution exists." GRAD [1963, 1, §1], referring to TRUESDELL's result, expressed this same fact as follows: "As observed by Hilbert, the quantities which arise in his theory are formally causal; their initial values uniquely determine their future. But it is a misinterpretation . . . to deduce from this that the fluid state itself is causal, even asymptotically."

Chapter XV

General Solution for the Pressures in Homo-energetic Affine Flows of a Gas of Maxwellian Molecules

(i) Affine flows in general

The reader who has mastered both the idea of the solution for homo-energetic simple shearing obtained in the preceding chapter and the general theory of homo-energetic affine flows in Section (v) of Chapter II will see at once that the latter applies straight off to the kinetic gas of Maxwellian molecules. The defining assumptions are

$$\text{div } \mathbf{P} = \mathbf{0}, \qquad \text{div } \mathbf{q} = 0; \tag{II.18}_r$$

$$\mathbf{u} = \mathbf{G}(t)\mathbf{r} + \mathbf{g}(t); \tag{II.41}_{1r}$$

ρ, ε, and $\mathbf{P}$ are functions of t alone; $\tag{II.42}_r$

$$\mathbf{b} = \text{const.} \tag{II.43}_r$$

When the body undergoing the flow consists of an ideal gas whose specific heats are constant and stand in the ratio $\tfrac{5}{3}$, the following conditions we have proved necessary and sufficient that the defining assumptions just set down be compatible with the field equations expressing balance of mass, momentum, and energy:

$$\rho = \rho(0) \exp\left(-\int_0^t E \, ds\right); \tag{II.44}_r$$

$$\mathbf{G} = [\mathbf{1} + t\mathbf{G}(0)]^{-1}\mathbf{G}(0),$$
$$\mathbf{g} = [\mathbf{1} + t\mathbf{G}(0)]^{-1}[\tfrac{1}{2}t^2\mathbf{G}(0)\mathbf{b} + t\mathbf{b} + \mathbf{g}(0)]; \tag{II.46}_r$$

$$\varepsilon = \varepsilon(0) \exp\left[-\tfrac{2}{3}\int_0^t \left(E + \frac{1}{p}\mathbf{P} \cdot \mathbf{E}\right) ds\right]. \tag{XV.1}$$

The last of these formulae results from (II.51) when $\gamma = \frac{5}{3}$. If by $-1/t_-$ and $-1/t_+$ we denote the greatest positive and least negative proper numbers of $G(0)$, the solution G, g given by (II.46) exists in the interval (t_-, t_+) and ceases to exist at t_- and t_+.

Further, in Chapter II we have introduced the definition

$$\tau \equiv \mu/p \qquad (II.52)_{1r}$$

and have shown that if μ is a linear function of ε alone, then

$$\tau = \tau(0) \exp\left(\int_0^t E\, ds\right). \qquad (II.54)_{2r}$$

In the preceding chapter we have seen how to use this basic corpus. The functions τ and μ in (II.52) we take, of course, as those given by (XIII.3) and (XIII.14) for a gas of Maxwellian molecules. We have seen reason to interpret this μ as the shear viscosity of the kinetic gas, but more important is the role that this τ plays in the differential equations satisfied by the moments, for example (XIII.4).

To apply these results to the kinetic theory, we first need to satisfy the essential conditions (II.18). Beyond that, we need to be sure that P is a function of t alone. If such is the case, we see from (XIII.2) that necessarily $M_{kma,a}$ is a function of t alone. If that function is 0, (XIII.2) uncouples from the system of moments and becomes a system restricting M as a function of t. That system is linear and homogeneous. Its coefficients are combinations of the components of G. From (II.46) we know that G is a continuous function in the interval (t_-, t_+). A standard theorem[1] on differential equations assures us that the system (XIII.2) has a unique solution corresponding to any prescribed initial value of $M(0)$, and that that solution exists in (t_-, t_+). We summarize the foregoing results, slightly rephrased, as the following **Fundamental Theorem on Homo-energetic Affine Flows**: *Let $\rho(0)$, $\varepsilon(0)$, $G(0)$, $g(0)$, and $P(0)$ be prescribed constants, and let $-1/t_-$ and $-1/t_+$, respectively, be the greatest positive and least negative proper numbers of $G(0)$. For a gas of Maxwellian molecules, corresponding to these initial values there is one and only one solution of the equations of balance and the equations of second moments such that*

$$\mathbf{u} = G(t)\mathbf{r} + \mathbf{g}(t); \qquad (II.41)_{1r}$$

$$\rho, \varepsilon, \text{ and } P \text{ are functions of } t \text{ alone}; \qquad (II.42)_r$$

$$M_{kma,a} = 0; \qquad (XV.2)$$

$$\mathbf{b} = \text{const.} \qquad (II.43)_r$$

---

[1] *E.g.* Lemma 1.1, Chapter IV of P. HARTMAN's *Ordinary Differential Equations*, New York, John Wiley & Sons, 1964.

(i) AFFINE FLOWS IN GENERAL

This solution exists in the interval (t_-, t_+); *at* $t = t_-$ *and at* $t = t_+$ *it ceases to exist.*

The special case of this theorem that results when

$$\|G_{km}\| = \begin{Vmatrix} 0 & K & 0 \\ 0 & 0 & 0 \\ 0 & 0 & 0 \end{Vmatrix} \tag{II.58}_{1r}$$

we worked out in full detail in the preceding chapter. The theorem as we state it renders explicit and generalizes some formal remarks of GALKIN[2], who noticed several further special cases of possible interest. Among them are the following, which GALKIN later went some way toward working out in detail. Each involves an arbitrary non-zero constant; in each we shall denote that constant by T.

Case 1. *Dilatation*[3]:

$$\|G_{km}(0)\| = T^{-1} \, \text{diag}(1, 1, 1). \tag{XV.3}$$

Case 2. *Extension*[4]:

$$\|G_{km}(0)\| = T^{-1} \, \text{diag}(1, 0, 0). \tag{XV.4}$$

Case 3. *Cylindrical dilatation*[5]:

$$\|G_{km}(0)\| = T^{-1} \, \text{diag}(1, 1, 0). \tag{XV.5}$$

The remainder of this chapter is devoted to elaboration of GALKIN's first and second examples. We select these for their intrinsic interest. They illustrate the very different kinds of behavior compatible with the kinetic theory. Notwithstanding that, they are mathematically simple and involve only elementary functions, something that is not true of the third example.

To what extent the exact solutions in the class here exhibited correspond to solutions of the Maxwell–Boltzmann equation is not yet established. If we assume that the moments ^3M, ^4M, ... are functions of t alone, the infinite system of equations for them (*cf.* Section (iii) of Chapter XVI) uncouples into an infinite sequence of finite systems of ordinary differential equations, one system for ^3M alone, the next one for ^4M alone, *etc.* We see from (XIII.5) that the system for ^3M is linear and homogeneous. We have illustrated the solution of this system twice already: for gross equilibrium in Chapter XIII, for simple shearing in Chapter XIV, and we shall illustrate it again below in connection with GALKIN's first example. However, a theorem to be proved in the following chapter shows that for ^4M, ^5M, ... the system is linear but inhomogeneous. The coefficients in the differential system for ^nM are components of G, as we may see by example from (XIII.5) and in general from (XIII.1). The theorem just mentioned shows that the inhomogeneous term is a bilinear function of the moments ^{n-1}M, ^{n-2}M, ..., which have been determined already and are differentiable functions of t. Thus the inhomogeneous term is certainly integrable. Accordingly, a unique solution ^nM corresponding to a prescribed initial value $^nM(0)$ exists in (t_-, t_+).

---

[2] GALKIN [1958].
[3] GALKIN [1966, §3].
[4] GALKIN [1964].
[5] GALKIN [1966, §5].

What is not clear is whether the functions $\mathbf{M}$, $^3\mathbf{M}$, ..., $^n\mathbf{M}$, ... so found correspond to a molecular density. As we have seen in Chapter IV, a further necessary condition is

$$\mathbf{n} \cdot \mathbf{Mn} > 0, \qquad (\text{IV}.7)_{2,4r}$$

but it is not clear that if this condition is satisfied at $t = 0$ it remains satisfied in (t_-, t_+). In Chapter XIV we satisfied it by imposing the requirement $A > 0$. In general, there are infinitely many further questions of this kind.

(ii) Homo-energetic dilatation

The analysis in Section (v) of Chapter II provides us with the following solution of the equations of balance of mass, momentum, and energy for a homo-energetic affine flow corresponding to the assumptions (XV.3), (XV.2), (II.42), and $\mathbf{b} = \mathbf{0}$:

$$\rho = \rho(0)\left(1 + \frac{t}{T}\right)^{-3},$$

$$\mathbf{u} = (t + T)^{-1}\mathbf{r}, \qquad (\text{II}.39)_r$$

$$\varepsilon = \varepsilon(0)\left(1 + \frac{t}{T}\right)^{-2};$$

$$p = p(0)\left(1 + \frac{t}{T}\right)^{-5}. \qquad (\text{II}.40)_r$$

These fields determine a unique Maxwellian density, and in Chapter X we have seen that that density is a solution of the Maxwell–Boltzmann equation. For it, of course, $\mathbf{P} = \mathbf{0}$ and $^3\mathbf{M} = \mathbf{0}$. For a gas of Maxwellian molecules we can go further and determine as follows the general solution for $\mathbf{P}$ and $^3\mathbf{M}$, on the assumption that they and also $^4\mathbf{M}$ are functions of t alone.

First we use (II.54)$_2$ to conclude that

$$\tau = \tau(0)\left(1 + \frac{t}{T}\right)^3. \qquad (\text{XV}.6)$$

Placing this formula and (II.39) into (XIII.4), we obtain the following differential system for the components of $\mathbf{P}$:

$$\dot{P}_{km} + \left(\frac{5}{t + T} + \frac{T^3}{\tau(0)(t + T)^3}\right)P_{km} = 0. \qquad (\text{XV}.7)$$

The general solution of this system is

$$P_{km} = P_{km}(0)\left(1 + \frac{t}{T}\right)^{-5} \exp\left[\frac{T^3}{2\tau(0)}\left(\frac{1}{(t + T)^2} - \frac{1}{T^2}\right)\right]. \qquad (\text{XV}.8)$$

(ii) DILATATION

Likewise, the general solution of (XIII.7) for this flow is

$$q_k = q_k(0)\left(1 + \frac{t}{T}\right)^{-6} \exp\left[\frac{T^3}{3\tau(0)}\left(\frac{1}{(t+T)^2} - \frac{1}{T^2}\right)\right],$$

$$P_{0|kmr} = P_{0|kmr}(0)\left(1 + \frac{t}{T}\right)^{-6} \exp\left[\frac{3T^3}{4\tau(0)}\left(\frac{1}{(t+T)^2} - \frac{1}{T^2}\right)\right]. \quad \text{(XV.9)}$$

The results (XV.8) and (XV.9) show that $\mathbf{P} \to \mathbf{0}$ and $^3\mathbf{M} \to \mathbf{0}$ as t approaches either ∞ or $-T$ according as T is positive or negative: As time proceeds, *the moments* $\mathbf{P}$ *and* $^3\mathbf{M}$ *approximate more and more closely the corresponding moments of the locally Maxwellian solution* F_M *determined by the gross condition of the flow.* Not only that, the limit of each moment equals the limit of the corresponding moment of F_M. This does not mean that F_M need approach in time a limit that is a molecular density whose moments have these limiting values. Indeed, by (X.8) and (X.9) we know that F_M does approach a limit, but that limit is not a molecular density at all.

We may describe these results in another way. Namely, *the principal solution of the general class of solutions for homo-energetic dilatation is* $\mathbf{P} = \mathbf{0}$, $^3\mathbf{M} = \mathbf{0}$, which, trivially, is grossly and momentally determined, just as in the case of MAXWELL's relaxation theorem. However, we see from (XV.8) and (XV.9) that in contrast with MAXWELL's solutions, the approach to $\mathbf{0}$ is *much slower than exponential*, being governed by some positive integral power of $(1 + t/T)^{-1}$. Also it does not depend on the molecular constants $\mathfrak{m}$ and $\mathfrak{g}$.

Because of (XIII.3)$_2$ we know that

$$\tau(0) = \frac{1}{3\mathfrak{a}}\sqrt{\frac{2\mathfrak{m}^3}{\mathfrak{g}}} \frac{1}{\rho(0)}. \quad \text{(XV.10)}$$

Therefore, if we hold t fixed and consider (XV.8) and (XV.9) in the limit as $\mathfrak{m} \to 0$, we find that $\mathbf{P} \to \mathbf{0}$ and $^3\mathbf{M} \to \mathbf{0}$, no matter what be the sign of T. Thus in this asymptotic process also *all solutions become consistent with the Stokes–Kirchhoff solution*, which is itself, of course, the same as the Euler–Hadamard solution of this problem.

All solutions corresponding to (II.39) are dissipationless, homochoric, and homo-energetic. Thus POISSON's law as expressed by (X.3) reduces to the statement that $p\rho^{-5/3} = \text{const.}$ in space and time, just as we have observed for the locally Maxwellian dilatations discussed in Chapter X. In the terms used in Sections (v)–(vii) of Chapter XI we may rephrase that statement as follows: $h_M = \text{const.} = h[F_U]$, F_U being the particular uniform Maxwellian density that corresponds to the initial principal moment $(\rho(0), \mathbf{0}, \varepsilon(0))$. It does not follow that $h \to h[F_U]$ or even that $h \to$ a limit. In Appendix B to Chapter XVII we shall mention an example for which $h \to$ a limit, but that limit is not h_M.

(iii) Homo-energetic extension, I. The general solution for the pressures

The analysis in Section (v) of Chapter II provides us with the following solution of the equations of balance of mass, momentum, and energy corresponding to the assumptions (XV.4), (XV.2), (II.42), and $\mathbf{b} = 0$:

$$\rho = \rho(0)\left(1 + \frac{t}{T}\right)^{-1}; \tag{II.67}_r$$

$$u_1 = (t + T)^{-1}(x_1 - x_0), \quad u_2 = 0, \quad u_3 = 0; \tag{XV.11}$$

$$\frac{\dot{\varepsilon}}{\varepsilon} = -(\gamma - 1)\frac{1 + P_{11}/p}{t + T}; \tag{II.70}_{2r}$$

$$\tau = \tau(0)\left(1 + \frac{t}{T}\right). \tag{II.75}_r$$

Again $\tau(0)$ for a gas of Maxwellian molecules is given by (XV.10). For our purposes it is more convenient to convert (II.70) into an equation for p, which follows at once from the above because $p = \frac{2}{3}\rho\varepsilon$ and $\gamma = \frac{5}{3}$:

$$(t + T)\dot{p} + \frac{5}{3}p + \frac{2}{3}P_{11} = 0. \tag{XV.12}$$

Again it is convenient to describe the solution in terms of the parameter T:

$$\mathsf{T} = \frac{4}{3}\frac{\tau(0)}{T}. \tag{II.76}_r$$

The sign of T is thus the sign of T.

We substitute (XV.11) and (II.75) into the equations of second moments (XIII.4) and so obtain the differential equations

$$(t + T)\dot{P}_{11} + \left(\frac{7}{3} + \frac{4}{3\mathsf{T}}\right)P_{11} + \frac{4}{3}p = 0,$$

$$(t + T)\dot{P}_{22} + \left(1 + \frac{4}{3\mathsf{T}}\right)P_{22} - \frac{2}{3}P_{11} - \frac{2}{3}p = 0,$$

$$(t + T)\dot{P}_{12} + \left(2 + \frac{4}{3\mathsf{T}}\right)P_{12} = 0, \tag{XV.13}$$

$$(t + T)\dot{P}_{13} + \left(2 + \frac{4}{3\mathsf{T}}\right)P_{13} = 0,$$

$$(t + T)\dot{P}_{23} + \left(1 + \frac{4}{3\mathsf{T}}\right)P_{23} = 0.$$

(iii) EXTENSION: GENERAL SOLUTION

A simple combination of these equations gives us

$$(t + T)(P_{22} + \tfrac{1}{2}P_{11})^{\cdot} + \left(1 + \frac{4}{3T}\right)(P_{22} + \tfrac{1}{2}P_{11}) = 0. \quad \text{(XV.14)}$$

The system formed by (XV.12) and (XV.13) suffices to determine p and $\mathbf{P}$. We may integrate (XV.13)$_{3,4,5}$ and (XV.14) at once:

$$P_{12} = P_{12}(0)\left(1 + \frac{t}{T}\right)^{-[2+(4/3)/T]},$$

$$P_{13} = P_{13}(0)\left(1 + \frac{t}{T}\right)^{-[2+(4/3)/T]},$$

$$P_{23} = P_{23}(0)\left(1 + \frac{t}{T}\right)^{-[1+(4/3)/T]}, \quad \text{(XV.15)}$$

$$P_{22} + \tfrac{1}{2}P_{11} = (P_{22}(0) + \tfrac{1}{2}P_{11}(0))\left(1 + \frac{t}{T}\right)^{-[1+(4/3)/T]}$$

The equations (XV.12) and (XV.13)$_1$ form a second-order system for p and P_{11}. In order to find the general solution of this system, GALKIN introduced the variable η defined as follows:

$$\eta \equiv \log\left(1 + \frac{t}{T}\right), \quad \text{(XV.16)}$$

so that for a function $f(\eta)$

$$f' \equiv \frac{df}{d\eta} = (t + T)\dot{f}, \quad \text{(XV.17)}$$

and then p and P_{11}, considered now as functions of η, must satisfy the simpler second-order system

$$p' + \tfrac{5}{3}p + \tfrac{2}{3}P_{11} = 0, \qquad P'_{11} + \left(\frac{7}{3} + \frac{4}{3T}\right)P_{11} + \tfrac{4}{3}p = 0. \quad \text{(XV.18)}$$

Since these are linear equations with constant coefficients, there is a solution proportional to $\exp(\chi\eta)$, χ being a root of the characteristic equation

$$\chi^2 + 4\chi\left(1 + \frac{1}{3T}\right) + 3 + \frac{20}{9T} = 0. \quad \text{(XV.19)}$$

The roots R and S are determined as follows:

$$R = -\frac{2(\tfrac{3}{4}T + \tfrac{5}{9})}{T + \tfrac{1}{3} + \sqrt{(T + \tfrac{1}{3})^2 - (\tfrac{3}{4}T^2 + \tfrac{5}{9}T)}},$$

$$S = -4\left(1 + \frac{1}{3T}\right) - R. \quad \text{(XV.20)}$$

R and S are real and distinct, no matter whether T be positive or negative. R is always negative. As given by the right-hand side of $(XV.20)_1$, it depends smoothly on T, and as $T \to 0$, it approaches the one and only root of the asymptotic form of $(XV.19)$ as $T \to \infty$, namely, $\frac{4}{3}\chi + \frac{20}{9} = 0$, the corresponding root being $R = -\frac{5}{3}$. As T varies from $-\infty$ to $+\infty$, R increases slowly and monotonically from -3 to -1. The root S is not defined at $T = 0$. As T varies from $-\infty$ to 0, S increases monotonically from -1 to $+\infty$; as T varies from 0 to $+\infty$, S increases from $-\infty$ to -3. When T is positive, R is the greater of the two roots; when T is negative, it is the smaller of them. These roots are shown in Figure XV.1.

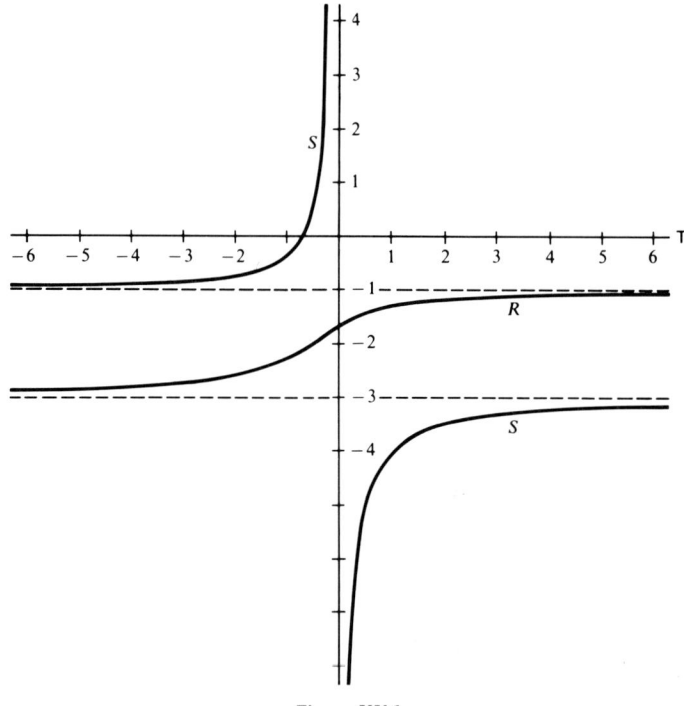

Figure XV.1

We may express the general solution for p and P_{11} as follows:

$$\frac{p}{p(0)} = A\left(1 + \frac{t}{T}\right)^R + (1 - A)\left(1 + \frac{t}{T}\right)^S,$$

$$\frac{P_{11}}{p(0)} = -\frac{3R + 5}{2} A\left(1 + \frac{t}{T}\right)^R - \frac{3S + 5}{2}(1 - A)\left(1 + \frac{t}{T}\right)^S, \quad (XV.21)$$

in which

$$A = \frac{S + \frac{5}{3} + \frac{2}{3}\frac{P_{11}(0)}{p(0)}}{S - R}.\qquad (XV.22)$$

This solution behaves differently from that for dilatations. *If* T *is negative, the magnitudes of the pressures generally approach* ∞ *at some finite time; if* T *is positive, the pressures approach* 0 *as* $t \to \infty$. Again the rate of relaxation to 0 is much slower than exponential, although for this solution, in contrast with that for dilatations, it is influenced by the molecular constants.

This solution is distinguished from dilatation in that it, like simple shearing, dissipates energy. Indeed, in obtaining (II.70) we have made use of the fact that

$$P_{ab}E_{ab} = P_{11}E = \frac{P_{11}}{t+T},\qquad (XV.23)$$

and so the rate of dissipation approaches ∞ or relaxes to 0 with the solution **P** itself.

(iv) *Homo-energetic extension, II. The principal solutions*

We shall see now that the solutions for homo-energetic extension are similar in many respects to the solutions for homo-energetic simple shearing[6]. They exhibit in certain cases dominant and damped parts, and from these we shall find principal solutions and corresponding gross determiners. We begin by investigating more closely the way the general solution behaves as time goes on.

Considering P_{12}, P_{13}, and P_{23} first, we ask for conditions under which each of these functions relaxes to zero as t increases. It is not difficult to see from (XV.15) that all do if T > 0. If T < 0, then $1 + t/T \to 0$ as $t \to -T$, and so a glance at (XV.15) shows that these three functions approach 0 as $t \to -T$ if both $1 + \frac{4}{3}/T$ and $2 + \frac{4}{3}/T$ are negative, that is, if $-\frac{2}{3} < T < 0$. If $-\frac{4}{3} < T \leq -\frac{2}{3}$, however, the functions P_{12} and P_{13} generally approach $\pm\infty$ as $t \to -T$, while P_{23} relaxes to zero. If $T \leq -\frac{4}{3}$, all three functions generally grow without bound as t approaches $-T$. Even more important than this is the fact that if $T \leq -\frac{2}{3}$, the effect of the initial values $P_{12}(0)$ and $P_{13}(0)$ upon the solution is not wiped out in time, as it is if $-\frac{2}{3} < T$. Thus, if we are to find any grossly determined solutions for homo-energetic extension, we must restrict our attention to cases in which $T > -\frac{2}{3}$.

---

[6] From this point onward we present results not to be found in GALKIN's papers.

To discuss the behavior of the normal pressures, it is convenient to write them out explicitly. Using (XV.21), (XV.15)$_4$, and (III.10)$_2$, we find that

$$\frac{M_{11}}{p(0)} = \left(1 - \frac{3R+5}{2}\right)A\left(1 + \frac{t}{T}\right)^R + \left(1 - \frac{3S+5}{2}\right)(1-A)\left(1 + \frac{t}{T}\right)^S,$$

$$\frac{M_{22}}{p(0)} = \left(1 + \frac{3R+5}{4}\right)A\left(1 + \frac{t}{T}\right)^R + \left(1 + \frac{3S+5}{4}\right)(1-A)\left(1 + \frac{t}{T}\right)^S$$
$$+ \left(\frac{P_{22}(0) + \frac{1}{2}P_{11}(0)}{p(0)}\right)\left(1 + \frac{t}{T}\right)^{-[1+(4/3)/T]}, \qquad \text{(XV.24)}$$

$$\frac{M_{33}}{p(0)} = \left(1 + \frac{3R+5}{4}\right)A\left(1 + \frac{t}{T}\right)^R + \left(1 + \frac{3S+5}{4}\right)(1-A)\left(1 + \frac{t}{T}\right)^S$$
$$- \left(\frac{P_{22}(0) + \frac{1}{2}P_{11}(0)}{p(0)}\right)\left(1 + \frac{t}{T}\right)^{-[1+(4/3)/T]}.$$

For analysis we divide these solutions into three classes.

Class 1. $A < 0$. Suppose first that $T > 0$. Then $0 > R > S$, and we see from (XV.21)$_1$ that of the two terms that make up p, the first dominates as $t \to \infty$. Next suppose that $T < 0$. The root R is still negative, but in this case $R < S$. From (XV.21)$_1$ we see that the first of the two terms in p again dominates, now as $t \to -T$. Since $A < 0$, this first term is negative in either case, and so all solutions in this case ultimately contradict the requirement $p > 0$. Hence Class 1 is empty: $A \geq 0$.

We now consider the effect of the requirement (IV.7), which makes M_{11}, M_{22}, and M_{33} positive always. In particular $P_{11}(0) + p(0) > 0$, and so it follows from (XV.22) that

$$(S - R)A > S + 1. \qquad \text{(XV.25)}$$

Referring to our discussion above of the roots (XV.20), we see that if $T > 0$, then $S + 1 < 0$ and $S - R < 0$. Accordingly

$$A < \frac{S+1}{S-R} \quad \text{if} \quad T > 0. \qquad \text{(XV.26)}$$

On the other hand, if $T < 0$, then $S + 1 > 0$ and $S - R > 0$, so

$$A > \frac{S+1}{S-R} \quad \text{if} \quad T < 0. \qquad \text{(XV.27)}$$

In both cases the ratio $(S + 1)/(S - R)$ is positive. Thus the latter case excludes the possibility $A = 0$. We are left, then, with the following two classes of solutions.

Class 2. $A = 0$ and $T > 0$. When the initial values of p and P_{11} are such that $A = 0$, namely

$$\frac{2}{3}\frac{P_{11}(0) + p(0)}{p(0)} = -(S + 1), \qquad \text{(XV.28)}$$

the first terms on the right-hand sides of (XV.24) are null. In this case, as we now show, the normal pressure M_{33} always contradicts the general requirement $M_{33} > 0$. Let us note first that with $p(0)$ and $P_{11}(0)$ satisfying (XV.28),

$$\frac{P_{22}(0) + \tfrac{1}{2}P_{11}(0)}{p(0)} = \frac{M_{22}(0)}{p(0)} - \tfrac{3}{4}(S + 3). \qquad \text{(XV.29)}$$

Since $T > 0$, we know that $S < -3$, and so the second term on the right-hand side here is positive. Since $M_{22}(0)/p(0) > 0$, we conclude that the third term which makes up M_{33} is always negative. Moreover, since $S < -3$, we see that $1 + (3S + 5)/4 < 0$, and so the second term making up M_{33} is also negative. The first term is null in this case, so $M_{33} < 0$ always. Thus Class 2 is empty: $A > 0$.

Class 3. $0 < A < (S + 1)/(S - R)$ and $T > 0$, or $A > (S + 1)/(S - R)$ and $-\tfrac{2}{3} < T < 0$. Since $R > -3$, we easily see from (XV.20)$_2$ that

$$-\left(1 + \frac{4}{3T}\right) > S \qquad \text{(XV.30)}$$

no matter what be the sign of T. In particular, then,

$$-\left(1 + \frac{4}{3T}\right) > S > R \qquad \text{if } T < 0. \qquad \text{(XV.31)}$$

Next we rewrite (XV.19) in the form

$$\left(\chi + 1 + \frac{4}{3T}\right)^2 + 2\left(\chi + 1 + \frac{4}{3T}\right)\left(1 - \frac{2}{3T}\right) - \frac{16}{9T} = 0, \qquad \text{(XV.32)}$$

from which it follows that for each root χ the number $\chi + 1 + \tfrac{4}{3}T^{-1}$ is either positive for all positive T or negative for all positive T. By (XV.30) this quantity is negative for the root S, so

$$R > -\left(1 + \frac{4}{3T}\right) > S \qquad \text{if } T > 0. \qquad \text{(XV.33)}$$

From this last inequality we see that of the two terms which make up p or the three terms which make up M_{11}, M_{22}, and M_{33}, in each case the first dominates as $t \to \infty$. If $T < 0$, we see from (XV.31) that the first term again dominates, this time as $t \to -T$. We may summarize this behavior by saying that the solutions of Class 3 have the *dominant part*

$$\frac{p^D}{p(0)} = A\left(1 + \frac{t}{T}\right)^R,$$

$$\frac{P^D_{11}}{p(0)} = -\frac{3R + 5}{2} A\left(1 + \frac{t}{T}\right)^R, \qquad \text{(XV.34)}$$

$$P^D_{22} = P^D_{33} = -\tfrac{1}{2} P^D_{11},$$

$$P^D_{12} = P^D_{13} = P^D_{23} = 0.$$

It is not difficult to see that the dominant forms (XV.34) are grossly determined. We note first that the components of $\mathbf{P}^D$ are proportional to the corresponding dominant form p^D for the pressure, the constants of proportionality being determined by the roots R and S. According to (XV.20) these roots are determined by T. Finally, from (II.76), (XV.10), (II.66), and (II.68) we see that

$$\mathsf{T} = \frac{4}{9\mathtt{a}} \sqrt{\frac{2\mathtt{m}^3}{\mathtt{g}} \frac{E(t)}{\rho(t)}}. \qquad \text{(XV.35)}$$

Thus T is a grossly determined parameter. If we denote by $o(1)$ any quantity which approaches 0 as t approaches ∞ or $-T$ according as T is positive or negative, we may infer from the preceding results two theorems that summarize the position of the general solutions:

First Main Theorem on homo-energetic extension. *In reference to the limits $t \to \infty$ or $t \to -T$, whichever is appropriate, a solution for which $A > 0$ and either $\mathsf{T} > 0$ or $-\tfrac{2}{3} < \mathsf{T} < 0$ satisfies the asymptotic relation*

$$P_{km}(t, \mathbf{x}) = P^G_{km}(\rho(t, \mathbf{x}), \text{grad } \rho(t, \mathbf{x}))[1 + o(1)], \qquad \text{(XV.36)}$$

P^G *being the* gross determiner:

$$P^G_{11} = -\frac{3R + 5}{2} p, \qquad P^G_{22} = \frac{3R + 5}{4} p, \qquad P^G_{33} = \frac{3R + 5}{4} p,$$

$$P^G_{12} = P^G_{13} = P^G_{23} = 0. \qquad \text{(XV.37)}$$

The solutions for which $\mathsf{T} \leq -\tfrac{2}{3}$ do not all have grossly determined asymptotic forms.

In the context of the theorem just stated we should recall that when $\mathsf{T} < 0$ it is not the tension number, which in this book the symbol T usually denotes. Rather, as we have stated shortly after (IV.13), the tension number is then $-\tfrac{1}{2}\mathsf{T}$. Thus solutions that fail to have a grossly determined asymptotic form correspond to tension numbers not less than $\tfrac{1}{3}$.

(v) STOKES–KIRCHHOFF APPROXIMATION

Just as we did for homo-energetic simple shearing, here also we may show that for certain particular solutions of the family the error term in (XV.36) is null. Indeed, we simply choose the initial conditions such as to annul the damped part of the solution. From (XV.36) this means that we choose $\mathbf{P}(0, \mathbf{x}) = \mathbf{P}^G(\rho(0, \mathbf{x}), \operatorname{grad} \rho(0, \mathbf{x}))$. The solutions which result from this choice of initial values will be called the *principal solutions* for homo-energetic extension. We present them and describe their relation to general solutions in the following

Second Main Theorem on homo-energetic extension. *Any solution for which $A > 0$ and either $-\frac{2}{3} < \mathsf{T} < 0$ or $\mathsf{T} > 0$ approaches asymptotically a principal solution:*

$$M_{km}(t, \mathbf{x}) = M^P_{km}(t, \mathbf{x})[1 + o(1)], \qquad (\text{XV}.38)$$

this principal solution being determined by the initial values $\mathbf{M}(0, \mathbf{x})$ as follows:

$$p^P = p^P(0)\left(1 + \frac{t}{T}\right)^R, \qquad p^P(0) \equiv p(0)\frac{S + \dfrac{5}{3} + \dfrac{2}{3}\dfrac{P_{11}(0)}{p(0)}}{S - R},$$

$$P^P_{11} = -\frac{3R + 5}{2}p^P, \qquad P^P_{22} = \frac{3R + 5}{4}p^P, \qquad P^P_{33} = \frac{3R + 5}{4}p^P, \qquad (\text{XV}.39)$$

$$P^P_{12} = P^P_{13} = P^P_{23} = 0.$$

Of course the principal solutions are grossly and momentally determined. That is, if $\boldsymbol{\rho}^P$ is the principal moment which corresponds to ρ, $\mathbf{u}$, and p^P through (III.15) and (VI.23), then

$$P^P_{km}(t, \mathbf{x}) = P^G_{km}(\boldsymbol{\rho}^P(t, \mathbf{x}), \operatorname{grad} \boldsymbol{\rho}^P(t, \mathbf{x})). \qquad (\text{XV}.40)$$

The particular principal solution that makes (XIV.43)$_1$ valid is specified by an initial condition analogous to (XIV.46), namely,

$$p^P(0) = Ap(0). \qquad (\text{XV}.41)$$

(v) *Homo-energetic extension, III. Asymptotic status of the Stokes–Kirchhoff solution*

In Chapter II we presented the solution for homo-energetic extension according to the Stokes–Kirchhoff theory. As the general remarks in Chapter IV indicate, if we are to find any agreement between it and the corresponding solution in the kinetic theory we should look only at circumstances which make T small. We shall investigate now the asymptotic form of the solution in the kinetic theory as $\mathsf{T} \to 0$.

Using either (XV.19) or (XV.20)$_1$, we can show that the root R has an expansion in powers of T:

$$R = -\tfrac{5}{3} + \tfrac{2}{3}T - \tfrac{1}{3}T^2 + O(T^3) \quad \text{as} \quad T \to 0. \tag{XV.42}$$

This expansion converges if $|T| < \tfrac{2}{3}$ and diverges otherwise. By placing (XV.42) into (XV.20)$_2$ we obtain the following expansion for the second root S:

$$S = -\frac{4}{3T} - \frac{7}{3} - \frac{2}{3}T + O(T^2) \quad \text{as} \quad T \to 0. \tag{XV.43}$$

This expansion converges if and only if $0 < |T| < \tfrac{2}{3}$.

Upon comparing (II.79) with (XV.34)$_1$, we might seek to identify the number R in the Stokes–Kirchhoff solution with one of the roots R and S. From (II.78) we see that this root must be R since it should be a smooth function of T near 0. In fact, if we specialize (II.78) to a gas for which $\gamma = \tfrac{5}{3}$, we obtain precisely the first two terms on the right-hand side of (XV.42), and so the number R in the Stokes–Kirchhoff theory is recovered here for small values of T. Moreover, from (XV.42) we easily find that

$$\frac{3R + 5}{2} = T - \tfrac{1}{2}T^2 + O(T^3) \quad \text{as} \quad T \to 0, \tag{XV.44}$$

and we see from (II.80) and (XV.37), then, that the Stokes–Kirchhoff constitutive relations also are borne out fairly well for small values of T.

However, the solutions as functions of t generally fail to agree as t increases. From (II.79) and (XV.34)$_1$ we find that

$$\frac{p^{S-K}}{p^D} = \frac{1}{A}\left(1 + \frac{t}{T}\right)^{-(5/3)+(2/3)T-R}, \tag{XV.45}$$

R being given by (XV.20)$_1$. From (XV.42) we see that $R + \tfrac{5}{3} - \tfrac{2}{3}T$ is negative for small values of T. Thus, *if* T > 0, *the Stokes–Kirchhoff solution for the pressure becomes infinitely larger than the corresponding solution according to the kinetic theory as time proceeds; if* T < 0, *it becomes infinitely smaller.* This conclusion is similar to that which we reached for homo-energetic simple shearing. It is more striking in that it refers not only to the limit $t \to \infty$ (as indeed it does if T > 0) but also, if T < 0, to the limit as $t \to$ the *finite* time $-T$.

In reaching this conclusion we have compared the solutions according to both theories by first letting t be large or close to $-T$, according as $T > 0$ or $T < 0$, and then considering the forms of the solutions as T $\to 0$. As with simple shearing, however, we shall see now that all solutions for extensions according to the kinetic theory agree well with the Stokes–Kirchhoff solution if we hold t and T fixed and consider the forms of the solutions for successively lighter gases: $\mathfrak{m} \to 0$. In view of (II.76) and (XV.10), results in this limit are the same as those in the limit as T $\to 0$. Of course we fix both $\mathfrak{g}$ and the initial values $\rho(0)$

(vi) RETROSPECT ON AFFINE FLOWS

and $\mathbf{M}(0)$ in considering this limit. From (XV.22), (XV.42), and (XV.43) we find that as $T \to 0$

$$A = 1 - \frac{1}{2}\frac{P_{11}(0)}{p(0)}T + O(T^2). \qquad (XV.46)$$

Therefore,

$$\frac{p}{p^{S-K}} = A\left(1 + \frac{t}{T}\right)^{R+(5/3)-(2/3)T} + (1-A)\left(1 + \frac{t}{T}\right)^{S+(5/3)-(2/3)T},$$

$$= (1 + O(T))\left(1 + \frac{t}{T}\right)^{O(T^2)} + T\left(\frac{1}{2}\frac{P_{11}(0)}{p(0)} + O(T)\right) \qquad (XV.47)$$

$$\times \left(1 + \frac{t}{T}\right)^{-4/(3T)-(2/3)+O(T)} \quad \text{as} \quad T \to 0.$$

It is not difficult to show[7] that if $t > 0$, then, no matter what be the sign of T,

$$\lim_{T \to 0}\left(1 + \frac{t}{T}\right)^{O(T^2)} = 1,$$

$$\lim_{T \to 0} T\left(1 + \frac{t}{T}\right)^{-4/(3T)} = 0. \qquad (XV.48)$$

From (XV.47) and (XV.48) we conclude that for each fixed time t

$$\frac{p(t)}{p^{S-K}(t)} \to 1 \quad \text{as} \quad T \to 0: \qquad (XV.49)$$

If we hold the initial conditions fixed, *in the limit of successively lighter gases the Stokes–Kirchhoff theory and the kinetic theory deliver asymptotically the same solutions at any fixed time.* This same fact was reflected in both other exact solutions the details of which we calculated earlier.

(vi) Retrospect

The three solutions we have worked out and analysed in this chapter and the preceding one illustrate three altogether different kinds of behavior compatible with the kinetic theory.

---

[7] For $(XV.48)_2$ we note that $1 + t/T > 1$ if $T > 0$, while $1 + t/T < 1$ if $T < 0$. Thus

$$T\left(1 + \frac{t}{T}\right)^{-4/(3T)} = T \exp\left(-\frac{4}{3}\left|\frac{1}{T}\log\left(1 + \frac{t}{T}\right)\right|\right),$$

and clearly the right-hand side here approaches 0 as $T \to 0$.

(a) *Simple shearing.* This is a steady flow which can be thought of as produced by the continued action of contact forces. The tension number T has the range $[0, \infty)$. There is no trend to equilibrium. As $t \to \infty$, the general results become consistent with NOLL's theory of the simple fluid. The flow is dissipative, and the rate of dissipation increases to ∞ as $t \to \infty$. The rates of approach are exponential.

(b) *Dilatation.* These flows are not dissipative. Always T $= 0$. There is one and only one locally Maxwellian solution F_M compatible with the flow. If the flow is an expansion, the pressures and the flux of energy become approximately consistent with those for F_M as t becomes large. However, $\rho(t) \to 0$ and $\varepsilon(t) \to 0$, so F_M itself does not tend to some F_U. If the flow is a condensation, the solution exists only in the interval $(-\infty, -T)$, and the corresponding $F_M \to$ a constant, not a molecular density, as $t \to -T$. The pressures and the flux of energy tend to values compatible with F_M, although now $\rho(t) \to \infty$ and $\varepsilon(t) \to \infty$. The rates of approach are much slower than exponential. There is no trend to equilibrium in either case.

(c) *Extension.* These flows are dissipative. The range of T is $[0, \infty)$. If the flow is a condensation, the results are much like those for a contracting dilatation. If the flow is an expansion, the velocity and the rate of dissipation tend to zero, and so do the components of **P**. The rates of approach are much slower than exponential. Since also $\rho \to 0$, there can be no trend to equilibrium.

If we consider limits as $\mathfrak{m} \to 0$ at a fixed value of t, with $\mathfrak{g}$ and initial conditions held fixed, *all three flows lead to results asymptotically compatible with their counterparts in the Stokes–Kirchhoff theory.* This fact lends further support to our interpretation of μ as being the shear viscosity in the sense of fluid mechanics.

Part E

The System of Equations for the Moments

Chapter XVI

The General System of Equations for the Moments in a Gas of Maxwellian Molecules. Ikenberry's Theorem on the Structure of Collisions Integrals

(i) *Explicit collisions integrals for a gas of Maxwellian molecules*

In Chapters XIII, XIV, and XV we presented some special classes of solutions for the moments of orders 0, 1, 2, and 3 in a gas of Maxwellian molecules. To obtain these we substituted into the equation of transfer (XIII.1) various special choices of the function g in order to get equations satisfied by the moments, (XIII.2) and (XIII.5) being particular examples of them. We then obtained the moments by solving those equations. We now ask whether a similar procedure can be used to determine the relative moments $^n\mathbf{M}$ when $n \geq 4$. An equation satisfied by $^n\mathbf{M}$ is found by putting $c_{k_1} \cdots c_{k_n}$ for g in (XIII.1). Using (III.19) and the fact that $g_{,m} = \partial_{c_m} g$, we obtain

$$\dot{M}_{k_1 \ldots k_n} + M_{k_1 \ldots k_n} E + \sum_{r=1}^{n} M_{k_1 \ldots k_{r-1} a k_{r+1} \ldots k_n} u_{k_r,a}$$

$$- \sum_{r=1}^{n} \frac{1}{\rho} M_{k_1 \ldots k_{r-1} k_{r+1} \ldots k_n} M_{k_r a, a} + M_{k_1 \ldots k_n a, a} = m\bar{\mathbb{C}} c_{k_1} \cdots c_{k_n}.$$

(XVI.1)

Before we can use this equation to determine $^n\mathbf{M}$, we must evaluate the collisions integral $\bar{\mathbb{C}} c_{k_1} \cdots c_{k_n}$. For the simplest cases, $n = 2$ and $n = 3$, we carried out this calculation in Chapter XII, and we obtained the results given by (XII.24) and (XII.25). However, the manipulations were elaborate; for polynomials of degree greater than 3 they become essentially more difficult, even for

Maxwellian molecules. MAXWELL himself[1] attempted to evaluate $\bar{\mathbb{C}} c_k c_m c_r c_s$, but because of its algebraic complexity he did not succeed. Nevertheless, the mathematical results needed for calculating such collisions integrals have been developed. For a gas of Maxwellian molecules they lead to relatively simple results, as we shall see now by exhibiting an explicit formula for $\bar{\mathbb{C}} c_{k_1} \cdots c_{k_n}$ once and for all. We begin by effecting some transformations valid for a much broader class of molecules.

If $\mathbf{f}$ is a vector, let the components of the n-fold tensor product $\mathbf{f} \otimes \cdots \otimes \mathbf{f}$ be denoted as follows: $f_n \equiv f_{k_1} f_{k_2} \cdots f_{k_n}$. In this notation, n is a multi-index representing a block of n indices $k_1 \ldots k_n$. In Appendix B to this chapter we discuss multi-indices in detail and outline some conventions relating to them which we shall use in this chapter and some later ones. In terms of this notation we wish to calculate the components $\bar{\mathbb{C}} c_n$; equivalently, we wish to calculate the scalar $S_n \bar{\mathbb{C}} c_n$ for all n^{th}-order symmetric tensors $\mathbf{S}$ (cf. (XVIB.1)). The latter expression is simpler to work with since it does not contain any free indices. To simplify the analysis still further, we consider only the case in which $\mathbf{S}$ is the n-fold tensor product $\mathbf{r} \otimes \cdots \otimes \mathbf{r}$ of a vector $\mathbf{r}$. With little change the analysis applies also to a general symmetric tensor $\mathbf{S}$. As an example of the simplifications which occur, it is not difficult to see that the following variant of the binomial theorem is valid for any three vectors $\mathbf{r}$, $\mathbf{f}$, and $\mathbf{g}$ (cf. (XVIB.5)$_1$):

$$r_n(f+g)_n = [r_a(f_a + g_a)]^n = \sum_{p=0}^{n} \binom{n}{p} (r_a f_a)^p (r_b g_b)^{n-p}; \quad \text{(XVI.2)}$$

for a general symmetric tensor $\mathbf{S}$ we have instead

$$S_{a_1 \ldots a_n}(f_{a_1} + g_{a_1}) \cdots (f_{a_n} + g_{a_n})$$
$$= S_{a_1 \ldots a_n} \sum_{p=0}^{n} \binom{n}{p} f_{a_1} \cdots f_{a_p} g_{a_{p+1}} \cdots g_{a_n}. \quad \text{(XVI.3)}$$

By repeated application of (XVI.2) we find that

$$r_n(f+g+h)_n = \sum_{p_1=0}^{n} \sum_{p_2=0}^{n-p_1} \binom{n}{p_1} \binom{n-p_1}{p_2} (r_a f_a)^{p_1} (r_b g_b)^{p_2} (r_c h_c)^{n-p_1-p_2}.$$
$$\text{(XVI.4)}$$

To evaluate $\bar{\mathbb{C}} g$, we use the form (VII.20)$_4$ of the total collisions operator so as to write

$$\bar{\mathbb{C}} g = \int F(\mathbf{v}) \int F(\mathbf{v}_*) \mathbb{O} g, \quad \text{(XVI.5)}$$

---

[1] MAXWELL [1879, Eq. (30)] was content to use as an approximation the reduction of fourth moments to combinations of second moments that is valid strictly when the molecular density is Maxwellian, namely, $\rho M_{kmrs} = M_{ks} M_{mr} + M_{ms} M_{kr} + M_{rs} M_{km}$, a relation we may read off by comparing (VIII.12)$_1$ with (VIII.9)$_2$. Cf. also the bracketed note MAXWELL added just before the Appendix.

(i) EXPLICIT COLLISIONS INTEGRALS

$\mathbb{O}$ being defined thus:

$$\mathbb{O}g \equiv w \int_0^{2\pi} \int_0^\infty (g' - g) r \, dr \, d\zeta. \tag{XVI.6}$$

It follows from (VII.7)$_2$ that $\mathbb{O}$ can also be written in the form

$$\mathbb{O}g = \int_0^{2\pi} \int_0^{\frac{1}{2}\pi} (g' - g) \mathbb{S}(\theta, w) \sin \theta \, d\theta \, d\zeta. \tag{XVI.7}$$

It is important to recall from the discussion following (VII.20) that for a given function g, the class of molecular densities F for which (VII.20)$_4$ is valid generally differs from the class for which $\bar{\mathbb{C}}g$ exists. Therefore, any conclusions which we might draw from (XVI.5) will hold only for this new class of functions, and this fact will be reflected in the statement of our final results.

The first step in evaluating $\bar{\mathbb{C}}c_n$ is to determine $\mathbb{O}c_n$. We note from (VI.38) and (VI.40) that

$$a_k(\theta, \zeta, \mathbf{w}) = \cos \theta \frac{w_k}{w} + \sin \theta \, n_k, \tag{XVI.8}$$

$\mathbf{n}$ being a unit vector orthogonal to $\mathbf{w}$. Subtracting $\mathbf{u}$ from each side of (VI.41)$_1$, then using (XVI.8) and the fact that $\mathbf{w} = \mathbf{c}_* - \mathbf{c}$, we may write $\mathbf{c}'$ as follows:

$$c'_k = c_k + \cos^2 \theta \, w_k + \sin \theta \cos \theta \, wn_k,$$
$$= \sin^2 \theta \, c_k + \cos^2 \theta \, c_{*k} + \sin \theta \cos \theta \, wn_k. \tag{XVI.9}$$

From (XVI.4) and (XVI.9)$_2$ we find that

$$r_n(c'_n - c_n)$$
$$= \sum_{p_1=0}^n \sum_{p_2=0}^{n-p_1} \binom{n}{p_1} \binom{n-p_1}{p_2} \sin^{2p_1} \theta \cos^{2p_2} \theta (w \sin \theta \cos \theta)^{n-p_1-p_2}$$
$$\times (r_b c_b)^{p_1} (r_d c_{*d})^{p_2} (r_e n_e)^{n-p_1-p_2} - (r_b c_b)^n, \tag{XVI.10}$$
$$= (\sin^{2n} \theta - 1)(r_b c_b)^n + \sum_{p_1=0}^{n-1} \sum_{p_2=0}^{n-p_1} \binom{n}{p_1} \binom{n-p_1}{p_2} \sin^{n+p_1-p_2} \theta$$
$$\times \cos^{n-p_1+p_2} \theta \, (r_b c_b)^{p_1} (r_d c_{*d})^{p_2} (wr_e n_e)^{n-p_1-p_2}.$$

Let us recall the definition

$$\mathbb{A}_{n|m}(w) \equiv 2\pi w \int_0^\infty \cos^n \theta \sin^m \theta \, r \, dr,$$
$$= 2\pi \int_0^{\frac{1}{2}\pi} \cos^n \theta \sin^{m+1} \theta \, \mathbb{S}(\theta, w) \, d\theta, \tag{XII.16}_r$$

and set

$$A_n(w) \equiv 2\pi \int_0^{\frac{1}{2}\pi} (\sin^{2n}\theta - 1) \sin\theta \, \mathbb{S}(\theta, w) \, d\theta,$$

$$= -\sum_{p=1}^{n} A_{2|2n-2p}(w). \tag{XVI.11}$$

If in addition we introduce the notation $\int_{n \perp w} \cdots$ for $\int_0^{2\pi} d\zeta \cdots$, then by applying $\mathbb{O}$, as given by (XVI.7), to the scalar $r_n c_n$ we conclude from (XVI.10)$_2$, (XII.16), and (XVI.11) that

$$r_n \mathbb{O} c_n = A_n (r_b c_b)^n + \sum_{p_1=0}^{n-1} \sum_{p_2=0}^{n-p_1} \binom{n}{p_1}\binom{n-p_1}{p_2} A_{n-p_1+p_2|n+p_1-p_2}$$

$$\times (r_b c_b)^{p_1} (r_d c_{*d})^{p_2} \left(\frac{1}{2\pi} \int_{n \perp w} (wr_e n_e)^{n-p_1-p_2} \right). \tag{XVI.12}$$

We now wish to evaluate the integral which appears on the right-hand side of (XVI.12). This step in the calculation of $\mathbb{O} c_n$ is the only one which is mathematically difficult, and so for the moment we simply assert that

$$\frac{1}{2\pi} \int_{n \perp w} (wr_e n_e)^s = \sum_{q_1=0}^{[\frac{1}{2}s]} \sum_{q_2=0}^{[\frac{1}{2}s]-q_1} a_{q_1}^s b_{q_2}^{s-2q_1} \lambda_{s-2q_1} r^{2(q_1+q_2)}$$

$$\times w^{2(q_1+q_2)} (r_e w_e)^{s-2q_1-2q_2}; \tag{XVI.13}$$

here

$$[u] \equiv \text{the greatest integer} \leq u,$$

$$a_q^s \equiv \frac{s!(2s-4q+1)!!}{(s-2q)!(2q)!!(2s-2q+1)!!},$$

$$b_q^s \equiv (-1)^q \frac{s!(2s-2q+1)!!(2s+1)}{(s-2q)!(2q)!!(2s+1)!!(2s-2q+1)!!}, \tag{XVI.14}$$

$$\lambda_s \equiv \begin{cases} 0 & \text{if } s \text{ is odd,} \\ (-1)^{\frac{1}{2}s} \dfrac{(s-1)!!}{s!!} & \text{if } s \text{ is even.} \end{cases}$$

The formula (XVI.13) expresses no more than a quadrature such as one finds in tables of integrals, although it is somewhat more complicated. It is valid for any two vectors $\mathbf{r}$ and $\mathbf{w}$, whether or not they have interpretations in the kinetic theory. We present a proof of it in Appendix A to this chapter. The special cases of (XVI.13) in which $s = 0, \ldots, 5$ are easy to write out explicitly, and for future reference we list them here:

(i) EXPLICIT COLLISIONS INTEGRALS

$$\frac{1}{2\pi}\int_{\mathbf{n}\perp\mathbf{w}} 1 = 1,$$

$$\frac{1}{2\pi}\int_{\mathbf{n}\perp\mathbf{w}} w n_k = 0,$$

$$\frac{1}{2\pi}\int_{\mathbf{n}\perp\mathbf{w}} w^2 n_k n_m = -\tfrac{1}{2}(w_k w_m - w^2 \delta_{km}),$$

$$\frac{1}{2\pi}\int_{\mathbf{n}\perp\mathbf{w}} w^3 n_k n_m n_r = 0,$$

$$\frac{1}{2\pi}\int_{\mathbf{n}\perp\mathbf{w}} w^4 n_k n_m n_r n_s = \tfrac{3}{8} w_k w_m w_r w_s - \tfrac{3}{4} w^2 w_{(k} w_m \delta_{rs)} + \tfrac{3}{8} w^4 \delta_{(km} \delta_{rs)},$$

$$\frac{1}{2\pi}\int_{\mathbf{n}\perp\mathbf{w}} w^5 n_k n_m n_r n_s n_t = 0;$$

(XVI.15)

as usual, parentheses around a set of s subscripts indicates the sum over the $s!$ permutations of the indices, divided by $s!$.

If we replace s by $n - p_1 - p_2$ in (XVI.13) and put the resulting formula into (XVI.12), we find that

$$r_n \mathbb{O} c_n = \mathbb{A}_n (r_b c_b)^n + \sum_{p_1=0}^{n} \sum_{p_2=0}^{n-p_1} \sum_{q_1=0}^{[\frac{1}{2}(n-p)]} \sum_{q_2=0}^{[\frac{1}{2}(n-p)]-q_1} \binom{n}{p_1}\binom{n-p_1}{p_2} \mathbb{A}_{n-p_1+p_2 | n+p_1-p_2}$$

$$\times a_{q_1}^{n-p} b_{q_2}^{n-p-2q_1} \lambda_{n-p-2q_1} r^{2q} (c^2 - 2c_b c_{*b} + c_*^2)^q \quad \text{(XVI.16)}$$

$$\times (r_d c_d)^{p_1} (r_e c_{*e})^{p_2} [r_f (c_{*f} - c_f)]^{n-p-2q},$$

$$p = p_1 + p_2, \quad q = q_1 + q_2.$$

Next we expand the terms

$$(c^2 - 2c_b c_{*b} + c_*^2)^q \quad \text{and} \quad [r_f(c_{*f} - c_f)]^{n-p-2q}.$$

For the first one we use the binomial theorem to obtain

$$(c^2 - 2c_b c_{*b} + c_*^2)^q = \sum_{m=0}^{q} \binom{q}{m} c^{2m} (c_*^2 - 2c_b c_{*b})^{q-m},$$

$$= \sum_{m_1=0}^{q} \sum_{m_2=0}^{q-m_1} \binom{q}{m_1}\binom{q-m_1}{m_2} c^{2m_1} c_*^{2m_2} (-2c_b c_{*b})^{q-m_1-m_2};$$

(XVI.17)

for the second expression, it follows from (XVI.2) that

$$[r_f(c_{*f} - c_f)]^{n-p-2q} = \sum_{a=0}^{n-p-2q} \binom{n-p-2q}{a} (r_f c_{*f})^a (-r_g c_g)^{n-p-2q-a}.$$

(XVI.18)

XVI. EQUATIONS FOR THE MOMENTS

Putting (XVI.17) and (XVI.18) into (XVI.16) and rearranging, we obtain

$$r_n \mathbb{O}c_n = \mathbb{A}_n (r_b c_b)^n + \Sigma^* \mathbb{A}_{n-p_1+p_2|n+p_1-p_2} r^{2q}$$
$$\times c_*^{2m_1} c_{**}^{2m_2}(c_b c_{*b})^{q-m_1-m_2}(r_d c_{*d})^{p_2+a}(r_e c_e)^{n-p_2-2q-a}, \quad \text{(XVI.19)}$$

the operator Σ^* being defined as follows:

$$\Sigma^* \cdots \equiv \sum_{p_1=0}^{n-1}\sum_{p_2=0}^{n-p_1}\sum_{q_1=0}^{[\frac{1}{2}(n-p)]}\sum_{q_2=0}^{[\frac{1}{2}(n-p)]-q_1}\sum_{m_1=0}^{q}\sum_{m_2=0}^{q-m_1}\sum_{a=0}^{n-p-2q}$$
$$\times \binom{n}{p_1}\binom{n-p_1}{p_2}\binom{q}{m_1}\binom{q-m_1}{m_2}\binom{n-p-2q}{a}$$
$$\times a_{q_1}^{n-p}b_{q_2}^{n-p-2q_1}\lambda_{n-p-2q_1}2^{q-m_1-m_2}(-1)^{n-p-m_1-m_2-a-q}\cdots. \quad \text{(XVI.20)}$$

Finally, since (XVI.19) holds for all vectors $\mathbf{r}$, the symmetric part of the coefficient of $r_{k_1} \cdots r_{k_n}$ on one side must equal that on the other side, and this gives the following formula for the components of the tensor $\mathbb{O}c_n$:

$$\mathbb{O}c_{k_1}\cdots c_{k_n} = \mathbb{A}_n c_{k_1}\cdots c_{k_n} + \Sigma^* \mathbb{A}_{n-p_1+p_2|n+p_1-p_2}$$
$$\times c_{b_1}c_{b_1}\cdots c_{b_{m_1}}c_{b_{m_1}}c_{d_1}\cdots c_{d_{q-m_1-m_2}}c_{(k_1}\cdots c_{k_{n-p_2-2q-a}}$$
$$\times \delta_{k_{n-p_2-2q-a+2}}\cdots \delta_{k_{n-p_2-a-1}k_{n-p_2-a}} \quad \text{(XVI.21)}$$
$$\times c_{*k_{n-p_2-a+1}}\cdots c_{*k_n)}c_{*d_1}\cdots c_{*d_{q-m_1-m_2}}c_{*e_1}c_{*e_1}\cdots c_{*e_{m_2}}c_{*e_{m_2}}.$$

In summary, (XVI.21) delivers $\mathbb{O}c_n$ *for any molecular model that renders convergent the integrals which define* $\mathbb{A}_n$ *and* $\mathbb{A}_{n-p_1+p_2|n+p_1-p_2}$. Because $\mathbb{A}_n$ and $\mathbb{A}_{n-p_1+p_2|n+p_1-p_2}$ are functions of the relative speed w, $\mathbb{O}c_n$ is generally a complicated function of $\mathbf{c}$ and $\mathbf{c}_*$. However, for one molecular model, the Maxwellian, the form of $\mathbb{O}c_n$ is particularly simple. Indeed, since $\mathbb{S}$ is a function of θ alone in a gas of such molecules, we see from (XII.16) and (XVI.11) that $\mathbb{A}_n$ and $\mathbb{A}_{n-p_1+p_2|n+p_1-p_2}$ are just constants. Thus the right-hand side of (XVI.21) becomes merely a polynomial in the components of $\mathbf{c}$ and $\mathbf{c}_*$. By placing this polynomial into (XVI.5) and evaluating the integrals with respect to $\mathbf{v}$ and $\mathbf{v}_*$, assuming these converge, we obtain the collisions integral $\overline{\mathbb{C}}c_n$.

Theorem 1. *In a gas of Maxwellian molecules, if F possesses moments up to those of order n and is such as to render* (XVI.5) *valid when* $g = c_n$, *then*

$$m\overline{\mathbb{C}}c_{k_1}\cdots c_{k_n} = \frac{1}{m}\mathbb{A}_n \rho M_{k_1\ldots k_n} + \Sigma^* \frac{1}{m}\mathbb{A}_{n-p_1+p_2|n+p_1-p_2}$$
$$\times M_{b_1b_1\ldots b_{m_1}b_{m_1}d_1\ldots d_{q-m_1-m_2}(k_1\ldots k_{n-p_2-2q-a}}\delta_{k_{n-p_2-2q-a+1}k_{n-p_2-2q-a+2}}$$
$$\cdots \delta_{k_{n-p_2-a-1}k_{n-p_2-a}}M_{k_{n-p_2-a+1}\ldots k_n)d_1\ldots d_{q-m_1-m_2}e_1e_1\ldots e_{m_2}e_{m_2}};$$

$$\text{(XVI.22)}$$

(i) EXPLICIT COLLISIONS INTEGRALS

Σ^* *is the summation operator* (XVI.20), *and* $\mathbb{A}_n$ *and* $\mathbb{A}_{n-p_1+p_2|n+p_1-p_2}$ *are constants determined explicitly by* $\mathfrak{m}$ *and* $\mathfrak{g}$ *through* (XII.16) *and* (XVI.11).

It is straightforward, though somewhat tedious, to evolve the sum explicitly for any desired value of n. For the collisions integrals of orders $0, \ldots, 5$ it gives us

$$\mathfrak{m}\bar{C}1 = 0,$$

$$\mathfrak{m}\bar{C}c_k = 0,$$

$$\mathfrak{m}\bar{C}c_k c_m = -3n\mathbb{A}_{2|2} P_{km},$$

$$\mathfrak{m}\bar{C}c_k c_m c_r = -\frac{n}{2}\mathbb{A}_{2|2}(9M_{kmr} - 6q_{(k}\delta_{mr)}),$$

$$\mathfrak{m}\bar{C}c_k c_m c_r c_s = \frac{n}{4}(35\mathbb{A}_{4|4} - 28\mathbb{A}_{2|2})M_{kmrs} - \frac{3n}{2}(5\mathbb{A}_{4|4} - 2\mathbb{A}_{2|2})M_{aa(km}\delta_{rs)}$$

$$+ \frac{3n}{4}\mathbb{A}_{4|4} M_{aabb}\delta_{(km}\delta_{rs)} + \frac{3n}{4\rho}(35\mathbb{A}_{4|4} - 4\mathbb{A}_{2|2})M_{(km}M_{rs)}$$

$$- \frac{15n}{\rho}\mathbb{A}_{4|4} M_{a(k}\delta_{mr}M_{s)a} - \frac{n}{2\rho}(45\mathbb{A}_{4|4} - 18\mathbb{A}_{2|2})pM_{(km}\delta_{rs)}$$

$$+ \frac{3n}{4\rho}\mathbb{A}_{4|4}(9p^2 + 2M_{ab}M_{ab})\delta_{(km}\delta_{rs)}, \qquad \text{(XVI.23)}$$

$$\mathfrak{m}\bar{C}c_k c_m c_r c_s c_t = \frac{5n}{8}(35\mathbb{A}_{4|4} - 16\mathbb{A}_{2|2})M_{kmrst}$$

$$- \frac{5n}{4}(15\mathbb{A}_{4|4} - 4\mathbb{A}_{2|2})M_{aa(kmr}\delta_{st)} + \frac{15n}{8}\mathbb{A}_{4|4} M_{aabb(k}\delta_{mr}\delta_{st)}$$

$$+ \frac{n}{4\rho}(175\mathbb{A}_{4|4} - 20\mathbb{A}_{2|2})M_{(kmr}M_{st)}$$

$$- \frac{15n}{4\rho}(15\mathbb{A}_{4|4} - 4\mathbb{A}_{2|2})pM_{(kmr}\delta_{st)}$$

$$+ \frac{15n}{2\rho}\mathbb{A}_{4|4} M_{ab}M_{ab(k}\delta_{mr}\delta_{st)}$$

$$+ \frac{45n}{4\rho}\mathbb{A}_{4|4}\, pM_{aa(k}\delta_{mr}\delta_{st)} + \frac{75n}{4\rho}\mathbb{A}_{4|4} M_{aa(k}\delta_{mr}M_{st)}$$

$$- \frac{15n}{2\rho}\mathbb{A}_{4|4} M_{aab}M_{b(k}\delta_{mr}\delta_{st)} - \frac{75n}{2\rho}\mathbb{A}_{4|4} M_{a(km}\delta_{rs}M_{t)a}.$$

Of course (XVI.23)$_{3,4}$ agree with (XII.24) and (XII.25) if we use the explicit formula (XII.18) for $\mathbb{A}_{2|2}$.

(ii) Ikenberry's theorem: The structure of collisions integrals

Since the two moments of F which appear in the summation Σ^* above are of orders

$$2m_1 + (q - m_1 - m_2) + (n - p_2 - 2q - a)$$

and

$$(p_2 + a) + (q - m_1 - m_2) + 2m_2,$$

respectively, and the sum of these two numbers is n, we may describe the general structure of $\bar{\mathbb{C}}$ for Maxwellian molecules as follows: The value of $\bar{\mathbb{C}}$ applied to a homogeneous polynomial of degree n in the components of $\mathbf{c}$ is a linear combination of the moments of order n plus a bilinear combination of the moments of lower order, the sum of the orders in each term being n. The coefficients in the bilinear combination are functions of $\mathfrak{m}$ and $\mathfrak{g}$ alone and are independent of F. This statement expresses the most important property of Maxwellian molecules: *We can evaluate the total effect of collisions upon a polynomial function of $\mathbf{c}$ directly in terms of expectations without first having to determine F.*

The general structure of $\bar{\mathbb{C}}$ just described, while suggested by the partial results of MAXWELL, was first revealed in a theorem discovered and proved by IKENBERRY[2]. Although he did not obtain the specific, general formula (XVI.22), his theorem, in addition to presenting the structure of $\bar{\mathbb{C}}$, describes an important property of the collisions integrals which is not apparent from Theorem 1. We now outline his results.

---

[2] IKENBERRY & TRUESDELL [1956, §§7 and 13].

As BOLTZMANN [1895] remarked, the notes MAXWELL [1879] added on the proofsheets of his last paper

> show evidently that he must have made a long and elaborate investigation on this subject a short time before his death, which, however, has not been published. I have treated the same subject by a different method, and have also found that many corrections of the equations of hydrodynamics can be derived from the theory of gases. It will be not easy, but perhaps not impossible, to test some of these differences by experiment. I have not yet published these results, because they do not agree in all respects with the results briefly announced by Maxwell, and the danger of falling into errors in this subject is great.
>
> With regard to this I beg the British Association to make efforts to ascertain if the manuscript of the investigation made by Maxwell on the application of spherical harmonics to the theory of gases is still in existence, and, if this manuscript should be lost, to encourage physicists to repeat these calculations.

BOLTZMANN's plea "to repeat these calculations" went unheeded, it seems, until IKENBERRY & TRUESDELL took up the work in 1954–1955. BOLTZMANN's warning about "the danger of falling into errors" was in the end confirmed, for his only published result of this kind, like MAXWELL's, is partly incorrect. *Cf.* our Footnote 1 to Chapter XIII, our discussion of (XXV.14), and our Footnote 27 to Chapter XXIV.

(ii) IKENBERRY'S POLYNOMIALS

Rather than consider $\bar{\mathbb{C}}c_n$ IKENBERRY chose to examine $\bar{\mathbb{C}}g$ for certain special polynomials g. These polynomials, which we denote by $\mathbf{Y}_s$ with components $Y_s = Y_{k_1 \ldots k_s}$, are defined[3] as follows. We obtain $\mathbf{Y}_s$ by subtracting from $c_{k_1} \cdots c_{k_s}$ that homogeneous symmetric polynomial of degree s in the components of $\mathbf{c}$ such as to annul the result of contracting the components of $\mathbf{Y}_s$ on any pair of indices. The first few $\mathbf{Y}_s$ are

$$Y(\mathbf{c}) \equiv 1,$$

$$Y_k(\mathbf{c}) \equiv c_k,$$

$$Y_{km}(\mathbf{c}) \equiv c_k c_m - \tfrac{1}{3} c^2 \, \delta_{km}, \qquad (\text{XVI}.24)$$

$$Y_{kmr}(\mathbf{c}) \equiv c_k c_m c_r - \tfrac{3}{5} c^2 c_{(k} \, \delta_{mr)},$$

$$Y_{kmrs}(\mathbf{c}) \equiv c_k c_m c_r c_s - \tfrac{6}{7} c^2 Y_{(km}(\mathbf{c}) \, \delta_{rs)} - \tfrac{1}{5} c^4 \, \delta_{(km} \, \delta_{rs)},$$

$$Y_{kmrst}(\mathbf{c}) \equiv c_k c_m c_r c_s c_t - \tfrac{10}{9} c^2 Y_{(kmr}(\mathbf{c}) \, \delta_{st)} - \tfrac{3}{7} c^4 Y_{(k}(\mathbf{c}) \, \delta_{mr} \, \delta_{st)}.$$

It is not difficult to derive the following general formula for the components of $\mathbf{Y}_s$:

$$Y_{k_1 \ldots k_s}(\mathbf{c}) = \sum_{q=0}^{[\tfrac{1}{2}s]} b_q^s c^{2q} c_{(k_1} \cdots c_{k_{s-2q}} \delta_{k_{s-2q+1} k_{s-2q+2}} \cdots \delta_{k_{s-1} k_s)}, \qquad (\text{XVI}.25)$$

b_q^s being given by $(\text{XVI}.14)_4$.

We define polynomials $\mathbf{Y}_{2r|s}$ as follows in terms of their components $Y_{2r|s}$:

$$Y_{2r|s}(\mathbf{c}) \equiv c^{2r} Y_s(\mathbf{c}). \qquad (\text{XVI}.26)$$

These polynomials form a complete set: Any symmetric polynomial can be expressed uniquely as a linear combination of them. For example,

$$c_{k_1} \cdots c_{k_s} = \sum_{q=0}^{[\tfrac{1}{2}s]} a_q^s \, Y_{2q|(k_1 \ldots k_{s-2q}}(\mathbf{c}) \, \delta_{k_{s-2q+1} k_{s-2q+2}} \cdots \delta_{k_{s-1} k_s)}, \qquad (\text{XVI}.27)$$

a_q^s being given by $(\text{XVI}.14)_3$.

A set of corrected proofsheets is included in the bound volume entitled *Kinetic Theory of Gases*, Vol. 1, in the Stokes Collection, presently suffering slow disintegration by shelf tides and occasional disappearance in the Johns Hopkins University Library. The corrections, some in STOKES's hand, are those mentioned in MAXWELL's letter of 21 August 1879 to STOKES in response to suggestions by KELVIN which STOKES had transmitted.

In 1970 Professor S. G. BRUSH kindly made available to TRUESDELL copies of such notes of MAXWELL on this subject as remain. They provide a basis for the remarks on spherical harmonics MAXWELL inserted on the proofsheets of his paper but do not go beyond them. They do not contain the details of his calculation of the "corrections" that BOLTZMANN questioned.

[3] IKENBERRY introduced these polynomials, the first few of which are displayed in (XVI.24), in his paper "A system of homogeneous spherical harmonics", *American Mathematical Monthly* **62**, 719–721 (1955); later he provided an improved development: "A system of homogeneous spherical harmonics", *Journal of Mathematical Analysis and its Applications* **3**, 355–357 (1961).

Corresponding to each polynomial $\mathbf{Y}_{2r|s}$ we define the *spherical moment* $\mathbf{P}_{2r|s}$ as follows:

$$P_{2r|s} \equiv \rho \overline{Y_{2r|s}} = \mathfrak{m} \int F Y_{2r|s}. \tag{XVI.28}$$

The sum $2r + s$ is the *order* of the spherical moment $\mathbf{P}_{2r|s}$. From (XVI.25) and (XVI.27) we see that the spherical moments $\mathbf{P}_{2r|s}$ and the relative moments $^{2r+s}\mathbf{M}$ may be expressed as linear combinations of each other. For example,

$$P_{0|km} = M_{km} - p\,\delta_{km} = P_{km},$$

$$P_{2|0} = 3p,$$

$$P_{0|kmr} = M_{kmr} - \tfrac{3}{5} M_{aa(k}\,\delta_{mr)},$$

$$P_{2|k} = M_{kaa} = 2q_k,$$

$$P_{0|kmrs} = M_{kmrs} - \tfrac{6}{7} M_{aa(km}\,\delta_{rs)} + \tfrac{3}{35} M_{aabb}\,\delta_{(km}\,\delta_{rs)}, \tag{XVI.29}$$

$$P_{2|km} = M_{kmaa} - \tfrac{1}{3} M_{aabb}\,\delta_{km},$$

$$P_{4|0} = M_{aabb},$$

$$P_{0|kmrst} = M_{kmrst} - \tfrac{10}{9} M_{aa(kmr}\,\delta_{st)} + \tfrac{5}{21} M_{aabb(k}\,\delta_{mr}\,\delta_{st)},$$

$$P_{2|kmr} = M_{aakmr} - \tfrac{3}{5} M_{aabb(k}\,\delta_{mr)},$$

$$P_{4|k} = M_{aabbk}.$$

Also, straightforward calculation shows that if F is Maxwellian

$$P_{2r|s} = P_{2r|s}^{(0)} \equiv \begin{cases} 0 & \text{if } s \ne 0, \\ (2r+1)!!\,\rho\left(\dfrac{p}{\rho}\right)^r & \text{if } s = 0. \end{cases} \tag{XVI.30}$$

We may now state

Ikenberry's Theorem. *In a gas of Maxwellian molecules, if F possesses moments of orders $1, \ldots, 2r + s$ and is such as to render* (XVI.5) *valid when* $g = Y_{2r|s}$, *then*

$$\mathfrak{m}\overline{\mathcal{C} Y_{2r|s}} = -\mathbb{C}_{2r|s} P_{2r|s} + Q_{2r|s}, \qquad 2r + s \geq 1; \tag{XVI.31}$$

$Q_{2r|s}$ *is a bilinear function of spherical moments the orders of which are positive numbers whose sum is* $2r + s$:

$$Q_{2r|s} = \sum \mathbb{C}_{r_1 r_2 | s_1 s_2} P_{2r_1|s_1} P_{2r_2|s_2}, \tag{XVI.32}$$

$$2r_1 + s_1 \geq 2r_2 + s_2 > 0, \qquad 2r_1 + s_1 + 2r_2 + s_2 = 2r + s;$$

(ii) IKENBERRY'S THEOREM 247

the tensorial coefficient $\mathbb{C}_{r_1 r_2 | s_1 s_2 s}$ is a function of $\mathfrak{m}$ and $\mathfrak{g}$ alone, and the scalar coefficient $\mathbb{C}_{2r|s}$ is given as follows:

$$\mathbb{C}_{2r|s} = 4\pi n \sqrt{\frac{\mathfrak{g}}{2\mathfrak{m}}} \int_0^\infty [1 - \cos^{2r+s}\theta \, P_s(\cos\theta) - \sin^{2r+s}\theta \, P_s(\sin\theta)] z \, dz,$$
(XVI.33)

in which P_s denotes the Legendre polynomial of order s, and θ is defined as a function of z by (XII.6).

Of course, in view of the general formula (XVI.22) for the collisions integrals we know that each of the coefficients $\mathbb{C}_{2r|s}$ and $\mathbb{C}_{r_1 r_2 | s_1 s_2 s}$ is a linear combination of the coefficients $\mathbb{A}_n$ and $\mathbb{A}_{n|m}$. For example, from (XVI.33), (XVI.11), and (XII.16) we find that

$$\mathbb{C}_{0|2} = 3n\mathbb{A}_{2|2},$$
$$\mathbb{C}_{0|4} = 7n(\mathbb{A}_{2|2} - \tfrac{5}{4}\mathbb{A}_{4|4}).$$
(XVI.34)

The values of $-\tfrac{3}{5}\tau \, \mathbb{C}_{2r|s}$ have been tabulated by numerical integration[4] of (XVI.33) for all values of r and s such that $2r + s \leq 36$.

To see the difference between the characterizations of $\bar{\mathbb{C}}$ given by Theorem 1 and by IKENBERRY's theorem, let us examine the terms on the right-hand side of (XVI.22) that are linear in moments of order n. One such term is $(1/\mathfrak{m})\mathbb{A}_n \rho M_n$. In addition, the expression following Σ^* will be linear in n^{th} moments if $p_2 + a = 0$, $q = m_1 + m_2$, and $m_2 = 0$;

$$M_{k_{n-p_2-a+1} \cdots k_n d_1 \cdots d_{q-m_1-m_2} e_1 e_1 \cdots e_{m_2} e_{m_2}}$$

is nothing other than ρ in this case. It will be linear also if $n - p_2 - 2q - a = 0$, $q = m_1 + m_2$, and $m_1 = 0$, since then

$$M_{b_1 b_1 \cdots b_{m_1} b_{m_1} d_1 \cdots d_{q-m_1-m_2} k_1 \cdots k_{n-p_2-2q-a}}$$

is simply ρ. Collecting these three contributions, we obtain the expression

$$\frac{1}{\mathfrak{m}} (\mathbb{A}_n + \mathbb{A}_{2n|0}) \rho M_{k_1 \ldots k_n} + \sum_{p=0}^{n-1} \sum_{q_1=0}^{[\frac{1}{2}(n-p)]} \sum_{q_2=0}^{[\frac{1}{2}(n-p)]-q_1} \binom{n}{p} a_{q_1}^{n-p} b_{q_2}^{n-p-2q_1} \lambda_{n-p-2q_1}$$

$$\times \frac{1}{\mathfrak{m}} (\mathbb{A}_{n+p|n-p} + (-1)^{n-p} \mathbb{A}_{n-p|n+p}) \rho M_{b_1 b_1 \ldots b_q b_q (k_1 \ldots k_{n-2q}}$$
(XVI.35)

$$\times \delta_{k_{n-2q+1} k_{n-2q+2}} \cdots \delta_{k_{n-1} k_n)}.$$

In contrast, if we isolate from the right-hand side of (XVI.31) the moments of highest order, namely those of order $2r + s$, we obtain not a linear combination like (XVI.35) but a single moment of order $2r + s$:

$$-\mathbb{C}_{2r|s} P_{2r|k_1 \ldots k_s}.$$
(XVI.36)

---

[4] ALTERMAN, FRANKOWSKI, & PEKERIS [1962, Table 1]. *Cf.* also WANG CHANG & UHLENBECK [1953, Appendix II].

XVI. EQUATIONS FOR THE MOMENTS

The difference may be seen explicitly if we compare (XVI.23) with the following list[5] of values of $\bar{\mathbb{C}}Y_{2r|s}$ that follows from (XVI.23), (XVI.24), and (XVI.29):

$$\mathfrak{m}\bar{\mathbb{C}} Y_{0|0} = 0,$$

$$\mathfrak{m}\bar{\mathbb{C}} Y_{0|k} = 0,$$

$$\mathfrak{m}\bar{\mathbb{C}} Y_{2|0} = 0,$$

$$\mathfrak{m}\bar{\mathbb{C}} Y_{0|km} = -3n\mathbb{A}_{2|2} P_{km},$$

$$\mathfrak{m}\bar{\mathbb{C}} Y_{2|k} = -4n\mathbb{A}_{2|2} q_{k},$$

$$\mathfrak{m}\bar{\mathbb{C}} Y_{0|kmr} = -\frac{9n}{2} \mathbb{A}_{2|2} P_{0|kmr},$$

$$\mathfrak{m}\bar{\mathbb{C}} Y_{4|0} = -\frac{2n}{\rho} \mathbb{A}_{2|2}(\rho P_{4|0} - 15p^2 + P_{ab} P_{ab}),$$

$$\mathfrak{m}\bar{\mathbb{C}} Y_{2|km} = -\frac{7n}{2\rho} \mathbb{A}_{2|2}(\rho P_{2|km} - p P_{km} + \tfrac{4}{7} P_{\{ka} P_{am\}}),$$

$$\mathfrak{m}\bar{\mathbb{C}} Y_{0|kmrs} = -\frac{7n}{4} (4\mathbb{A}_{2|2} - 5\mathbb{A}_{4|4}) P_{0|kmrs} \qquad \text{(XVI.37)}$$
$$- \frac{3n}{4\rho} (4\mathbb{A}_{2|2} - 35\mathbb{A}_{4|4}) P_{\{km} P_{rs\}},$$

$$\mathfrak{m}\bar{\mathbb{C}} Y_{4|k} = -\frac{3n}{\rho} \mathbb{A}_{2|2}(\rho P_{4|k} - \tfrac{28}{3} p q_{k} + \tfrac{2}{3} P_{0|kab} P_{ab} + \tfrac{28}{15} q_{a} P_{ka}),$$

$$\mathfrak{m}\bar{\mathbb{C}} Y_{2|kmr} = -\frac{n}{2} (11\mathbb{A}_{2|2} - 10\mathbb{A}_{4|4}) P_{2|kmr} + \frac{9n}{\rho} (\mathbb{A}_{2|2} - 5\mathbb{A}_{4|4}) p P_{0|kmr}$$
$$- \frac{3n}{2\rho} (2\mathbb{A}_{2|2} + 5\mathbb{A}_{4|4}) P_{\{ak} P_{0|mra\}}$$
$$- \frac{27n}{5\rho} (\mathbb{A}_{2|2} - 10\mathbb{A}_{4|4}) q_{\{k} P_{mr\}},$$

$$\mathfrak{m}\bar{\mathbb{C}} Y_{0|kmrst} = -\frac{5n}{8} (16\mathbb{A}_{2|2} - 35\mathbb{A}_{4|4}) P_{0|kmrst}$$
$$- \frac{5n}{4\rho} (4\mathbb{A}_{2|2} - 35\mathbb{A}_{4|4}) P_{0|\{kmr} P_{st\}}.$$

[5] All items but the last were given by IKENBERRY & TRUESDELL [1956, §8], except for corrected numerical coefficients in $\mathfrak{m}\bar{\mathbb{C}} Y_{4|k}$ provided by STREET [1960, Appendix B]. The coefficients $\mathfrak{B}_{k}$ used by IKENBERRY & TRUESDELL are expressed as follows in terms of the coefficients $\mathbb{A}_{m|n}$:

$$\mathfrak{B}_{2} = 3\mathbb{A}_{2|2}, \qquad \mathfrak{B}_{4} = 10\mathbb{A}_{2|2} - 35\mathbb{A}_{4|4}.$$

For shorter writing, braces around a set of indices indicate the totally symmetric traceless part, formed after all repeated indices have been summed. For example,

$$A_{\{km\}} \equiv A_{(km)} - \tfrac{1}{3} A_{aa}\, \delta_{km},$$
$$A_{\{ka} B_{am\}} \equiv \tfrac{1}{2} A_{ka} B_{am} + \tfrac{1}{2} A_{ma} B_{ak} - \tfrac{1}{3} A_{ab} B_{ba}\, \delta_{km}, \qquad \text{(XVI.38)}$$
$$A_{\{kmr\}} \equiv A_{(kmr)} - \tfrac{1}{15}[(A_{aak} + A_{aka} + A_{kaa})\, \delta_{mr}$$
$$+ (A_{aam} + A_{ama} + A_{maa})\, \delta_{rk} + (A_{aar} + A_{ara} + A_{raa})\, \delta_{km}].$$

The values of $\mathbb{A}_{2|2}$ and $\mathbb{A}_{4|4}$ determined through use of the formulae given in Footnote 5 and the numerical values of $\mathfrak{B}_2$ and $\mathfrak{B}_4$ calculated for IKENBERRY & TRUESDELL and given above in (XII.20) and Footnote 5 to Chapter XII are as follows[6]:

$$\mathbb{A}_{2|2} = \sqrt{\frac{\mathfrak{g}}{2\mathfrak{m}}} \times 1.3703\ldots, \qquad \mathbb{A}_{4|4} = \sqrt{\frac{\mathfrak{g}}{2\mathfrak{m}}} \times 0.111\ldots. \qquad \text{(XVI.39)}$$

MAXWELL's methods would have enabled him to obtain (XVI.22). In following them, IKENBERRY & TRUESDELL found that difficulties arose at the very stage at which MAXWELL himself abandoned his attempts after having calculated a few terms: $n = 4$. The collisions integrals, indeed, could be calculated, but it was not at all clear that the result could be inverted so as to obtain an explicit formula for $^4\mathbf{M}$ in terms of them. Rather than prove the algebraic inversion possible, IKENBERRY introduced the spherical moments $\mathbf{P}_{2r|s}$ and through use of them not only effected that inversion but also evaluated the coefficient $\mathbb{c}_{2r|s}$ explicitly. As we shall see in Chapter XVIII, and then again in Chapter XXV, the fact that (XVI.36) is a scalar multiple of the spherical moment $\mathbf{P}_{2r|s}$ alone leads to results which would be difficult to extract from (XVI.22), although that formula certainly implies them.

Since the general structure of $\bar{\mathbb{C}}$ has already been described in Theorem 1, all there remains to prove in IKENBERRY's theorem is the special form (XVI.36) of that part of $\bar{\mathbb{C}} \mathbf{Y}_{2r|s}$ which is linear in the spherical moments of order $2r + s$. We do so in Appendix A to this chapter.

(iii) The general system of equations for the moments

One of the most important illustrations of both Theorem 1 and IKENBERRY's theorem comes when we substitute (XVI.22) into (XVI.1). The result is the ***General System of Equations for the Moments***, an infinite system of differential equations into which F enters only implicitly, that is, only through its

---

[6] The table provided by ALTERMAN, FRANKOWSKI, & PEKERIS [1962] yields the ratio $\mathbb{A}_{4|4}/\mathbb{A}_{2|2} = 0.15778\ldots$, nearly twice as large as what (XVI.39) makes it. Unfortunately their results do not include a determination of $\mathbb{A}_{2|2}$.

moments. The member of this system which results from combining (XVI.1) and (XVI.22) is referred to as the *equation for the moment* $M_{k_1 \ldots k_n}$. Of course, (XIII.4) and (XIII.7)$_1$ are particular members of the system, the former being the equation for the moment P_{km}; the latter, that for q_k. One feature of the general system of equations for the moments should be noted. From (XVI.1) we see that $^{n+1}\mathbf{M}$ appears in the equations for $^n\mathbf{M}$. Thus the equations which govern the evolution in time of the nth moments do not become equations for the nth moments alone until the $n + 1^{st}$ moments are known. Because of this *forward coupling*, the system of equations for moments is in general of infinite order, indissolubly linked.

Since the polynomials $\mathbf{Y}_{2r|s}$ form a complete set, an alternative arrangement of the equations for the moments is obtained by choosing $\mathbf{Y}_{2r|s}$ for the function g in the general equation of transfer (IX.10). A somewhat more convenient result is obtained by use of the variant (IX.13) of the equation of transfer; it gives us the system

$$\mathfrak{L} Y_{2r|s} = \mathfrak{m}\bar{\mathbf{C}}\, Y_{2r|s}. \tag{XVI.40}$$

Bearing in mind the caution described following (IX.13), we may say that *the system (XVI.40) with the equation of balance of linear momentum (IX.4) adjoined is equivalent to the general system of equations for the moments*. For some purposes (XVI.40) is easier to use than the system arising from (XVI.1) and (XVI.22). The right-hand side of (XVI.40) is given, of course, by IKENBERRY's theorem (XVI.31), special cases of which are set out in (XVI.37). Specific expressions for the quantities $\mathfrak{L} Y_{2r|s}$ may be determined from (IX.14) by no more than routine calculations. For the moments of lowest orders they are as follows:

$$\mathfrak{L} Y_{0|0} = \dot{\rho} + \rho E,$$

$$\mathfrak{L} Y_{0|k} = 0,$$

$$\mathfrak{L} Y_{2|0} = 3\dot{p} + 5pE + 2P_{ab} E_{ab} + 2q_{a,a},$$

$$\mathfrak{L} Y_{0|km} = \dot{P}_{km} + P_{km} E + 2P_{\{ak} u_{m,a\}} + 2pE_{km} + P_{0|kma,a} + \tfrac{4}{5} q_{\{k,m\}},$$

$$\mathfrak{L} Y_{2|k} = 2\dot{q}_k + \tfrac{10}{3} q_k E + 2P_{0|kab} E_{ab} + 2q_a u_{k,a} + \tfrac{8}{5} q_a E_{ka}$$
$$- \frac{5p}{\rho} M_{ka,a} - \frac{2}{\rho} P_{ka} M_{ab,b} + P_{2|ka,a} + \tfrac{1}{3} P_{4|0,k},$$

$$\mathfrak{L} Y_{0|kmr} = \dot{P}_{0|kmr} + P_{0|kmr} E + 3P_{0|\{akm} u_{r,a\}} + \tfrac{12}{5} q_{\{k} E_{mr\}}$$
$$- \frac{3}{\rho} P_{\{km} M_{ra,a\}} + P_{0|kmra,a} + \tfrac{3}{7} P_{2|\{km,r\}},$$

$$\mathfrak{L} Y_{4|0} = \dot{P}_{4|0} + \tfrac{7}{3} P_{4|0} E + 4P_{2|ab} E_{ab} - \frac{8}{\rho} q_a M_{ab,b} + P_{4|a,a},$$

$$\mathfrak{L} Y_{2|km} = \dot{P}_{2|km} + \tfrac{5}{3} P_{2|km} E + 2 P_{0|kmab} E_{ab} + 2 P_{2|\{ak} u_{m,a\}}$$

$$+ \tfrac{8}{7} P_{2|\{ak} E_{ma\}} + \tfrac{14}{15} P_{4|0} E_{km} - \frac{2}{\rho} P_{0|kma} M_{ab,b}$$

$$- \frac{28}{5\rho} q_{\{k} M_{ma,a\}} + P_{2|kma,a} + \tfrac{2}{5} P_{4|\{k,m\}}, \qquad \text{(XVI.41)}$$

$$\mathfrak{L} Y_{0|kmrs} = \dot{P}_{0|kmrs} + P_{0|kmrs} E + 4 P_{0|\{akmr} u_{s,a\}} + \tfrac{12}{7} P_{2|\{km} E_{rs\}}$$

$$- \frac{4}{\rho} P_{0|\{kmr} M_{sa,a\}} + P_{0|kmrsa,a} + \tfrac{4}{9} P_{2|\{kmr,s\}},$$

$$\mathfrak{L} Y_{4|k} = \dot{P}_{4|k} + \tfrac{7}{3} P_{4|k} E + P_{4|a} u_{k,a} + 4 P_{2|kab} E_{ab} + \tfrac{8}{5} P_{4|a} E_{ak}$$

$$- \frac{7}{3\rho} P_{4|0} M_{ka,a} - \frac{4}{\rho} P_{2|ka} M_{ab,b} + P_{4|ka,a} + \tfrac{1}{3} P_{6|0,k},$$

$$\mathfrak{L} Y_{2|kmr} = \dot{P}_{2|kmr} + \tfrac{5}{3} P_{2|kmr} E + 2 P_{0|kmrab} E_{ab} + \tfrac{4}{3} P_{2|\{akm} E_{ra\}}$$

$$+ 3 P_{2|\{akm} u_{r,a\}} + \tfrac{54}{35} P_{4|\{k} E_{mr\}} - \frac{2}{\rho} P_{0|kmra} M_{ab,b}$$

$$- \frac{27}{7\rho} P_{2|\{km} M_{ra,a\}} + P_{2|kmra,a} + \tfrac{3}{7} P_{4|\{km,r\}}.$$

These values of $\mathfrak{L}$ and the list of collisions integrals (XVI.37) suffice to determine all equations for moments of orders 0, 1, 2, 3, and 4, and for a suitably arranged subset of those of order 5. Alternatively, an equivalent system is obtained from the list (XVI.23) and the appropriate special cases of (XVI.1).

It is natural to ask now whether one might be able to extend the exact solutions presented in Chapters XIII, XIV, and XV for a gas of Maxwellian molecules by solving the remaining equations in this general system for the moments. We shall examine this question more closely in Chapter XVIII. Before that, we insert a chapter explaining a different approach to the calculation of collisions integrals, which although purely formal and not yet rigorously established, may be applied to a broad class of molecular models, not just the Maxwellian.

Appendix A

Integration Formulae and the Proof of Ikenberry's Theorem

We turn now to the verification of (XVI.13). We shall use certain properties of spherical harmonics Y_q^m considered as functions defined on the unit sphere in three dimensions. Namely, there are matrices $(T_{mk}^{(q)}(\mathbf{R}))$ such that for all unit vectors **n** and rotations **R**

$$Y_q^m(\mathbf{R}^{-1}\mathbf{n}) = \sum_{k=-q}^{q} T_{mk}^{(q)}(\mathbf{R}) Y_q^k(\mathbf{n}), \quad m = -q, -q+1, \ldots, q. \quad \text{(XVIA.1)}$$

Moreover, to within a scalar factor, (XVIA.1) defines the spherical harmonics[1]; that is, if S_q^m, $m = -q, -q+1, \ldots, q$, is any set of functions defined on the unit sphere such that

$$S_q^m(\mathbf{R}^{-1}\mathbf{n}) = \sum_{k=-q}^{q} T_{mk}^{(q)}(\mathbf{R}) S_q^k(\mathbf{n}), \quad m = -q, -q+1, \ldots, q, \quad \text{(XVIA.2)}$$

for all **R** and **n**, $(T_{mk}^{(q)}(\mathbf{R}))$ being the matrix that appears in (XVIA.1), then for some constant λ_q

$$S_q^m(\mathbf{n}) = \lambda_q Y_q^m(\mathbf{n}), \quad m = -q, -q+1, \ldots, q. \quad \text{(XVIA.3)}$$

We begin the proof of (XVI.13) by considering the following result[2]:

Lemma 1. *If*

$$X_q^m(\hat{\mathbf{n}}) \equiv \frac{1}{2\pi} \int_{\mathbf{n} \perp \hat{\mathbf{n}}} Y_q^m(\mathbf{n}), \quad m = -q, -q+1, \ldots, q, \quad \text{(XVIA.4)}$$

for all unit vectors $\hat{\mathbf{n}}$, there is a constant λ_q such that

$$X_q^m = \lambda_q Y_q^m, \quad m = -q, -q+1, \ldots, q. \quad \text{(XVIA.5)}$$

Proof. For any rotation **R**,

$$\begin{aligned} X_q^m(\mathbf{R}^{-1}\hat{\mathbf{n}}) &= \frac{1}{2\pi} \int_{\mathbf{n} \perp \mathbf{R}^{-1}\hat{\mathbf{n}}} Y_q^m(\mathbf{n}), \\ &= \frac{1}{2\pi} \int_{\mathbf{R}\mathbf{n} \perp \hat{\mathbf{n}}} Y_q^m(\mathbf{n}), \quad \text{(XVIA.6)} \\ &= \frac{1}{2\pi} \int_{\mathbf{n} \perp \hat{\mathbf{n}}} Y_q^m(\mathbf{R}^{-1}\mathbf{n}). \end{aligned}$$

---

[1] KANIEL [1977, Lemma 3.1].

[2] MAXWELL [1879, bracketed additions after §§14(1), 14(5), and 14(15)] introduced spherical harmonics so as to calculate collisions integrals. His results include Lemma 1 and (XVIA.11). IKENBERRY & TRUESDELL [1956, §7] stated that MAXWELL's "cryptic remarks" led them to conjecture the theorem which IKENBERRY then proceeded to prove. The proof of Lemma 1 we give here is due to KANIEL [1977, Lemma 3.2].

By (XVIA.1),

$$X_q^m(\mathbf{R}^{-1}\hat{\mathbf{n}}) = \frac{1}{2\pi}\int_{\mathbf{n}\perp\hat{\mathbf{n}}} \sum_{k=-q}^{q} T_{mk}^{(q)}(\mathbf{R})Y_q^k(\mathbf{n}),$$

$$= \sum_{k=-q}^{q} T_{mk}^{(q)}(\mathbf{R})X_q^k(\hat{\mathbf{n}}). \qquad \text{(XVIA.7)}$$

We conclude from the general properties of spherical harmonics described following (XVIA.1) that (XVIA.5) is valid. △

To determine[3] the constants λ_q, let us set $m = 0$ and $\hat{\mathbf{n}} = (0, 0, 1)$ in (XVIA.4) and (XVIA.5) and so obtain

$$\frac{1}{2\pi}\int_0^{2\pi} Y_q^0(\cos\zeta, \sin\zeta, 0)\, d\zeta = \lambda_q Y_q^0(0, 0, 1). \qquad \text{(XVIA.8)}$$

But

$$Y_q^0(n_1, n_2, n_3) = P_q(n_3), \qquad \text{(XVIA.9)}$$

P_q being the qth Legendre polynomial:

$$P_q(z) = \frac{1}{2^q q!} \frac{d^q}{dz^q} (z^2 - 1)^q. \qquad \text{(XVIA.10)}$$

From (XVIA.9) we see that (XVIA.8) reduces to

$$P_q(0) = \lambda_q P_q(1). \qquad \text{(XVIA.11)}$$

Finally, by expanding the right-hand side of (XVIA.10), we find that

$$P_q(1) = 1,$$

$$P_q(0) = \begin{cases} 0 & \text{if } q \text{ is odd,} \\ (-1)^{\frac{1}{2}q}\dfrac{(q-1)!!}{q!!} & \text{if } q \text{ is even.} \end{cases} \qquad \text{(XVIA.12)}$$

By combining (XVIA.11) and (XVIA.12) we find that λ_q is given by $(\text{XVI.14})_{4,5}$.

The polynomials $\mathbf{Y}_s$ introduced by IKENBERRY have the important property of being harmonic functions. Since, as is shown by the general formula (XVI.25), these polynomials are homogeneous of degree s, the functions $\mathbf{Y}_s(\mathbf{n})$ lie in the subspace spanned by the spherical harmonics $Y_s^m(\mathbf{n})$, $m = -s, -s+1, \ldots, s$. Thus, since λ_s does not depend upon m, we can express the results of the preceding lemma in the equivalent form

$$\frac{1}{2\pi}\int_{\mathbf{n}\perp\hat{\mathbf{n}}} \mathbf{Y}_s(\mathbf{n}) = \lambda_s \mathbf{Y}_s(\hat{\mathbf{n}}). \qquad \text{(XVIA.13)}$$

---

[3] KANIEL [1977, Lemma 3.3].

If we multiply each side by w^s and set $\mathbf{w} = w\hat{\mathbf{n}}$, we obtain

$$\frac{1}{2\pi}\int_{\mathbf{n}\perp\mathbf{w}} w^s Y_s(\mathbf{n}) = \lambda_s Y_s(\mathbf{w}). \tag{XVIA.14}$$

From (XVIA.14), (XVI.25), and (XVI.27) we find that

$$\frac{1}{2\pi}\int_{\mathbf{n}\perp\mathbf{w}} (wr_e n_e)^s = \sum_{q=0}^{[\frac{1}{2}s]} a_q^s r^{2q} w^{2q} \frac{1}{2\pi} r_{k_1} \cdots r_{k_{s-2q}} \int_{\mathbf{n}\perp\mathbf{w}} w^{s-2q} Y_{k_1\ldots k_{s-2q}}(\mathbf{n}),$$

$$= \sum_{q=0}^{[\frac{1}{2}s]} a_q^s r^{2q} w^{2q} \lambda_{s-2q} r_{k_1} \cdots r_{k_{s-2q}} Y_{k_1\ldots k_{s-2q}}(\mathbf{w}),$$

$$= \sum_{q_1=0}^{[\frac{1}{2}s]} \sum_{q_2=0}^{[\frac{1}{2}s]-q_1} a_{q_1}^s b_{q_2}^{s-2q_1} \lambda_{s-2q_1} r^{2(q_1+q_2)}$$

$$\times w^{2(q_1+q_2)}(r_e w_e)^{s-2q_1-2q_2}, \tag{XVIA.15}$$

and this completes the proof of (XVI.13).

We turn now to the proof of IKENBERRY's theorem. We use again the properties of spherical harmonics outlined at the beginning of this appendix.

As we remarked after (XVI.39), all we need prove is that the part of $\bar{\mathbb{C}}Y_{2r|s}$ which is linear in the moments of order $2r + s$ is given by (XVI.36). For an equivalent statement of this problem, let us recall that if g is a polynomial of degree m in the components of $\mathbf{c}$, then in a gas of Maxwellian molecules $\mathbb{O}g$ is a homogeneous polynomial of degree m in the components of $\mathbf{c}$ and $\mathbf{c}_*$. If we denote its value by $\mathbb{O}g(\mathbf{c}, \mathbf{c}_*)$ and introduce an operator $\mathbb{L}$ defined as follows:

$$\mathbb{L}g(\mathbf{c}) \equiv \mathbb{O}g(\mathbf{c}, \mathbf{0}) + \mathbb{O}g(\mathbf{0}, \mathbf{c}), \tag{XVIA.16}$$

then clearly the part of $\bar{\mathbb{C}}g$ which is linear in the moments of order n is simply $\int F(\mathbf{v}) \int F(\mathbf{v}_*) \mathbb{L}g(\mathbf{c})$. Thus what we wish to show is that

$$\mathbb{L} Y_{2r|s} = -\frac{1}{n} \mathbb{C}_{2r|s} Y_{2r|s} \tag{XVIA.17}$$

and that $\mathbb{C}_{2r|s}$ is given by (XVI.33).

If

$$\int\int_{\mathbf{n}\perp\mathbf{w}} \cdots \equiv \int_0^{2\pi} \int_0^{\frac{1}{2}\pi} \mathbb{S}(\theta) \sin\theta\, d\theta\, d\zeta \cdots, \tag{XVIA.18}$$

$\mathbb{S}$ being the scattering factor for Maxwellian molecules, then from (XVI.7) and (XVI.9)$_2$ we find that

$$\mathbb{O}g(\mathbf{c}, \mathbf{c}_*) = \int\int_{\mathbf{n}\perp(\mathbf{c}_*-\mathbf{c})} [g(\sin^2\theta\, \mathbf{c} + \cos^2\theta\, \mathbf{c}_* + \sin\theta\cos\theta\, |\mathbf{c}_* - \mathbf{c}|\mathbf{n}) - g(\mathbf{c})]. \tag{XVIA.19}$$

APPENDIX A. PROOF OF IKENBERRY'S THEOREM

By placing (XVIA.19) into (XVIA.16) we obtain the following explicit formula for the operator $\mathbb{L}$:

$$\mathbb{L}g(\mathbf{c}) = \int\!\!\int_{\mathbf{n} \perp \mathbf{c}} [g(\sin^2 \theta \, \mathbf{c} + \sin \theta \cos \theta \, c\mathbf{n}) + g(\cos^2 \theta \, \mathbf{c}$$

$$+ \sin \theta \cos \theta \, c\mathbf{n}) - g(\mathbf{c})]. \qquad \text{(XVIA.20)}$$

We may interpret (XVIA.17) as stating that $\mathbf{Y}_{2r|s}$ is a proper function of the operator $\mathbb{L}$. In the next lemma we exhibit a large class of proper functions of $\mathbb{L}$, and thereafter we show that $\mathbf{Y}_{2r|s}$ lies in the space spanned by the elements of this class.

Lemma 2. *If* $\mathbf{c} = c\hat{\mathbf{n}}$ *and* $|\hat{\mathbf{n}}| = 1$, *and if*

$$Y_{mk}^q(\mathbf{c}) \equiv c^m Y_{m-2k}^q(\hat{\mathbf{n}}), \qquad q = -m + 2k, \, -m + 2k + 1, \ldots, m - 2k,$$
$$\text{(XVIA.21)}$$

then there are constants $_m\lambda_k$ *such that*

$$\mathbb{L}Y_{mk}^q = {_m\lambda_k} Y_{mk}^q. \qquad \text{(XVIA.22)}$$

Proof. Because $\mathbb{L}Y_{mk}^q$ is homogeneous of degree m, it is sufficient to prove that there are constants $_m\lambda_k$ such that

$$\mathbb{L}Y_{mk}^q(\hat{\mathbf{n}}) = {_m\lambda_k} Y_{m-2k}^q(\hat{\mathbf{n}}) \qquad \text{(XVIA.23)}$$

for all unit vectors $\hat{\mathbf{n}}$. From (XVIA.20) we have

$$\mathbb{L}Y_{mk}^q(\hat{\mathbf{n}}) = \int\!\!\int_{\mathbf{n} \perp \hat{\mathbf{n}}} [Y_{mk}^q(\sin^2 \theta \, \hat{\mathbf{n}} + \sin \theta \cos \theta \, \mathbf{n})$$

$$+ Y_{mk}^q(\cos^2 \theta \, \hat{\mathbf{n}} + \sin \theta \cos \theta \, \mathbf{n}) - Y_{mk}^q(\hat{\mathbf{n}})]. \qquad \text{(XVIA.24)}$$

Since $\mathbf{n}$ is orthogonal to $\hat{\mathbf{n}}$, there are unit vectors $\mathbf{n}_1$ and $\mathbf{n}_2$ such that

$$\sin^2 \theta \, \hat{\mathbf{n}} + \sin \theta \cos \theta \, \mathbf{n} = \sin \theta \, \mathbf{n}_1,$$

$$\cos^2 \theta \, \hat{\mathbf{n}} + \sin \theta \cos \theta \, \mathbf{n} = \cos \theta \, \mathbf{n}_2. \qquad \text{(XVIA.25)}$$

Use of (XVIA.21) and (XVIA.25) shows that (XVIA.24) becomes

$$\mathbb{L}Y_{mk}^q(\hat{\mathbf{n}}) = \int\!\!\int_{\mathbf{n} \perp \hat{\mathbf{n}}} [\sin^m \theta \, Y_{m-2k}^q(\mathbf{n}_1) + \cos^m \theta \, Y_{m-2k}^q(\mathbf{n}_2) - Y_{m-2k}^q(\hat{\mathbf{n}})].$$

$$\text{(XVIA.26)}$$

Next, if **R** is any rotation, then from (XVIA.24) we have

$$\mathbb{L}Y_{mk}^q(\mathbf{R}^{-1}\hat{\mathbf{n}}) = \int\int_{\mathbf{n}\perp\mathbf{R}^{-1}\hat{\mathbf{n}}} [Y_{mk}^q(\sin^2\theta\,\mathbf{R}^{-1}\hat{\mathbf{n}} + \sin\theta\cos\theta\,\mathbf{n})$$
$$+ Y_{mk}^q(\cos^2\theta\,\mathbf{R}^{-1}\hat{\mathbf{n}} + \sin\theta\cos\theta\,\mathbf{n}) - Y_{mk}^q(\mathbf{R}^{-1}\hat{\mathbf{n}})],$$

$$= \int\int_{\mathbf{n}\perp\hat{\mathbf{n}}} [Y_{mk}^q(\sin^2\theta\,\mathbf{R}^{-1}\hat{\mathbf{n}} + \sin\theta\cos\theta\,\mathbf{R}^{-1}\mathbf{n})$$
$$+ Y_{mk}^q(\cos^2\theta\,\mathbf{R}^{-1}\hat{\mathbf{n}} + \sin\theta\cos\theta\,\mathbf{R}^{-1}\mathbf{n}) - Y_{mk}^q(\mathbf{R}^{-1}\hat{\mathbf{n}})],$$

$$= \int\int_{\mathbf{n}\perp\hat{\mathbf{n}}} [\sin^m\theta\,Y_{m-2k}^q(\mathbf{R}^{-1}\mathbf{n}_1) + \cos^m\theta\,Y_{m-2k}^q(\mathbf{R}^{-1}\mathbf{n}_2)$$
$$- Y_{m-2k}^q(\mathbf{R}^{-1}\hat{\mathbf{n}})]. \qquad \text{(XVIA.27)}$$

By (XVIA.1), (XVIA.26), and (XVIA.27)$_3$, the functions Y_{mk}^q satisfy the relation

$$\mathbb{L}Y_{mk}^q(\mathbf{R}^{-1}\hat{\mathbf{n}}) = \sum_{p=-(m-2k)}^{m-2k} T_{qp}^{(m-2k)}(\mathbf{R})\mathbb{L}Y_{mk}^p(\hat{\mathbf{n}}) \qquad \text{(XVIA.28)}$$

for all $\hat{\mathbf{n}}$ and **R**. From the comments following (XVIA.1) we find there are constants $_m\lambda_k$ such that (XVIA.23) holds for all unit vectors $\hat{\mathbf{n}}$. △

In order to compute[4] the numbers $_m\lambda_k$, let us choose $\hat{\mathbf{n}} = (0, 0, 1)$. Since **n** is perpendicular to $\hat{\mathbf{n}}$, we may set $\mathbf{n} = (\cos\zeta, \sin\zeta, 0)$, and then we see from (XVIA.25) that

$$\mathbf{n}_1 = (\cos\zeta\cos\theta, \sin\zeta\cos\theta, \sin\theta),$$
$$\mathbf{n}_2 = (\cos\zeta\sin\theta, \sin\zeta\sin\theta, \cos\theta). \qquad \text{(XVIA.29)}$$

If in addition we choose $q = 0$, we see from (XVIA.22) and (XVIA.26) that

$$\int\int_{\mathbf{n}\perp\hat{\mathbf{n}}} [\sin^m\theta\,Y_{m-2k}^0(\cos\zeta\cos\theta, \sin\zeta\cos\theta, \sin\theta)$$
$$+ \cos^m\theta\,Y_{m-2k}^0(\cos\zeta\sin\theta, \sin\zeta\sin\theta, \cos\theta) - Y_{m-2k}^0(0, 0, 1)]$$
$$= {_m\lambda_k}\,Y_{m-2k}^0(0, 0, 1). \qquad \text{(XVIA.30)}$$

In view of (XVIA.9), (XVIA.12)$_1$, and (XVIA.18) this last result becomes

$${_m\lambda_k} = 2\pi\int_0^{\frac{1}{2}\pi} \mathbb{S}(\theta)\sin\theta\,[\sin^m\theta\,P_{m-2k}(\sin\theta) + \cos^m\theta\,P_{m-2k}(\cos\theta) - 1]\,d\theta. \qquad \text{(XVIA.31)}$$

---

[4] KANIEL [1977, Lemma 4.4].

APPENDIX A. PROOF OF IKENBERRY'S THEOREM

We complete the proof of IKENBERRY's theorem by making the following simple observation. As we noted previously, the functions $\mathbf{Y}_s(\mathbf{n})$ lie in the subspace spanned by the spherical harmonics $Y_s^q(\mathbf{n})$, $q = -s, -s+1, \ldots, s$. Because of (XVI.26) this implies that the polynomials $\mathbf{Y}_{2r|s}(\mathbf{c})$ lie in the subspace spanned by the polynomials $c^{2r+s} Y_s^q(\hat{\mathbf{n}})$, $q = -s, -s+1, \ldots, s$, where $\mathbf{c} = c\hat{\mathbf{n}}$, $|\hat{\mathbf{n}}| = 1$. Thus, since the constants $_m\lambda_k$ appearing in (XVIA.22) are independent of the index q, we may replace Y_{mk}^q in (XVIA.22) by $\mathbf{Y}_{2r|s}$ provided we set $m = 2r + s$, $k = r$:

$$\mathbb{L}\mathbf{Y}_{2r|s} = {}_{2r+s}\lambda_r \, \mathbf{Y}_{2r|s}. \tag{XVIA.32}$$

This completes the proof of IKENBERRY's theorem, for (XVIA.32) is nothing other than (XVIA.17) with $\mathbb{c}_{2r|s}$ given by[5]

$$\mathbb{c}_{2r|s} = -n \,_{2r+s}\lambda_r,$$

$$= 2\pi n \int_0^{\frac{1}{2}\pi} \mathbb{S}(\theta) \sin\theta \left[1 - \sin^{2r+s}\theta \, P_s(\sin\theta) - \cos^{2r+s}\theta \, P_s(\cos\theta)\right] d\theta,$$

$$= 4\pi n \sqrt{\frac{\mathscr{G}}{2\mathfrak{m}}} \int_0^\infty \left[1 - \sin^{2r+s}\theta \, P_s(\sin\theta) - \cos^{2r+s}\theta \, P_s(\cos\theta)\right] z \, dz;$$

(XVIA.33)

in the last step we have used (XII.9) with $\mathbb{k} = 5$.

---

[5] The proof of IKENBERRY's theorem published by IKENBERRY & TRUESDELL [1956, §13] rests partly upon geometric considerations and partly upon properties of Legendre polynomials which few students today will find familiar.

Lemma 2 is due in statement to KANIEL [1977, Lemma 4.3], but our proof is different from his. He proves the result first for Maxwellian pseudomolecules, which we have defined at the end of Chapter XII, and thereafter states that a well-defined limit procedure proves it valid also for Maxwellian molecules. The latter step in his proof we have not been able to justify from his somewhat brief description of the limit procedure. The proof we have provided here, while adapted from his, does not have recourse at any step to pseudomolecules.

KANIEL [1977, §1] claims to have "extended" IKENBERRY's results to any molecular model in which $\mathbb{S}$ is a function of θ alone, which he calls "generalized Maxwellian molecules". This claim, however, seems somewhat strained since IKENBERRY's proof makes no use of the specific form of $\mathbb{S}(\theta)$, and the corresponding generalization of $\mathbb{c}_{2r|s}$, namely, (XVIA.33)$_2$, does appear in the work of IKENBERRY & TRUESDELL [1956, Eq. (7.4)$_2$].

The straightforward proof of IKENBERRY's theorem we have given here has two distinct parts:

1. The *general* determination of the collisions integrals of all polynomials, for a gas of Maxwellian molecules. This determination, presented as Theorem 1, we believe to be new. In the only difficult step of the proof, expressed by (XVI.13), we have made use of MAXWELL's theorem expressed here as Lemma 1 and of KANIEL's evaluation of λ_k.

2. *Evaluation* of the leading coefficient when the polynomial is $\mathbf{Y}_{2r|s}$. To obtain it, we have applied another of KANIEL's results concerning spherical harmonics, here stated as Lemma 2, and have used KANIEL's evaluation of $_m\lambda_k$.

Appendix B

Multi-indices

In a number of chapters we abbreviate some formulae by using multi-indices, and here we introduce the notation and conventions we shall use.

A block of n indices $k_1, \ldots, k_n$ will be denoted by the symbol n, and we call n a multi-index. If **A** is an nth-order tensor, its components will be written collectively as A_n:

$$A_n \equiv A_{k_1 \ldots k_n}. \tag{XVIB.1}$$

The n-fold tensor product $\mathbf{f} \otimes \cdots \otimes \mathbf{f}$ of a vector $\mathbf{f}$, and the nth spatial gradient $\text{grad}^n \phi$ of a field ϕ are both examples of nth-order tensors, and for these (XVIB.1) reduces to the respective forms

$$f_n = f_{k_1} \cdots f_{k_n}, \qquad \phi_{,n} = \phi_{,k_1 \ldots k_n}. \tag{XVIB.2}$$

Sometimes we shall view a large block of indices as made up of a number of smaller blocks, each denoted by its own multi-index. For example, if **B** is a tensor of order $n + m$, and its components have indices $k_1, \ldots, k_n, p_1, \ldots, p_m$, we may write these components collectively as

$$B_{\text{nm}} \equiv B_{k_1 \ldots k_n p_1 \ldots p_m}, \tag{XVIB.3}$$

n consisting in the first n indices, m consisting in the last m.

Two general conventions concerning multi-indices will be useful. First, if $n = 0$, then the block of indices represented by n is empty. To encompass this possibility we adopt the convention that

$$A_n \equiv 1 \quad \text{if} \quad n = 0. \tag{XVIB.4}$$

Second, when a multi-index is repeated in an expression, we assume that a summation is implied over all indices which the repeated multi-index represents. For example, if **g** is a second vector, then

$$\begin{aligned} f_n g_n &= f_{a_1} \cdots f_{a_n} g_{a_1} \cdots g_{a_n} = (f_a g_a)^n, \\ A_n \phi_{,n} &= A_{a_1 \ldots a_n} \phi_{,a_1 \ldots a_n}, \\ f_n B_{\text{nm}} &= f_{a_1} \cdots f_{a_n} B_{a_1 \ldots a_n p_1 \ldots p_m}. \end{aligned} \tag{XVIB.5}$$

If n consists in indices $k_1, \ldots, k_n$, by the *multi-index decomposition*

$$\text{n} = \text{pq} \tag{XVIB.6}$$

we mean that p is the multi-index that represents the first p indices of n, namely, $k_1, \ldots, k_p$, and q is the multi-index consisting in the remaining indices $k_{p+1}, \ldots, k_n$. Necessarily, then, $q = n - p$. In the special case in which $p = 1$, so the block represented by p has one index, namely, k_1, we replace p by that index and so write (XVIB.6) in the form

$$n = k_1 q. \tag{XVIB.7}$$

More generally, then, for any tensor **A** of order $n + 1$ we may write the components of **A** in the form

$$A_{mn} \equiv A_{mk_1 \ldots k_n}. \tag{XVIB.8}$$

We may also interpret (XVIB.6) as a rule for forming one multi-index from two others. In general we may set

$$n = p_1 \cdots p_s, \tag{XVIB.9}$$

and n is now the multi-index whose first p_1 members are the indices represented by p_1, whose next p_2 members are the indices represented by p_2, *etc.*

Chapter XVII

Grad's Formal Evaluation of Collisions Integrals, and His Method of Approximating the Initial-value Problem

In a gas of Maxwellian molecules IKENBERRY's theorem and further explicit formulae stated and proved in the preceding chapter deliver the collisions integral of any polynomial in the components of c as an exhibited bilinear function of certain moments of the molecular density. Using this result, we were able to outflank the Maxwell–Boltzmann equation itself by working directly with the equation of transfer (XIII.1) for polynomial functions g. The exact results for Maxwellian molecules presented in Chapters XIII, XIV, and XV all depend essentially on this fact. Any attempt to work with the equation of transfer for other molecular models immediately faces the problem of evaluating the collisions integral $\bar{C}g$. Although no proved solution of this problem is presently known, GRAD[1] invented for quite general gases and for a broad class of molecular densities a formal method of calculating such collisions integrals as infinite series in the moments.

(i) Grad's expansion and equations of transfer for the Hermite coefficients

GRAD makes use of a system of three-dimensional Hermite polynomials $\mathbf{H}_s$. In terms of a vector $\mathbf{a}$, $\mathbf{H}_s(\mathbf{a})$ is a function of $\mathbf{a}$ whose values are components of symmetric tensors: $\mathbf{H}_s(\mathbf{a}) = H_{k_1 \ldots k_s}(\mathbf{a})$, each of these component functions being a polynomial of degree s in the components of $\mathbf{a}$, as is readily seen from the definition

$$H_{k_1 \ldots k_s}(\mathbf{a}) \equiv (-1)^s e^{\frac{1}{2}a^2} \partial_{a_{k_1}} \cdots \partial_{a_{k_s}} e^{-\frac{1}{2}a^2}. \tag{XVII.1}$$

[1] GRAD [1949, §4].

The first few of these are
$$H(\mathbf{a}) \equiv 1,$$
$$H_k(\mathbf{a}) \equiv a_k,$$
$$H_{km}(\mathbf{a}) \equiv a_k a_m - \delta_{km}, \qquad \text{(XVII.2)}$$
$$H_{kmr}(\mathbf{a}) \equiv a_k a_m a_r - 3a_{(k} \delta_{mr)},$$
$$H_{kmrs}(\mathbf{a}) \equiv a_k a_m a_r a_s - 6a_{(k} a_m \delta_{rs)} + 3 \delta_{(km} \delta_{rs)}.$$

The Hermite polynomials[2] are orthogonal with respect to the weight function $(2\pi)^{-\frac{3}{2}} \exp(-\frac{1}{2}a^2)$. That is, if $\delta_{(k_1 \ldots k_r)(p_1 \ldots p_r)}$ represents the sum of the $r!$ distinct products $\delta_{m_1 q_1} \cdots \delta_{m_r q_r}$ in which $(m_1 \ldots m_r)$ is a permutation of $(k_1 \ldots k_r)$ and $(q_1 \ldots q_r)$ is a permutation of $(p_1 \ldots p_r)$, then

$$\int_{\mathscr{V}} \frac{1}{(2\pi)^{\frac{3}{2}}} e^{-\frac{1}{2}a^2} H_{k_1 \ldots k_r} H_{p_1 \ldots p_s} \, d\mathbf{a} = \begin{cases} 0 & \text{if } r \neq s, \\ \delta_{(k_1 \ldots k_r)(p_1 \ldots p_r)} & \text{if } r = s. \end{cases} \qquad \text{(XVII.3)}$$

We shall refer to the functions $\exp(-\frac{1}{4}a^2)H_s$ as *Hermite functions*. In the space of functions which are square-integrable over $\mathscr{V}$, the Hermite functions are *complete* in the following sense: If f satisfies the condition

$$\int_{\mathscr{V}} (f(\mathbf{a}))^2 \, d\mathbf{a} < \infty, \qquad \text{(XVII.4)}$$

then f has a unique expansion

$$f(\mathbf{a}) = e^{-\frac{1}{4}a^2} \sum_{s=0}^{\infty} A_{b_1 \ldots b_s} H_{b_1 \ldots b_s}(\mathbf{a}), \qquad \text{(XVII.5)}$$

and this expansion converges in the mean[3] to the function f.

GRAD considers only such molecular densities F as can be expanded in a series of Hermite functions. In order to set down this expansion, we let ρ, $\mathbf{u}$, and ε be the density, gross velocity, and energetic, respectively, that correspond to F. Introducing the dimensionless random velocity

$$\kappa_k \equiv \sqrt{\frac{3}{2\varepsilon}} c_k, \qquad \text{(XVII.6)}$$

---

[2] These polynomials were introduced and studied by GRAD in his "Note on N-Dimensional Hermite Polynomials", *Communications on Pure and Applied Mathematics* **2**, 325–330 (1949), in which he describes many of their properties, some of which we simply list here.

[3] This means that
$$\lim_{N \to \infty} \int_{\mathscr{V}} \left(f - e^{-\frac{1}{4}a^2} \sum_{s=0}^{N} A_{b_1 \ldots b_s} H_{b_1 \ldots b_s} \right)^2 d\mathbf{a} = 0;$$
the expansion (XVII.5) need not converge pointwise.

(i) GRAD'S EXPANSION 263

$\mathbf{c}$ being the random velocity $\mathbf{v} - \mathbf{u}$, we suppose that $\exp(\frac{1}{4}\kappa^2)F$ be square-integrable:

$$\int_{\mathscr{V}} e^{\frac{1}{2}\kappa^2} [F(t, \mathbf{x}, \mathbf{u}(t, \mathbf{x}) + \kappa\sqrt{\tfrac{2}{3}\varepsilon(t, \mathbf{x})})]^2 \, d\kappa < \infty \qquad \text{(XVII.7)}$$

for each t and $\mathbf{x}$. Then $\exp(\frac{1}{4}\kappa^2)F$ may be expanded in the Hermite functions $\exp(-\frac{1}{4}\kappa^2)H_s(\kappa)$ as indicated generally by (XVII.5). That is, there are tensor-valued functions $A_s(t, \mathbf{x})$ such that

$$F(t, \mathbf{x}, \mathbf{v}) = e^{-\frac{1}{2}\kappa^2} \sum_{s=0}^{\infty} A_s(t, \mathbf{x}) H_s(\kappa). \qquad \text{(XVII.8)}$$

Throughout this chapter we use multi-indices and the conventions relating to them as presented in Appendix B to Chapter XVI. In view of (XVII.6), an equivalent expansion of F is easily seen to be

$$F(t, \mathbf{x}, \mathbf{v}) = F_M(t, \mathbf{x}, \mathbf{v}) \sum_{s=0}^{\infty} \frac{1}{s!} A_s(t, \mathbf{x}) H_s(\kappa), \qquad \text{(XVII.9)}$$

the tensors A_s here generally being different from those appearing in (XVII.8), and F_M being the Maxwellian density which has the same principal moment as F:

$$F_M = \frac{\rho}{\mathfrak{m}(\frac{4}{3}\pi\varepsilon)^{\frac{3}{2}}} e^{-\frac{3}{4}c^2/\varepsilon}. \qquad \text{(XVII.10)}$$

The infinite sum in (XVII.9) represents $1 + G$ in the decomposition (VIII.24). The coefficients A_s in (XVII.9) may be called the *Hermite coefficients* of F.

The Hermite coefficients have a simple representation in terms of F. Since H_s is a symmetric tensor, we see from (XVII.9) that we may assume A_s to be symmetric also. If we multiply each side of (XVII.9) by $H_r(\kappa)$ and then integrate with respect to κ, we find that

$$\int_{\mathscr{V}} H_{k_1 \ldots k_r}(\kappa) F \, d\kappa = \frac{\rho}{\mathfrak{m}(\frac{2}{3}\varepsilon)^{\frac{3}{2}}} \sum_{s=0}^{\infty} \frac{1}{s!} A_{a_1 \ldots a_s} \int_{\mathscr{V}} \frac{e^{-\frac{1}{2}\kappa^2}}{(2\pi)^{\frac{3}{2}}} H_{a_1 \ldots a_s} H_{k_1 \ldots k_r} \, d\kappa. \qquad \text{(XVII.11)}$$

Because of (XVII.3) the integral on the right-hand side is zero when $s \neq r$; when $s = r$, it is non-zero only for the $r!$ sets of indices $a_1, \ldots, a_r$ that are permutations of $k_1, \ldots, k_r$. Therefore, because of the symmetry of A_r and the orthogonality conditions (XVII.3), we obtain

$$A_r = \frac{\mathfrak{m}}{\rho} (\tfrac{2}{3}\varepsilon)^{\frac{3}{2}} \int_{\mathscr{V}} H_r(\kappa) F \, d\kappa, \qquad \text{(XVII.12)}$$

and if we use (XVII.6) to change the variable of integration here from κ to $\mathbf{v}$, we find that A_r is simply the average of $H_r(\kappa)$:

$$A_r = \overline{H_r(\kappa)}. \qquad \text{(XVII.13)}$$

Thus the Hermite coefficients of F are the expectations of the corresponding Hermite polynomials. Of course, it follows directly from the uniqueness of the expansion (XVII.9) that the Hermite coefficients of F_M, other than $\mathbf{A}_0$, are null.

Since the components of $\mathbf{H}_r(\mathbf{\kappa})$ are polynomials in the components of $\mathbf{\kappa}$, and since $\mathbf{\kappa}$ is proportional to the random velocity $\mathbf{c}$, it follows from (III.19) and (XVII.13) that $\rho\mathbf{A}_r$ is a linear combination of the relative moments $^n\mathbf{M}$ with coefficients depending upon ε. Moreover, since the Hermite polynomials form a complete set, it follows from (III.19) that each $^n\mathbf{M}$ is a linear combination of the functions $\rho\mathbf{A}_r$. In this sense the $\mathbf{A}_r$ are simply an alternative set of moments of F, just as were the spherical moments $\mathbf{P}_{2r|s}$ introduced in Chapter XVI. In order to interpret results, it is useful to express the first few of the functions $\mathbf{A}_r$ in terms of the relative moments. From (XVII.2)$_1$ and (XVII.13) we see that the scalar field $\mathbf{A}_0$ has the single component

$$A = 1; \tag{XVII.14}$$

from (XVII.2)$_2$ and (III.8), that the vector field $\mathbf{A}_1$ has the components

$$A_k = 0; \tag{XVII.15}$$

from (XVII.2)$_3$, (III.9), (III.10), and (III.15), that the second-order tensor field $\mathbf{A}_2$ has the components

$$A_{km} = \frac{1}{p} M_{km} - \delta_{km} = \frac{1}{p} P_{km}; \tag{XVII.16}$$

and in a similar manner that

$$A_{kmr} = \frac{1}{p\sqrt{\frac{2}{3}\varepsilon}} M_{kmr},$$

$$A_{kmrs} = \frac{1}{\frac{2}{3}p\varepsilon} M_{kmrs} - \frac{6}{p} P_{(km} \delta_{rs)} - 3\delta_{(km}\delta_{rs)}. \tag{XVII.17}$$

To determine fields $\mathbf{A}_s(t, \mathbf{x})$ that render (XVII.9) a solution of the Maxwell–Boltzmann equation, we appeal to the equation of transfer (IX.12)$_2$. First we note that if $g = g(\mathbf{\kappa})$, then from (XVII.6)

$$\partial_t g = -\left(\sqrt{\frac{3}{2\varepsilon}}\,\partial_t u_a + \frac{1}{2\varepsilon}\kappa_a\,\partial_t \varepsilon\right) g_{,a},$$

$$\partial_{x_k} g = -\left(\sqrt{\frac{3}{2\varepsilon}}\,u_{a,k} + \frac{1}{2\varepsilon}\kappa_a \varepsilon_{,k}\right) g_{,a}, \tag{XVII.18}$$

$$\partial_{v_k} g = \sqrt{\frac{3}{2\varepsilon}}\,g_{,k}.$$

(i) GRAD'S EXPANSION

By placing these expressions into (IX.12)$_2$ and then using (IX.4) and (IX.5) to eliminate $\dot{\mathbf{u}}$ and $\dot{\varepsilon}$ we obtain an equation of transfer for functions of $\mathbf{\kappa}$ alone:

$$\rho \dot{g} + \left(\rho\sqrt{\frac{2}{3}\varepsilon}\,\overline{\kappa_a g}\right)_{,a} - \sqrt{\frac{3}{2\varepsilon}}\,\overline{g_{,a} M_{ab,b}} - \frac{1}{2\varepsilon}\overline{\kappa_a g_{,a}}(M_{bc}u_{b,c} + q_{b,b})$$

$$+ \rho u_{a,b}\overline{\kappa_b g_{,a}} + \frac{1}{3}\rho\sqrt{\frac{3}{2\varepsilon}}\,\varepsilon_{,b}\overline{\kappa_b \kappa_a g_{,a}} = \mathfrak{m}\bar{C}g. \qquad\text{(XVII.19)}$$

We wish to set $g = H_r$ in this equation, but first we require averages of the functions $\kappa_k H_r$, $H_{r,k}$, $\kappa_a H_{r,a}$, $\kappa_m H_{r,k}$, and $\kappa_m \kappa_a H_{r,a}$. Because these functions are polynomials in the components of $\mathbf{\kappa}$, we can express each as a linear combination of Hermite polynomials. We write the first two in the general form

$$\kappa_k H_r = \sum_{s=0}^{r+1} \Gamma_{rs}^k H_s,$$

$$H_{r,k} = \sum_{s=0}^{r-1} \Lambda_{rs}^k H_s, \qquad\text{(XVII.20)}$$

the coefficients Γ_{rs}^k and Λ_{rs}^k being pure numbers. For example, directly from (XVII.2) we find that[4]

$$\kappa_k H_{mr} = H_{kmr} + \delta_{km} H_r + \delta_{kr} H_m,$$

$$H_{km,r} = \delta_{rk} H_m + \delta_{rm} H_k. \qquad\text{(XVII.21)}$$

Simple combinations of (XVII.20) yield expressions for the remaining functions:

$$\kappa_a H_{r,a} = \sum_{s=0}^{r-1}\sum_{p=0}^{s+1} \Lambda_{rs}^a \Gamma_{sp}^a H_p,$$

$$\kappa_m H_{r,k} = \sum_{s=0}^{r-1}\sum_{p=0}^{s+1} \Lambda_{rs}^k \Gamma_{sp}^m H_p, \qquad\text{(XVII.22)}$$

$$\kappa_m \kappa_a H_{r,a} = \sum_{s=0}^{r-1}\sum_{p=0}^{s+1}\sum_{q=0}^{p+1} \Lambda_{rs}^a \Gamma_{sp}^a \Gamma_{pq}^m H_q.$$

---

[4] GRAD (see Footnote 2) has derived the following explicit form of (XVII.20):

$$\kappa_k H_{k_1 \ldots k_r} = H_{kk_1 \ldots k_r} + \sum_{p=1}^{r} \delta_{kk_p} H_{k_1 \ldots k_{p-1}k_{p+1}\ldots k_r},$$

$$H_{k_1 \ldots k_r, k} = \sum_{p=1}^{r} \delta_{kk_p} H_{k_1 \ldots k_{p-1}k_{p+1}\ldots k_r}.$$

If we set $g = H_r$ in (XVII.19) and then use (XVII.13), (XVII.20), and (XVII.22), we obtain

$$\rho \dot{A}_r + \sum_{s=0}^{r+1} \Gamma_{rs}^a \left(\rho \sqrt{\tfrac{2}{3} \varepsilon} A_s \right)_{,a} - \sqrt{\tfrac{3}{2\varepsilon}} M_{ab,b} \sum_{s=0}^{r-1} \Lambda_{rs}^a A_s$$
$$- \frac{1}{2\varepsilon}(M_{bc}u_{b,c} + q_{b,b}) \sum_{s=0}^{r-1} \sum_{p=0}^{s+1} \Lambda_{rs}^a \Gamma_{sp}^a A_p + \rho \sum_{s=0}^{r-1} \sum_{p=0}^{s+1} \Lambda_{rs}^a \Gamma_{sp}^b u_{a,b} A_p$$
$$+ \tfrac{1}{3}\rho \sqrt{\tfrac{3}{2\varepsilon}} \sum_{s=0}^{r-1} \sum_{p=0}^{s+1} \sum_{q=0}^{p+1} \Lambda_{rs}^a \Gamma_{sp}^a \Gamma_{pq}^b \varepsilon_{,b} A_q = \mathbb{\bar{C}} H_r. \qquad \text{(XVII.23)}$$

It remains to evaluate the collisions integral on the right-hand side. To this end GRAD formally substitutes the expansion (XVII.9) of F into (XII.13)$_1$ and expands the result:

$$\mathbb{\bar{C}} H_r = \frac{1}{2} \iint F_M F_{M*} \sum_{m,n=0}^{\infty} \frac{1}{m!\,n!} A_m A_n H_m H_{n*} \mathbb{B} H_r,$$
$$= \frac{\rho^2}{2\mathfrak{m}^2} \sum_{m,n=0}^{\infty} \mathbb{A}_{rmn} A_m A_n, \qquad \text{(XVII.24)}$$

in which

$$\mathbb{A}_{rmn}(\varepsilon) \equiv \frac{1}{(2\pi)^3 m!\,n!} \int_{\mathscr{V}} \int_{\mathscr{V}} \int_{\mathscr{O}} e^{-\tfrac{1}{2}\kappa^2} e^{-\tfrac{1}{2}\kappa_*^2} H_m H_{n*}$$
$$\times (H'_{r*} + H'_r - H_{r*} - H_r) \mathbb{S}(\theta, \omega\sqrt{\tfrac{2}{3\varepsilon}})\, d\kappa\, d\kappa_*. \qquad \text{(XVII.25)}$$

Here we have used (VII.7) and have set $\kappa_* \equiv \sqrt{\tfrac{3}{2}/\varepsilon}\,c_*$, $\omega \equiv \sqrt{\tfrac{3}{2}/\varepsilon}\,w$. The result of using (XVII.24)$_2$ to express the right-hand side of (XVII.23) is GRAD's *Equation of Transfer for the Expectations A_r of the Hermite Polynomials H_r*.

GRAD[5] has described a general procedure for expressing the tensors $\mathbb{A}_{rmn}$ in terms of certain scalar functions of ε alone. The main ingredients that are needed in his procedure have appeared in previous chapters, but the calculations are long. Here we shall rest content to indicate the main steps involved and then list some specimens of the calculation.

For a function g of $\mathbf{c}$, we see from (XII.13)$_2$ that $\mathbb{B}g$ is a function of $\mathbf{c}$ and $\mathbf{c}_*$. Equivalently, if g is a function of κ, then $\mathbb{B}g$ is a function of κ and κ_*. Denoting the value of this function by $\mathbb{B}g(\kappa, \kappa_*)$, we may write (XVII.25) in the explicit form

$$\mathbb{A}_{rmn} = \frac{1}{(2\pi)^3 m!\,n!} \int_{\mathscr{V}} \int_{\mathscr{V}} e^{-\tfrac{1}{2}\kappa^2} e^{-\tfrac{1}{2}\kappa_*^2} H_m(\kappa) H_n(\kappa_*) \mathbb{B} H_r(\kappa, \kappa_*)\, d\kappa\, d\kappa_*.$$
$$\text{(XVII.26)}$$

---

[5] GRAD [1949, Appendix 3].

(i) FORMAL COLLISIONS INTEGRALS

We introduce now the new dimensionless vector $\boldsymbol{\xi} \equiv \boldsymbol{\kappa}_* + \boldsymbol{\kappa}$ and recall that $\boldsymbol{\omega} = \boldsymbol{\kappa}_* - \boldsymbol{\kappa}$. The following list contains simple identities involving $\boldsymbol{\xi}$, $\boldsymbol{\omega}$, $\boldsymbol{\kappa}$, and $\boldsymbol{\kappa}_*$:

$$\kappa_k = \tfrac{1}{2}(\xi_k - \omega_k), \qquad \kappa_{*k} = \tfrac{1}{2}(\xi_k + \omega_k),$$

$$\kappa^2 + \kappa_*^2 = \tfrac{1}{2}(\xi^2 + \omega^2), \qquad \text{(XVII.27)}$$

$$\left| \frac{\partial(\boldsymbol{\kappa}, \boldsymbol{\kappa}_*)}{\partial(\boldsymbol{\xi}, \boldsymbol{\omega})} \right| = \frac{1}{8}.$$

Using these, we find that $\mathbb{A}_{rmn}$ may be expressed in the form

$$\mathbb{A}_{rmn} = \frac{1}{(2\pi)^3 8 m! n!} \int_{\mathcal{V}} \int_{\mathcal{V}} e^{-\frac{1}{4}\xi^2} e^{-\frac{1}{4}\omega^2} H_m(\tfrac{1}{2}\boldsymbol{\xi} - \tfrac{1}{2}\boldsymbol{\omega}) H_n(\tfrac{1}{2}\boldsymbol{\xi} + \tfrac{1}{2}\boldsymbol{\omega})$$

$$\times \mathbb{B} H_r(\tfrac{1}{2}\boldsymbol{\xi} - \tfrac{1}{2}\boldsymbol{\omega}, \tfrac{1}{2}\boldsymbol{\xi} + \tfrac{1}{2}\boldsymbol{\omega}) \, d\boldsymbol{\xi} \, d\boldsymbol{\omega}. \qquad \text{(XVII.28)}$$

Before we can evaluate the two integrals here, we need an explicit expression for $\mathbb{B}H_r$. This we may obtain from some of the results in Chapter XVI. First let us recall the property (VI.68) of the encounter operator $\mathbb{E}$: If the initial asymptotic velocities $\mathbf{v}$ and $\mathbf{v}_*$ for an encounter are interchanged, so also are the final asymptotic velocities $\mathbf{v}'$ and $\mathbf{v}'_*$. Using this fact, we see from (XVI.6) that the value $\mathbb{O}g(\mathbf{c}_*, \mathbf{c})$ of $\mathbb{O}g$ is given by the corresponding integral of $w(g'_* - g_*)$. Hence

$$\mathbb{B}g(\mathbf{c}, \mathbf{c}_*) = \mathbb{O}g(\mathbf{c}, \mathbf{c}_*) + \mathbb{O}g(\mathbf{c}_*, \mathbf{c}). \qquad \text{(XVII.29)}$$

But we already have an explicit formula for $\mathbb{O}c_n$, namely (XVI.21), and so by placing it in (XVII.29) we obtain a corresponding formula for $\mathbb{B}c_n$. From this formula and (XVII.2) we obtain $\mathbb{B}H_r$. For example,

$$\mathbb{B}H = 0,$$

$$\mathbb{B}H_k = 0,$$

$$\mathbb{B}H_{km} = -3\mathbb{A}_{2|2}(\omega\sqrt{\tfrac{2}{3}\varepsilon})(\omega_k \omega_m - \tfrac{1}{3}\omega^2 \delta_{km}),$$

$$\mathbb{B}H_{kmr} = -\tfrac{9}{2}\mathbb{A}_{2|2}(\omega\sqrt{\tfrac{2}{3}\varepsilon})(\xi_{(k}\omega_m \omega_{r)} - \tfrac{1}{3}\omega^2 \xi_{(k}\delta_{mr)}),$$

$$\mathbb{B}H_{kmrs} = -\tfrac{9}{2}\mathbb{A}_{2|2}(\omega\sqrt{\tfrac{2}{3}\varepsilon})(\xi_{(k}\xi_m \omega_r \omega_{s)} - \tfrac{1}{3}\omega^2 \xi_{(k}\xi_m \delta_{rs)}$$

$$- 4\omega_{(k}\omega_m \delta_{rs)} + \tfrac{4}{3}\omega^2 \delta_{(km}\delta_{rs)}) \qquad \text{(XVII.30)}$$

$$+ \tfrac{1}{4}(35\mathbb{A}_{4|4}(\omega\sqrt{\tfrac{2}{3}\varepsilon}) - 10\mathbb{A}_{2|2}(\omega\sqrt{\tfrac{2}{3}\varepsilon}))\omega_k \omega_m \omega_r \omega_s$$

$$- \tfrac{3}{2}(5\mathbb{A}_{4|4}(\omega\sqrt{\tfrac{2}{3}\varepsilon}) - \mathbb{A}_{2|2}(\omega\sqrt{\tfrac{2}{3}\varepsilon}))\omega^2 \omega_{(k}\omega_m \delta_{rs)}$$

$$+ \tfrac{3}{4}\mathbb{A}_{4|4}(\omega\sqrt{\tfrac{2}{3}\varepsilon})\omega^4 \delta_{(km}\delta_{rs)}.$$

We see from (XVI.21), (XVII.29), and (XVII.2) that in general $\mathbb{B}H_r(\tfrac{1}{2}\xi - \tfrac{1}{2}\omega, \tfrac{1}{2}\xi + \tfrac{1}{2}\omega)$ may be written as a linear combination each term of which is a product of three expressions, one of these being a polynomial in the components of ξ, another being a polynomial in the components of ω, and the last being a sum of coefficients $\mathbb{A}_{n|m}(\omega\sqrt{\tfrac{2}{3}\varepsilon})$ defined by (XII.16). The explicit formulae (XVII.30) illustrate this fact. Of course, because H_s is a polynomial, the same conclusion applies also to the more complicated quantity $H_m(\tfrac{1}{2}\xi - \tfrac{1}{2}\omega)H_n(\tfrac{1}{2}\xi + \tfrac{1}{2}\omega)\mathbb{B}H_r(\tfrac{1}{2}\xi - \tfrac{1}{2}\omega, \tfrac{1}{2}\xi + \tfrac{1}{2}\omega)$ appearing in (XVII.28). Since ξ enters this sum only as a polynomial, the integral with respect to it in each term may be evaluated immediately. Since the functions $\mathbb{A}_{n|m}$ depend upon ω only through its magnitude ω, the integral with respect to ω in each term may be reduced to one with respect to ω, and the resulting terms may then be expressed as linear combinations of the new coefficients:

$$\mathbb{A}_{p|n|m}(\varepsilon) \equiv \frac{p!}{2\sqrt{\pi}(2p+1)!} \int_0^\infty e^{-\tfrac{1}{4}\omega^2} \omega^{2p+2} \mathbb{A}_{n|m}(\omega\sqrt{\tfrac{2}{3}\varepsilon})\, d\omega. \qquad \text{(XVII.31)}$$

Each $\mathbb{A}_{n|m}$, and hence also each $\mathbb{A}_{p|n|m}$ is determined by the molecular model through the scattering factor $\mathbb{S}$. Since $\mathbb{A}_{n|m} > 0$, also $\mathbb{A}_{p|n|m} > 0$. For ideal spheres[6] it follows from (XII.17) that

$$\mathbb{A}_{p|2|2}(\varepsilon) = \frac{p!(p+1)!\, 2^{2p+1}}{3(2p+1)!}\sqrt{\tfrac{2}{3}\pi\varepsilon}\, \mathbb{d}^2, \qquad \text{(XVII.32)}$$

while for inverse $\mathbb{k}^{\text{th}}$-power molecules it follows from (XII.18) that

$$\mathbb{A}_{p|2|2}(\varepsilon) = \frac{p!}{2\sqrt{\pi}(2p+1)!}\frac{\mathbb{a}}{2}\left(\frac{2\mathfrak{g}}{\mathfrak{m}}\right)^{2/(\mathbb{k}-1)}\left(\frac{2}{3}\varepsilon\right)^{\tfrac{1}{2}(\mathbb{k}-5)/(\mathbb{k}-1)}$$

$$\times \int_0^\infty e^{-\tfrac{1}{4}\omega^2}\omega^{2p+2+[(\mathbb{k}-5)/(\mathbb{k}-1)]}\, d\omega, \qquad \text{(XVII.33)}$$

$$= \frac{2^{2p+[(\mathbb{k}-5)/(\mathbb{k}-1)]}p!\,\mathbb{a}}{\sqrt{\pi}(2p+1)!}\left(\frac{2\mathfrak{g}}{\mathfrak{m}}\right)^{2/(\mathbb{k}-1)}\Gamma\!\left(p + 2\frac{\mathbb{k}-2}{\mathbb{k}-1}\right)\left(\frac{2}{3}\varepsilon\right)^{\tfrac{1}{2}(\mathbb{k}-5)/(\mathbb{k}-1)}$$

The number $\mathbb{a}$, which is given by (XII.19), depends upon $\mathbb{k}$. The definition of $\mathbb{A}_{p|n|m}$ has been so arranged that when $\mathbb{A}_{n|m}$ is a constant, as it is for example for Maxwellian molecules, then $\mathbb{A}_{p|n|m} = \mathbb{A}_{n|m}$ for all p. When $m = n = 2$, this can be verified directly by use of (XVII.33)$_2$.

---

[6] GRAD [1949, Eqs. (A3.54) and (A3.49)]. GRAD's coefficients B_1 and $B_1^{(p)}$ are related as follows to those employed here:

$$2B_1 = \mathbb{A}_{2|2}, \qquad 2B_1^{(p)} = \mathbb{A}_{p|2|2}.$$

(i) FORMAL COLLISIONS INTEGRALS

The longest step in this procedure is to write

$$H_m(\tfrac{1}{2}\xi - \tfrac{1}{2}\omega)H_n(\tfrac{1}{2}\xi + \tfrac{1}{2}\omega)\mathbb{B}H_r(\tfrac{1}{2}\xi - \tfrac{1}{2}\omega, \tfrac{1}{2}\xi + \tfrac{1}{2}\omega)$$

as a sum of terms of the form described previously and then to effect the integrations with respect to ξ and ω. This step is particularly tedious when m, n, and r are large. The simplest cases of this calculation, those in which $r = 0, 1, \ldots, 4$, follow by use of (XVII.30) and lead to the following list[7] of *collisions integrals for general molecular models*:

$$m\bar{C}H = 0,$$

$$m\bar{C}H_k = 0,$$

$$m\bar{C}H_{km} = -3n\rho\mathbb{A}_{2|2|2}A_{km} - \tfrac{3}{2}n\rho(\mathbb{A}_{3|2|2} - \mathbb{A}_{2|2|2})A_{\{ak}A_{ma\}}$$
$$+ \tfrac{3}{16}n\rho(\mathbb{A}_{4|2|2} - 2\mathbb{A}_{3|2|2} + \mathbb{A}_{2|2|2})(2A_{\{kma}A_{abb\}}$$
$$+ 2A_{\{abk}A_{mab\}} + A_{\{aak}A_{mbb\}}) + Q_{km},$$

$$m\bar{C}H_{kmr} = -\tfrac{9}{2}n\rho\mathbb{A}_{2|2|2}(A_{kmr} - \tfrac{1}{3}A_{aa(k}\delta_{mr)}) + \tfrac{3}{8}n\rho(\mathbb{A}_{3|2|2} - \mathbb{A}_{2|2|2})$$
$$\times (3A_{(km}A_{r)aa} - 6A_{a(k}A_{mr)a} + 4A_{ab}A_{ab(k}\delta_{mr)}$$
$$- 4A_{aab}A_{b(k}\delta_{mr)}) + Q_{kmr}, \qquad (XVII.34)$$

$$m\bar{C}H_{kmrs} = \tfrac{1}{4}n\rho(35\mathbb{A}_{4|4|4} - 10\mathbb{A}_{4|2|2} - 18\mathbb{A}_{2|2|2})A_{kmrs}$$
$$- \tfrac{3}{4}n\rho(10\mathbb{A}_{4|4|4} + \mathbb{A}_{4|2|2} - 6\mathbb{A}_{3|2|2} + \mathbb{A}_{2|2|2})A_{aa(km}\delta_{rs)}$$
$$+ \tfrac{3}{4}n\rho(\mathbb{A}_{4|4|4} + \mathbb{A}_{4|2|2} - 2\mathbb{A}_{3|2|2} + \mathbb{A}_{2|2|2})A_{aabb}\delta_{(km}\delta_{rs)}$$
$$+ \tfrac{3}{4}n\rho(35\mathbb{A}_{4|4|4} - 10\mathbb{A}_{4|2|2} + 6\mathbb{A}_{2|2|2})A_{(km}A_{rs)}$$
$$- \tfrac{3}{2}n\rho(10\mathbb{A}_{4|4|4} + \mathbb{A}_{4|2|2} - 6\mathbb{A}_{3|2|2} + 5\mathbb{A}_{2|2|2})A_{a(k}\delta_{mr}A_{s)a}$$
$$+ \tfrac{3}{2}n\rho(\mathbb{A}_{4|4|4} + \mathbb{A}_{4|2|2} - 2\mathbb{A}_{3|2|2} + \mathbb{A}_{2|2|2})$$
$$\times A_{ab}A_{ab}\delta_{(km}\delta_{rs)} + Q_{kmrs};$$

Q_{km} and Q_{kmr} represent those parts of the infinite sum (XVII.24)$_2$ for $m\bar{C}H_{km}$ and $m\bar{C}H_{kmr}$, respectively, that involve Hermite coefficients A_s for which $s \geq 4$, and Q_{kmrs} represents the part of the infinite sum for $m\bar{C}H_{kmrs}$ that involves all terms in (XVII.24)$_2$ for which $m + n \neq 4$.

The equations of transfer for A_2 and A_3 may be written out explicitly. First we use (XVII.21) and similar relations to find the coefficients Γ_{rs}^k and Λ_{rs}^k, and then we substitute these into the left-hand side of (XVII.23). The right-hand

---

[7] GRAD [1949; Eqs. (A3.56), (A3.57)], with the exception of (XVII.34)$_5$, which we believe to be new.

side is given, of course, by (XVII.34)$_{3,4}$. We obtain then the two equations[8]:

$$\rho \dot{A}_{km} + \left(\rho \sqrt{\frac{2}{3}\varepsilon}\, A_{akm}\right)_{,a} - \frac{1}{\varepsilon}(M_{ab}u_{a,b} + q_{a,a})(A_{km} + \delta_{km})$$

$$+ 2\rho A_{a(k}u_{m),a} + 2\rho u_{(k,m)} + \frac{2}{3}\rho\sqrt{\frac{3}{2\varepsilon}}\,\varepsilon_{,a}A_{akm}$$

$$= -3n\rho \mathbb{A}_{2|2|2} A_{km} - \tfrac{3}{2}n\rho(\mathbb{A}_{3|2|2} - \mathbb{A}_{2|2|2})A_{\{ak}A_{ma\}}$$

$$+ \tfrac{3}{16}n\rho(\mathbb{A}_{4|2|2} - 2\mathbb{A}_{3|2|2} + \mathbb{A}_{2|2|2}) \qquad\qquad\text{(XVII.35)}$$

$$\times (2A_{\{kma}A_{abb\}} + 2A_{\{kab}A_{mab\}} + A_{\{kaa}A_{mbb\}}) + Q_{km},$$

$$\rho \dot{A}_{kmr} + \left(\rho \sqrt{\frac{2}{3}\varepsilon}\, A_{akmr}\right)_{,a} + 3\left(\rho\sqrt{\frac{2}{3}\varepsilon}\, A_{(km}\right)_{,r)} - 3\sqrt{\frac{3}{2\varepsilon}}\, A_{(km}M_{r)a,a}$$

$$- \frac{3}{2\varepsilon}(M_{ab}u_{a,b} + q_{a,a})A_{kmr} + 3\rho A_{a(km}u_{r),a}$$

$$+ \rho\sqrt{\frac{3}{2\varepsilon}}\,(\varepsilon_{,a}A_{akmr} + 3A_{(km}\varepsilon_{,r)} + 2\varepsilon_{,a}A_{a(k}\delta_{mr)} + \varepsilon_{,(k}\delta_{mr)})$$

$$= -\tfrac{9}{2}n\rho\mathbb{A}_{2|2|2}(A_{kmr} - \tfrac{1}{3}A_{aa(k}\delta_{mr)}) + \tfrac{3}{8}n\rho(\mathbb{A}_{3|2|2} - \mathbb{A}_{2|2|2})$$

$$\times (3A_{(km}A_{r)aa} - 6A_{a(k}A_{mr)a} + 4A_{ab}A_{ab(k}\delta_{mr)}$$

$$- 4A_{aab}A_{b(k}\delta_{mr)}) + Q_{kmr}.$$

The equations of transfer for $A = 1$, $A_k = 0$, and $A_{aa} = 0$ are satisfied identically. The usual equations of balance, (IX.3)–(IX.5), may be written in terms of $\mathbb{A}_2$ and $\mathbb{A}_3$ by use of (XVII.16) and (XVII.17).

(ii) Contrast and comparison of Grad's formal expansion with Ikenberry's theorem

Because the Hermite polynomials $\mathbf{H}_r$ form a complete set, we may use (XVII.24)$_2$ to evaluate $\bar{C}g$ formally for any polynomial g. The result, generally, is an infinite series, the coefficients of which, being linear combinations of the $\mathbb{A}_{rmn}$, are determined from the kinetic constitutive relations through use of (XVII.25). Of course GRAD's result applies to a gas of Maxwellian molecules as a special case. For such a gas IKENBERRY's theorem also delivers the collisions integrals $\bar{C}g$. It is important to note that these two results, namely, IKENBERRY's theorem and the specialization of GRAD's formula (XVII.24)$_2$ to Maxwellian molecules, are not equivalent. First, GRAD considers only such molecular density functions F as may be represented in mean by a series of Hermite functions.

---

[8] GRAD [1949, Eqs. (A3.31), (A3.46), (A3.56), and (A3.57)].

(ii) CONTRAST WITH IKENBERRY'S THEOREM

Because of (XVII.13), such molecular densities must possess moments of all orders. In contrast, IKENBERRY's theorem is valid, as its statement in Chapter XVI shows, for a much broader class of functions F, indeed for ones which need possess only a finite number of moments. Secondly, GRAD's results are only formal since (XVII.24)$_1$ involves the product of two Hermite series, neither of which need converge pointwise. In contrast to this, IKENBERRY's theorem involves no series; it is exact, rigorously proved in the preceding chapter. GRAD's assertion if specialized to Maxwellian molecules should follow from IKENBERRY's theorem, without the restrictions laid down at the start in GRAD's arguments. Thus for Maxwellian molecules we should expect the formulae for $\bar{C}g$ obtained by GRAD and IKENBERRY to agree. Such is the case. Indeed, (XVII.24)$_2$ should reduce to a finite sum over those products $A_m A_n$ for which $m + n = r$, and secondly the coefficients in this sum should be constants. This second conjecture is certainly true, since $\mathbb{S} = \mathbb{S}(\theta)$ for Maxwellian molecules, and so (XVII.25) shows that each $\mathbb{A}_{rmn}$ is independent of ε. In order to see that the first conjecture is true, we require an alternative form of the coefficient $\mathbb{A}_{rmn}$. If

$$F_r \equiv \frac{1}{(\frac{4}{3}\pi\varepsilon)^{\frac{3}{2}}} e^{-\frac{1}{2}\kappa^2} H_r(\kappa), \tag{XVII.36}$$

then from (XVII.6), (XVII.25), and (VII.20)$_3$ we find that

$$\mathbb{A}_{rmn} = \frac{1}{m!n!} \bar{\mathbb{C}}_{F_m, F_n} H_r(\kappa). \tag{XVII.37}$$

If we use the form of $\bar{\mathbb{C}}_{G,H}$ given by (VII.15)$_2$, it follows from (VI.2) that $\mathbb{A}_{rmn}$ is given also by the formula

$$\mathbb{A}_{rmn} = \frac{1}{(2\pi)^3 m!n!} \int_{\mathscr{V}} \int_{\mathscr{V}} \int_{\mathscr{O}} e^{-\frac{1}{2}\kappa^2} e^{-\frac{1}{2}\kappa_*^2} H_r(H'_m H'_{n*} + H'_{m*} H'_n$$

$$- H_m H_{n*} - H_{m*} H_n) \mathbb{S}(\theta, \omega \sqrt{\tfrac{2}{3}\varepsilon}) \, d\kappa \, d\kappa_*. \tag{XVII.38}$$

A simple application of (XVI.13) shows that for Maxwellian molecules

$$\int_{\mathscr{O}} (H'_r + H'_{r*} - H_r - H_{r*}) \mathbb{S}(\theta) \tag{XVII.39}$$

is a polynomial[9] of degree r in the components of κ and κ_*. By (XVII.25) and the orthogonality conditions (XVII.3) we see that $\mathbb{A}_{rmn}$ vanishes if $m + n > r$.

---

[9] GRAD [1949, §4] noted this fact, as well as the corresponding result for (XVII.40), but he supplied no proof.

Similarly

$$\int_\Omega (H'_m H'_{n*} + H'_{m*} H'_n - H_m H_{n*} - H_{m*} H_n) \mathbb{S}(\theta) \qquad \text{(XVII.40)}$$

is a polynomial of degree $m + n$ in the components of κ and κ_*, and so (XVII.38) and the orthogonality conditions (XVII.3) show that $\mathbb{A}_{rmn}$ is zero when $r > m + n$. Thus (XVII.24)$_2$ when specialized to Maxwellian molecules reduces to a finite sum over those terms for which $r = m + n$, the coefficients in this sum being constants.

(iii) Grad's method of truncation. His 13-moment system and his 20-moment system

For general molecular models, the sum on the right-hand side of (XVII.24)$_2$ may be expected to be infinite, and as a result the explicit equation of transfer for A_r is expressed in terms of an infinite series in fields $A_s, s = 1, 2, \dots$. For this reason it would be difficult to use the equations of transfer to determine the functions A_r in general. To overcome this difficulty, GRAD has set down a formal method[10] of calculating successive approximations to the function F in which the equations of transfer for the A_r are replaced by equations which are much simpler. His method may be described in four steps. First he truncates the expansion (XVII.9) for F, retaining only a finite number of terms. Let us denote by $\hat{F}$ the finite sum which results and remark that it is completely determined by a finite number of the functions A_r. In order to illustrate the procedure, let us retain only those terms in (XVII.9) for which $s \leq 3$. Then, by means of (XVII.2), (XVII.6), (XVII.16)$_2$, and (XVII.17)$_1$ we can write $\hat{F}$ in the explicit form

$$\hat{F} = F_M\left(1 + \frac{3}{4p\varepsilon} P_{ab} c_a c_b + \frac{3}{8p\varepsilon^2} M_{abd}(c_a c_b c_d - 2\varepsilon c_{(a} \delta_{bd)})\right). \qquad \text{(XVII.41)}$$

As we remarked just after (VIII.36), the function $\hat{F}$ cannot be non-negative for all values of $\mathbf{c}$ unless it reduces to F_M. Thus it is not a molecular density, but of course it may approximate one in various ways. If we think of $\hat{F}$ in such a context, the gross behavior of the gas is described completely by the fields ρ, $\mathbf{u}$, ε, $\mathbf{P}$, and $^3\mathbf{M}$, the components of these being twenty in number. Whatever be the $\hat{F}$ considered, the second step is to calculate fields $\hat{A}_r$ from $\hat{F}$ just as the fields A_r are determined by F through (XVII.13). For example, when

---

[10] GRAD [1949, §§4 and 5].

(iii) GRAD'S METHOD OF TRUNCATION 273

F is given by (XVII.41) we find that

$$\hat{A} = 1,$$
$$\hat{A}_k = 0,$$
$$\hat{A}_{km} = \frac{1}{p} P_{km},$$ (XVII.42)
$$\hat{A}_{kmr} = \frac{1}{p\sqrt{\frac{2}{3}\varepsilon}} M_{kmr},$$
$$\hat{A}_r = 0, \quad r \geq 4,$$

the last statement here being a consequence of the orthogonality conditions (XVII.3), while the others follow from (XVII.14)–(XVII.17)$_1$ since $\hat{F}$ is simply a particular example of the expansion (XVII.9). Third, GRAD replaces A_r by $\hat{A}_r$ in each of the equations of transfer and then retains only the equations of transfer for the fields which determine $\hat{F}$. In the special case (XVII.41), which we are considering as illustration, we retain the equations of balance (IX.3), (IX.4), and (IX.5) as well as equation (XVII.35) for the second and third moments. Because of (XVII.42)$_5$ the infinite sums represented by Q_{km} and Q_{kmr} in the latter equations vanish identically, and the fields A_{kmrs} on the left-hand side of (XVII.35)$_2$ are zero. Fourth and finally, GRAD introduces a simplification of the collisions integrals $\bar{C}H_r$. He reasons that if F is to be approximated fairly closely by F_M, then it is reasonable to expect the components of each tensor in the list $\hat{A}_2, \hat{A}_3, \ldots$ to be significantly smaller than those of the tensors preceding it. Therefore, if $\hat{F}$ is found by dropping from the general expansion for F all terms whose order exceeds a given integer N, all products $A_m A_n$ in which $m + n > N$ may be expected to be small compared to those in which $m + n \leq N$. Accordingly GRAD simply casts away from each $\bar{C}H_r$ all the former terms. In this way he obtains a system of differential equations for the fields which determine $\hat{F}$. In the special case (XVII.41) we take $N = 3$, and so we must cast away from (XVII.34)$_3$ and (XVII.34)$_4$ all terms which are quadratic in the fields A_s. The appropriate equations for this case may be read off directly from (XVII.35); when expressed in terms of ρ, $\mathbf{u}$, ε, $\mathbf{P}$, and $^3\mathbf{M}$ by means of (XVII.16)$_2$ and (XVII.17)$_1$, these become

$$\dot{P}_{km} + P_{km} E - \tfrac{2}{3} P_{ab} E_{ab} \delta_{km} + 2 P_{a(k} u_{m),a} + 2p E_{km}$$
$$+ (M_{akm} - \tfrac{2}{3} q_a \delta_{km})_{,a} = -3n \mathbb{A}_{2|2|2} P_{km},$$

$$\dot{M}_{kmr} + M_{kmr} E + 2\varepsilon P_{(km,r)} - \frac{3}{\rho} P_{(km} M_{r)a,a} + 3 M_{a(km} u_{r),a}$$ (XVII.43)
$$+ 2P_{(km} \varepsilon_{,r)} + 2\varepsilon_{,a} P_{a(k} \delta_{mr)} + 2p\varepsilon_{,(k} \delta_{mr)}$$
$$= -\tfrac{9}{2} n \mathbb{A}_{2|2|2} (M_{kmr} - \tfrac{2}{3} q_{(k} \delta_{mr)}).$$

These two equations, when taken together with the equations of balance (IX.3), (IX.4), and (IX.5), constitute GRAD's 20-*moment system*.

A second example of this method of truncation comes from retaining only those terms in (XVII.9) which involve the 13 fields $\rho, u_k, \varepsilon, P_{km}$, and q_k, these being the fields of principal interest in gas flows. First let us introduce the polynomials

$$H_k^{(3)}(\kappa) \equiv \kappa_k(\kappa^2 - 5),$$
$$H_{0|kmr}^{(3)}(\kappa) \equiv \kappa_k \kappa_m \kappa_r - \tfrac{3}{5}\kappa^2 \kappa_{(k} \delta_{mr)}. \quad \text{(XVII.44)}$$

It is a straightforward matter to show that

$$H_k^{(3)} = H_{kaa},$$
$$H_{0|kmr}^{(3)} = H_{kmr} - \tfrac{3}{5} H_{aa(k} \delta_{mr)}, \quad \text{(XVII.45)}$$
$$H_{kmr} = H_{0|kmr}^{(3)} + \tfrac{3}{5} H_{(k}^{(3)} \delta_{mr)};$$

thus each member of one of the two sets (XVII.44) and H_{kmr} is a linear combination of the members of the other. Moreover, from (XVII.45) we can write the third-order terms in the expansion (XVII.9) in the alternative form

$$A_{abc} H_{abc} = A_{0|abc}^{(3)} H_{0|abc}^{(3)} + \tfrac{3}{5} A_a^{(3)} H_a^{(3)}, \quad \text{(XVII.46)}$$

in which

$$A_{0|kmr}^{(3)} \equiv A_{kmr} - \tfrac{3}{5} A_{aa(k} \delta_{mr)},$$
$$A_k^{(3)} \equiv A_{kaa}. \quad \text{(XVII.47)}$$

Directly from (XVII.17)$_1$ we can express $A_{0|kmr}^{(3)}$ and $A_k^{(3)}$ in terms of the moments of F as follows:

$$A_{0|kmr}^{(3)} = \frac{1}{p\sqrt{\tfrac{2}{3}\varepsilon}} P_{0|kmr},$$
$$A_k^{(3)} = \frac{2}{p\sqrt{\tfrac{2}{3}\varepsilon}} q_k, \quad \text{(XVII.48)}$$

$P_{0|kmr}$ being defined by (XIII.6).

We now proceed to drop all terms in (XVII.9) for which $s \geq 4$, and also $A_{0|kmr}^{(3)}$ in the third-order terms (XVII.46). By use of (XVII.6), (XVII.16)$_2$, and (XVII.48)$_2$ we obtain the following $\hat{F}$:

$$\hat{F} = F_M \left[1 + \frac{3}{4p\varepsilon} P_{ab} c_a c_b - \frac{3}{2p\varepsilon} q_a c_a \left(1 - \frac{3c^2}{10\varepsilon}\right) \right]. \quad \text{(XVII.49)}$$

The expression in brackets here we denoted in Chapter VIII by $1 + G_G$, and in Section (iii) of that chapter we examined the degree to which the averages with respect to $\hat{F}$ approximate those with respect to F. Next we calculate the functions $\hat{A}_r$ that correspond to $\hat{F}$, and these are given again by (XVII.42), with the exception of (XVII.42)$_4$ which is now replaced by

$$\hat{A}_{kmr} = \frac{6}{5p\sqrt{\frac{2}{3}\varepsilon}} q_{(k} \delta_{mr)}. \tag{XVII.50}$$

This can be seen from (XVII.47) and (XVII.48)$_2$ by replacing A_r by $\hat{A}_r$ and setting $A^{(3)}_{0|kmr}$ equal to zero. Next we consider the equation of transfer (XVII.35)$_1$ for A_{km} and the equation of transfer for $A^{(3)}_k$ obtained from (XVII.35)$_2$ by contracting on m and r. In these equations we replace A_r by $\hat{A}_r$ as calculated above, and on the right-hand sides drop those products $\hat{A}_m \hat{A}_n$ in which $m + n > 3$. Equivalently we may simply contract (XVII.43)$_2$ on m and r, and whenever M_{kmr} appears in this equation and in (XVII.43), we replace it by the value

$$M_{kmr} = \tfrac{6}{5} q_{(k} \delta_{mr)}, \tag{XVII.51}$$

which follows from (XVII.17)$_1$ and (XVII.50). The foregoing arguments paraphrase GRAD's. In terms of IKENBERRY's spherical moments the statement (XVII.51) is just the result of setting $P_{0|kmr} = 0$ in (XVI.29)$_4$. With this simplifying assumption the 20-moment system (XVII.43) reduces to

$$\begin{aligned}
&\dot{P}_{km} + P_{km} E - \tfrac{2}{3} P_{ab} E_{ab} \delta_{km} + 2 P_{a(k} u_{m),a} + 2p E_{km} \\
&\quad + \tfrac{4}{5}(q_{(k,m)} - \tfrac{1}{3} q_{u,u} \delta_{km}) = -3n \mathbb{A}_{2|2|2} P_{km}, \\
&\dot{q}_k + \tfrac{7}{5} q_k E + \tfrac{2}{5}\varepsilon P_{ka,a} - \frac{1}{\rho} P_{ka} M_{ab,b} + \tfrac{7}{5} q_a u_{k,a} \\
&\quad + \tfrac{2}{5} q_a u_{a,k} + \tfrac{7}{3} P_{ka} \varepsilon_{,a} + \tfrac{5}{3} p \varepsilon_{,k} = -2n \mathbb{A}_{2|2|2} q_k,
\end{aligned} \tag{XVII.52}$$

and these, together with (IX.3), (IX.4), and (IX.5), constitute GRAD's *13-moment system*. These are field equations, each of which is of first order in t. In terms of continuum mechanics as described in Chapter I, we see that GRAD's *13-moment system describes a continuum theory of fluids of the rate type*.

However, (XVII.52)$_1$ is not of the form (I.28). First, it delivers $\dot{\mathbf{P}}$ as a function not only of $\mathbf{P}, \mathbf{E}, \mathbf{W}, \rho$, and ε, but also of $\mathbf{q}$. Second, the terms involving the spin $\mathbf{W}$ are of the form $\dot{\mathbf{P}} + \mathbf{WP} - \mathbf{PW}$; thus, even apart from the dependence of $\dot{\mathbf{P}}$ upon $\mathbf{q}$, GRAD's approximation (XVII.52)$_1$ is not of the form (I.28). Of course all the steps both approximate and exact that led GRAD to his 13-moment system presume that a solution of the Maxwell-Boltzmann equation is being considered. Such being the case, MAXWELL's consistency theorem renders the variables entering (XVII.52)$_1$ dependent upon each other through (IX.3), (IX.4), and (IX.5). The relation (XVII.52)$_1$ as it stands, with variables taken as independent, is one of *infinitely many possible extensions* of the results GRAD has inferred. That particular extension does not satisfy the principle of material frame-indifference for materials of the rate type. Whether there be an extension such as to satisfy that principle, is not presently known. As the results of GRAD's process are intentionally approximate, not exact, the question is of scant importance. An arbitrarily close approximation need not share the invariance of the quantity it approximates. For example, the power series for $\sin x$ converges for all x, but no partial sum is periodic, and the Riemann upper and lower sums to approximate the area of a plane figure are not generally invariant under rotation of the axes used to define them.

As far as this general method of truncation is concerned, it is felt that the density function $\hat{F}$ will approximate a solution F, the approximation improving as the number of terms which are included in $\hat{F}$ increases, but, as far as we know, no reasoning to this effect has ever been brought forward. This conjecture is at least suspect due to the fact that GRAD simply casts away those terms on the right-hand side of (XVII.24)$_2$ for which $m + n > N$, his justification of this step being itself somewhat dubious[11]. However there is one molecular model, namely, a gas of Maxwellian molecules, for which the terms that GRAD drops are identically zero. That is because for such a gas the coefficients $\mathbb{A}_{rmn}$ vanish unless $m + n = r$, as we have shown above. We can say somewhat more in this case. If, as we have assumed previously, F is truncated by dropping all terms whose orders exceed some integer N, so $\hat{F}$ is given by

$$\hat{F} = F_M \sum_{s=0}^{N} \frac{1}{s!} A_s H_s, \qquad \text{(XVII.53)}$$

then the orthogonality conditions (XVII.3) show that

$$\hat{A}_r = \begin{cases} A_r & \text{if } r \leq N, \\ 0 & \text{if } r > N. \end{cases} \qquad \text{(XVII.54)}$$

Thus

$$\bar{\mathbb{C}}_{\hat{F}} H_r = \bar{\mathbb{C}}_F H_r \quad \text{if } r \leq N, \qquad \text{(XVII.55)}$$

again because $\mathbb{A}_{rmn}$ vanishes unless $m + n = r$. This means that the right-hand sides of the equations for $\hat{A}_r$, $r \leq N$, are exact. Moreover, since the functions $\hat{A}_r$, $r > N$, appear only in the equation of transfer for $\hat{A}_N$, the left-hand sides of the equations of transfer for $\hat{A}_r$, $r < N$, remain unchanged when A_s is replaced by $\hat{A}_s$. Thus, for a gas of Maxwell molecules, *when F is truncated after terms of order N, the equations for $\hat{A}_0, \ldots, \hat{A}_{N-1}$ are exact. The equation for $\hat{A}_N$ is in general not exact.* For example, equation (XVII.43)$_1$ for the pressures in GRAD's 20-moment system is precisely the same as the exact equation (XIII.4), which was obtained in principle by MAXWELL, but (XVII.43)$_2$ differs from the exact equation (XIII.5).

GRAD's method provides a sequence that approximates formally the exact, infinite system of equations for moments. Each approximation is a finite system of partial differential equations, each member of which is of the first order in the time. The further we carry the approximation, the higher becomes the order of the system of differential equations.

A system of equations, each of which is of first order in the time, suggests an initial-value problem. The higher is the order of the system, the greater the

---

[11] The reasoning of GRAD [1949, Appendix 5] to justify his 13-moment system is much like that we have described in Footnote 7 to Chapter VIII, and so his argument suffers from the objections we have noted there.

number of initial values we expect to be able to prescribe. For example, GRAD's 13-moment system (IX.3), (IX.4), (IX.5), and (XVII.52) suggests that in order to obtain a unique solution we should have to prescribe the initial values of the thirteen moments involved: the components of $\rho(0, \mathbf{x})$, $\mathbf{u}(0, \mathbf{x})$, $\varepsilon(0, \mathbf{x})$, $\mathbf{P}(0, \mathbf{x})$, and $\mathbf{q}(0, \mathbf{x})$ if $\mathbf{x}$ lies in the part of space we wish to consider. Similarly, GRAD's 20-moment system suggests that we prescribe twenty initial values.

GRAD's method is not one of successive approximations in the usual sense. It does not determine the pressure tensor and energy flux in terms of gross fields. All it produces is equations to be solved for them, not solutions of problems concerning them. If we have the solution to some problem formulated with respect to the 13-moment system, and if we decide for some reason that it is not accurate enough, GRAD's framework does not show us a way to use that solution as a basis for improvement. It does not even provide the data necessary to formulate a corresponding problem according to the 20-moment system. Thus GRAD's method cannot justly be regarded as a method for solving the Maxwell–Boltzmann equation; rather, it is a systematic, rational procedure for *formally approximating the kinetic theory by a sequence of theories of continuum mechanics*. Some of the higher moments appear in GRAD's approximating theories. They are not of interest in themselves but rather are parameters which spring up there as a nuisance and have to be eliminated. The aim is to determine $\mathbf{P}$ and $\mathbf{q}$ for a gas flow of some particular kind. In order to do so, we have to expect either to prescribe the initial values of certain higher moments or to prove that the effect of those values upon $\mathbf{P}$ and $\mathbf{q}$ is negligible.

In this sense GRAD's results have deservedly received much attention. There is a literature on the 13-moment equations, studied as an end in themselves, and some work has been done on the 20-moment equations in the same spirit. Some exact solutions of these particular equations of continuum mechanics have even been labelled incorrectly "exact solutions" according to the kinetic theory itself.

(iv) *Comparison of solutions of Grad's systems with corresponding exact solutions for shearing*

Insight into the accuracy of GRAD's successive systems of equations for moments can be gained by comparing the results they deliver in a special problem with the exact solutions for that same problem. Since we have presented in Chapter XIV the exact solution for homo-energetic simple shearing in a gas of Maxwellian molecules, let us see what GRAD's method, suitably specialized to Maxwellian molecules, tells us about such a flow.

First we consider the 13-moment equations. Since the third moment $^3\mathbf{M}$ is a function of time alone, GRAD's equation (XVII.52)$_1$ for the pressures is exact, and so his equations deliver just the same system of pressures as that which follows from the exact solution. The same, however, is not true of the energy flux. We can read off GRAD's equations for them from (XIV.27) by setting $P_{0|kmr}$ equal to zero and retaining only the equations containing $\dot{q}$, these being

$$\tau \dot{q}_1 + \tfrac{2}{3} q_1 + \tfrac{7}{5} T q_2 = 0,$$
$$\tau \dot{q}_2 + \tfrac{2}{3} q_2 + \tfrac{2}{5} T q_1 = 0, \qquad (XVII.56)$$
$$\tau \dot{q}_3 + \tfrac{2}{3} q_3 = 0.$$

The general solution of these is[12]

$$q_1 = \left(\tfrac{1}{2} q_1(0) - \tfrac{\sqrt{14}}{4} q_2(0)\right) e^{R_1 t/\tau} + \left(\tfrac{1}{2} q_1(0) + \tfrac{\sqrt{14}}{4} q_2(0)\right) e^{R_2 t/\tau},$$

$$q_2 = \left(-\tfrac{1}{\sqrt{14}} q_1(0) + \tfrac{1}{2} q_2(0)\right) e^{R_1 t/\tau} + \left(\tfrac{1}{\sqrt{14}} q_1(0) + \tfrac{1}{2} q_2(0)\right) e^{R_2 t/\tau},$$

$$q_3 = q_3(0) e^{-2t/3\tau}, \qquad (XVII.57)$$

the factors R_1 and R_2 being given as follows:

$$R_1 = -\tfrac{2}{3} + \tfrac{\sqrt{14}}{5} T, \qquad R_2 = -\tfrac{2}{3} - \tfrac{\sqrt{14}}{5} T. \qquad (XVII.58)$$

We see that R_2 is always negative, while as T increases, R_1 is negative at first, vanishes when $T = 10/(3\sqrt{14})$, and thereafter is positive. Therefore, q_1 and q_2 are unstable when $T > 10/(3\sqrt{14})$. This critical value is slightly larger than the corresponding value $\sqrt{\tfrac{18}{31}}$ for which the exact solutions for q_1 and q_2 are unstable. Moreover, (XVII.57)$_3$ fails to predict any instability[13] in the component q_3, while we know from the exact solution that if $T > 3/\sqrt{2}$, then $|q_3(t)| \to \infty$ as $t \to \infty$ unless $q_3(0) = 0$.

If we go now to the 20-moment equations, we find that for homo-energetic simple shearing they are just the same as the exact equations.

(v) *The relaxation theorem for Grad's 13-moment system. Grad's derivation of Enskog's first approximation to the viscosity and the Maxwell number*

Many problems that are presently beyond our grasp if approached through the kinetic theory become amenable to solution if that theory is replaced by one of GRAD's finite systems of partial differential equations. A complete survey

---

[12] Eqs. (XVII.57)$_{1,2}$ correct their counterparts in the paper of TRUESDELL [1956, Eqs. (51.25)$_{2,3}$].

[13] TRUESDELL [1956, §51].

of these solutions is outside the scope of this book. One of them, however, is of such great interest in itself that we record it now.

In Chapter XIII we presented MAXWELL's solution for the pressure deviator and the energy flux of a gas of Maxwellian molecules in a grossly homogeneous condition. For general molecular models we can obtain an approximate solution of this same problem by using GRAD's 13-moment system. This system implies that ρ, $\mathbf{u}$, and ε are constant, while (XVII.52) reduces to the same two ordinary differential equations as appear in MAXWELL's solution,

$$\tau^{(1)} \dot{P}_{km} = -P_{km},$$
$$\tfrac{3}{2}\tau^{(1)} \dot{q}_k = -q_k, \qquad \text{(XVII.59)}$$

and

$$\tau^{(1)} = \frac{1}{3n\mathbb{A}_{2|2|2}}. \qquad \text{(XVII.60)}$$

Of course $\tau^{(1)}$ reduces for a gas of Maxwellian molecules to the time of relaxation τ obtained in Chapter XIII, namely, (XIII.3)$_1$. Thus GRAD's 13-moment system makes the pressure deviator and the energy flux for a gas of general molecules decay according to MAXWELL's formulae (XIII.10) except that the time of relaxation depends upon the kinetic constitutive relations. In particular, since $\mathbb{A}_{2|2|2}$ is a function of ε, from (XVII.60) we see that the rates at which the pressure and energy flux relax to their values in kinetic equilibrium depend not only upon ρ but also upon ε, in different ways for different molecular models.

Although, if $\mathbb{k} \neq 5$, these results refer only to solutions of GRAD's equations and not strictly to the kinetic theory, they do suggest the way in which the essential phenomena of relaxation, discovered by MAXWELL for a gas of Maxwellian molecules and to this day rigorously demonstrated only for such a gas, may occur approximately (in a sense not yet made precise) when a more general molecular model is used.

If we are willing to accept GRAD's 13-moment system, we may use it to determine the viscosity and the Maxwell number of the gas by the procedure MAXWELL devised for a gas of Maxwellian molecules, which we presented in Section (iv) of Chapter XIII. Looking back at that chapter, we can see from (XIII.14)$_2$ that the value $\mu^{(1)}$ so obtained as an approximation to μ is[14]

$$\mu^{(1)} = \frac{2}{9} \frac{\mathbb{m}\varepsilon}{\mathbb{A}_{2|2|2}}; \qquad \text{(XVII.61)}$$

the value $\mathbf{M}^{(1)}$ of the Maxwell number is

$$\mathbf{M}^{(1)} = \tfrac{5}{2}. \qquad \text{(XVII.62)}$$

[14] GRAD [1949, Eqs. (5.30), (5.31), (5.32)$_1$, and (5.33)$_1$].

For ideal spheres we may use (XVII.32) and so reduce (XVII.61) to[15]

$$\mu^{(1)} = \frac{5}{24}\sqrt{\frac{3}{2\pi}}\frac{m\sqrt{\varepsilon}}{d^2}, \qquad \text{(XVII.63)}$$

while for inverse k^{th}-power molecules we see from (XVII.33)$_2$ that

$$\mu^{(1)} = \frac{5}{6}\frac{m\sqrt{\pi}}{a\Gamma\left(2\frac{2k-3}{k-1}\right)}\left(\frac{m}{2g}\right)^{2/(k-1)}\varepsilon\left(\frac{3}{8\varepsilon}\right)^{\frac{1}{2}(k-5)/(k-1)},$$

$$= \frac{5}{12}\sqrt{\frac{3}{2}}\left(\frac{4}{3}\right)^{2/(k-1)}\frac{\sqrt{\pi}}{a\Gamma\left(2\frac{2k-3}{k-1}\right)}m\sqrt{\varepsilon}\left(\frac{m\varepsilon}{g}\right)^{2/(k-1)}. \qquad \text{(XVII.64)}$$

We see immediately that the way $\mu^{(1)}$ depends upon m, d, g, and ε in (XVII.63) and (XVII.64)$_2$ conforms with (IV.6). We should recall also that the considerations of Chapter IV do not suffice to prove μ and M independent of ρ. That conclusion does follow here, along with the numerical factors of proportionality for $\mu^{(1)}$ and $M^{(1)}$. Those quantities as delivered by (XVII.61) and (XVII.62) reduce for a gas of Maxwellian molecules to the quantities we have denoted by μ and M, respectively. In this sense (XVII.61) and (XVII.62) are exact for such a gas. For other molecular models their accuracy is difficult to assess directly. As we shall see in Chapter XXIV, there is a method of defining precisely what is meant by the viscosity and the Maxwell number for a gas specified by general kinetic constitutive equations. Once this shall have been done, explicit formulae for μ and M are delivered by a formal procedure. These formulae presume that two integral equations have been solved. CHAPMAN & COWLING[16], following ENSKOG[17], have set up a process of approximation for effecting their solution. GRAD's results (XVII.61) and (XVII.62) agree with what the first step in that process delivers[18]. For that reason we have attached the superscript 1 to them here. We present a direct derivation of ENSKOG's first approximations in Section (xii) of Chapter XXIV.

[15] In terms of the temperature θ the result (XVII.63) may be expressed as follows:

$$\mu^{(1)} = \frac{5}{16}\frac{1}{\sqrt{\pi}}\frac{\sqrt{km\theta}}{d^2}.$$

The celebrated formula MAXWELL obtained from his first kinetic theory was of this kind with $\frac{5}{16}$ replaced by the much smaller factor $\frac{2}{3}/\pi$; cf. TRUESDELL [1975, §8]. The value most modern elementary treatments obtain through a different argument replaces $\frac{5}{16}$ by $1/\pi$, very nearly the same; cf. CHAPMAN & COWLING [1939, §6.2].

[16] CHAPMAN & COWLING [1939, §§7.5–7.52] [1970, ibid.].

[17] ENSKOG [1917, leading terms in the expansions (92) and (93)]. CHAPMAN & COWLING [1970, Historical Note after §7.52] state that results given by CHAPMAN [1916, *1* & *2*] are "essentially equivalent" to ENSKOG's.

[18] Cf. CHAPMAN & COWLING [1939, Eqs. 7.51.13 and 7.52.7] [1970, Eqs. 7.51.14 and 7.52.8].

Among the many contributions GRAD has made to the kinetic theory, his formal expansion (XVII.24) may turn out to be the most important. If, following his suggestion, we cast away the terms involving Hermite coefficients the sum of whose orders differs from r, then his series reduces to a bilinear function of just the same kind as that provided for Maxwellian molecules by IKENBERRY's theorem. To illustrate this fact, we may simply inspect (XVII.34) when the terms Q_{km}, Q_{kmr}, and Q_{kmrs} are neglected. *Henceforth we shall mean the term "in* GRAD's *approximation" to describe use of this truncation*, and we shall distinguish by superscript 1 the results that follow from it. The reader is not to confuse such results with those that follow from GRAD's further assumption (XVII.41), from which he obtained his 20-moment system. In Chapter XXV we shall see that use of GRAD's approximation enables us to extend almost effortlessly to general molecular models a vast quantity of statements previously thought limited to a gas of Maxwellian molecules. GRAD's (XVII.60), (XVII.61), and (XVII.62) are the simplest specimens of what to expect.

Before exploiting GRAD's approximation we shall complete our survey of results that are formally exact. We begin with results that are rigorously proved as well.

Appendix A

Conversion Formulae

In order to facilitate conversions between Hermite polynomials and IKENBERRY's polynomials, we record here formulae which express one set in terms of the other for the polynomials of orders 2, 3, and 4. They follow directly from (XVI.24), (XVI.26), and (XVII.2). In each, the Hermite polynomials are evaluated at the dimensionless random velocity κ, and IKENBERRY's polynomials are evaluated at $\mathbf{c}$, the two arguments being related through (XVII.6).

$$H_{km} = \frac{3}{2\varepsilon} Y_{0|km} + \left(\frac{c^2}{2\varepsilon} - 1\right)\delta_{km},$$

$$H_{kmr} = \left(\frac{3}{2\varepsilon}\right)^{\frac{3}{2}} Y_{0|kmr} - 3\sqrt{\frac{3}{2\varepsilon}}\left(1 - \frac{3c^2}{10\varepsilon}\right)c_{(k}\,\delta_{mr)},$$

$$H_{kmrs} = \left(\frac{3}{2\varepsilon}\right)^{2}\left(Y_{0|kmrs} + \tfrac{6}{7}Y_{2|(km}\,\delta_{rs)} + \tfrac{1}{5}Y_{4|0}\,\delta_{(km}\,\delta_{rs)}\right)$$

$$\qquad - \frac{9}{\varepsilon} Y_{0|(km}\,\delta_{rs)} + 3\left(1 - \frac{c^2}{\varepsilon}\right)\delta_{(km}\,\delta_{rs)};$$

(XVIIA.1)

$$Y_{2|0} = \tfrac{2}{3}\varepsilon(H_{aa} + 3),$$

$$Y_{0|km} = \tfrac{2}{3}\varepsilon H_{\{km\}},$$

$$Y_{2|k} = (\tfrac{2}{3}\varepsilon)^{\frac{3}{2}}(H_{kaa} + 5H_k),$$

$$Y_{0|kmr} = (\tfrac{2}{3}\varepsilon)^{\frac{3}{2}} H_{\{kmr\}}, \qquad \text{(XVIIA.2)}$$

$$Y_{4|0} = (\tfrac{2}{3}\varepsilon)^{2}(H_{aabb} + 10H_{aa} + 15),$$

$$Y_{2|km} = (\tfrac{2}{3}\varepsilon)^{2}(H_{\{kmaa\}} + 7H_{\{km\}}),$$

$$Y_{0|kmrs} = (\tfrac{2}{3}\varepsilon)^{2} H_{\{kmrs\}}.$$

If we integrate each of these equations with respect to **v**, we obtain, by use of (XVI.28), (XVII.13), (XVII.16), and (XVII.17), corresponding formulae relating the Hermite coefficients $\mathbf{A}_p$ and the spherical moments $\mathbf{P}_{2r|s}$. These formulae are

$$A_{km} = \frac{1}{p} P_{km},$$

$$A_{kmr} = \frac{1}{p\sqrt{\tfrac{2}{3}\varepsilon}} (P_{0|kmr} + \tfrac{6}{5} q_{(k} \delta_{mr)}), \qquad \text{(XVIIA.3)}$$

$$A_{kmrs} = \frac{1}{\tfrac{2}{3}p\varepsilon}(P_{0|kmrs} + \tfrac{6}{7} P_{2|(km} \delta_{rs)} + \tfrac{1}{5} P_{4|0} \delta_{(km} \delta_{rs)})$$

$$\qquad - \frac{6}{p} P_{(km} \delta_{rs)} - 3 \delta_{(km} \delta_{rs)};$$

$$P_{2|0} = 3p,$$

$$P_{km} = p A_{km},$$

$$P_{2|k} = p\sqrt{\tfrac{2}{3}\varepsilon}\, A_{kaa},$$

$$P_{0|kmr} = p\sqrt{\tfrac{2}{3}\varepsilon}\, A_{\{kmr\}}, \qquad \text{(XVIIA.4)}$$

$$P_{4|0} = \tfrac{2}{3}p\varepsilon(A_{aabb} + 15),$$

$$P_{2|km} = \tfrac{2}{3}p\varepsilon(A_{\{kmaa\}} + 7A_{km}),$$

$$P_{0|kmrs} = \tfrac{2}{3}p\varepsilon A_{\{kmrs\}}.$$

As an example of the use of these formulae, we may convert the list (XVII.34) of collisions integrals of Hermite polynomials into an equivalent list of collisions integrals for IKENBERRY's polynomials. For example, by use of

(XVIIA.2)$_1$ we see that $\mathfrak{m}\bar{\mathbb{C}}Y_{0|km} = \frac{2}{3}\varepsilon\mathfrak{m}\bar{\mathbb{C}}H_{\{km\}}$. Then we use the list (XVIIA.3) to express the right-hand side of (XVII.34)$_3$ in terms of spherical moments, and from the result we obtain $\mathfrak{m}\bar{\mathbb{C}}Y_{0|km}$ directly. This procedure leads to the following list of *collisions integrals for the spherical harmonics for general molecular models*:

$$\mathfrak{m}\bar{\mathbb{C}}Y_{0|0} = 0,$$

$$\mathfrak{m}\bar{\mathbb{C}}Y_{0|k} = 0,$$

$$\mathfrak{m}\bar{\mathbb{C}}Y_{2|0} = 0,$$

$$\mathfrak{m}\bar{\mathbb{C}}Y_{0|km} = -3n\mathbb{A}_{2|2|2}P_{km} + R_{0|km},$$

$$\mathfrak{m}\bar{\mathbb{C}}Y_{2|k} = -4n\mathbb{A}_{2|2|2}q_k + R_{2|k},$$

$$\mathfrak{m}\bar{\mathbb{C}}Y_{0|kmr} = -\frac{9n}{2}\mathbb{A}_{2|2|2}P_{0|kmr} + R_{0|kmr},$$

$$\mathfrak{m}\bar{\mathbb{C}}Y_{4|0} = -\frac{2n}{\rho}\mathbb{A}_{2|2|2}(\rho P_{4|0} - 15p^2 + P_{ab}P_{ab}) + R_{4|0},$$

(XVIIA.5)

$$\mathfrak{m}\bar{\mathbb{C}}Y_{2|km} = -\frac{n}{8}(27\mathbb{A}_{4|2|2} - 42\mathbb{A}_{3|2|2} + 43\mathbb{A}_{2|2|2})P_{2|km}$$

$$-\frac{n}{4\rho}(27\mathbb{A}_{4|2|2} - 42\mathbb{A}_{3|2|2} + 23\mathbb{A}_{2|2|2})P_{\{ak}P_{ma\}}$$

$$+\frac{7n}{8\rho}(27\mathbb{A}_{4|2|2} - 42\mathbb{A}_{3|2|2} + 19\mathbb{A}_{2|2|2})pP_{km} + R_{2|km},$$

$$\mathfrak{m}\bar{\mathbb{C}}Y_{0|kmrs} = \frac{n}{4}(35\mathbb{A}_{4|4|4} - 10\mathbb{A}_{4|2|2} - 18\mathbb{A}_{2|2|2})P_{0|kmrs}$$

$$+\frac{3n}{4\rho}(35\mathbb{A}_{4|4|4} - 10\mathbb{A}_{4|2|2} + 6\mathbb{A}_{2|2|2})P_{\{km}P_{rs\}} + R_{0|kmrs}.$$

To shorten the calculation, we have retained explicitly only those summands in (XVII.34) which, in terms of the general formula (XVII.24)$_2$, correspond to the choice $m + n = r$. All other summands have been absorbed into the remainders $R_{2r|s}$, each of which is in general an infinite sum. IKENBERRY's theorem tells us that *for a gas of Maxwellian molecules each $R_{2r|s}$ is null*. Indeed, if we set each $R_{2r|s}$ equal to zero in (XVIIA.5), and then set $\mathbb{A}_{p|n|m} = \mathbb{A}_{n|m}$ for all values of p, we recover immediately the list of collisions integrals (XVI.37) which illustrates IKENBERRY's theorem. Moreover, IKENBERRY's theorem renders the resulting formulae rigorously proved, not merely formal.

Appendix B

Exact Solutions of the Maxwell–Boltzmann Equation for a Gas of Maxwellian Molecules

In GRAD's method of truncation a molecular density F is approximated by a function $\hat{F}$ which is the product of a Maxwellian density F_M and a polynomial in the components of $\mathbf{c}$. We have recorded two approximations of this type in (XVII.41) and (XVII.49). In each of these F_M is the unique Maxwellian density whose principal moment is the same as that of F, and the coefficients of the polynomial are certain moments of F. To determine these moments, GRAD uses the equations of transfer corresponding to them and simply disregards the equations of transfer for all other moments. Because of this the function $\hat{F}$ is only an *approximate solution* of the Maxwell–Boltzmann equation. If, however, we could choose the moments determining $\hat{F}$ such as to satisfy also the equations of transfer GRAD chose to disregard, then $\hat{F}$ would actually be an *exact solution*. Since there are infinitely many equations for moments and only a finite number of unknown functions determining $\hat{F}$, in general we do not expect to be able to do this. However, in certain special circumstances it is possible to find[1] exact solutions in precisely this way.

Consider a molecular density function F having the form

$$F = F_M S_N, \qquad \text{(XVIIB.1)}$$

S_N being the Nth partial sum of GRAD's expansion (XVII.9):

$$S_N = \sum_{s=0}^{N} \frac{1}{s!} A_s H_s. \qquad \text{(XVIIB.2)}$$

In contrast with GRAD's assumptions we do not require F_M to have the same principal moment as F. Instead we let the fields of density, gross velocity, and energetic for F_M be arbitrary functions δ, $\mathbf{v}$, and ξ, respectively. In order that F shall be integrable over $\mathscr{V}$, we assume that $\xi > 0$. When we wish to emphasize that F_M is determined by δ, $\mathbf{v}$, and ξ, we shall write $F_M(\delta, \mathbf{v}, \xi)$ in place of F_M.

The function F given by (XVIIB.1) is determined, then, by δ, $\mathbf{v}$, ξ, and the coefficients $A_0, \ldots, A_N$. In order to decide, for a given choice of these fields, whether F satisfies the Maxwell–Boltzmann equation, we must first calculate $\mathcal{D}F$ and $\mathcal{C}F$. By use of (XVII.10) the former of these can be evaluated explicitly:

$$\mathcal{D}F = F_M \mathcal{D}S_N + S_N \mathcal{D}F_M,$$

$$= F_M \mathcal{D}S_N + F_M S_N \left[\frac{1}{\delta}(\partial_t \delta + v_a \delta_{,a}) + \left(\frac{3}{4\xi^2} |\mathbf{v} - \mathbf{u}|^2 - \frac{3}{2\xi} \right)(\partial_t \xi + v_a \xi_{,a}) \right.$$

$$\left. + \frac{3}{2\xi}(v_b - v_b)(\partial_t v_b + v_a v_{b,a}) - \frac{3}{2\xi}(v_a - v_a)b_a \right]. \qquad \text{(XVIIB.3)}$$

---

[1] MUNCASTER [1979].

We shall not need the specific details of this formula. We have recorded it so as to note the structure of $\mathfrak{D}F$ as a function of $\mathbf{v}$. Since S_N is a polynomial of degree N in the components of $\mathbf{v}$, we see that $F_M^{-1}\mathfrak{D}(F_M S_N)$ is *a polynomial in $\mathbf{v}$ of degree $N + 3$*. Moreover, since the term of degree $N + 3$ in (XVIIB.3)$_2$ is multiplied by grad ξ, we have the somewhat more special result that *if ξ is independent of $\mathbf{x}$, then $F_M^{-1}\mathfrak{D}(F_M S_N)$ is a polynomial in $\mathbf{v}$ of degree $N + 2$.*

We now wish to draw similar conclusions about $\mathbb{C}F$. To do so, let us recall that the Hermite functions $\exp(-\tfrac{1}{4}a^2)\mathbf{H}_r(\mathbf{a})$ form a complete orthogonal set in the space of functions which are square-integrable over $\mathscr{V}$. Equivalently the Hermite polynomials $\mathbf{H}_r$ form a complete orthogonal set with respect to the inner product

$$(f, g) \equiv \int F_M \, fg, \qquad \text{(XVIIB.4)}$$

the orthogonality relations being expressed by (XVII.3). Using (XVII.24)$_2$, we may write the Fourier coefficient of the function $F_M^{-1}\mathbb{C}(F_M S_N)$ with respect to $\mathbf{H}_r$ in the form

$$\begin{aligned}(\mathbf{H}_r, F_M^{-1}\mathbb{C}(F_M S_N)) &= \int \mathbf{H}_r \, \mathbb{C}(F_M S_N), \\ &= \bar{\mathbb{C}}_{F_M S_N} \mathbf{H}_r, \qquad \text{(XVIIB.5)} \\ &= \frac{\rho^2}{2\mathfrak{m}^2} \sum_{m,n=0}^{N} \mathbb{A}_{rmn} A_m A_n.\end{aligned}$$

The calculations which led us to (XVII.24)$_2$ were only formal since the double sum appearing there was in general infinite. In our use of it here, however, each of these sums is finite. For general molecular models we know little about the coefficients $\mathbb{A}_{rmn}$ and so there is little that we can conclude from (XVIIB.5)$_3$. However, for a gas of Maxwellian molecules we showed in Section (ii) of this chapter that $\mathbb{A}_{rmn}$ vanishes if $r \neq m + n$. In this case the sum on the right-hand side of (XVIIB.5)$_3$ vanishes whenever $r > 2N$. Therefore $\mathbb{C}F$ has an Hermite expansion of the form (XVII.9), one which contains only the first $2N$ terms. Equivalently, *in a gas of Maxwellian molecules $F_M^{-1}\mathbb{C}(F_M S_N)$ is a polynomial in $\mathbf{v}$ of degree $2N$.*

Now let us combine this result with the one stated just after (XVIIB.3). We see that for the Maxwellian gas the Maxwell–Boltzmann equation, when evaluated at a function F of the form (XVIIB.1), reduces to an equality between two polynomials in the components of $\mathbf{v}$, one of degree $N + 3$ and the other of degree $2N$. Therefore F will be a solution if and only if the Fourier coefficients of each side of this equation agree, and we need only consider the Fourier coefficients with respect to $\mathbf{H}_r$ when $r = 0, \ldots, \max(N + 3, 2N)$. We may state this result somewhat differently as a **Theorem**. *In a gas of Maxwellian molecules, the function $F_M(\delta, \mathbf{v}, \xi)S_N$ satisfies the Maxwell–Boltzmann equation if and only if*

$\xi > 0$ *and the moments of* $F_M S_N$ *satisfy the equations of moments of orders* $0, \ldots,$ $\max(N + 3, 2N)$. Thus, while in general there are infinitely many equations for moments which we must satisfy, for the very special circumstances of a gas of Maxwellian molecules and a molecular density of the form (XVIIB.1) we need only satisfy a finite number of those equations. Once having done so, we find that all the others are automatically satisfied. It is this fact which makes the search for exact solutions of the type (XVIIB.1) feasible, at least for Maxwellian molecules.

Using the second result stated after (XVIIB.3), we obtain the **Corollary**. *If* ξ *is a positive function of t alone, then* $F_M(\delta, \mathbf{v}, \xi)S_N$ *is a solution of the Maxwell–Boltzmann equation if and only if its moments satisfy the equations of moments of orders* $0, \ldots, \max(N + 2, 2N)$.

In applying the theorem or its corollary to the problem of finding exact solutions, we face two difficulties. The first is the fact that for a function of the form (XVIIB.1) we must determine the functions δ, $\mathbf{v}$, ξ, $\mathbf{A}_0$, $\mathbf{A}_1$, ..., and $\mathbf{A}_N$, and the number of moment equations these must satisfy is $\max(N + 3, 2N)$. As a result the system governing these functions is highly over-determined, particularly when N is large, and so in general it will have no solution. The smaller the value of N, of course, the more likely a solution will exist. Second, the value of N is restricted, in a practical sense, by the number of equations of moments we have at our disposal. From (XVI.1) and the list of collisions integrals (XVI.23) we obtain those of orders $0, \ldots, 5$. Since $\max(N + 3, 2N) = 5$ if and only if $N = 2$, the presently available list of collisions integrals restricts us to polynomials of degree 2. For larger values of N the number of equations of moments we should need increases sharply. If $N = 3$, the function F is not non-negative, so the next interesting case is $N = 4$. For it we should need the equations of moments of orders $0, \ldots, 8$.

To illustrate how the theorem may be applied, let us look for exact solutions having the particularly simple form

$$F = F_M(\delta, \mathbf{v}, \xi)(\alpha + \beta|\mathbf{v} - \mathbf{v}|^2). \qquad \text{(XVIIB.6)}$$

All moments of F are determined by the five fields δ, $\mathbf{v}$, ξ, α, and β. For example, the fields of density, gross velocity, and energetic corresponding to F are given by

$$\rho = (\alpha + 2\xi\beta)\delta, \qquad u_k = v_k, \qquad \varepsilon = (\alpha + \tfrac{10}{3}\xi\beta)\frac{\xi\delta}{\rho}. \qquad \text{(XVIIB.7)}$$

If we replace δ, $\mathbf{v}$, α, and β in (XVIIB.6) by their values in terms of ρ, $\mathbf{u}$, and ε, which result from solving (XVIIB.7), we may write F in the more explicit form

$$F = \frac{\rho}{m\left(\tfrac{4}{3}\pi\xi\right)^{\frac{3}{2}}} e^{-\tfrac{3}{4}c^2/\xi}\left[1 - \frac{3}{2}\left(\frac{\varepsilon}{\xi} - 1\right) + \frac{3}{4\xi}\left(\frac{\varepsilon}{\xi} - 1\right)c^2\right], \qquad \text{(XVIIB.8)}$$

c being the random velocity **v** − **u**. Unless ξ is a grossly determined field, this molecular density is not grossly determined. To exhibit the other moments of F, it is convenient to work in terms of the spherical moments $\mathbf{P}_{2r|s}$. Since the Maxwellian density appearing in (XVIIB.8) is $F_M(\rho, \mathbf{u}, \xi)$, we see from (XVI.30) that the spherical moments of F are given by

$$P_{2r|s} = \left[1 - \frac{3}{2}\left(\frac{\varepsilon}{\xi} - 1\right)\right] P^{(0)}_{2r|s}(\rho, \mathbf{u}, \xi) + \frac{3}{4\xi}\left(\frac{\varepsilon}{\xi} - 1\right) P^{(0)}_{2(r+1)|s}(\rho, \mathbf{u}, \xi),$$

$$= \begin{cases} 0 & \text{if } s \neq 0, \\ (2r+1)!!\,\rho(\tfrac{2}{3}\xi)^r\left[1 + r\left(\frac{\varepsilon}{\xi} - 1\right)\right] & \text{if } s = 0. \end{cases} \qquad \text{(XVIIB.9)}$$

In particular, all spherical moments with indices vanish, and **M**, **q**, and $P_{4|0}$ are given explicitly by

$$M_{km} = p\,\delta_{km} = \tfrac{2}{3}\rho\varepsilon\,\delta_{km},$$

$$q_k = 0, \qquad \text{(XVIIB.10)}$$

$$P_{4|0} = \tfrac{20}{3}\rho\xi^2\left[1 + 2\left(\frac{\varepsilon}{\xi} - 1\right)\right].$$

We specialize still further now by choosing ξ to be a positive function of t alone. Since $\max(N + 2, 2N) = 4$ when $N = 2$, we see from the corollary that F will satisfy the Maxwell–Boltzmann equation if and only if ρ, **u**, ε, and ξ satisfy the equations of moments of orders $0, \ldots, 4$. We shall use these equations in the forms expressed by the equation of balance of linear momentum (IX.4) and the result of substituting $(XVI.37)_{1-9}$ and $(XVI.41)_{1-9}$ into the equation of transfer (XVI.40) for the spherical harmonics. When the equations are specialized by letting $\mathbf{P}_{2r|s}$ have the form $(XVIIB.9)_{2,3}$, they reduce to the system

$$\dot{\rho} + \rho E = 0,$$

$$\rho \dot{u}_k + p_{,k} = \rho b_k,$$

$$\dot{p} + \tfrac{5}{3} p E = 0,$$

$$E_{km} = 0, \qquad \text{(XVIIB.11)}$$

$$-\frac{5p}{2\rho} p_{,k} + \frac{1}{6} P_{4|0,k} = 0,$$

$$\dot{P}_{4|0} + \frac{7}{3} P_{4|0} E = -\frac{2}{\mathfrak{m}} \mathbb{A}_{2|2}(\rho P_{4|0} - 15p^2).$$

So as to exhibit a simple class of solutions of this system, let us suppose that $\mathbf{b} = \mathbf{0}$. Then the first four equations are satisfied by the particular fields

$$\rho = \rho(0)\left(1 + \frac{t}{T}\right)^{-3},$$

$$\mathbf{u} = (t + T)^{-1}\mathbf{r}, \qquad (\text{II}.39)_r$$

$$\varepsilon = \varepsilon(0)\left(1 + \frac{t}{T}\right)^{-2},$$

which we proved in Chapter X to be the fields of density, gross velocity, and energetic of a locally Maxwellian solution. Moreover (XVIIB.11)$_5$ tells us that $P_{4|0}$ is a function of t alone, and using (XVIIB.10)$_4$ we may write (XVIIB.11)$_6$ as the ordinary differential equation

$$\dot{\sigma} = -\frac{1}{\mathfrak{m}}\mathbb{A}_{2|2}\rho\sigma, \qquad \sigma \equiv \frac{\zeta}{\varepsilon} - 1. \qquad (\text{XVIIB}.12)$$

Since $\mathbb{A}_{2|2}$ is a constant and ρ is given explicitly by (II.39)$_1$, this equation can be solved:

$$\frac{\zeta}{\varepsilon} - 1 = \left(\frac{\zeta(0)}{\varepsilon(0)} - 1\right)\exp\left[\frac{\mathbb{A}_{2|2}T^3\rho(0)}{2\mathfrak{m}}\left(\frac{1}{(t+T)^2} - \frac{1}{T^2}\right)\right],$$

$$= \left(\frac{\zeta(0)}{\varepsilon(0)} - 1\right)\exp\left[\frac{T^3}{6\tau(0)}\left(\frac{1}{(t+T)^2} - \frac{1}{T^2}\right)\right], \qquad (\text{XVIIB}.13)$$

τ being the time of relaxation for Maxwellian molecules, namely, (XIII.3). Since ε is known, this equation determines ζ.

There are two more conditions to satisfy, namely, ζ must be positive and F must be non-negative. By (XVIIB.8) such will be the case if

$$0 \leq \frac{\varepsilon}{\zeta} - 1 \leq \frac{2}{3} \quad \text{when} \quad t \geq 0. \qquad (\text{XVIIB}.14)$$

If we consider now only the case in which $T > 0$, so the solution exists for all time, we see from (XVIIB.13) that the function $\zeta/\varepsilon - 1$ is monotone, increasing if $\zeta(0)/\varepsilon(0) \leq 1$ and decreasing if $\zeta(0)/\varepsilon(0) \geq 1$. Using this fact, we may reduce (XVIIB.14) to the following simple inequality for the initial values:

$$\frac{3}{5} \leq \frac{\zeta(0)}{\varepsilon(0)} \leq 1. \qquad (\text{XVIIB}.15)$$

In summary, then, we see that *when ρ, $\mathbf{u}$, ε, and ζ are given by (XVIIB.13) and (II.39), when $\rho(0) > 0$, $\varepsilon(0) > 0$, $T > 0$, and when $\zeta(0)$ satisfies (XVIIB.15), the function F given by (XVIIB.8) is an exact non-negative solution of the Maxwell–Boltzmann equation.* In view of (II.39)$_2$, all these solutions depend upon $\mathbf{x}$. As far as we are aware, this is the only class of explicit, exact, place-dependent solutions presently known, beyond, of course, the locally Maxwellian ones.

It is not difficult to analyse the asymptotic behavior of these solutions as $t \to \infty$. Directly from (XVIIB.13)$_2$ we see that

$$\frac{\xi}{\varepsilon} - 1 \to \left(\frac{\xi(0)}{\varepsilon(0)} - 1\right) \exp\left(-\frac{T}{6\tau(0)}\right) \quad \text{as} \quad t \to \infty, \qquad \text{(XVIIB.16)}$$

and so in general the term in (XVIIB.8) multiplying c^2 does not vanish as time proceeds. Moreover, $\xi \to 0$ as $t \to \infty$. Hence $F \to 0$. Thus *F does not approach a Maxwellian density in the course of time: There is no approach to equilibrium.* In drawing this conclusion we have considered the behavior of F as $t \to \infty$ for fixed values of **x** and **v**. For a weak rather than pointwise limit we may look at the moments of F, but for these the conclusion is no better. Indeed, by (XVIIB.16) and (XVIIB.9)$_3$ we see that each function $\rho^{-1}(\tfrac{2}{3}\varepsilon)^{-r}P_{2r|0}$ has a well-defined limit as $t \to \infty$, but in general this limit is not 1. Thus, *asymptotically as $t \to \infty$ the spherical moments of F do not approach the spherical moments of a Maxwellian density.* It is particularly remarkable that this F and a locally Maxwellian solution correspond to *exactly* the *same* gross fields ρ, **u**, ε, namely (II.39), and the *same* **b**, namely **0**; they have also the *same* **P** and **q**, namely null fields, as the Euler-Hadamard and Stokes-Kirchhoff theories provide in correspondence with (II.39) when **b** = **0**. Thus this F and the corresponding locally Maxwellian one are *grossly indistinguishable.* Their difference reflects itself in the higher moments $P_{2r|0}$, $r \geq 2$. Contrary to widespread prejudice, examples of which we have quoted in Footnote 14 to Chapter XIV, F does *not* approach a grossly determined molecular density, even though there is one that corresponds *exactly* to the same gross condition. In the note added at the end of this chapter we shall see that there are in fact infinitely many solutions F for which these same conclusions hold.

We may use these solutions to test some of the old claims about the H-theorem. Since (XVIIB.8) is an even function of **c**, for all choices of ξ we conclude that **s** = **0**; thus *the formal narrow H-theorem holds*: $\dot{h} < 0$ unless $\xi = \varepsilon$. By calculating h explicitly we can show that it does exist and is a continuously differentiable function of $\rho\xi^{-\frac{3}{2}}$ and $\varepsilon\xi^{-1}$, so whenever these two quantities are continuously differentiable functions of t, so also is h, and thus the qualification "formal" may be removed. The asymptotic behavior exhibited in (XVIIB.16) adds to the example given in Chapter X another to show that *the narrow H-theorem does not imply a trend to equilibrium.* Indeed, when we use the principal moment (II.39) and so reduce (XVIIB.11) to (XVIIB.13), we associate with each solution in the present class a unique equilibrium solution, the F_U determined by the principal moment $(\rho(0), \mathbf{0}, \varepsilon(0))$. We know from the results in Section (vi) of Chapter XI that h_M = const. = $h[F_U]$. Because $h \geq h_M$, again we conclude that

$$h[F] \downarrow \text{a limit} \geq h[F_U]. \qquad \text{(XI.64)}_r$$

Dr. MARIO PITTERI has shown us analysis of the behavior of the h that corresponds to a particular choice of $\xi(0)/\varepsilon(0)$ satisfying (XVIIB.15). For that choice he concludes that $\lim h > h[F_U]$. Thus the caution we expressed in Chapter XI, especially in Sections (v) and (vi), regarding the old style of argument in these matters is supported by an explicit counterexample.

There is, however, a sense in which the common claims may be supported. If we hold t, $\mathbf{x}$, and $\mathbf{v}$ fixed and consider lighter and lighter gases, $\mathfrak{m} \to 0$, we see from (XIII.3) that $\tau(0) \to 0$, and so in this same limit $\xi/\varepsilon \to 1$. Therefore,

$$\frac{F}{F_M(\rho, \mathbf{u}, \varepsilon)} \to 1 \quad \text{as} \quad \mathfrak{m} \to 0. \tag{XVIIB.17}$$

For fixed t, $\mathbf{x}$, and $\mathbf{v}$, the exact solutions (XVIIB.8) *approach asymptotically as $\mathfrak{m} \to 0$ the unique locally Maxwellian solution determined by the principal moment* (II.39).

Another class of exact solutions arising in the same way is found if we replace (II.39) by the choice

$$\rho = \text{const.}, \quad \mathbf{u} = \text{const.}, \quad \varepsilon = \text{const.} \tag{XVIIB.18}$$

If also $\mathbf{b} = \mathbf{0}$ and ξ is a function of t alone, (XVIIB.11)$_{1-5}$ are satisfied, and (XVIIB.11)$_6$ reduces again to (XVIIB.12). Since ρ and ε are both constant, the solution of this differential equation is now

$$\begin{aligned}\frac{\xi}{\varepsilon} - 1 &= \left(\frac{\xi(0)}{\varepsilon} - 1\right)\exp\left(-\frac{\mathbb{A}_{2|2}\rho}{\mathfrak{m}} t\right), \\ &= \left(\frac{\xi(0)}{\varepsilon} - 1\right)\exp\left(-\frac{t}{3\tau}\right).\end{aligned} \tag{XVIIB.19}$$

To satisfy the condition (XVIIB.14) requiring that ξ be positive and F be non-negative, we proceed just as we did in the preceding case and obtain (XVIIB.15) again. Thus, *when ρ, $\mathbf{u}$, ε, and ξ are given by* (XVIIB.18) *and* (XVIIB.19), *when $\rho > 0$, $\varepsilon > 0$, and when $\xi(0)$ satisfies* (XVIIB.15), *the function F given by* (XVIIB.8) *is an exact non-negative solution of the Maxwell-Boltzmann equation*. This F was first obtained by KROOK & WU[2]. In contrast with those of the preceding class, all these solutions are spatially homogeneous. They appear to be the only explicit, exact, spatially homogeneous solutions presently known aside from the *uniform* Maxwellian ones.

From (XVIIB.19) we see that $\xi/\varepsilon \to 1$ exponentially as $t \to \infty$. *Thus, for each fixed $\mathbf{v}$, F approaches in the course of time the unique Maxwellian density F_U*

---

[2] The method of KROOK & WU [1976] [1977, Eqs. (83) and (84)] is less straightforward than that used here. In addition they implicitly considered only pseudomolecules (*cf.* their passage from Eq. (14) to Eq. (15) of their paper of 1977), and at that only Maxwellian ones (*cf.* their Eq. (1)). Our analysis shows that the molecular density they obtained is a solution also for a gas of Maxwellian molecules.

APPENDIX B. MUNCASTER'S EXACT SOLUTIONS

with principal moment (XVIIB.18). *In other words, these solutions exhibit a trend to kinetic equilibrium*. By (XVIIB.9)$_2$ the moments of F also approach their values in kinetic equilibrium. This last conclusion we will confirm and extend to general spatially homogeneous solutions in Section (ii) of Chapter XVIII. If now we hold t and $\mathbf{v}$ fixed and let $\mathfrak{m} \to 0$, (XIII.3) shows that $\tau \to 0$ and so $\xi/\varepsilon \to 1$. Thus (XVIIB.17) applies: *For fixed t and $\mathbf{v}$, the exact solutions* (XVIIB.8) *specialized by* (XVIIB.19) *approach asymptotically as* $\mathfrak{m} \to 0$ *the unique locally Maxwellian solution determined by the principal moment* (XVIIB.18). Thus KROOK & WU's *solutions conform with the standard claims*.

Note added in proof. Let $F^M(t, \mathbf{x}, \mathbf{v})$ denote MUNCASTER's solution as given by (XVIIB.8), (II.39), and (XVIIB.13), and let $F^{K-U}(t, \mathbf{c})$ denote KROOK & WU's solution, considered as a function of t and the random velocity $\mathbf{c}$, as given by (XVIIB.8), (XVIIB.18), and (XVIIB.19). Then a simple calculation shows that

$$F^M(t, \mathbf{x}, \mathbf{v}) = F^{K-U}\left(-\frac{T^3}{2}\left[\frac{1}{(t+T)^2} - \frac{1}{T^2}\right], \left(1 + \frac{t}{T}\right)[\mathbf{v} - (t+T)^{-1}(\mathbf{x} - \mathbf{x}_0)]\right). \quad \text{(XVIIB.20)}$$

This relation follows also as a special case from a general transformation proved by NIKOL'SKII which generates a solution for dilatations directly from any given spatially homogeneous solution: *For a gas of inverse* $\mathbb{k}^{th}$*-power molecules,* $\mathbb{k} \neq \frac{7}{3}$*, set*

$$Z(t) = \frac{(\mathbb{k}-1)T}{(7-3\mathbb{k})}\left[\left(1 + \frac{t}{T}\right)^{[(7-3\mathbb{k})/(\mathbb{k}-1)]} - 1\right], \quad \text{(XVIIB.21)}$$

T *being an arbitrary constant other than* 0*, and let the body force* $\mathbf{b}$ *be null. If* $F^O(t, \mathbf{c})$ *is any spatially homogeneous solution of the Maxwell-Boltzmann equation for such a gas, considered as a function of time t and the random velocity* $\mathbf{c}$*, then*

$$F = F^O\left(Z(t), \left(1 + \frac{t}{T}\right)[\mathbf{v} - (t+T)^{-1}(\mathbf{x} - \mathbf{x}_0)]\right) \quad \text{(XVIIB.22)}$$

is a solution for the same gas. For proof we refer the reader to the paper of NIKOL'SKII[3]. Each solution of this kind has principal moment (II.39), and so *each corresponds to a homo-energetic dilatation*. Directly from (III.19) we see that the relative moment $^n\mathbf{M}$ of F is given in terms of the relative moment $^n\mathbf{M}^O$ of F^O by

$$^n\mathbf{M}(t) = \left(1 + \frac{t}{T}\right)^{-(n+3)} {}^n\mathbf{M}^O(Z(t)), \quad \text{(XVIIB.23)}$$

and in the particular case $\mathbb{k} = 5$ giving a Maxwellian gas this formula allows us to obtain GALKIN's solutions (XV.8) and (XV.9) directly from the solutions (XIII.10) and (XIII.11) expressing MAXWELL's relaxation theorem. The behavior of the higher moments corresponding to (XVIIB.22) may be calculated similarly by applying (XVIIB.23) to TRUESDELL's general relaxation theorem (XVIII.4) for a grossly homogeneous condition of a Maxwellian gas.

[3] NIKOL'SKII [1963, *2*]. *Cf.* also the corresponding result for ideal spheres obtained by NIKOL'SKII [1963, *1*].

Part F

Existence, Uniqueness, and Qualitative Behavior

Chapter XVIII

Existence Theory for the General Initial-value Problem. Part I: Molecules with Intermolecular Forces of Infinite Range

(i) *Prolegomena to existence theory*

We first began to study the Maxwell–Boltzmann equation, and properties which its solutions must have, in Chapter IX. MAXWELL's consistency theorem, ENSKOG's equation of transfer, and BOLTZMANN's H-theorem were some of the results we found. Implicit in our derivation of them was the assumption that there *were* solutions, ones possessing various degrees of regularity and integrability. To justify this assumption, though only generally, we turn now to the theory of existence and the related problems it encompasses.

Theorems on existence, uniqueness, and asymptotic behavior serve a number of purposes in any domain subjected to mathematical analysis. At a very naive level the bare existence of solutions reflects in some way the reasonableness of the basic axioms which underlie the subject. For us these are the equation of evolution (III.38) and MAXWELL's choice (VII.2) of the collisions operator $\mathbb{C}$. At a somewhat higher level we gain from such theorems insight into the variety of solutions the theory may deliver, a variety reflected by technical properties such as regularity and integrability and their relations to the underlying data of specific problems. With this information we may justify, or at least place on a firmer footing, many of the formal calculations which have gone before. Finally, at the deepest level we learn from such theorems the qualitative behavior of broad classes of solutions. Qualitative properties may be suggested by formal results, such as the formal H-theorems we have presented in Chapter XI, or by exact solutions, such as those we have presented in Chapters XIII–XV, or by both, but ultimately they can be ascertained only from keen study of solutions demonstrated to exist. The theorems we shall present in this part of the book will serve us in all these ways.

We may divide existence theorems into two broad classes: first, theorems which apply to a gas occupying all of space—the pure initial-value problem, and second, theorems which apply to a gas occupying a bounded region—the initial-value problem with boundary conditions. Any treatment of the second class requires, as a prerequisite, a comprehensive discussion of boundary conditions for the molecular density function F. In this book, however, we have chosen to develop the subject of boundary conditions only insofar as it relates to the formal H-theorems, which we have presented in Chapter XI. Accordingly we shall limit ourselves here[1] to the pure initial-value problem:

$$\mathfrak{D}F = \partial_t F + \mathbf{v} \cdot \partial_{\mathbf{x}} F + \mathbf{b} \cdot \partial_{\mathbf{v}} F = \int\!\!\int_{\mathscr{S}} w(F'F'_* - FF_*) = \mathbb{C}F, \quad (\text{IX.1})_r$$

$$F|_{t=t_0} = F_0, \quad (\text{V.13})_r$$

for functions F and F_0 defined on all of the point space $\mathscr{E}$. As we shall soon see, even with this restriction we are left with a rich body of results.

We divide our discussion of the initial-value problem into three parts. The first concerns a gas of molecules whose intermolecular forces extend to ∞; it is the subject of the following sections of this chapter. For such molecules only one existence theorem is known, and in several regards its scope is narrow. First, it applies only to a gas of Maxwellian molecules; second, it concerns solutions of the infinite system of equations for the moments, not of the Maxwell–Boltzmann equation itself; third, it deals only with spatially homogeneous solutions of that system. However, in exchange for these limitations we obtain, by no more than elementary calculations, not only proofs of existence and uniqueness, not only proof of a strict trend to equilibrium, but even estimates of the rate at which that trend proceeds.

The second and third parts of our discussion will deal with a gas of molecules with a cut-off[2]: the former, presented in Chapter XX, concerns place-dependent solutions; the latter, in Chapter XXI, spatially homogeneous solutions. In each case we begin by surveying the course of development of the corresponding part of existence theory, and thereafter we present in detail the most conclusive results that are presently known. In both cases the theorems we prove apply to all spherically symmetric molecules with a cut-off. This means that

(a) The *cross-section* $\mathscr{S}$ is an open disk, its radius d being any positive number;

---

[1] A survey of results up to 1973 on the initial-value problem with boundary conditions may be found in the review article on existence theory by GUIRAUD [1973]; more recent results appear in the work of UKAI [1974] [1976], SHIZUTA [1979, *1 & 2*], NISHIDA & IMAI [1977], SHIZUTA & ASANO [1977], KANIEL & SHINBROT [1978].

[2] MUNCASTER acknowledges with gratitude the criticism his teachers P. HARTMAN and A. MENIKOFF provided for an early version of this part of the work.

(b) The *encounter operator* $\mathbb{E}: \mathscr{V} \times \mathscr{V} \times \mathscr{S} \to \mathscr{V} \times \mathscr{V} \times \mathscr{S}$ is any piecewise continuously differentiable function such that

$$\mathbb{E} \circ \mathbb{E} = \text{identity}; \tag{VI.65}_r$$

$$\mathbb{M} \circ \mathbb{E} = \mathbb{M}, \quad \mathbb{K} \circ \mathbb{E} = \mathbb{K}; \tag{VI.66}_r$$

$$\mathbb{X} \circ \mathbb{E} = \mathbb{E} \circ \mathbb{X}; \tag{VI.68}_r$$

$$\frac{r'}{r} \left| \frac{\partial(\mathbf{v}', \mathbf{v}'_*, r', \zeta')}{\partial(\mathbf{v}, \mathbf{v}_*, r, \zeta)} \right| = 1; \tag{VI.70}_r$$

$\mathbb{M}$, $\mathbb{K}$, $\mathbb{X}$, and the primed variables are defined by (VI.67)$_1$, (VI.67)$_2$, (VI.69), and (VI.64), respectively. We have not included (VI.75) in our list defining $\mathbb{E}$, for no use is made of it in proofs of existence.

(ii) *Spatially homogeneous solutions for a gas of Maxwellian molecules: Existence, uniqueness, and the trend to equilibrium*

As we showed in Chapter XVI, in a gas of Maxwellian molecules the Maxwell–Boltzmann equation gives rise to a general system of differential equations for the moments of the molecular density. Looking back now at the classes of solutions presented earlier for the moments of orders 0, 1, 2, and 3, one corresponding to gross homogeneity (Chapter XIII) and the others to homo-energetic simple shearing (Chapter XIV), dilatation (Chapter XV), and extension (Chapter XV), we may ask whether these solutions can be completed by solving the remaining equations in the general system for the moments. The answer is positive, as we have shown at the end of Section (i) of Chapter XV. We now verify this fact in detail for a grossly homogeneous condition[3].

To this end we use the equation of transfer for a function of **c** alone:

$$(\rho \bar{g})^{\cdot} + \rho \bar{g} E + \rho \overline{c_a g}_{,b} u_{b,a} - \overline{g_{,a}} M_{ab,b} + (\rho \overline{c_a g})_{,a} = \mathfrak{m}\bar{C}g, \tag{XIII.1}_r$$

and in it we take as g the polynomial $Y_{2r|s}$. In a grossly homogeneous condition all terms on the left-hand side are null except the first; by use of (XVI.28) and IKENBERRY's theorem (XVI.31) we reduce (XIII.1) to

$$\dot{P}_{2r|s} + c_{2r|s} P_{2r|s} = Q_{2r|s}. \tag{XVIII.1}$$

The superimposed dot now indicates the ordinary derivative with respect to t. The coefficient $c_{2r|s}$ is given as follows:

$$c_{2r|s} = 4\pi n \sqrt{\frac{g}{2\mathfrak{m}}} \int_0^\infty [1 - \cos^{2r+s} \theta \, P_s(\cos \theta) - \sin^{2r+s} \theta \, P_s(\sin \theta)] z \, dz. \tag{XVI.33}_r$$

---

[3] TRUESDELL [1956, Chapter VI]. RAY [1979] has noticed a counterpart of our Theorem 1 for Maxwellian pseudomolecules defined by a constant scattering factor $\mathbb{S}$.

Not only is the system of equations no longer coupled in the sense of increasing n, as we knew would be the case, but also it reduces to a set of ordinary linear differential equations for individual spherical moments $\mathbf{P}_{2r|s}$, provided all spherical moments of order lower than $2r + s$ shall have been determined already. The condition stated in the last clause follows from (XVI.32).

A uniform Maxwellian density is a solution of the Maxwell–Boltzmann equation. Hence its moments, which are constant, satisfy (XVIII.1). If we denote by $\mathbf{P}_{2r|s}^{(0)}$ and $\mathbf{Q}_{2r|s}^{(0)}$ the functions $\mathbf{P}_{2r|s}$ and $\mathbf{Q}_{2r|s}$ corresponding to a Maxwellian density, then $\mathbf{P}_{2r|s}^{(0)}$ is given by (XVI.30), and (XVIII.1) shows that

$$\mathbf{Q}_{2r|s}^{(0)} = \mathbb{C}_{2r|s}\,\mathbf{P}_{2r|s}^{(0)}. \tag{XVIII.2}$$

It is clear that (XVIII.1) has a unique solution for all r and s because it is an ordinary linear differential equation of first order with a differentiable right-hand side. Since we are interested not only in existence and uniqueness but also in estimating the rates of decay of the spherical moments $\mathbf{P}_{2r|s}$ for kinetic equilibrium, we shall give a formal proof by induction based upon explicit integration. For a general value of $2r + s$ we assert that the general solution of (XVIII.1) is

$$\mathbf{P}_{2r|s} = \sum_{k=1}^{N} \exp(-\lambda_{2r|s|k}\,t)\mathbf{B}_{2r|s|k}(t) + \mathbf{P}_{2r|s}^{(0)}; \tag{XVIII.3}$$

here $\mathbf{B}_{2r|s|k}$ is a polynomial function of t with coefficients which are functions of ρ, ε, $\mathfrak{m}$, and $\mathfrak{g}$; the integer N, the number of summands, depends upon r and s; and the decay factor $\lambda_{2r|s|k} > 0$. Obviously MAXWELL's results (XIII.10) and (XIII.11) are of this kind, so we know already that the assertion is valid when $2r + s = 2$ or 3. Now we fix $2r + s$ and suppose that (XVIII.3) has been established for all moments of order less than $2r + s$. Then $\mathbf{Q}_{2r|s}$ is a known function (in fact analytic), and we may integrate (XVIII.1):

$$\mathbf{P}_{2r|s} = \exp(-\mathbb{C}_{2r|s}\,t)\left[\mathbf{K}_{2r|s} + \int_0^t \exp(\mathbb{C}_{2r|s}\,u)\mathbf{Q}_{2r|s}(u)\,du\right], \tag{XVIII.4}$$

$\mathbf{K}_{2r|s}$ being a constant of integration. Because of (XVI.32) and the hypothesis of induction, $\mathbf{Q}_{2r|s}$ is given by

$$\mathbf{Q}_{2r|s} = \sum \mathbb{C}_{r_1 r_2 | s_1 s_2} \left[\sum_{k_1=1}^{N_1} \mathbf{B}_{2r_1|s_1|k_1}(t)\exp(-\lambda_{2r_1|s_1|k_1}t) + \mathbf{P}_{2r_1|s_1}^{(0)}\right]$$
$$\times \left[\sum_{k_2=1}^{N_2} \mathbf{B}_{2r_2|s_2|k_2}(t)\exp(-\lambda_{2r_2|s_2|k_2}t) + \mathbf{P}_{2r_2|s_2}^{(0)}\right]. \tag{XVIII.5}$$

Multiplying out the terms shows that there are polynomials $\mathbf{C}_{2r|s|k}$, positive constants $\mu_{2r|s|k}$, and a positive integer R such that

$$\mathbf{Q}_{2r|s} = \sum_{k=1}^{R} \exp(-\mu_{2r|s|k}\,t)\mathbf{C}_{2r|s|k}(t) + \mathbf{Q}_{2r|s}^{(0)}. \tag{XVIII.6}$$

We may now put (XVIII.2) into (XVIII.6), then put the result into (XVIII.4), then carry out the quadrature. Thus we obtain

$$\mathbf{P}_{2r|s} = \exp(-\mathbb{C}_{2r|s}t)\mathbf{K}_{2r|s}$$
$$+ \sum_{k=1}^{R} \exp(-\mu_{2r|s|k}t)\mathbf{B}_{2r|s|k}(t) + \mathbf{P}_{2r|s}^{(0)}. \quad \text{(XVIII.7)}$$

Since this result is of the form (XVIII.3), the induction is complete. △

Thus we have proved our first major result:

Theorem 1. *For a gas of Maxwellian molecules, the infinite system of equations for moments which are compatible with gross rest possesses a unique solution for each choice of the initial values. The solution exists for all time and tends exponentially to its value for kinetic equilibrium.*

Though the proof of this theorem is elementary, it not only has delivered the existence and uniqueness of solutions but also has demonstrated the trend to equilibrium for such a gas. Since the theorem considers only the infinite system of equations for the moments, we must view it in the context of the Maxwell-Boltzmann equation itself as delivering only a solution "weak" in one of the various senses of the word. Alternatively we may interpret the theorem as asserting the following necessary condition for solutions of (IX.1): *For a gas of Maxwellian molecules, assume that the moments up to order n of a molecular density compatible with gross rest exist and are differentiable functions of time. Then these moments are in fact analytic functions of time; they may assume arbitrary initial values*[4], *by which they are determined uniquely; and they tend exponentially to their values for kinetic equilibrium.*

In Appendix B to Chapter XVII we have seen that the particular spatially homogeneous solution F found by KROOK & WU itself approaches exponentially the corresponding solution F_U for kinetic equilibrium.

(iii) *Estimate of the rates of approach to equilibrium*

The formulae used to prove Theorem 1 enable us to estimate the several moments' rates of approach to equilibrium. MAXWELL's results (XIII.10) and (XIII.11) do not suggest what to expect, since the logarithmic decrements for $^3\mathbf{M}$ are $\frac{2}{3}\tau$ and $\frac{3}{2}\tau$, one of which is greater and one of which is less than the time

---

[4] As far as the differential equations are concerned, the initial values are arbitrary. The fact that moments are indeed moments of an essentially positive molecular density imposes on the initial values the inequalities

$$n_a M^{(0)}_{ab} n_b \geq 0, \qquad n_a M^{(0)}_{abbc} n_c \geq 0, \ldots,$$

the first of which we have discussed in Section (iv) of Chapter IV.

of relaxation τ for **P**. We might wonder whether as the order of the moment grows greater, the logarithmic decrement might also grow greater. We prove now that such is not the case.

By (XVIII.7) the logarithmic decrement of one term in the solution is $(\mathbb{C}_{2r|s})^{-1}$. To estimate the others, we wish to find the smallest $\lambda_{2r|s|k}$ for given r and s, since as $t \to \infty$ the term in which this one appears ultimately dominates the sum in (XVIII.3). We shall prove by induction that

$$\min_{k} \lambda_{2r|s|k} \geqq \tfrac{2}{3}/\tau \quad \text{if} \quad 2r + s \geqq 3. \tag{XVIII.8}$$

As we have said already, MAXWELL's results (XIII.10)$_2$ and (XIII.11) are equivalent to

$$\tau\lambda_{2|1|1} = \tfrac{2}{3}, \qquad \tau\lambda_{0|3|1} = \tfrac{3}{2}, \tag{XVIII.9}$$

and a fairly easy explicit calculation based on IKENBERRY's theorem yields for the fourth moments[5]

$$\begin{aligned}
\tau\lambda_{4|0|1} &= \tfrac{2}{3}, & \tau\lambda_{4|0|2} &= 2, & & \\
\tau\lambda_{2|2|1} &= \tfrac{7}{6}, & \tau\lambda_{2|2|2} &= 1, & \tau\lambda_{2|2|3} &= 2, \\
\tau\lambda_{0|4|1} &= 2.096\ldots, & \tau\lambda_{0|4|2} &= 2. & &
\end{aligned} \tag{XVIII.10}$$

Thus the results for the third and fourth moments conform with (XVIII.8). Now fixing r and s, we see from (XVIII.3) and its expanded form (XVIII.7) that

$$\min_{k} \lambda_{2r|s|k} = \min\left(\mathbb{C}_{2r|s}, \min_{k} \mu_{2r|s|k}\right). \tag{XVIII.11}$$

The coefficients $\mu_{2r|s|k}$ are obtained by multiplying out the terms in (XVIII.5), so each of them is of one of the following forms:

$$\mu_{2r|s|k} = \begin{cases} \lambda_{2r_1|s_1|k_1} + \lambda_{2r_2|s_2|k_2}, \\ \lambda_{2r_1|s_1|k_1}, \\ \lambda_{2r_2|s_2|k_2}, \end{cases} \tag{XVIII.12}$$

---

[5] The cryptic note about "harmonics of the fourth and sixth orders" which MAXWELL [1879] added in proof, just before the appendix, may indicate that he did not calculate correctly the seven times that we list here in (XVIII.10). It seems that MAXWELL, working while moribund, convinced himself there were only a few distinct times of relaxation; on the contrary, the analysis of this chapter indicates that there are infinitely many.

The value of $\tau\lambda_{0|4|1}$ given above is that which follows from the fact that $\lambda_{0|4|1} = \mathbb{C}_{0|4}$, the determination of $\mathbb{C}_{0|4}$ through (XVI.34)$_2$, and the numerical evaluations (XVI.39), which are based on calculations done for IKENBERRY & TRUESDELL in 1956. The later and doubtless more accurate evaluation by ALTERMAN, FRANKOWSKI, & PEKERIS [1962, Table 1] yields $\tau\lambda_{0|4|1} = 1.8731\ldots$. On the proof at hand this difference has no effect.

and $2r_1 + s_1 < 2r + s$, $2r_2 + s_2 < 2r + s$, $2r_1 + s_1 + 2r_2 + s_2 = 2r + s$. The hypothesis of induction is that (XVIII.8) holds for all orders less than $2r + s$. Thus (XVIII.12) shows that

$$\mu_{2r|s|k} \geq \tfrac{2}{3}/\tau. \tag{XVIII.13}$$

It follows from (XVIII.11) that

$$\min_k \lambda_{2r|s|k} \geq \min(\mathbb{C}_{2r|s}, \tfrac{2}{3}/\tau). \tag{XVIII.14}$$

To obtain an estimate of $\mathbb{C}_{2r|s}$, let us extend the definition of it as follows: For any positive q,

$$\mathbb{C}_{q|0} \equiv 4\pi n \sqrt{\frac{g}{2m}} \int_0^\infty \{1 - \cos^q \theta - \sin^q \theta\} z \, dz. \tag{XVIII.15}$$

We note from (XVI.33) that

$$\mathbb{C}_{2r|s} \geq \mathbb{C}_{2r+s|0}, \tag{XVIII.16}$$

because $|P_s(z)| \leq 1$ if $|z| \leq 1$. Now in (XVIII.15) the integrand is an increasing function of q for each θ in $(0, \tfrac{1}{2}\pi)$. Hence

$$\mathbb{C}_{q_1|0} > \mathbb{C}_{q_2|0} \quad \text{if} \quad q_1 > q_2. \tag{XVIII.17}$$

Therefore[6], if $2r + s \geq 4$,

$$\mathbb{C}_{2r|s} \geq \mathbb{C}_{4|0} = \tfrac{2}{3}/\tau, \tag{XVIII.18}$$

the value of $\mathbb{C}_{4|0}$ having been obtained by explicit calculation. We now appeal to (XVIII.14) and (XVIII.18) and so complete the proof of (XVIII.8) by induction. Since $\tfrac{3}{2}\tau$ is the logarithmic decrement of $\mathbf{q}$, we can express the result as the following estimate of the rate of approach to equilibrium:

Theorem 2. *In a gas of Maxwellian molecules, among all moments of a molecular density consistent with a grossly homogeneous condition the slowest approach to the equilibrium value is achieved by the energy flux.*

In Appendix B to Chapter XVII we have seen that the particular spatially homogeneous solution F found by KROOK & WU approaches the corresponding F_U exponentially with logarithmic decrement 3τ; that is, just twice as slowly as $\mathbf{q}$ relaxes.

In the prologue we declared we should not take up informal approximations and linearizations, but one result connected therewith is so immediate here that we will mention it. A popular linearized theory replaces the operator $\mathbb{C}$ by a linear one. In a gas of Maxwellian molecules, the linearization amounts to casting away the bilinear function $\mathbf{Q}_{2r|s}$ in IKENBERRY's theorem (XVI.31).

[6] IKENBERRY & TRUESDELL [1956, §14].

The factor $c_{2r|s}$ then becomes[7] what is often called an "eigenvalue". At the same time, the counterpart of the relaxation theorem we have just proved then becomes an immediate and trivial consequence of the estimate (XVIII.18). However, the rates of approach to equilibrium of some of the moments are not given correctly by the linearized theory. Many of the exponential terms are lost altogether by the linearization. To see this fact, it suffices to consider the list (XVIII.10) of values of logarithmic decrements for the fourth moments. On each line, only the first term appears in the linearized theory, which thus provides only three decrements instead of seven. Moreover, for two of the three kinds of moments of order 4, the decrements provided by the linearized theory are not the largest. Thus, in general, the approach to equilibrium according to the linearized theory is faster than according to MAXWELL'S. Of course no such discrepancy arises for the moments of orders 2 and 3, since for them the linearized equation is exact, as MAXWELL showed in effect.

The existence theorem given in this chapter is the only one presently known in the kinetic theory for gases whose molecules are subject to intermolecular forces of infinite range. It rests heavily, indeed essentially, on IKENBERRY'S theorem and thus is strictly limited to Maxwellian molecules. Moreover, it is an existence theorem for the general system of equations for the moments.

Perhaps the results of this chapter can be extended easily to the system of equations for moments that corresponds to GRAD'S approximation for arbitrary spherically symmetric molecules (Chapter XVII), because that system has just the same form as (XVIII.1). A counterpart of Theorem 1 follows at once if we can prove that $c_{2r|s} > 0$ for general molecules. A counterpart of Theorem 2 follows if we can prove a counterpart of (XVIII.18). Of course results of this kind fall far short of an existence theorem for the Maxwell-Boltzmann equation.

For the subclass of pseudomolecules he calls "generalized Maxwellian molecules" (see the end of Chapter XII) KANIEL[8] has proved counterparts of our Theorem 1 and of some other results in this chapter.

(iv) Retrospect

We have chosen to classify the results presented in this chapter as expressing an existence theorem. In first publishing them TRUESDELL preferred to regard them as constituting an explicit solution. This explicit solution is the last we present in this book. Here is a table of all those we have found in print:

---

[7] The factor $c_{2r|s}$ appeared in the work of WANG CHANG & UHLENBECK [1952, Appendix III] on the linearized theory. They noticed the inequality (XVIII.18) and asserted that $c_{2r|s} \to \infty$ as $r \to \infty$. IKENBERRY & TRUESDELL [1956, §14] stated they could not fully justify WANG CHANG & UHLENBECK's argument, but they were able to demonstrate a strict lower bound from which WANG CHANG & UHLENBECK's assertion follows:

$$\frac{c_{q|0}}{2\pi n \sqrt{g/m}} > (1 - e^{-a}) \cot \sqrt[4]{\frac{512a}{9q}} - \sqrt{\pi} \frac{\Gamma(\tfrac{1}{2}q - \tfrac{1}{2})}{\Gamma(\tfrac{1}{2}q)},$$

a being any number such that $0 < a \leq \tfrac{9}{512}(\tfrac{1}{2}\pi)^4 q$. They asserted that they had found also an upper bound of the same order as $q \to \infty$, as well as an asymptotic form of that order.

[8] KANIEL [1977, §§3–5]. In regard to KANIEL's claim to include Maxwellian molecules as a special case the reader should note our comments in Footnote 5 to Appendix A to Chapter XVI.

(iv) RETROSPECT

Name	Chapter in which treated	Quantity exhibited	Molecular model
Locally Maxwellian	X	F	all
Krook & Wu's solution	XVII, Appendix B	F	Maxwellian
Muncaster's solutions	XVII, Appendix B	F	Maxwellian
Homo-energetic affine flows			
in general	XV, Section (i)	^2M	Maxwellian
gross rest	XVIII, Sections (ii) & (iii)	$^2M, {}^3M, {}^4M, \ldots$	Maxwellian
shearing	XIV	$^2M, {}^3M$	Maxwellian
dilatation	XV, Section (ii)	$^2M, {}^3M$	Maxwellian
extension	XV, Section (iii)	^2M	Maxwellian
cylindrical dilatation	XV, Section (i) (but not developed in detail)	^2M	Maxwellian

Before we turn to existence theory for the Maxwell–Boltzmann equation itself, we shall pick up and resolve certain questions relating to the domain of $\mathbb{C}$ and the convergence of integrals.

Chapter XIX

Convergence Theorems and the Domain of the Collisions Operator

It may seem odd that we have not yet confronted the important question, for what functions F, G, and H do $\mathbb{C}F$ and $\mathbb{C}(G, H)$ exist? The reason is that we did not wish to bury the conceptual structure of the kinetic theory in merely ancillary and often technical analyses of the convergence of integrals. With few exceptions our steps have been easily justifiable in the following sense: If the integrals from which we began converged, so did all those occurring in the manipulations derived from them. One case which cannot be so justified we noted in Section (iv) of Chapter VII. Namely, we wished to pass from the form

$$\bar{\mathbb{C}}_{G,H}g = \tfrac{1}{8}\iint\int_{\mathscr{S}} w(g + g_* - g' - g'_*)(G'H'_* + G'_*H' - GH_* - G_*H)$$

$$(\text{VII.20})_{2r}$$

of the total collisions operator to the form

$$\bar{\mathbb{C}}_{G,H}g = \tfrac{1}{4}\iint\int_{\mathscr{S}} w(g' + g'_* - g - g_*)(GH_* + G_*H), \qquad (\text{VII.20})_{3r}$$

and thence to

$$\bar{\mathbb{C}}_{G,H}g = \tfrac{1}{2}\iint\int_{\mathscr{S}} w(g' - g)(GH_* + G_*H). \qquad (\text{VII.20})_{4r}$$

Unfortunately, the convergence of the integrals $(\text{VII.20})_{3,4}$ does not generally follow from the convergence of the integral that defines $\bar{\mathbb{C}}_{G,H}g$:

$$\bar{\mathbb{C}}_{G,H}g = \tfrac{1}{2}\iint\int_{\mathscr{S}} wg(G'H'_* + G'_*H' - GH_* - G_*H). \qquad (\text{VII.15})_{2r}$$

Our objective in this chapter is to resolve in general terms some of the problems raised earlier concerning the convergence of integrals. In particular we shall exhibit some broad spaces of functions for F, G, and H for which $\mathbb{C}F$

and $\mathbb{C}(G, H)$ exist, and we shall establish the transformations (VII.17)–(VII.20) rigorously for a variety of functions g. At the same time we shall lay the foundations for the existence theorems which we shall present in Chapters XX and XXI.

(i) Preliminaries

We begin by recording some simple results which interrelate $\mathbb{C}(G, H)$, $\bar{\mathbb{C}}_{G,H}\, g$, and the transformations (VII.17)–(VII.20). These results do not require detailed information about G, H, and g, and for that reason they will be valuable later.

Let $\mathcal{L}^1(\mathcal{X})$ denote the set of functions which are Lebesgue integrable on a set $\mathcal{X}$. Then we have

Lemma 1. *Let G, H, and g be such that $wg(G'H'_* + G'_*H' - GH_* - G_*H) \in \mathcal{L}^1(\mathscr{V} \times \mathscr{V} \times \mathscr{S})$. Then $g\mathbb{C}(G, H) \in \mathcal{L}^1(\mathscr{V})$. In particular, if the integral defining $\bar{\mathbb{C}}_{G,H}1$ converges absolutely, then (G, H) lies in the domain of $\mathbb{C}(\,,\,)$, and $\mathbb{C}(G, H)$ is integrable.*

Proof. Since $wg(G'H'_* + G'_*H' - GH_* - G_*H) \in \mathcal{L}^1(\mathscr{V} \times \mathscr{V} \times \mathscr{S})$, we conclude from Fubini's theorem[1] that the function[2] $(\mathbf{v}_*, r, \zeta) \mapsto wg(G'H'_* + G'_*H' - GH_* - G_*H)(\mathbf{v}, \mathbf{v}_*, r, \zeta)$ is integrable on $\mathscr{V} \times \mathscr{S}$ for almost all $\mathbf{v} \in \mathscr{V}$, and its integral $\iint_{\mathscr{S}} wg(G'H'_* + G'_*H' - GH_* - G_*H) = g\mathbb{C}(G, H)$ lies in $\mathcal{L}^1(\mathscr{V})$. △

This lemma tells us that if $\bar{\mathbb{C}}_{G,H}\, g$ exists, then so does $g\mathbb{C}(G, H)$. By use of it in the particular case when $g \equiv 1$ we may find functions in the domain of $\mathbb{C}(\,,\,)$ by finding functions for which (VII.15)$_2$ converges. We shall see some simple applications of this result in Section (iii).

Next we return to the formal transformations (VII.17)–(VII.20) and record some conditions sufficient to justify them rigorously.

Lemma 2. *Let G, H, and g be such that $wg(G'H'_* + G'_*H' - GH_* - G_*H) \in \mathcal{L}^1(\mathscr{V} \times \mathscr{V} \times \mathscr{S})$ and $w(g' - g)(GH_* + G_*H) \in \mathcal{L}^1(\mathscr{V} \times \mathscr{V} \times \mathscr{S})$. Then $\bar{\mathbb{C}}_{G,H}\, g$ exists and is given by (VII.17)–(VII.20).*

Proof. First we show that (VII.20)$_4$ implies (VII.20)$_3$. By (VI.62) and (VI.54)$_{2,3}$ the transformation $(\mathbf{v}, \mathbf{v}_*, r, \zeta) \mapsto (\mathbf{v}_*, \mathbf{v}, r_*, \zeta_*)$ is piecewise continuously differentiable, it has a piecewise continuously differentiable inverse, and

---

[1] E.g. Chapter 12, Theorem 19, of H. L. ROYDEN, *Real Analysis*, London, Macmillan, 1968.
[2] Throughout this chapter we suppress the dependence of F, G, H, and g on t and $\mathbf{x}$.

its Jacobian is 1. Thus, by (VI.68) and a theorem[3] in measure theory on the change of variables in an integral, $w(g' - g)(GH_* + G_*H) \in \mathscr{L}^1(\mathscr{V} \times \mathscr{V} \times \mathscr{S})$ implies that $w(g'_* - g_*)(G_*H + GH_*) \in \mathscr{L}^1(\mathscr{V} \times \mathscr{V} \times \mathscr{S})$, and moreover the two functions have the same integral. In particular, then, we find $w(g' + g'_* - g - g_*)(G_*H + GH_*) \in \mathscr{L}^1(\mathscr{V} \times \mathscr{V} \times \mathscr{S})$, and (VII.20)$_3$ follows from (VII.20)$_4$.

To obtain (VII.20)$_2$ from (VII.20)$_3$, we apply the same procedure again, this time making the change of variables $(\mathbf{v}, \mathbf{v}_*, r, \zeta) \mapsto (\mathbf{v}', \mathbf{v}'_*, r', \zeta')$ and using (VI.65), (VI.70), and (VI.45)$_2$. The remaining forms of $\bar{\mathbb{C}}_{G,H}g$ follow from the assumption that $wg(G'H'_* + G'_*H' - GH_* - G_*H) \in \mathscr{L}^1(\mathscr{V} \times \mathscr{V} \times \mathscr{S})$ by the steps outlined after (VII.16). △

This lemma tells us little more about the transformations (VII.17)-(VII.20) than we may already have suspected: If the first and last forms of $\bar{\mathbb{C}}_{G,H}g$, namely (VII.15)$_2$ and (VII.20)$_4$, are valid, so also are the steps which connect them. Its importance, however, lies in the fact that by use of it we reduce the problem of finding functions G, H, and g for which (VII.17)-(VII.20) are valid to that of finding functions for which only (VII.15)$_2$ and (VII.20)$_4$ need exist. That is, *we wish to find functions G, H, and g for which $wg(G'H'_* + G'_*H' - GH_* - G_*H)$ and $w(g' - g)(GH_* + G_*H)$ are integrable on $\mathscr{V} \times \mathscr{V} \times \mathscr{S}$.* It is this problem that we consider in the remainder of the chapter. Not only will its solution yield functions for which (VII.17)-(VII.20) are valid, but by Lemma 1 we shall obtain also directly from it functions which lie in the domains of $\mathbb{C}(\ ,\ )$ and $\mathbb{C}$.

To shorten the analysis, we shall actually consider a simpler problem. Namely, we shall seek to find functions F and g for which $wg(F'F'_* - FF_*)$ and $w(g' - g)FF_*$ are integrable on $\mathscr{V} \times \mathscr{V} \times \mathscr{S}$. This is the special case of our problem in which $F = G = H$, and we shall be able to generalize its solution, by inspection, to the full problem.

(ii) *Restrictions on the growth of the integrand*

When, in the preceding chapters, we made use of the transformations (VII.17)-(VII.20), the function g was either a polynomial in the components of $\mathbf{v}$, as in Chapters XVI and XVII, or it was $\log F$, as in the discussion of BOLTZMANN's H-theorem. Setting aside for the moment the second case, we treat the first by considering measurable functions g which satisfy the growth condition

$$|g| \leq \alpha(1 + v)^n, \qquad \text{(XIX.1)}$$

---

[3] *E.g.* Chapter XVI, §2 of S. LANG, *Analysis II*, Reading, *etc.*, Addison-Wesley, 1969.

α and n being constants, and $\alpha > 0$, $n \geq 0$. Sometimes we shall be more restrictive and consider continuously differentiable functions g which not only satisfy (XIX.1) but also have continuous derivatives which satisfy the growth condition

$$|\partial_{v_k} g| \leq \beta_k (1 + v)^{n-1} \tag{XIX.2}$$

for some non-negative constants β_k. Any polynomial of degree n in the components of $\mathbf{v}$ satisfies all these conditions. By replacing $\mathbf{v}$ by $\mathbf{v}_*$ in (XIX.1) and (XIX.2) we obtain similar restrictions on g_*. Growth conditions which g' and g'_* satisfy in consequence of (XIX.1) and (XIX.2) can be derived in the following way. From (VI.33)$_1$ and the inequality $w \leq v + v_*$ we see that

$$1 + v' = 1 + |\mathbf{v} + (\mathbf{w} \cdot \boldsymbol{\alpha})\boldsymbol{\alpha}|,$$

$$\leq 1 + v + w, \tag{XIX.3}$$

$$\leq 2(1 + v)(1 + v_*).$$

If we combine this result with (XIX.1) and (XIX.2), we obtain

$$|g'| \leq 2^n \alpha (1 + v)^n (1 + v_*)^n,$$
$$|(\partial_{v_k} g)'| \leq 2^{n-1} \beta_k (1 + v)^{n-1} (1 + v_*)^{n-1}. \tag{XIX.4}$$

Similar conditions apply to g'_*: they are derived by interchanging $\mathbf{v}$ and $\mathbf{v}_*$, and hence also v' and v'_*, in (XIX.4).

For the choice $g = \log F$ the problem is more complicated, and in fact for it we know of no analysis which establishes (VII.17)–(VII.20) in any generality. This choice of g is used only in proofs of the H-theorem, and such proofs as have been constructed so far resort to an indirect approach. This is certainly true of the proof we present in Section (v) of Chapter XXI. Without going into details, we remark that there we reduce the problem to applying the Boltzmann monotonicity theorem (VIII.5) only to measurable functions F satisfying inequalities of the form

$$ae^{-b(1+v^2)} \leq F \leq k, \tag{XIX.5}$$

k, a, and b being positive constants. Directly from these we obtain the simple bound

$$|\log F| \leq \max(|\log k|, b + |\log a|)(1 + v)^2. \tag{XIX.6}$$

Since this is a special case of (XIX.1), henceforth we shall consider the choice $g = \log F$ only in so far as it falls within the scope of our analysis for (XIX.1) and (XIX.2) alone.

(iii) Convergence theorems

MAXWELL's collisions operator $\mathbb{C}$ is determined by two kinetic constitutive quantities: the encounter operator $\mathbb{E}$ and the cross-section $\mathscr{S}$. Different choices of these two quantities define different collisions operators, and therefore we may expect that conditions under which (VII.17)-(VII.20) are valid will vary with this choice. We shall treat molecules with a cut-off and molecules with intermolecular forces of infinite range separately[4]. For the former, $\mathscr{S}$ being a disk in $\mathscr{R}^2$, we shall find sufficient conditions for convergence much more general than those we shall obtain afterward for intermolecular forces for which $\mathscr{S} = \mathscr{P}$. We may also expect that considerations of this kind may restrict the possible choices of $\mathbb{E}$ and $\mathscr{S}$, that is, restrict the class of molecular models for which the kinetic theory itself can be formulated definitely. Some such restrictions will be suggested by our analysis of the case in which $\mathscr{S} = \mathscr{P}$.

Molecules with a cut-off. For molecules with a cut-off we take g to be a measurable function satisfying (XIX.1). In this case it is straightforward to derive restrictions on F which ensure that $wg(F'F'_* - FF_*)$ and $w(g' - g)FF_*$ both lie in $\mathscr{L}^1(\mathscr{V} \times \mathscr{V} \times \mathscr{S})$. We note first that if F is measurable, then $wgFF_*$ is a product of four measurable functions and so is measurable on the product space $\mathscr{V} \times \mathscr{V} \times \mathscr{S}$. Next, by (XIX.1) and the inequality

$$w \leq v + v_* \leq (1 + v)(1 + v_*), \qquad \text{(XIX.7)}$$

we see that

$$|wgFF_*| \leq \alpha w(1 + v)^n |F| |F_*|,$$
$$\leq \alpha(1 + v)^{n+1}(1 + v_*)|F| |F_*|. \qquad \text{(XIX.8)}$$

Since the function on the right-hand side here is independent of the position $\mathbf{x}^e$ in the plane $\mathscr{P}$, a necessary condition for this function to be integrable is that $\mathscr{S}$ be bounded. Since we here assume such to be the case, $(XIX.8)_2$ shows that $wgFF_*$ is integrable in the product space provided $(1 + v)^{n+1}F$ be integrable. If n is an integer, this condition implies that the moments of F of all orders not exceeding $n + 1$ exist. Let us consider now the function $wg'FF_*$. Since g is measurable and $\mathbb{E}$ is piecewise continuously differentiable, g' is measurable in the product space $\mathscr{V} \times \mathscr{V} \times \mathscr{S}$. Therefore $wg'FF_*$ is also measurable, and, in view of the growth condition $(XIX.4)_1$ on g', (XIX.8) can be replaced by

$$|wg'FF_*| \leq 2^n \alpha(1 + v)^{n+1}(1 + v_*)^{n+1}|F||F_*|. \qquad \text{(XIX.9)}$$

---

[4] The theorems we present here for molecules with a cut-off are used in most papers dealing with existence theory for such molecules, although they are rarely, if ever, stated explicitly. For molecules with intermolecular forces of infinite range our results are new. In fact, in the common literature of the kinetic theory justifications of formal steps for them are as rare as hen's teeth. We have found only two major works in which the corresponding questions of convergence of collisions integrals are so much as mentioned: CHAPMAN & COWLING [1939, §3.6], GRAD [1958, §15].

This shows that $wg'FF_*$ is integrable if $(1 + v)^{n+1}F$ is integrable. Now we pass from $wg'FF_*$ to $wgF'F'_*$ by applying the transformation $(\mathbf{v}, \mathbf{v}_*, r, \zeta) \mapsto (\mathbf{v}', \mathbf{v}'_*, r', \zeta')$. Since this transformation is piecewise continuously differentiable, its inverse is also piecewise continuously differentiable, and since by (VI.70) its Jacobian is 1, we conclude[5] that $wgF'F'_*$ is integrable on $\mathscr{V} \times \mathscr{V} \times \mathscr{S}$ under this same condition on F. Collecting these three results, we conclude that *if $(1 + v)^{n+1}F$ lies in $\mathscr{L}^1(\mathscr{V})$ then $wg(F'F'_* - FF_*)$ and $w(g' - g)FF_*$ lie in $\mathscr{L}^1(\mathscr{V} \times \mathscr{V} \times \mathscr{S})$*.

The preceding analysis is easily generalized to show that if $(1 + v)^{n+1}G$ and $(1 + v)^{n+1}H$ are integrable, so also are the functions $wg(G'H'_* + G'_*H' - GH_* - G_*H)$ and $w(g' - g)(GH_* + G_*H)$. From Lemma 2, then, we obtain immediately the first main result:

Theorem 1. *Let $\mathscr{S}$ be a disk. If $(1 + v)^{n+1}G \in \mathscr{L}^1(\mathscr{V})$, $(1 + v)^{n+1}H \in \mathscr{L}^1(\mathscr{V})$, and if g is a measurable function satisfying (XIX.1), then $\bar{\mathbb{C}}_{G,H}g$ exists and is given by (VII.17)–(VII.20).*

If we combine this theorem with Lemma 1 we obtain our second main conclusion:

Theorem 2. *Let $\mathscr{S}$ be a disk. If $(1 + v)G \in \mathscr{L}^1(\mathscr{V})$ and $(1 + v)H \in \mathscr{L}^1(\mathscr{V})$, then $\mathbb{C}(G, H)$ exists and is integrable. In particular, if $(1 + v)F \in \mathscr{L}^1(\mathscr{V})$, then $\mathbb{C}F$ exists and is integrable.*

In our analysis leading to these results we actually showed that both $wgFF_*$ and $wgF'F'_*$ were integrable. Because of this we have for Theorem 2 the

Corollary. $\mathbb{F}F$ *and* $\mathbb{P}F$ *exist, and*

$$\mathbb{C}F = -(\mathbb{F}F)F + \mathbb{P}F. \tag{VII.5}_r$$

Here $\mathbb{F}$ and $\mathbb{P}$ are are the frequency operator and partial collisions operator, respectively, as defined by (VII.4) and (VII.6).

For molecules with a cut-off, then, no restrictions need be placed on our choice of $\mathbb{E}$ and $\mathscr{S}$. This fact and the decomposition (VII.5) provide two reasons why mathematicians find molecules with a cut-off particularly attractive. As we shall see in Chapters XX and XXI, all presently known existence theorems for the Maxwell–Boltzmann equation, with the exception of the one presented in Chapter XVIII, refer to this class of molecules.

Theorem 1, Theorem 2, and the Corollary constitute our main results on convergence of collisions integrals for molecules with a cut-off. There is, however, one other which we shall need in Chapter XXI. First let us note that if for g we choose $(1 + v)^n$ and then apply Theorem 1 and Lemma 1, we find that $(1 + v)^{n+1}F \in \mathscr{L}^1(\mathscr{V})$ implies $(1 + v)^n \mathbb{C}F \in \mathscr{L}^1(\mathscr{V})$. Thus the moments of

---
[5] See the reference cited in Footnote 3.

(iii) FORCES OF INFINITE RANGE

order n of $\mathbb{C}F$ exist whenever those of F of order $n + 1$ exist. The result we shall need is the relation[6] between the moments of F and of $\mathbb{C}F$ which we record in

Theorem 3. Let $\mathscr{S}$ be a disk. If $n \geq 2$, there is a positive constant $C(n)$ such that if $(1 + v)^{n+1}F$ is non-negative and integrable, then

$$\int (1 + v^2)^{\frac{1}{2}n} \mathbb{C}F \leq C(n)\pi \mathrm{d}^2 \left(\int (1 + v^2)^{\frac{1}{2}n} F \right) \left(\int (1 + v^2) F \right). \quad \text{(XIX.10)}$$

Proof. We begin from the inequality

$$a^s + b^s \leq (a + b)^s \leq a^s + b^s + C(2s)(a^\theta b^{s-\theta} + a^{s-\theta} b^\theta), \quad \text{(XIX.11)}$$

which is valid when $a, b \geq 0$, $s \geq 1$, and $0 \leq \theta \leq 1$. The number $C(2s)$ can be chosen independently of a, b, and θ. From this inequality and (VI.2) we find that

$$(1 + v'^2)^s + (1 + v'^2_*)^s - (1 + v^2)^s - (1 + v^2_*)^s$$
$$\leq C(2s)[(1 + v^2_*)^\theta (1 + v^2)^{s-\theta} + (1 + v^2_*)^{s-\theta} (1 + v^2)^\theta]. \quad \text{(XIX.12)}$$

Since $(1 + v)^{n+1}F$ is integrable and since the function $g = (1 + v^2)^{\frac{1}{2}n}$ satisfies (XIX.1), we see from Theorem 1 that (VII.17)–(VII.20), when specialized by setting $G = H = F$, are valid. If we set $s = \frac{1}{2}n$ in (XIX.12) and then use the form of $\bar{\mathbb{C}}_F g$ implied by (VII.20)$_3$, we obtain

$$\int (1 + v^2)^{\frac{1}{2}n} \mathbb{C}F \leq \tfrac{1}{2} C(n) \pi \mathrm{d}^2 \iint w[(1 + v^2_*)^\theta (1 + v^2)^{\frac{1}{2}n-\theta}$$
$$+ (1 + v^2_*)^{\frac{1}{2}n-\theta}(1 + v^2)^\theta] F F_*. \quad \text{(XIX.13)}$$

Now (XIX.10) follows from (XIX.13) by use of the inequality

$$w \leq (1 + v^2)^{\frac{1}{2}} (1 + v^2_*)^{\frac{1}{2}}, \quad \text{(XIX.14)}$$

and the special choice $\theta = \frac{1}{2}$. △

Molecules with intermolecular forces of infinite range. In the preceding analysis for molecules with a cut-off the fact that $\mathbb{C}$ satisfied (VII.5) was essential. If the intermolecular forces have infinite range, so that $\mathscr{S} = \mathscr{P}$, such a decomposition is not possible; in order to obtain positive results, we must undertake a more subtle analysis, making use of such cancellation of infinities as the differences $F'F'_* - FF_*$ and $g' - g$ may provide[7].

---

[6] POVZNER [1962, §2].

[7] FINKELSTEIN [1965] has derived some alternative forms of the collisions operator and the bilinear form which take account of such cancellations. Some of the forms of these we derive here are of the same kind.

We shall restrict our attention in the analysis for this case to encounter operators $\mathbb{E}$ which arise from a two-body problem of analytical dynamics. We have described these operators in Chapter VI in our third solution of the encounter problem. For them we may use the form of the collisions operator given in (VII.10). In view of (VII.7) we may restate the problem to be considered here as that of finding functions g and F for which $\mathbb{S}g(F'F'_* - FF_*)$ and $\mathbb{S}(g' - g)FF_*$ are integrable on $\mathscr{V} \times \mathscr{V} \times \triangle$ with respect to the measure $\sin \theta \, d\theta \, d\zeta \, d\mathbf{v}_* \, d\mathbf{v}$. We begin by deriving some convenient forms of the integrals of these two functions.

From (VI.41) and (VI.40)$_2$ we see that $\mathbf{v}'$ and $\mathbf{v}'_*$ are functions of $\mathbf{v}$, $\mathbf{v}_*$, θ, and ζ. Suppressing for the moment the variables $\mathbf{v}$, $\mathbf{v}_*$, and ζ, we shall emphasize the dependence upon θ by writing $\mathbf{v}'(\theta)$ and $\mathbf{v}'_*(\theta)$ in place of $\mathbf{v}'$ and $\mathbf{v}'_*$. In particular it follows from (VI.40)$_2$ and (VI.41) that $\mathbf{v}'(\tfrac{1}{2}\pi) = \mathbf{v}$ and $\mathbf{v}'_*(\tfrac{1}{2}\pi) = \mathbf{v}_*$. Similarly, if H is any function of $\mathbf{v}$, we shall write $H'(\theta)$ and $H'_*(\theta)$ in place of H' and H'_*. If F is continuously differentiable, we can use (VI.41) and the chain rule to write

$$F'(\theta)F'_*(\theta) - FF_* = F'(\theta)F'_*(\theta) - F'(\tfrac{1}{2}\pi)F'_*(\tfrac{1}{2}\pi),$$

$$= -\int_\theta^{\frac{1}{2}\pi} \frac{d}{d\phi}(F'(\phi)F'_*(\phi)) \, d\phi, \qquad (\text{XIX.15})$$

$$= -\int_\theta^{\frac{1}{2}\pi} [F'_*(\phi)(\partial_\mathbf{v} F)'(\phi) - F'(\phi)(\partial_\mathbf{v} F)'_*(\phi)] \cdot \gamma(\phi, \zeta, \mathbf{w}) \, d\phi,$$

γ being given by

$$\gamma(\phi, \zeta, \mathbf{w}) \equiv \partial_\phi[(\mathbf{w} \cdot \mathbf{a}(\phi, \zeta, \mathbf{w}))\mathbf{a}(\phi, \zeta, \mathbf{w})]. \qquad (\text{XIX.16})$$

If g is continuously differentiable, we similarly obtain

$$g'(\theta) - g = g'(\theta) - g'(\tfrac{1}{2}\pi),$$

$$= -\int_\theta^{\frac{1}{2}\pi} (\partial_\mathbf{v} g)'(\phi) \cdot \gamma(\phi, \zeta, \mathbf{w}) \, d\phi. \qquad (\text{XIX.17})$$

Thus

$$\iint\int_\triangle \mathbb{S}g(F'F'_* - FF_*) = -\iint\int_0^{2\pi}\int_0^{\frac{1}{2}\pi}\int_\theta^{\frac{1}{2}\pi} \mathbb{S}(\theta, \mathbf{w})g[F'_*(\partial_\mathbf{v} F)' - F'(\partial_\mathbf{v} F)'_*](\phi)$$

$$\cdot \gamma(\phi, \zeta, \mathbf{w}) \sin \theta \, d\phi \, d\theta \, d\zeta, \qquad (\text{XIX.18})$$

$$\iint\int_\triangle \mathbb{S}(g' - g)FF_* = -\iint\int_0^{2\pi}\int_0^{\frac{1}{2}\pi}\int_\theta^{\frac{1}{2}\pi} \mathbb{S}(\theta, \mathbf{w})(\partial_\mathbf{v} g)'(\phi)$$

$$\cdot \gamma(\phi, \zeta, \mathbf{w})FF_* \sin \theta \, d\phi \, d\theta \, d\zeta.$$

(iii) FORCES OF INFINITE RANGE 313

We shall be able to analyse the integrability of $\mathbb{S}(g'-g)FF_*$ directly from (XIX.18)$_2$, but for $\mathbb{S}g(F'F'_* - FF_*)$ we require another equivalent form of (XIX.18)$_1$. First we interchange the orders of integration with respect to θ and ϕ. Since

$$\int_0^{\frac{1}{2}\pi} d\theta \int_\theta^{\frac{1}{2}\pi} d\phi \cdots = \int_0^{\frac{1}{2}\pi} d\phi \int_0^\phi d\theta \cdots, \qquad \text{(XIX.19)}$$

we obtain

$$\iint\int_\Omega \mathbb{S}g(F'F'_* - FF_*) = -\iint\int_0^{2\pi}\int_0^{\frac{1}{2}\pi} g[F'_*(\partial_v F)' - F'(\partial_v F)'_*](\phi)$$

$$\cdot \gamma(\phi, \zeta, \mathbf{w}) \left\{ \int_0^\phi \mathbb{S}(\theta, w) \sin\theta \, d\theta \right\} d\phi \, d\zeta.$$

(XIX.20)

Now we make the change of variables $(\mathbf{v}, \mathbf{v}_*, \phi, \zeta) \mapsto (\mathbf{v}', \mathbf{v}'_*, \phi', \zeta') \equiv \mathbb{E}(\mathbf{v}, \mathbf{v}_*, \phi, \zeta)$. By (VI.46) and (VI.45)$_2$ we know that $\phi' = \phi$ and $w' = w$, so the expression in braces in (XIX.20) is unaltered by this change. Moreover, we recall that $\mathbb{E}$ is piecewise continuously differentiable, that it is its own inverse, and that the transformation it defines has Jacobian 1. From these facts we find that (XIX.20) becomes

$$\iint\int_\Omega \mathbb{S}g(F'F'_* - FF_*) = \iint\int_0^{2\pi}\int_0^{\frac{1}{2}\pi} g'(\phi)[F_* \partial_v F - F(\partial_v F)_*]$$

$$\cdot \gamma(\phi', \zeta', \mathbf{w}') \left\{ \int_0^\phi \mathbb{S}(\theta, w) \sin\theta \, d\theta \right\} d\phi \, d\zeta,$$

$$= \iint\int_0^{2\pi}\int_0^{\frac{1}{2}\pi}\int_\theta^{\frac{1}{2}\pi} \mathbb{S}(\theta, w)g'(\phi)[F_* \partial_v F - F(\partial_v F)_*]$$

$$\cdot \gamma(\phi', \zeta', \mathbf{w}') \sin\theta \, d\phi \, d\theta \, d\zeta, \qquad \text{(XIX.21)}$$

where in the last step we have applied (XIX.19) again. To analyse the integrability of $\mathbb{S}g(F'F'_* - FF_*)$, we shall use (XIX.21)$_2$. It is important to note that the various conversions and transformations which have led us to (XIX.21)$_2$ are justifiable in the following sense: If the integral in (XIX.21)$_2$ converges absolutely, then Fubini's theorem and the theorem on the change of variables in an integral deliver the results (XIX.21)$_1$, (XIX.20), and (XIX.18)$_1$. Thus we have reduced our basic problem to that of finding continuously differentiable functions g and F for which the integrals on the right-hand sides of (XIX.18)$_2$ and (XIX.21)$_2$ converge absolutely.

From the definition (VI.40)$_2$ of $\mathbf{a}$ we find that
$$[\mathbf{w} \cdot \mathbf{a}(\phi, \zeta, \mathbf{w})]\mathbf{a}(\phi, \zeta, \mathbf{w}) = \tfrac{1}{2}[(\cos 2\phi + 1)\mathbf{w} - \sin 2\phi \cos \zeta \, w\boldsymbol{\xi}$$
$$- \sin 2\phi \sin \zeta \, w\boldsymbol{\eta}], \qquad \text{(XIX.22)}$$
and also that
$$\cos \zeta \, \boldsymbol{\xi} + \sin \zeta \, \boldsymbol{\eta} = \frac{1}{\sin \phi}\left(\frac{\cos \phi}{w}\mathbf{w} - \mathbf{a}(\phi, \zeta, \mathbf{w})\right). \qquad \text{(XIX.23)}$$

Therefore, from (XIX.16) we obtain
$$\boldsymbol{\gamma}(\phi, \zeta, \mathbf{w}) = -w\left(\frac{\sin 2\phi}{w}\mathbf{w} + \cos 2\phi \cos \zeta \, \boldsymbol{\xi} + \cos 2\phi \sin \zeta \, \boldsymbol{\eta}\right),$$
$$= -\frac{\cos \phi}{\sin \phi}\mathbf{w} + \frac{\cos 2\phi}{\sin \phi}w\mathbf{a}(\phi, \zeta, \mathbf{w}). \qquad \text{(XIX.24)}$$

From (VI.43), (VI.45)$_{1,2}$, and (VI.46) we now find that
$$\boldsymbol{\gamma}(\phi', \zeta', \mathbf{w}') = -\frac{\cos \phi}{\sin \phi}\mathbf{w} + \frac{1}{\sin \phi}w\mathbf{a}(\phi, \zeta, \mathbf{w}),$$
$$= -w(\cos \zeta \, \boldsymbol{\xi} + \sin \zeta \, \boldsymbol{\eta}). \qquad \text{(XIX.25)}$$

Since the expressions in brackets on the right-hand sides of (XIX.24)$_1$ and (XIX.25)$_2$ are unit vectors, we see that
$$|\boldsymbol{\gamma}(\phi, \zeta, \mathbf{w})| \leqq w, \qquad |\boldsymbol{\gamma}(\phi', \zeta', \mathbf{w}')| \leqq w. \qquad \text{(XIX.26)}$$

Therefore, if g satisfies the growth restrictions (XIX.1) and (XIX.2), by using (XIX.4) and (XIX.26) we prove that
$$|\mathbb{S}(\theta, w)g'(\phi)[F_* \, \partial_v F - F(\partial_v F)_*] \cdot \boldsymbol{\gamma}(\phi', \zeta', \mathbf{w}')\sin \theta|$$
$$\leqq 2^n \alpha \sum_{k=1}^{3} (1 + v)^n (1 + v_*)^n [|F_*| \, |\partial_{v_k} F| + |F| \, |(\partial_{v_k} F)_*|]$$
$$\times \mathbb{S}(\theta, w)w \sin \theta, \qquad \text{(XIX.27)}$$
$$|\mathbb{S}(\theta, w)(\partial_v g)'(\phi) \cdot \boldsymbol{\gamma}(\phi, \zeta, \mathbf{w})FF_* \sin \theta|$$
$$\leqq 2^{n-1}\left(\sum_{k=1}^{3} \beta_k\right)(1 + v)^{n-1}(1 + v_*)^{n-1}|F| \, |F_*| \mathbb{S}(\theta, w)w \sin \theta.$$

If we integrate the right-hand sides of these two inequalities with respect to ϕ, ζ, and θ, we find that a sufficient condition for convergence is
$$\int_0^{\frac{1}{2}\pi} (\tfrac{1}{2}\pi - \theta)\mathbb{S}(\theta, w)w \sin \theta \, d\theta < \infty. \qquad \text{(XIX.28)}$$

Namely, $\mathbb{S}$ satisfies the condition that $(\tfrac{1}{2}\pi - \theta)\sin\theta\, \mathbb{S}(\theta, w)$ be integrable with respect to θ. Since $\mathbb{S}$ is determined by the molecular model, we see that in order to obtain further information from (XIX.27) we must restrict our choice of kinetic constitutive quantities. We should note, however, that (XIX.27) provides us only with *sufficient* conditions for the convergence of (XIX.18)$_2$ and (XIX.21)$_2$. Thus these two integrals may converge also for some scattering factors that violate (XIX.28).

For the purposes of this book we may safely restrict attention to molecular models such that for some positive constant C

$$\int_0^{\tfrac{1}{2}\pi} (\tfrac{1}{2}\pi - \theta)\mathbb{S}(\theta, w) w \sin\theta\, d\theta \leq C(1 + v)^2(1 + v_*)^2. \qquad \text{(XIX.29)}$$

In Section (iv) of this chapter we prove that (XIX.29) is satisfied for a gas of inverse $\mathbb{k}^{\text{th}}$-power molecules provided that $\mathbb{k} > 3$, and so any consequences of (XIX.29) will be valid for such a gas. However, these particular molecules constitute only a subclass of all possible ones for which (XIX.29) holds, and so it is preferable first to draw some conclusions that follow from (XIX.29) without further specialization. Indeed, from (XIX.27) and (XIX.29) we see that the integrals of the left-hand sides of (XIX.27)$_{1,2}$ may be bounded by expressions of the form

$$\text{const.}\left(\int (1 + v)^{n+2}|F|\right)\left(\sum_{k=1}^{3} \int (1 + v)^{n+2}|\partial_{v_k} F|\right), \qquad \text{(XIX.30)}$$

and

$$\text{const.}\left(\int (1 + v)^{n+1}|F|\right)^2, \qquad \text{(XIX.31)}$$

respectively. From these we find that if $(1 + v)^{n+2}F \in \mathscr{L}^1(\mathscr{V})$ and $(1 + v)^{n+2}\partial_{v_k} F \in \mathscr{L}^1(\mathscr{V})$, $k = 1, 2, 3$, the integrals on the right-hand sides of (XIX.18)$_{1,2}$ converge absolutely.

These conclusions are easily extended to the more general problem concerning $\mathbb{S}g(G'H'_* + G'_*H' - GH_* - G_*H)$ and $\mathbb{S}(g' - g)(GH_* + G_*H)$, and they lead, by Lemma 2, to our next main result:

Theorem 4. *Let $\mathbb{S}$ satisfy (XIX.29), and let g be any continuously differentiable function satisfying (XIX.1) and (XIX.2). If G and H are continuously differentiable, and if $(1 + v)^{n+2}K \in \mathscr{L}^1(\mathscr{V})$ when $K = G, H, \partial_{v_k} G$, or $\partial_{v_m} H$, then $\bar{\mathbb{C}}_{G,H}\, g$ exists and is given by (VII.17)–(VII.20).*

From this theorem and Lemma 1 we obtain immediately our final result concerning convergence.

Theorem 5. *Let $\mathbb{S}$ satisfy (XIX.29). If G and H are continuously differentiable, and if $(1 + v)^2 K \in \mathscr{L}^1(\mathscr{V})$ when $K = G, H, \partial_{v_k} G$, or $\partial_{v_m} H$, then $\mathbb{C}(G, H)$ exists and is integrable. In particular, if $(1 + v)^2 F$ and $(1 + v)^2\, \partial_{v_k} F$ are integrable, then $\mathbb{C}F$ exists and is integrable.*

(iv) Inverse k^{th}-power molecules

The convergence theorems we have obtained for molecules with a cut-off are independent of the kinetic constitutive quantities $\mathbb{E}$ and $\mathscr{S}$. In particular, they apply to a gas of ideal spheres. For molecules with intermolecular forces of infinite range, our theorems apply only to those whose scattering factors $\mathbb{S}$ satisfy the inequality (XIX.29). Therefore, to conclude our study of convergence, we shall decide whether (XIX.29) is satisfied by the one class of molecular forces of infinite range which we have considered in this book, namely inverse k^{th}-power molecules. For them we summarize our results in

Theorem 6. *For inverse k^{th}-power molecules, (XIX.29) is satisfied when $k > 3$.*

Proof. From (XII.9)$_3$ we find that

$$(\tfrac{1}{2}\pi - \theta)\mathbb{S}(\theta, w)w \sin\theta \, d\theta = \left(\frac{2\mathfrak{g}}{\mathfrak{m}}\right)^{2/(k-1)} w^{(2k-6)/(k-1)}(\tfrac{1}{2}\pi - \theta)z \, dz, \tag{XIX.32}$$

θ being given as a function of z alone by (XII.6). Recall that $\delta = D(z)$ is the unique positive root of (XII.5). It is convenient to change variables in (XIX.32) from z to δ. By solving (XII.5) for z we exhibit this transformation explicitly:

$$z = \left(\frac{2}{k-1}\right)^{1/(k-1)} \delta(1 - \delta^2)^{-1/(k-1)}, \tag{XIX.33}$$

and so (XIX.32) becomes

$$(\tfrac{1}{2}\pi - \theta)\mathbb{S}(\theta, w)w \sin\theta \, d\theta = \left(\frac{4\mathfrak{g}}{\mathfrak{m}(k-1)}\right)^{2/(k-1)} w^{(2k-6)/(k-1)} f(\delta) \, d\delta, \tag{XIX.34}$$

$$f(\delta) \equiv (\tfrac{1}{2}\pi - \theta)\,\delta\!\left(1 - \delta^2 + \frac{2}{k-1}\delta^2\right)(1 - \delta^2)^{-[2/(k-1)]-1}.$$

The function f depends, of course, upon the parameter k. If we substitute (XIX.33) into (XII.6) and then change the variable of integration from u to u/δ, we obtain the following formula for θ as a function of δ^2:

$$\theta = \int_0^1 \left[\left(\frac{1}{\delta^2} - 1\right)(1 - u^{k-1}) + 1 - u^2\right]^{-\frac{1}{2}} du. \tag{XIX.35}$$

Finally we note that as δ varies from 0 to 1, θ varies from 0 to $\tfrac{1}{2}\pi$. This fact follows directly from (XIX.35). Therefore, using (XIX.34), we may restate

(XIX.28) as the requirement that f be integrable on $(0, 1)$, and (XIX.29) may be replaced by

$$w^{(2k-6)/(k-1)} \int_0^1 f(\delta) \, d\delta \leq \text{const.}(1+v)^2(1+v_*)^2. \tag{XIX.36}$$

Let us assume now that $k > 3$. This means that $(2k-6)/(k-1)$ never exceeds 2, and so (XIX.36) follows from $(XIX.7)_2$ provided we can show that f is integrable on $(0, 1)$.

From (XIX.35) we note that θ, considered as a function of δ^2, is continuous on $[0, 1]$, provided we assume that $\theta(0) \equiv 0$. Hence from $(XIX.34)_2$ we see that f is bounded and continuous on $[0, \lambda]$ if $\lambda < 1$, and so f is integrable on such intervals. Let us choose and fix now a specific λ and show that f is integrable on $(\lambda, 1)$. Using (XIX.35), we can show that θ is differentiable with respect to δ^2 in the open interval $(0, 1)$, and so the mean-value theorem provides a number $\xi \in (\lambda^2, 1)$ such that

$$\tfrac{1}{2}\pi - \theta = \theta(1) - \theta(\delta^2) = \theta'(\xi)(1 - \delta^2) \tag{XIX.37}$$

if $\delta^2 \in (\lambda^2, 1)$. We can determine the derivative of θ directly from (XIX.35):

$$\theta'(\delta^2) = \frac{1}{2\delta^4} \int_0^1 (1 - u^{k-1}) \left[\left(\frac{1}{\delta^2} - 1 \right)(1 - u^{k-1}) + 1 - u^2 \right]^{-\frac{3}{2}} du. \tag{XIX.38}$$

Since $\xi \in (\lambda^2, 1)$ and $k > 3$,

$$0 \leq \theta'(\xi) \leq \frac{1}{2\lambda^4} \int_0^1 \frac{1 - u^{k-1}}{(1 - u^2)^{\frac{3}{2}}} du < \infty, \tag{XIX.39}$$

and so there is a constant C_λ such that

$$0 \leq \tfrac{1}{2}\pi - \theta \leq C_\lambda(1 - \delta^2) \quad \text{if} \quad \lambda^2 < \delta^2 < 1. \tag{XIX.40}$$

Thus, if $\lambda^2 < \delta^2 < 1$,

$$0 \leq f(\delta) \leq (\tfrac{1}{2}\pi - \theta)\left(1 + \frac{2}{k-1}\right)\delta(1 - \delta^2)^{-[2/(k-1)]-1},$$

$$\leq \left(1 + \frac{2}{k-1}\right) C_\lambda \, \delta(1 - \delta^2)^{-2/(k-1)}. \tag{XIX.41}$$

Since the last expression here is integrable on $(\lambda, 1)$ whenever $k > 3$, we conclude that f is integrable on $(0, 1)$. △

Chapter XX

Existence Theory for the General Initial-value Problem. Part II: Place-dependent Solutions for Molecules with a Cut-off

(i) *Integral forms of the Maxwell–Boltzmann equation*

Each of the existence theorems which we shall describe in this chapter begins from an integral equation which is formally equivalent to (IX.1) and (V.13). One such equation we have presented already in Section (iii) of Chapter V:

$$F = \mathfrak{R}_{t-t_0} F_0 + \int_{t_0}^{t} \mathfrak{R}_{t-s} \mathbb{C} F \, ds; \qquad (V.12)_r$$

each function here is presumed evaluated at $(t, \mathbf{x}, \mathbf{v})$.

To obtain a second one, we recall that for molecules with a cut-off

$$\mathbb{C} F = -(\mathbb{F} F) F + \mathbb{P} F, \qquad (VII.5)_r$$

$\mathbb{F}$ and $\mathbb{P}$ being the frequency operator and the partial collisions operator, respectively, which were introduced in Chapter VII. In view of (VII.5), the Maxwell–Boltzmann equation can be written as

$$\mathfrak{D} F + (\mathbb{F} F) F = \mathbb{P} F. \qquad (XX.1)$$

Applying the retrogressor $\mathfrak{R}_{t-s}$ to each side here and then using (III.33), (V.4), and (V.10), we obtain

$$\frac{d}{ds} \mathfrak{R}_{t-s} F + (\mathfrak{R}_{t-s} \mathbb{F} F) \mathfrak{R}_{t-s} F = \mathfrak{R}_{t-s} \mathbb{P} F. \qquad (XX.2)$$

If we regard $\mathfrak{R}_{t-s} \mathbb{F} F$ and $\mathfrak{R}_{t-s} \mathbb{P} F$ for the moment as known functions of s, we may interpret (XX.2) as a linear ordinary differential equation of first order for

$\Re_{t-s} F$. By solving this differential equation we obtain a second integral form of (IX.1):

$$F = \mathsf{J} F, \tag{XX.3}$$

J being defined as follows:

$$\mathsf{J} F \equiv (\Re_{t-t_0} F_0) \exp\left(-\int_{t_0}^{t} \Re_{t-s} \mathbb{F} F \, ds\right)$$

$$+ \int_{t_0}^{t} (\Re_{t-r} \mathbb{P} F) \exp\left(-\int_{r}^{t} \Re_{t-s} \mathbb{F} F \, ds\right) dr; \tag{XX.4}$$

in deriving this we have used (III.34) and (III.30), and again each function is presumed evaluated at $(t, \mathbf{x}, \mathbf{v})$.

In Section (iii) of Chapter V we have referred to the solution of (V.12) as a weak solution. By our general agreement there we shall apply this same term to the solutions of (XX.3). Under very weak conditions these solutions are also mild solutions. That is, their mild derivatives as defined by (V.1) exist and are given by (V.14). For (V.12) this result was established following (V.14), and its proof required the function $s \mapsto \Re_{t-s} \mathbb{C} F(t, \mathbf{x}, \mathbf{v})$ to be continuous. The corresponding proof for (XX.3) requires that $s \mapsto \Re_{t-s} \mathbb{F} F(t, \mathbf{x}, \mathbf{v})$ and $s \mapsto \Re_{t-s} \mathbb{P} F(t, \mathbf{x}, \mathbf{v})$ both be continuous. The details of the proof are given in the Corollary in Section (v) of this chapter.

(ii) Survey of possibly place-dependent solutions

The first existence theorem for solutions that may depend upon place was obtained by GRAD[1]. He supposed the body force to be null, and he worked only with Maxwellian pseudomolecules. Assuming the initial value F_0 to be bounded by a uniform Maxwellian density F_U, he used (V.12) to establish the existence and uniqueness of weak solutions in a finite interval of time. He proved also that these were mild solutions, but he did not show that they were non-negative.

Subsequently, GRAD[2] obtained some more conclusive results. Restricting attention again to null body force, he considered the full class of molecules with

---

[1] GRAD [1958, §20].
[2] GRAD [1965, 2, §3]. GRAD [1965, 1, §7] proved a similar result for solutions which are periodic in space by reducing the initial-value problem to an initial-boundary-value problem in which the gas is confined in a rectangular box with ideally smooth walls (cf. Eq. (XI.19)).

a *cut-off hard potential*[3], which he defined by the conditions

$$\mathbb{S}(\theta, w) < \alpha(w + w^{\delta-1}) \cos \theta, \quad 0 \leq \theta \leq \tfrac{1}{2}\pi, \tag{XX.5}$$

$$\int_0^{\frac{1}{2}\pi} \mathbb{S}(\theta, w) \sin \theta \, d\theta > \frac{\beta w}{1 + w},$$

for constants α, β, and δ such that $\alpha > 0$, $\beta > 0$, and $0 < \delta < 1$. He looked for solutions F of the Maxwell–Boltzmann equation which were close to a uniform Maxwellian density F_U, close in the sense that $\|F - F_U\|$ was small, the norm $\|\ \|$ being defined by

$$\|F\| \equiv \max_{0 \leq s \leq 3} \max_{v \in \mathscr{V}} (1 + v^2)^{\frac{3}{2}} \left\{ \int_{\mathscr{E}} F_U^{-1} |\partial_x^s F|^2 \, dx \right\}^{\frac{1}{2}}. \tag{XX.6}$$

He showed that there were positive constants σ_1 and σ_2 such that for each smooth function F_0 satisfying the conditions $F_0 \geq 0$ and $\|F_0 - F_U\| < \infty$ a unique, smooth, non-negative solution F existed when

$$0 \leq t - t_0 \leq \sigma_1 \|F_0 - F_U\|^{-1} - \sigma_2.$$

Clearly we can make this interval of existence as long as we wish simply by choosing F_0 sufficiently close to the uniform Maxwellian density F_U. In Chapter XXI we shall see that spatially homogeneous solutions generally exist for all time. GRAD's result supports this fact since it says roughly that a solution which is almost spatially homogeneous and almost Maxwellian exists for a long time.

In regard to GRAD's theorem it is natural to ask whether the interval of existence need be finite. That is, is the finiteness a product of the method of proof, or is it inherent to solutions? This question has been answered recently by UKAI[4]. By extending the analysis initiated by GRAD, he proved that the F corresponding to an F_0 sufficiently close to F_U exists for all time and tends to F_U as $t \to \infty$. This is the strongest trend to equilibrium yet established. More recently, some further extensions and generalizations have been obtained by SHIZUTA[5] and NISHIDA & IMAI[6].

---

[3] Even though (XX.5) contains no explicit reference to molecules with a cut-off, it can hold only for these. In order to see why this is so, let us recall the definition (VII.8) of $\mathbb{S}$ and consider the simple case in which R is an increasing function of θ. Ideal spheres and inverse $\Bbbk^{\text{th}}$-power molecules both have this property. It follows, then, from (VII.8) that

$$\int_0^\theta \sin \sigma \, \mathbb{S}(\sigma, w) \, d\sigma = \tfrac{1}{2} w R^2(\theta, w).$$

If the molecules do not have a cut-off, by (VI.83)$_1$ the right-hand side approaches ∞ as $\theta \to \tfrac{1}{2}\pi$. We conclude that $\theta \mapsto \mathbb{S}(\theta, w)$ is unbounded for each w, and this possibility is clearly ruled out by (XX.5)$_1$. GRAD has remarked, however, that (XX.5) is satisfied by a broad class of molecules with a cut-off, a class which includes ideal spheres and molecules whose intermolecular force *within their spheres of action* is that of an inverse $\Bbbk^{\text{th}}$-power molecule, $\Bbbk \geq 5$.

[4] UKAI [1974].
[5] SHIZUTA [1979, *1* & *2*].
[6] NISHIDA & IMAI [1977].

UKAI, SHIZUTA, and NISHIDA & IMAI have all used the same general method of proof, the beginnings of which appear in the theorems of GRAD. We shall not present the details of this method here, for it involves a deep study of an appropriate linearized problem, and in this book we have chosen not to take up linearizations. However, because of the extent and breadth of these results, and their subsequent importance in the development of existence theory, we describe briefly in general terms how such proofs proceed. One begins by setting $F = F_U + F_U^{\frac{1}{2}} G$ in (IX.1), thereby replacing the problem for F by an equivalent one for G. Note that $F \to F_U$ if and only if $G \to 0$ as $t \to \infty$. The linearized problem, that obtained by casting away all terms that are not linear in G, is analysed using the theory of linear semigroups, and a global existence and uniqueness theorem is proved for it, for a broad class of initial values G_0. Thereafter a detailed spectral analysis of this linearized problem reveals that the solution G decays to 0 as $t \to \infty$, and certain estimates of the rate of decay may be derived. These estimates vary from one proof to the next, but one which will serve for these remarks is

$$\|G(t)\| \leq e^{-\nu(t-t_0)} \|G_0\|, \tag{XX.7}$$

ν being a positive constant and $\|\ \|$ being some norm on functions of $\mathbf{x}$ and $\mathbf{v}$. In order now to treat the non-linear problem for G, a space of the type

$$\mathscr{F} \equiv \left\{ G \,\Big|\, \sup_{t \geq t_0} e^{\nu(t-t_0)} \|G(t)\| < \infty \right\} \tag{XX.8}$$

is introduced. The elements of $\mathscr{F}$ are defined whenever $t \geq t_0$, and each approaches 0 as $t \to \infty$. By (XX.7) $\mathscr{F}$ contains the solutions of the linearized problem. Considering the non-linear problem as a perturbation of the linear one, we may use perturbation techniques to prove the existence and uniqueness of solutions lying in $\mathscr{F}$ for the full problem for G, at least in some neighborhood of $G = 0$. By the choice of $\mathscr{F}$ these solutions automatically exist for all time, they tend to equilibrium, and the decay estimate (XX.7) is valid.

Working in a somewhat different direction, GLIKSON[7] obtained a very general local existence theorem, one whose beginnings may be found in GRAD's first theorem. Using the integral equation (XX.3), GLIKSON was able to treat the general initial-value problem for the complete class of molecules with a cut-off and for a large class of body forces **b**. He proved the existence and uniqueness of weak solutions in all of space and in a finite interval of time. His solutions are non-negative and continuous, and, as we shall show later, they are actually mild solutions. With respect to smoothness, however, the results for cut-off hard potentials outlined earlier are the strongest yet obtained, being the *only* ones presently available that deliver possibly place-dependent solutions of the Maxwell–Boltzmann equation itself. GLIKSON's work is interesting not only for the theorem which it delivers but also for the method of proof. On a theoretical level it is nothing more than an application of one of the most elementary results from functional analysis: the principle of contraction mappings. Practically, however, it is a good example of the difficulties which sometimes confront him who would apply that principle to a concrete problem. The remainder of this chapter is devoted to a precise statement of GLIKSON's theorem and the details of its proof.

---

[7] GLIKSON [1972, *1* & *2*]. We are indebted to Professor GLIKSON for his comments on a preliminary draught of this chapter.

(iii) A class of body forces

We begin by defining the class of body forces to be considered. Let $\mathscr{D}$ be the set of functions

$$\mathbf{b} = \mathbf{b}(\mathbf{x}) \qquad\qquad (\text{III.21})_r$$

such that

(a) the initial-value problem described by (III.22) and (III.23) has a unique solution[8] for all $\mathbf{x}$ and $\mathbf{v}$ and for all s in some interval $[t_0, t_\mathbf{b})$, t_0 being the fixed initial time for the initial-value problem.

(b) $\mathbf{b}$ has a potential U,

$$\mathbf{b} = -\mathrm{grad}\, U, \qquad U = U(\mathbf{x}), \qquad\qquad (\text{II.15})_r$$

and this potential is bounded from below; since U is determined only to within an additive constant, we may take it as having a positive lower bound:

$$\inf_{\mathbf{x}\in\mathscr{E}} U(\mathbf{x}) > 0; \qquad\qquad (\text{XX.9})$$

in what follows we shall assume always that U is so selected.

In Chapter III, where we first introduced the retrogressors $\mathbf{r}_s$ and $\mathfrak{R}_s$, we assumed the body force $\mathbf{b}$ to be such that each was defined for all s. In this chapter we shall weaken that assumption by considering any $\mathbf{b} \in \mathscr{D}$. From (III.25) we can see that $\mathbf{r}_{t-s}(t, \mathbf{x}, \mathbf{v})$ is well defined when $t_0 \leq s < t_\mathbf{b}$, and so the values $\mathsf{J}F(t, \mathbf{x}, \mathbf{v})$ are well defined when $t_0 \leq t < t_\mathbf{b}$. Henceforth we shall consider only times in this interval, and it is only for such times that we shall seek a solution of (XX.3).

(iv) Preliminary estimates

If

$$E_0 \equiv \alpha_0(v^2 + 2U), \qquad\qquad (\text{XX.10})$$

α_0 being a positive constant, then E_0 is proportional to the energy of a molecule the motion of which is governed by the equations (III.22) and (II.15). According to classical dynamics this energy is constant along the trajectory of a molecule subject to the force $\mathbf{b}$, and therefore E_0, considered as a function of t,

---

[8] Examples are furnished by those functions $\mathbf{b}$ which are uniformly Lipschitz-continuous in all of space. This follows by applying to each bounded neighborhood of a fixed place $\mathbf{x}$ the standard existence and uniqueness theorem for ordinary differential equations, *e.g.* Chapter II, Theorem 11, of HARTMAN's book cited in Footnote 1 to Chapter XV.

x, and v, is unaffected by retrogression. This same result holds for any function G of E_0 alone:

$$\Re_s G(E_0) = G(E_0). \tag{XX.11}$$

Let us set

$$E \equiv E_0[1 - \beta(t - t_0)], \qquad \alpha \equiv \alpha_0[1 - \beta(t - t_0)], \tag{XX.12}$$

β being a positive constant. Clearly E_0 and α_0 are the values of E and α, respectively, at the initial time t_0. Let t_1 be so chosen that

$$\alpha(t) > 0 \quad \text{if} \quad t_0 \leq t < t_1 \leq t_b. \tag{XX.13}$$

Henceforth we shall consider only times in the interval $[t_0, t_1]$.

The functions e^{-E_0} and e^{-E} will be central to our discussion, and so we examine now some of their properties. From the definition (VII.2) of $\mathbb{C}$ and the law of conservation of energy (VI.2) it is plain that these two functions are unaffected by collisions:

$$\mathbb{C}e^{-E_0} = \mathbb{C}e^{-E} = 0. \tag{XX.14}$$

In view of the decomposition (VII.5) of $\mathbb{C}$, an equivalent statement of these same properties is

$$\mathbb{P}e^{-E_0} = e^{-E_0}\mathbb{F}e^{-E_0},$$
$$\mathbb{P}e^{-E} = e^{-E}\mathbb{F}e^{-E}. \tag{XX.15}$$

There are certain bounds on $\mathbb{F}e^{-E}$ and related quantities which will be useful later. Let us note first from (XX.9) that E_0 is positive:

$$E_0 \geq 2\alpha_0 \inf_{\mathbf{x} \in \mathscr{E}} U(\mathbf{x}) > 0. \tag{XX.16}$$

Let us set

$$\inf U \equiv \inf_{\mathbf{x} \in \mathscr{E}} U(\mathbf{x}),$$

$$\inf \alpha \equiv \inf_{t_0 \leq t \leq t_1} \alpha(t) = \alpha(t_1), \tag{XX.17}$$

$$a \equiv \int_{\mathscr{S}} 1 = \pi \mathrm{d}^2.$$

Lemma 1.

$$\mathbb{F}e^{-E} \leq a e^{-2 \inf U \times \inf \alpha} \Phi(E_0), \qquad t_0 \leq t \leq t_1, \tag{XX.18}$$

Φ being defined by

$$\Phi(E_0) \equiv \left(\frac{\pi}{\inf \alpha}\right)^{\frac{3}{2}} \left(\frac{E_0}{\alpha_0} - 2 \inf U\right)^{\frac{1}{2}} + \frac{2\pi}{(\inf \alpha)^2}. \tag{XX.19}$$

Proof. From the inequality $w \leq v + v_*$ and the definition (VII.4) of $\mathbb{F}$ we see that

$$\mathbb{F}e^{-E} \leq \int\int_{\mathcal{S}} (v + v_*)e^{-E_*},$$

$$= ae^{-2U\alpha}\left(v\int e^{-\alpha v_*^2} + \int v_* e^{-\alpha v_*^2}\right). \qquad \text{(XX.20)}$$

We can evaluate the two integrals on the right-hand side of this inequality:

$$\int e^{-\alpha v_*^2} = 4\pi \int_0^\infty v_*^2 e^{-\alpha v_*^2}\, dv_* = (\pi/\alpha)^{3/2},$$

$$\int v_* e^{-\alpha v_*^2} = 4\pi \int_0^\infty v_*^3 e^{-\alpha v_*^2}\, dv_* = 2\pi/\alpha^2. \qquad \text{(XX.21)}$$

Also, by solving (XX.10) for v we obtain

$$v = \left(\frac{E_0}{\alpha_0} - 2U\right)^{1/2}. \qquad \text{(XX.22)}$$

Putting (XX.21) and (XX.22) into (XX.20)$_2$, we get

$$\mathbb{F}e^{-E} \leq ae^{-2U\alpha}\left[\left(\frac{\pi}{\alpha}\right)^{3/2}\left(\frac{E_0}{\alpha_0} - 2U\right)^{1/2} + \frac{2\pi}{\alpha^2}\right]; \qquad \text{(XX.23)}$$

using (XX.16), we can replace U and α by their lower bounds and so obtain (XX.18). $\triangle$

Lemma 2. *There are positive functions B_k, $k = 1, 2, 3$, defined on $[t_0, t_1]$, such that for all $\mathbf{x}$ and $\mathbf{v}$ and for all $t \in [t_0, \tau]$, $t_0 < \tau \leq t_1$,*

$$e^E \int_{t_0}^t \mathfrak{R}_{t-s}(e^{-E}\mathbb{F}e^{-E})\, ds \leq B_1(\tau),$$

$$a(t - t_0)\Phi(E_0)e^E \int_{t_0}^t \mathfrak{R}_{t-s}(e^{-E}\mathbb{F}e^{-E})\, ds \leq B_2(\tau), \qquad \text{(XX.24)}$$

$$a(t - t_0)\Phi(E_0)e^{-\beta E_0(t - t_0)} \leq B_3(\tau),$$

and

$$B_k(\tau) \to 0 \quad \text{as} \quad \tau \to t_0, \quad k = 1, 2, 3. \qquad \text{(XX.25)}$$

Proof. The properties of the retrogressor described in (III.33) and (XX.11) allow us to apply Lemma 1 and so obtain the inequalities

$$\int_{t_0}^{t} \mathfrak{R}_{t-s}(e^{-E}\mathbb{F}e^{-E}) \, ds = \int_{t_0}^{t} (\mathfrak{R}_{t-s}e^{-E})(\mathfrak{R}_{t-s}\mathbb{F}e^{-E}) \, ds,$$

$$\leq \int_{t_0}^{t} (\mathfrak{R}_{t-s}e^{-E}) a e^{-2 \inf U \times \inf \alpha} \mathfrak{R}_{t-s}\Phi(E_0) \, ds,$$

$$= ae^{-2 \inf U \times \inf \alpha}\Phi(E_0) \int_{t_0}^{t} \mathfrak{R}_{t-s}e^{-E} \, ds. \quad (XX.26)$$

Since E_0 is unaffected by retrogression, (III.31) shows that

$$\mathfrak{R}_{t-s}(e^{\beta E_0(t-t_0)}) = e^{\beta E_0(s-t_0)}. \quad (XX.27)$$

Thus, using the properties (III.33) and (XX.11) of the retrogressor again, we deduce from $(XX.12)_1$ the relations

$$\int_{t_0}^{t} \mathfrak{R}_{t-s}e^{-E} \, ds = \int_{t_0}^{t} \mathfrak{R}_{t-s}(e^{-E_0}e^{\beta E_0(t-t_0)}) \, ds,$$

$$= e^{-E_0} \int_{t_0}^{t} e^{\beta E_0(s-t_0)} \, ds, \quad (XX.28)$$

$$= e^{-E_0} \frac{e^{\beta E_0(t-t_0)} - 1}{\beta E_0}.$$

By combining (XX.26) and (XX.28) we obtain the estimates

$$e^{E} \int_{t_0}^{t} \mathfrak{R}_{t-s}(e^{-E}\mathbb{F}e^{-E}) \, ds \leq ae^{-2 \inf U \times \inf \alpha} \Phi(E_0) \frac{1 - e^{-\beta E_0(\tau - t_0)}}{\beta E_0},$$

$$a(t - t_0)\Phi(E_0)e^{E} \int_{t_0}^{t} \mathfrak{R}_{t-s}(e^{-E}\mathbb{F}e^{-E}) \, ds \leq a^2 e^{-2 \inf U \times \inf \alpha} \frac{\Phi(E_0)^2}{\beta E_0}(\tau - t_0).$$

$$(XX.29)$$

It follows from (XX.19) and (XX.16) that $\Phi(E_0)/\sqrt{\beta E_0}$ is a bounded function of E_0, say

$$\Phi(E_0)/\sqrt{\beta E_0} \leq C_1, \quad (XX.30)$$

C_1 being independent of t. By comparing $(XX.24)_2$ with $(XX.29)_2$ and (XX.30) we see that a function B_2 of the desired kind is given by

$$B_2(\tau) \equiv a^2 e^{-2 \inf U \times \inf \alpha} C_1^2(\tau - t_0). \quad (XX.31)$$

We may also write

$$\Phi(E_0)\frac{1-e^{-\beta E_0(\tau-t_0)}}{\beta E_0} = \frac{\Phi(E_0)}{\sqrt{\beta E_0}}\frac{1-e^{-\beta E_0(\tau-t_0)}}{\sqrt{\beta E_0(\tau-t_0)}}\sqrt{\tau-t_0},$$

$$\leq C_1 C_2 \sqrt{\tau-t_0}, \qquad (XX.32)$$

C_2 being the supremum of the function $y \mapsto (1-e^{-y^2})/y$ when $y > 0$. Inspection of $(XX.24)_1$, $(XX.29)_1$, and $(XX.32)$ shows that a function B_1 of the desired kind is given by

$$B_1(\tau) \equiv ae^{-2\inf U \times \inf \alpha} C_1 C_2 \sqrt{\tau-t_0}. \qquad (XX.33)$$

Finally, from (XX.30)

$$a(t-t_0)\Phi(E_0)e^{-\beta E_0(t-t_0)} = a\frac{\Phi(E_0)}{\sqrt{\beta E_0}}\sqrt{\beta E_0(t-t_0)}e^{-\beta E_0(t-t_0)}\sqrt{t-t_0},$$

$$\leq aC_1 C_3 \sqrt{\tau-t_0}, \qquad (XX.34)$$

C_3 being the supremum of the function $y \mapsto ye^{-y^2}$ when $y \geq 0$. Thus a function B_3 of the desired kind is given by

$$B_3(\tau) \equiv aC_1 C_3 \sqrt{\tau-t_0}. \qquad \triangle \qquad (XX.35)$$

(v) *Glikson's theorem*

The proof of GLIKSON's existence theorem applies the principle of contraction mappings to (XX.3). As a first step towards this, we introduce function spaces for F_0 and F. Let ξ and δ be constants such that

$$0 < \xi - \delta < \xi. \qquad (XX.36)$$

If we use $\mathscr{C}(\mathscr{X})$ to denote the set of continuous functions on a set $\mathscr{X}$, then the function space $\mathscr{F}_0$ for the initial values will be the following:

$$\mathscr{F}_0 \equiv \{F_0 \in \mathscr{C}(\mathscr{E} \times \mathscr{V}) | 0 \leq F_0 \leq (\xi-\delta)e^{-E_0}\}. \qquad (XX.37)$$

That is, we shall consider only those non-negative initial values that are continuous and bounded by a specific Maxwellian density. For each τ in $(t_0, t_1]$ we introduce also the following definitions:

$$\mathscr{F}_\tau \equiv \{F \in \mathscr{C}([t_0, \tau] \times \mathscr{E} \times \mathscr{V}) | 0 \leq F \leq \xi e^{-E}\},$$

$$\rho_\tau(F_1, F_2) \equiv \sup_{[t_0,\tau] \times \mathscr{E} \times \mathscr{V}} (|F_1 - F_2|e^E), \qquad F_1, F_2 \in \mathscr{F}_\tau. \qquad (XX.38)$$

It is straightforward to show that the pair $(\mathscr{F}_\tau, \rho_\tau)$ forms a metric space for each such τ. Moreover, this metric space is complete since it is isometric, by the map $F \leftrightarrow Fe^E$, to the set of non-negative continuous functions on $[t_0, \tau] \times \mathscr{E} \times \mathscr{V}$ which are bounded by the constant ξ, this being a complete metric space also when equipped with the supremum norm.

The essential idea in the proof of the existence theorem is standard. We must show that there is a number ε in the interval $(0, 1)$ and that there are times t_2 and t_3, both greater than t_0, such that

$$\mathbb{J}: \mathscr{F}_\tau \to \mathscr{F}_\tau \quad \text{if} \quad t_0 \leq \tau \leq t_2, \tag{XX.39}$$

$$\rho_\tau(\mathbb{J}F_1, \mathbb{J}F_2) \leq \varepsilon \rho_\tau(F_1, F_2) \quad \text{for all} \quad F_1, F_2 \in \mathscr{F}_\tau \quad \text{if} \quad t_0 \leq \tau \leq t_3.$$

If $T \equiv \min(t_2, t_3)$, it follows from (XX.39) that $\mathbb{J}$ is a contraction mapping on the complete metric space $\mathscr{F}_T$, and by the principle of contraction mappings[9] we conclude that (XX.3) has a unique solution F in $\mathscr{F}_T$. We proceed now to the details.

Theorem[10]. *In a gas of molecules with a cut-off, to each body force $\mathbf{b} \in \mathscr{D}$ corresponds a number T such that there is a unique solution $F \in \mathscr{F}_T$ of (XX.3) for each $F_0 \in \mathscr{F}_0$.*

Proof. We show first that there is a number t_2 such that (XX.39)$_1$ is satisfied. If $F \in \mathscr{F}_\tau$ for some τ, then F is non-negative. Since F_0 is also non-negative, we may deduce from the definition (XX.4) of $\mathbb{J}$ the inequality

$$0 \leq \mathbb{J}F \leq \mathfrak{R}_{t-t_0}F_0 + \int_{t_0}^{t} \mathfrak{R}_{t-r}\mathbb{P}F \, dr. \tag{XX.40}$$

From the upper bound on F_0 indicated in (XX.37), and from the properties of the retrogressor which are set forth by (III.33) and (XX.11), we find that

$$\mathfrak{R}_{t-t_0}F_0 \leq (\xi - \delta)\mathfrak{R}_{t-t_0}e^{-E_0},$$
$$= (\xi - \delta)e^{-E_0}, \tag{XX.41}$$
$$\leq (\xi - \delta)e^{-E}.$$

According to (XX.15)$_2$ and the upper bound on F indicated in (XX.38)$_1$

$$\mathbb{P}F \leq \xi^2 \mathbb{P}e^{-E} = \xi^2 e^{-E} \mathbb{F}e^{-E}. \tag{XX.42}$$

---

[9] E.g. Chapter VI, §1, of LANG's book cited in Footnote 3 to Chapter XIX.

[10] GLIKSON [1972, *1*, Theorem 3] [1972, *2*, Theorem 3(a)]. In the interests of clarity and completeness we have reorganized GLIKSON's proof extensively, and we have filled in many details absent from his proof.

(v) GLIKSON'S THEOREM

Placing (XX.41)$_3$ and (XX.42)$_2$ into (XX.40), we obtain

$$0 \leq \mathsf{J}F \leq \left[\xi - \delta + \xi^2 e^E \int_{t_0}^t \mathfrak{R}_{t-\tau}(e^{-E}\mathbb{F}e^{-E})\, d\tau \right] e^{-E}. \quad (\text{XX}.43)$$

If we use Lemma 2 now, we can replace the preceding inequality by

$$0 \leq \mathsf{J}F \leq [\xi - \delta + \xi^2 B_1(\tau)]e^{-E}. \quad (\text{XX}.44)$$

By (XX.25) we know that $B_1(\tau) > 0$ and that $B_1(\tau) \to 0$ as $\tau \to t_0$, and so there is a number t_2 greater than t_0 such that

$$\xi - \delta + \xi^2 B_1(t_2) \leq \xi. \quad (\text{XX}.45)$$

Thus, if $t_0 \leq \tau \leq t_2$ and if $F \in \mathscr{F}_\tau$, then

$$0 \leq \mathsf{J}F \leq \xi e^{-E}. \quad (\text{XX}.46)$$

According to (XX.46) and the definition (XX.38)$_1$ of $\mathscr{F}_\tau$, $\mathsf{J}F$ will lie in $\mathscr{F}_\tau$ provided it be continuous. To prove that $\mathsf{J}F$ is continuous, let us recall that the solution $\chi[t, \mathbf{x}, \mathbf{v}]$ of (III.22) and (III.23) is unique. This means that[11] it is continuous in all of its arguments, and so from (III.31) we see that the retrogressor maps continuous functions onto continuous functions. Considering $\mathbb{F}F$ now, we see from (XX.38)$_1$, (XX.10), (XX.12), and the inequality $w \leq v + v_*$ that

$$wF_* \leq \xi w e^{-\alpha(v_*^2 + 2U)} \leq \xi(v + v_*)e^{-\inf \alpha \times (v_*^2 + 2\inf U)}. \quad (\text{XX}.47)$$

Since wF_* is continuous and since for any bounded set of velocities $\mathbf{v}$ the right-hand side of (XX.47)$_2$ is an integrable function of $\mathbf{v}_*$, it follows from a standard theorem[12] in measure theory that $\int \int_{\mathscr{S}} wF_*$ is also continuous, that is, $\mathbb{F}F$ is continuous. Similarly, since (VI.2) implies that $E' + E'_* = E + E_*$, for $wF'F'_*$ we have the bounds

$$\begin{aligned} wF'F'_* &\leq \xi^2 w e^{-(E' + E_*')}, \\ &= \xi^2 w e^{-(E + E_*)}, \\ &\leq \xi^2(v + v_*)e^{-\inf \alpha \times (v_*^2 + 2\inf U)}. \end{aligned} \quad (\text{XX}.48)$$

Thus by the same reasoning we see that $\mathbb{P}F$ is continuous. Finally, we can see from the definition (XX.4) that $\mathsf{J}F$ is also continuous, and this completes the proof of (XX.39)$_1$.

---

[11] E.g. Chapter V, Theorem 2.1 of HARTMAN's book cited in Footnote 1 to Chapter XV.
[12] E.g. Chapter XIV, Lemma 1, of LANG's book cited in Footnote 3 to Chapter XIX.

To prove (XX.39)$_2$, we begin with the estimates

$$|\mathsf{J}F_1 - \mathsf{J}F_2| \leq (\mathfrak{R}_{t-t_0}F_0)\left|\exp\left(-\int_{t_0}^t \mathfrak{R}_{t-s}\mathbb{F}F_1\,ds\right)\right.$$

$$\left.- \exp\left(-\int_{t_0}^t \mathfrak{R}_{t-s}\mathbb{F}F_2\,ds\right)\right|$$

$$+ \int_{t_0}^t \left|(\mathfrak{R}_{t-r}\mathbb{P}F_1)\exp\left(-\int_r^t \mathfrak{R}_{t-s}\mathbb{F}F_1\,ds\right)\right.$$

$$\left.- (\mathfrak{R}_{t-r}\mathbb{P}F_2)\exp\left(-\int_r^t \mathfrak{R}_{t-s}\mathbb{F}F_2\,ds\right)\right|dr,$$

$$\leq J_1 + J_2 + J_3, \tag{XX.49}$$

the quantities J_1, J_2, and J_3 being defined as follows:

$$J_1 \equiv (\mathfrak{R}_{t-t_0}F_0)\left|\exp\left(-\int_{t_0}^t \mathfrak{R}_{t-s}\mathbb{F}F_1\,ds\right) - \exp\left(-\int_{t_0}^t \mathfrak{R}_{t-s}\mathbb{F}F_2\,ds\right)\right|,$$

$$J_2 \equiv \int_{t_0}^t (\mathfrak{R}_{t-r}|\mathbb{P}F_1 - \mathbb{P}F_2|)\exp\left(-\int_r^t \mathfrak{R}_{t-s}\mathbb{F}F_1\,ds\right)dr, \tag{XX.50}$$

$$J_3 \equiv \int_{t_0}^t (\mathfrak{R}_{t-r}\mathbb{P}F_2)\left|\exp\left(-\int_r^t \mathfrak{R}_{t-s}\mathbb{F}F_1\,ds\right) - \exp\left(-\int_r^t \mathfrak{R}_{t-s}\mathbb{F}F_2\,ds\right)\right|dr.$$

These estimates follow from the identity $ab - cd = a(b - c) + c(a - d)$ and the properties of the retrogressor set forth in (III.33). Because $y \mapsto y + e^{-y}$ is an increasing function when $y \geq 0$,

$$|e^{-y_1} - e^{-y_2}| \leq |y_1 - y_2| \quad \text{if} \quad y_1, y_2 \geq 0. \tag{XX.51}$$

This inequality allows us to conclude that

$$J_1 \leq (\mathfrak{R}_{t-t_0}F_0)\int_{t_0}^t \mathfrak{R}_{t-s}\mathbb{F}|F_1 - F_2|\,ds,$$

$$J_2 \leq \int_{t_0}^t \mathfrak{R}_{t-r}|\mathbb{P}F_1 - \mathbb{P}F_2|\,dr, \tag{XX.52}$$

$$J_3 \leq \int_{t_0}^t (\mathfrak{R}_{t-r}\mathbb{P}F_2)\int_r^t \mathfrak{R}_{t-s}\mathbb{F}|F_1 - F_2|\,ds\,dr.$$

The preceding estimates are valid for quite general functions. If $F_0 \in \mathscr{F}_0$ and $F_1, F_2 \in \mathscr{F}_\tau$, then we have also the bounds

$$0 \leq F_0 \leq (\xi - \delta)e^{-E_0}, \quad 0 \leq F_1 \leq \xi e^{-E}, \quad 0 \leq F_2 \leq \xi e^{-E}, \tag{XX.53}$$

$$|F_1 - F_2| \leq \rho_\tau(F_1, F_2)e^{-E}.$$

Using these in (XX.52)$_{1,3}$, we obtain

$$J_1 \leq (\xi - \delta)e^{-E_0}\rho_\tau(F_1, F_2) \int_{t_0}^t \mathfrak{R}_{t-s}\mathbb{F}e^{-E}\, ds,$$
$$J_3 \leq \xi^2 \rho_\tau(F_1, F_2) \int_{t_0}^t \mathfrak{R}_{t-r}\mathbb{P}e^{-E}\, dr \int_{t_0}^t \mathfrak{R}_{t-s}\mathbb{F}e^{-E}\, ds. \quad (XX.54)$$

We can use the inequality proved in Lemma 1 to bound $\mathbb{F}e^{-E}$ here, and (XX.15)$_2$ may be used to eliminate the operator $\mathbb{P}$. When this is done, we get the new estimates

$$J_1 \leq (\xi - \delta) a e^{-2 \inf U \times \inf \alpha} \rho_\tau(F_1, F_2)(t - t_0)e^{-E_0}\Phi(E_0),$$
$$J_3 \leq \xi^2 a e^{-2 \inf U \times \inf \alpha} \rho_\tau(F_1, F_2)(t - t_0)\Phi(E_0) \int_{t_0}^t \mathfrak{R}_{t-s}(e^{-E}\mathbb{F}e^{-E})\, ds, \quad (XX.55)$$

and if we multiply each side of these by e^E and use the inequalities (XX.24)$_{2,3}$ from Lemma 2, we obtain

$$J_1 e^E \leq (\xi - \delta)e^{-2 \inf U \times \inf \alpha} B_3(\tau)\rho_\tau(F_1, F_2),$$
$$J_3 e^E \leq \xi^2 e^{-2 \inf U \times \inf \alpha} B_2(\tau)\rho_\tau(F_1, F_2). \quad (XX.56)$$

Now let us examine J_2. From (XX.53), (XX.15)$_2$, and the definition (VII.6) of $\mathbb{P}$ we have

$$|\mathbb{P}F_1 - \mathbb{P}F_2| \leq \int\int_{\mathscr{S}} w|F'_1 F'_{1*} - F'_2 F'_{2*}|,$$
$$\leq \int\int_{\mathscr{S}} w(F'_1|F'_{1*} - F'_{2*}| + F'_{2*}|F'_1 - F'_2|),$$
$$\leq 2\xi \rho_\tau(F_1, F_2) \int\int_{\mathscr{S}} w e^{-E'}e^{-E'_*}, \quad (XX.57)$$
$$= 2\xi \rho_\tau(F_1, F_2)\mathbb{P}e^{-E},$$
$$= 2\xi \rho_\tau(F_1, F_2)e^{-E}\mathbb{F}e^{-E}.$$

If we place this into (XX.52)$_2$, we get

$$J_2 \leq 2\xi \rho_\tau(F_1, F_2) \int_{t_0}^t \mathfrak{R}_{t-r}(e^{-E}\mathbb{F}e^{-E})\, dr; \quad (XX.58)$$

multiplying each side by e^E and then using (XX.24)$_1$, we obtain

$$J_2 e^E \leq 2\xi B_1(\tau)\rho_\tau(F_1, F_2). \quad (XX.59)$$

It follows now from (XX.49)$_2$, (XX.56), and (XX.59) that

$$|\mathsf{J}F_1 - \mathsf{J}F_2|e^E \leq \{2\xi B_1(\tau) + e^{-2 \inf U \times \inf \alpha}[(\xi - \delta)B_3(\tau) + \xi^2 B_2(\tau)]\}\rho_\tau(F_1, F_2). \quad (XX.60)$$

In view of (XX.25) the coefficient of $\rho_\tau(F_1, F_2)$ on the right-hand side here tends to zero as $\tau \to t_0$. Thus for each ε in the interval $(0, 1)$ there is a number t_3 greater than t_0 such that this coefficient is less than ε when $t_0 \leq \tau \leq t_3$. Therefore

$$|\mathsf{J}F_1 - \mathsf{J}F_2|e^E \leq \varepsilon\rho_\tau(F_1, F_2) \qquad \text{when } t_0 \leq \tau \leq t_3. \qquad (\text{XX.61})$$

If we take the supremum now on each side over all x and v and over all $t \in [t_0, \tau]$, we obtain (XX.39)$_2$.

Finally, if we set $T = \min(t_2, t_3)$ and apply the principle of contraction mappings to $\mathsf{J} \colon \mathscr{F}_T \to \mathscr{F}_T$, we complete the proof. △

With regard to regularity of GLIKSON's solutions, we have the

Corollary. *F is a mild solution of the Maxwell–Boltzmann equation:*

$$\mathfrak{d}F = \mathbb{C}F. \qquad (\text{V}.14)_r$$

This corollary is an immediate consequence of (XX.3), (VII.5), and the identity

$$\mathfrak{d}\mathsf{J}G = -(\mathbb{F}G)\mathsf{J}G + \mathbb{P}G, \qquad (\text{XX}.62)$$

which is valid for all G in $\mathscr{F}_\tau$. To prove (XX.62), let us recall that in the proof of the theorem we have shown that for any $G \in \mathscr{F}_\tau$, $\mathbb{F}G$ and $\mathbb{P}G$ are continuous. By (III.31), then, the function $s \mapsto \mathfrak{R}_{t-s}\mathbb{F}G$ is continuous, and so we may replace G by $\mathbb{F}G$ in (V.8) to obtain

$$\mathfrak{d}\int_r^t \mathfrak{R}_{t-s}\mathbb{F}G \, ds = \mathbb{F}G. \qquad (\text{XX}.63)$$

Because $\mathbb{F}G$ and $\mathbb{P}G$ are continuous, we see from (XX.63) that the function $r \mapsto (\mathfrak{R}_{t-r}\mathbb{P}G)\exp(-\int_r^t \mathfrak{R}_{t-s}\mathbb{F}G \, ds)$ is continuous, and for each fixed r it is differentiable along extrinsic trajectories. Thus we may denote this function by H_r, apply the identity (V.6), and so obtain

$$\mathfrak{d}\int_{t_0}^t (\mathfrak{R}_{t-r}\mathbb{P}G)\exp\left(-\int_r^t \mathfrak{R}_{t-s}\mathbb{F}G \, ds\right) dr$$

$$= (\mathfrak{R}_{t-t}\mathbb{P}G)\exp\left(-\int_t^t \mathfrak{R}_{t-s}\mathbb{F}G \, ds\right) \qquad (\text{XX}.64)$$

$$- \int_{t_0}^t (\mathfrak{R}_{t-r}\mathbb{P}G)\exp\left(-\int_r^t \mathfrak{R}_{t-s}\mathbb{F}G \, ds\right) \mathfrak{d}\int_r^t \mathfrak{R}_{t-u}\mathbb{F}G \, du \, dr,$$

$$= \mathbb{P}G - (\mathbb{F}G)\int_{t_0}^t (\mathfrak{R}_{t-r}\mathbb{P}G)\exp\left(-\int_r^t \mathfrak{R}_{t-s}\mathbb{F}G \, ds\right) dr.$$

(v) GLIKSON'S THEOREM

We conclude from (V.3), (XX.63), and (XX.64) that $\mathsf{J}G$ itself is differentiable along the trajectories, and that its mild derivative is given by

$$\begin{aligned}
\mathsf{d}\mathsf{J}G = {} & (\mathsf{d}\mathfrak{R}_{t-t_0}F_0)\exp\left(-\int_{t_0}^{t}\mathfrak{R}_{t-s}\mathsf{F}G\,ds\right) \\
& - (\mathfrak{R}_{t-t_0}F_0)\exp\left(-\int_{t_0}^{t}\mathfrak{R}_{t-s}\mathsf{F}G\,ds\right)\mathsf{d}\int_{t_0}^{t}\mathfrak{R}_{t-u}\mathsf{F}G\,du \\
& + \mathsf{d}\int_{t_0}^{t}(\mathfrak{R}_{t-r}\mathbb{P}G)\exp\left(-\int_{r}^{t}\mathfrak{R}_{t-s}\mathsf{F}G\,ds\right)dr, \qquad\text{(XX.65)} \\
= {} & -(\mathsf{F}G)\bigg[(\mathfrak{R}_{t-t_0}F_0)\exp\left(-\int_{t_0}^{t}\mathfrak{R}_{t-s}\mathsf{F}G\,ds\right) \\
& + \int_{t_0}^{t}(\mathfrak{R}_{t-r}\mathbb{P}G)\exp\left(-\int_{r}^{t}\mathfrak{R}_{t-s}\mathsf{F}G\,ds\right)dr\bigg] + \mathbb{P}G,
\end{aligned}$$

this being nothing more than (XX.62). △

Chapter XXI

Existence Theory for the General Initial-value Problem. Part III: Spatially Homogeneous Solutions for Molecules with a Cut-off

(i) Survey of spatially homogeneous solutions

We turn now to the initial-value problem for the equation

$$\partial_t F = \mathbb{C} F, \qquad \text{(XXI.1)}$$

to which the Maxwell–Boltzmann equation reduces when $\mathbf{b} = 0$ and $\varGamma$ is assumed independent of place. Such solutions are of no interest for the description of gas flows because they all correspond to gross free equilibrium. Indeed, as we have shown in Case 1, Section (i) of Chapter XI, for all these solutions

$$\rho = \text{const.}, \qquad \mathbf{u} = \mathbf{0}, \qquad \varepsilon = \text{const.}, \qquad \text{(XXI.2)}$$

provided that these quantities exist and be differentiable. Recalling the definitions of these quantities as moments of F, we can write (XXI.2) in the form

$$\int F = \int F_0, \qquad \int F\mathbf{v} = \int F_0 \mathbf{v}, \qquad \int Fv^2 = \int F_0 v^2, \qquad \text{(XXI.3)}$$

F_0 being the initial value of F. That is, F and F_0 have the same principal moment. In view of the Boltzmann–Gronwall theorem, we may express this fact also in the form

$$\int SF = \int SF_0 \qquad \text{for all summational invariants } S. \qquad \text{(XXI.4)}$$

The interest of the solutions independent of place lies in the light they cast upon the difference between gross equilibrium and kinetic equilibrium. Recalling that the principal moment of a Maxwellian density defines that density uniquely, we can see one thing at once from (XXI.3): The principal moment of any solution of (XXI.1) being at all times the principal moment of one and only one uniform Maxwellian density F_U, *there is one and only one such F_U that F could approach as $t \to \infty$.* Therefore, with a knowledge of no more than the initial value F_0 we know exactly which Maxwellian density a solution will approach *if it approaches one at all*. This fact helps to explain why it has been possible in some cases to establish rigorously and estimate an ultra-narrow trend to equilibrium.

When it comes to rigorous work, of course we cannot use the relations (XXI.3) straight off, for they have been shown to follow from the assumption that a smooth solution exists. Rather, (XXI.3) must be proved as a by-product of an existence theorem.

Henceforth in this chapter inequalities involving measurable functions will be understood to hold "almost everywhere"; of course, when the functions in question are continuous, this restriction may be dropped. Similarly, "non-negative" shall mean "non-negative almost everywhere".

The earliest results concerning existence of spatially homogeneous solutions are due to CARLEMAN[1]. He considered a gas consisting of ideal spheres and looked for solutions which depended upon **v** only through its magnitude v. By choosing continuous initial values satisfying the conditions

$$0 \leq F_0(v) \leq \frac{\alpha}{(1+v)^\kappa}, \qquad \kappa > 6, \qquad (\text{XXI.5})$$

for some constant α, he proved that a solution of (XXI.1) existed for all time and was unique. His solutions are continuous functions of v; they, too, satisfy the inequality (XXI.5), but with a different constant α. Since $(1+v)^{-\kappa} \in \mathscr{L}^1(\mathscr{V})$ if $\kappa \geq 4$, we can see in addition from (XXI.5) that the principal moments of CARLEMAN's solutions exist for all time. Finally, CARLEMAN proved a rigorous ultra-narrow H-theorem and established the trend to equilibrium for his solutions in the sense of uniform convergence.

Subsequent work on the initial-value problem sought to establish existence and uniqueness for a class of initial values and solutions broader than those considered by CARLEMAN. MORGENSTERN[2], improving certain partial results due to WILD[3], proved an existence theorem for a gas consisting of Maxwellian pseudomolecules. His results are actually valid for the much broader class of all pseudomolecules. For the initial value F_0 he took any non-negative function in $\mathscr{L}^1(\mathscr{V})$, and he proved that a non-negative weak solution existed for all time and was unique. We apply the term "weak" because he considered an integral equation which is formally equivalent to (XXI.1). MORGENSTERN's proof is elegant, clear, and brief. It has exerted influence on many later studies.

---

[1] CARLEMAN [1933].
[2] MORGENSTERN [1954].
[3] WILD [1951].

The third existence theorem for spatially homogeneous solutions is that of TRUESDELL for the equations of moments of a gas of Maxwellian molecules. We have presented it in Chapter XVIII.

The results of CARLEMAN, MORGENSTERN, and TRUESDELL are quite restrictive in that they apply only to special types of molecules. The next existence theorem is due to POVZNER[4], and in this respect his results are much broader. POVZNER considered the full class of molecules with a cut-off and proved the existence and uniqueness of a non-negative weak solution for all time, provided only that the initial value F_0 be non-negative and be such that $(1 + v)^4 F_0 \in \mathscr{L}^1(\mathscr{V})$.

While POVZNER's theorem goes a long way toward solving the initial-value problem for (XXI.1), even it is restrictive in certain regards. First, it delivers only weak solutions. Secondly, the requirement $(1 + v)^4 F_0 \in \mathscr{L}^1(\mathscr{V})$ implies that the fields ρ, $\mathbf{u}$, ε, $\mathbf{P}$, $^3\mathbf{M}$, and $^4\mathbf{M}$ must exist initially, while for the purposes of describing a gas flow by means of the kinetic theory it is more natural to require only that ρ, $\mathbf{u}$, ε, and possibly also h exist initially. A decade later ARKERYD[5] in an extensive study of existence theory for spatially homogeneous solutions came close to removing these drawbacks. We may state his main results as follows:

Theorem 1. *In a gas of molecules with a cut-off, if*

$$F_0 \geq 0, \quad F_0 \log F_0 \in \mathscr{L}^1(\mathscr{V}), \qquad (XXI.6)_{1,2}$$

and if for some number κ, $\kappa \geq 2$,

$$(1 + v)^\kappa F_0 \in \mathscr{L}^1(\mathscr{V}), \qquad (XXI.6)_3$$

then there is a non-negative solution of

$$\frac{dF}{dt} = \mathbb{C}F, \quad t > 0, \quad F\Big|_{t=0} = F_0, \qquad (XXI.7)$$

such that

$$(1 + v)^\kappa F \in \mathscr{L}^1(\mathscr{V}) \qquad (XXI.8)$$

and $(XXI.3)_{1,2}$ are satisfied. If $\kappa > 2$, then $(XXI.3)_3$ is also satisfied. If $\kappa \geq 4$, then the solution is unique, and $(XXI.6)_2$ may be omitted from the hypotheses for existence. If in addition $(XXI.6)_2$ is satisfied, then

$$F \log F \in \mathscr{L}^1(\mathscr{V}), \qquad (XXI.9)$$

and the ultra-narrow H-theorem is valid. If $\kappa > 4$, then the ultra-narrow trend to equilibrium holds in the sense of weak convergence in $\mathscr{L}^1(\mathscr{V})$.

---

[4] POVZNER [1962].
[5] ARKERYD [1972, *1* & *2*].

At approximately the time when ARKERYD's results appeared, DI BLASIO[6] proved a new existence theorem for pseudomolecules. Though her results extend somewhat those of MORGENSTERN, their main importance lies in a subsequent study[7] in which DI BLASIO generalized her analysis to the full class of molecules with a cut-off. For initial values F_0 satisfying (XXI.6)$_{1,3}$ when $\kappa = 4$, he proved that there is a unique, non-negative, differentiable solution of (XXI.7), and that it satisfies (XXI.3)$_{1,3}$. While her results are superseded by those of ARKERYD, her method of proof has independent interest since it draws upon results from the relatively new field of non-linear semigroups.

ARKERYD has given two different proofs of existence, one supposing that all three conditions (XXI.6) are satisfied as stated, the other, only that the first and third hold but that κ is at least as large as 4. In the remainder of this chapter we present the details of the second of these two proofs. It is constructive in nature, and with it we shall be able to present also proofs of the ultra-narrow H-theorem and the ultra-narrow trend to equilibrium.

As the notation in (XXI.7)$_1$ suggests, ARKERYD's theorem delivers solutions as $\mathscr{L}^1(\mathscr{V})$-valued functions of time:

$$F: \mathscr{R}^+ \to \mathscr{L}^1(\mathscr{V}), \qquad (\text{XXI.10})$$

$\mathscr{R}^+$ being the set of non-negative real numbers. In contrast, (XXI.1) is a partial differential equation for a function

$$F: \mathscr{R}^+ \times \mathscr{V} \to \mathscr{R}. \qquad (\text{XXI.11})$$

This distinction may seem unimportant, but in connection with regularity it leads to differentiation in two different senses. Indeed, for the function in (XXI.10) derivatives have meaning only in the sense of the calculus in Banach spaces, while for (XXI.11) derivatives with respect to time are interpreted as partial derivatives in the sense of classical differential calculus.

In the kinetic theory, use of the latter type of derivative is at present unavoidable in many calculations. We should like, then, to complete ARKERYD's results by showing that his solutions are also differentiable in the sense of (XXI.11). Though this will not be true for all functions F, it is true for the solutions delivered by ARKERYD's theorem. There are two ways in which we might prove this. The first would be to use the results already proved by ARKERYD and construct from them a direct proof[8]. Alternatively, we might adjust ARKERYD's proof itself so as to obtain directly functions of the type (XXI.11). In this book we shall take the second of these two courses. Our proof follows very closely the presentation of ARKERYD, the main difference being our use not of an $\mathscr{L}^1(\mathscr{V})$ setting but rather of one based upon measurable functions on $\mathscr{R}^+ \times \mathscr{V}$.

(ii) General results on existence and regularity

We begin by changing from the initial-value problem for (XXI.1) to a related problem. If we multiply (XXI.4) by αSF, α being a constant, and add the result to (XXI.1), we obtain the new equation

$$\partial_t F + \beta SF = \mathbb{Q}F, \qquad (\text{XXI.12})$$

---

[6] DI BLASIO [1972].
[7] DI BLASIO [1974].
[8] We are indebted to Professor ARKERYD for indicating how this may be done and for his valuable comments on an earlier draught of this chapter.

in which

$$\beta \equiv \alpha \int SF_0, \qquad \mathbb{Q}F \equiv \mathbb{C}F + \alpha SF \int SF. \qquad (\text{XXI.13})$$

Any solution of (XXI.1) will also be a solution of (XXI.12). Conversely, if F satisfies both (XXI.4) and (XXI.12) for some summational invariant S, then we can reverse the steps leading to (XXI.12) and conclude that F satisfies also (XXI.1). Therefore, if an existence theorem for (XXI.1) is to be based upon the initial-value problem for (XXI.12), the proof is not complete until we show that also (XXI.4) is satisfied. This observation will be important at a number of stages in our analysis.

From (XXI.12) we derive the following integral equation, which any solution of (XXI.12) and (XXI.7)$_3$ must satisfy:

$$F(t, \mathbf{v}) = F_0(\mathbf{v})e^{-\beta S(\mathbf{v})t} + \int_0^t e^{-\beta S(\mathbf{v})(t-s)} \mathbb{Q}F(s, \mathbf{v})\, ds. \qquad (\text{XXI.14})$$

Later we shall have occasion to consider several equations having the same structure as (XXI.14), and certain results about existence will be required for all of them before we can ultimately consider (XXI.12). Therefore we begin now by considering a slightly more general problem.

Let $\mathcal{M}(\mathcal{X})$ be the space of extended real-valued Lebesgue-measurable functions on a measurable set $\mathcal{X} \subset \mathcal{R}^N$, and let $\mathcal{M}^+(\mathcal{X})$ be the set of non-negative functions in $\mathcal{M}(\mathcal{X})$. We shall consider the integral equation

$$F(t, \mathbf{v}) = F_0(\mathbf{v})e^{-g(\mathbf{v})t} + \int_0^t e^{-g(\mathbf{v})(t-s)} AF(s, \mathbf{v})\, ds, \qquad (\text{XXI.15})$$

in which F_0, g, and A have the following properties:

(A1) $F_0, g \in \mathcal{M}^+(\mathcal{V})$,
(A2) $A: \mathcal{M}^+(\mathcal{R}^+ \times \mathcal{V}) \to \mathcal{M}^+(\mathcal{R}^+ \times \mathcal{V})$,
(A3) $F \geq G \geq 0 \Rightarrow AF \geq AG$,
(A4) If $\{F^n\}$ is an increasing sequence in $\mathcal{M}^+(\mathcal{R}^+ \times \mathcal{V})$ with pointwise limit F, then $AF^n \to AF$ pointwise as $n \to \infty$.

We shall denote (XXI.15) by the ordered triple (F_0, g, A); by convention we shall use this notation only for such F_0, g, and A as enjoy properties (A1)–(A4).

It is rather easy to show that there is a measurable solution of (F_0, g, A). One simply introduces an approximating sequence $\{F^n\}$ by means of the recurrence formulae

$$F^1 \equiv 0, \qquad F^{n+1} \equiv F_0 e^{-gt} + \int_0^t e^{-g(t-s)} AF^n\, ds, \qquad n \geq 1, \qquad (\text{XXI.16})$$

and then shows that this sequence converges to a solution.

Lemma 1. $\{F^n\}$ is an increasing sequence in $\mathcal{M}^+(\mathcal{R}^+ \times \mathcal{V})$, and it converges pointwise to a function $F \in \mathcal{M}^+(\mathcal{R}^+ \times \mathcal{V})$ which satisfies (F_0, g, A).

Proof. We shall use induction on n to show that $F^n \in \mathcal{M}^+(\mathcal{R}^+ \times \mathcal{V})$ if $n = 1, 2, \ldots$. Trivially $F^1 \in \mathcal{M}^+(\mathcal{R}^+ \times \mathcal{V})$. Suppose now that $F^n \in \mathcal{M}^+(\mathcal{R}^+ \times \mathcal{V})$. Property (A2) shows that $AF^n \in \mathcal{M}^+(\mathcal{R}^+ \times \mathcal{V})$. We can write (XXI.16)$_2$ in the form

$$F^{n+1} = F_0 e^{-gt} + \int_{\mathcal{R}} \chi(s, t) e^{-g(t-s)} AF^n \, ds, \qquad \text{(XXI.17)}$$

χ being the characteristic function of the set $\{(s, t) \mid 0 \leq s \leq t\}$. But the integrand on the right-hand side here is non-negative, and moreover being a product of measurable functions it is measurable in the product space of the variables $(t, s, \mathbf{v})$. By Tonelli's theorem[9] the integral of such a function with respect to s is a function in $\mathcal{M}^+(\mathcal{R}^+ \times \mathcal{V})$. It follows, then, from (XXI.17) that $F^{n+1} \in \mathcal{M}^+(\mathcal{R}^+ \times \mathcal{V})$, and this completes the induction.

From (XXI.16) we find that

$$F^2 - F^1 \geq 0, \qquad F^{n+1} - F^n = \int_0^t e^{-g(t-s)} (AF^n - AF^{n-1}) \, ds \qquad \text{if } n \geq 2. \qquad \text{(XXI.18)}$$

Thus (XXI.18)$_2$ and property (A3) show that for n given and fixed, $F^n \leq F^{n+1}$ if $F^{n-1} \leq F^n$. By (XXI.18)$_1$, then, induction shows that $F^n \leq F^{n+1}$, $n \geq 1$. Therefore $\{F^n\}$ is an increasing sequence in $\mathcal{M}^+(\mathcal{R}^+ \times \mathcal{V})$. By the monotone convergence theorem[10] the pointwise limit F of this sequence also lies in $\mathcal{M}^+(\mathcal{R}^+ \times \mathcal{V})$. If for fixed $\mathbf{v}$ we let $n \to \infty$ on each side of (XXI.16)$_2$, it follows from property (A4) and the monotone convergence theorem that F satisfies (F_0, g, A). △

The function F delivered by Lemma 1 will henceforth be called the *iterative solution* of (F_0, g, A). It has the important property expressed by the following

Corollary. If $G \in \mathcal{M}^+(\mathcal{R}^+ \times \mathcal{V})$ is any solution of (F_0, g, A), then

$$F \leq G. \qquad \text{(XXI.19)}$$

Proof. If we set $F = G$ in (XXI.15) and then subtract from it (XXI.16)$_2$, we find that

$$G - F^{n+1} = \int_0^t e^{-g(t-s)} (AG - AF^n) \, ds, \qquad n \geq 1. \qquad \text{(XXI.20)}$$

Because A is monotone in the sense (A3), (XXI.20) shows that $F^{n+1} \leq G$ if $F^n \leq G$. Since G is non-negative, $G - F^1 = G \geq 0$, and it follows by induction that $F^n \leq G$ if $n = 1, 2, \ldots$. △

---

[9] E.g. Chapter 12, Theorem 20, of ROYDEN's book cited in Footnote 1 to Chapter XIX.
[10] E.g. Theorem 1.26 of RUDIN's book cited in Footnote 15 to Chapter XI.

(ii) GENERAL PRELIMINARIES

We shall wish to compare different equations of the type (XXI.15), and so by the statement

$$(F_{01}, g_1, A_1) \leq (F_{02}, g_2, A_2), \tag{XXI.21}$$

we shall mean that

$$F_{01} \leq F_{02}, \quad g_1 \geq g_2, \quad A_1 G \leq A_2 G \quad \text{for all} \quad G \in \mathcal{M}^+(\mathcal{R}^+ \times \mathcal{V}). \tag{XXI.22}$$

The ordering (XXI.21) implies that the corresponding iterative solutions are also ordered:

Lemma 2. *If F_1 and F_2 are the iterative solutions of (F_{01}, g_1, A_1) and (F_{02}, g_2, A_2), respectively, then (XXI.21) implies that*

$$F_1 \leq F_2. \tag{XXI.23}$$

Proof. Let $\{F_1^n\}$ and $\{F_2^n\}$ be the sequences delivered by (XXI.16) for the two equations. We shall use induction on n to show that $F_1^n \leq F_2^n$ when $n \geq 1$, and (XXI.23) will then follow by letting n approach ∞. We note first that $F_1^1 \leq F_2^1$ since both F_1^1 and F_2^1 are zero. Assume now that $F_1^n \leq F_2^n$. From (XXI.16)$_2$ we find that

$$\begin{aligned}F_2^{n+1} - F_1^{n+1} &= F_{02} e^{-g_2 t} - F_{01} e^{-g_1 t} \\ &\quad + \int_0^t (e^{-g_2(t-s)} A_2 F_2^n - e^{-g_1(t-s)} A_1 F_1^n)\, ds, \\ &= (F_{02} - F_{01}) e^{-g_2 t} + F_{01}(e^{-g_2 t} - e^{-g_1 t}) \\ &\quad + \int_0^t \{(e^{-g_2(t-s)} - e^{-g_1(t-s)}) A_2 F_2^n \\ &\quad + e^{-g_1(t-s)}[(A_2 F_2^n - A_2 F_1^n) + (A_2 F_1^n - A_1 F_1^n)]\}\, ds.\end{aligned} \tag{XXI.24}$$

Of the five differences on the right-hand side here, the first three are non-negative in virtue of (XXI.22)$_{1,2}$. Also $A_2 F_1^n \geq A_1 F_1^n$ by (XXI.22)$_3$, and since $F_1^n \leq F_2^n$ and A_2 is monotone in the sense (A3), it follows that $A_2 F_1^n \leq A_2 F_2^n$. Hence the fourth and fifth terms in (XXI.24)$_2$ are non-negative, and so $F_1^{n+1} \leq F_2^{n+1}$. This completes the induction. △

We consider now sequences $\{(F_{0m}, g_m, A_m)\}$ of integral equations, and we shall use the notation

$$(F_{0m}, g_m, A_m) \to (F_0, g, A) \quad \text{as} \quad m \to \infty \tag{XXI.25}$$

as an abbreviation for

$$\left.\begin{aligned}F_{0m} &\to F_0 \\ g_m &\to g \\ A_m G &\to AG\end{aligned}\right\} \text{pointwise as } m \to \infty \quad \text{for all} \quad G \in \mathcal{M}^+(\mathcal{R}^+ \times \mathcal{V}). \tag{XXI.26}$$

In view of the ordering of equations defined by (XXI.21), it is clear what is meant by an increasing sequence of such equations. For these sequences we have the important result expressed by

Lemma 3. *Let $\{(F_{0m}, g_m, A_m)\}$ be an increasing sequence satisfying (XXI.25), and let F_m be the iterative solution of (F_{0m}, g_m, A_m). Then $\{F_m\}$ is an increasing sequence in $\mathcal{M}^+(\mathcal{R}^+ \times \mathcal{V})$, and it converges to a function $F \in \mathcal{M}^+(\mathcal{R}^+ \times \mathcal{V})$ that satisfies (F_0, g, A).*

Proof. Since

$$(F_{0m}, g_m, A_m) \leq (F_{0m+1}, g_{m+1}, A_{m+1}) \quad \text{when} \quad m \geq 1, \quad \text{(XXI.27)}$$

it follows from Lemma 2 that $\{F_m\}$ is an increasing sequence in $\mathcal{M}^+(\mathcal{R}^+ \times \mathcal{V})$. By the monotone convergence theorem the pointwise limit F of this sequence lies in $\mathcal{M}^+(\mathcal{R}^+ \times \mathcal{V})$. All that remains to be shown is that F satisfies (F_0, g, A). Now, F_m satisfies the integral equation (XXI.15):

$$F_m = F_{0m} e^{-g_m t} + \int_0^t e^{-g_m(t-s)} A_m F_m \, ds. \quad \text{(XXI.28)}$$

If we write

$$A_{m+1} F_{m+1} - A_m F_m = (A_{m+1} F_{m+1} - A_{m+1} F_m) + (A_{m+1} F_m - A_m F_m) \quad \text{(XXI.29)}$$

and note that each group of terms on the right-hand side is non-negative, the first being so because A_{m+1} is monotone and the second being so in virtue of (XXI.27) and (XXI.22)$_3$, we see that $\{A_m F_m\}$ is increasing. Therefore this sequence converges pointwise in $\mathcal{M}^+(\mathcal{R}^+ \times \mathcal{V})$, and we may easily show that its limit is AF. To do so, we use (XXI.22)$_3$ and the fact that A_m is monotone for each m to write

$$A_m F_n \leq A_m F_m \leq A_p F_m \quad \text{if} \quad n \leq m \leq p. \quad \text{(XXI.30)}$$

If we let p, m, and n approach ∞, in that order, then it follows from (XXI.26)$_3$ and property (A4) that

$$AF \leq \lim_{m \to \infty} A_m F_m \leq AF, \quad \text{(XXI.31)}$$

so that $\{A_m F_m\}$ is an increasing sequence converging pointwise to AF. Finally we let m approach ∞ in (XXI.28) and conclude from the monotone convergence theorem that F satisfies (F_0, g, A). △.

We now have two solutions of the equation (F_0, g, A) appearing in this lemma, one being the function F and the other being the iterative solution delivered by Lemma 1. It is a simple matter to show that these two solutions are equal:

Corollary. *F is the iterative solution of (F_0, g, A).*

(ii) GENERAL PRELIMINARIES

Proof. Since $\{(F_{0m}, g_m, A_m)\}$ is an increasing sequence,

$$(F_{0m}, g_m, A_m) \leq (F_0, g, A) \quad \text{if} \quad m \geq 1. \tag{XXI.32}$$

Therefore, if $\hat{F}$ is the iterative solution of (F_0, g, A), Lemma 2 shows that $F_m \leq \hat{F}$ if $m \geq 1$, and in the limit as $m \to \infty$ this becomes $F \leq \hat{F}$. But the corollary to Lemma 1 states that $\hat{F} \leq F$, and so $F = \hat{F}$. △

The preceding results concern measurable solutions of (F_0, g, A), and we shall go on now to discuss their regularity. For this we need a special type of continuity and differentiability. We shall say that $G \in \mathcal{M}^+(\mathcal{R}^+ \times \mathcal{V})$ is *t-continuous* if for almost all $\mathbf{v} \in \mathcal{V}$, $t \mapsto G(t, \mathbf{v})$ is continuous when $t \geq 0$. Similarly we shall say that G is *t-differentiable* if for almost all $\mathbf{v} \in \mathcal{V}$, $t \mapsto G(t, \mathbf{v})$ is differentiable when $t > 0$. We begin by giving conditions on a measurable solution of (F_0, g, A) sufficient that it be *t*-continuous.

Lemma 4. Let $F \in \mathcal{M}^+(\mathcal{R}^+ \times \mathcal{V})$ be any solution of (F_0, g, A), and suppose that $F_0 \in \mathcal{L}^1(\mathcal{V})$, that $g(\mathbf{v})$ is finite for all $\mathbf{v}$, and that $F(t, \cdot) \in \mathcal{L}^1(\mathcal{V})$ when $t > 0$. Then F is *t*-continuous.

Proof. The functions $F_0, F(1, \cdot), F(2, \cdot), F(3, \cdot), \ldots$ are all integrable. Therefore, if we let $\mathcal{N} \subset \mathcal{V}$ be the set of points $\mathbf{v}$ for which the values of all these functions are finite, $\mathcal{N}$ will differ from $\mathcal{V}$ by at most a set of measure zero. We shall show that if $\mathbf{v} \in \mathcal{N}$, then $t \mapsto F(t, \mathbf{v})$ is continuous when $t > 0$. Choose $\mathbf{v} \in \mathcal{N}$, and let it be fixed throughout the remainder of the proof. Since F satisfies (XXI.15), we see that to prove $t \mapsto F(t, \mathbf{v})$ is continuous, it suffices to show that $s \mapsto e^{g(\mathbf{v})s} AF(s, \mathbf{v})$ is integrable[11] on each finite interval $[0, T]$. Choose any integer N such that $T \leq N$. Then

$$\int_0^T e^{g(\mathbf{v})s} AF(s, \mathbf{v}) \, ds \leq \int_0^N e^{g(\mathbf{v})s} AF(s, \mathbf{v}) \, ds. \tag{XXI.33}$$

But if we set $t = N$ in (XXI.15), we find that

$$e^{g(\mathbf{v})N} F(N, \mathbf{v}) = F_0(\mathbf{v}) + \int_0^N e^{g(\mathbf{v})s} AF(s, \mathbf{v}) \, ds, \tag{XXI.34}$$

and a simple combination of (XXI.33) and (XXI.34) gives us

$$\int_0^T e^{g(\mathbf{v})s} AF(s, \mathbf{v}) \, ds \leq e^{g(\mathbf{v})N} F(N, \mathbf{v}) - F_0(\mathbf{v}). \tag{XXI.35}$$

Since $\mathbf{v} \in \mathcal{N}$ and $g(\mathbf{v})$ is finite, the right-hand side here is finite, and so $s \mapsto e^{g(\mathbf{v})s} AF(s, \mathbf{v})$ is integrable on $[0, T]$. △

---

[11] *E.g.* Chapter 5, Lemma 6, of ROYDEN's book cited in Footnote 1 to Chapter XIX.

Let us suppose now that F is a solution of (F_0, g, A) and also that AF is t-continuous. Then each expression on the right-hand side of (XXI.15) is t-differentiable, and hence F is t-differentiable and satisfies the differential equation

$$\partial_t F + gF = AF \qquad \text{(XXI.36)}$$

when $t > 0$. Since (XXI.36) is an equality between functions in $\mathscr{L}^1(\mathscr{V})$, it need only be satisfied for almost all $\mathbf{v} \in \mathscr{V}$. This is the reason that t-continuity and t-differentiability are natural concepts when discussing solutions of (XXI.36). What we have just shown is that to prove F t-differentiable, it suffices to prove AF t-continuous. Therefore we shall now examine conditions under which a solution of (F_0, g, A) has the latter property. It does not seem feasible to do so without descending to somewhat specific operators A. Thus, with a view toward our objective in the kinetic theory, we shall restrict our attention henceforth to operators A having the general form

$$AF \equiv \int\!\!\int_{\mathscr{S}} (K_1(\mathbf{v}, \mathbf{v}_*) F' F'_* + K_2(\mathbf{v}, \mathbf{v}_*) F F_*), \qquad \text{(XXI.37)}$$

K_1 and K_2 being non-negative measurable functions satisfying the following requirements:

(A5) K_1 is a collisional invariant:

$$K_1(\mathbf{v}', \mathbf{v}'_*) = K_1(\mathbf{v}, \mathbf{v}_*); \qquad \text{(XXI.38)}$$

(A6) there is a positive constant ξ and a non-negative integer n such that

$$K_J(\mathbf{v}, \mathbf{v}_*) \leq \xi(1 + v^2)^n (1 + v_*^2)^n, \qquad J = 1, 2. \qquad \text{(XXI.39)}$$

In order to be sure that the results we have obtained already for the equation (F_0, g, A) are valid when A has this special form, it is important to establish

Lemma 5. *Any operator A of the form* (XXI.37) *satisfying* (A5) *and* (A6) *satisfies also* (A2)–(A4).

Proof. Suppose that $F \in \mathscr{M}^+(\mathscr{R}^+ \times \mathscr{V})$. If

$$G(t, \mathbf{v}, \mathbf{v}_*) \equiv F(t, \mathbf{v}) F(t, \mathbf{v}_*), \qquad \text{(XXI.40)}$$

then $G \in \mathscr{M}^+(\mathscr{R}^+ \times \mathscr{V} \times \mathscr{V})$, and so, trivially, $G \in \mathscr{M}^+(\mathscr{R}^+ \times \mathscr{V} \times \mathscr{V} \times \mathscr{S})$. Using (XXI.40), we can write the integrand in (XXI.37) in the form

$$K_1 F' F'_* + K_2 F F_* = K_1 (G \circ \mathsf{E}) + K_2 G. \qquad \text{(XXI.41)}$$

Since the product of measurable functions is measurable and the composition of a measurable function with one which is piecewise continuously differentiable and whose inverse is also piecewise continuously differentiable is again a measurable function, we see that the left-hand member of (XXI.41) lies in

$\mathcal{M}^+(\mathcal{R}^+ \times \mathcal{V} \times \mathcal{V} \times \mathcal{S})$. If we integrate $K_1 F'F'_* + K_2 FF_*$ over $\mathcal{V} \times \mathcal{S}$, Tonelli's theorem shows that the resulting function, namely AF, lies in $\mathcal{M}^+(\mathcal{R}^+ \times \mathcal{V})$, and this completes the proof of property (A2).

If F and G both lie in $\mathcal{M}^+(\mathcal{R}^+ \times \mathcal{V})$, and moreover $G \leq F$, then

$$K_1 G'G'_* + K_2 GG_* \leq K_1 F'F'_* + K_2 FF_*. \tag{XXI.42}$$

Therefore $AG \leq AF$, and so property (A3) is satisfied.

Let $\{F^n\}$ be an increasing sequence in $\mathcal{M}^+(\mathcal{R}^+ \times \mathcal{V})$, and let $F \in \mathcal{M}^+(\mathcal{R}^+ \times \mathcal{V})$ be its pointwise limit. If we replace G by F^{n-1} and F by F^n in (XXI.42), we see that $\{K_1 F^{n'} F^{n'}_* + K_2 F^n F^n_*\}$ is an increasing sequence in $\mathcal{M}^+(\mathcal{R}^+ \times \mathcal{V} \times \mathcal{V} \times \mathcal{S})$. Thus, for each $(t, \mathbf{v}) \in \mathcal{R}^+ \times \mathcal{V}$, this same sequence is increasing in $\mathcal{M}^+(\mathcal{V} \times \mathcal{S})$ with pointwise limit $K_1 F'F'_* + K_2 FF_*$. By use of the monotone convergence theorem we see that

$$\int\int_{\mathcal{S}} (K_1 F^{n'} F^{n'}_* + K_2 F^n F^n_*) \to \int\int_{\mathcal{S}} (K_1 F'F'_* + K_2 FF_*) \quad \text{as} \quad n \to \infty \tag{XXI.43}$$

for each $(t, \mathbf{v}) \in \mathcal{R}^+ \times \mathcal{V}$. But this is nothing more than property (A4). △

Our final general result on regularity of solutions may be stated as follows.

Lemma 6. *Let $F \in \mathcal{M}^+(\mathcal{R}^+ \times \mathcal{V})$ be any solution of (F_0, q, A), and suppose A has the form* (XXI.37), *that it satisfies* (A5) *and* (A6), *and that $g(\mathbf{v})$ is finite for all $\mathbf{v}$. If for each positive T there is a constant C_T such that*

$$\|(1 + v^2)^{2n} F(t, \mathbf{v})\| \leq C_T \quad \text{if} \quad 0 \leq t \leq T, \tag{XXI.44}$$

then F is t-differentiable and satisfies (XXI.36).

Henceforth in this chapter $\| \ \|$ denotes the norm in $\mathcal{L}^1(\mathcal{V})$.

Proof. First we show that on each finite interval $[0, T]$, $(1 + v^2)^n F$ can be bounded by a non-negative function in $\mathcal{L}^1(\mathcal{V})$ which is independent of t. Since F satisfies (XXI.15),

$$F \leq F_0 + \int_0^t AF \, ds. \tag{XXI.45}$$

Therefore

$$(1 + v^2)^n F(t, \mathbf{v}) \leq G_T(\mathbf{v}) \quad \text{if} \quad 0 \leq t \leq T, \tag{XXI.46}$$

G_T being given by

$$G_T(\mathbf{v}) \equiv (1 + v^2)^n F_0(\mathbf{v}) + \int_0^T (1 + v^2)^n AF(s, \mathbf{v}) \, ds. \tag{XXI.47}$$

Clearly G_T is a bound of the type indicated above, provided only that it lie in $\mathscr{L}^1(\mathscr{V})$. In order to see that it does so, we calculate its norm:

$$\|G_T\| = \|(1+v^2)^n F_0\| + \int_0^T \left[\int (1+v^2)^n AF\right] ds, \qquad \text{(XXI.48)}$$

the integrals with respect to s and $\mathbf{v}$ on the right-hand side having been interchanged by appeal to Tonelli's theorem. By (XXI.37), (XXI.38), and (VI.70),

$$\int (1+v^2)^n AF = \iint\int_{\mathscr{S}} (1+v^2)^n K_1 F'F'_* + \iint\int_{\mathscr{S}} (1+v^2)^n K_2 FF_*,$$
$$\qquad \text{(XXI.49)}$$
$$= \iint\int_{\mathscr{S}} [(1+v'^2)^n K_1 + (1+v^2)^n K_2] FF_*.$$

Since, in view of (VI.2),

$$1+v'^2 \leq 1+v'^2 + v'^2_* = 1+v^2 + v^2_* \leq (1+v^2)(1+v^2_*), \quad \text{(XXI.50)}$$

we see from (XXI.39) that

$$(1+v'^2)^n K_1 + (1+v^2)^n K_2 \leq (1+v^2)^n (1+v^2_*)^n (K_1 + K_2),$$
$$\leq 2\xi(1+v^2)^{2n}(1+v^2_*)^{2n}. \qquad \text{(XXI.51)}$$

Placing this result into (XXI.49)$_2$, we obtain

$$\int (1+v^2)^n AF \leq 2\xi\pi d^2 \|(1+v^2)^{2n} F\|^2, \qquad \text{(XXI.52)}$$

and so from (XXI.44), (XXI.48), and (XXI.52),

$$\|G_T\| \leq \|(1+v^2)^n F_0\| + 2\xi T\pi d^2 C_T^2. \qquad \text{(XXI.53)}$$

Since $(1+v^2)^n \leq (1+v^2)^{2n}$, we see from (XXI.44) that $(1+v^2)^n F_0 \in \mathscr{L}^1(\mathscr{V})$, and so we conclude from (XXI.53) that $G_T \in \mathscr{L}^1(\mathscr{V})$.

Next we show that $K_1 F'F'_* + K_2 FF_*$, the integrand in (XXI.37), is bounded on each finite interval $[0, T]$ by a non-negative function in $\mathscr{L}^1(\mathscr{V} \times \mathscr{V} \times \mathscr{S})$ which is independent of t. Indeed, by (XXI.38), (XXI.39), and (XXI.46)

$$K_1(\mathbf{v}, \mathbf{v}_*) F'F'_* + K_2(\mathbf{v}, \mathbf{v}_*) FF_* = K_1(\mathbf{v}', \mathbf{v}'_*) F'F'_* + K_2(\mathbf{v}, \mathbf{v}_*) FF_*,$$
$$\leq \xi(1+v'^2)^n(1+v'^2_*)^n F'F'_*$$
$$\quad + \xi(1+v^2)^n(1+v^2_*)^n FF_*, \qquad \text{(XXI.54)}$$
$$\leq H_T \quad \text{if} \quad 0 \leq t \leq T,$$

H_T being the function

$$H_T \equiv \xi(G'_T G'_{T*} + G_T G_{T*}). \qquad \text{(XXI.55)}$$

Moreover,

$$\iint \int_{\mathscr{S}} H_T = 2\xi \pi \mathbb{d}^2 \|G_T\|^2 < \infty, \qquad (XXI.56)$$

and so $H_T \in \mathscr{L}^1(\mathscr{V} \times \mathscr{V} \times \mathscr{S})$.

Next we prove that AF is t-continuous. By (XXI.44) and Lemma 4, F is t-continuous. Moreover, by Fubini's theorem[12], $H_T(\mathbf{v}, \cdot, \cdot) \in \mathscr{L}^1(\mathscr{V} \times \mathscr{S})$ for almost all $\mathbf{v} \in \mathscr{V}$. Hence, for almost all $\mathbf{v} \in \mathscr{V}$, the integrand in (XXI.37) is a continuous function of t for almost all $(\mathbf{v}_*, \mathbf{r}^e) \in \mathscr{V} \times \mathscr{S}$, and it is bounded independently of t on $[0, T]$ by an integrable function. It follows from a theorem[13] in measure theory that AF is continuous in $[0, T]$ for almost all $\mathbf{v} \in \mathscr{V}$. Since T was chosen arbitrarily, AF is t-continuous.

Finally, since AF is t-continuous, the right-hand side of (XXI.15) is t-differentiable. This proves F to be t-differentiable. By differentiating each side of (XXI.15) with respect to t we obtain (XXI.36). △

(iii) A modified collisions operator and its properties

Let v be a positive constant, and define

$$\mathbb{C}_v F \equiv \iint \int_{\mathscr{S}} \min(v, w)(F' F'_* - F F_*). \qquad (XXI.57)$$

As this modified collisions operator will be of central importance in the proof of the existence theorem, we shall take a moment to describe some of its properties. Because $\mathbb{C}_v$ and $\mathbb{C}$ are so closely related in structure, we shall see that they have many properties in common.

Let us note first that if we replace w in (VII.15)$_2$ by any measurable function $k(\mathbf{v}, \mathbf{v}_*)$ which is unaffected either by interchange of $\mathbf{v}$ and $\mathbf{v}_*$ or by interchange of $(\mathbf{v}, \mathbf{v}_*)$ and $(\mathbf{v}', \mathbf{v}'_*)$, then the calculations which lead to (VII.20) can still be performed provided that the integrals which appear converge. In particular we can choose $k = \min(v, w)$ and conclude that

$$\int g \mathbb{C}_v F = \tfrac{1}{4} \iint \int_{\mathscr{S}} \min(v, w)(g + g_* - g' - g'_*)(F' F'_* - F F_*). \qquad (XXI.58)$$

---

[12] See the reference to ROYDEN's book cited in Footnote 1 to Chapter XIX.

[13] Specifically, if $f(x, y)$ is continuous with respect to $x \in \mathscr{X}$ for almost all $y \in \mathscr{Y}$ and is integrable with respect to y for each $x \in \mathscr{X}$, and if there is an integrable function $g(y)$ such that $|f(x, y)| \leq g(y)$ for all $x \in \mathscr{X}$, then $\int_{\mathscr{Y}} f(x, y)\,dy$ is continuous at all $x \in \mathscr{X}$. E.g. Chapter XIV, Lemma 1, of LANG's book cited in Footnote 3 to Chapter XIX.

Therefore, *for any summational invariant S,*

$$\int SC_v F = 0 \quad \text{for all} \quad F, \tag{XXI.59}$$

and this result parallels (VII.23) for $\mathbb{C}$. As a result of this property, if we consider the initial-value problem described by

$$\partial_t F = \mathbb{C}_v F, \tag{XXI.60}$$

we can transform it, just as we did (XXI.1), into the formally equivalent problem

$$\partial_t F + \beta SF = \mathbb{Q}_v F, \quad \int SF = \int SF_0, \tag{XXI.61}$$

in which β remains given by (XXI.13)$_1$, and

$$\mathbb{Q}_v F \equiv \mathbb{C}_v F + \alpha SF \int SF. \tag{XXI.62}$$

The corresponding results for $\mathbb{C}$ are outlined in (XXI.4), (XXI.12), and (XXI.13).

Next we show that the domain of $\mathbb{C}_v$ differs from that of $\mathbb{C}$. If we proceed to determine the domain of $\mathbb{C}_v$ as we did in Chapter XIX for $\mathbb{C}$, we may replace (XIX.7) by the simpler inequality $\min(v, w) \leq v$, and therefore also we may replace (XIX.8) and (XIX.9), respectively, by

$$\begin{aligned}|\min(v, w)gFF_*| &\leq \alpha v(1+v)^n|F||F_*|, \\ |\min(v, w)g'FF_*| &\leq 2^n \alpha v(1+v)^n(1+v_*)^n|F||F_*|.\end{aligned} \tag{XXI.63}$$

We find that *if $\mathscr{S}$ is a disk and g satisfies (XIX.1), then (XXI.58) will be absolutely convergent if $(1 + v)^n F$ is integrable.* Moreover, in regard to the domain of $\mathbb{C}_v$ we find that *if F is integrable, then $\mathbb{C}_v F$ exists and is also integrable.* The corresponding result for $\mathbb{C}$ required that $(1 + v)F$ be integrable.

One last result which we shall need is POVZNER's inequality (XIX.10), but now with $\mathbb{C}_v$ replacing $\mathbb{C}$. To prove this, we proceed just as in Chapter XIX except that now we replace (XIX.14) by

$$\min(v, w) \leq (1 + v^2)^{\frac{1}{2}}(1 + v_*^2)^{\frac{1}{2}}. \tag{XXI.64}$$

We find that *if $F \geq 0$ and if $(1 + v)^n F$ is integrable, then*

$$\int (1 + v^2)^{\frac{1}{2}n} \mathbb{C}_v F \leq C(n)\pi d^2 \left(\int (1 + v^2)^{\frac{1}{2}n} F\right)\left(\int (1 + v^2) F\right), \tag{XXI.65}$$

$C(n)$ *being a positive constant determined by n but independent of v.* That $C(n)$ is indeed the same for all v, follows from the fact that the right-hand side of (XXI.64) is independent of v. This property of $C(n)$ will be important in the proof of the existence theorem.

(iv) An existence theorem for spatially homogeneous solutions

As a step toward obtaining solutions of the initial-value problem (XXI.1), we shall give first a complete solution of the corresponding problem (XXI.60). Thereby we shall construct a sequence $\{F_m\}$ whose m^{th} member is the unique solution of (XXI.60) when $v = m$. Since $\min(m, w) \to w$ as $m \to \infty$, one might conjecture that the sequence $\{F_m\}$ converged to a function F satisfying (XXI.1). This conjecture is in fact true, but such proofs of it as we have seen require that an existence theorem for F be proved beforehand[14]. As we shall see now, however, a proof of existence and uniqueness can be constructed using a slight variation of this idea.

Theorem[15] **2.** *For a gas of molecules with a cut-off, if*

$$F_0 \geq 0, \qquad (\text{XXI.6})_{1r}$$

$$(1 + v^2)^2 F_0 \in \mathscr{L}^1(\mathscr{V}), \qquad (\text{XXI.66})$$

then there is a non-negative t-differentiable function $F \in \mathscr{M}^+(\mathscr{R}^+ \times \mathscr{V})$ which satisfies (XXI.1), (XXI.3), *and* (XXI.7)$_3$ *and for which, corresponding to each positive T, there is a constant C_T such that*

$$\|(1 + v^2)^2 F(t, \mathbf{v})\| \leq C_T, \qquad 0 \leq t \leq T. \qquad (\text{XXI.67})$$

There is only one solution with all these properties.

Proof. *Step 1. Proof of existence for* (XXI.60). Consider the integral equation

$$F = F_0 e^{-v\|F_0\|t} + \int_0^t e^{-v\|F_0\|(t-s)} \mathbb{Q}_v F \, ds. \qquad (\text{XXI.68})$$

This is simply an integrated version of (XXI.61)$_1$ in which $S = 1$ and $\alpha = v$. For convenience we take $\int_{\mathscr{S}} 1 = 1$ in the remainder of the proof. In effect, we multiply (XXI.1), (XXI.7)$_3$, and (XXI.60) by πd^2 and thenceforth denote $\pi d^2 F$ by F and $\pi d^2 F_0$ by F_0. This convention lets us write $\mathbb{Q}_v$ in the more explicit form

$$\mathbb{Q}_v F = \int \int_{\mathscr{S}} \min(v, w) F' F'_* + \int \int_{\mathscr{S}} (v - \min(v, w)) F F_*. \qquad (\text{XXI.69})$$

---

[14] We are indebted to Professor ARKERYD for this observation.
[15] ARKERYD [1972, 2, Theorem 1.1]. The proof we give here is ARKERYD's, suitably adjusted, of course, to deliver solutions of the type (XXI.11) rather than ones represented by (XXI.10). In addition, for clarity and completeness we have filled in many of the steps left unstated in ARKERYD's proof and improved upon others. In doing so we are led to set down separately Lemmas 1–6, not all of which appear in ARKERYD's work.

We see then that $\mathbb{Q}_v$ has the form (XXI.37) with

$$K_1 \equiv \min(v, w), \qquad K_2 \equiv v - \min(v, w). \qquad \text{(XXI.70)}$$

The functions K_1 and K_2 are non-negative and measurable. Moreover, (XXI.38) is satisfied in virtue of (VI.45)$_2$, and since K_1 and K_2 are bounded, (XXI.39) is satisfied when $n = 0$. We conclude by Lemma 5 that $\mathbb{Q}_v$ has the general properties (A2)–(A4), and so we may now write (XXI.68) as $(F_0, v\|F_0\|, \mathbb{Q}_v)$.

Let $\{F^n\}$ be the sequence of iterates for $(F_0, v\|F_0\|, \mathbb{Q}_v)$ defined by (XXI.16). By induction we shall prove that

$$\|F^n\| \leq \|F_0\| \qquad \text{if} \quad n \geq 1. \qquad \text{(XXI.71)}$$

Since $F^1 = 0$, this inequality is certainly satisfied if $n = 1$. Assume that it holds for some fixed n. From (XXI.16)$_2$ we see that

$$\|F^{n+1}\| = \|F_0\| e^{-v\|F_0\|t} + \int_0^t e^{-v\|F_0\|(t-s)} \|\mathbb{Q}_v F^n\| \, ds. \qquad \text{(XXI.72)}$$

To obtain this we have used Tonelli's theorem to interchange the order of integration in the last term on the right-hand side. Since (XXI.59) is valid when $S = 1$, we find from (XXI.62) that

$$\|\mathbb{Q}_v F\| = v\|F\|^2 \qquad \text{(XXI.73)}$$

for any function F, and so by (XXI.72), (XXI.73), and the hypothesis of induction

$$\|F^{n+1}\| \leq \|F_0\| e^{-v\|F_0\|t} + \int_0^t e^{-v\|F_0\|(t-s)} v\|F_0\|^2 \, ds,$$

$$= \|F_0\| \left(e^{-v\|F_0\|t} + v\|F_0\| \int_0^t e^{-v\|F_0\|(t-s)} \, ds \right), \qquad \text{(XXI.74)}$$

$$= \|F_0\|.$$

This completes the proof of (XXI.71).

According to Lemma 1, $\{F^n\}$ is an increasing sequence which converges pointwise to a function $F \in \mathcal{M}^+(\mathcal{R}^+ \times \mathcal{V})$ satisfying $(F_0, v\|F_0\|, \mathbb{Q}_v)$. Thus, by (XXI.71) and the monotone convergence theorem, we find that

$$\|F\| \leq \|F_0\|. \qquad \text{(XXI.75)}$$

If we recall that we may choose $n = 0$ in (XXI.39), we see that the hypotheses of Lemma 6 are satisfied, since by (XXI.75) we may set $C_T = \|F_0\|$ in (XXI.44) for all T. Therefore, F is t-differentiable and satisfies

$$\partial_t F + v\|F_0\| F = \mathbb{Q}_v F. \qquad \text{(XXI.76)}$$

(iv) PROOF OF EXISTENCE THEOREM 351

We show now that

$$\|F\| = \|F_0\|. \tag{XXI.77}$$

This is nothing more that (XXI.3)$_1$. To prove it, we integrate (XXI.76) first with respect to t, then with respect to $\mathbf{v}$, and thereafter apply Tonelli's theorem. We obtain

$$\|F\| - \|F_0\| = \int_0^t \int (\mathbb{Q}_v F - v\|F_0\|F) \, ds. \tag{XXI.78}$$

Using (XXI.73), we can rewrite this as

$$\|F\| - \|F_0\| = \int_0^t v\|F\|(\|F\| - \|F_0\|) \, ds. \tag{XXI.79}$$

We may interpret this as a Volterra integral equation for the unknown $\|F\| - \|F_0\|$. The kernel $v\|F\|$ is measurable; (XXI.75) shows that it has a bound independent of t. Therefore[16] (XXI.79) has a unique solution, and since zero is a solution, we obtain (XXI.77).

It follows now from (XXI.76), (XXI.77), and the definition of $\mathbb{Q}_v$ that F is a solution of the initial-value problem (XXI.60), and so this first step in the proof is complete.

Step 2. *Bounds for the moments and proof of the corresponding first-integrals* (XXI.3) *for* (XXI.60). We derive now some properties of the solution we have found. We have shown already that $\|F\|$ is bounded independently of t; in fact we have proved a stronger statement, namely (XXI.77), which is another way of asserting the first-integral (XXI.3)$_1$. We go on now to show that $(1 + v^2)F \in \mathscr{L}^1(\mathscr{V})$ and that

$$\|(1 + v^2)F\| = \|(1 + v^2)F_0\|. \tag{XXI.80}$$

This latter statement is equivalent, by (XXI.77), to the first-integral (XXI.3)$_3$. According to (XXI.16)$_2$ and Tonelli's theorem, the iterate F^n satisfies

$$\|(1 + v^2)F^{n+1}\| = \|(1 + v^2)F_0\|e^{-v\|F_0\|t} \\ + \int_0^t e^{-v\|F_0\|(t-s)}\|(1 + v^2)\mathbb{Q}_v F^n\| \, ds. \tag{XXI.81}$$

Since $1 + v^2$ is a summational invariant, by taking 1 for S in (XXI.59) and (XXI.62) we see that

$$\|(1 + v^2)\mathbb{Q}_v F\| = v\|F\| \, \|(1 + v^2)F\| \tag{XXI.82}$$

---

[16] *E.g.* Chapter I of H. WIDOM, *Lectures on Integral Equations*, New York, *etc.*, Van Nostrand Reinhold, 1969.

for general functions F, and so we can rewrite (XXI.81) as

$$\|(1 + v^2)F^{n+1}\| = \|(1 + v^2)F_0\|e^{-v\|F_0\|t} + \int_0^t e^{-v\|F_0\|(t-s)}v\|F^n\|\,\|(1 + v^2)F^n\|\,ds,$$

$$= \|(1 + v^2)F_0\|\left(e^{-v\|F_0\|t} + v\|F_0\|\int_0^t e^{-v\|F_0\|(t-s)}\,ds\right)$$

$$+ \int_0^t e^{-v\|F_0\|(t-s)}v(\|F^n\|\,\|(1 + v^2)F^n\| - \|F_0\|\,\|(1 + v^2)F_0\|)\,ds,$$

$$= \|(1 + v^2)F_0\| + \int_0^t e^{-v\|F_0\|(t-s)}v\|F^n\|(\|(1 + v^2)F^n\| - \|(1 + v^2)F_0\|)\,ds$$

$$+ \int_0^t e^{-v\|F_0\|(t-s)}v\|(1 + v^2)F_0\|(\|F^n\| - \|F_0\|)\,ds. \quad \text{(XXI.83)}$$

In view of (XXI.71) it follows by induction that

$$\|(1 + v^2)F^n\| \leq \|(1 + v^2)F_0\|, \qquad n \geq 1, \quad \text{(XXI.84)}$$

and since the sequence $\{F^n\}$ is increasing and converges pointwise to F, it follows from (XXI.84) and the monotone convergence theorem that

$$\|(1 + v^2)F\| \leq \|(1 + v^2)F_0\|. \quad \text{(XXI.85)}$$

Thus $(1 + v^2)F \in \mathscr{L}^1(\mathscr{V})$ if $t > 0$. Using this fact now, we multiply each side of (XXI.76) by $1 + v^2$, then integrate first with respect to t and thereafter with respect to $\mathbf{v}$ so as to obtain

$$\|(1 + v^2)F\| - \|(1 + v^2)F_0\| = \int_0^t \left[\int (1 + v^2)(\mathbb{Q}_v F - v\|F_0\|F)\right] ds. \quad \text{(XXI.86)}$$

In view of (XXI.82), (XXI.86) has the more explicit form

$$\|(1 + v^2)F\| - \|(1 + v^2)F_0\| = \int_0^t v\|(1 + v^2)F\|(\|F\| - \|F_0\|)\,ds. \quad \text{(XXI.87)}$$

We see immediately from (XXI.87) and (XXI.77) that (XXI.80) holds.

Since $|v_k| \leq 1 + v^2$, we know immediately that $Fv_k \in \mathscr{L}^1(\mathscr{V})$ for each k. By an argument parallel to that which led us to (XXI.87) we find that (XXI.87) remains valid if in it we replace $\|(1 + v^2)F\|$ by $\int F\mathbf{v}$ and $\|(1 + v^2)F_0\|$ by $\int F_0\mathbf{v}$. Hence we conclude that the first-integral (XXI.3)$_2$ is valid.

Another important property of F is the fact that $(1 + v^2)^2 F \in \mathscr{L}^1(\mathscr{V})$, and in particular that for each positive T there is a positive constant C_T, *independent of* $\mathbf{v}$, such that

$$\|(1 + v^2)^2 F\| \leq C_T \qquad \text{when} \quad 0 \leq t \leq T. \quad \text{(XXI.88)}$$

This inequality is similar to (XXI.75) and (XXI.85) in that it gives bounds independent of t to the moments of F, but it is much more difficult to establish. The reason is that in the proofs of (XXI.75) and (XXI.85) we appealed to (XXI.59) with $S = 1$ and $S = 1 + v^2$. When $S = (1 + v^2)^2$ we no longer have (XXI.59), and so we appeal to a somewhat less sharp estimate, namely POVZNER's inequality (XXI.65) with $n = 4$. The details of the proof, being somewhat technical, are presented in an appendix to this chapter.

Step 3. *Proof of existence for* (XXI.1) *and proof of the first-integrals* (XXI.3). We turn now to the initial-value problem for (XXI.1). Consider the operators $\mathbb{Q}$, $\hat{\mathbb{Q}}_m$, and $\check{\mathbb{Q}}_m$ defined as follows:

$$\mathbb{Q}F \equiv \iint_{\mathscr{S}} wF'F'_* + \iint_{\mathscr{S}} [\alpha(1 + v^2)(1 + v_*^2) - w]FF_*,$$

$$\hat{\mathbb{Q}}_m F \equiv \iint_{\mathscr{S}} \min(m, w)F'F'_* + \iint_{\mathscr{S}} [\alpha(1 + v^2)(1 + v_*^2) - w]FF_*,$$

$$m = 1, 2, \ldots,$$

$$\check{\mathbb{Q}}_m F \equiv \iint_{\mathscr{S}} \min(m, w)F'F'_* + \iint_{\mathscr{S}} [\alpha(1 + v^2)(1 + v_*^2) - \min(m, w)]FF_*,$$

$$m = 1, 2, \ldots,$$

(XXI.89)

α being any positive constant such that $\alpha \geq 1$. Each of these operators has the form (XXI.37) for certain non-negative measurable functions K_1 and K_2. Moreover, (XXI.38) is satisfied in each case due to (VI.45)$_2$, and in view of the inequalities

$$\min(m, w) \leq w \leq (1 + v^2)(1 + v_*^2),$$
$$1 \leq \alpha,$$

(XXI.90)

(XXI.39) is also satisfied when $n = 1$. By Lemma 5, then, $\mathbb{Q}$, $\hat{\mathbb{Q}}_m$, and $\check{\mathbb{Q}}_m$ all have the general properties (A2)–(A4). Therefore we may consider the integral equation $(F_0, g, \mathbb{Q})$, and the sequences of integral equations $\{(F_0, g, \hat{\mathbb{Q}}_m)\}$ and $\{(F_0, g, \check{\mathbb{Q}}_m)\}$, g being the function

$$g \equiv \alpha(1 + v^2)\|(1 + v^2)F_0\|. \qquad (XXI.91)$$

Directly from (XXI.89)$_{2,3}$ we find that if $G \in \mathscr{M}^+(\mathscr{R}^+ \times \mathscr{V})$, then

$$\hat{\mathbb{Q}}_m G \leq \check{\mathbb{Q}}_m G,$$
$$\hat{\mathbb{Q}}_{m_1} G \leq \hat{\mathbb{Q}}_{m_2} G \qquad \text{if} \quad m_1 \leq m_2, \qquad (XXI.92)$$

while from (XXI.89)$_{1,2}$ and the monotone convergence theorem,

$$\hat{\mathbb{Q}}_m G \to \mathbb{Q}G \qquad \text{pointwise as} \quad m \to \infty. \qquad (XXI.93)$$

We can restate these properties in terms of the integral equations by means of (XXI.21), (XXI.22), (XXI.25), and (XXI.26):

$$(F_0, g, \hat{\mathbb{Q}}_m) \leq (F_0, g, \check{\mathbb{Q}}_m),$$
$$(F_0, g, \hat{\mathbb{Q}}_{m_1}) \leq (F_0, g, \hat{\mathbb{Q}}_{m_2}) \quad \text{if} \quad m_1 \leq m_2, \tag{XXI.94}$$
$$(F_0, g, \hat{\mathbb{Q}}_m) \to (F_0, g, \mathbb{Q}) \quad \text{as} \quad m \to \infty.$$

Let $\hat{F}_m$ and $\check{F}_m$ be the iterative solutions of $(F_0, g, \hat{\mathbb{Q}}_m)$ and $(F_0, g, \check{\mathbb{Q}}_m)$, respectively. From (XXI.94)$_1$ and Lemma 2 we have

$$\hat{F}_m \leq \check{F}_m \quad \text{when} \quad m \geq 1. \tag{XXI.95}$$

Now we let F_v be the solution of (XXI.60) obtained in the first step of the proof and consider the sequence $\{F_m\}$ whose m^{th} member is the solution F_v when $v = m$. Since (XXI.80) is nothing more than (XXI.61)$_2$ with $S = 1 + v^2$, we see that F_m satisfies the differential equation (XXI.61)$_1$ for this choice of S. In view of (XXI.62), (XXI.89)$_3$, and (XXI.91), this differential equation is

$$\partial_t F + gF = \check{\mathbb{Q}}_m F, \tag{XXI.96}$$

and so F_m is also a solution of $(F_0, g, \check{\mathbb{Q}}_m)$. By the corollary to Lemma 1 we conclude that

$$\check{F}_m \leq F_m \quad \text{when} \quad m \geq 1. \tag{XXI.97}$$

Combining (XXI.95) with (XXI.97), we see that

$$\hat{F}_m \leq F_m \quad \text{when} \quad m \geq 1. \tag{XXI.98}$$

Moreover, since each function F_m satisfies (XXI.88), and we know that the constant C_T appearing there is independent of v, we conclude that

$$\|(1 + v^2)^2 \hat{F}_m\| \leq C_T \quad \text{when} \quad 0 \leq t \leq T, \tag{XXI.99}$$

C_T being independent of m.

Finally, let F be the iterative solution of $(F_0, g, \mathbb{Q})$. By (XXI.94)$_{2,3}$ and Lemma 3 and its corollary, we see that $\{\hat{F}_m\}$ is an increasing sequence and

$$\hat{F}_m \to F \quad \text{pointwise as} \quad m \to \infty. \tag{XXI.100}$$

Using (XXI.99) and the fact that C_T is independent of m, we see by the monotone convergence theorem that F satisfies (XXI.67). The hypotheses of Lemma 6 are now satisfied, and so F is t-differentiable and satisfies the differential equation

$$\partial_t F + gF = \mathbb{Q}F. \tag{XXI.101}$$

In view of (XXI.89)$_1$, (XXI.91), and the definition of the collisions operator $\mathbb{C}$ we can write this as

$$\partial_t F = \mathbb{C}F + \alpha(1 + v^2)F(\|(1 + v^2)F\| - \|(1 + v^2)F_0\|), \tag{XXI.102}$$

and it is clear that F will satisfy (XXI.1) if first we can show that it satisfies (XXI.80).

We prove (XXI.80) now as a special case of the more general fact that if G is any function in $\mathcal{M}^+(\mathcal{R}^+ \times \mathcal{V})$ which satisfies $(XXI.7)_3$, (XXI.101), and (XXI.67), then

$$\|(1 + v^2)G\| = \|(1 + v^2)F_0\|. \qquad (XXI.103)$$

In order to see this, we multiply (XXI.101) by $1 + v^2$, and then integrate first with respect to t and then with respect to $\mathbf{v}$. By use of Tonelli's theorem we obtain

$$\|(1 + v^2)G\| - \|(1 + v^2)F_0\| = \int_0^t \left[\int (1 + v^2)(\mathbb{Q}G - gG) \right] ds. \qquad (XXI.104)$$

Using (VII.23) with $S = 1 + v^2$, we may simplify the right-hand side and so find that

$$\|(1 + v^2)G\| - \|(1 + v^2)F_0\|$$
$$= \int \alpha \|(1 + v^2)^2 G\| (\|(1 + v^2)G\| - \|(1 + v^2)F_0\|) \, ds. \qquad (XXI.105)$$

As before, we interpret this as a Volterra integral equation for the function $\|(1 + v^2)G\| - \|(1 + v^2)F_0\|$. The kernel $\alpha\|(1 + v^2)^2 G\|$ is measurable; (XXI.67) shows that it has an upper bound independent of t on each interval $[0, T]$. We conclude that this integral equation has a unique solution, and since zero is a solution, we obtain (XXI.103).

From (XXI.102) and (XXI.80) we see that F satisfies (XXI.1). This completes the proof of existence. We have shown in the course of it that (XXI.80) is satisfied. By an argument parallel to that which gave us (XXI.87) we may show that (XXI.77) is satisfied. These two conclusions together amount to $(XXI.3)_{1,3}$. In just the same way we may obtain $(XXI.3)_2$ also.

Step 4. *Proof of uniqueness.* Suppose that $G \in \mathcal{M}^+(\mathcal{R}^+ \times \mathcal{V})$ also satisfies the hypotheses of the theorem. This means that G satisfies (XXI.67) and (XXI.101); as the preceding steps of the proof indicate, G satisfies also (XXI.103) and the integral equation $(F_0, g, \mathbb{Q})$. Since F is the iterative solution of this equation, the corollary to Lemma 1 shows that

$$F \leq G, \qquad (XXI.106)$$

and so by (XXI.80) and (XXI.103)

$$\|(1 + v^2)(G - F)\| = \|(1 + v^2)G\| - \|(1 + v^2)F\|,$$
$$= \|(1 + v^2)F_0\| - \|(1 + v^2)F_0\|, \qquad (XXI.107)$$
$$= 0.$$

Thus F and G are equal almost everywhere. △

(v) Proof of the ultra-narrow H-theorem

We turn now to the proof of the H-theorem for the solutions exhibited in the preceding section. Generally speaking, any proof of this theorem is effected in three steps. If $\log^+ F$ and $\log^- F$ denote the positive and negative parts, respectively, of $\log F$, one proves, assuming that $F_0 \log F_0 \in \mathscr{L}^1(\mathscr{V})$, first that $F \log^- F \in \mathscr{L}^1(\mathscr{V})$ when $t > 0$, then that $F \log^+ F \in \mathscr{L}^1(\mathscr{V})$ when $t > 0$, and finally that

$$h[F](t_2) \leq h[F](t_1) \qquad \text{when} \quad t_2 \geq t_1. \qquad (XI.62)_r$$

We may dispense with the first of these immediately. We begin with the simple inequality

$$-y + x \log y \leq x \log x \qquad \text{if} \quad y > 0, \quad x \geq 0. \qquad (XXI.108)$$

When $y \leq 1$, the left-hand side is negative, whence we conclude that

$$x \log^- x \leq y - x \log y \qquad \text{if} \quad 0 < y \leq 1, \quad x \geq 0. \qquad (XXI.109)$$

If we set $y = \exp(-v^2)$ and $x = F$, we obtain

$$F \log^- F \leq e^{-v^2} + v^2 F. \qquad (XXI.110)$$

It follows from this inequality that *if $F \geq 0$ and $(1 + v^2)^n F \in \mathscr{L}^1(\mathscr{V})$, then $(1 + v^2)^{n-1} F \log^- F \in \mathscr{L}^1(\mathscr{V})$*. This result simply expresses a certain property of functions in $\mathscr{L}^1(\mathscr{V})$. If we apply it to the solutions we have found for (XXI.1) and note that for them $(1 + v^2)^2 F \in \mathscr{L}^1(\mathscr{V})$, we conclude immediately that $F \log^- F \in \mathscr{L}^1(\mathscr{V})$. It is not even necessary yet to assume that $F_0 \log F_0 \in \mathscr{L}^1(\mathscr{V})$.

Following ARKERYD, we set down a sequence of new equations, ones for which the H-theorem may be established quite easily and whose solutions tend in a certain sense to our solution of (XXI.1). This sequence of new equations is

$$\partial_t G + g_p G = \mathbb{Q}^k G, \qquad G\bigg|_{t=0} = F_{0km}; \qquad (XXI.111)$$

we assume that F_0 is a given function such that

$$F_0 \geq 0, \qquad F_0 \log F_0 \in \mathscr{L}^1(\mathscr{V}); \qquad (XXI.6)_{1,2r}$$

$$(1 + v^2)^2 F_0 \in \mathscr{L}^1(\mathscr{V}), \qquad (XXI.66)_r$$

and we lay down the definitions

$$\mathbb{Q}^k G \equiv \min(\mathbb{Q}G, k) \qquad \text{for all} \quad G \in \mathscr{M}^+(\mathscr{V}),$$
$$g_p \equiv \alpha(1 + v^2)\|(1 + v^2)F_{0p}\|, \qquad F_{0km} \equiv \min(F_{0m}, k), \qquad (XXI.112)$$

in which

$$F_{0m} \equiv F_0 + (1/m)e^{-v^2}. \qquad (XXI.113)$$

(v) ULTRA-NARROW H-THEOREM   357

Here α is a positive constant, and $\mathbb{Q}$ is given by $(XXI.89)_1$.

First we show that (XXI.111) has a solution. Let F_{kmp} be the iterative solution of $(F_{0km}, g_p, \mathbb{Q}^k)$, the integral equation corresponding to (XXI.111). From (XXI.112) we can show that

$$(F_{0k_1m_1}, g_{p_1}, \mathbb{Q}^{k_1}) \leq (F_{0k_2m_2}, g_{p_2}, \mathbb{Q}^{k_2}) \quad \text{if} \quad k_1 \leq k_2, m_2 \leq m_1, p_1 \leq p_2,$$

$$(F_{0km}, g_p, \mathbb{Q}^k) \to (F_{0m}, g_p, \mathbb{Q}) \quad \text{as} \quad k \to \infty. \quad (XXI.114)$$

Using the first of these and Lemma 2 we find that

$$F_{k_1m_1p_1} \leq F_{k_2m_2p_2} \quad (XXI.115)$$

for the range of indices indicated in $(XXI.114)_1$. We also conclude from (XXI.114) that the sequence $\{(F_{0km}, g_p, \mathbb{Q}^k)\}_{k=1}^{\infty}$ is increasing and that its limit is $(F_{0m}, g_p, \mathbb{Q})$. By Lemma 3 and its corollary, $\{F_{kmp}\}_{k=1}^{\infty}$ is increasing, and its pointwise limit is the iterative solution, say $F_{mp} \in \mathcal{M}^+(\mathcal{R}^+ \times \mathcal{V})$, of $(F_{0m}, g_p, \mathbb{Q})$. By taking the limit as $k_2 \to \infty$ and then as $k_1 \to \infty$ in (XXI.115) we obtain

$$F_{m_1p_1} \leq F_{m_2p_2} \quad \text{if} \quad m_2 \leq m_1, \quad p_1 \leq p_2. \quad (XXI.116)$$

Combining (XXI.115) and (XXI.116), we see that

$$F_{kmp} \leq F_{mp} \leq F_{pp} \quad \text{if} \quad p \leq m. \quad (XXI.117)$$

Thus for fixed p the function F_{pp} is an upper bound for all of the functions F_{kmp}. We know already that F_{pp} is the iterative solution of $(F_{0p}, g_p, \mathbb{Q})$. However, this equation is nothing more than the integral form of (XXI.1) with initial value F_{0p} instead of F_0. From $(XXI.6)_{1,2}$ and (XXI.66) it follows that $F_{0p} \geq 0$ and $(1 + v^2)^2 F_{0p} \in \mathcal{L}^1(\mathcal{V})$, and so by the existence theorem presented earlier, F_{pp} is t-differentiable, $(1 + v^2)^2 F_{pp} \in \mathcal{L}^1(\mathcal{V})$, and F_{pp} satisfies

$$\|(1 + v^2)F_{pp}\| = \|(1 + v^2)F_{0p}\| \quad (XXI.118)$$

and the inequality (XXI.67). In view of (XXI.117) we see that each function F_{kmp} satisfies the inequality (XXI.44) with $n = 1$, and so by Lemma 6, F_{kmp} is t-differentiable and satisfies (XXI.111).

Next we indicate in what sense the functions F_{kmp} tend to the solution of (XXI.1). Let us look once again at the special solution F_{pp}. We saw earlier that it satisfied (XXI.67) with a suitable constant C_T. In general this constant depends upon p, but it is possible to show that it need not. For this we need the detailed form of (XXI.67) proved in the appendix to this chapter, namely (XXIA.15). Since $F_{pp}(0) = F_{0p}$, that form becomes in this case

$$\|(1 + v^2)^2 F_{pp}\| \leq 2^N \|(1 + v^2)^2 F_{0p}\| \quad \text{if} \quad 0 \leq t \leq \frac{N}{2C(4)\|(1 + v^2)F_{0p}\|},$$

$$(XXI.119)$$

N being an arbitrary positive integer. As $F_{0p} \leq F_{01}$, we may replace F_{0p} on the right-hand sides of both these inequalities by F_{01} and conclude, using (XXI.117), that for each positive T there is a constant C_T *not depending upon k, m, or p*, such that

$$\|(1+v^2)^2 F_{kmp}\| \leq C_T \quad \text{if} \quad 0 \leq t \leq T. \tag{XXI.120}$$

Recall now that $\{F_{kmp}\}_{k=1}^\infty$ is increasing and converges pointwise to F_{mp}. According to (XXI.116), $\{F_{mp}\}_{m=p}^\infty$ is a decreasing sequence of non-negative measurable functions, and hence it has a pointwise limit $F_p \in \mathcal{M}^+(\mathcal{R}^+ \times \mathcal{V})$. Since $(F_{0m}, g_p, \mathbb{Q}) \to (F_0, g_p, \mathbb{Q})$ as $m \to \infty$ and since F_{mp} is bounded independently of m by $F_{pp} \in \mathcal{L}^1(\mathcal{V})$, we may show using the dominated convergence theorem[17] that F_p satisfies $(F_0, g_p, \mathbb{Q})$. Next we let $m_1 \to \infty$ and $m_2 \to \infty$ in (XXI.116) and obtain

$$F_{p_1} \leq F_{p_2} \quad \text{if} \quad p_1 \leq p_2. \tag{XXI.121}$$

We see that $\{F_p\}_{p=1}^\infty$ is increasing and so converges pointwise to a function $F \in \mathcal{M}^+(\mathcal{R}^+ \times \mathcal{V})$. Moreover we note that $(F_0, g_p, \mathbb{Q}) \to (F_0, g, \mathbb{Q})$ as $p \to \infty$, the function g being given by (XXI.91). It follows then from the monotone convergence theorem that F satisfies $(F_0, g, \mathbb{Q})$. That is, F is a solution of the integral form of (XXI.1). Finally we note that F is t-differentiable and satisfies (XXI.1). For we may let k, m, and p approach ∞ in (XXI.120) and use Fatou's Lemma[18] so as to conclude that $(1+v^2)^2 F \in \mathcal{L}^1(\mathcal{V})$. We then apply Lemma 6, and this completes the proof. By uniqueness, F is the solution of (XXI.1) given by the existence theorem.

Finally we indicate how the special choice of the system (XXI.111) permits us to establish quite easily the Boltzmann monotonicity theorem for the solutions F_{kmp}. We know that F_{kmp} satisfies $(F_{0km}, g_p, \mathbb{Q}^k)$, that is,

$$F_{kmp} = F_{0km} e^{-g_p t} + \int_0^t e^{-g_p(t-s)} \mathbb{Q}^k F_{kmp} \, ds. \tag{XXI.122}$$

From this equation and the definitions (XXI.112) we obtain two simple bounds. Choose α such that $g_p \geq 1$ when $p \geq 1$. Then

$$F_{kmp} \leq k\left(e^{-g_p t} + \int_0^t e^{-g_p(t-s)} \, ds\right),$$

$$\leq k,$$

$$F_{kmp} \geq F_{0km} e^{-g_p t}, \tag{XXI.123}$$

$$\geq (1/m) e^{-v^2} e^{-t\alpha(1+v^2)\|(1+v^2) F_{0p}\|},$$

$$\geq (1/m) e^{-(1+v^2)(1+\alpha t\|(1+v^2) F_{01}\|)},$$

$$\geq (1/m) e^{-c_1(1+v^2)} \quad \text{if} \quad 0 \leq t \leq T,$$

---

[17] *E.g.* Theorem 1.34 in RUDIN's book cited in Footnote 15 to Chapter XI.
[18] *E.g.* Theorem 1.28 of RUDIN's book cited in Footnote 15 to Chapter XI.

(v) ULTRA-NARROW H-THEOREM

in which c_1 is a positive constant determined by T but independent of $k, m,$ and p. These show us immediately that $F_{kmp} \log F_{kmp} \in \mathscr{L}^1(\mathscr{V})$. More important, however, is the fact that, by (XIX.5) and (XIX.6), on any finite interval $[0, T]$

$$|\log F_{kmp}| \leq \text{const.}(1 + v)^2. \tag{XXI.124}$$

Since this is of the form (XIX.6) and $(1 + v^2)^2 F_{kmp} \in \mathscr{L}^1(\mathscr{V})$, the transformations (VII.17)–(VII.20) are valid, and so the Boltzmann monotonicity theorem holds:

$$\bar{\mathbb{C}}_{F_{kmp}} \log F_{kmp} \leq 0. \tag{XXI.125}$$

It is this inequality that we shall use in proving the H-theorem for F. We proceed now to the details.

Theorem[19] **3.** *Let F_0 satisfy* (XXI.6)$_{1,2}$ *and* (XXI.66), *and let F be the corresponding unique solution of* (XXI.1). *Then $F \log F \in \mathscr{L}^1(\mathscr{V})$ when $t \geq 0$, and* (XI.62) *is satisfied.*

Proof. Let us note first a simple observation. If we can show that $F \log F \in \mathscr{L}^1(\mathscr{V})$ when $t \geq 0$ and that for each positive T

$$h[F](T) \leq h[F](0), \tag{XXI.126}$$

then (XI.62) follows, for since $F \log F \in \mathscr{L}^1(\mathscr{V})$ we may choose $F(t_1)$ as a new initial value, and the hypothesis (XXI.6)$_2$ is still satisfied. Applying (XXI.126) with $T = t_2 - t_1$, we obtain (XI.62). Hence our problem reduces to proving $F \log F \in \mathscr{L}^1(\mathscr{V})$ and (XXI.126).

We may write

$$F_{kmp} \log F_{kmp} \bigg|_{t=T} - F_{kmp} \log F_{kmp} \bigg|_{t=0}$$

$$= \int_0^T \partial_t (F_{kmp} \log F_{kmp}) \, dt,$$

$$= \int_0^T (1 + \log F_{kmp}) \, \partial_t F_{kmp} \, dt, \tag{XXI.127}$$

$$= \int_0^T (\mathbb{Q}^k F_{kmp} - g_p F_{kmp}) \log F_{kmp} \, dt + \int_0^T (\mathbb{Q}^k F_{kmp} - g_p F_{kmp}) \, dt.$$

---

[19] ARKERYD [1972, 2, Theorem 1.3].

We note that

$$(\mathbb{Q}^k F_{kmp} - g_p F_{kmp}) \log F_{kmp}$$
$$= (\mathbb{Q}^k F_{kmp} - g_p F_{kmp})(\log^+ F_{kmp} - \log^- F_{kmp}),$$
$$\leq (\mathbb{Q} F_{kmp} - g_p F_{kmp}) \log^+ F_{kmp} - (\mathbb{Q}^k F_{kmp} - g_p F_{kmp}) \log^- F_{kmp},$$
$$= (\mathbb{Q} F_{kmp} - g_p F_{kmp}) \log F_{kmp} + (\mathbb{Q} F_{kmp} - \mathbb{Q}^k F_{kmp}) \log^- F_{kmp}.$$
(XXI.128)

Placing this into (XXI.127), integrating with respect to **v**, and then using Tonelli's theorem, we obtain

$$h[F_{kmp}](T) \leq h[F_{kmp}](0) + \frac{1}{n} \int_0^T (A_1 + A_2 + A_3) \, dt, \quad \text{(XXI.129)}$$

in which

$$A_1 \equiv \int (\mathbb{Q} F_{kmp} - g_p F_{kmp}) \log F_{kmp},$$

$$A_2 \equiv \int (\mathbb{Q} F_{kmp} - g_p F_{kmp}), \quad \text{(XXI.130)}$$

$$A_3 \equiv \int_{\mathbb{Q} F_{kmp} > k} (\mathbb{Q} F_{kmp} - k) \log^- F_{kmp} \, d\mathbf{v}.$$

We shall examine separately the behavior of A_1, A_2, and A_3.

Consider first A_3. Using (XXI.117), (XXI.123)$_6$, and the monotonicity of $\mathbb{Q}$, we find that

$$0 \leq \int_0^T A_3 \, dt \leq \int_0^T \left[\int_{\mathbb{Q} F_{pp} > k} (\mathbb{Q} F_{pp} - k) \log^- F_{kmp} \, d\mathbf{v} \right] dt,$$
$$\leq \int_0^T \left[\int \chi (\mathbb{Q} F_{pp} - k) \left(c_1(1 + v^2) - \log \frac{1}{m} \right) \right] dt, \quad \text{(XXI.131)}$$

χ being the characteristic function of the set $\{\mathbf{v} \mid \mathbb{Q} F_{pp} > k\}$. Since the constant c_1 does not depend upon k, the integrand here is bounded independently of k by the non-negative integrable function $[c_1(1 + v^2) + \log m] \mathbb{Q} F_{pp}$. Moreover, the integrand tends to zero almost everywhere as $k \to \infty$. By the dominated convergence theorem we conclude that

$$\lim_{k \to \infty} \int_0^T A_3 \, dt = 0. \quad \text{(XXI.132)}$$

Now consider A_2. Using (XXI.89)$_1$ and (XXI.112)$_2$, we see that

$$A_2 = \int \mathbb{C} F_{kmp} + \alpha \|(1 + v^2) F_{kmp}\| [\|(1 + v^2) F_{kmp}\| - \|(1 + v^2) F_{0p}\|].$$
(XXI.133)

(v) ULTRA-NARROW H-THEOREM 361

We know by (VII.23) that the first term is null. Moreover, by (XXI.117) and (XXI.118) we see that the second term is negative. Hence $A_2 \leq 0$, and so we may simply omit it from (XXI.129).

Finally we consider A_1. Using (XXI.89)$_1$ and (XXI.112)$_2$ again, we find that

$$A_1 = \int (\log F_{kmp})\mathbb{C}F_{kmp} + \alpha \left[\int \cdot (1+v^2)F_{kmp} \log F_{kmp} \right]$$
$$\times [\|(1+v^2)F_{kmp}\| - \|(1+v^2)F_{0p}\|],$$
$$\leq \bar{C}_{F_{kmp}} \log F_{kmp} - \alpha \|(1+v^2)F_{kmp} \log^- F_{kmp}\| \quad \text{(XXI.134)}$$
$$\times [\|(1+v^2)F_{kmp}\| - \|(1+v^2)F_{0p}\|].$$

We may discard the first term here because of (XXI.125). Next, using (XXI.110) and (XXI.120), we see that

$$\|(1+v^2)F_{kmp} \log^- F_{kmp}\| \leq \|(1+v^2)e^{-v^2}\| + \|(1+v^2)^2 F_{kmp}\|,$$
$$\leq \|(1+v^2)e^{-v^2}\| + C_T \quad \text{if} \quad 0 \leq t \leq T.$$
$$\text{(XXI.135)}$$

Thus $\|(1+v^2)F_{kmp} \log^- F_{kmp}\|$ is bounded on any finite interval of time by a constant that is independent of k, m, and p. Furthermore, we know that $F_{0p} - F_{kmp}$ is bounded above, independently of k and m, by $F_{0p} \in \mathscr{L}^1(\mathscr{V})$, and below, independently of k and m, by $-F_{pp} \in \mathscr{L}^1(\mathscr{V})$, and so by the dominated convergence theorem

$$\lim_{m \to \infty} \lim_{k \to \infty} [\|(1+v^2)F_{0p}\| - \|(1+v^2)F_{kmp}\|]$$
$$= \|(1+v^2)F_{0p}\| - \|(1+v^2)F_p\|, \quad \text{(XXI.136)}$$
$$= [\|(1+v^2)F_0\| - \|(1+v^2)F\|] + \frac{1}{p}\|(1+v^2)e^{-v^2}\| + \|(1+v^2)(F-F_p)\|.$$

Since $\{F_p\}_{p=1}^\infty$ is increasing and converges pointwise to F, we conclude from the dominated convergence theorem that the last term here approaches 0 as $p \to \infty$. The second-from-last term certainly approaches 0 as $p \to \infty$. The first term is null by (XXI.80). Hence

$$\lim_{p \to \infty} \lim_{m \to \infty} \lim_{k \to \infty} \int_0^T A_1 \, dt = 0. \quad \text{(XXI.137)}$$

Now let us examine the behavior of $h[F_{kmp}](0)$. Using (XXI.110), we see that

$$F_{kmp} \log^- F_{kmp} \bigg|_{t=0} \leq e^{-v^2} + v^2 F_{kmp} \bigg|_{t=0},$$
$$= e^{-v^2} + v^2 F_{0km}, \quad \text{(XXI.138)}$$
$$\leq e^{-v^2} + v^2(F_0 + e^{-v^2}).$$

We note also that

$$F_{kmp} \log^+ F_{kmp}\bigg|_{t=0} = F_{0km} \log^+ F_{0km},$$

$$\leq \left(F_0 + \frac{1}{m}e^{-v^2}\right) \log^+ \left(F_0 + \frac{1}{m}e^{-v^2}\right). \quad \text{(XXI.139)}$$

The function $x \mapsto x \log^+ x$ is convex when $x \geq 0$, and therefore

$$(x + y) \log^+(x + y) \leq x \log^+ 2x + y \log^+ 2y. \quad \text{(XXI.140)}$$

We conclude from (XXI.139) and (XXI.140) that

$$F_{kmp} \log^+ F_{kmp}\bigg|_{t=0} \leq F_0 \log 2 + F_0 \log^+ F_0 \quad \text{if } m \geq 2. \quad \text{(XXI.141)}$$

We see by (XXI.138) and (XXI.141) that $|F_{kmp} \log F_{kmp}|_{t=0}|$ is bounded independently of k, m, and p by a function in $\mathcal{L}^1(\mathcal{V})$. We also know that $F_{kmp}|_{t=0} \to F_0$ as $k \to \infty$, $m \to \infty$, and $p \to \infty$. From the dominated convergence theorem we conclude that

$$\lim_{p \to \infty} \lim_{m \to \infty} \lim_{k \to \infty} h[F_{kmp}](0) = h[F](0). \quad \text{(XXI.142)}$$

We are now ready to use (XXI.129). Let us set $h^+[F] \equiv \overline{\log^+ F}$ and $h^-[F] \equiv \overline{\log^- F}$ and rewrite (XXI.129) in the form

$$h^+[F_{kmp}](T) \leq h^-[F_{kmp}](T) + h[F_{kmp}](0) + \frac{1}{n}\int_0^T A_1 \, dt + \frac{1}{n}\int_0^T A_3 \, dt. \quad \text{(XXI.143)}$$

Using (XXI.110) and (XXI.117), we see that

$$F_{kmp} \log^- F_{kmp} \leq e^{-v^2} + v^2 F_{kmp},$$

$$\leq e^{-v^2} + v^2 F_{pp}. \quad \text{(XXI.144)}$$

We have bounded $F_{kmp} \log^- F_{kmp}$ by a function in $\mathcal{L}^1(\mathcal{V})$ that is independent of k and m. Hence we may let $k \to \infty$ in (XXI.143) using Fatou's lemma on the left-hand side and the dominated convergence theorem on the right-hand side. In view of (XXI.132) we obtain

$$h^+[F_{mp}](T) \leq h^-[F_{mp}](T) + \lim_{k \to \infty} h[F_{kmp}](0) + \frac{1}{n} \lim_{k \to \infty} \int_0^T A_1 \, dt. \quad \text{(XXI.145)}$$

Hence $F_{mp} \log^+ F_{mp} \in \mathscr{L}^1(\mathscr{V})$. We may apply the same procedure now as $m \to \infty$, and we obtain

$$h^+[F_p](T) \leq h^-[F_p](T) + \lim_{m \to \infty} \lim_{k \to \infty} h[F_{kmp}](0) + \frac{1}{n} \lim_{m \to \infty} \lim_{k \to \infty} \int_0^T A_1 \, dt.$$
(XXI.146)

Thus $F_p \log^+ F_p \in \mathscr{L}^1(\mathscr{V})$. Now we observe that since the sequence $\{F_p\}_{p=1}^\infty$ is increasing and converges pointwise to F, the inequality (XXI.110) implies that

$$F_p \log^- F_p \leq e^{-v^2} + v^2 F_p,$$
$$\leq e^{-v^2} + v^2 F,$$
(XXI.147)

and therefore we have bounded $F_p \log^- F_p$ by a function in $\mathscr{L}^1(\mathscr{V})$ that is independent of p. Hence, applying once again Fatou's lemma and the dominated convergence theorem, we obtain in view of (XXI.137) and (XXI.142)

$$h^+[F](T) \leq h^-[F](T) + h[F](0).$$
(XXI.148)

We conclude that $F \log^+ F \in \mathscr{L}^1(\mathscr{V})$ and (XXI.126) is satisfied. △

(vi) *Proof of the ultra-narrow trend to equilibrium*

Finally we examine the asymptotic behavior of solutions of (XXI.1) in order to establish the trend to equilibrium:

$$F \to F_U \quad \text{as} \quad t \to \infty.$$
(XI.61)$_r$

Here F_U is the unique uniform Maxwellian density that has the same principal moment as the initial value F_0. We know already, because of the first integrals (XXI.3), that if F is to approach some Maxwellian density in time, that density must be F_U.

We have purposely left unspecified the precise sense of the limit in (XI.61). It may be a type of uniform convergence, or possibly convergence in $\mathscr{L}^1(\mathscr{V})$, or perhaps something much weaker. This ambiguity reflects our present lack of knowledge about the trend to equilibrium. We now present ARKERYD's proof of (XI.61) interpreted as convergent in the sense of the weak topology in $\mathscr{L}^1(\mathscr{V})$. There are a number of ways in which to define the weak topology, but for the purposes of this discussion we shall say that a sequence $\{F_n\} \subset \mathscr{L}^1(\mathscr{V})$ converges weakly as $n \to \infty$ to $F \in \mathscr{L}^1(\mathscr{V})$, and we shall write $F_n \rightharpoonup F$, if[20] $\{\|F_n\|\}$

---

[20] *E.g.* Chapter V, Section 1 of K. YOSIDA, *Functional Analysis*, Berlin, *etc.*, Springer-Verlag, 1974.

is a bounded set, and for all measurable sets $\mathcal{U} \subset \mathcal{V}$,

$$\lim_{n\to\infty} \int_{\mathcal{U}} F_n \, d\mathbf{v} < \infty. \tag{XXI.149}$$

It is well known that if $\{F_n\}$ converges to F in $\mathscr{L}^1(\mathcal{V})$ then it also converges weakly, but in general the converse is false.

Several reasons, some of them reasons of expediency and one of them a reason based upon the desired physical interpretation, show why weak convergence is useful in a first proof of (XI.61). The first is the fact that if the convergence in (XI.61) is strong, say a type of uniform convergence or convergence in $\mathscr{L}^1(\mathcal{V})$, then it will also be weak. On the other hand, because weak convergence is not very restrictive, it is reasonable to expect that if (XI.61) is valid in any sense, that sense will include weak convergence. Thirdly, since weak convergence does not imply strong convergence, a weak version of (XI.61) should be easier to prove than a strong version. Next we note that since in the kinetic theory we are primarily interested in moments of the molecular density, a type of convergence that is based upon averages of F would seem appropriate. In view of (XXI.149) weak convergence has this property. Finally we remark that we wish to prove the trend to equilibrium for a very broad class of initial values, say for all those satisfying the requirements $(XXI.6)_{1,2}$ and (XXI.66) of the H-theorem. Because the corresponding class of solutions is also large, only for a weak type of convergence may we expect (XI.61) to hold *in one sense for all members* of that class. Of course a stronger type of convergence may hold for certain subclasses. For example, if we restrict attention to a gas of ideal spheres and consider F_0 belonging to the class of initial values that are continuous functions of $v = |\mathbf{v}|$ and that satisfy (XXI.5) when $\kappa \geq 8$, then by CARLEMAN's existence theorem and ARKERYD's theorem of uniqueness we know that ARKERYD's solutions are actually continuous functions of v and, as CARLEMAN showed, (XI.61) is valid in the sense of uniform convergence.

Despite the extent of our analysis in the preceding sections a number of results are still needed in preparation for the proof of the trend to equilibrium. We present these results in the next three lemmas. However, in contrast with our preceding analysis the proofs of these lemmas make use of somewhat deeper properties of functions in $\mathscr{L}^1(\mathcal{V})$ than those we have presented so far, and so we shall simply omit the proofs. For the details we refer the interested reader to the papers of ARKERYD.

As the first step in establishing an approach to equilibrium we must be able to show that the solution F actually exhibits some trend in time, even if only in the sense of weak convergence. ARKERYD noted that such information could be obtained within the framework of $\mathscr{L}^1(\mathcal{V})$ from little more than the first integrals (XXI.3) and the existence of $h[F]$. He stated this result as follows[21].

---

[21] ARKERYD [1972, 2, Corollary 3.4].

Lemma 7. *Let F be a solution of* (XXI.1) *whose initial value F_0 satisfies the conditions*

$$F_0 \geq 0, \qquad F_0 \log F_0 \in \mathscr{L}^1(\mathscr{V}), \qquad (1+v)^\kappa F_0 \in \mathscr{L}^1(\mathscr{V}), \qquad \text{(XXI.6)}_r$$

for some κ greater than 4. Then the sets

$$\mathscr{A}_k \equiv \{(1+v^k)F \,|\, t \geq 0\} \subset \mathscr{L}^1(\mathscr{V}), \qquad k = 0, 1, 2, \qquad \text{(XXI.150)}$$

are weakly relatively compact.

To interpret this result, let us note that if a set $\mathscr{A} \subset \mathscr{L}^1(\mathscr{V})$ is weakly relatively compact, then[22] given any sequence $\{F_n\} \subset \mathscr{A}$ we may extract a subsequence $\{F_{n_v}\}$ such that F_{n_v} converges weakly as $v \to \infty$ to an element of $\mathscr{L}^1(\mathscr{V})$. In the light of this property of weakly relatively compact sets, the preceding lemma tells us something, however little, about the trend of solutions in time.

Given any solution F of (XXI.1) whose initial value satisfies (XXI.6), by use of the preceding lemma we may construct many weakly convergent sequences $\{F(t_n)\}$. In order to deduce properties of the limits of these sequences, we require a form of the H-theorem that is somewhat stronger than just the inequality (XI.62). More precisely, we must have some estimate of the difference $h[F](t_2) - h[F](t_1)$ in terms of F. Formally we see from (XI.1) and (XI.2) that this difference is simply $(1/n) \int_{t_1}^{t_2} \bar{\mathbb{C}}_F \log F \, dt$. ARKERYD showed[23] that a result of this type could be established once the solutions were proved positive.

Lemma 8. *Let F be a solution of* (XXI.1) *whose initial value F_0 satisfies* (XXI.6) *for some κ not less than 4 and is such that $\|F_0\| > 0$. Then*

$$F(t, \mathbf{v}) > 0 \qquad \text{(XXI.151)}$$

for all positive t and for almost all $\mathbf{v} \in \mathscr{V}$, and (XI.62) *may be replaced by*

$$h[F](t_2) \leq h[F](t_1) + \frac{1}{n} \int_{t_1}^{t_2} \bar{\mathbb{C}}_F \log F \, dt. \qquad \text{(XXI.152)}$$

The only difference between (XXI.152) and the formal result mentioned previously is the fact that here we have an inequality rather than an equality. The inequality will suffice, however, for proof of the trend to equilibrium.

In order to show that a given solution F approaches F_U, if indeed it does approach something, we must make use of part (VIII.6) of the Boltzmann monotonicity theorem. At the same time we wish to work with weakly convergent sequences $\{F(t_n)\}$ in $\mathscr{L}^1(\mathscr{V})$. In order to interconnect these two ideas we must show that we may pass to the limit in $\bar{\mathbb{C}}_{F(t_n)} \log F(t_n)$ as $n \to \infty$, knowing only that $F(t_n)$ converges weakly. By adapting an argument of CARLEMAN's, ARKERYD showed[24] that one could pass to this limit in an important case.

---

[22] E.g. Theorem 8.12.1 of R. E. EDWARDS, *Functional Analysis*, New York *etc.*, Holt, Rinehart and Winston, 1965.

[23] ARKERYD [1972, 2, Theorem 2.3 and Eq. (2.10)].

[24] ARKERYD [1972, 2, Theorem 3.6].

Lemma 9. *Let F be a solution of* (XXI.1) *whose initial value F_0 satisfies* (XXI.6) *for some κ greater than 4. Let $\{t_n\}$ be a sequence of times for which*

$$F(t_n) \rightharpoonup \psi \qquad \text{as} \quad n \to \infty,$$
$$\bar{\mathbb{C}}_{F(t_n)} \log F(t_n) \to 0 \qquad \text{a.e. as} \quad n \to \infty. \qquad \text{(XXI.153)}$$

Then

$$\bar{\mathbb{C}}_\psi \log \psi = 0. \qquad \text{(XXI.154)}$$

With these three lemmas as preparation we may now present the proof of the ultra-narrow trend to equilibrium.

Theorem[25] 4. *Let F be a solution of* (XXI.1) *whose initial value F_0 satisfies* (XXI.6) *for some κ greater than 4. Then*

$$F(t) \rightharpoonup F_U \qquad \text{as} \quad t \to \infty. \qquad \text{(XXI.155)}$$

Proof. We know already that $h[F](t)$ is a decreasing function of t. By (XI.63) it is bounded from below by $h[F_U]$ and so it tends to a limit as $t \to \infty$. We conclude then from (XXI.152) and the Boltzmann monotonicity theorem that there is a sequence of times $\{t_n\}$ for which (XXI.153)$_2$ is satisfied. By Lemma 7 we may extract a subsequence $\{t_{n_\nu}\}$ such that

$$F(t_{n_\nu}) \rightharpoonup \psi \qquad \text{as} \quad \nu \to \infty \qquad \text{(XXI.156)}$$

for some function $\psi \in \mathscr{L}^1(\mathscr{V})$. By extracting subsequences once again we may ensure that $SF(t_{n_\nu})$ converges weakly to $S\psi$ for all summational invariants S. Hence

$$\int S\psi = \int SF_0. \qquad \text{(XXI.157)}$$

Applying Lemma 9 to (XXI.153)$_2$ and (XXI.156), we find that ψ satisfies (XXI.154), and so by (VIII.6) ψ is a Maxwellian density. In view of (XXI.157) we obtain

$$\psi = F_U. \qquad \text{(XXI.158)}$$

Now consider the behavior of $F(t)$ as $t \to \infty$. If the subsequence $\{t_{n_\nu}\}$ in (XXI.156) has a finite limit $\hat{t}$ as $\nu \to \infty$, then because F is a continuous function of t we see that $F(\hat{t}) = F_U$. By the uniqueness of solutions of (XXI.1) we find that $F(t) = F_U$, $t \geq \hat{t}$, and so we have finished the proof. If the limit of the sequence of times is infinite, let us suppose that F does not converge weakly to F_U. We conclude that there is a set $\mathcal{O}_1$ containing F_U and open in the weak topology in $\mathscr{L}^1(\mathscr{V})$, and a sequence of times $\{s_n\}$ such that $F(s_n)$ lies outside this set for all n. By using Lemma 7 again we may extract a certain subsequence,

---

[25] ARKERYD [1972, 2, Theorem 3.6].

(vi) TREND TO EQUILIBRIUM

which for convenience we also denote by $\{s_n\}$, with the following properties. First, s_n has the form $k_n + \kappa_n$ in which k_n is a positive integer and

$$0 \leq \kappa_n < 1, \quad \lim_{n \to \infty} \kappa_n \text{ exists}, \quad k_n < k_{n+1} \quad \text{for} \quad n = 1, 2, \ldots. \tag{XXI.159}$$

Secondly, there is a function $\hat{\psi} \in \mathscr{L}^1(\mathscr{V})$ such that

$$F(s_n) \to \hat{\psi} \neq F_U \quad \text{as} \quad n \to \infty. \tag{XXI.160}$$

Now let us perturb this sequence slightly by varying the numbers κ_n. In order to do this without moving too far from $\hat{\psi}$, we use the fact that $F(t)$ is uniformly continuous[26] in t. This implies that if $\mathcal{O}_2$ is a weakly open neighborhood of the origin in $\mathscr{L}^1(\mathscr{V})$ there is a positive number τ_0 such that

$$F(t) - F(t + \tau) \in \mathcal{O}_2 \quad \text{when} \quad t > 0, \quad |\tau| \leq \tau_0. \tag{XXI.161}$$

Without loss of generality we assume that $\mathcal{O}_1$ and $\hat{\psi} + \mathcal{O}_2$ are disjoint.

Next we note from (XXI.152) and the fact that $h[F](t)$ is bounded from below that

$$\int_{k_n}^{k_n+1} \bar{\mathbb{C}}_{F(t)} \log F(t) \, dt \to 0 \quad \text{a.e. as} \quad n \to \infty. \tag{XXI.162}$$

This implies that

$$\bar{\mathbb{C}}_{F(t+k_n)} \log F(t + k_n) \to 0 \quad \text{a.e. as} \quad n \to \infty \tag{XXI.163}$$

for almost all $t \in [0, 1]$. Let us choose and fix t in this interval such that (XXI.163) holds and also

$$|t - \kappa_n| < \tau_0 \tag{XXI.164}$$

---

[26] This fact follows directly from (XXI.1) and (XXI.80). Indeed, we may write

$$\|F(t_2) - F(t_1)\| \leq \int_{t_1}^{t_2} \|CF\| \, dt,$$

and since $w = w'$ and $w \leq (1 + v^2)(1 + v_*^2)$, we have the bounds

$$\|CF\| \leq \iint\int_{\mathscr{S}} w'F'F'_* + \iint\int_{\mathscr{S}} wFF_*,$$

$$\leq 2\|(1 + v^2)F\|^2,$$

$$= 2\|(1 + v^2)F_0\|^2.$$

Thus

$$\|F(t_2) - F(t_1)\| \leq \text{const.}(t_2 - t_1).$$

for all sufficiently large n. This is possible because of $(XXI.159)_{1,2}$. Because of (XXI.161) the points of the sequence $\{F(t + k_n)\}$ lie in $\hat{\psi} + \mathcal{O}_2$ for all sufficiently large n. Hence they do not approach F_U. On the other hand, since (XXI.163) applies we may use the argument leading to (XXI.158) and so conclude that there is a subsequence $\{t + k_{n_\nu}\}$ for which

$$F(t + k_{n_\nu}) \to F_U \quad \text{as} \quad \nu \to \infty. \tag{XXI.165}$$

This is a contradiction since $F(t + k_n)$ does not approach F_U. Hence we obtain (XXI.155). △

Appendix A

Estimation of Fourth Moments

We turn now to the proof of (XXI.88). Let us begin by looking more closely at the iterates F^n defined by (XXI.16). If $\|F_0\| = 0$, then from (XXI.77) we see that $\|F\| = 0$, so $F = 0$ almost everywhere in $\mathscr{V}$. In such a case (XXI.88) is satisfied trivially for any constant C_T. If $\|F_0\| \neq 0$, then we can use (XXI.16) to calculate the iterates, the first few of these being

$$F^1 = 0,$$

$$F^2 = F_0 e^{-\nu \|F_0\| t}, \tag{XXIA.1}$$

$$F^3 = F_0 e^{-\nu \|F_0\| t} + \frac{\mathcal{Q}_\nu F_0}{\nu \|F_0\|} (e^{-\nu \|F_0\| t} - e^{-2\nu \|F_0\| t}).$$

Continuing this calculation, we can show by induction that each iterate is a finite sum of terms of the form $\phi(t)\psi(\mathbf{v})$, the functions ϕ being differentiable and ψ being integrable. We conclude that F^n is t-differentiable; if we differentiate each side of $(XXI.16)_2$ with respect to t, we obtain

$$\partial_t F^{n+1} + \nu \|F_0\| F^{n+1} = \mathcal{Q}_\nu F^n. \tag{XXIA.2}$$

Let τ be a positive number such that

$$\xi \equiv \tau C(4) \|(1 + v^2)F_0\| < 1, \tag{XXIA.3}$$

$C(n)$ being the positive constant that appears in (XXI.65). We shall use induction on n to prove that for any fixed positive integer N

$$\|(1 + v^2)^2 F^n\|$$

$$\leq \frac{1}{1 - \xi} \|(1 + v^2)^2 F^n((N-1)\tau, \mathbf{v})\| \quad \text{when} \quad (N-1)\tau \leq t \leq N\tau. \tag{XXIA.4}$$

Since $F^1 = 0$, this result is certainly true if $n = 1$. Assume it is true for some fixed integer n. Let us multiply (XXIA.2) by $(1 + v^2)^2$ and then integrate first with respect to t from $(N - 1)\tau$ to t and then with respect to $\mathbf{v}$. Using Tonelli's theorem, we obtain

$$\|(1 + v^2)^2 F^{n+1}(t, \mathbf{v})\| = \|(1 + v^2)^2 F^{n+1}((N - 1)\tau, \mathbf{v})\|$$
$$+ \int_{(N-1)\tau}^{t} \left[\int (1 + v^2)^2 (\mathbb{Q}_v F^n - v\|F_0\|F^{n+1}) \right] ds. \quad \text{(XXIA.5)}$$

But, from (XXI.62) with $S = 1$ and $\alpha = v$, we can write

$$\int (1 + v^2)^2 (\mathbb{Q}_v F^n - v\|F_0\|F^{n+1})$$
$$= \int (1 + v^2)^2 \mathbb{C}_v F^n + \int (1 + v^2)^2 v(\|F^n\|F^n - \|F_0\|F^{n+1}), \quad \text{(XXIA.6)}$$
$$= \int (1 + v^2)^2 \mathbb{C}_v F^n + \int (1 + v^2)^2 v \|F^n\|(F^n - F^{n+1})$$
$$+ \int (1 + v^2)^2 v F^{n+1}(\|F^n\| - \|F_0\|).$$

By (XXI.71) the last integral is negative, while the next to last integral is also negative since $F^n \leq F^{n+1}$. In order to estimate the first term, we appeal to POVZNER's inequality (XXI.65) with $n = 4$. In view of (XXI.84) and the hypothesis of induction, and recalling that $\int_{\mathscr{S}} 1 = \pi d^2 = 1$, we obtain

$$\int (1 + v^2)^2 \mathbb{C}_v F^n \leq C(4)\|(1 + v^2)^2 F^n\| \, \|(1 + v^2)F^n\|,$$
$$\leq \frac{C(4)}{1 - \xi} \|(1 + v^2)^2 F^n((N - 1)\tau, \mathbf{v})\| \, \|(1 + v^2)F_0\| \quad \text{(XXIA.7)}$$

when $(N - 1)\tau \leq t \leq N\tau$. Since $F^n \leq F^{n+1}$,

$$\int (1 + v^2)^2 \mathbb{C}_v F^n \leq \frac{1}{\tau} \frac{\xi}{1 - \xi} \|(1 + v^2)^2 F^{n+1}((N - 1)\tau, \mathbf{v})\| \quad \text{(XXIA.8)}$$

when $(N - 1)\tau \leq t \leq N\tau$. By combining (XXIA.5), (XXIA.6), and (XXIA.8) we find that

$$\|(1 + v^2)^2 F^{n+1}(t, \mathbf{v})\| \leq \|(1 + v^2)^2 F^{n+1}((N - 1)\tau, \mathbf{v})\| \left(1 + \frac{\xi}{1 - \xi}\right) \quad \text{(XXIA.9)}$$

when $(N - 1)\tau \leq t \leq N\tau$, and since $1 + \xi/(1 - \xi) = 1/(1 - \xi)$, we obtain (XXIA.4).

If we set $N = 1, 2, \ldots$ in (XXIA.4), we obtain

$$\|(1 + v^2)^2 F^n(\tau, \mathbf{v})\| \leq \frac{1}{1 - \xi} \|(1 + v^2)^2 F_0\|,$$

$$\|(1 + v^2)^2 F^n(2\tau, \mathbf{v})\| \leq \frac{1}{1 - \xi} \|(1 + v^2)^2 F^n(\tau, \mathbf{v})\|,$$

$$\vdots \qquad \text{(XXIA.10)}$$

$$\|(1 + v^2)^2 F^n((N-1)\tau, \mathbf{v})\| \leq \frac{1}{1 - \xi} \|(1 + v^2)^2 F^n((N-2)\tau, \mathbf{v})\|.$$

By combining these, we find that

$$\|(1 + v^2)^2 F^n((N-1)\tau, \mathbf{v})\| \leq \frac{1}{(1 - \xi)^{N-1}} \|(1 + v^2)^2 F_0\|, \quad \text{(XXIA.11)}$$

and using this in (XXIA.4) we obtain

$$\|(1 + v^2)^2 F^n(t, \mathbf{v})\| \leq \frac{1}{(1 - \xi)^N} \|(1 + v^2)^2 F_0\| \quad \text{when} \quad (N-1)\tau \leq t \leq N\tau.$$

$$\text{(XXIA.12)}$$

Finally, since $\xi < 1$, we see that $1/(1 - \xi) > 1$, and so we conclude from (XXIA.12) that

$$\|(1 + v^2)^2 F^n(t, \mathbf{v})\| \leq \frac{1}{(1 - \xi)^N} \|(1 + v^2)^2 F_0\| \quad \text{when} \quad 0 \leq t \leq N\tau.$$

$$\text{(XXIA.13)}$$

Now we let n approach ∞ and use the monotone convergence theorem to conclude that

$$\|(1 + v^2)^2 F(t, \mathbf{v})\| \leq \frac{1}{(1 - \xi)^N} \|(1 + v^2)^2 F_0\| \quad \text{when} \quad 0 \leq t \leq N\tau.$$

$$\text{(XXIA.14)}$$

Therefore we may choose N large enough that $T \leq N\tau$ and then choose $C_T \equiv (1 - \xi)^{-N} \|(1 + v^2)^2 F_0\|$. This will give us (XXI.88). To complete the proof we must be sure that C_T is independent of v. But v can enter the formula for C_T only through $C(4)$, and by the statement of POVZNER's inequality (XXI.65), $C(n)$ is independent of v. $\triangle$

APPENDIX A. ESTIMATION OF FOURTH MOMENTS

We may obtain a more explicit form of (XXIA.14) by giving ξ some convenient value. For example, if we set $\xi = \frac{1}{2}$ and use (XXIA.3)$_1$ to eliminate τ, we see that (XXIA.14) becomes

$$\|(1 + v^2)^2 F(t, \mathbf{v})\| \leq 2^N \|(1 + v^2)^2 F_0\| \quad \text{when} \quad 0 \leq t \leq \frac{N}{2C(4)\|(1 + v^2)F_0\|}.$$

(XXIA.15)

Part G

Grossly and Momentally Determined Solutions and the Iterative Procedures of the Kinetic Theory

Chapter XXII

Hilbert's Formal Iterative Procedure for Calculating Gas-dynamic Solutions. The Assertion of Gross Causality. The Hilbert Mapping

In Chapters IV, XIII, XIV, and XV we have shown both by general reasoning and by exhibiting solutions for special gas flows that for a *typical* solution of the Maxwell–Boltzmann equation the fields **P** and **q** are *not* grossly determined. In Chapter X we have seen that the solutions constituting one special class, namely, the locally Maxwellian ones, *are* grossly determined; of course, so are the corresponding **P** and **q**, both of them being in fact null for each member of the class. In Chapters XIII, XIV, and XV we have seen that the **P** and **q** corresponding to certain other solutions *are* momentally determined and are *not null*; they are also grossly determined at least when $\mathbf{b} = \mathbf{0}$. The importance of momentally or grossly determined solutions is obvious, for it is only they, if not directly then at least through some limiting process, that permit us even to frame such questions as whether the kinetic gas has a viscosity and a Maxwell number. We have shown in Chapters XIII and XIV some ways in which these quantities may be defined and calculated.

(i) *Hilbert's formal iterative procedure*

In a celebrated memoir, second only to the works of MAXWELL and BOLTZMANN themselves in its importance and influence upon later research into the kinetic theory, HILBERT[1] was the first to recognize and assert that "the most general task" of the kinetic theory was to determine solutions of a *special kind*.

---

[1] HILBERT [1912]. In presenting the formal results obtained in or implied by this paper we have made good use of the exposition of GRAD [1958, §§23–24].

He sought[2] molecular densities corresponding to "a gas in a stable state of motion", which he explained as follows: "F is to be finite for all t and remain continuous." A power series in $t - t_0$, he wrote, "would contradict the above condition of stability," because it could be expected to converge only if $|t - t_0|$ were sufficiently small. So as to obviate solutions lacking this "stability", HILBERT chose to consider a class of molecular densities $(t, \mathbf{x}, \mathbf{v}) \mapsto F(t, \mathbf{x}, \mathbf{v}; \delta)$ depending upon a parameter δ in such a way as to render $F - F^{(0)}/\delta$ an analytic function of δ, $F^{(0)}$ being some function of t, $\mathbf{x}$, and $\mathbf{v}$ only. Thus HILBERT sought to determine functions $F^{(n)}$ such that

$$F(t, \mathbf{x}, \mathbf{v}; \delta) = \frac{1}{\delta} F^{(0)}(t, \mathbf{x}, \mathbf{v}) + F^{(1)}(t, \mathbf{x}, \mathbf{v}) + \delta F^{(2)}(t, \mathbf{x}, \mathbf{v}) + \cdots,$$

$$= \sum_{n=0}^{\infty} \delta^{n-1} F^{(n)}(t, \mathbf{x}, \mathbf{v}). \tag{XXII.1}$$

Although HILBERT gave no further reason for making this strange assumption[3], he claimed[4] it to be "*the most general expression for the Maxwellian fundamental function of a gas in a stable state of motion*" and "*the mathematical expression of the stability of the state of motion of the gas* ...".

HILBERT proceeded with a master's hand to determine the severe restrictions that his assumption (XXII.1) entailed. Although he considered only ideal spheres, his development is easily extended to arbitrary kinetic constitutive quantities[5]. HILBERT's analysis is purely formal; even for the existence of the general term $F^{(n)}$ he gave only a partial proof; even presuming the existence of that term, he made no attempt to prove that the series (XXII.1) converged. In recent years one of his steps has been justified, but the rest of his theory remains today in just the state where he left it, nearly seventy years ago.

Any solution G of the Maxwell-Boltzmann equation may be represented in the form (XXII.1): We need only choose $F^{(0)} = 0$, $F^{(1)} = G$, $F^{(2)} = F^{(3)} = \ldots = 0$. HILBERT ruled out this possibility by assuming implicitly that

$$F^{(0)} \neq 0, \tag{XXII.2}$$

and it is the combination of this assumption with (XXII.1) that makes the class of solutions that he obtained a very special one indeed. The sign of F is the

---

[2] HILBERT [1912, pp. 564, 565, 576].

[3] Physicists and chemists often interpret δ as a measure of the mean free path and call (XXII.1) an expansion of F for small mean free path. However, this interpretation is somewhat obscure (*cf.* Footnote 5 to Chapter VII), and we shall not pursue it further. We are free to make any special choice of δ we please. For example, we may choose δ as $\mathfrak{m}$ and then take the results as $\mathfrak{m} \to 0$ to concern lighter and lighter gases of a given kind. However, since we need not identify δ, in this book we will not do so.

[4] HILBERT [1912, pp. 565, 576].

[5] A remark of LUNN [1913] may refer to this fact or to the far more difficult matter described in Footnote 2 to Chapter XXIII.

(i) HILBERT'S FORMAL PROCEDURE 377

same as the sign of $F^{(0)}/\delta$, at least for sufficiently small δ. For naturalness of interpretation we shall assume that $F^{(0)} > 0$; consequently in interpreting HILBERT's expansion (XXII.1) we shall always assume that $\delta > 0$, even though, of course, when δ is interior to the interval of convergence the series converges also for $-\delta$.

If we substitute (XXII.1) into the Maxwell–Boltzmann equation (IX.1) and use the fact that $\mathbb{C}$ is bilinear, we may express each side of the resulting equation as a power series in δ. By equating coefficients of like powers of δ we obtain HILBERT's *iterative system*:

$$\mathbb{C}(F^{(0)}, F^{(0)}) = 0,$$

$$2\mathbb{C}(F^{(0)}, F^{(1)}) = \mathfrak{D}F^{(0)}, \tag{XXII.3}$$

$$2\mathbb{C}(F^{(0)}, F^{(n)}) = \mathfrak{D}F^{(n-1)} - \sum_{m=1}^{n-1} \mathbb{C}(F^{(m)}, F^{(n-m)}) \quad \text{if } n \geq 2.$$

The first member of this system asserts that $\mathbb{C}F^{(0)} = 0$. By (VIII.7) and (XXII.2) we conclude that $F^{(0)}$ is *Maxwellian, its density and energetic being any positive scalar fields $\rho^{(0)}$ and $\varepsilon^{(0)}$, and its gross velocity $\mathbf{u}^{(0)}$ being an arbitrary vector field.* The conditions (XXII.2) and (XXII.3)$_1$ place no further restrictions upon $F^{(0)}$. With regard to the remaining members of (XXII.3), let us note that since $\mathbb{C}(\ ,\ )$ is bilinear and $F^{(0)}$ does not vanish, $\mathbb{C}(F^{(0)}, \cdot)$ is a non-trivial linear integral operator. Thus, when $F^{(0)}$ is specified, (XXII.3)$_2$ becomes a linear integral equation for the function $F^{(1)}$. More generally, for any given functions $F^{(0)}, \ldots, F^{(n-1)}$, the right-hand side of (XXII.3)$_3$ is a known function. Accordingly (XXII.3)$_3$ is nothing more than a linear integral equation for $F^{(n)}$, the integral operator in this equation being again $\mathbb{C}(F^{(0)}, \cdot)$. It appears, then, that the problem of determining the terms $F^{(n)}$ is simply one of solving an iterative system of linear integral equations. However, upon closer examination of $\mathbb{C}(F^{(0)}, \cdot)$ and its properties we shall see that the problem is actually much more difficult.

We begin by obtaining necessary conditions that the iterative system (XXII.3) have a solution. Suppose that there be a particular solution $F_P^{(n)}$ of each member of the system of integral equations (XXII.3)$_{2,3}$. Since these equations are linear, the difference $D^{(n)}$ of two such particular solutions must satisfy the homogeneous equation

$$\mathbb{C}(F^{(0)}, D^{(n)}) = 0, \tag{XXII.4}$$

and so the general solution of the system (XXII.3)$_{2,3}$ is formed by adding to $F_P^{(n)}$ the general solution of this homogeneous equation. If in (VII.20)$_2$ we replace G by $F^{(0)}$ and H by $F^{(0)}g$ and then use the fact that any Maxwellian density $F^{(0)}$ satisfies the relation $F_*^{(0)'}F^{(0)'} = F_*^{(0)}F^{(0)}$, we find that

$$\int g\mathbb{C}(F^{(0)}, F^{(0)}g) = -\tfrac{1}{8} \iiint_{\mathscr{S}} w F_*^{(0)} F^{(0)} (g'_* + g' - g_* - g)^2. \tag{XXII.5}$$

Therefore $D^{(n)}$ cannot satisfy (XXII.4) unless it be the product of $F^{(0)}$ and a summational invariant. The Boltzmann–Gronwall theorem, stated and proved in Chapter VI, asserts that every measurable summational invariant is a linear combination of the components of the principal invariant **I**, defined by (VI.20), and so the general solution of (XXII.4) is

$$D^{(n)} = \sum_{\alpha=0}^{4} C_\alpha^{(n)} I_\alpha F^{(0)}. \tag{XXII.6}$$

Therefore, the general solution of the system $(\text{XXII.3})_{2,3}$ has the form

$$F^{(n)} = F_P^{(n)} + \sum_{\alpha=0}^{4} C_\alpha^{(n)} I_\alpha F^{(0)}, \tag{XXII.7}$$

$\mathbf{C}^{(n)}$ being an arbitrary function of t and $\mathbf{x}$.

If some solution $F^{(n)}$ of the system of integral equations is known, we may easily calculate a field $\mathbf{C}^{(n)}$ such as to express this solution in terms of $F^{(0)}$ and the particular solution $F_P^{(n)}$ in the form (XXII.7). To do so, we multiply (XXII.7) by I_β and then integrate with respect to $\mathbf{v}$. The result is

$$\sum_{\alpha=0}^{4} A_{\beta\alpha} C_\alpha^{(n)} = \int I_\beta(F^{(n)} - F_P^{(n)}), \qquad \beta = 0, 1, \ldots, 4, \tag{XXII.8}$$

A being given by

$$A_{\beta\alpha} \equiv \int I_\beta I_\alpha F^{(0)}. \tag{XXII.9}$$

Since $F^{(0)}$ is a Maxwellian density, we may calculate **A** explicitly and thereby display its properties. One is

$$\det \mathbf{A} \neq 0. \tag{XXII.10}$$

Thus (XXII.8), interpreted as a system of five linear equations for the five fields $C_\alpha^{(n)}$, $\alpha = 0, 1, \ldots, 4$, has a unique solution. In terms of this solution, $F^{(n)}$ is given by (XXII.7). We may express this simple result in the following form. *If the system $(\text{XXII.3})_{2,3}$ has a solution $F_P^{(n)}$, then, for any function $\rho^{(n)}$ there is a unique solution $F^{(n)}$ of this same system having $\rho^{(n)}$ as its principal moment.* Indeed, (XXII.10) allows us to choose the field $\mathbf{C}^{(n)}$ as one that satisfies the conditions

$$\sum_{\alpha=0}^{4} A_{\beta\alpha} C_\alpha^{(n)} = \frac{1}{\mathfrak{m}} \rho_\beta^{(n)} - \int I_\beta F_P^{(n)}, \tag{XXII.11}$$

define $F^{(n)}$ by (XXII.7), and then note by subtracting (XXII.11) from (XXII.8) that $\boldsymbol{\rho}^{(n)} = \mathfrak{m} \int F^{(n)} \mathbf{I}$. In particular there is a unique solution whose principal moment vanishes. We may assume, without loss of generality, that this solution is $F_P^{(n)}$:

$$\int I_\alpha F_P^{(n)} = 0, \qquad \alpha = 0, 1, \ldots, 4. \tag{XXII.12}$$

(i) HILBERT'S FORMAL PROCEDURE

According to (VII.22), $\int \mathbb{C}(G, H)\mathbf{I} = \mathbf{0}$ for general functions G and H. If we multiply (XXII.3)$_{2,3}$ by I_α, integrate with respect to $\mathbf{v}$, and then use the property of $\mathbb{C}$ we have just recalled, we obtain another necessary condition for the existence of a solution, namely

$$\int I_\alpha \mathfrak{D} F^{(n)} = 0 \quad \text{if} \quad n \geq 0, \quad \alpha = 0, 1, \ldots, 4. \tag{XXII.13}$$

Recalling the notation (VI.20) and the theorem proved in Section (iv) of Chapter V, we may state this result as follows: *the fields ρ, $\mathbf{u}$, ε, p, $\mathbf{P}$, and $\mathbf{q}$ defined from $F^{(n)}$ satisfy the field equations of continuum mechanics*, namely, (I.1), (I.2), and (I.4). When $n = 0$, those field equations have the especially simple forms

$$\dot{\rho}^{(0)} = -\rho^{(0)} u^{(0)}_{a,a},$$

$$\rho^{(0)} \dot{u}^{(0)}_k = -p^{(0)}_{,k} + \rho^{(0)} b_k = -\tfrac{2}{3}(\rho^{(0)} \varepsilon^{(0)})_{,k} + \rho^{(0)} b_k, \tag{XXII.14}$$

$$\rho^{(0)} \dot{\varepsilon}^{(0)} = -p^{(0)} u^{(0)}_{a,a} = \tfrac{2}{3} \dot{\rho}^{(0)} \varepsilon^{(0)},$$

$\rho^{(0)}$, $\mathbf{u}^{(0)}$, and $\varepsilon^{(0)}$ being respectively the mass density, gross velocity, and energetic that define the Maxwellian molecular density $F^{(0)}$. The superimposed dot here denotes the material derivative with respect to the velocity field $\mathbf{u}^{(0)}$. Thus *the 0th term $F^{(0)}$ in* HILBERT'S *procedure must correspond to a flow of an* EULER-HADAMARD *gas with constant specific heats whose ratio is $\tfrac{5}{3}$*, and, conversely, any such flow with everywhere positive ρ and ε determines uniquely one admissible $F^{(0)}$. The starting point of HILBERT'S procedure thus requires the solution of some problem of classical gas dynamics. It may be any such problem we please. We must note that smooth solutions of the initial-value problem of gas dynamics generally fail to exist for more than some finite interval of time, and this interval may be very small.

As we have just seen, when $n = 0$ the necessary condition (XXII.13) is equivalent to a system of differential equations. The same is true for greater values of n, for if we substitute (XXII.7) into (XXII.13) and then set

$$B^k_{\alpha\beta} \equiv \int v_k I_\alpha I_\beta F^{(0)}, \quad C_{\alpha\beta} \equiv \int I_\alpha \mathfrak{D}(I_\beta F^{(0)}), \quad d^{(n)}_\alpha \equiv \int I_\alpha \mathfrak{D} F^{(n)}_P, \tag{XXII.15}$$

we obtain

$$\sum_{\beta=0}^{4} (A_{\alpha\beta} \partial_t C^{(n)}_\beta + B^k_{\alpha\beta} C^{(n)}_{\beta,k} + C_{\alpha\beta} C^{(n)}_\beta) + d^{(n)}_\alpha = 0,$$

$$\alpha = 0, 1, \ldots, 4, \quad n = 1, 2, \ldots. \tag{XXII.16}$$

These are HILBERT'S *field equations*. They provide restrictions on $F^{(0)}$, $F^{(n)}_P$, and $C^{(n)}$ which must be satisfied if there is to be a solution of the integral equations (XXII.3).

Anyone familiar with modern continuum mechanics will be keenly disappointed by HILBERT's field equations for positive values of n. Such a person will expect that those equations describe some natural hierarchy of continuum theories of fluids. Certainly he will expect that when $n = 1$ they will be the differential equations of the Stokes-Kirchhoff theory. Nothing of the sort. They are obscure in mechanics; they are likewise obscure in analysis; it is not even clear that they possess solutions.

(ii) *Proof of effectiveness*

Before interpreting HILBERT's field equations we turn to conditions sufficient that a solution of the iterative system exist. That is, aside from the question whether HILBERT's series (XXII.1) converge or not, and, if so, whether it converge to a solution, we may ask simply *whether or not the procedure be effective* in the sense that the iterates to which it refers actually exist. We have shown that if there is a solution of the iterative system $(XXII.3)_{2,3}$, it will be given by (XXII.7), $F_P^{(n)}$ being the unique solution of $(XXII.3)_{2,3}$ that satisfies the integral condition (XXII.12), and the field $C^{(n)}$ being such as to satisfy HILBERT's field equations (XXII.16). Proceeding by induction, then, we suppose that $F^{(0)}, \ldots, F^{(k-1)}$ be known, and hence that (XXII.12) and (XXII.16) be valid when $n = 1, \ldots, k - 1$. We ask first whether $F_P^{(k)}$ exists. That is, we seek a solution of the equations

$$C(F^{(0)}, F_P^{(k)}) = H^{(k)}, \qquad \int I_\alpha F_P^{(k)} = 0, \qquad \alpha = 0, 1, \ldots, 4, \quad (XXII.17)$$

$H^{(k)}$ being given in terms of $F^{(0)}, \ldots, F^{(k-1)}$ by

$$H^{(k)} \equiv \begin{cases} \frac{1}{2}\mathfrak{D}F^{(0)} & \text{if } k = 1, \\ \frac{1}{2}\mathfrak{D}F^{(k-1)} - \frac{1}{2}\sum_{m=1}^{k-1} C(F^{(m)}, F^{(k-m)}) & \text{if } k > 1. \end{cases} \quad (XXII.18)$$

In view of (VII.22) and (XXII.13), $H^{(k)}$ satisfies

$$\int I_\alpha H^{(k)} = 0. \qquad (XXII.19)$$

The first stage in the proof of effectiveness is thus to prove that $F_P^{(k)}$ exists. Of the several difficult problems of analysis HILBERT's scheme presents, this is the only one he himself attempted to solve. He considered only a gas of ideal spheres, and even then he obtained only a partial solution. Because mathematicians in recent years have grown interested in operators such as that which appears in $(XXII.17)_1$, results more inclusive than his are now available. We state them here without proof.

(ii) PROOF OF EFFECTIVENESS

Let $\mathscr{L}_M^2(\mathscr{V})$ denote the space of functions which are square-integrable on $\mathscr{V}$ with weight $(F^{(0)})^{-1}$, and consider the set $\mathscr{L}_M^2(\mathscr{V})^\perp$ of functions $F \in \mathscr{L}_M^2(\mathscr{V})$ such that

$$\int I_\alpha F = 0, \qquad \alpha = 0, 1, \ldots, 4. \tag{XXII.20}$$

If we denote by $\mathbb{L}$ the restriction of $\mathbb{C}(F^{(0)}, \cdot)$ to $\mathscr{L}_M^2(\mathscr{V})^\perp$, then

$$\mathbb{L}: \mathscr{L}_M^2(\mathscr{V})^\perp \to \mathscr{L}_M^2(\mathscr{V})^\perp, \tag{XXII.21}$$

and so in view of (XXII.17)$_2$ and (XXII.19) we seek a solution of

$$\mathbb{L} F_P^{(k)} = H^{(k)}. \tag{XXII.22}$$

That this equation has a solution for many common molecular models, is ensured by the following **Theorem**[6]. *For a gas consisting of either*

(a) *molecules with a cut-off hard potential, or*
(b) *inverse* $\mathbb{k}^{th}$-*power molecules for which* $\mathbb{k} > 3$,

the operator $\mathbb{L}$ *is invertible.* We see, then, that (XXII.17) does have a solution, at least for some types of molecules. From the necessary conditions derived earlier we know that there is only one such solution.

---

[6] Part (a) of this theorem is due to GRAD [1963, 2, §§II and IV]; part (b), in the generality asserted here, to PAO [1974, 1 & 2]. PAO's proof is quite intricate, making use of abstract properties of pseudo-differential operators, but to our knowledge it is the first analytical study of any kind that treats molecules subject to forces of infinite range other than Maxwellian ones. For molecules with a cut-off, all proofs we have seen proceed in two steps. In the first, one shows that $\mathbb{C}(F^{(0)}, \cdot)$ can be written in the form

$$\mathbb{C}(F^{(0)}, G)(\mathbf{v}) = -\lambda(v) \frac{G(\mathbf{v})}{F^{(0)}(\mathbf{v})} + \int_{\mathscr{V}} k(\mathbf{v}, \mathbf{v}_*) \sqrt{\lambda(v)\lambda(v_*)} \frac{G(\mathbf{v}_*)}{F^{(0)}(\mathbf{v}_*)} d\mathbf{v}_*,$$

λ being a positive function of v, and k being symmetric. The variables t and $\mathbf{x}$ have been suppressed here, although all functions, including λ and k, generally depend upon them. HILBERT [1912, Eqs. (18), (23)] by a staggering calculation established this form of $\mathbb{C}(F^{(0)}, \cdot)$ for a gas of ideal spheres; GRAD [1963, 2, §II] and CHAPMAN & COWLING [1939, §7.6], for general molecules with a cut-off. The last reference is typical of many studies in the kinetic theory in that it blurs the class of kinetic constitutive equations to which its results apply, often giving the false impression of having considered all possible molecular models. Indeed, in the middle of their proof CHAPMAN & COWLING lay down the assumption (their Eq. 7.6.6 and the line of text preceding it) that $\mathscr{S}$ is finite, but they do not mention this restriction in §7.13 when they apply the result to justify use of "the theory of integral equations". Also their appeal to pp. 99 and 129 of COURANT & HILBERT's *Methoden der mathematischen Physik* **1**, 2nd ed., Springer-Verlag, Berlin, 1931, is a mere feint, for the passage they cite refers to integration over a square, and its results cannot be extended by "analytical convenience" to integration over $\mathscr{V} \times \mathscr{V}$. Cf. Chapter XIX.

Once the basic expression of $\mathbb{C}(F^{(0)}, \cdot)$ in terms of a symmetric kernel has been established, we may set

$$g \equiv \sqrt{\lambda}\, G/F^{(0)}$$

(*continued*)

Since we have now a particular solution $F_P^{(k)}$, we may apply the remaining necessary conditions set forth above. In particular, the general solution of (XXII.3)$_{2,3}$ has the form (XXII.7), and HILBERT's field equations (XXII.16) are all that remain to be satisfied. Since only the field $C^{(n)}$ in (XXII.7) is still arbitrary, we might hope to prove that by choosing this field appropriately we could satisfy (XXII.16). Let us explore this possibility further.

From (XXII.9) and (XXII.15)$_{1,2}$ the fields **A**, **B**^{k}, and **C** are determined uniquely by the Maxwellian density $F^{(0)}$ which is presumed known. Indeed,

and write that expression in the form

$$\frac{1}{\sqrt{\lambda}} \mathbb{C}\left(F^{(0)}, \frac{1}{\sqrt{\lambda}} F^{(0)} g\right) = -g + \mathbb{K} g,$$

$\mathbb{K}$ being the integral operator

$$\mathbb{K} g(\mathbf{v}) \equiv \int_{\mathscr{V}} k(\mathbf{v}, \mathbf{v}_*) g(\mathbf{v}_*) \, d\mathbf{v}_*.$$

Since the function k is symmetric, $\mathbb{K}$ is formally a self-adjoint operator.

The second step in the proof of invertibility is a proof that $\mathbb{K}$, or some power of $\mathbb{K}$, is completely continuous. In this regard HILBERT [1912, p. 571] simply noted that the kernel k had "a first-order infinity" at $\mathbf{v} = \mathbf{v}_*$ and concluded that the theory of linear integral equations could be applied. His ideas on this subject, as reflected in the two paragraphs at the end of Chapter 15 of his *Grundzüge einer allgemeinen Theorie der linearen Integralgleichungen*, Teubner, Leipzig & Berlin, 1912, do not seem to bear out this conclusion. Indeed, HILBERT's student HECKE [1922, p. 274] noticed that the kernel k did not satisfy the requirements then known to be sufficient for $\mathbb{K}$ to be completely continuous, so that "it was uncertain whether the classical theory would apply to Hilbert's equation." HECKE knew that a condition sufficient for the abstract theory to apply was that some power of $\mathbb{K}$ be completely continuous; he knew also that such would be the case if the corresponding iterated kernel were square-integrable. HECKE [1922, §1] went on to prove, for a gas of ideal spheres, that the fifth iterated kernel was square-integrable, thereby completing the gap left by HILBERT. A second proof for ideal spheres has been given by CARLEMAN [1957, Part 2, Chapter 1, §2].

In order to complete the general proof we appeal to the Riesz-Schauder theory of completely continuous operators, as presented, for example, in Chapter X of YOSIDA's book cited in Footnote 20 to Chapter XXI. Since $\mathbb{K}$ is a self-adjoint completely continuous operator, that theory tells us that the integral equation

$$-g + \mathbb{K} g = h$$

has a solution if and only if h is orthogonal, in the $\mathscr{L}^2(\mathscr{V})$ inner product, to all solutions of the homogeneous equation. But this means that

$$\mathbb{C}(F^{(0)}, F) = H$$

has a solution if and only if

$$\int I_\alpha H = 0, \qquad \alpha = 0, 1, \ldots 4.$$

In view of the domain and range indicated in (XXII.21), this last result is equivalent to the invertibility of $\mathbb{L}$.

they can easily be calculated, but here we do not need them. The field $\mathbf{d}^{(k)}$, given by (XXII.15)$_3$, is determined uniquely by $F_P^{(k)}$, or equivalently, in view of the existence theorem established above, by $F^{(0)}, \ldots, F^{(k-1)}$, these functions being presumed known also. Thus, according to the hypothesis of induction, the four fields $\mathbf{A}$, $\mathbf{B}^k$, $\mathbf{C}$, and $\mathbf{d}^{(k)}$ are known at the kth stage. HILBERT's field equations, then, form a system of partial differential equations for the field $\mathbf{C}^{(k)}$. In view of (XXII.10) we may solve these equations for $\partial_t \mathbf{C}^{(k)}$, and thus it is reasonable to set up an initial-value problem. HILBERT simply assumed that a unique solution $\mathbf{C}^{(k)}(t, \mathbf{x})$ of this problem would exist, at least for t sufficiently close to the initial time. If we accept this assumption, we see that the kth member of (XXII.3) has a unique solution, and the induction is complete.

Although some steps in this proof of effectiveness have been established rigorously, some remain either purely formal or absent altogether. It is not difficult to see what these are. First, on the assumption that $F^{(0)}, \ldots, F^{(k-1)}$ were known, we attempted to find $F_P^{(k)}$ such as to satisfy (XXII.17), and this we were able to do by applying the aforementioned theorem. However, in doing so we assumed that $F^{(k-1)}$ was differentiable and that $C(F^{(m)}, F^{(k-m)})$ existed for $m = 1, \ldots, k-1$ in order that we might form $H^{(k)}$ as given by (XXII.18)$_2$. In addition we assumed implicitly that $H^{(k)}$ was in $\mathscr{L}_M^2(\mathscr{V})$ so that the theorem would apply. In order to account for these steps in the proof we should expand the hypothesis of induction to include them and then prove that the solution $F^{(k)}$ which results also has these properties. Before we can do this, we must examine again the steps delivering $F^{(k)}$, using now such an expanded hypothesis of induction. The preceding theorem tells us that (XXII.17) has a solution $F_P^{(k)}$ lying in $\mathscr{L}_M^2(\mathscr{V})$. The next step in finding $F^{(k)}$ involves HILBERT's field equations, and in order to write down these equations we must calculate the fields $\mathbf{A}$, $\mathbf{B}^k$, $\mathbf{C}$, and $\mathbf{d}^{(k)}$. As we can see from (XXII.15)$_3$, we must prove first that $F_P^{(k)}$ is differentiable and that $I_\alpha \mathfrak{D} F_P^{(k)}$ is integrable. Even if this were possible, we should still have before us the task of proving that HILBERT's field equations have a solution. Once this were done, to complete the induction we should have to show that $H^{(k+1)}$ exists and lies in $\mathscr{L}_M^2(\mathscr{V})$.

These gaps in the analysis, though clear and well defined, have escaped the attention of mathematicians ever since HILBERT first published his results, and they remain to this day[7]. Even so, one further step is required before HILBERT's

---

[7] One reason for this is that HILBERT's results are too often interpreted as being mathematically rigorous. CHAPMAN & COWLING [1939, Historical Summary, §10] wrote that HILBERT "approached the subject from the standpoint of the pure mathematician, laying stress on the necessity of a proof that a solution of Boltzmann's equation actually exists—a point that the physicist is content to assume, the more readily because he knows in how many respects such an equation as that of Boltzmann represents but an approximation to the conditions of the natural problem. Hilbert reformulated Maxwell's idea of successive approximation to f, and showed that the solution of Boltzmann's equation could be effected by solving an infinite series of linear integral

(continued)

procedure may be considered complete. That is a proof that the series (XXII.1) formed from the functions $F^{(n)}$ converges, at least for small values of δ. HILBERT simply assumed this to be so, and then proceeded to derive certain properties of the function it defined. Indeed, convergence may be the most difficult aspect to investigate since it would require certain information from preceding stages of the procedure. For example, even if the initial-value problem for (XXII.16) has a solution in some finite interval $[0, t_n]$, it might happen that $t_n \to 0$ as $n \to \infty$, and the existence of a limit function F would then be unlikely. As we know of no analysis bearing on these points, we can do no more now than follow HILBERT's example and look at the properties of his series solutions, presuming they exist.

(iii) Hilbert's assertion of gross causality

To find a term in HILBERT's expansion (XXII.1), we must solve a system of partial differential equations. At the 0th step this system is (XXII.14), while at all subsequent stages of the procedure it is HILBERT's field equations (XXII.16). Interpreting each of these systems as defining an initial-value problem, we must specify the initial values of the fields $\rho^{(0)}$ and $\mathbf{C}^{(n)}$. Our choice of these fields is restricted by the conditions $\rho^{(0)} > 0$ and $\varepsilon^{(0)} > 0$ set down in the result stated after (XXII.3), but otherwise it is arbitrary.

We may describe this choice of initial values in another way. By combining (XXII.11) and (XXII.12) we obtain the following expression for the principal

---

equations of the second kind." These statements may be found repeated in almost the same words in the recollections of CHAPMAN [1967, p. 8].

However ridiculous it would be to rest content, as CHAPMAN & COWLING seem to have expressed themselves ready to be, with equations (but an approximation!) that provide no solutions and hence no information at all, HILBERT himself did much to obscure his own work. His title, "Begründung der kinetischen Gastheorie", is equivocal in German and has no precise equivalent in other common languages. "Begründung" derives from a verb meaning to found, establish, start, substantiate, give arguments in support of, give grounds for doing, justify. Certainly HILBERT's paper does none of these things, but such a title gives some Begründung to the physicists' blank failure to grasp what HILBERT was doing.

Not only that, HILBERT [1912, p. 562] opened his paper with the claim "... *I show that it is a certain linear integral equation of the second kind with symmetric kernel that forms the mathematical foundation* [Grundlage] *of the kinetic theory of gases, and that unless we investigate that equation by the modern methods of the theory of integral equations, it is impossible to establish the theory of gases systematically.*" When a great mathematician claims he has done something, lesser men can be forgiven for believing him. In fact, far from proving anything at all to be *impossible*, HILBERT did not prove any mathematical status for his brilliantly conceived though purely formal procedure.

But confusion did not stop here. Mathematicians, especially German ones, were long wont to state that HILBERT had settled "all the fundamental problems of the kinetic theory", so that nothing remained but "applications".

moment of $F^{(n)}$ in terms of the field $\mathbf{C}^{(n)}$:

$$\rho_\alpha^{(n)} = \mathfrak{m} \sum_{\beta=0}^{4} A_{\alpha\beta} C_\beta^{(n)}. \qquad (\text{XXII.23})$$

In view of (XXII.10), then, to prescribe the initial value $\mathbf{C}^{(n)}(0, \cdot)$ is equivalent to prescribing the initial value $\boldsymbol{\rho}^{(n)}(0, \cdot)$. Also, we may use (XXII.1) to express the principal moment $\boldsymbol{\rho}$ of F in terms of the fields $\boldsymbol{\rho}^{(n)}$ by the series

$$\rho_\alpha(t, \mathbf{x}; \delta) = \sum_{n=0}^{\infty} \delta^{n-1} \rho_\alpha^{(n)}(t, \mathbf{x}). \qquad (\text{XXII.24})$$

Therefore, to prescribe the initial values $\boldsymbol{\rho}^{(n)}(0, \cdot)$, $n \geq 0$, is the same thing as to prescribe the series $\boldsymbol{\rho}(0, \cdot; \delta)$. By combining these last two facts we find that *in HILBERT's procedure the series $\boldsymbol{\rho}(0, \cdot; \delta)$ may be chosen arbitrarily, subject only to the requirements that it converge for sufficiently small δ and that $\rho^{(0)}$ and $\varepsilon^{(0)}$ be positive.*

One of the interesting features of HILBERT's procedure is the way the choice of the initial value $\boldsymbol{\rho}(0, \cdot; \delta)$ affects the solutions delivered by the procedure. To see how it does so, let us look back at the equations $F^{(n)}$ must satisfy. Since $F^{(0)}$ is a Maxwellian density, it is determined uniquely by its principal moment $\boldsymbol{\rho}^{(0)}$. Also, this principal moment is determined uniquely by its initial value $\boldsymbol{\rho}^{(0)}(0, \cdot)$ for each body force $\mathbf{b}$ through the initial-value problem (XXII.14), at least according to HILBERT's formal reasoning. Thus $F^{(0)}$ is determined completely by $\boldsymbol{\rho}^{(0)}(0, \cdot)$ and $\mathbf{b}$. More generally now, $F^{(n)}$ is given in terms of $F_P^{(n)}$, $F^{(0)}$, and $\mathbf{C}^{(n)}$ by (XXII.7). But we know already that $F_P^{(n)}$ is determined *uniquely* by $F^{(0)}$, ..., $F^{(n-1)}$. To find the field $\mathbf{C}^{(n)}$, we must solve HILBERT's field equations (XXII.16). According to HILBERT's formal reasoning again, this field is determined uniquely by its initial value $\mathbf{C}^{(n)}(0, \cdot)$ and the various coefficients appearing in (XXII.16). But, as (XXII.9) and (XXII.15) show, these coefficients are known once $F^{(0)}$, $F_P^{(n)}$, and $\mathbf{b}$ be known, the body force entering explicitly through the operator $\mathfrak{D}$. Thus $F^{(n)}$ is determined by $F^{(0)}$, ..., $F^{(n-1)}$, $\mathbf{C}^{(n)}(0, \cdot)$, and $\mathbf{b}$. But by the same reasoning $F^{(n-1)}$ is known once $F^{(0)}$, ..., $F^{(n-2)}$, $\mathbf{C}^{(n-1)}(0, \cdot)$, and $\mathbf{b}$ be known, and so $F^{(n)}$ is determined by $F^{(0)}$, ..., $F^{(n-2)}$, $\mathbf{C}^{(n-1)}(0, \cdot)$, $\mathbf{C}^{(n)}(0, \cdot)$, and $\mathbf{b}$. Continuing this process, we see that $F^{(n)}$ is determined uniquely by the initial values $\boldsymbol{\rho}^{(0)}(0, \cdot)$ and $\mathbf{C}^{(k)}(0, \cdot)$, $k = 1, \ldots, n$, and by the body force $\mathbf{b}$. Since these initial values are determined by the series $\boldsymbol{\rho}(0, \cdot; \delta)$ through (XXII.23) and (XXII.24), and since the $F^{(n)}$ determine $F(t, \mathbf{x}, \mathbf{v}; \delta)$ through (XXII.1), we arrive at

Hilbert's Assertion of Gross Causality. *For a given body force $\mathbf{b}$, each solution delivered by HILBERT's procedure is determined uniquely by the series expansions of the initial values of its fields of density, gross velocity, and energetic.* We have used the term "assertion" here to recall that the proof, since it is based upon HILBERT's procedure, is merely formal. All of its results must be viewed against the background of HILBERT's characterizing hypotheses (XXII.1) and (XXII.2).

HILBERT's solutions, then, are causal: their values at any positive time at which they exist are determined by the values they had at earlier times, indeed, by their values at the initial time $t = 0$. More important, however, is the fact that they are grossly causal: their values when $t > 0$ are determined by the gross condition of the gas when $t = 0$. We shall refer to solutions with this property as *gas-dynamic solutions*. In Chapter VIII we defined a grossly determined solution as one whose values at time t are completely determined by the gross condition of the gas at that same time. It is natural to ask whether HILBERT's solutions are also grossly determined. We shall soon see that such is indeed the case. To prepare for the analysis, we now look more closely[8] at HILBERT's series solutions.

(iv) Properties of Hilbert's formal solutions. The Hilbert mapping

We may interpret HILBERT's assertion of gross causality as stating that there is a unique function $\mathbb{H}_S$ which maps series expansions $\sum_{n=0}^{\infty} \delta^{n-1} \boldsymbol{\rho}^{(n)}(0, \cdot)$ onto series solutions $\sum_{n=0}^{\infty} \delta^{n-1} F^{(n)}$ of the Maxwell–Boltzmann equation. More specifically, there is a unique function $\mathbb{H}_S$ of t, $\mathbf{x}$, $\mathbf{v}$, δ, time-independent fields $\boldsymbol{\phi}^{(n)}$, $n = 0, 1, \ldots$, and the body force $\mathbf{b}$ such that

(a) $$F(t, \mathbf{x}, \mathbf{v}; \delta) \equiv \mathbb{H}_S(t, \mathbf{x}, \mathbf{v}; \delta, \boldsymbol{\phi}^{(0)}, \boldsymbol{\phi}^{(1)}, \ldots, \mathbf{b}) \qquad \text{(XXII.25)}$$

is a solution of the Maxwell–Boltzmann equation with body force $\mathbf{b}$ for all sufficiently small δ;

(b) there is a positive Maxwellian density $F^{(0)}$ such that for small values of δ, $F - F^{(0)}/\delta$ equals a convergent power series in δ;

(c) for all δ near 0, all $\mathbf{x}$, and all fields $\mathbf{b}$, $\boldsymbol{\phi}^{(n)}$, $n = 0, 1, \ldots$,

$$m \int I_\alpha \mathbb{H}_S(0, \mathbf{x}, \mathbf{v}; \delta, \boldsymbol{\phi}^{(0)}, \boldsymbol{\phi}^{(1)}, \ldots, \mathbf{b}) = \sum_{n=0}^{\infty} \delta^{n-1} \phi_\alpha^{(n)}(\mathbf{x}), \qquad \alpha = 0, 1, \ldots, 4.$$

(XXII.26)

From this last formula we see that $\boldsymbol{\phi}^{(n)}$ is nothing more than the initial value $\boldsymbol{\rho}^{(n)}(0, \cdot)$ of the principal moment of $F^{(n)}$. Of course, this means that the field $\boldsymbol{\phi}^{(0)}$ must be such that $\rho^{(0)}$ and $\varepsilon^{(0)}$ are positive, and we shall assume so without further mention.

The function $\mathbb{H}_S$ defines, then, HILBERT's series solutions since its values are precisely those. To state a property of such solutions is equivalent to stating a

---

[8] The results we present in the remainder of this chapter, with the exception of Lemma 1, we regard as new. Lemma 1 is due, in content though not in the form stated here, to GRAD [1958, §23].

property of $\mathbb{H}_S$. Every such property must follow from (a), (b), and (c) above since these define $\mathbb{H}_S$. Of course, in view of these three conditions we know that F has the form (XXII.1), the terms $F^{(n)}$ being determined by the iterative system (XXII.3). Thus we may, if we wish, use the iterative procedure to deduce properties of $\mathbb{H}_S$. For the most part we shall find it more convenient to use (a), (b), and (c) directly.

Unfortunately these three defining conditions do not tell us directly enough about $\mathbb{H}_S$ to justify the formal calculations which we shall present. What we need is certain technical information about $\mathbb{H}_S$, information relating to its radius of convergence as a series and its smoothness as a function. Since it is technical properties that we require, we shall rest content here simply to list those which we require for our analysis.

Interval of convergence. For each given set of fields $\boldsymbol{\phi}^{(n)}$, $n \geq 0$, the series defining $\mathbb{H}_S(t, \mathbf{x}, \mathbf{v}; \delta, \boldsymbol{\phi}^{(0)}, \boldsymbol{\phi}^{(1)}, \ldots, \mathbf{b})$ is presumed to converge for all δ in some interval of convergence $(0, r)$. The number r will in general depend upon the choice of these fields. Since the precise nature of the dependence is presently unknown, we adopt the following hypotheses as being sufficient to justify the formal steps we shall go through:

(d) if $\sum_{n=0}^{\infty} \delta^n \boldsymbol{\phi}^{(n)}$ converges for all δ, so also does the corresponding series expansion of $\mathbb{H}_S$.

As we shall see later, this assumption is in fact true for the few explicit series solutions that are known. *In what follows we shall consider only those sets of fields $\boldsymbol{\phi}^{(n)}$, $n \geq 0$, whose associated series converge on $(0, \infty)$.*

Regularity. We assume that

(e) if $\sum_{n=0}^{\infty} \delta^n \boldsymbol{\phi}^{(n)}$ converges for all δ, then the function

$$F(t, \mathbf{x}, \mathbf{v}; \delta) = \mathbb{H}_S\left(t, \mathbf{x}, \mathbf{v}; \delta, \sum_{n=0}^{\infty} \delta^n \boldsymbol{\phi}^{(n)}, \mathbf{0}, \mathbf{0}, \ldots, \mathbf{b}\right) \qquad \text{(XXII.27)}$$

is a series of the form (XXII.1) satisfying (XXII.2), and this series also converges for all δ.

This statement holds formally if $\mathbb{H}_S$ is an analytic function of its argument $\boldsymbol{\phi}^{(0)}$.

One of the main problems in the kinetic theory is to find solutions of the Maxwell–Boltzmann equation. In this respect HILBERT's series solutions are useful because each such series gives us not just one but rather a one-parameter family of solutions. If we have two distinct series solutions, we have two one-parameter families. However, because two such expansions, even though they be different, may represent one and the same function of t, $\mathbf{x}$, and $\mathbf{v}$ when their respective parameters are given fixed values, the corresponding families of solutions will in general have some members in common. GRAD[9] has shown how one can avoid many such repetitions. His results may be summarized by an identity satisfied by $\mathbb{H}_S$, which we now present as

Lemma 1.

$$\mathbb{H}_S(t, \mathbf{x}, \mathbf{v}; \delta, \boldsymbol{\phi}^{(0)}, \boldsymbol{\phi}^{(1)}, \ldots, \mathbf{b}) = \mathbb{H}_S\left(t, \mathbf{x}, \mathbf{v}; \delta, \sum_{n=0}^{\infty} \delta^n \boldsymbol{\phi}^{(n)}, \mathbf{0}, \ldots, \mathbf{b}\right). \quad \text{(XXII.28)}$$

The mapping $\mathbb{H}_S$ is determined by its restriction to arguments such that $\boldsymbol{\phi}^{(1)} = \mathbf{0}$, $\boldsymbol{\phi}^{(2)} = \mathbf{0}, \ldots$.

Proof. Consider the function F given by (XXII.27). By condition (e) it is a series of the type HILBERT considered. Next we note that for each δ it is a solution of the Maxwell–Boltzmann equation. Indeed, if δ_0 is any positive number, then $F(t, \mathbf{x}, \mathbf{v}; \delta_0)$ is simply the value at $\delta = \delta_0$ of the series solution $\mathbb{H}_S(t, \mathbf{x}, \mathbf{v}; \delta, \sum_{n=0}^{\infty} \delta_0^n \boldsymbol{\phi}^{(n)}, \mathbf{0}, \ldots, \mathbf{b})$. Thus, by HILBERT's assertion of gross causality, F is determined through (XXII.25) by the initial value of its principal moment. According to (XXII.26) this initial value is given by

$$\mathfrak{m} \int I_\alpha \mathbb{H}_S\left(0, \mathbf{x}, \mathbf{v}; \delta, \sum_{n=0}^{\infty} \delta^n \boldsymbol{\phi}^{(n)}, \mathbf{0}, \ldots, \mathbf{b}\right) = \frac{1}{\delta} \sum_{n=0}^{\infty} \delta^n \phi_\alpha^{(n)}(\mathbf{x}),$$

$$= \sum_{n=0}^{\infty} \delta^{n-1} \phi_\alpha^{(n)}(\mathbf{x}), \quad \text{(XXII.29)}$$

$$\alpha = 0, 1, \ldots, 4.$$

In view of (XXII.26), we obtain (XXII.28). △

We may phrase GRAD's conclusion in another way: In applying HILBERT's procedure it is sufficient to choose $\mathbf{C}^{(n)}(0, \cdot) = \mathbf{0}$, $n = 1, 2, \ldots$. This last result tells us that in exploring the properties of $\mathbb{H}_S$ we need only examine the function $\mathbb{H}_S(t, \mathbf{x}, \mathbf{v}; \delta, \boldsymbol{\phi}^{(0)}, \mathbf{0}, \ldots, \mathbf{b})$. We derive now three important properties of this function, which we state as

Lemma 2. *For each time-independent field $\boldsymbol{\phi}$ the function $\mathbb{H}_S(t, \mathbf{x}, \mathbf{v}; \delta, \delta\boldsymbol{\phi}, \mathbf{0}, \ldots, \mathbf{b})$*

(a) *is independent of δ,*
(b) *is a solution of the Maxwell–Boltzmann equation,*
(c) *has principal moment $\boldsymbol{\phi}$ initially.*

Proof. Let a field $\boldsymbol{\phi}$ be given, and consider the functions $F_1(t, \mathbf{x}, \mathbf{v}; \delta) \equiv \mathbb{H}_S(t, \mathbf{x}, \mathbf{v}; \delta, \delta_1 \boldsymbol{\phi}, \mathbf{0}, \ldots, \mathbf{b})$ and $F_2(t, \mathbf{x}, \mathbf{v}; \delta) \equiv \mathbb{H}_S(t, \mathbf{x}, \mathbf{v}; \delta\delta_2/\delta_1, \delta_2\boldsymbol{\phi}, \mathbf{0}, \ldots, \mathbf{b})$, δ_1 and δ_2 being positive numbers. From property (a) of $\mathbb{H}_S$ we know

---

[9] GRAD [1949, §30] [1958, §23]. GRAD presumed the body force to be null, but trivially his result is valid in the more general form given here.

(iv) THE HILBERT MAPPING   389

that both of these functions are solutions of the Maxwell–Boltzmann equation for all δ. From property (b), each is a convergent series of the form (XXII.1) satisfying (XXII.2). Thus, by HILBERT's assertion of gross causality, each is determined through (XXII.25) by the initial value of its principal moment. From (XXII.26) we see that these initial values are given by

$$m \int I_\alpha \mathbb{H}_S(0, \mathbf{x}, \mathbf{v}; \delta, \delta_1 \boldsymbol{\phi}, 0, \ldots, \mathbf{b}) = \frac{1}{\delta} \delta_1 \phi_\alpha(\mathbf{x}),$$

$$m \int I_\alpha \mathbb{H}_S\left(0, \mathbf{x}, \mathbf{v}; \frac{\delta \delta_2}{\delta_1}, \delta_2 \boldsymbol{\phi}, 0, \ldots, \mathbf{b}\right) = \frac{\delta_1}{\delta \delta_2} \delta_2 \phi_\alpha(\mathbf{x}), \quad \text{(XXII.30)}$$

$$= \frac{1}{\delta} \delta_1 \phi_\alpha(\mathbf{x}).$$

Since they are the same for both solutions, we conclude from the uniqueness of $\mathbb{H}_S$ that the two solutions are equal: $F_1 = F_2$. Since this is true in particular when $\delta = \delta_1$, we find that

$$\mathbb{H}_S(t, \mathbf{x}, \mathbf{v}; \delta_1, \delta_1 \boldsymbol{\phi}, 0, \ldots, \mathbf{b}) = \mathbb{H}_S(t, \mathbf{x}, \mathbf{v}; \delta_2, \delta_2 \boldsymbol{\phi}, 0, \ldots, \mathbf{b}). \quad \text{(XXII.31)}$$

This completes the proof of (a). Conclusion (b) follows directly from the fact that the left-hand side of (XXII.31) is the value at $\delta = \delta_1$ of the series solution F_1. Conclusion (c) follows directly from (XXII.30)$_1$ by setting $\delta_1 = \delta$. △

The conclusions of these two lemmas may be summarized compactly in terms of the *Hilbert mapping* $\mathbb{H}$ defined by

$$\mathbb{H}(t, \mathbf{x}, \mathbf{v}; \boldsymbol{\phi}, \mathbf{b}) \equiv \mathbb{H}_S(t, \mathbf{x}, \mathbf{v}; \delta, \delta \boldsymbol{\phi}, 0, \ldots, \mathbf{b}). \quad \text{(XXII.32)}$$

According to Lemma 2, $\mathbb{H}(\cdot, \cdot, \cdot; \boldsymbol{\phi}, \mathbf{b})$ *is a solution of the Maxwell–Boltzmann equation which is determined by the initial value of its principal moment*

$$m \int I_\alpha \mathbb{H}(0, \mathbf{x}, \mathbf{v}; \boldsymbol{\phi}, \mathbf{b}) = \phi_\alpha(\mathbf{x}). \quad \text{(XXII.33)}$$

The solutions delivered by the Hilbert mapping, then, are gas-dynamic solutions. If we set $\boldsymbol{\phi} = \sum_{n=0}^{\infty} \delta^{n-1} \boldsymbol{\phi}^{(n)}$ in (XXII.32) and then compare this equation with (XXII.28), we obtain the following result connecting $\mathbb{H}$ and $\mathbb{H}_S$:

$$\mathbb{H}_S(t, \mathbf{x}, \mathbf{v}; \delta, \boldsymbol{\phi}^{(0)}, \boldsymbol{\phi}^{(1)}, \ldots, \mathbf{b}) = \mathbb{H}\left(t, \mathbf{x}, \mathbf{v}; \sum_{n=0}^{\infty} \delta^{n-1} \boldsymbol{\phi}^{(n)}, \mathbf{b}\right). \quad \text{(XXII.34)}$$

Recalling that $\boldsymbol{\phi}^{(n)}$ is simply the initial value $\boldsymbol{\rho}^{(n)}(0, \cdot)$, we obtain from (XXII.24), (XXII.25), and (XXII.34) the **Generating Formula**

$$F(t, \mathbf{x}, \mathbf{v}; \delta) = \mathbb{H}(t, \mathbf{x}, \mathbf{v}; \boldsymbol{\rho}(0, \cdot; \delta), \mathbf{b}) \quad \text{(XXII.35)}$$

from which flow many important properties of HILBERT's solutions. Indeed, we see immediately that the Hilbert mapping ℍ *determines all of* HILBERT's *solutions*. From the definition of ℍ we know that the converse of this result is valid also. Concerning the structure of the series solutions we see from this formula that HILBERT's *solutions depend upon* δ *and* $\rho^{(n)}(0, \cdot)$, $n = 0, 1, \ldots$, *only through* the sum of the series $\rho(0, \cdot; \delta)$. Once $\rho(0, \cdot; \delta)$ has been chosen, we simply place it into ℍ so as to obtain the corresponding series solution. However, if we desire only to obtain solutions of the Maxwell–Boltzmann equation, there is no compelling reason to evaluate ℍ at a series expansion. Indeed, we know from the properties of ℍ that any time-independent field $\rho(0, \cdot)$ will do. *It is the Hilbert mapping itself that is fundamental.* We do not mean to say that the series solutions serve no purpose. To see their importance, we need only restate in terms of ℍ the conditions (a), (b), and (c) defining ℍ$_S$. We find that ℍ is that function of t, **x**, **v**, and time-independent fields **ϕ** and **b** such that

(a)′ $$F(t, \mathbf{x}, \mathbf{v}) \equiv \mathbb{H}(t, \mathbf{x}, \mathbf{v}; \boldsymbol{\phi}, \mathbf{b}) \tag{XXII.36}$$

is a solution of the Maxwell–Boltzmann equation with body force **b**;

(b)′ for each convergent series expansion $\sum_{n=0}^{\infty} \delta^n \boldsymbol{\phi}^{(n)}$, $\mathbb{H}(t, \mathbf{x}, \mathbf{v}; \sum_{n=0}^{\infty} \delta^{n-1} \boldsymbol{\phi}^{(n)}, \mathbf{b})$ has the form (XXII.1), in which (XXII.2) holds;

(c)′ the field **ϕ** is the initial value of the principal moment of F.

There is only one function with these three properties. From (b)′ we see that the series solutions play only an auxiliary role. Once we have exploited the requirement (b)′, we may dispense with them. *The parameter δ is nothing more than a dummy variable*, something like the variable x of the function $f(x) \equiv x - x$. Despite what some studies in the kinetic theory might lead us to expect, this dummy appears to have no special status, at least insofar as we may trust the formal calculations presented here. After having used δ as a tool to formulate the iterative scheme, we might as well give it some convenient value.

(v) *Locally Maxwellian solutions*

In Chapter X we delimited and displayed all locally Maxwellian solutions. Our analysis included constructive proof that if ρ, **u**, and ε are the fields of density, gross velocity, and energetic, respectively, of a locally Maxwellian solution, these fields are determined uniquely by their own initial values and the body force **b**. Since ρ, **u**, and ε determine exactly one Maxwellian density, we conclude that *the locally Maxwellian solutions are gas-dynamic solutions.* Thus, it is natural to ask whether these be values of the Hilbert mapping ℍ, and if so, to ask what information they give us concerning HILBERT's series solutions.

We have seen in Chapter X that in order for a Maxwellian density to be a solution of the Maxwell–Boltzmann equation, the body force **b** must be of a

(v) LOCALLY MAXWELLIAN SOLUTIONS 391

special type. Specifically, **b** must have the form given by (II.28) and (II.29), suitably specialized to the case in which $\gamma = \frac{5}{3}$. It is reasonable to expect that if $\mathbb{H}$ delivers a locally Maxwellian solution, it will do so only when **b** has this form. In the remainder of this section we shall consider without further mention only body forces of this kind.

We ask now whether there is a solution of the iterative system (XXII.3) in which $F^{(1)}$, $F^{(2)}$, ... all vanish. We can easily see that if such is the case the iterative system reduces to

$$\mathbb{C}(F^{(0)}, F^{(0)}) = 0, \qquad \mathfrak{D}F^{(0)} = 0. \qquad \text{(XXII.37)}$$

These conditions state that $F^{(0)}$ must be a locally Maxwellian solution. Because of the form of the body force being considered, they are satisfied by every such solution. Thus HILBERT's procedure does indeed give us all the locally Maxwellian solutions, although it does so, according to (XXII.1), in the form $F = F^{(0)}/\delta$. More specifically we have shown that *if the fields $\rho^{(0)}$, $\mathbf{u}^{(0)}$, and $\varepsilon^{(0)}$ define a locally Maxwellian solution subject to the body force* **b**, *and if* $\boldsymbol{\phi}^{(0)} \equiv \rho^{(0)}(0, \cdot)$, *then*

$$\mathbb{H}_S(t, \mathbf{x}, \mathbf{v}; \delta, \boldsymbol{\phi}^{(0)}, 0, \ldots, \mathbf{b}) = \frac{1}{\delta} \frac{\rho^{(0)}}{\mathfrak{m}(\frac{4}{3}\pi\varepsilon^{(0)})^{\frac{3}{2}}} \exp\left(-\frac{3}{4\varepsilon^{(0)}} |\mathbf{v} - \mathbf{u}^{(0)}|^2\right).$$
(XXII.38)

Of course, $\rho^{(0)}$, $\mathbf{u}^{(0)}$, and $\varepsilon^{(0)}$ are certain functions of t, $\mathbf{x}$, and $\boldsymbol{\phi}^{(0)}$ specified by the general characterization of such solutions presented in Chapter X.

The preceding result is altogether natural, and we should be disappointed if HILBERT's procedure failed to reconfirm the status of the locally Maxwellian solutions. However, it shows that HILBERT's expressed reason for introducing his series (XXII.1) is questionable. As we have seen in Chapter X, a locally Maxwellian solution need not exist for all time, and even if it does exist for all time, it may approach as $t \to \infty$ a limit that is not a molecular density. Thus the one class of functions $F^{(0)}/\delta$ for which HILBERT's procedure can be proved to converge fails to provide only solutions F that satisfy his condition of "stability". This fact does not diminish the importance of HILBERT's method. That importance, as displayed by the function $\mathbb{H}$ and its properties, rests on ideas altogether different from those HILBERT adduced to suggest his starting point.

It is a simple matter now to find the corresponding values of $\mathbb{H}$. Since the right-hand side of (XXII.32) is independent of δ, we may set $\delta = 1$ in both (XXII.32) and (XXII.38). Comparing the two equations that result, we find that the restriction of the Hilbert mapping to locally Maxwellian solutions is

$$\mathbb{H}(t, \mathbf{x}, \mathbf{v}; \boldsymbol{\phi}, \mathbf{b}) = \frac{\rho}{\mathfrak{m}(\frac{4}{3}\pi\varepsilon)^{\frac{3}{2}}} \exp\left(-\frac{3}{4\varepsilon} |\mathbf{v} - \mathbf{u}|^2\right). \qquad \text{(XXII.39)}$$

That is, *if a body force* **b** *and a field* $\rho(0, \cdot)$ *are consistent with a certain locally Maxwellian solution, that solution is* $\mathbb{H}(t, \mathbf{x}, \mathbf{v}; \rho(0, \cdot), \mathbf{b})$.

We have seen already that if we replace the argument $\boldsymbol{\phi}$ in $\mathbb{H}$ by a convergent series $\sum_{n=0}^{\infty} \delta^{n-1} \boldsymbol{\phi}^{(n)}$, we obtain one of HILBERT's series solutions. If we choose the fields $\boldsymbol{\phi}^{(n)}$ such that for all δ the expansion $\sum_{n=0}^{\infty} \delta^{n-1} \boldsymbol{\phi}^{(n)}$ is the initial value of the principal moment of a locally Maxwellian solution, we may use (XXII.39) to exhibit explicitly some of HILBERT's series solutions. We shall develop a simple example in which this procedure is easy to carry out. Let the body force be null and the field of density be uniform. We have shown in Chapter X that all locally Maxwellian solutions consistent with this possibility are homo-energetic dilatations, ρ, $\mathbf{u}$, and ε being given explicitly by (II.39). Placing these fields into (XXII.39), we obtain the somewhat more explicit value of $\mathbb{H}$

$$\mathbb{H}(t, \mathbf{x}, \mathbf{v}; \boldsymbol{\phi}, 0) = \frac{\rho(0)}{\mathfrak{m}(\tfrac{4}{3}\pi\varepsilon(0))^{\tfrac{3}{2}}} \exp\left(-\frac{3}{4\varepsilon(0)} (t+T)^2 |\mathbf{v} - (t+T)^{-1}\mathbf{r}|^2\right);$$
(XXII.40)

in view of (II.39)$_2$, T is determined by the initial value of the expansion E, and hence by the initial value of $\mathbf{u}$, as follows:

$$T = 3/E(0). \tag{XXII.41}$$

Also, from (VI.23) we see that

$$\phi_\alpha(\mathbf{x}) = \begin{cases} \rho(0) & \text{if } \alpha = 0, \\ \rho(0) u_\alpha(0, \mathbf{x}) & \text{if } \alpha = 1, 2, 3, \\ \rho(0)(2\varepsilon(0) + u^2(0, \mathbf{x})) & \text{if } \alpha = 4. \end{cases} \tag{XXII.42}$$

Clearly, one way to make the field $\boldsymbol{\phi}$ equal an expansion of the form $\sum_{n=0}^{\infty} \delta^{n-1} \boldsymbol{\phi}^{(n)}$ is to choose $\mathbf{u}(0, \mathbf{x})$ and $\varepsilon(0)$ independent of δ and replace $\rho(0)$ by a series

$$\rho(0; \delta) = \sum_{n=0}^{\infty} \delta^{n-1} \rho^{(n)}(0). \tag{XXII.43}$$

Thus, by replacing $\rho(0)$ by $\rho(0; \delta)$ in (XXII.40) we obtain from (XXII.35) the series solution

$$F(t, \mathbf{x}, \mathbf{v}; \delta) = \sum_{n=0}^{\infty} \delta^{n-1} \frac{\rho^{(n)}(0)}{\mathfrak{m}(\tfrac{4}{3}\pi\varepsilon(0))^{\tfrac{3}{2}}} \exp\left(-\frac{3}{4\varepsilon(0)} (t+T)^2 |\mathbf{v} - (t+T)^{-1}\mathbf{r}|^2\right).$$
(XXII.44)

This solution is uninteresting since for it each of the terms $F^{(n)}$ is a Maxwellian density. Indeed, each is a locally Maxwellian solution. However, it is not difficult to exhibit series solutions which are not of this kind. If we replace $\rho(0)$ by $\rho(0; \delta)$, let $\mathbf{u}(0, \mathbf{x})$ be independent of δ again, and this time replace $\varepsilon(0)$ by an analytic function of δ, we see from (XXII.42) that $\boldsymbol{\phi}$ becomes a series of the form $\sum_{n=0}^{\infty} \delta^{n-1} \boldsymbol{\phi}^{(n)}$. Thus we may make this same replacement in (XXII.40) and so obtain another of HILBERT's solutions. As a simple example of what happens,

(v) LOCALLY MAXWELLIAN SOLUTIONS 393

let us replace $\rho(0)$ by $\rho(0)/\delta$ and $\varepsilon(0)$ by $\varepsilon(0)(1 + \delta)^{-2}$. The resulting series solution, the interval of convergence of which is easily seen to be $(0, \infty)$, is obtained from (XXII.35):

$$F(t, \mathbf{x}, \mathbf{v}; \delta) = \frac{1}{\delta} \frac{\rho(0)(1 + \delta)^3}{\mathfrak{m}(\tfrac{4}{3}\pi\varepsilon(0))^{\tfrac{3}{2}}} \exp\left(-\frac{3(1 + \delta)^2}{4\varepsilon(0)}(t + T)^2 |\mathbf{v} - (t + T)^{-1}\mathbf{r}|^2\right),$$

$$= \frac{1}{\delta} F^{(0)} + F^{(1)} + \delta F^{(2)} + \cdots, \qquad \text{(XXII.45)}$$

and

$$F^{(0)} = \frac{\rho(0)}{\mathfrak{m}(\tfrac{4}{3}\pi\varepsilon(0))^{\tfrac{3}{2}}} \exp\left(-\frac{3}{4\varepsilon(0)}(t + T)^2 |\mathbf{v} - (t + T)^{-1}\mathbf{r}|^2\right),$$

$$F^{(1)} = \left\{-\frac{3}{2\varepsilon(0)}(t + T)^2 |\mathbf{v} - (t + T)^{-1}\mathbf{r}|^2 + 3\right\} F^{(0)}, \qquad \text{(XXII.46)}$$

$$F^{(2)} = \left\{2\left[-\frac{3}{4\varepsilon(0)}(t + T)^2 |\mathbf{v} - (t + T)^{-1}\mathbf{r}|^2 + \frac{7}{4}\right]^2 - \frac{25}{8}\right\} F^{(0)}, \ldots.$$

In this case each term is a product of $F^{(0)}$ and a polynomial in the components of $\mathbf{v}$. No $F^{(n)}$ except $F^{(0)}$ is a locally Maxwellian density, yet the series (XXII.45) converges to such a density. In this example, then, HILBERT's process begins from the locally Maxwellian solution (XXII.46)$_1$ and converges to another locally Maxwellian solution, namely, (XXII.45)$_1$.

Beyond the locally Maxwellian solutions, the only other explicit molecular densities presently known to satisfy the Maxwell-Boltzmann equation are those of MUNCASTER and of KROOK & WU, both of which we have developed in Appendix B to Chapter XVII. Their status with regard to HILBERT's formal series solutions is yet to be investigated.

These two explicit examples give us some idea of the diversity of HILBERT's series solutions. Depending upon the choice of the constants $\rho^{(n)}(0)$ in (XXII.43), we can easily arrange that the solution (XXII.44) shall have only a finite number of non-zero terms, or, if we so desire, that none of its terms shall vanish. For a broad class of choices of $\rho(0; \delta)$ the interval of convergence of this solution will be infinite, while for other choices we can arrange that its interval of convergence be as small as we please. However, despite this variety of possibilities, the series (XXII.44) always defines a locally Maxwellian solution, a fact that is described concisely by the Hilbert mapping H in the one simple formula (XXII.40). The second example (XXII.45) shows how various each of the terms $F^{(n)}$ can be in one of HILBERT's solutions. However, at the same time it shows us that no matter what these terms are, it is the sum of the series (XXII.45)$_2$ that is important. Indeed, the form of the terms (XXII.46) might lead us to expect a new and unusual solution of the Maxwell-Boltzmann equation, while we know in fact that when placed in the series (XXII.45)$_2$ they generate nothing more than a locally Maxwellian solution.

(vi) Proof that Hilbert's solutions are grossly determined

We may rewrite the solutions given by $\mathbb{H}$ in the form

$$F(t, \mathbf{x}, \mathbf{v}) = \mathbb{H}(t, \mathbf{x}, \mathbf{v}; \sigma(0, \cdot)), \qquad \text{(XXII.47)}$$

σ being the gross condition of the gas as defined by (VIII.68). Our objective now is to prove that these solutions are grossly determined, that is, that there is a gross determiner G for which they may be expressed in the form (VIII.69). To do so, we introduce a new notation for $\mathbb{H}$ that is slightly more explicit than (XXII.47). Let the gas occupy all of space. The point space $\mathscr{E}$ is then a 3-dimensional Euclidean space. We denote by $\mathscr{W}$ the translation space of $\mathscr{E}$. This is the 3-dimensional vector space of all displacements. Thus if $\mathbf{x}$ lies in $\mathscr{E}$ and $\mathbf{r}$ lies in $\mathscr{W}$, then $\mathbf{x} + \mathbf{r}$ lies in $\mathscr{E}$ also. Let $\mathbf{x}_0$ be a fixed place. Then $\mathscr{E}$ is simply the set of all places $\mathbf{x}_0 + \mathbf{r}$, $\mathbf{r}$ being any vector in $\mathscr{W}$. Since $\mathbb{H}$ depends upon the values of $\sigma(0, \cdot)$ at all places, we may view it as a function of $\sigma(0, \mathbf{x}_0 + \mathbf{r})$ for all $\mathbf{r} \in \mathscr{W}$. This leads us to set down the following alternative form of Hilbert's solutions:

$$F(t, \mathbf{x}, \mathbf{v}) = \underset{\mathbf{r}}{\mathbb{H}}(t, \mathbf{x}, \mathbf{v}; \sigma(0, \mathbf{x}_0 + \mathbf{r})), \qquad \text{(XXII.48)}$$

$\mathbf{r}$ being nothing more here than a dummy variable that takes on all values in $\mathscr{W}$, and the fixed place $\mathbf{x}_0$ being any one we choose. This notation, although apparently more cumbersome, is useful, and in fact we have already used an extended version of it to describe the mappings appearing in (I.24). The mapping $\mathbb{H}$ in (XXII.48) will in general be different for different choices of the place $\mathbf{x}_0$, but we will not denote this dependence explicitly since some $\mathbf{x}_0$ is assumed to be chosen and fixed throughout the discussion.

To prove that the solutions (XXII.48) are grossly determined, we proceed as follows. Let s be a real number, and let $\mathbf{h}$ be a vector. By direct calculation we can show that if F is a solution of the Maxwell–Boltzmann equation, then so also is the function G defined as follows:

$$G(t, \mathbf{x}, \mathbf{v}) \equiv F(t + s, \mathbf{x} + \mathbf{h}, \mathbf{v}). \qquad \text{(XXII.49)}$$

Let σ be the gross condition of the gas described by F. It follows from (XXII.49) and (IX.1) that the gross condition ξ of the gas described by G is just

$$\xi(t, \mathbf{x}) = \sigma(t + s, \mathbf{x} + \mathbf{h}). \qquad \text{(XXII.50)}$$

If in addition F is one of Hilbert's series solutions, then by replacing t by $t + s$ and $\mathbf{x}$ by $\mathbf{x} + \mathbf{h}$ in (XXII.1) we find that G is one also. By Hilbert's assertion of gross causality both F and G are determined uniquely by the initial value of the gross condition of the gas they describe. This means that F is the value of $\mathbb{H}$ given by (XXII.48) and G is the value of $\mathbb{H}$ given by

$$G(t, \mathbf{x}, \mathbf{v}) = \underset{\mathbf{r}}{\mathbb{H}}(t, \mathbf{x}, \mathbf{v}; \xi(0, \mathbf{x}_0 + \mathbf{r})). \qquad \text{(XXII.51)}$$

It follows from (XXII.48), (XXII.49), (XXII.50), and (XXII.51) that $\mathbb{H}$ satisfies the identity

$$\underset{r}{\mathbb{H}}(t, \mathbf{x}, \mathbf{v}; \sigma(s, \mathbf{x}_0 + \mathbf{h} + \mathbf{r})) = \underset{r}{\mathbb{H}}(t + s, \mathbf{x} + \mathbf{h}, \mathbf{v}; \sigma(0, \mathbf{x}_0 + \mathbf{r})) \quad \text{(XXII.52)}$$

for all choices of t, $\mathbf{x}$, $\mathbf{v}$, s, $\mathbf{h}$, and σ. Let us replace t, $\mathbf{x}$, s, and $\mathbf{h}$ now by 0, $\mathbf{x}_0$, t, and $\mathbf{x} - \mathbf{x}_0$, respectively. This gives us

$$\underset{r}{\mathbb{H}}(0, \mathbf{x}_0, \mathbf{v}; \sigma(t, \mathbf{x} + \mathbf{r})) = \underset{r}{\mathbb{H}}(t, \mathbf{x}, \mathbf{v}; \sigma(0, \mathbf{x}_0 + \mathbf{r})). \quad \text{(XXII.53)}$$

Thus, if we define $\mathbb{G}(\mathbf{v}; \sigma)$ to be the function $\mathbb{H}(0, \mathbf{x}_0, \mathbf{v}; \sigma)$, it follows from (XXII.48) and (XXII.53) that

$$F(t, \mathbf{x}, \mathbf{v}) = \underset{r}{\mathbb{G}}(\mathbf{v}; \sigma(t, \mathbf{x} + \mathbf{r})). \quad \text{(XXII.54)}$$

As this is simply an alternative form of (VIII.69), we have proved that *the solutions delivered by the Hilbert mapping are grossly determined.* Since HILBERT's series solutions are determined by the Hilbert mapping through (XXII.35), they, too, are grossly determined.

It is important to note that the field ρ appearing in (XXII.54) through the gross condition $\sigma = (\rho, \mathbf{b})$ is *not* arbitrary. We have assumed implicitly from the outset that ρ is the principal moment of F, and since F is a solution of the Maxwell–Boltzmann equation, MAXWELL's consistency theorem tells us that its moments must satisfy the equations of balance (IX.3)–(IX.5). We shall have more to say on this point in the next chapter. Here we add just one more remark: That ρ is the principal moment of F, imposes upon $\mathbb{G}$ the five further conditions:

$$\rho_\alpha(t, \mathbf{x}) = \mathfrak{m} \int I_\alpha \underset{r}{\mathbb{G}}(\mathbf{v}; \rho(t, \mathbf{x} + \mathbf{r}), \mathbf{b}(\mathbf{x} + \mathbf{r})), \quad \alpha = 0, 1, \ldots, 4, \quad \text{(XXII.55)}$$

for all ρ and $\mathbf{b}$.

(vii) *Retrospect*

We have seen that the "stability" HILBERT used to motivate his definition of a special class of solutions cannot generally hold for all the solutions his formal results seem to deliver. Indeed, classical gas dynamics gives us no reason to expect that gas flows should exist for all time.

Although, as far as proof is concerned, only one of the several major gaps in HILBERT's reasoning has been closed by later researches, HILBERT's paper remains a monument of the subject. It was HILBERT who first stated outright that *only a special class of solutions* can describe gas flows in the terms appropriate to gas dynamics; who first specified it as his objective to calculate directly *just those solutions*; who first conceived a *systematic procedure*, if only a formal one, for approximating such solutions; and who first showed, though again only formally, that a suitable limiting process led to a *solution determined uniquely by the initial value, which may be assigned at will, of its own principal moment.*

Chapter XXIII

Grossly Determined Solutions. The Equations of Gross Determinism

We have seen again and again that if we wish to obtain from the kinetic theory relations resembling the constitutive equations of continuum mechanics, we should turn our attention away from general solutions of the Maxwell-Boltzmann equation in favor of a class of special ones, the *grossly determined*—at least, this is what the exact solutions presented in Chapters XIII, XIV, and XV suggest. HILBERT's gas-dynamic solutions, presented and dissected in the preceding chapter, constitute such a class, for as we have shown already the solutions delivered by his procedure are grossly determined, provided we may trust the purely formal analysis which today is all that is available on the subject. Though HILBERT gave no evidence of knowing this fact, he thought his solutions would be valuable in determining material properties and gross flows. Indeed, in reference to his formal procedure, he claimed[1] that "calculation of the second approximation would deliver not only a proof of the second theorem on heat and Boltzmann's expression for the entropy of the gas but also the equations of motion with account taken for internal friction and the conduction of heat; the viscosity and heat conductivity appear as numbers which can be calculated numerically through the solution of certain integral equations." These claims he did not support. They remain disputed[2].

---

[1] HILBERT [1912, pp. 576–577].

[2] It is widely believed in the circles of physics and aeronautics that HILBERT's method of approximating special solutions of the Maxwell-Boltzmann equation does not lend itself to calculation of μ and M. BOGUSLAWSKI [1915], introducing approximations HILBERT would have disdained, reached some unsatisfactory conclusions, whereupon the physicists pronounced HILBERT's method defective. Indeed, CHAPMAN & COWLING [1939, Historical Summary, §10] wrote that "Hilbert, however, did not succeed in obtaining a solution for f in a convenient form, because of his treatment of the term $\partial f/\partial t$ in Boltzmann's equation. An example of the defect in Hilbert's solution is afforded by a research founded upon it, by Boguslawski, ..." CHAPMAN in his recollections [1967, p. 8] reaffirmed this judgment.

(continued)

We shall now study grossly determined solutions directly; we seek to learn what they tell us about gross relations pertaining to gas flows. We begin by asking the following question, one which is only too often forgotten: *Why in the kinetic theory do we wish to determine gross relations?* To answer this, we shall take one step backward and examine again the concept of a gas-dynamic solution.

(i)   Gas-dynamic solutions. The importance of grossly determined solutions

In connection with HILBERT's formal iterative procedure we defined a gas-dynamic solution of the Maxwell-Boltzmann equation as one that is determined completely by the initial value of its own principal moment. In doing so we have extended to the kinetic theory the familiar expectation that for the equations of gas dynamics a unique solution will exist, at least in a short interval of time, for each choice of the initial values of the fields of density, velocity, and energetic. HILBERT's procedure supports the existence of an extensive class of these solutions, and so it is reasonable to set down the

Hypothesis Regarding Existence.   *There is a non-negative function* $\mathbb{F}$ *such that*

$$F(t, \mathbf{x}, \mathbf{v}) = \mathbb{F}_{\mathbf{r}}(t, \mathbf{x}, \mathbf{v}; \boldsymbol{\phi}(\mathbf{x}_0 + \mathbf{r})) \qquad\text{(XXIII.1)}$$

On the other hand, hardly had HILBERT's paper appeared but LUNN [1913] claimed results summarized as follows:

> The second part of the paper shows that if the kernel be thought of as expanded in a series of Legendre polynomials in cos δ, and the known functions in the integral equations in series of surface harmonics in the velocity space, then the corresponding expansions of the unknown functions are determined by the solution of certain Fredholm equations in the one-dimensional realm of the absolute value of the velocity.
>
> The first term of the power series was shown by Hilbert to be the well-known Maxwell formula. The general form of the second term is here found to contain, besides the additive solutions of the homogeneous equations, only harmonics of the first and second orders, corresponding respectively to thermal gradient and viscous stress, and leading to a computation of the coefficients of thermal conduction and viscosity. The value of the latter coefficient given by Meyer, and that of the former resulting from one of Tait's equations, which seems to be little known but fits experimental conditions better than Meyer's, both prove to be only the first terms in series resulting from solution of the integral equations by iteration.

Unfortunately this work was never published; in 1952 LUNN's surviving colleagues at the University of Chicago reported to TRUESDELL that they could find no trace of it. Perhaps a hint toward the kind of analysis it contained may be read in the cryptic remarks of HECKE [1922, Introduction and end of §3]; on the other hand, LUNN may have formulated something akin to ENSKOG's process, which we shall discuss in the following chapter.

is a solution of the Maxwell–Boltzmann equation, and

$$\phi_\alpha(\mathbf{x}) = \mathfrak{m} \int I_\alpha \mathbb{F}(0, \mathbf{x}, \mathbf{v}; \boldsymbol{\phi}(\mathbf{x}_0 + \mathbf{r})), \qquad \alpha = 0, 1, \ldots, 4, \quad \text{(XXIII.2)}$$

for all functions $\boldsymbol{\phi} = (\phi_\alpha)$ *such that* $\phi_0 > 0$, $\phi_0 \phi_4 > \phi_a \phi_a$.

If $\boldsymbol{\rho}$ is the principal moment of F, then from (XXIII.2) and (VI.21)$_2$ we see that $\boldsymbol{\phi}$ is the initial value of $\boldsymbol{\rho}$:

$$\phi_\alpha(\mathbf{x}) = \rho_\alpha(0, \mathbf{x}). \qquad \text{(XXIII.3)}$$

The formula (XXIII.2) simply tells us how to obtain this initial value from the solution F. The formula (XXIII.1) tells us how we may use $\boldsymbol{\phi}$ so as to retrieve the solution from whence it came. From (XXIII.3) and (VI.23) we see that the inequalities imposed here on $\boldsymbol{\phi}$ ensure that $\rho(0, \cdot)$ and $\varepsilon(0, \cdot)$ be positive. This must be the case if the value of $\mathbb{F}$ is to be a molecular density.

Just as we have noted in connection with (XXII.48), the mapping $\mathbb{F}$ in (XXIII.1) depends upon the choice of $\mathbf{x}_0$. However, since we assume here that some $\mathbf{x}_0$ has been chosen and kept fixed during our discussion, we leave the dependence of $\mathbb{F}$ upon $\mathbf{x}_0$ undenoted.

HILBERT's procedure tells us that we should expect one function $\mathbb{F}$ for each choice of the body force $\mathbf{b}$, and we could include this fact here by adding $\mathbf{b}$ to the list of arguments of $\mathbb{F}$. However, rather than do so at once, we at first restrict attention to a gas subject to the action of its intermolecular forces alone: $\mathbf{b} = \mathbf{0}$. Simplifications, both conceptual and formal, are thereby gained in our discussion of grossly determined solutions; even so, as we shall see in the following chapter, long and tedious calculations are needed to complete the work in detail. In Section (vii) of this chapter and in Section (x) of the next we shall indicate briefly how to restore $\mathbf{b}$ and what alteration results when we do so.

It is one thing to state that there are certain special solutions of the Maxwell–Boltzmann equation, but it is quite another to determine them. If we substitute (XXIII.1) directly into (IX.1), we learn nothing. Clearly we need to require more of the function $\mathbb{F}$. Let us look once more at continuum gas dynamics. A familiar property of gas flows subject to null body force allows us to obtain from any one many others simply by adding to each place $\mathbf{x}$ a constant vector $\mathbf{h}$ and by adding to each time t a constant s. That is, if $\boldsymbol{\rho}$ is a gas flow, so also is $\boldsymbol{\rho}(\cdot + s, \cdot + \mathbf{h})$ for each choice of the constants s and $\mathbf{h}$. This property can easily be verified in both the Euler–Hadamard and the Stokes–Kirchhoff theories, and it is true in many others. We search for gas-dynamic solutions in the kinetic theory because we expect that their fields of principal moments might correspond with gas flows according to continuum theories. Thus it is reasonable to set down the following second condition:

Hypothesis Regarding Invariance. *If F is the value of $\mathbb{F}$ at some field $\boldsymbol{\phi}$, so also is $F(\cdot + s, \cdot + \mathbf{h}, \cdot)$ for all constants s and $\mathbf{h}$.*

If $\mathscr{G}$ denotes the set of molecular densities that are values of $\mathbb{F}$, we may restate this hypothesis in words by saying that $\mathscr{G}$ is mapped into itself by the transformations[3] $F \mapsto F(\cdot + s, \cdot + \mathbf{h}, \cdot)$. The fact that a set of solutions is reproduced by certain transformations implies that it is special. However, this requirement is not enough, in itself, to give us a class of gas-dynamic solutions. Indeed, the set consisting in all solutions of the Maxwell–Boltzmann equation is such a class, as is also the set consisting in all uniform Maxwellian densities. However, both these examples fail to satisfy the hypothesis regarding existence, the former containing too many solutions and the latter containing too few. It is the combination of these two hypotheses that is important. Any class of solutions which is compatible with both we call a *class of gas-dynamic solutions*. Our hypothesis here is that one exists. If there is more than one, then each will be defined by a different function $\mathbb{F}$ according to (XXIII.1). In Section (vii) of Chapter XXIV we shall find evidence that $\mathbb{F}$ may be unique.

HILBERT's *solutions lie in the set $\mathscr{G}$.* This follows from two simple observations. The first is that HILBERT's assertion of gross causality, when specialized to the case in which $\mathbf{b} = \mathbf{0}$, implies that his solutions satisfy the first hypothesis. Next we note that if in (XXII.1) we replace t by $t + s$ and $\mathbf{x}$ by $\mathbf{x} + \mathbf{h}$, from one of the series that HILBERT considered we obtain another. This fact implies that the second hypothesis is also satisfied. Indeed, by comparing (XXIII.2) with (XXII.26) we see that the solutions HILBERT obtained are precisely those in the set $\mathscr{G}$ for which ϕ is a series $\sum_{n=0}^{\infty} \delta^{n-1} \phi^{(n)}$.

Mathematically the two hypotheses we have set up tell us that for each field ϕ and for each set of constants s and $\mathbf{h}$ there is a field ξ such that

$$\mathbb{F}_{\mathbf{r}}(t + s, \mathbf{x} + \mathbf{h}, \mathbf{v}; \phi(\mathbf{x}_0 + \mathbf{r})) = \mathbb{F}_{\mathbf{r}}(t, \mathbf{x}, \mathbf{v}; \xi(\mathbf{x}_0 + \mathbf{r})). \qquad \text{(XXIII.4)}$$

If we multiply each side of this equation by $\mathfrak{m} I_\alpha$, integrate with respect to $\mathbf{v}$, and then evaluate at $t = 0$, we see from (XXIII.2) that the right-hand side reduces to $\xi_\alpha(\mathbf{x})$. Thus

$$\begin{aligned}\xi_\alpha(\mathbf{x}) &= \mathfrak{m} \int I_\alpha \mathbb{F}_{\mathbf{r}}(s, \mathbf{x} + \mathbf{h}, \mathbf{v}; \phi(\mathbf{x}_0 + \mathbf{r})), \\ &= \rho_\alpha(s, \mathbf{x} + \mathbf{h}),\end{aligned} \qquad \text{(XXIII.5)}$$

ρ being the principal moment of $\mathbb{F}(\cdot\,; \phi)$. Placing this formula for ξ back into (XXIII.4), we obtain

$$\mathbb{F}_{\mathbf{r}}(t + s, \mathbf{x} + \mathbf{h}, \mathbf{v}; \phi(\mathbf{x}_0 + \mathbf{r})) = \mathbb{F}_{\mathbf{r}}(t, \mathbf{x}, \mathbf{v}; \rho(s, \mathbf{x}_0 + \mathbf{h} + \mathbf{r})). \qquad \text{(XXIII.6)}$$

---

[3] Or more abstractly, $\mathscr{G}$ is *invariant* with respect to the group of transformations $F \mapsto F(\cdot + s, \cdot + \mathbf{h}, \cdot)$.

Since this formula must hold for all choices of t, $\mathbf{x}$, s, and $\mathbf{h}$, we may replace these variables by 0, $\mathbf{x}_0$, t, and $\mathbf{x} - \mathbf{x}_0$, respectively, and so obtain

$$\underset{r}{\mathbb{F}}(t, \mathbf{x}, \mathbf{v}; \boldsymbol{\phi}(\mathbf{x}_0 + \mathbf{r})) = \underset{r}{\mathbb{F}}(0, \mathbf{x}_0, \mathbf{v}; \boldsymbol{\rho}(t, \mathbf{x} + \mathbf{r})). \tag{XXIII.7}$$

Therefore, if we define $\mathbb{G}(\mathbf{v}; \boldsymbol{\phi})$ to be $\mathbb{F}(0, \mathbf{x}_0, \mathbf{v}; \boldsymbol{\phi})$, we see from (XXIII.7) and (XXIII.1) that

$$F(t, \mathbf{x}, \mathbf{v}) = \underset{r}{\mathbb{G}}(\mathbf{v}; \boldsymbol{\rho}(t, \mathbf{x} + \mathbf{r})): \tag{XXIII.8}$$

Gas-dynamic solutions are grossly determined. This result points out very clearly the importance of grossly determined solutions in any attempt to recover from the kinetic theory gas flows that resemble those found in continuum gas dynamics. It is a consequence of the two hypotheses set down above, hypotheses motivated by simple properties that we should expect of gas flows in general.

(ii) *Methods of determining gas flows*

We return now to the question of why we wish to calculate gross relations. It is important to recall that *the primary objective in the kinetic theory is to determine gas flows.* One method of doing so is to find solutions of the Maxwell–Boltzmann equation. The fields ρ, $\mathbf{u}$, and ε determined by a solution constitute a gas flow according to the kinetic theory. However, not only is it difficult to find a solution, but also any one solution determines much more than ρ, $\mathbf{u}$, and ε and hence contains *more information than we need or wish.* Thus we seek other methods of finding ρ, $\mathbf{u}$, and ε. The examples in Chapters XIII, XIV, XV, and Appendix B to Chapter XVII, which are the only explicit solutions we have, suggest that many solutions yield one and the same gas flow, and that among all those solutions the grossly determined ones enjoy a special importance. Of course, it is a trivial property of such solutions that their fields $\mathbf{P}$ and $\mathbf{q}$ are also grossly determined:

$$P_{km}(t, \mathbf{x}) = \underset{r}{\mathfrak{P}}_{km}(\boldsymbol{\rho}(t, \mathbf{x} + \mathbf{r})),$$

$$q_k(t, \mathbf{x}) = \underset{r}{\mathfrak{q}}_k(\boldsymbol{\rho}(t, \mathbf{x} + \mathbf{r})), \tag{XXIII.9}$$

these being nothing more than (VIII.71) specialized by setting $\mathbf{b} = \mathbf{0}$. A second method, then, of determining gas flows, perhaps an easier one, would be *to calculate the gross determiners $\mathfrak{P}$ and $\mathfrak{q}$ of grossly determined solutions directly, without first determining solutions of the Maxwell–Boltzmann equation.* Once having obtained these gross determiners, we can place them into the equations of balance (IX.3)–(IX.5) so as to obtain *differential equations of motion*. We may write these equations of motion in the explicit forms

$$\dot{\rho} + \rho E = 0,$$

$$\rho \dot{u}_k = -p_{,k} - \left(\underset{\mathbf{r}}{\mathfrak{P}}_{ka}(\rho(t, \mathbf{x} + \mathbf{r}))\right)_{,a}, \qquad \text{(XXIII.10)}$$

$$\rho \dot{\varepsilon} = -pE - \underset{\mathbf{r}}{\mathfrak{P}}_{ab}(\rho(t, \mathbf{x} + \mathbf{r}))E_{ab} - \left(\underset{\mathbf{r}}{\mathfrak{q}}_a(\rho(t, \mathbf{x} + \mathbf{r}))\right)_{,a}.$$

This is just the way we use constitutive relations in continuum mechanics; it is what MAXWELL himself attempted to do.

Of the two methods just described, HILBERT's is an example of the first. Its objective is to calculate solutions of the Maxwell–Boltzmann equation. The fact that the solutions so obtained are grossly determined is only incidental to the method. Since HILBERT's procedure delivers gas flows first and only afterward infers from them the existence of $\mathfrak{P}$ and $\mathfrak{q}$, *in it there is no reason* to calculate these gross determiners, for they add nothing to what is known already. HILBERT's analysis suggests that grossly determined solutions exist in abundance, but it does not provide a useful way to ascertain the gross determiners. If we are to find a method of the second kind, we require a different theory of grossly determined solutions, one which from the outset treats on an equal footing the mapping G and the field ρ appearing in (XXIII.8).

The essential step in such a theory is clear: *If G is known*, (XXIII.10) *describes an initial-value problem for ρ, $\mathbf{u}$, and ε. By the hypothesis regarding existence and MAXWELL's consistency theorem this problem has a unique solution for each initial value ϕ. Each solution of this problem defines a gas flow. The value of G at an arbitrary field ρ not satisfying* (XXIII.10) *is not a solution of* (IX.1). If we can determine G without first solving the Maxwell–Boltzmann equation— or, better, if we can first determine $\mathfrak{P}$ and $\mathfrak{q}$—we may *thereafter* find gas flows in the kinetic theory by solving an initial-value problem similar to the one that arises in continuum mechanics. *This is the reason that we wish to learn about gross relations.*

This fact might suggest that (XXIII.9) be identified with the constitutive relations of some continuum theory of fluids, and several students of the kinetic theory have thought such an identification possible. It is not. The reason has been given already in Chapter IX in connection with MAXWELL's consistency theorem. In continuum mechanics the arguments of constitutive functions are arbitrary vector fields or tensor fields, subject only to conditions of smoothness or algebraic class; on the contrary, the field ρ appearing in (XXIII.9) is by no means arbitrary. We know *a priori* that this field is restricted by the differential system (XXIII.10); hence if we vary one component of ρ, we generally vary the others also. Of course (XXIII.9) might be *restrictions* of constitutive relations, restrictions to arguments ρ such as to satisfy (XXIII.10). If so, we must recall that many different functions have a common restriction.

Whether $\mathfrak{P}$ and $\mathfrak{q}$ can or cannot be identified with quantities familiar to us from continuum thermomechanics, is one question. However, whether the kinetic theory is able to provide conditions sufficient to determine $\mathfrak{P}$ and $\mathfrak{q}$, possibly uniquely, is quite another, and we should take care not to confuse the two. G, $\mathfrak{P}$, and $\mathfrak{q}$ are functions of a time-independent field ϕ, and the hypothesis regarding existence ensures that apart from certain conditions of positivity we may choose ϕ arbitrarily. This fact provides some hope that equations governing the *mappings* G, $\mathfrak{P}$, and $\mathfrak{q}$ may indeed arise from the kinetic theory, and shortly we shall see that at least formally they do.

(iii) *The Maxwell–Boltzmann equation for grossly determined solutions*

To find G, we must use the Maxwell–Boltzmann equation. Thus we must examine first the special form that that equation takes for the solutions (XXIII.8). To this end we rewrite the grossly determined solutions in a notation that lends itself better to analysis.

The arguments of G are a molecular velocity $\mathbf{v}$ and a function $\psi = (\psi_\alpha)$ defined on the translation space $\mathscr{W}$. In (XXIII.8) G is evaluated at a different ψ for each t and $\mathbf{x}$, namely $\psi(\mathbf{r}) = \rho(t, \mathbf{x} + \mathbf{r})$. Many functions in our analysis are formed by shifting the argument $\mathbf{x}$ of a field in this way, so we introduce a simple notation for them in terms of an operator $\mathfrak{S}$ we call the *shifter*. For a general field ϕ we define as follows a field we call its *shift* $\mathfrak{S}\phi$:

$$\mathfrak{S}\phi(t, \mathbf{x}) \equiv \phi(t, \mathbf{x} + \cdot). \qquad \text{(XXIII.11)}$$

Thus, for each t and $\mathbf{x}$, $\mathfrak{S}\phi(t, \mathbf{x})$ is a function defined on $\mathscr{W}$. If ϕ is real-valued, $\mathfrak{S}\phi(t, \mathbf{x})$ is a real-valued function on $\mathscr{W}$. In terms of the shifter we may write (XXIII.8) in either of the two equivalent forms

$$\begin{aligned} F(t, \mathbf{x}, \mathbf{v}) &= \mathop{\mathsf{G}}_{\mathbf{r}}(\mathbf{v}; \mathfrak{S}\rho(t, \mathbf{x})(\mathbf{r})), \\ &= \mathsf{G}(\mathbf{v}; \mathfrak{S}\rho(t, \mathbf{x})). \end{aligned} \qquad \text{(XXIII.12)}$$

In the calculations we present in this chapter and the next one the variable $\mathbf{v}$ plays only a secondary role. Therefore, henceforth we shall suppress $\mathbf{v}$ whenever it appears as an argument of F or G, and we shall interpret F as a function of t and $\mathbf{x}$ alone, G as a function of ψ alone. This allows us to write the grossly determined solutions F as the composite functions

$$F = \mathsf{G} \circ \mathfrak{S}\rho. \qquad \text{(XXIII.13)}$$

To calculate the derivatives of F with respect to t and $\mathbf{x}$, we shall use the chain rule. Before doing so we pause briefly to explain what we shall mean by the derivatives of G and $\mathfrak{S}\rho$.

We assume that the domain of G is an open subset $\mathcal{O}$ of a Banach space $\mathscr{F}$. In the appendix to this chapter we present the parts of the theory of Banach spaces that we shall need here and in Chapter XXIV. A typical element ψ of $\mathscr{F}$ is a function defined on $\mathscr{W}$. As we shall see later, the constant functions on $\mathscr{W}$ play a special role in our analysis, so we shall assume that $\mathcal{O}$ contains them. We assume also that G *is differentiable on* $\mathcal{O}$. This means (see (XXIIIA.1), (XXIIIA.2), and (XXIIIA.3)) that if $\psi \in \mathcal{O}$ then for all ξ near $\mathbf{0}$ in $\mathscr{F}$ the mapping G has the expansion

$$\mathsf{G}(\psi + \xi) = \mathsf{G}(\psi) + \partial_\psi \mathsf{G}(\psi)[\xi] + \chi(\psi, \xi)\|\xi\|, \qquad \text{(XXIII.14)}$$

in which $\partial_\psi \mathsf{G}(\psi)$ is the derivative of G at ψ and $\|\ \|$ is the norm in $\mathscr{F}$. For each fixed ψ in $\mathcal{O}$ the derivative $\partial_\psi \mathsf{G}(\psi)[\cdot]$ is a continuous linear function on $\mathscr{F}$, and

$$\lim_{\|\xi\| \to 0} |\chi(\psi, \xi)| = 0. \qquad \text{(XXIII.15)}$$

Next we consider $\mathfrak{S}\rho$. By definition $\mathfrak{S}\rho$ is a field whose values are functions ψ defined on $\mathscr{W}$. Henceforth we assume that ρ has been so chosen that these values be in $\mathscr{F}$. Then $\mathfrak{S}\rho$ is differentiable at $(t, \mathbf{x})$ if (see (XXIIIA.1), (XXIIIA.2), (XXIIIA.3), and (XXIIIA.9)) for all s and $\mathbf{h}$ near zero, each component of $\mathfrak{S}\rho(t + s, \mathbf{x} + \mathbf{h})$ has an expansion

$$\mathfrak{S}\rho_\alpha(t + s, \mathbf{x} + \mathbf{h}) = \mathfrak{S}\rho_\alpha(t, \mathbf{x}) + \partial_t \mathfrak{S}\rho_\alpha(t, \mathbf{x})s + \partial_{x_a} \mathfrak{S}\rho_\alpha(t, \mathbf{x})h_a$$
$$+ \chi_\alpha(t, \mathbf{x}, s, \mathbf{h})|(s, \mathbf{h})|; \qquad \text{(XXIII.16)}$$

$\partial_t \mathfrak{S}\rho(t, \mathbf{x})$ and $\partial_{x_k} \mathfrak{S}\rho(t, \mathbf{x})$ are the partial derivatives of $\mathfrak{S}\rho$ at $(t, \mathbf{x})$, and $|(s, \mathbf{h})| \equiv |s| + |\mathbf{h}|$. Each of these derivatives is an element of $\mathscr{F}$, and

$$\lim_{|(s,\mathbf{h})| \to 0} \|\chi(t, \mathbf{x}, s, \mathbf{h})\| = 0. \qquad \text{(XXIII.17)}$$

These statements concern the differentiability of $\mathfrak{S}\rho$, not that of ρ itself. However, if ρ is differentiable, we may easily relate its partial derivatives to those of $\mathfrak{S}\rho$. By differentiating formally on each side of (XXIII.11) and then replacing ϕ with ρ we obtain

$$\partial_t \mathfrak{S}\rho_\alpha(t, \mathbf{x}) = \partial_t \rho_\alpha(t, \mathbf{x} + \cdot) = \mathfrak{S}\, \partial_t \rho_\alpha(t, \mathbf{x}),$$
$$\partial_{x_k} \mathfrak{S}\rho_\alpha(t, \mathbf{x}) = \partial_{x_k} \rho_\alpha(t, \mathbf{x} + \cdot) = \mathfrak{S}\, \partial_{x_k} \rho_\alpha(t, \mathbf{x}). \qquad \text{(XXIII.18)}$$

We hasten to emphasize that this result is only formal. To establish it rigorously, we should have to show that $\mathfrak{S}\rho$ had an expansion of the form (XXIII.16) in which $\partial_t \mathfrak{S}\rho$ and $\partial_{x_k} \mathfrak{S}\rho$ had these special forms. Nevertheless, this formal argument suggests what to expect, at least for fields ρ which are themselves smooth.

Let ρ be any smooth field for which $\mathfrak{S}\rho: \mathscr{I} \times \mathscr{E} \to \mathcal{O}$, $\mathscr{I}$ being an open interval of times containing 0. Since $\mathcal{O}$ is the domain of G, for such fields the composite function $F = \mathsf{G} \circ \mathfrak{S}\rho$ is well defined. We assume also that $\mathfrak{S}\rho$ is differentiable and that its derivatives are related to those of ρ by (XXIII.18).

(iv)   EQUATIONS OF GROSS DETERMINISM

Since $\mathbb{G}$ is by assumption differentiable, so also is F, and by the chain rule (see (XXIIIA.18)) we obtain

$$\partial_t F = \partial_\psi \mathbb{G}(\mathfrak{S}\rho)[\mathfrak{S}\,\partial_t\,\rho], \qquad \partial_{x_k} F = \partial_\psi \mathbb{G}(\mathfrak{S}\rho)[\mathfrak{S}\,\partial_{x_k}\rho]. \qquad \text{(XXIII.19)}$$

Substituting these into (IX.1), suitably specialized to the case in which $\mathbf{b} = \mathbf{0}$, and using the fact that $\partial_\psi \mathbb{G}(\psi)[\,\cdot\,]$ is a linear operator, we obtain finally the form of *the Maxwell–Boltzmann equation appropriate to grossly determined solutions*:

$$\partial_\psi \mathbb{G}(\mathfrak{S}\rho)[\mathfrak{S}\,\partial_t\,\rho + v_a\,\mathfrak{S}\,\partial_{x_a}\rho] = \mathbb{C}\mathbb{G}(\mathfrak{S}\rho). \qquad \text{(XXIII.20)}$$

Each term in this equation is given explicitly in terms of ρ and $\mathbb{G}$ and their derivatives. Such must be the case if we wish to deduce properties not only of the combination of ρ and $\mathbb{G}$ appearing in (XXIII.8) but also of ρ and $\mathbb{G}$ separately.

(iv)   The equations of gross determinism and properties of gross determiners

We wish to find equations governing the gross determiner $\mathbb{G}$, equations that are independent of what the field ρ must be in order for (XXIII.8) to satisfy the Maxwell–Boltzmann equation. Since ρ is determined completely, once $\mathbb{G}$ is known, by the equations of motion (XXIII.10), we must remove the information contained in those equations from the Maxwell–Boltzmann equation (XXIII.20) that is appropriate to grossly determined solutions. To reveal how this may be done, let us rewrite (XXIII.10) in a more convenient form. Since the equations of balance (IX.3)–(IX.5) were derived in the kinetic theory by taking for g in the equation of transfer (IX.2) a summational invariant, we may derive equivalent forms of these equations by letting g be the principal invariants I_α in turn. Setting $\mathbf{b} = \mathbf{0}$ and using (VI.22), we obtain

$$\partial_t \rho_\alpha = -(\mathfrak{F}_{\alpha a}(\mathfrak{S}\rho))_{,a}, \qquad \text{(XXIII.21)}$$

$\mathfrak{F}_k$ being defined by

$$\mathfrak{F}_{\alpha k}(\mathfrak{S}\rho) \equiv \mathfrak{m} \int I_\alpha v_k\, \mathbb{G}(\mathfrak{S}\rho). \qquad \text{(XXIII.22)}$$

Of course we have already used here the fact that the molecular density F is grossly determined. We can easily interchange (XXIII.10) and (XXIII.21) by means of (VI.23) and the formula

$$\mathfrak{F}_{\alpha k} = \begin{cases} \rho u_k & \text{if } \alpha = 0, \\ \rho u_\alpha u_k + p\,\delta_{\alpha k} + \mathfrak{P}_{\alpha k} & \text{if } \alpha = 1, 2, 3, \\ \rho(2\varepsilon + u^2)u_k + 2pu_k + 2\mathfrak{P}_{ka}u_a + 2q_k & \text{if } \alpha = 4. \end{cases} \qquad \text{(XXIII.23)}$$

We may look upon (XXIII.21) as a formula giving the time derivative $\partial_t \rho$ in terms of $\mathbb{G}$ and ρ. This time derivative occurs also in the Maxwell-Boltzmann equation (XXIII.20) governing grossly determined solutions. If we eliminate $\partial_t \rho$ between them, we obtain

$$\partial_\psi \mathbb{G}(\mathfrak{S}\rho)[-\mathfrak{S}(\mathfrak{F}_a(\mathfrak{S}\rho))_{,a} + v_a \mathfrak{S} \partial_{x_a}\rho] = \mathbb{C}\mathbb{G}(\mathfrak{S}\rho). \qquad \text{(XXIII.24)}$$

This equation holds for all t for which ρ is defined. In particular it holds when $t = 0$, and then, in view of (XXIII.3) and the fact that no time derivatives are present, it becomes

$$\partial_\psi \mathbb{G}(\mathfrak{S}\phi)[-\mathfrak{S}(\mathfrak{F}_a(\mathfrak{S}\phi))_{,a} + v_a \mathfrak{S} \partial_{x_a}\phi] = \mathbb{C}\mathbb{G}(\mathfrak{S}\phi). \qquad \text{(XXIII.25)}$$

In addition, we know that $\mathbb{G}$ satisfies

$$\phi_\alpha = \mathfrak{m} \int I_\alpha \mathbb{G}(\mathfrak{S}\phi), \qquad \alpha = 0, 1, \ldots, 4, \qquad \text{(XXIII.26)}$$

this being a direct consequence of (XXIII.2) and the definition of $\mathbb{G}$ in terms of $\mathbb{F}$ given after (XXIII.7).

Our objective is to determine $\mathbb{G}$ without first finding the principal moment ρ of the solution (XXIII.8). In this respect we note that (XXIII.25) and (XXIII.26) involve only $\mathbb{G}$ and the field ϕ. Since by the hypothesis of existence in Section (i) the field ϕ is arbitrary, (XXIII.25) and (XXIII.26) give us conditions on $\mathbb{G}$ that are independent of what ρ might be. We shall call these two conditions the **Equations of Gross Determinism**[4]. *They provide restrictions on the gross determiner $\mathbb{G}$ alone.*

We ask now whether $\mathbb{G}$ must satisfy any conditions other than the equations of gross determinism, and of course the requirement that it be non-negative. Suppose, then, that $\mathbb{G}$ does satisfy these conditions, and let ρ be a solution of the equations of motion (XXIII.10), the gross determiners $\mathfrak{P}$ and $\mathfrak{q}$ being those determined by the solution $\mathbb{G}$. Since (XXIII.25) and (XXIII.26) are valid for all initial values ϕ, and since for each fixed t the field $\rho(t, \cdot)$ may be interpreted as the initial value of a new field, we may replace ϕ in (XXIII.25) by $\rho(t, \cdot)$ and so obtain (XXIII.24). It is a simple matter now to reverse the steps leading to (XXIII.24) and conclude that (XXIII.20) is satisfied also. Thus, the function F given by (XXIII.8) is a non-negative solution of the Maxwell-Boltzmann equation. *The equations of gross determinism and the restriction $\mathbb{G} \geq 0$ are necessary and sufficient conditions for $\mathbb{G}$ to be a gross determiner.* It is these conditions alone that $\mathbb{G}$ must satisfy. They are explicit mathematical statements about the gross determiner, and from these alone flow the properties it must have. *A fortiori* it is from these conditions alone that the gross determiners $\mathfrak{P}$ and $\mathfrak{q}$ must follow.

The first of the equations of gross determinism, namely (XXIII.25), is very similar in structure to the Maxwell-Boltzmann equation. Indeed, both are

---

[4] MUNCASTER [1975, Chapter VII].

integro-differential equations in which the right-hand side is a value of the collisions operator $\mathbb{C}$ and the left-hand side is the value of a differential operator. We may expect, therefore, that many of the properties which we have described previously for the Maxwell–Boltzmann equation will have analogues for the equations of gross determinism. We shall reserve our discussion of these until Section (vii), in which we extend our analysis so as to take account of the body force $\mathbf{b}$. There are some conclusions, however, that we may deduce now. First suppose that ϕ is independent of place. Recall that the open set $\mathcal{O}$ contains all fields of this type. Then $\mathfrak{F}_{a,a}$ and $\phi_{,k}$ both vanish, and so, because $\partial_\psi \mathsf{G}(\psi)[\,\cdot\,]$ is a linear operator, the left-hand side of (XXIII.25) reduces to zero. In view of (VIII.7) and (XXIII.26) we conclude that *if G is any solution of the equations of gross determinism, and if ϕ is a constant field, then $\mathsf{G}(\mathfrak{S}\phi)$ is the unique Maxwellian density whose principal moment is ϕ.* This is a universal property of all gross determiners, one which is independent of the molecular model of the gas. It is important to note that this result does not give us a solution of the equations of gross determinism. Rather, it tells us the explicit form of all solutions G on a very special subset of their domain, namely that consisting in the constant fields.

Second, let us recall that the locally Maxwellian solutions are grossly determined. Thus we may take for F any one of these for which $\mathbf{b} = \mathbf{0}$, and the steps leading from (IX.1) to (XXIII.25) and (XXIII.26) still apply. We conclude, then, that if $\mathcal{A}$ is the set of initial values ϕ of the principal moments of all locally Maxwellian solutions for which $\mathbf{b} = \mathbf{0}$, *the equations of gross determinism are satisfied on $\mathcal{A} \cap \mathcal{O}$ if the restriction of G to $\mathcal{A} \cap \mathcal{O}$ is a Maxwellian density.* This result points out the status of locally Maxwellian solutions in the theory of gross determiners, but in contrast with the universal property obtained previously its value seems obscure because it does not tell us that the restriction of G to $\mathcal{A} \cap \mathcal{O}$ *must* be a Maxwellian density.

The locally Maxwellian solutions are of scant interest. To learn about more important solutions directly is presently very difficult. Consequently we might explore the possibility of approximating the values of G. To do so by an iterative procedure will be our objective in Chapter XXIV.

(v) *Principles of local action and the domain of the gross determiner*

We have said very little about the Banach space $\mathscr{F}$ and the open subset $\mathcal{O} \subset \mathscr{F}$ on which gross determiners are defined. The reason is that the steps presented so far in our analysis do not require us to make any special choice of them. This choice represents, then, an extra variable in the theory, one we may adjust so as to draw the concept of a gross determiner even closer to its intended objective: a kinetic analogue of constitutive relations in continuum thermomechanics. To see how this may be done, we look again at gas dynamics.

Our interest in studying grossly determined solutions lies in the hope that they may yield gas flows in the kinetic theory that are similar to ones found in continuum mechanics. These hopes are strengthened by the fact that ρ is to be calculated, just as in continuum mechanics, by solving the equations of motion (XXIII.10). However this fact, in itself, is not enough. We should not expect any agreement with continuum theories unless the gross determiners $\mathfrak{P}$ and $\mathbf{q}$ resemble constitutive relations *not only in form but also in properties*. In Chapter I we have stated general principles now commonly accepted in continuum thermomechanics, principles which serve to dictate some properties we should like constitutive relations to have. The *principle of determinism* is one of these. An analogue of it that might be applied in the kinetic theory is: *at a place* $\mathbf{x}$ *and a time* t *the values of the pressure deviator* $\mathbf{P}$ *and the energy flux vector* $\mathbf{q}$ *are determined by the history of the principal moment* ρ *at all places in a neighborhood of* $\mathbf{x}$. The formula (XXIII.9) shows that this is indeed the case for grossly determined solutions, though the history of ρ enters only through its present value. In addition to the principle of determinism we have the *principle of local action*. An analogue of it for the kinetic theory might be: *two gas flows that agree in a neighborhood of the place* $\mathbf{x}$ *give rise to the same values of* $\mathbf{P}$ *and* $\mathbf{q}$ *at* $\mathbf{x}$. It is this principle which we shall use as our guideline in choosing $\mathscr{F}$ and $\mathscr{O}$.

In continuum thermomechanics COLEMAN & NOLL have shown[5] how some principles of local and non-local action may be interpreted mathematically as demanding that a constitutive functional be continuous on a suitable Banach space. This suggests that one might impose upon $\mathfrak{P}$ and $\mathbf{q}$ some requirement of local action by choosing $\mathscr{F}$ and $\mathscr{O}$ appropriately. Unfortunately there is no general rule for doing so. We have to proceed by examples. Therefore, in this section we present one choice of $\mathscr{F}$ and $\mathscr{O}$ that implies a special type of local action for $\mathfrak{P}$ and $\mathbf{q}$. There may be many others[6].

Let $\mathscr{C}(\mathscr{W})$ be the set of real-valued continuous functions on $\mathscr{W}$. It is well known that $\mathscr{C}(\mathscr{W})$ is a vector space if for each ψ and ξ in $\mathscr{C}(\mathscr{W})$ and each α in $\mathscr{R}$ we define $\psi + \xi$ and $\alpha\psi$ pointwise:

$$(\psi + \xi)(\mathbf{r}) \equiv \psi(\mathbf{r}) + \xi(\mathbf{r}), \qquad (\alpha\psi)(\mathbf{r}) \equiv \alpha\psi(\mathbf{r}). \tag{XXIII.27}$$

For each positive integer m let H_m be a real-valued function on $[0, \infty)$ such that

(H1) H_m is positive and continuous;
(H2) $r^m H_m(r) \to 0$ as $r \to \infty$.

We could, for example, choose H_m to be $r \mapsto (1 + r)^{-m-1}$. The function $r \mapsto \exp(-r)$ is a possible choice of H_m for all values of m. Let us denote by $\mathscr{C}_m(\mathscr{W})$ the set of all $\psi \in \mathscr{C}(\mathscr{W})$ for which the number

$$\|\psi\|_m \equiv \sup_{\mathbf{r} \in \mathscr{W}} |\psi(\mathbf{r})| H_m(|\mathbf{r}|) \tag{XXIII.28}$$

is finite. That is, $\mathscr{C}_m(\mathscr{W})$ is the set of continuous functions on $\mathscr{W}$ whose rate of growth as $|\mathbf{r}| \to \infty$ is no greater than that of $H_m(|\mathbf{r}|)^{-1}$. By property (H2) we know that the polynomials of degree m are examples of functions in $\mathscr{C}_m(\mathscr{W})$. The mapping $\|\ \|_m$ is a norm on this space, and with respect to it we have

Lemma 1. *$\mathscr{C}_m(\mathscr{W})$ is a Banach space.*

---

[5] B. D. COLEMAN & W. NOLL, "An approximation theorem for functionals, with applications in continuum mechanics," *Archive for Rational Mechanics and Analysis* **6**, 355–370 (1960).

[6] Whether or not our choice be reasonable, is a question that must ultimately be decided by an analytical study of the equations of gross determinism. However, the sets $\mathscr{F}$ and $\mathscr{O}$ that we choose in this book play more than only a superficial role. The iterative procedure that we shall present in Chapter XXIV for approximating gross determiners depends on this choice, and the fact that that procedure happens to be an effective one (*cf.* Section (vii) of Chapter XXIV) suggests that our choice may be reasonable.

(v) PRINCIPLES OF LOCAL ACTION

Proof. It is well known[7] that the set $\mathscr{C}_\infty(\mathscr{W})$ of bounded continuous functions on $\mathscr{W}$, when equipped with the supremum norm

$$\|\psi\|_\infty \equiv \sup_{\mathbf{r} \in \mathscr{W}} |\psi(\mathbf{r})|, \tag{XXIII.29}$$

is a Banach space. The function Φ defined by

$$\Phi(\psi)(\mathbf{r}) \equiv \psi(\mathbf{r}) H_m(|\mathbf{r}|) \tag{XXIII.30}$$

is an invertible linear transformation from $\mathscr{C}_m(\mathscr{W})$ onto $\mathscr{C}_\infty(\mathscr{W})$ which preserves norms: $\|\Phi(\psi)\|_\infty = \|\psi\|_m$. Thus $\mathscr{C}_m(\mathscr{W})$ is isometric to $\mathscr{C}_\infty(\mathscr{W})$. That $\mathscr{C}_m(\mathscr{W})$ is a Banach space, now follows from the fact[8] that a normed vector space that is isometric to a Banach space is itself a Banach space. △

We now choose for $\mathscr{F}$ the set of functions $\boldsymbol{\psi} = (\psi_\alpha)$ each of whose components ψ_α lies in $\mathscr{C}_m(\mathscr{W})$:

$$\mathscr{F} \equiv (\mathscr{C}_m(\mathscr{W}))^5. \tag{XXIII.31}$$

The norm $\|\ \|$ in $\mathscr{F}$ is defined in terms of the norm $\|\ \|_m$ as follows:

$$\|\boldsymbol{\psi}\| \equiv \sum_{\alpha=0}^{4} \|\psi_\alpha\|_m. \tag{XXIII.32}$$

In choosing the open set $\mathcal{O} \subset \mathscr{F}$ we must bear in mind the inequalities set out in the hypothesis of existence in Section (i): $\phi_0 > 0$, $\phi_0 \phi_4 > \phi_a \phi_a$. A natural choice for $\mathcal{O}$, then, is the set $\mathcal{O}_\infty$ defined as follows:

$$\mathcal{O}_\infty \equiv \{\boldsymbol{\psi} \in \mathscr{F} \mid \psi_0 > 0, \psi_0 \psi_4 > \psi_a \psi_a\}. \tag{XXIII.33}$$

Unfortunately this set is not open, but by enlarging it slightly we obtain many open sets.

Lemma 2. *For each positive number d the set*

$$\mathcal{O}_d \equiv \{\boldsymbol{\psi} \in \mathscr{F} \mid \psi_0(\mathbf{r}) > 0, \psi_0(\mathbf{r})\psi_4(\mathbf{r}) > \psi_a(\mathbf{r})\psi_a(\mathbf{r}) \text{ if } |\mathbf{r}| \leq d\} \tag{XXIII.34}$$

is open in $\mathscr{F}$.

Proof. Choose $\boldsymbol{\psi} \in \mathcal{O}_d$, and let $\mathscr{B}_\delta(\boldsymbol{\psi})$ be the open ball of radius δ centered upon $\boldsymbol{\psi}$:

$$\mathscr{B}_\delta(\boldsymbol{\psi}) \equiv \{\boldsymbol{\xi} \in \mathscr{F} \mid \|\boldsymbol{\psi} - \boldsymbol{\xi}\| < \delta\}. \tag{XXIII.35}$$

We must prove that $\mathscr{B}_\delta(\boldsymbol{\psi}) \subset \mathcal{O}_d$ for some δ. For any $\boldsymbol{\xi} \in \mathscr{B}_\delta(\boldsymbol{\psi})$ it follows from (XXIII.28) and (XXIII.32) that

$$|\psi_\alpha(\mathbf{r}) - \xi_\alpha(\mathbf{r})| < \frac{\delta}{H_m(|\mathbf{r}|)}, \quad \alpha = 0, 1, \ldots, 4. \tag{XXIII.36}$$

In particular

$$\xi_0 > \psi_0 - \frac{\delta}{H_m}, \tag{XXIII.37}$$

---

[7] *E.g.* Chapter VII of J. A. DIEUDONNÉ, *Foundations of Modern Analysis*, Academic Press, New York, 1969.

[8] *E.g.* §14 of Chapter III of the reference cited in Footnote 7.

and also

$$\xi_0\xi_4 - \xi_a\xi_a = [(\xi_0 - \psi_0) + \psi_0][(\xi_4 - \psi_4) + \psi_4] - [(\xi_a - \psi_a) + \psi_a][(\xi_a - \psi_a) + \psi_a],$$
$$= \psi_0\psi_4 - \psi_a\psi_a + (\xi_0 - \psi_0)(\xi_4 - \psi_4) + \psi_0(\xi_4 - \psi_4)$$
$$+ \psi_4(\xi_0 - \psi_0) - (\xi_a - \psi_a)(\xi_a - \psi_a) - 2\psi_a(\xi_a - \psi_a), \tag{XXIII.38}$$
$$> \psi_0\psi_4 - \psi_a\psi_a - \frac{\delta}{H_m}\left(\psi_0 + \psi_4 + 2\sum_{k=1}^{3}|\psi_k|\right) - 2\frac{\delta^2}{H_m^2}.$$

Recall now that H_m and ψ are continuous, H_m is positive, and ψ satisfies the inequalities (XXIII.34). Thus there are positive constants $\varepsilon_1, \varepsilon_2, C_1$, and C_2 such that if $|\mathbf{r}| \leq d$

$$\psi_0(\mathbf{r}) \geq \varepsilon_1, \quad \psi_0(\mathbf{r})\psi_4(\mathbf{r}) - \psi_a(\mathbf{r})\psi_a(\mathbf{r}) \geq \varepsilon_2,$$
$$\frac{1}{H_m(|\mathbf{r}|)} \leq C_1, \quad \frac{1}{H_m(|\mathbf{r}|)}\left(\psi_0(\mathbf{r}) + \psi_4(\mathbf{r}) + 2\sum_{k=1}^{3}|\psi_k(\mathbf{r})|\right) \leq C_2. \tag{XXIII.39}$$

From (XXIII.39), (XXIII.38)$_3$, and (XXIII.37) we see that if $|\mathbf{r}| \leq d$ then

$$\xi_0(\mathbf{r}) > \varepsilon_1 - \delta C_1, \quad \xi_0(\mathbf{r})\xi_4(\mathbf{r}) - \xi_a(\mathbf{r})\xi_a(\mathbf{r}) > \varepsilon_2 - \delta C_2 - 2\delta^2 C_1^2. \tag{XXIII.40}$$

The right-hand sides of these two inequalities are determined by ψ, H_m, and δ alone. Moreover, for sufficiently small δ they are positive. We conclude that ξ obeys the inequalities in (XXIII.34) provided that it lie in a sufficiently small open ball about ψ. That is, $\mathcal{O}_d$ is an open set. $\triangle$

Each of the sets $\mathcal{O}_d$ is a possible choice for the set $\mathcal{O}$ on which G is defined. Henceforth we shall assume that a positive number d has been chosen and that $\mathcal{O} = \mathcal{O}_d$. We should note that, in keeping with our general assumptions about $\mathscr{F}$, the special choice (XXIII.31) is large enough to contain those fields ψ which are constant on $\mathscr{W}$. Likewise, the open set $\mathcal{O} = \mathcal{O}_d$ contains the constant fields, at least those which satisfy the physically appropriate inequalities $\psi_0 > 0$, $\psi_0\psi_4 > \psi_a\psi_a$.

Reviewing our choice of the Banach space $\mathscr{F}$, we note that since G is differentiable on $\mathcal{O}$, it is also continuous there. Roughly speaking this means that if ψ^1 and ψ^2 both lie in $\mathcal{O}$ and $\|\psi^1 - \psi^2\|$ is small, then $|\mathsf{G}(\psi^1) - \mathsf{G}(\psi^2)|$ is small also. Next let us note what it means for $\|\psi^1 - \psi^2\|$ to be small. From the definition (XXIII.28) of $\|\ \|_m$ we see that if ψ^1 and ψ^2 lie in $\mathscr{C}_m(\mathscr{W})$ and if $\|\psi^1 - \psi^2\|_m$ is small, then $|\psi^1(\mathbf{r}) - \psi^2(\mathbf{r})|H_m(|\mathbf{r}|)$ is small for all $\mathbf{r} \in \mathscr{W}$. But since $|\mathbf{r}|^m H_m(|\mathbf{r}|) \to 0$ as $|\mathbf{r}| \to \infty$, the number $\|\psi^1 - \psi^2\|_m$ may be small even when the values $\psi^1(\mathbf{r})$ and $\psi^2(\mathbf{r})$ differ greatly for large values of $|\mathbf{r}|$. For example, if $\psi^1 - \psi^2$ is a polynomial of degree m in the components of $\mathbf{r}$, then $\|\psi^1 - \psi^2\|_m$ is finite even though $|\psi^1(\mathbf{r}) - \psi^2(\mathbf{r})| \to \infty$ as $|\mathbf{r}| \to \infty$. The continuity of G, when viewed in terms of this property of $\|\ \|_m$, tells us that if the values of ψ^1 and ψ^2 differ only slightly in a neighborhood of $\mathbf{r} = 0$, then the difference of the corresponding values of G is small also. That is, G *is influenced only slightly by the values of* ψ *at points far from* $\mathbf{0}$. In fact, the function H_m that defines $\|\ \|_m$ is sometimes called an *influence function*[9] since it alone determines the relative influence upon the value of G of the values of ψ in different neighborhoods of $\mathbf{r} = \mathbf{0}$.

It is useful to summarize these ideas in terms of the gross determiners $\mathfrak{P}$ and $\mathbf{q}$. Roughly speaking, we may say that with the choice made above for $\mathscr{F}$, continuity of G implies that *the values of* $\mathbf{P}$ *and* $\mathbf{q}$ *at a place* $\mathbf{x}$ *are influenced primarily by the values of* $\mathbf{\rho}$ *at places in a neighborhood of* $\mathbf{x}$ *and only slightly, if at all, by the values of* $\mathbf{\rho}$ *at places far from* $\mathbf{x}$. To require G to be continuous in this sense is the form of the principle of local action that we shall impose upon $\mathfrak{P}$ and $\mathbf{q}$ in this book. Of course there may be many other types of local action that one might use in place of it.

---

[9] The concept of an influence function has been used extensively in continuum mechanics to describe materials with fading memory. A discussion of materials of this type may be found in §38 of the book by TRUESDELL & NOLL cited in Footnote 4 to Chapter I.

Next let us consider our choice of the open set $\mathcal{O}$. We note from (XXIII.34) that there are elements in $\mathcal{O}$ which do not satisfy the inequalities $\psi_0 > 0$ and $\psi_0 \psi_4 > \psi_a \psi_a$ for all $\mathbf{r} \in \mathscr{W}$. Consequently such functions need not lie in the domain of $\mathbf{G}$ as set down in the hypothesis of existence. Indeed, by assuming that $\mathbf{G}$ is defined on all of $\mathcal{O}$ we actually *extend* the domain of $\mathbf{G}$ to a broader class of functions, one whose elements need satisfy these inequalities only for $\mathbf{r}$ near $\mathbf{0}$. We have done this so as to apply the differential calculus in Banach spaces. However, we do not mean to say by this that we shall allow $\mathbf{G}$ to deliver solutions F through the formula (XXIII.8) for which the corresponding density and energetic are not always positive. We are simply assuming that $\mathbf{G}$ delivers a solution in this way *only* when its argument ψ lies in the subset $\mathcal{O}_\infty$ of $\mathcal{O}_d$ even though $\mathbf{G}$ be defined on all of $\mathcal{O}_d$.

We should note two more things. First, even though we have extended the domain of $\mathbf{G}$ from the set of functions $\mathcal{O}_\infty$ to all of $\mathcal{O}_d$, this extension is realistic: Since the value of $\mathbf{G}$ is determined mainly by the values of ψ near $\mathbf{r} = \mathbf{0}$, it should be sufficient to consider arguments ψ for which $\psi_0 > 0$ and $\psi_0 \psi_4 > \psi_a \psi_a$ in some neighborhood of $\mathbf{r} = \mathbf{0}$. The elements of $\mathcal{O}_d$ are precisely of this type. Second, for[10] no value of d is $\mathcal{O}_\infty$ dense in $\mathcal{O}_d$, so in general an extension of $\mathbf{G}$ to all of $\mathcal{O}_d$ will not be unique unless additional conditions are imposed on $\mathbf{G}$. Our hope here is that such additional conditions will consist in the equations of gross determinism, applied now on $\mathcal{O}_d$ rather than $\mathcal{O}_\infty$. If on $\mathcal{O}_d$ they have a unique solution, then the corresponding extension will also be unique.

(vi) *A space of functions for the principal moment*

In our derivation of the equations of gross determinism we restricted attention to a certain class of fields $\boldsymbol{\rho}$. It is not difficult to draw from our analysis conditions which define this class. In order to obtain (XXIII.19) and thence the Maxwell–Boltzmann equation in the special form (XXIII.20), we required that:

(a) for all places $\mathbf{x}$ and for all times t in an interval $\mathscr{I}$ containing 0, $\mathfrak{S}\boldsymbol{\rho}(t, \mathbf{x})$ lies in the domain $\mathcal{O}$ of $\mathbf{G}$;

(b) $\mathfrak{S}\boldsymbol{\rho}$ is differentiable, and its derivatives are given by (XXIII.18);

in all of Section (iv) we considered only gas flows:

(c) $\boldsymbol{\rho}$ satisfies the equations of motion (XXIII.10) in some neighborhood of $\mathbf{x}$ for all $t \in \mathscr{I}$;

(d) $$\rho_0 > 0, \qquad \rho_0 \rho_4 > \rho_a \rho_a. \qquad \text{(XXIII.41)}$$

To complete the derivation, we should like to show that such a class of fields exists. Unfortunately we do not know at present how to do so, for that would involve proving the existence of solutions of the equations of motion. This step is complicated by the fact that the gross determiners $\mathfrak{P}$ and $\mathbf{q}$ appearing in (XXIII.10) are defined in terms of $\mathbf{G}$, and so we should have to establish (c) for all functions $\mathbf{G}$ in some broad class, one large enough to contain solutions of the equations of gross determinism. We can, however, go part way toward establishing (a)–(d) by exhibiting a function space $\mathscr{M}$ for the existence problem (c) whose elements $\boldsymbol{\rho}$ satisfy (a), (b), and (d). There are two reasons why it is important to do this. First, it is not clear at present what properties of $\boldsymbol{\rho}$ will ensure the differentiability of $\mathfrak{S}\boldsymbol{\rho}$ and the validity of (XXIII.18). Second, our choice of $\mathscr{F}$ and $\mathcal{O}$ will have some influence on the properties of $\boldsymbol{\rho}$ through (a) and (b), but it is not clear what this influence will be. For convenience we shall say that a set $\mathscr{M}$ is *compatible* with $\mathscr{F}$ and $\mathcal{O}$ if each $\boldsymbol{\rho} \in \mathscr{M}$ satisfies (a), (b), and (d). To clarify these two points, at least by example, our aim here will be to exhibit explicitly one set $\mathscr{M}$ which is compatible with the sets $\mathscr{F} = (\mathscr{C}_m(\mathscr{W}))^5$ and $\mathcal{O} = \mathcal{O}_d$ considered in Section (v).

---

[10] We are indebted to Dr. C. DAVINI for this observation.

Let $\mathscr{C}^p(\mathscr{E})$ be the set of real-valued functions on $\mathscr{E}$ that are p times continuously differentiable. For each integer m, $m \geq p$, let $\mathscr{C}^p_m(\mathscr{E})$ be the set of all $\phi \in \mathscr{C}^p(\mathscr{E})$ for which

$$\sup_{|\mathbf{y}| \geq 1} \frac{|\partial_{y_{k_1}} \cdots \partial_{y_{k_q}} \phi(\mathbf{y})|}{|\mathbf{y}|^{m-q}} < \infty \quad \text{when} \quad \begin{cases} q = 0, 1, \ldots, p, \\ k_1, \ldots, k_q = 1, 2, 3. \end{cases} \quad \text{(XXIII.42)}$$

That is, $\mathscr{C}^p_m(\mathscr{E})$ consists in functions ϕ whose rate of growth as $|\mathbf{y}| \to \infty$ is no greater than that of $|\mathbf{y}|^m$, while the rate of growth of their q^{th} derivatives is no greater than that of $|\mathbf{y}|^{m-q}$. A simple but important example of such a function is provided by any polynomial of degree less than or equal to m in the components of $\mathbf{y}$.

So as to be able to consider fields $\boldsymbol{\rho}$ that depend upon both place and time, we let $\mathscr{I} \subset \mathscr{T}$ be any open interval containing 0 and denote by $\mathscr{C}^p(\bar{\mathscr{I}} \times \mathscr{E})$ the set of real-valued continuous functions on $\bar{\mathscr{I}} \times \mathscr{E}$ that are p times continuously differentiable on $\mathscr{I} \times \mathscr{E}$, $\bar{\mathscr{I}}$ being the closure of $\mathscr{I}$. For each integer m, $m \geq p$, we set

$$\mathscr{C}^p_m(\bar{\mathscr{I}} \times \mathscr{E}) \equiv \{ \phi \in \mathscr{C}^p(\bar{\mathscr{I}} \times \mathscr{E}) \,|\, \partial^n_t \phi(t, \cdot) \in \mathscr{C}^{p-n}_m(\mathscr{E})$$

for all $t \in \bar{\mathscr{I}}$, $n = 0, \ldots, p\}$. (XXIII.43)

Thus, if $\phi \in \mathscr{C}^p_m(\bar{\mathscr{I}} \times \mathscr{E})$, then $\phi(t, \cdot) \in \mathscr{C}^p_m(\mathscr{E})$ for each $t \in \bar{\mathscr{I}}$, and the effect of applying q spatial differentiations to ϕ is to reduce the maximum allowable growth rate at ∞ from that of $|\mathbf{y}|^m$ to that of $|\mathbf{y}|^{m-q}$. However, the rates of growth of ϕ and its spatial derivatives need not change when they are differentiated repeatedly with respect to time. Examples of such functions are the polynomials in $\mathbf{y}$ of degree less than m, the coefficients of which are smooth functions on $\bar{\mathscr{I}}$.

We now prove the following.

Lemma 1. *If* $\boldsymbol{\rho} \in (\mathscr{C}^0_m(\bar{\mathscr{I}} \times \mathscr{E}))^5$, *then* $\mathfrak{S}\boldsymbol{\rho}(t, \mathbf{x}) \in \mathscr{F}$ *for all* $t \in \bar{\mathscr{I}}$ *and for all* $\mathbf{x} \in \mathscr{E}$.

Proof. Choose $\rho_\alpha \in \mathscr{C}^0_m(\bar{\mathscr{I}} \times \mathscr{E})$. Then by (XXIII.42) and (XXIII.43) the rate of growth of $\mathfrak{S}\rho_\alpha(t, \mathbf{x})(\mathbf{r}) \equiv \rho_\alpha(t, \mathbf{x} + \mathbf{r})$ as $|\mathbf{x} + \mathbf{r}| \to \infty$ is no greater than that of $|\mathbf{x} + \mathbf{r}|^m$. Therefore if we fix $t \in \bar{\mathscr{I}}$ and $\mathbf{x} \in \mathscr{E}$, the rate of growth of $\mathfrak{S}\rho_\alpha(t, \mathbf{x})(\mathbf{r})$ as $|\mathbf{r}| \to \infty$ is no greater than that of $|\mathbf{r}|^m$. In view of (XXIII.28) and property (H2), this implies that $\|\mathfrak{S}\rho_\alpha(t, \mathbf{x})\|_m$ is finite. Therefore $\mathfrak{S}\boldsymbol{\rho}(t, \mathbf{x}) \in \mathscr{F}$. $\triangle$

Next we prove that $\mathfrak{S}\boldsymbol{\rho}$ is differentiable when $\boldsymbol{\rho}$ is sufficiently smooth and satisfies certain restrictions on its growth at ∞.

Lemma 2. *If* $\boldsymbol{\rho} \in (\mathscr{C}^2_m(\bar{\mathscr{I}} \times \mathscr{E}))^5$, *then* $\mathfrak{S}\boldsymbol{\rho}$ *is differentiable on* $\mathscr{I} \times \mathscr{E}$, *and its partial derivatives are given by* (XXIII.18).

Proof. Let $\boldsymbol{\rho}$ be twice continuously differentiable on $\mathscr{I} \times \mathscr{E}$, and set $\mathbf{z} = (t, \mathbf{x})$, $\mathbf{k} = (s, \mathbf{h})$. Then the components ρ_α of $\boldsymbol{\rho}$ have the finite Taylor expansions

$$\rho_\alpha(\mathbf{z} + \mathbf{k}) = \rho_\alpha(\mathbf{z}) + \sum_{p=0}^{3} \partial_{z_p} \rho_\alpha(\mathbf{z}) k_p + \sum_{p=0}^{3} R_{\alpha p}(\mathbf{z}, \mathbf{k}) k_p, \quad \text{(XXIII.44)}$$

in which

$$R_{\alpha p}(\mathbf{z}, \mathbf{k}) \equiv \frac{1}{2} \sum_{q=0}^{3} \partial_{z_p} \partial_{z_q} \rho_\alpha(\mathbf{z} + \theta \mathbf{k}) k_q, \quad \text{(XXIII.45)}$$

and θ is some number in the interval $(0, 1)$. If we replace $\mathbf{x}$ now by $\mathbf{x} + \mathbf{r}$ and use the definition (XXIII.11) of the shifter, (XXIII.44) becomes

$$\mathfrak{S}\rho_\alpha(t + s, \mathbf{x} + \mathbf{h}) = \mathfrak{S}\rho_\alpha(t, \mathbf{x}) + \mathfrak{S} \partial_t \rho_\alpha(t, \mathbf{x}) s + \mathfrak{S} \partial_{x_a} \rho_\alpha(t, \mathbf{x}) h_a$$
$$+ R_{\alpha 0}(t, \mathbf{x} + \cdot, s, \mathbf{h}) s + R_{\alpha a}(t, \mathbf{x} + \cdot, s, \mathbf{h}) h_a. \quad \text{(XXIII.46)}$$

Upon comparing (XXIII.46) with (XXIII.16) and (XXIII.17) we find that in order for $\mathfrak{S}\rho$ to be differentiable on $\mathscr{I} \times \mathscr{E}$ and for its partial derivatives to be given by (XXIII.18), it will suffice to prove that

$$\lim_{|(s,\mathbf{h})|\to 0} \|\mathbf{R}_p(t, \mathbf{x} + \cdot, s, \mathbf{h})\| = 0 \quad \text{when} \quad p = 0, 1, 2, 3. \qquad (\text{XXIII.47})$$

Suppose now that $\rho_\alpha \in \mathscr{C}_m^2(\mathscr{I} \times \mathscr{E})$. Then as $|\mathbf{x}| \to \infty$ the rates of growth of $\partial_t^2 \rho_\alpha(t, \mathbf{x})$, $\partial_t \partial_{x_k} \rho_\alpha(t, \mathbf{x})$, and $\partial_{x_k} \partial_{x_n} \rho_\alpha(t, \mathbf{x})$ are no greater than those of $|\mathbf{x}|^m$, $|\mathbf{x}|^{m-1}$, and $|\mathbf{x}|^{m-2}$, respectively. Therefore there are constants C_1 and C_2 such that

$$|\partial_{z_p} \partial_{z_q} \rho_\alpha(t, \mathbf{x})| \leq C_1 |\mathbf{x}|^m + C_2 \quad \text{for all} \quad \mathbf{x} \in \mathscr{E}, \text{ and } p, q = 0, 1, 2, 3. \qquad (\text{XXIII.48})$$

Since $\partial_{z_p} \partial_{z_q} \rho_\alpha$ is a continuous function of t on the closed interval $\bar{\mathscr{I}}$, we may take C_1 and C_2 to be independent of t. Thus

$$\begin{aligned}
&\|R_{\alpha p}(t, \mathbf{x} + \cdot, s, \mathbf{h})\|_m \\
&= \sup_{\mathbf{r} \in \mathscr{W}} |R_{\alpha p}(t, \mathbf{x} + \mathbf{r}, s, \mathbf{h})| H_m(|\mathbf{r}|), \\
&= \sup_{\mathbf{r} \in \mathscr{W}} \left| \sum_{q=0}^{3} \partial_{z_p} \partial_{z_q} \rho_\alpha(t + \theta s, \mathbf{x} + \mathbf{r} + \theta \mathbf{h}) k_q \right| H_m(|\mathbf{r}|), \\
&\leq 4 \left(C_1 \sup_{\mathbf{r} \in \mathscr{W}} |\mathbf{x} + \mathbf{r} + \theta \mathbf{h}|^m H_m(|\mathbf{r}|) + C_2 \sup_{\mathbf{r} \in \mathscr{W}} H_m(|\mathbf{r}|) \right) |(s, \mathbf{h})|.
\end{aligned} \qquad (\text{XXIII.49})$$

Since $|\mathbf{r}|^m H_m(|\mathbf{r}|) \to 0$ as $|\mathbf{r}| \to \infty$, we see that

$$\|R_{\alpha p}(t, \mathbf{x} + \cdot, s, \mathbf{h})\|_m \leq \text{const.} |(s, \mathbf{h})|, \qquad (\text{XXIII.50})$$

and so from (XXIII.50) and (XXIII.32) it follows that $\mathbf{R}_p$ satisfies (XXIII.47). $\triangle$

A space of functions that are compatible with $\mathscr{F}$ and $\mathcal{O}$ can now be described. This space is defined in the following theorem, the proof of which follows from the preceding two lemmas.

Theorem. *Let $\mathscr{M}$ be the set of all fields $\boldsymbol{\rho} \in (\mathscr{C}_m^2(\mathscr{I} \times \mathscr{E}))^5$ that satisfy the inequalities (XXIII.41). Then the elements of $\mathscr{M}$ are compatible with $\mathscr{F} = (\mathscr{C}_m(\mathscr{W}))^5$ and $\mathcal{O} = \mathcal{O}_d$.*

(vii) Gross determiners depending upon the body force. The generalized equations of gross determinism and the equation of transfer for gross determiners

So far we have examined only a gas subject to a null body force. We indicate now how our analysis may be altered so as to take account of a **b** that does not vanish. Generalizing (XXIII.8), we examine a class of solutions determined by the gross condition $\boldsymbol{\sigma} = (\boldsymbol{\rho}, \mathbf{b})$ of the gas:

$$F(t, \mathbf{x}, \mathbf{v}) = \mathbb{Z}(\mathbf{v}; \boldsymbol{\sigma}(t, \mathbf{x} + \mathbf{r})). \qquad (\text{XXIII.51})$$

We require the members of this class to satisfy the Maxwell–Boltzmann equation with body force **b**, and as before they have principal moment **ρ**. That is, for all **ρ**,

$$\rho_\alpha(t, \mathbf{x}) = \mathfrak{m} \int I_\alpha \, \mathbb{Z}(\mathbf{v}; \sigma(t, \mathbf{x} + \mathbf{r})), \qquad \alpha = 0, 1, \ldots, 4. \quad \text{(XXIII.52)}$$

Just as before, the gross determiner is a function of a molecular velocity **v** and a vector-valued function **χ** defined on $\mathscr{W}$. This time, however, the second argument has eight components: the first five correspond to principal moments and are denoted collectively by **ψ**; the last three correspond to body forces and are denoted collectively by **f**. Thus $\chi = (\psi, \mathbf{f})$, and in (XXIII.51) $\mathbb{Z}$ is evaluated at $\chi(\mathbf{r}) = \sigma(t, \mathbf{x} + \mathbf{r}) = (\rho(t, \mathbf{x} + \mathbf{r}), \mathbf{b}(\mathbf{x} + \mathbf{r}))$. Again suppressing **v** when it appears as an argument and using the shifter $\mathfrak{S}$, we may write (XXIII.51) as the composite function

$$F = \mathbb{Z} \circ \mathfrak{S}\sigma. \quad \text{(XXIII.53)}$$

Assuming that $\mathbb{Z}$ and $\mathfrak{S}\sigma$ are smooth functions, we may use the chain rule to calculate the derivatives of F in terms of those of $\mathbb{Z}$ and $\mathfrak{S}\sigma$. Since **b** does not depend upon time, we obtain instead of (XXIII.19) the formulae

$$\partial_t F = \partial_\chi \mathbb{Z}(\mathfrak{S}\sigma)[\mathfrak{S}\,\partial_t\sigma] = \partial_\psi \mathbb{Z}(\mathfrak{S}\sigma)[\mathfrak{S}\,\partial_t\rho],$$

$$\partial_{x_k} F = \partial_\chi \mathbb{Z}(\mathfrak{S}\sigma)[\mathfrak{S}\,\partial_{x_k}\sigma] = \partial_\psi \mathbb{Z}(\mathfrak{S}\sigma)[\mathfrak{S}\,\partial_{x_k}\rho] + \partial_\mathbf{f} \mathbb{Z}(\mathfrak{S}\sigma)[\mathfrak{S}\,\partial_{x_k}\mathbf{b}], \quad \text{(XXIII.54)}$$

$$\partial_{v_k} F = \partial_{v_k} \mathbb{Z}(\mathfrak{S}\sigma).$$

Placing these into (IX.1), we obtain *the Maxwell–Boltzmann equation appropriate to general grossly determined solutions*:

$$\partial_\psi \mathbb{Z}(\mathfrak{S}\sigma)[\mathfrak{S}\,\partial_t\rho + v_a\,\mathfrak{S}\,\partial_{x_a}\rho] + \partial_\mathbf{f} \mathbb{Z}(\mathfrak{S}\sigma)[v_a\,\mathfrak{S}\,\partial_{x_a}\mathbf{b}] + b_a\,\partial_{v_a} \mathbb{Z}(\mathfrak{S}\sigma) = \mathbb{C}\mathbb{Z}(\mathfrak{S}\sigma). \quad \text{(XXIII.55)}$$

To derive for $\mathbb{Z}$ conditions analogous to the equations of gross determinism (XXIII.25) and (XXIII.26) which govern $\mathbb{G}$, we again *eliminate time derivatives*, this time between (XXIII.55) and the equations of balance. When account is taken of a general body force, the equations of balance change from (XXIII.21) to

$$\partial_t \rho_\alpha = -(\mathfrak{F}_{\alpha a}(\mathfrak{S}\sigma))_{,a} - \mathfrak{B}_\alpha(\mathfrak{S}\sigma), \quad \text{(XXIII.56)}$$

in which $\mathfrak{F}_k$ and $\mathfrak{B}$ are given by

$$\mathfrak{F}_{\alpha k}(\mathfrak{S}\sigma) \equiv \mathfrak{m} \int I_\alpha v_k \mathbb{Z}(\mathfrak{S}\sigma), \qquad \mathfrak{B}_\alpha(\mathfrak{S}\sigma) \equiv \mathfrak{m} b_a \int I_\alpha \, \partial_{v_a} \mathbb{Z}(\mathfrak{S}\sigma). \quad \text{(XXIII.57)}$$

In order to derive these we simply multiply (IX.1) by $\mathfrak{m} I_\alpha$ and then integrate with respect to **v**. Comparing (XXIII.57)$_1$ with (XXIII.22), we see that $\mathfrak{F}_k$ is again given by (XXIII.23), provided we interpret $\mathfrak{P}$ and **q** there as functions of

(vii) THE GOVERNING EQUATIONS 415

σ. If Z vanishes sufficiently quickly as $v \to \infty$, we may write $\mathfrak{B}$ in the more explicit form

$$\mathfrak{B}_\alpha = \begin{cases} 0 & \text{if } \alpha = 0, \\ -\rho b_\alpha & \text{if } \alpha = 1, 2, 3, \\ -2\rho u_a b_a & \text{if } \alpha = 4. \end{cases} \qquad \text{(XXIII.58)}$$

Placing (XXIII.56) into (XXIII.55), setting $t = 0$ in the equation that results and also in (XXIII.52), and then using (XXIII.3), we obtain

$$\mathbb{D}Z(\mathfrak{S}\sigma) = \mathbb{C}Z(\mathfrak{S}\sigma),$$

$$\phi_\alpha = \mathfrak{m} \int I_\alpha Z(\mathfrak{S}\sigma), \qquad \alpha = 0, 1, \ldots, 4, \qquad \text{(XXIII.59)}$$

σ denoting the pair $(\phi, \mathbf{b})$ now rather than $(\rho, \mathbf{b})$, and $\mathbb{D}$ being the operator

$$\mathbb{D}Z(\mathfrak{S}\sigma) \equiv \partial_\psi Z(\mathfrak{S}\sigma)[-\mathfrak{S}(\mathfrak{F}_a(\mathfrak{S}\sigma))_{,a} - \mathfrak{S}\mathfrak{B}(\mathfrak{S}\sigma) + v_a \mathfrak{S}\phi_{,a}]$$
$$+ \partial_\mathfrak{f} Z(\mathfrak{S}\sigma)[v_a \mathfrak{S}\mathbf{b}_{,a}] + b_a \partial_{v_a} Z(\mathfrak{S}\sigma). \qquad \text{(XXIII.60)}$$

We call these the *Generalized Equations of Gross Determinism*. They extend (XXIII.25) and (XXIII.26) to the case in which the gross determiner depends upon $\mathbf{b}$.

If we set $\mathbf{b} = \mathbf{0}$ in (XXIII.60), the third term on the right-hand side vanishes, and in addition $\mathfrak{B} = \mathbf{0}$. The derivative $\partial_\mathfrak{f} Z(\mathfrak{S}\rho, \mathbf{0})[\cdot]$ is in general not null, but in (XXIII.60) its argument is $v_a \mathfrak{S} \mathbf{b}_{,a}$ and so this term also vanishes when $\mathbf{b} = \mathbf{0}$. Thus (XXIII.59) reduce in this case to (XXIII.25) and (XXIII.26), whence we find that

$$Z(\psi, \mathbf{0}) = G(\psi), \qquad \text{(XXIII.61)}$$

G being the gross determiner that we have examined previously in the chapter. This fact, which should not be surprising, allows us to use directly in analysing Z the properties of G that we have already found. In particular we conclude that *when ϕ is uniform and $\mathbf{b} = \mathbf{0}$, the value of Z is a Maxwellian density*:

$$Z(\phi, \mathbf{0}) = F_M. \qquad \text{(XXIII.62)}$$

Here F_M is the unique Maxwellian density with principal moment ϕ. This is a universal property of all solutions of (XXIII.59), independent of the molecular model.

The formal similarity between (XXIII.59)$_1$ and (IX.1) suggests that we may be able to derive for the generalized equations of gross determinism analogues of some of the results which we have derived already for the Maxwell–Boltzmann equation. We now do so. In Section (iv) of Chapter V we have calculated the principal moment of $\mathfrak{D}F$ and have recorded the result in (V.23). From it and (VII.23) we obtained MAXWELL's consistency theorem: The moments of a solution of the Maxwell–Boltzmann equation satisfy the equations of balance (IX.3)–(IX.5). Let us derive a corresponding result for (XXIII.59). First we differentiate each side of (XXIII.59)$_2$ with respect to ϕ. A simple way to do this is to replace ϕ by $\phi + \lambda\xi$, then differentiate each side with respect to λ, and finally set $\lambda = 0$. We obtain

$$\xi_\alpha = \mathfrak{m} \int I_\alpha \partial_\psi Z[\mathfrak{S}\xi], \qquad (\text{XXIII}.63)$$

the argument $\mathfrak{S}\sigma$ of $\partial_\psi Z$ being suppressed. Then by use of (XXIII.60), (XXIII.54)$_4$, (XXIII.57), and (XXIII.63) we find

$$\int I_\alpha \mathbb{D}Z = -\int I_\alpha \partial_\psi Z[\mathfrak{S}\mathfrak{F}_{a,a} + \mathfrak{S}\mathfrak{B}] + \int I_\alpha v_a \partial_{x_a} Z + \int I_\alpha b_a \partial_{v_a} Z,$$

$$= -\frac{1}{\mathfrak{m}}\mathfrak{F}_{\alpha a,a} - \frac{1}{\mathfrak{m}}\mathfrak{B}_\alpha + \partial_{x_a}\!\left(\int I_\alpha v_a Z\right) + b_a \int I_\alpha \partial_{v_a} Z, \qquad (\text{XXIII}.64)$$

$$= 0.$$

If Z satisfies (XXIII.59)$_2$, *then the principal moment of* $\mathbb{D}Z$ *is null.* This result differs sharply from the corresponding one for $\mathfrak{D}F$. It tells us that if we multiply each side of (XXIII.59)$_1$ by I_α and integrate with respect to **v**, thereby deriving conditions on Z analogous to the equations of balance for F, we obtain nothing more than $\mathbf{0} = \mathbf{0}$. This result, however, should not be surprising. Indeed, it simply confirms the fact that in deriving the equations of gross determinism we have removed from (IX.1) the information contained in the equations of balance.

Let us suppress the field σ now whenever it appears as an argument of Z. Next we note, in view of the chain rule, that the two terms in $\mathbb{D}Z$ which contain **v** explicitly combine to give us $v_a \partial_{x_a} Z$. This observation allows us to write $\mathbb{D}$ in the somewhat shorter form

$$\mathbb{D}Z = -\partial_\psi Z[\mathfrak{S}\mathfrak{F}_{a,a} + \mathfrak{S}\mathfrak{B}] + v_a \partial_{x_a} Z + b_a \partial_{v_a} Z. \qquad (\text{XXIII}.65)$$

It is important to note, as we easily see from (XXIII.57), that $\mathfrak{F}_k$ and $\mathfrak{B}$ are themselves functions of Z. Thus $\mathbb{D}$ is not a linear operator, nor is it a purely differential operator.

If we express Z in terms of a different though equivalent set of arguments, we obtain an alternative form of $\mathbb{D}Z$. Beginning with the argument of Z in the form (ψ, **f**), we shall define new functions δ, **v**, ξ, and ϖ in terms of ψ just as ρ, **u**, ε, and p are defined in terms of ϕ. In view of (VI.23) and (III.15), this means that

$$\delta \equiv \psi_0, \qquad v_k \equiv \psi_k/\psi_0, \qquad k = 1, 2, 3,$$
$$\xi \equiv (\psi_0\psi_4 - \psi_a\psi_a)/2\psi_0^2, \qquad \varpi \equiv 2\,\delta\xi/3. \qquad (\text{XXIII}.66)$$

Given any gross determiner which is expressed as a function of (ψ, **f**), we shall denote by a superimposed caret the corresponding function of (δ, **v**, ξ, **f**), and by a superimposed caret inverted the corresponding function of (δ, **v**, ϖ, **f**).

(vii) EQUATIONS OF TRANSFER

Thus
$$Z(\psi, \mathbf{f}) = \hat{Z}(\delta, \mathbf{v}, \xi, \mathbf{f}) = \check{Z}(\delta, \mathbf{v}, \varpi, \mathbf{f}), \qquad \text{(XXIII.67)}$$
or, in terms of a general set of shifted arguments,
$$Z(\mathfrak{S}\phi, \mathfrak{S}\mathbf{b}) = \hat{Z}(\mathfrak{S}\rho, \mathfrak{S}\mathbf{u}, \mathfrak{S}\varepsilon, \mathfrak{S}\mathbf{b}) = \check{Z}(\mathfrak{S}\rho, \mathfrak{S}\mathbf{u}, \mathfrak{S}p, \mathfrak{S}\mathbf{b}). \qquad \text{(XXIII.68)}$$
To obtain expressions for $\mathbb{D}\hat{Z}$ and $\mathbb{D}\check{Z}$ from that given above for $\mathbb{D}Z$, we need only make the change of variables (XXIII.66) in (XXIII.65) and then use (XXIII.23) and (XXIII.58). However, there is a simpler way to obtain the same result. Considering $\mathbb{D}\hat{Z}$ first, we use the chain rule to write $\mathfrak{D}\hat{Z}$ as follows:
$$\mathfrak{D}\hat{Z} = \partial_\delta \hat{Z}[\mathfrak{S} \, \partial_t \rho] + \partial_{v_a} \hat{Z}[\mathfrak{S} \, \partial_t u_a] + \partial_\xi \hat{Z}[\mathfrak{S} \, \partial_t \varepsilon] + v_a \, \partial_{x_a} \hat{Z} + b_a \, \partial_{v_a} \hat{Z}. \qquad \text{(XXIII.69)}$$
This is simply an alternative form of the left-hand side of (XXIII.55). Thus we may eliminate time derivatives directly between (XXIII.69) and the equations of balance (IX.3)–(IX.5), and then set $t = 0$ as before. If we denote each of the fields ρ, $\mathbf{u}$, ε and its initial value both by the same symbol, then by use of (I.6) we find that
$$\mathbb{D}\hat{Z} = -\partial_\delta \hat{Z}[\mathfrak{S}(u_a \rho_{,a} + \rho E)] - \partial_{v_a} \hat{Z}\left[\mathfrak{S}\left(u_b u_{a,b} + \frac{1}{\rho} p_{,a} + \frac{1}{\rho} \hat{\mathfrak{P}}_{ab,b} - b_a\right)\right]$$
$$- \partial_\xi \hat{Z}\left[\mathfrak{S}\left(u_a \varepsilon_{,a} + \frac{1}{\rho} pE + \frac{1}{\rho} \hat{\mathfrak{P}}_{ab} E_{ab} + \frac{1}{\rho} \hat{q}_{a,a}\right)\right] + v_a \, \partial_{x_a} \hat{Z} + b_a \, \partial_{v_a} \hat{Z}. \qquad \text{(XXIII.70)}$$
In precisely the same way we obtain the following expression for $\mathbb{D}$ applied to $\check{Z}$:
$$\mathbb{D}\check{Z} = -\partial_\delta \check{Z}[\mathfrak{S}(u_a \rho_{,a} + \rho E)] - \partial_{v_a} \check{Z}\left[\mathfrak{S}\left(u_b u_{a,b} + \frac{1}{\rho} p_{,a} + \frac{1}{\rho} \hat{\mathfrak{P}}_{ab,b} - b_a\right)\right]$$
$$- \partial_\varpi \check{Z}[\mathfrak{S}(u_a p_{,a} + \tfrac{5}{3} pE + \tfrac{2}{3} \hat{\mathfrak{P}}_{ab} E_{ab} + \tfrac{2}{3} \hat{q}_{a,a})] + v_a \partial_{x_a} \check{Z} + b_a \partial_{v_a} \check{Z}. \qquad \text{(XXIII.71)}$$
The terms $\partial_{x_m} Z$, $\partial_{x_m} \hat{Z}$, and $\partial_{x_m} \check{Z}$ in each of (XXIII.65), (XXIII.70), and (XXIII.71) are to be calculated by use of the chain rule, bearing in mind the appropriate set of arguments in each case:
$$\partial_{x_m} Z = \partial_{\psi_a} Z[\mathfrak{S}\phi_{a,m}] + \partial_{f_a} Z[\mathfrak{S}b_{a,m}],$$
$$\partial_{x_m} \hat{Z} = \partial_\delta \hat{Z}[\mathfrak{S}\rho_{,m}] + \partial_{v_a} \hat{Z}[\mathfrak{S}u_{a,m}] + \partial_\xi \hat{Z}[\mathfrak{S}\varepsilon_{,m}] + \partial_{f_a} \hat{Z}[\mathfrak{S}b_{a,m}], \qquad \text{(XXIII.72)}$$
$$\partial_{x_m} \check{Z} = \partial_\delta \check{Z}[\mathfrak{S}\rho_{,m}] + \partial_{v_a} \check{Z}[\mathfrak{S}u_{a,m}] + \partial_\varpi \check{Z}[\mathfrak{S}p_{,m}] + \partial_{f_a} \check{Z}[\mathfrak{S}b_{a,m}].$$

It is possible to derive from the equations of gross determinism an analogue of the equation of transfer (IX.2). First let us note that for any function g of $\mathbf{c}$

alone, the expectation $\rho\bar{g}$, when calculated using a gross determiner $\mathbb{Z}$ is grossly determined:

$$\rho\bar{g} = \mathfrak{m} \int g\mathbb{Z}(\mathfrak{S}\sigma). \tag{XXIII.73}$$

We shall not use a special symbol to denote the gross determiner of $\rho\bar{g}$, but rather we shall interpret $\rho\bar{g}$ itself as the gross determiner. If we multiply (XXIII.59)$_1$ by g and then integrate with respect to $\mathbf{v}$, we obtain an equation of transfer which relates various gross determiners. There is a simple way to obtain this equation directly from the equation of transfer (IX.13) which we have derived already. If we specialize that equation by choosing F to be grossly determined, then by means of (I.6) and the chain rule we may write the material derivative appearing there in the form

$$(\rho\bar{g})^{\cdot} = \partial_{\delta}(\rho\bar{g})[\mathfrak{S}\,\partial_t\rho] + \partial_{v_a}(\rho\bar{g})[\mathfrak{S}\,\partial_t u_a] + \partial_{\varpi}(\rho\bar{g})[\mathfrak{S}\,\partial_t p] + u_a(\rho\bar{g})_{,a}. \tag{XXIII.74}$$

Here, as in the remainder of the chapter, all gross determiners will be interpreted as functions of δ, $\mathbf{v}$, ϖ, and $\mathbf{f}$, and so we shall drop from the notation the superimposed inverted caret introduced earlier. Now we place (XXIII.74) into (IX.13), eliminate time derivatives using the equations of balance (IX.3)–(IX.5), and finally set $t = 0$. This procedure gives us the *Equation of Transfer for Gross Determiners*:

$$-\partial_{\delta}(\rho\bar{g})[\mathfrak{S}(u_a\rho_{,a} + \rho E)] - \partial_{v_a}(\rho\bar{g})\left[\mathfrak{S}\left(u_b u_{a,b} + \frac{1}{\rho}p_{,a} + \frac{1}{\rho}\mathfrak{P}_{ab,b} - b_a\right)\right]$$
$$- \partial_{\varpi}(\rho\bar{g})[\mathfrak{S}(u_a p_{,a} + \tfrac{5}{3}pE + \tfrac{2}{3}\mathfrak{P}_{ab}E_{ab} + \tfrac{2}{3}q_{a,a})] + u_a(\rho\bar{g})_{,a}$$
$$+ \rho\bar{g}E + \overline{\rho c_a g}_{,b} u_{b,a} - \overline{g}_{,a}(p_{,a} + \mathfrak{P}_{ab,b}) + (\overline{\rho c_a g})_{,a} = \mathfrak{m}\bar{C}g. \tag{XXIII.75}$$

Each of the expectations $\rho\bar{g}$, $\overline{\rho c_k g}_{,m}$, $\rho\bar{g}_{,k}$, and $\overline{\rho c_k g}$ is, of course, a gross determiner, and so the two gradients $(\rho\bar{g})_{,m}$ and $(\overline{\rho c_a g})_{,a}$ are to be calculated using the chain rule just as is illustrated in (XXIII.72)$_3$. As an example of the use of (XXIII.75) let us specialize it by choosing g to be either $c_k c_m - \tfrac{1}{3}c^2\,\delta_{km}$, $\tfrac{1}{2}c^2 c_k$, or $c_k c_m c_r - \tfrac{3}{5}c^2 c_{(k}\,\delta_{mr)}$. This will give us analogues of the equations of transfer for P_{km}, q_k, and $P_{0|kmr}$ recorded as (XIII.4) and (XIII.7), but now they will relate the corresponding gross determiners $\mathfrak{P}_{km}$, q_k, and $\mathfrak{P}_{0|kmr}$. Either by specializing (XXIII.75) in this way, or by eliminating time derivatives in (XIII.4) and (XIII.7) directly, we obtain

$$-\partial_{\delta}\mathfrak{P}_{km}[\mathfrak{S}(u_a\rho_{,a} + \rho E)] - \partial_{v_a}\mathfrak{P}_{km}\left[\mathfrak{S}\left(u_b u_{a,b} + \frac{1}{\rho}p_{,a} + \frac{1}{\rho}\mathfrak{P}_{ab,b} - b_a\right)\right]$$
$$- \partial_{\varpi}\mathfrak{P}_{km}[\mathfrak{S}(u_a p_{,a} + \tfrac{5}{3}pE + \tfrac{2}{3}\mathfrak{P}_{ab}E_{ab} + \tfrac{2}{3}q_{a,a})] + u_a\,\mathfrak{P}_{km,a} + \mathfrak{P}_{km}E$$
$$+ 2\mathfrak{P}_{(ak}u_{m,a)} + 2pE_{km} + \mathfrak{P}_{0|kma,a} + \tfrac{4}{5}q_{(k,m)} = \mathfrak{m}\bar{C}c_k c_m,$$

$$-\partial_\delta q_k[\mathfrak{S}(u_a\rho_{,a} + \rho E)] - \partial_{v_a} q_k\left[\mathfrak{S}\left(u_b u_{a,b} + \frac{1}{\rho}p_{,a} + \frac{1}{\rho}\mathfrak{P}_{ab,b} - b_a\right)\right]$$

$$- \partial_\varpi q_k[\mathfrak{S}(u_a p_{,a} + \tfrac{5}{3}pE + \tfrac{2}{3}\mathfrak{P}_{ab} E_{ab} + \tfrac{2}{3}q_{a,a})] + u_a q_{k,a} + \tfrac{5}{3}q_k E$$

$$+ \mathfrak{P}_{0|kab} E_{ab} + q_a u_{k,a} + \tfrac{4}{5}q_a E_{ka} - \frac{5p}{2\rho}p_{,k} - \frac{5p}{2\rho}\mathfrak{P}_{ka,a} - \frac{1}{\rho}\mathfrak{P}_{ka}p_{,a}$$

$$- \frac{1}{\rho}\mathfrak{P}_{ka}\mathfrak{P}_{ab,b} + \tfrac{1}{2}\mathfrak{M}_{kaab,b} = \tfrac{1}{2}\mathfrak{m}\bar{\mathbb{C}}c^2 c_k, \qquad \text{(XXIII.76)}$$

$$-\partial_\delta \mathfrak{P}_{0|kmr}[\mathfrak{S}(u_a\rho_{,a} + \rho E)] - \partial_{v_a} \mathfrak{P}_{0|kmr}\left[\mathfrak{S}\left(u_b u_{a,b} + \frac{1}{\rho}p_{,a} + \frac{1}{\rho}\mathfrak{P}_{ab,b} - b_a\right)\right]$$

$$- \partial_\varpi \mathfrak{P}_{0|kmr}[\mathfrak{S}(u_a p_{,a} + \tfrac{5}{3}pE + \tfrac{2}{3}\mathfrak{P}_{ab} E_{ab} + \tfrac{2}{3}q_{a,a})]$$

$$+ u_a \mathfrak{P}_{0|kmr,a} + \mathfrak{P}_{0|kmr} E + \mathfrak{P}_{0|a(km} u_{r),a} - \tfrac{6}{5}\mathfrak{P}_{0|ab(k}\,\delta_{mr)} E_{ab}$$

$$+ \tfrac{12}{5}q_{(k} E_{mr)} - \tfrac{24}{25}q_a E_{a(k}\,\delta_{mr)} - \frac{3}{\rho}\mathfrak{P}_{(km}p_{,r)} - \frac{3}{\rho}\mathfrak{P}_{(km}\mathfrak{P}_{r)a,a}$$

$$+ \frac{6}{5\rho}\mathfrak{P}_{a(k}\,\delta_{mr)}(p_{,a} + \mathfrak{P}_{ab,b}) + \mathfrak{M}_{kmra,a} - \tfrac{3}{5}\delta_{(km}\mathfrak{M}_{r)aab,b}$$

$$= \mathfrak{m}\bar{\mathbb{C}}(c_k c_m c_r - \tfrac{3}{5}c^2 c_{(k}\,\delta_{mr)}),$$

$\mathfrak{M}$ being the gross determiner of the fourth moments $^4\mathbf{M}$:

$$\mathfrak{M}_{kmrs} \equiv \mathfrak{m} \int c_k c_m c_r c_s\, \mathbb{Z}. \qquad \text{(XXIII.77)}$$

We have expressed the right-hand sides of these three equations as collisions integrals, for in this form they are valid, just as (XXIII.75) is, for general molecular models. If we specialize to a gas of Maxwellian molecules, we see from the right-hand sides of (XIII.4) and (XIII.7) that these collisions integrals, when evaluated using a grossly determined molecular density $\mathbb{Z}$, become $-\mathfrak{P}_{km}/\tau$, $-2\mathfrak{q}_k/3\tau$, and $-3\mathfrak{P}_{0|kmr}/2\tau$, respectively.

Our ultimate objective in studying gross determiners is to determine gas flows by solving the equations of motion. As we can see from (XXIII.10), to do this we need only the gross determiners $\mathfrak{P}$ and $\mathfrak{q}$. In general we cannot expect to find these without first producing a solution $\mathbb{Z}$ of the equations of gross determinism. In some exceptional cases, however, we may be able to use the equations of transfer (XXIII.76)$_{1,2}$ directly. It is to one of these exceptional cases that we now turn our attention.

(viii) Gross determinism for affine flows

The first of the generalized equations of gross determinism, namely (XXIII.59)$_1$, is an integro-differential equation set in a Banach space. As such it presents formidable problems of analysis. The equations of transfer (XXIII.76), even though they do not involve the molecular velocity **v**, are no better. One way to simplify each of these equations is to turn away from general fields **ϕ** and **b** and consider instead special ones. That is, we might seek to determine $\mathbb{Z}$ on a subset of its domain. We have already presented one specimen of this type of specialization: On the set of constant fields of the form $\sigma = (\boldsymbol{\phi}, \mathbf{0})$, $\mathbb{Z}$ is a Maxwellian density. Generalizing this result, we consider now the class of affine flows and determine the special forms that our general results take when specialized to these.

To be specific, we wish to consider the restriction of $\mathbb{Z}$ to the set of fields of the form

$$\delta(\mathbf{r}) = \rho, \quad v_k(\mathbf{r}) = G_{ka}r_a + g_k, \quad \varpi(\mathbf{r}) = p, \quad f_k(\mathbf{r}) = 0, \quad \text{(XXIII.78)}$$

ρ, **G**, **g**, and p being constants. As in the preceding section, all gross determiners here will be interpreted as functions of δ, **v**, ϖ, and **f**, and the superimposed inverted caret in (XXIII.67)$_2$ will be suppressed. Of course, the restriction of any gross determiner to the specific fields (XXIII.78) is simply a function of ρ, **G**, **g**, and p themselves. In particular, let us define as follows *reduced functions* Z, **P**, **q**, $P_{0|3}$:

$$Z(\rho, \mathbf{G}, \mathbf{g}, p) \equiv \underset{\mathbf{r}}{\mathbb{Z}}(\rho, \mathbf{Gr} + \mathbf{g}, p, \mathbf{0}),$$

$$P_{km}(\rho, \mathbf{G}, \mathbf{g}, p) \equiv \underset{\mathbf{r}}{\mathfrak{P}}_{km}(\rho, \mathbf{Gr} + \mathbf{g}, p, \mathbf{0}),$$

$$q_k(\rho, \mathbf{G}, \mathbf{g}, p) \equiv \underset{\mathbf{r}}{\mathfrak{q}}_k(\rho, \mathbf{Gr} + \mathbf{g}, p, \mathbf{0}),$$

$$P_{0|kmr}(\rho, \mathbf{G}, \mathbf{g}, p) \equiv \underset{\mathbf{r}}{\mathfrak{P}}_{0|kmr}(\rho, \mathbf{Gr} + \mathbf{g}, p, \mathbf{0}).$$

(XXIII.79)

Since we are considering here only gross determiners, there is no confusion in using the symbols P_{km}, q_k, and $P_{0|kmr}$ to denote gross determiners rather than fields.

The restrictions (XXIII.59)$_2$ on $\mathbb{Z}$ imply corresponding restrictions on the reduced function Z. Evaluating each side of (XXIII.59)$_2$ at $\mathbf{x} = \mathbf{0}$, setting

$\psi(\mathbf{r}) \equiv \phi(\mathbf{r})$, and using (XXIII.66) and (XXIII.78), we may write these restrictions in the form

$$\rho = \mathfrak{m} \int Z(\rho, \mathbf{G}, \mathbf{g}, p),$$

$$g_k = \frac{\mathfrak{m}}{\rho} \int v_k Z(\rho, \mathbf{G}, \mathbf{g}, p), \qquad \text{(XXIII.80)}$$

$$p = \frac{\mathfrak{m}}{3} \int c^2 Z(\rho, \mathbf{G}, \mathbf{g}, p).$$

By the second of these, the random velocity of $Z(\rho, \mathbf{G}, \mathbf{g}, p)$ appearing in the third is given by $\mathbf{c} = \mathbf{v} - \mathbf{g}$.

Let us relate these reduced functions now to the values of the corresponding gross determiners in affine flows. As (XXIII.78)$_4$ indicates, we will consider only the case in which $\mathbf{b} = \mathbf{0}$. From (II.42) we see that the fields of density and pressure in such flows are independent of $\mathbf{x}$ and the gross velocity is given by (II.41). Of course, in the theory of gross determiners we are only interested in the initial values of these fields, and it is convenient to use the same symbols to denote these initial values. Thus we take $\rho(\mathbf{x}) = \rho$, $\mathbf{u}(\mathbf{x}) = \mathbf{G}(\mathbf{x} - \mathbf{x}_0) + \mathbf{g}$, and $p(\mathbf{x}) = p$, in which ρ, p, $\mathbf{G}$, and $\mathbf{g}$ are constants. From (XXIII.11) we find that

$$\mathfrak{S}\rho(\mathbf{x})(\mathbf{r}) = \rho,$$
$$\mathfrak{S}u_k(\mathbf{x})(\mathbf{r}) = G_{ka} r_a + [G_{ka}(x_a - x_{0a}) + g_k], \qquad \text{(XXIII.81)}$$
$$\mathfrak{S}p(\mathbf{x})(\mathbf{r}) = p.$$

Since for fixed $\mathbf{x}$ each of these functions has the form (XXIII.78) with $\mathbf{g}$ replaced by $\mathbf{G}(\mathbf{x} - \mathbf{x}_0) + \mathbf{g}$, the values of $\mathbb{Z}$ in affine flow are determined as follows by the reduced function Z:

$$\mathbb{Z}(\mathfrak{S}\rho(\mathbf{x}), \mathfrak{S}\mathbf{u}(\mathbf{x}), \mathfrak{S}p(\mathbf{x}), 0) = Z(\rho, \mathbf{G}, \mathbf{G}(\mathbf{x} - \mathbf{x}_0) + \mathbf{g}, p). \qquad \text{(XXIII.82)}$$

Of course, each gross determiner is related to its corresponding reduced function in exactly this way.

We wish to specialize (XXIII.59) and (XXIII.76) to the fields given by (XXIII.78). Before we do this, let us look at the gross determiner for the pressures:

$$\mathfrak{P}_{km}(\mathfrak{S}\rho(\mathbf{x}), \mathfrak{S}\mathbf{u}(\mathbf{x}), \mathfrak{S}p(\mathbf{x}), 0)$$
$$= P_{km}(\rho, \mathbf{G}, \mathbf{G}(\mathbf{x} - \mathbf{x}_0) + \mathbf{g}, p), \qquad \text{(XXIII.83)}$$
$$= \mathfrak{m} \int (c_k c_m - \tfrac{1}{3}c^2 \, \delta_{km}) Z(\rho, \mathbf{G}, \mathbf{G}(\mathbf{x} - \mathbf{x}_0) + \mathbf{g}, p).$$

From (II.42) we see that in an affine flow the field of pressures must be independent of **x**. By (XXIII.83)$_1$ it will be so if the gross determiner **P** is independent of **g**. A glance at (XXIII.83)$_2$ shows us one way in which this might happen, namely, if Z should depend upon **v** and **g** only through the random velocity $\mathbf{c} = \mathbf{v} - \mathbf{g}$:

$$Z(\mathbf{v};\rho, \mathbf{G}, \mathbf{g}, p) = Z(\mathbf{v} - \mathbf{g}; \rho, \mathbf{G}, \mathbf{0}, p) \equiv Z^G(\mathbf{c}; \rho, \mathbf{G}, p), \quad \text{(XXIII.84)}$$

say; in this expression we have restored in the notation the variable **v**, which we have suppressed up to now. This would imply that

$$P_{km}(\rho, \mathbf{G}, \mathbf{g}, p) = \mathfrak{m} \int (c_k c_m - \tfrac{1}{3}c^2\,\delta_{km}) Z^G(\mathbf{c}; \rho, \mathbf{G}, p) \quad \text{(XXIII.85)}$$

which is independent of **g** and so the value of $\mathfrak{P}_{km}$ in (XXIII.83) would be independent of **x**. Of course the same conclusion applies to the expectation of any function of **c** alone. In particular the values of q_k, $\mathfrak{P}_{0|kmr}$, and $\mathfrak{M}_{kmrs}$ would be independent of **x**. So as to remain consistent with (II.42), we shall assume henceforth that Z does have the form (XXIII.84) for some function Z^G. Whether or not such an assumption be consistent with the equations of gross determinism is another matter, and we shall return to discuss it somewhat later.

Using (XXIII.79)$_1$ and (XXIII.84), we may express the derivatives of $\mathbb{Z}$ directly in terms of those of Z^G. This may be done by means of a simple rule which we illustrate in the case of the derivative with respect to **v**:

$$\partial_\mathbf{v} \underset{\mathbf{r}}{\mathbb{Z}}(\mathbf{v}; \rho, \mathbf{G}\mathbf{r} + \mathbf{g}, p, 0)[\mathbf{K}\mathbf{r} + \mathbf{k}] = \frac{d}{d\lambda}\underset{\mathbf{r}}{\mathbb{Z}}(\mathbf{v}; \rho, \mathbf{G}\mathbf{r} + \mathbf{g} + \lambda(\mathbf{K}\mathbf{r} + \mathbf{k}), p, 0)\big|_{\lambda=0},$$

$$= \frac{d}{d\lambda} Z^G(\mathbf{v} - \mathbf{g} - \lambda\mathbf{k}; \rho, \mathbf{G} + \lambda\mathbf{K}, p)\big|_{\lambda=0}, \quad \text{(XXIII.86)}$$

$$= -k_a\,\partial_{c_a} Z^G(\mathbf{c}; \rho, \mathbf{G}, p) + K_{ab}\,\partial_{G_{ab}} Z^G(\mathbf{c}; \rho, \mathbf{G}, p),$$

K and **k** being arbitrary constants. In precisely the same way we obtain

$$\partial_\delta \underset{\mathbf{r}}{\mathbb{Z}}(\mathbf{v}; \rho, \mathbf{G}\mathbf{r} + \mathbf{g}, p, 0)[\chi] = \chi\,\partial_\rho Z^G(\mathbf{c}; \rho, \mathbf{G}, p),$$

$$\partial_\varpi \underset{\mathbf{r}}{\mathbb{Z}}(\mathbf{v}; \rho, \mathbf{G}\mathbf{r} + \mathbf{g}, p, 0)[\eta] = \eta\,\partial_p Z^G(\mathbf{c}; \rho, \mathbf{G}, p),$$

(XXIII.87)

for any constants χ and η. Now we may specialize (XXIII.59)$_1$ to affine flows. Since ρ, p, E, **E**, $\mathfrak{P}$, and **q** are independent of **x**, by appeal to (XXIII.87) we simplify as follows the first and third terms on the right-hand side of (XXIII.71):

$$-\partial_\delta \mathbb{Z}[\mathfrak{S}(u_a \rho_{,a} + \rho E)] = -\rho E\,\partial_\rho Z^G,$$

$$-\partial_\varpi \mathbb{Z}[\mathfrak{S}(u_a p_{,a} + \tfrac{5}{3}pE + \tfrac{2}{3}\mathfrak{P}_{ab}E_{ab} + \tfrac{2}{3}q_{a,a})] = -(\tfrac{5}{3}pE + \tfrac{2}{3}P_{ab}E_{ab})\,\partial_p Z^G.$$

(XXIII.88)

Using (XXIII.86), we find that the second term in (XXIII.71) becomes

$$-\partial_{v_a} \mathbb{Z}\left[\mathfrak{S}\left(u_b u_{a,b} + \frac{1}{\rho} p_{,a} + \frac{1}{\rho} \mathfrak{P}_{ab,b} - b_a\right)\right]$$

$$= -\partial_{v_a} \mathbb{Z}[G_{ab} G_{bc} r_c + G_{ab}(G_{bc}(x_c - x_{0c}) + g_b)], \qquad \text{(XXIII.89)}$$

$$= G_{ab}[G_{bc}(x_c - x_{0c}) + g_b] \partial_{c_a} Z^G - G_{ab} G_{bc} \partial_{G_{ac}} Z^G.$$

Since $\mathbf{b} = \mathbf{0}$, the last term in (XXIII.71) is null, and from (XXIII.72)$_3$ and (XXIII.86) we see that the second-from-last term becomes

$$v_a \partial_{x_a} \mathbb{Z} = v_a \partial_{v_b} \mathbb{Z}[G_{ba}],$$

$$= -v_a G_{ba} \partial_{c_b} Z^G, \qquad \text{(XXIII.90)}$$

$$= -c_a G_{ba} \partial_{c_b} Z^G - G_{ba}[G_{ac}(x_c - x_{0c}) + g_a] \partial_{c_b} Z^G.$$

Combining these results, we obtain finally the *equations of gross determinism for affine flows*:

$$-\rho E \partial_\rho Z^G - G_{ab} G_{bc} \partial_{G_{ac}} Z^G - (\tfrac{5}{3} p E + \tfrac{2}{3} P_{ab} E_{ab}) \partial_p Z^G - c_a G_{ba} \partial_{c_b} Z^G = \mathbb{C} Z^G,$$

$$\rho = \mathfrak{m} \int Z^G(\mathbf{c}; \rho, \mathbf{G}, p), \qquad 0 = \mathfrak{m} \int c_k Z^G(\mathbf{c}; \rho, \mathbf{G}, p), \qquad \text{(XXIII.91)}$$

$$p = \frac{\mathfrak{m}}{3} \int c^2 Z^G(\mathbf{c}; \rho, \mathbf{G}, p),$$

E and $\mathbf{E}$ being given in terms of $\mathbf{G}$ by (II.41) and (I.5)$_{1,3}$. It is important to note that the variable $\mathbf{g}$ does not appear explicitly in these equations. This means that we have indeed obtained equations for a function Z^G of $\mathbf{c}$, ρ, $\mathbf{G}$, and p alone, and so, in retrospect, our assumption (XXIII.84) is certainly consistent with the equations of gross determinism.

It is also possible to specialize the equations of transfer (XXIII.76) to affine flows, and in fact we have already prepared the calculation needed to do this. Indeed, the formulae (XXIII.88) and (XXIII.89) are valid for any gross determiner $\mathbb{Z}$, regardless of whether it happens to depend upon the molecular velocity $\mathbf{v}$. Thus, we may simply replace $\mathbb{F}$ and Z^G in them by $\mathfrak{P}$ and $\mathbf{P}$, respectively. Of course $\mathbf{P}$ does not depend upon $\mathbf{c}$, so some terms vanish, but the results are valid nevertheless. Using these and recalling that the values of $\mathfrak{P}_{km}$, q_k, and $\mathfrak{P}_{0|kmr}$ in affine flows do not depend upon $\mathbf{x}$, we find that (XXIII.76)$_1$ reduces to

$$-\rho E \partial_\rho P_{km} - G_{ab} G_{bc} \partial_{G_{ac}} P_{km} - (\tfrac{5}{3} p E + \tfrac{2}{3} P_{ab} E_{ab}) \partial_p P_{km}$$
$$+ P_{km} E + 2 P_{\{ak} G_{ma\}} + 2 p E_{km} = -P_{km}/\tau. \qquad \text{(XXIII.92)}$$

In precisely the same way, (XXIII.76)$_{2,3}$ reduce to

$$-\rho E \, \partial_p q_k - G_{ab} G_{bc} \, \partial_{G_{ac}} q_k - (\tfrac{5}{3}pE + \tfrac{2}{3}P_{ab}E_{ab}) \, \partial_p q_k$$
$$+ \tfrac{5}{3}q_k E + P_{0|kab}E_{ab} + q_a G_{ka} + \tfrac{4}{5}q_a E_{ka} = -2q_k/3\tau,$$
$$-\rho E \, \partial_p P_{0|kmr} - G_{ab}G_{bc} \, \partial_{G_{ac}} P_{0|kmr} - (\tfrac{5}{3}pE + \tfrac{2}{3}P_{ab}E_{ab}) \, \partial_p P_{0|kmr} \quad \text{(XXIII.93)}$$
$$+ P_{0|kmr}E + P_{0|a(km}G_{r)a} - \tfrac{6}{5}P_{0|ab(k} \, \delta_{mr)}E_{ab} + \tfrac{12}{5}q_{(k}E_{mr)}$$
$$- \tfrac{24}{25}q_a E_{a(k} \, \delta_{mr)} = -3P_{0|kmr}/2\tau.$$

Since our analysis of affine flows in Chapter XV concerned only a gas of Maxwellian molecules, we have specialized the equations of transfer here to such a gas as indicated following (XXIII.77). In affine flows, then, the equation of transfer for $\mathfrak{P}_{km}$ reduces to an equation for P_{km} alone, independent of the other gross determiners. Similarly the equations of transfer for q_k and $\mathfrak{P}_{0|kmr}$ reduce to equations for q_k and $P_{0|kmr}$ alone. By solving these equations we may find the gross determiners $\mathfrak{P}$ and $\mathfrak{q}$ in affine flow. There is much we can learn from these equations, but we shall defer our discussion of them until later. Indeed, we shall meet (XXIII.92) again in Chapter XXVI, and at that point we shall look at it in detail and actually display some values of its solution.

Appendix A

Calculus in Banach Spaces

The differential calculus in Banach spaces is a well known subject, and it may be found in a number of standard texts on analysis[1]. For completeness we present here the results from this subject which are used in Chapters XXIII and XXIV. We presume only some familiarity with the calculus of functions of several real variables and with the definition of a Banach space.

The notion of the derivative of a function f at a point x is usually associated with the existence of a certain linear approximation to $f(y) - f(x)$ when y is near x. When the domain and range of f are subsets of Banach spaces, this notion can be formalized as follows. Let $\mathcal{M}_1$ and $\mathcal{M}_2$ be Banach spaces with norms $\|\ \|_1$ and $\|\ \|_2$, respectively. Suppose that $\mathcal{U}$ is an open subset of $\mathcal{M}_1$, and let us consider a function $f\colon \mathcal{U} \to \mathcal{M}_2$. We say that f is *differentiable at* $x \in \mathcal{U}$ if there is a function ψ and a continuous linear function λ such that for all h near 0 in $\mathcal{M}_1$

$$f(x + h) = f(x) + \lambda(h) + \psi(h)\|h\|_1, \quad \text{(XXIIIA.1)}$$

---

[1] E.g. Chapter V of the book by LANG cited in Footnote 3 to Chapter XIX or Chapter VIII of the book by DIEUDONNÉ cited in Footnote 7 to the foregoing chapter.

ψ having the property
$$\lim_{\|h\|_1 \to 0} \|\psi(h)\|_2 = 0. \qquad (\text{XXIIIA.2})$$
We shall write $\lambda(h)$ in the form
$$\lambda(h) = \partial_x f(x)[h] \qquad (\text{XXIIIA.3})$$
to emphasize that it is *the derivative of f at x* and that it is a *continuous linear function of h*. We say that f is *differentiable on $\mathcal{U}$* if it is differentiable at all points $x \in \mathcal{U}$. Of course ψ is generally a function of x as well as of h, but for briefer writing we let this dependence remain silent.

In order to bring the above definition of a derivative closer to its classical special case, it is useful to introduce a certain convention when $\mathcal{M}_1$ is $\mathcal{R}$, the set of real numbers. Then, since h is real and $\partial_x f(x)[\cdot]$ is linear,
$$\partial_x f(x)[h] = h\, \partial_x f(x)[1]. \qquad (\text{XXIIIA.4})$$
The quantity $\partial_x f(x)[1]$ is simply some vector in $\mathcal{M}_2$, and we shall denote it by
$$\partial_x f(x) \equiv \partial_x f(x)[1]. \qquad (\text{XXIIIA.5})$$
Thus (XXIIIA.1) becomes
$$f(x+h) = f(x) + h\, \partial_x f(x) + \psi(h)|h|. \qquad (\text{XXIIIA.6})$$
In this special case it is traditional to give the name "derivative of f at x" to the vector $\partial_x f(x)$ rather than to the function $\partial_x f(x)[\cdot]$. If also $\mathcal{M}_2 = \mathcal{R}$, then $\partial_x f(x)$ is simply a real number, and in view of (XXIIIA.2) it is given by
$$\partial_x f(x) = \lim_{h \to 0} \frac{f(x+h) - f(x)}{h}. \qquad (\text{XXIIIA.7})$$
Therefore, in terms of the convention introduced above, the classical special definition of the derivative agrees with the more general one given here.

If $\mathcal{M}_1$ is the product of n Banach spaces, say $\mathcal{M}_1 = \mathcal{M}_{11} \times \cdots \times \mathcal{M}_{1n}$, and if we write a typical point of $\mathcal{M}_1$ in the form $x = (x_1, \ldots, x_n)$, we may define partial derivatives of f just as we do classically. Thus $\partial_{x_k} f(x)[\cdot]$ is defined as the derivative of f considered as a function of x_k, the remaining variables $x_1, \ldots, x_{k-1}, x_{k+1}, \ldots, x_n$ being held fixed. The derivative $\partial_x f(x)[\cdot]$ is related to these partial derivatives by
$$\partial_x f(x)[(h_1, \ldots, h_n)] = \sum_{k=1}^n \partial_{x_k} f(x)[h_k]. \qquad (\text{XXIIIA.8})$$
If $\mathcal{M}_{11} = \ldots = \mathcal{M}_{1n} = \mathcal{R}$, then we may apply the convention introduced previously to each partial derivative and so from (XXIIIA.4) and (XXIIIA.5) write (XXIIIA.8) in the equivalent form
$$\partial_x f(x)[(h_1, \ldots, h_n)] = \sum_{k=1}^n h_k\, \partial_{x_k} f(x). \qquad (\text{XXIIIA.9})$$

Note that each quantity $\partial_{x_k} f(x)$ is now some vector from $\mathcal{M}_{1k}$.

Most of the elementary properties of the derivative of a real-valued function of a real variable carry over to this more general setting. As an illustration we shall examine the form of the chain rule for differentiating composite functions. Let $\mathcal{U}_1$ be an open subset of $\mathcal{M}_1$, and suppose $f: \mathcal{U}_1 \to \mathcal{M}_2$ be differentiable at $x \in \mathcal{U}_1$. Also suppose that $\mathcal{U}_2$ be an open subset of $\mathcal{M}_2$ containing $f(x)$ and that $g: \mathcal{U}_2 \to \mathcal{M}_3$ be differentiable at $y = f(x)$, $\mathcal{M}_3$ being a Banach space with norm $\|\ \|_3$. We wish to show that $g \circ f$ is differentiable at x and also to calculate its derivative in terms of the derivatives of f and g. Since g is differentiable at y, we know that for k near 0 in $\mathcal{M}_2$

$$g(y + k) = g(y) + \partial_y g(y)[k] + \chi(k)\|k\|_2, \qquad \text{(XXIIIA.10)}$$

and χ satisfies the condition

$$\lim_{\|k\|_2 \to 0} \|\chi(k)\|_3 = 0. \qquad \text{(XXIIIA.11)}$$

From (XXIIIA.1) we see that

$$g \circ f(x + h) = g(f(x + h)),$$
$$= g(y + k), \qquad \text{(XXIIIA.12)}$$

in which

$$k \equiv \partial_x f(x)[h] + \psi(h)\|h\|_1. \qquad \text{(XXIIIA.13)}$$

Applying (XXIIIA.10) now, we obtain

$$g \circ f(x + h) = g \circ f(x) + \partial_y g(f(x))[\partial_x f(x)[h] + \psi(h)\|h\|_1]$$
$$+ \chi(\partial_x f(x)[h] + \psi(h)\|h\|_1)\|\partial_x f(x)[h] + \psi(h)\|h\|_1\|_2,$$
$$= g \circ f(x) + \partial_y g(f(x))[\partial_x f(x)[h]] + \phi(h)\|h\|_1; \qquad \text{(XXIIIA.14)}$$

the function ϕ is given by

$$\phi(h) = \|h\|_1^{-1}\{\partial_y g(f(x))[\psi(h)\|h\|_1]$$
$$+ \chi(\partial_x f(x)[h] + \psi(h)\|h\|_1)\|\partial_x f(x)[h] + \psi(h)\|h\|_1\|_2\}. \qquad \text{(XXIIIA.15)}$$

It follows from (XXIIIA.2) and (XXIIIA.11) that ϕ satisfies

$$\lim_{\|h\|_1 \to 0} \|\phi(h)\|_3 = 0. \qquad \text{(XXIIIA.16)}$$

By comparing (XXIIIA.14)$_2$ and (XXIIIA.16) with (XXIIIA.1), (XXIIIA.2), and (XXIIIA.3), we see that $g \circ f$ is differentiable at x, and its derivative is given in terms of $\partial_x f$ and $\partial_y g$ by the *chain rule*:

$$\partial_x (g \circ f)(x)[h] = \partial_y g(f(x))[\partial_x f(x)[h]]. \qquad \text{(XXIIIA.17)}$$

If $\mathcal{M}_1 = \mathcal{R}$, we may apply our previous convention to both g and $g \circ f$. Using (XXIIIA.4) and (XXIIIA.5), we may then write (XXIIIA.17) in the equivalent form

$$\partial_x(g \circ f)(x) = \partial_y g(f(x))[\partial_x f(x)]. \qquad \text{(XXIIIA.18)}$$

If in addition $\mathcal{M}_2 = \mathcal{M}_3 = \mathcal{R}$, we may apply our convention to g also and so write this last formula as

$$\partial_x(g \circ f)(x) = \partial_y g(f(x)) \, \partial_x f(x), \qquad \text{(XXIIIA.19)}$$

the usual form of the chain rule from the calculus of functions of one variable. The formula (XXIIIA.17) is the generalization of it appropriate to mappings between Banach spaces.

If a function $f: \mathcal{R} \to \mathcal{R}$ is N times differentiable, we may use Taylor's theorem to approximate $f(x+h) - f(x)$ when h is small by a polynomial of degree N in the variable h, the coefficients in this polynomial being given easily in terms of the derivatives of f at x. Using this idea, we can extend (XXIIIA.1) so as to define derivatives of second and higher orders of a general mapping on Banach spaces. Specifically, let $\mathcal{M}_1$ and $\mathcal{M}_2$ be Banach spaces, and let $\mathcal{U} \subset \mathcal{M}_1$ be open. Then we say that $f: \mathcal{U} \to \mathcal{M}_2$ is N times differentiable at $x \in \mathcal{U}$ if for all h near 0, $f(x+h)$ may be expanded in the form

$$f(x+h) = f(x) + \sum_{n=1}^{N} \frac{1}{n!} \partial_x^n f(x)[h, \ldots, h] + \psi(h)\|h\|_1^N, \qquad \text{(XXIIIA.20)}$$

in which $\partial_x^n f(x)[\cdot, \ldots, \cdot]$ is a continuous linear function of each of its n arguments, and the function ψ satisfies (XXIIIA.2). We call $\partial_x^n f(x)$ the nth derivative of f at x. If f is N times differentiable at all points $x \in \mathcal{U}$, then we say that f is N times differentiable on $\mathcal{U}$. It is not difficult to show that if f is N times differentiable on $\mathcal{U}$, $\partial_x^m f$ is $N-m$ times differentiable if $m = 1, \ldots, N$, just as we should expect.

If $\mathcal{M}_1 = \mathcal{R}$, it follows from the linearity of $\partial_x^n f(x)[\cdot, \ldots, \cdot]$ that for real numbers $h_1, \ldots, h_n$

$$\partial_x^n f(x)[h_1, \ldots, h_n] = h_1 \cdots h_n \, \partial_x^n f(x)[1, \ldots, 1]. \qquad \text{(XXIIIA.21)}$$

This suggests that we extend the convention introduced previously by setting

$$\partial_x^n f(x) \equiv \partial_x^n f(x)[1, \ldots, 1], \qquad \text{(XXIIIA.22)}$$

and calling the vector $\partial_x^n f(x) \in \mathcal{M}_2$ the nth derivative of f at x. Doing so, we find that (XXIIIA.20) reduces to the somewhat more familiar form

$$f(x+h) = f(x) + \sum_{n=1}^{N} \frac{h^n}{n!} \partial_x^n f(x) + |h|^N \psi(h). \qquad \text{(XXIIIA.23)}$$

When $\mathcal{M}_2 = \mathcal{R}$ also, this last formula is nothing more than the classical Taylor expansion of f.

Chapter XXIV

The Method of Stretched Fields for Approximating Gross Determiners. Use of It to Obtain the Results of Enskog's Procedure

HILBERT's procedure has served as a prototype for many calculations in the kinetic theory that are designed to produce results comparable with constitutive relations of continuum mechanics. It deals from its very outset with series expansions and approximations, the significance of which is difficult to assess. It aims to find solutions of the Maxwell–Boltzmann equation and only afterwards leads, as if by accident, to equations that have all the appearances of constitutive relations of continuum mechanics. The theory of gross determiners we have presented in the preceding chapter is different. It is an exact theory. The equations of gross determinism, which express the conditions that govern gross determiners, we derived without appeal to approximations of any kind. Moreover, we dealt with gross determiners themselves and not with such solutions of the Maxwell–Boltzmann equation as may or may not correspond with them.

However, no matter how precise a theory may be, the ultimate test of it lies in the results it delivers. When the equations underlying a theory are complicated, to approximate may be the only way to obtain something concrete. Therefore we shall look now to approximating solutions of the equations of gross determinism. We reiterate the caution that gross determiners as they stand are not solutions of the Maxwell–Boltzmann equation; likewise the value F of a gross determiner $\mathfrak{G}$ at an arbitrary field ρ is not a solution of the Maxwell–Boltzmann equation. Rather, such an F is a solution if and only if ρ satisfies the equations of motion (XXIII.10), the gross determiners $\mathfrak{P}$ and $\mathfrak{q}$ being those corresponding to $\mathfrak{G}$.

To motivate the iteration we shall use, we begin by surveying some standard formulae in the kinetic theory that we might obtain and generalize by such a procedure.

(i) Enskog's procedure

Much of the early work in the kinetic theory focused upon calculating the viscosity and the Maxwell number of a gas. These quantities were obtained through use of the beginnings of some procedure designed to deliver approximate solutions of the Maxwell–Boltzmann equation. The best known of these is ENSKOG's.

The literature of the kinetic theory refers fluently to this procedure as "the Chapman–Enskog process". After examining the sources we conclude that this term misrepresents as equivalent CHAPMAN's work and ENSKOG's. It confuses *two different problems*, each of which has given rise to methods of approximation of grossly determined functions:

Problem 1. To derive formulae giving μ and $M\mu$ as functionals of the kinetic constitutive quantities such that

$$P_{km} \approx -2\mu E_{km}, \qquad q_k \approx -M\mu \varepsilon_{,k}, \qquad \text{(XXIV.1)}$$

when ρ is nearly uniform.

Problem 2. To derive general series expansions for **P** and **q** as functionals of ρ, and to give necessary and sufficient conditions which uniquely determine the coefficients in these expansions.

We can illustrate these two problems perfectly in terms of some calculations which we have presented already and some which we shall provide later. In Chapter XIII we have repeated MAXWELL's rough derivation of (XXIV.1) and the formulae (XIII.14) and (XIII.18) which he therewith obtained for μ and M, respectively. For a gas of Maxwellian molecules these constitute a solution of Problem 1. On the surface MAXWELL's argument does not seem to reflect an iterative procedure at all but rather to advance some crude approximations tailored to deliver equations like (XXIV.1). However, IKENBERRY & TRUESDELL invented an iteration which delivers **P** and **q** at each stage as explicitly calculated functionals of ρ and which embeds MAXWELL's results as its first stage. This iteration constitutes a solution of Problem 2, once again for Maxwellian molecules. Because it renders systematic some of the ideas MAXWELL seems to have employed in his own later attempts to extend (XXIV.1), they named their iteration "Maxwellian". In Chapter XXV we present the details of their procedure as well as some variants of it; specific references to MAXWELL's work will be found there, too.

These illustrations show us that a solution of Problem 1 need not entail a solution of Problem 2; necessarily, of course, solution of the latter contains a solution of the former.

(i) ENSKOG'S PROCEDURE

Until fairly recently Problem 1 was uppermost in nearly all studies by physicists[1] and chemists. This was certainly true of the work of CHAPMAN. It is clear from his original papers[2] that his only aim was to decide upon definitions of μ and Mμ which would yield (XXIV.1), thereby giving a solution of Problem 1. This he did. It is equally clear, however, that he did not solve Problem 2 or even suggest a method for solving it[3].

ENSKOG[4] also solved Problem 1, but he did so by deriving (XXIV.1) as the first non-zero terms in a general iterative procedure which represented **P** and **q** as functionals of **ρ**. Thus he solved Problem 2 first. Certain formulae for μ and Mμ appeared in this solution. These formulae, in one form, are the same as those derived by CHAPMAN, or essentially so[5]. In this sense one may justly assign the attributory "Chapman-Enskog" to these formulae. However, it is a misnomer when applied to the general *iterative procedure*. If CHAPMAN had any method for approaching Problem 2 in general, certainly the steps in the

---

[1] BOLTZMANN [1880] [1881] filled over 150 pages with an attack on this problem for a gas of ideal spheres. He concluded on a note of despair. *Cf.* Footnotes 27 and 30 to this chapter.

[2] CHAPMAN [1916, *2*]. The summary by CHAPMAN [1916, *1*] adds nothing.

[3] Very crudely, CHAPMAN's solution of Problem 1 resembles in some respects the one given by MAXWELL. He represented his molecular density f in the form $F_M(1 + F)$, F being presumed small. He substituted this expression into the equation of transfer (IX.12), then neglected terms which he considered small. The procedure seems to consist in replacing f by F_M in the left-hand side, a fact which CHAPMAN [1916, *2*, §3(A)] described as follows: "In taking mean values of functions of U, V, W [respectively c_1, c_2, c_3 in our notation] ... we shall neglect the part F in the velocity-distribution function f, in cases where the mean value is to be differentiated or multiplied by a small factor, since the resulting error is only of the second order." In retrospect, then, we may expect that any general procedure which gives rise at its first stage to CHAPMAN's calculation will resemble more closely Maxwellian iteration than the procedure proposed by ENSKOG. This expectation is strengthened by the work of [LENNARD-]JONES [1923], in which CHAPMAN's calculation is carried one step further so as to deliver second approximations to **P** and **q**. However, so far as we know, no such general procedure has ever been put forward.

(There are a number of steps in CHAPMAN's analysis which would be difficult to justify mathematically. We point out two of these in Footnotes 18 and 31 to this chapter.)

[4] ENSKOG [1917].

[5] CHAPMAN & COWLING [1939] make the difference clear. Abandoning CHAPMAN's original calculation altogether, they present in their Chapter 7 a solution of Problem 1 based on expansions in terms of Sonine polynomials, which had been introduced into the kinetic theory by BURNETT [1935]. For its historical interest they present the corresponding calculation of ENSKOG in their Appendix A. In his recollections CHAPMAN [1967, pp. 9-10] writes, "unknown to each other, Enskog and I were trying to improve our first imperfect attacks upon the problem of determining F and the transport coefficients for a gas composed of molecules of general spherically symmetric type", and later ENSKOG [1921] found "a number of numerical and other oversights in my work that had escaped my notice. Apart from this, our results were identical." Also, "As I slowly read [Enskog's dissertation], I was astonished to find how differently, using Boltzmann's equation, he had arrived at almost exactly the same results as mine." This statement makes no sense for Problem 2, since CHAPMAN published nothing concerning a scheme of iteration; it easily applies to Problem 1, since their methods of determining μ and Mμ are indeed different. *Cf.* Footnote 29 to this chapter.

analysis he published are so meagre and so arbitrary as to leave obscure what it may have been. The expansion often called "Chapman-Enskog" seems to be due, in all its insecurity, to ENSKOG alone, and in their book CHAPMAN & COWLING[6] attribute it to him.

ENSKOG's iterative procedure is special in that, like HILBERT's, it delivers particular rather than general solutions of the Maxwell-Boltzmann equation, namely, grossly determined ones. However, ENSKOG's procedure is special for a further reason, indeed, the reason that underlies its very importance and distinguishes it from HILBERT's procedure: At each stage it delivers the pressure deviator and the energy flux vector by formulae resembling gross determiners.

The 0th iterates are

$$P_{km}^{\langle 0 \rangle} = 0,$$
$$q_k^{\langle 0 \rangle} = 0; \qquad \text{(XXIV.2)}$$

these statements are formally the same as the constitutive relations of an Euler-Hadamard fluid. The first iterates are formally the same as the constitutive relations of a Stokes-Kirchhoff fluid:

$$P_{km}^{\langle 1 \rangle} = -2\mu E_{km},$$
$$q_k^{\langle 1 \rangle} = -M\mu\varepsilon_{,k}. \qquad \text{(XXIV.3)}$$

ENSKOG obtained particular expressions for μ and M, which we shall give as our equations (XXIV.91). The second iterates are much more complicated:

$$P_{km}^{\langle 2 \rangle} = -2\mu E_{km} + \varpi_1 \frac{\mu^2}{p} EE_{km} + \varpi_2 \frac{\mu^2}{p}\left[b_{\{k,m\}} - \left(\frac{1}{\rho}p_{,\{k}\right)_{,m\}}\right.$$
$$\left. - u_{\{a,k}u_{m,a\}} - 2u_{\{a,k}E_{ma\}}\right] + \varpi_3 \frac{\mu^2}{\rho\varepsilon}\varepsilon_{,\{km\}}$$
$$+ \varpi_4 \frac{\mu^2}{p\rho\varepsilon}p_{,\{k}\varepsilon_{,m\}} + \varpi_5 \frac{\mu^2}{\rho\varepsilon^2}\varepsilon_{,\{k}\varepsilon_{,m\}} + \varpi_6 \frac{\mu^2}{p}E_{\{ak}E_{ma\}}, \quad \text{(XXIV.4)}$$

---

[6] CHAPMAN & COWLING [1939, Chapter 7]. The familiar remark that CHAPMAN began from MAXWELL's equations of transfer while ENSKOG began from BOLTZMANN's equation is true, but it merely illustrates a difference of tradition: British and teutonic. The method of ENSKOG could have been applied also to CHAPMAN's starting point. In his recollections CHAPMAN [1967, p. 12] writes, "I had ..., at the outset, decided to develop the book along Enskog's lines rather than along those that I had followed. I felt that my way did a good engineer's job, but that Enskog's was more elegant, artistic and logical. Cowling and I tried to make his arguments easier to follow than they were in his dissertation." This remark can refer only to Problem 1.

Although ENSKOG [1917, Footnotes 14, 28, 33, 35] mentions HILBERT's work only vaguely, and in such a way as to emphasize his departure from it, any informed reader easily discerns HILBERT's influence. In ENSKOG's earlier attempt [1911, 1] there is no trace of the theory of integral equations, but his final treatment rests essentially if imprecisely upon that theory. On the contrary, CHAPMAN [1967, p. 8] tells us that he bought HILBERT's book but did not find it helpful. To the end of his life he continued to misrepresent HILBERT's work so crudely as to suggest that indeed he had no more than glanced at the paper in which it is presented.

(i) ENSKOG'S PROCEDURE

$$q_k^{\langle 2 \rangle} = -\mathsf{M}\mu\varepsilon_{,k} + \theta_1 \frac{\mu^2}{\rho\varepsilon} E\varepsilon_{,k} - \theta_2 \frac{\mu^2}{\rho\varepsilon} [\tfrac{2}{3}(\varepsilon E)_{,k} + 2\varepsilon_{,a} u_{a,k}]$$

$$+ \theta_3 \frac{\mu^2}{\rho p} E_{ka} p_{,a} + \theta_4 \frac{\mu^2}{\rho} E_{ka,a} + \theta_5 \frac{\mu^2}{\rho\varepsilon} E_{ka}\varepsilon_{,a}.$$

Braces surrounding a pair of free indices indicate the deviator of the symmetric part, formed after all repeated indices have been summed:

$$A_{\{km\}} \equiv A_{(km)} - \tfrac{1}{3} A_{aa} \delta_{km},$$

$$A_{\{ka} B_{am\}} \equiv \tfrac{1}{2} A_{ka} B_{am} + \tfrac{1}{2} A_{ma} B_{ak} - \tfrac{1}{3} A_{ab} B_{ba} \delta_{km}, \quad \text{etc.,}$$

(XVI.38)$_{1,2r}$

and the coefficients $\varpi_1, \varpi_2, \ldots, \theta_5$ are dimensionless functions of ε alone which depend upon the molecular model. The former result was first obtained by BURNETT[7] and is called ***Burnett's Formula***; the latter is due to CHAPMAN & COWLING. We have printed both of them here in CHAPMAN & COWLING's arrangement[8].

We see that (XXIV.4)$_1$ contains a term involving grad **b**. Since we at first consider only the case in which **b** = **0**, until we reach Section (x) of this chapter we shall refer to (XXIV.4) only when it, too, is so specialized.

ENSKOG's method applies to a general molecular model. It delivers μ and M as functions of the kinetic constitutive quantities $\mathfrak{m}$, $\mathbb{E}$, and $\mathscr{S}$ introduced in Chapter VII. We shall derive these in Section (ix) of this chapter. However, we shall not follow ENSKOG's procedure to get them, for it labors under vagueness as well as complexity. We know of no presentation of the details of it which is even formally explicit[9]. Nevertheless, in retrospect one thing is clear: ENSKOG's *objective was to approximate what we call gross determiners*. We attribute the indirectness and obscurity of his method to *his not having had at his disposal a general theory of gross determiners*. Indeed, to our knowledge the first general theory of this type is that which we have presented in the preceding chapter. The essential step in the analysis contained therein is *the elimination of time derivatives* between the equations of balance (XXIII.21) and the Maxwell-Boltzmann equation (XXIII.20) for grossly determined solutions. ENSKOG's

---

[7] BURNETT [1936]. Some of the terms had been obtained by [LENNARD-] JONES [1923]. *Cf.* also the pioneer work of MAXWELL, presented and discussed in Section (iv) of Chapter XXV.

[8] CHAPMAN & COWLING [1939, Eqs. 15.3.6, 15.41.1, 15.41.3].

[9] GRAD [1958, §25], who chose to follow ENSKOG in explaining the procedure, described the presentation of CHAPMAN & COWLING [1939] as "a completely *ad hoc* recipe for juggling terms in an expansion $f = f^{(0)} + f^{(1)} + \cdots$ in order to get results of the desired form ...". A profile of CHAPMAN published in *The Observer*, July 2, 1957 (reprinted on pp. 463–465 of S. G. BRUSH, *The Kind of Motion We Call Heat*, Volume 2, Amsterdam *etc.*, North-Holland Publishing Co., 1976) quotes him as saying the book was very heavy going, "like chewing glass". To us ENSKOG's own presentation seems to be, as CHAPMAN & COWLING imply, even more confusing.

procedure also eliminates time derivatives[10], but it does so at each stage of a general iterative procedure and so leads to unnecessary complication. By effecting this elimination at the very outset we have obtained a general theory of gross determiners, a theory which stands by itself, independent of any iterative procedure one might wish to use in calculating gross determiners. Because ENSKOG did not perceive this important step, he could only work with the Maxwell–Boltzmann equation itself.

In this respect ENSKOG's method is akin to HILBERT's. The difference, however, is that HILBERT sought solutions of the Maxwell–Boltzmann equation in the first place, while ENSKOG sought gross determiners. Indeed, the very reason the physicists have always advanced in favor of ENSKOG's treatment and against HILBERT's is that ENSKOG finally did obtain values for μ and M; that is, that he *did* seek and *did* find approximate gross determiners.

(ii) *The method of stretched fields*

We introduce now a new and direct method for obtaining iterative approximations to gross determiners. Among its products will be counterparts of (XXIV.2)–(XXIV.4)—namely, iterates $\mathfrak{P}^{(0)}$, $\mathfrak{q}^{(0)}$, $\mathfrak{P}^{(1)}$, $\mathfrak{q}^{(1)}$, $\mathfrak{P}^{(2)}$, $\mathfrak{q}^{(2)}$ to the gross determiners $\mathfrak{P}$ and $\mathfrak{q}$. These *approximate gross determiners will have exactly the same forms* as (XXIV.2)–(XXIV.4), but the principal moment ρ will be an arbitrary smooth field from the outset. Thus we shall obtain directly, with no detour through time-dependent solutions of the Maxwell–Boltzmann equation, the results that ENSKOG's process finally delivers. The relations between $\mathbf{P}^{\langle n \rangle}$ and $\mathfrak{P}^{(n)}$ and between $\mathbf{q}^{\langle n \rangle}$ and $\mathfrak{q}^{(n)}$ are clear: If $\mathbf{P}^{\langle n \rangle}$ and $\mathbf{q}^{\langle n \rangle}$ are given by (XXIV.2)–(XXIV.4), then $\mathbf{P}^{\langle n \rangle} = \mathfrak{P}^{(n)}(\rho)$ and $\mathbf{q}^{\langle n \rangle} = \mathfrak{q}^{(n)}(\rho)$, at least if $n = 0$, 1, or 2. We have not carried the new process further than $n = 2$, but merely formal labor would suffice to do so. Further comparison with ENSKOG's process would be impossible anyway, since its complete results for greater values of n are not available, so far as we know.

---

[10] In almost all iterative procedures designed to deliver approximate gross determiners, this elimination of time derivatives plays an essential and mysterious role. GRAD [1958, §24] used it to show how ENSKOG's results could be obtained from those of HILBERT. He writes "It is crucial that we do not set $t = 0$ too soon." We do not understand this statement. Indeed, the iteration we present here is applied directly to the equations of gross determinism, that is, to relations from which the time derivatives have been eliminated *exactly* by expressing them in terms of space derivatives *at the outset* and *at each and every time*.

For a gas of ideal spheres, PEKERIS [1955] may have perceived a reduction similar to GRAD's, but the steps in this part of his analysis are too meager for us to say for sure (an elimination of time derivatives appears nowhere in his work). His major conclusions concern the aspect of ENSKOG's procedure which we present in Section (xii) of this chapter (*cf.* Footnote 30 to this chapter).

As (XXIII.25) and (XXIII.26) indicate, gross determiners are functions of the initial value ϕ of the principal moment ρ. However, so as to permit comparisons of the type indicated above, in this chapter we will consider all gross determiners to be evaluated at a general time-dependent field ρ. The reader may, if he so wishes, replace ρ in all formulae by ϕ without affecting the analysis.

The basic idea is one MAXWELL himself[11] noticed in part: The right-hand sides of (XXIV.3) and (XXIV.4) suggest an expansion in powers of grad ρ, grad^2 ρ, ... For example, $\mathfrak{P}^{(1)} - \mathfrak{P}^{(0)}$ should be linear in grad ρ, while $\mathfrak{P}^{(2)} - \mathfrak{P}^{(1)}$ should be linear in grad^2 ρ and quadratic in grad ρ. If such an expansion of a function is justified, its results will be the more accurate the more nearly constant is the argument field ρ near the place $\mathbf{x}$ at which the expansion is effected. Accordingly, we introduce a special class of fields such as to render this idea precise.

Let $\mathbf{x}_0$ be a fixed place, and let δ be a positive number. From each function ϕ on $\mathscr{E}$ we may form another function $\overset{\delta}{\phi}$, also defined on $\mathscr{E}$, as follows:

---

[11] In surveying such approximations of this kind as he had obtained MAXWELL [1879, §(15)] wrote, "It appears from the investigation that the condition of the successful use of this method of approximation is that $l\, d/dh$ should be small, where d/dh denotes the differentiation with respect to a line drawn in any direction. In other words, the properties of the medium must not be sensibly different at points within a distance of each other, comparable with the 'mean free path' of a molecule." Although physicists are wont to echo this statement, its meaning is not clear. Formally, it is specific if we take for l some length characteristic of the problem, for example $\mathfrak{m}/[p(\tau^{(1)})^2]$, $\tau^{(1)}$ being given by (XVII.60). The concept of magnitude of a differential operator can be rendered precise in many ways, but these differ from each other, and the corresponding requirements that such a quantity be small lead to different criteria. For the special solutions discussed in Chapters XIV and XV a natural parameter for expansion was found to be T, the tension number, which was defined as a ratio of the time of relaxation τ to a time characteristic of the problem. To render this ratio small, it suffices to make τ small, other things being equal.

However, no single dimensionless number will control all derivatives. For example, the heat-transfer number H introduced in Footnote 7 to Chapter VIII could also be interpreted as a "Knudsen number", yet it and T are independent. In the particular problems we have seen to be controlled by T, also H = 0. If we look again at (XXIV.4) and try to interpret it in MAXWELL's terms, it suggests to us that not only $l\, d/dh$ but also $l^2\, d^2/dh^2$, $l^3\, d^3/dh^3$, ... are presumed small. In the special solutions discussed in Chapters XIV and XV all terms but the first are null, so T suffices there as a governing parameter. In more general problems, additional dimensionless parameters will be needed, not only H but also numbers defined in terms of higher derivatives.

The matter was discussed in phenomenological terms by TRUESDELL in §28 of "A new definition of a fluid. II. The Maxwellian fluid", *Journal de Mathématiques* (9) **30**, 111–158 (1951). If we transfer to the kinetic theory the suggestion he made there for continuum theories of fluids, we might regard the gross condition as given and consider lighter and lighter gases: $\mathfrak{m} \to 0$ while other molecular parameters and ρ are held fixed. This point of view conforms with the fact that iterates $\mathfrak{P}^{(n)}$ and $\mathbf{q}^{(n)}$ are *polynomials* in μ, as is exemplified by (XXIV.101). Because μ is a function of ε and molecular constants alone, it follows from (IV.2) and (more clearly) from (IV.5) that for a given gross condition $\mu \to 0$ if $\mathfrak{m} \to 0$. Considerations of this kind might render concrete the common vague claim of physicists that expressions such as (XXIV.4) are "asymptotic".

$$\overset{\delta}{\phi}(\mathbf{x}) \equiv \phi(\mathbf{x}_0 + \delta(\mathbf{x} - \mathbf{x}_0)). \tag{XXIV.5}$$

If ϕ is continuous, then for each place $\mathbf{x}$, $\overset{\delta}{\phi}(\mathbf{x})$ approaches the constant $\phi(\mathbf{x}_0)$ as $\delta \to 0$. If ϕ is smooth, then

$$\begin{aligned}\overset{\delta}{\phi}_{,k}(\mathbf{x}) &= \delta \phi_{,k}(\mathbf{x}_0 + \delta(\mathbf{x} - \mathbf{x}_0)), \\ \overset{\delta}{\phi}_{,km}(\mathbf{x}) &= \delta^2 \phi_{,km}(\mathbf{x}_0 + \delta(\mathbf{x} - \mathbf{x}_0)), \quad \text{etc.,}\end{aligned} \tag{XXIV.6}$$

so that as $\delta \to 0$ we see that $\overset{\delta}{\phi}(\mathbf{x})$ is $O(1)$, $\overset{\delta}{\phi}_{,k}(\mathbf{x})$ is $O(\delta)$, and $\overset{\delta}{\phi}_{,km}(\mathbf{x})$ and $\overset{\delta}{\phi}_{,k}(\mathbf{x})\overset{\delta}{\phi}_{,m}(\mathbf{x})$ are $O(\delta^2)$. This is the behavior of the field $\mathbf{\rho}$ that (XXIV.2)–(XXIV.4) suggest. The effect of choosing successively smaller values of δ in $\overset{\delta}{\phi}$ is illustrated in Figure XXIV.1. The parameter δ may be interpreted as introducing a change of scale in the point space $\mathscr{E}$, the effect of which is to stretch out the spatial variables radially from the fixed place $\mathbf{x}_0$ as δ becomes small. We call $\overset{\delta}{\phi}$ the *stretched field* corresponding to ϕ.

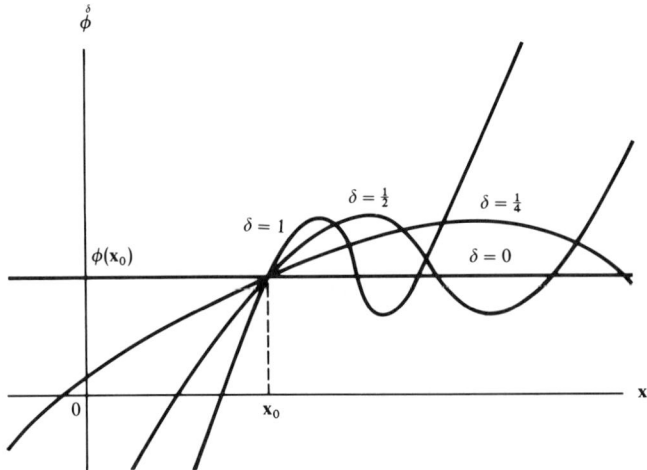

Figure XXIV.1

Using the shifter $\mathfrak{S}$ defined by (XXIII.11), we may summarize the general observations made above as follows. When δ is small, we expect that $\mathfrak{P}(\mathfrak{S}\overset{\delta}{\rho})$ and $\mathbf{q}(\mathfrak{S}\overset{\delta}{\rho})$ will be approximated by (XXIV.2)–(XXIV.4). Following this suggestion, we present in the remainder of this chapter a procedure for approximating gross determiners, the argument of which is a general stretched field $\overset{\delta}{\mathbf{\rho}}$. We call it the **Method of Stretched Fields**.

(iii) *The basic expansion of gross determiners*

We wish to approximate $G(\mathfrak{S}\overset{\delta}{\rho})$ when δ is small. To do so, we begin by considering a problem of pure analysis: expand a smooth functional of $\overset{\delta}{\rho}$ in a neighborhood of $\delta = 0$ as a series of functions of $\overset{\delta}{\rho}$ and its spatial derivatives. Although we denote the functional in question by G, and although when interpreting results we call it a gross determiner, in this and the next two sections of the chapter we make no use of the equations of gross determinism (XXIII.25) and (XXIII.26). G is an arbitrary mapping of the same kind.

One way to approximate $G(\mathfrak{S}\overset{\delta}{\rho})$ is by use of a finite Taylor expansion[12] in δ about $\delta = 0$. Unfortunately this leads to an expansion in terms of functions of ρ rather than $\overset{\delta}{\rho}$. To obtain one in which $\overset{\delta}{\rho}$ and its spatial derivatives appear, we proceed as follows.

Consider the function of two real numbers λ and δ defined by

$$g(\lambda, \delta) \equiv \underset{\mathbf{r}}{G}\left(\rho(t, \mathbf{x}_0 + \lambda(\mathbf{x} - \mathbf{x}_0) + \delta\mathbf{r})\right). \tag{XXIV.7}$$

Using (XXIV.5) and the definition (XXIII.11) of the shifter $\mathfrak{S}$, we see that

$$g(\delta, \delta) = \underset{\mathbf{r}}{G}\left(\rho(t, \mathbf{x}_0 + \delta(\mathbf{x} + \mathbf{r} - \mathbf{x}_0))\right) = G(\mathfrak{S}\overset{\delta}{\rho}(t, \mathbf{x})), \tag{XXIV.8}$$

so it is the value $g(\delta, \delta)$ that we wish to approximate. To do this by altering only the functional character of G, as recalled by the dummy variable $\mathbf{r}$ in the notation, we simply fix λ, expand $g(\lambda, \delta)$ in a finite Taylor sum about $\delta = 0$, and thereafter set $\lambda = \delta$:

$$\begin{aligned}g(\delta, \delta) = g(\delta, 0) + \partial_\delta g(\delta, 0)\,\delta + \partial_\delta^2 g(\delta, 0)\frac{\delta^2}{2!} \\ + \cdots + \partial_\delta^N g(\delta, 0)\frac{\delta^N}{N!} + o(\delta^N).\end{aligned} \tag{XXIV.9}$$

---

[12] Functional expansions obtained in this way were introduced in the context of continuum mechanics by B. D. COLEMAN, "On retardation theorems", *Archive for Rational Mechanics and Analysis* **43**, 1–23 (1971). His work grew from an earlier analysis by B. D. COLEMAN & W. NOLL, "An approximation theorem for functionals, with applications in continuum mechanics", *Archive for Rational Mechanics and Analysis* **6**, 355–370 (1960).

Rather than consider a stretched field, COLEMAN introduces the related concept of the retardation of a function. The δ-retardation of ψ is the function $\mathbf{r} \mapsto \psi(\delta\mathbf{r})$, and COLEMAN's "retardation theorem" establishes an expansion of $\underset{\mathbf{r}}{G}(\psi(\delta\mathbf{r}))$ as a series of functions of ψ and its spatial derivatives. Here we require an expansion of $G(\mathfrak{S}\overset{\delta}{\rho}(t, \mathbf{x})) = \underset{\mathbf{r}}{G}(\rho(t, \mathbf{x}_0 + \delta(\mathbf{x} - \mathbf{x}_0) + \delta\mathbf{r}))$. To obtain one from COLEMAN's results we could choose ψ in his analysis to be $\psi(\mathbf{r}) = \rho(t, \mathbf{x}_0 + \delta(\mathbf{x} - \mathbf{x}_0) + \mathbf{r})$. However, we should then effect δ-retardation upon a function which depends already on δ, and so there would be some difficulty in interpreting the expansion obtained. To resolve this difficulty, we have chosen here to derive the expansion anew using essentially COLEMAN's method, both extending and restricting his analysis to suit our applications to the kinetic theory.

This formula is the heart of the method of stretched fields. To make it explicit, we substitute g as defined by (XXIV.7) and then derive convenient forms of the coefficients $\partial_\delta^n g(\delta, 0)$. To simplify the steps in doing so, we hold t, $\mathbf{x}$, and λ fixed and set

$$\psi_\alpha(\mathbf{r}) \equiv \rho_\alpha(t, \mathbf{x}_0 + \lambda(\mathbf{x} - \mathbf{x}_0) + \mathbf{r}),$$
$$\overset{\delta}{\psi}_\alpha(\mathbf{r}) \equiv \psi_\alpha(\delta \mathbf{r}). \qquad \text{(XXIV.10)}$$

Then g becomes the composite function

$$g = \mathsf{G} \circ \overset{\delta}{\psi}, \qquad \text{(XXIV.11)}$$

and so we may calculate the coefficients $\partial_\delta^n g(\delta, 0)$ by means of the chain rule for differentiating composite functions. We proceed now to the details.

Let G be N times differentiable (cf. (XXIIIA.20)): If ψ lies in the domain $\mathcal{O} \subset \mathscr{F}$ of G, then for all ξ near $\mathbf{0}$

$$\mathsf{G}(\psi + \xi) = \mathsf{G}(\psi) + \sum_{n=1}^{N} \frac{1}{n!} \partial_\psi^n \mathsf{G}(\psi)[\xi, \ldots, \xi] + \chi(\psi, \xi) \|\xi\|^N, \qquad \text{(XXIV.12)}$$

$\| \ \|$ being the norm in $\mathscr{F}$ and $\partial_\psi^n \mathsf{G}(\psi)$ being the nth derivative of G at ψ. For each fixed ψ, $\partial_\psi^n \mathsf{G}(\psi)[\,\cdot\,, \ldots, \cdot\,]$ is a continuous linear function of each of its n arguments, and

$$\lim_{\|\xi\| \to 0} |\chi(\psi, \xi)| = 0. \qquad \text{(XXIII.15)}_r$$

Next let ρ be so chosen that the mapping $\delta \mapsto \overset{\delta}{\psi}$ is N times differentiable in a neighborhood of 0. In particular, then, we have (cf. (XXIIIA.23)) the finite Taylor expansion

$$\overset{\delta}{\psi}_\alpha = \overset{0}{\psi}_\alpha + \sum_{n=1}^{N} \frac{\delta^n}{n!} \left(\frac{d^n}{d\delta^n} \overset{\delta}{\psi}_\alpha \bigg|_{\delta=0} \right) + \delta^N \Psi_\alpha(\delta), \qquad \text{(XXIV.13)}$$

in which

$$\lim_{\delta \to 0} \|\Psi(\delta)\| = 0. \qquad \text{(XXIV.14)}$$

Suppose now that ρ is smooth. Then by (XXIV.10) the function $\delta \mapsto \overset{\delta}{\psi}(\mathbf{r})$ has a finite Taylor expansion about $\delta = 0$ for each fixed $\mathbf{r} \in \mathscr{W}$:

$$\overset{\delta}{\psi}_\alpha(\mathbf{r}) = \overset{0}{\psi}_\alpha(\mathbf{r}) + \sum_{n=1}^{N} \frac{\delta^n}{n!} \psi_{\alpha, a_1 \ldots a_n}(0) r_{a_1} \cdots r_{a_n} + \delta^N R_\alpha(\delta, \mathbf{r}),$$
$$\lim_{\delta \to 0} |R_\alpha(\delta, \mathbf{r})| = 0. \qquad \text{(XXIV.15)}$$

(iii) BASIC FUNCTIONAL EXPANSION

The formal similarity between (XXIV.15)$_1$ and the result of evaluating (XXIV.13) at $\mathbf{r} \in \mathscr{W}$ suggests that when ψ is smooth, then

$$\frac{d^n}{d\delta^n} \overset{\delta}{\psi}_\alpha \bigg|_{\delta=0} = \psi_{\beta,a_1\ldots a_n}(0) J^{a_1 \ldots a_n}_{\alpha\beta}, \qquad \text{(XXIV.16)}$$

$J^{p_1 \ldots p_n}_\beta$ being the vector-valued function on $\mathscr{W}$ with components $J^{p_1 \ldots p_n}_{\alpha\beta}$, $\alpha = 0, 1, \ldots, 4$, which are given by

$$J^{p_1 \ldots p_n}_{\alpha\beta}(\mathbf{r}) \equiv \begin{cases} r_{p_1} \cdots r_{p_n} & \text{if } \beta = \alpha, \\ 0 & \text{if } \beta \neq \alpha. \end{cases} \qquad \text{(XXIV.17)}$$

We have not proved (XXIV.16). To do so, we should have to show that (XXIV.14) is satisfied when $\boldsymbol{\Psi}(\delta) = \mathbf{R}(\delta, \cdot)$, a result that does not generally follow from the pointwise limit (XXIV.15)$_2$. Indeed, (XXIV.13) and (XXIV.14) are statements about the regularity of the $\mathscr{F}$-valued function $\delta \mapsto \overset{\delta}{\psi}$ while (XXIV.15) concerns the regularity, for each fixed $\mathbf{r} \in \mathscr{W}$, of the vector-valued function $\delta \mapsto \overset{\delta}{\psi}(\mathbf{r})$. Nevertheless, this formal analogy suggests that (XXIV.16) will be valid for a broad class of functions ψ, and we shall assume that it holds in the calculations to follow.

By a simple extension of the proof of Lemma 2 in Section (vi) of Chapter XXIII, we can exhibit a large class of functions for which (XXIV.16) is valid. Recalling that we chose $\mathfrak{G}$ to be defined on an open set in $\mathscr{F} = (\mathscr{C}_m(\mathscr{W}))^5$, we consider now integers m such that $m \geq N+1$. Then any field $\boldsymbol{\rho} \in (\mathscr{C}^{N+1}_m(\mathscr{I} \times \mathscr{E}))^5$ gives rise to a function ψ that satisfies (XXIV.16). To prove this, we write the remainder $\mathbf{R}$ in (XXIV.15)$_1$ explicitly in the form

$$R_\alpha(\delta, \mathbf{r}) = \frac{\overset{\delta}{\psi}_{\alpha,a_1 \ldots a_{N+1}}(\delta\theta \mathbf{r})}{(N+1)!} r_{a_1} \cdots r_{a_{N+1}}, \qquad \text{(XXIV.18)}$$

θ being some number in the interval $(0, 1)$. Then, by (XXIII.32) and (XXIII.28), we see that

$$\|\mathbf{R}(\delta, \cdot)\| = \sum_{\alpha=0}^{4} \sup_{\mathbf{r} \in \mathscr{W}} \left| \frac{\overset{\delta}{\psi}_{\alpha,a_1\ldots a_{N+1}}(\delta\theta\mathbf{r}) r_{a_1} \cdots r_{a_{N+1}}}{(N+1)!} \right| H_m(|\mathbf{r}|), $$
$$\leq \delta k \sup_{\mathbf{r} \in \mathscr{W}} \sup_{(\alpha, a_1, \ldots, a_{N+1})} |\psi_{\alpha,a_1\ldots a_{N+1}}(\delta\theta\mathbf{r})| \, |\mathbf{r}|^{N+1} H_m(|\mathbf{r}|), \qquad \text{(XXIV.19)}$$

k being a constant. Because $\rho_\alpha \in \mathscr{C}^{N+1}_m(\mathscr{I} \times \mathscr{E})$, we know that the rate of growth of $\psi_{\alpha,p_1 \ldots p_q}(\mathbf{r})$ as $|\mathbf{r}| \to \infty$ is no greater than that of $|\mathbf{r}|^{m-q}$, so there are constants C_1 and C_2 such that

$$|\psi_{\alpha,p_1 \ldots p_{N+1}}(\delta\theta\mathbf{r})| \leq C_1(\delta\theta|\mathbf{r}|)^{m-N-1} + C_2. \qquad \text{(XXIV.20)}$$

Hence, since $m \geq N+1$,

$$\|\mathbf{R}(\delta, \cdot)\| \leq \delta k C_1 (\delta\theta)^{m-N-1} \sup_{\mathbf{r} \in \mathscr{W}} |\mathbf{r}|^m H_m(|\mathbf{r}|) + \delta k C_2 \sup_{\mathbf{r} \in \mathscr{W}} |\mathbf{r}|^{N+1} H_m(|\mathbf{r}|),$$
$$\leq \text{const.} \, \delta. \qquad \text{(XXIV.21)}$$

Thus (XXIV.14) is satisfied when $\boldsymbol{\Psi}(\delta) = \mathbf{R}(\delta, \cdot)$, and hence (XXIV.16) follows. △

Since $\mathbb{G}$ and $\delta \mapsto \overset{\delta}{\psi}$ are N times differentiable, so also is the function $\delta \mapsto \mathbb{G} \circ \overset{\delta}{\psi}$. To calculate its successive derivatives we need a formula for the nth derivative of a composite function. When $n = 1$ such a formula is the chain rule

$$\frac{d}{d\delta} \mathbb{G} \circ \overset{\delta}{\psi} = \partial_\psi \mathbb{G}(\overset{\delta}{\psi}) \left[\frac{d}{d\delta} \overset{\delta}{\psi} \right]. \tag{XXIV.22}$$

The derivative of this equation is

$$\frac{d^2}{d\delta^2} \mathbb{G} \circ \overset{\delta}{\psi} = \partial_\psi^2 \mathbb{G}(\overset{\delta}{\psi}) \left[\frac{d}{d\delta} \overset{\delta}{\psi}, \frac{d}{d\delta} \overset{\delta}{\psi} \right] + \partial_\psi \mathbb{G}(\overset{\delta}{\psi}) \left[\frac{d^2}{d\delta^2} \overset{\delta}{\psi} \right]. \tag{XXIV.23}$$

The first term is obtained by holding $d\overset{\delta}{\psi}/d\delta$ fixed in (XXIV.22) and applying the chain rule. The second term comes from holding $\overset{\delta}{\psi}$ fixed and applying the chain rule once again. By repeated derivation of (XXIV.22) with respect to δ we obtain the following general formula:

$$\frac{d^n}{d\delta^n} \mathbb{G} \circ \overset{\delta}{\psi} = \sum_{p=1}^{n} \sum_{\substack{n_1 + \cdots + n_p = n \\ n_m > 0}} \frac{n!}{p! \, n_1! \cdots n_p!} \partial_\psi^p \mathbb{G}(\overset{\delta}{\psi}) \left[\frac{d^{n_1}}{d\delta^{n_1}} \overset{\delta}{\psi}, \ldots, \frac{d^{n_p}}{d\delta^{n_p}} \overset{\delta}{\psi} \right]. \tag{XXIV.24}$$

When $n = 1$ and $n = 2$, this formula reduces to (XXIV.22) and (XXIV.23), respectively, and by differentiating once more the reader may show by induction that it holds in general.

We may determine now the coefficients in (XXIV.9). We first evaluate each side of (XXIV.24) at $\delta = 0$ and then use the formula (XXIV.16) and the linearity of the derivatives of $\mathbb{G}$. This gives us

$$\partial_\delta^n g(\lambda, 0) = \sum_{p=1}^{n} \sum_{\substack{n_1 + \cdots + n_p = n \\ n_m > 0}} \frac{n!}{p! \, n_1! \cdots n_p!} \psi_{\alpha_1, n_1}(0) \cdots$$
$$\cdots \psi_{\alpha_p, n_p}(0) \, \partial_\psi^p \mathbb{G}(\overset{0}{\psi}) [J_{\alpha_1}^{n_1}, \ldots, J_{\alpha_p}^{n_p}]. \tag{XXIV.25}$$

In this formula and henceforth in the chapter we use multi-indices (cf. Appendix B of Chapter XVI). Next note from (XXIV.10) that

$$\overset{0}{\psi}_\alpha(\mathbf{r}) = \rho_\alpha(t, \mathbf{x}_0 + \lambda(\mathbf{x} - \mathbf{x}_0)) = \overset{\lambda}{\rho}_\alpha(t, \mathbf{x}),$$
$$\psi_{\alpha, n}(0) = \rho_{\alpha, n}(t, \mathbf{x}_0 + \lambda(\mathbf{x} - \mathbf{x}_0)) = \frac{1}{\lambda^n} \overset{\lambda}{\rho}_{\alpha, n}(t, \mathbf{x}), \tag{XXIV.26}$$

and so

$$\overset{0}{\psi}_\alpha \big|_{\lambda = \delta} = \overset{\delta}{\rho}_\alpha(t, \mathbf{x}), \tag{XXIV.27}$$

$$\delta^{n_1 + \cdots + n_p} \psi_{\alpha_1, n_1}(0) \cdots \psi_{\alpha_p, n_p}(0) \big|_{\lambda = \delta} = \overset{\delta}{\rho}_{\alpha_1, n_1}(t, \mathbf{x}) \cdots \overset{\delta}{\rho}_{\alpha_p, n_p}(t, \mathbf{x}).$$

Setting $\lambda = \delta$ in (XXIV.25), using these formulae, and placing the result into (XXIV.9), we obtain finally the **Basic Expansion of Gross Determiners**[13]:

$$\mathbb{G}(\mathfrak{S}\overset{\delta}{\rho}(t, \mathbf{x})) = \mathbb{G}(\overset{\delta}{\rho}(t, \mathbf{x}))$$

$$+ \sum_{n=1}^{N} \sum_{p=1}^{n} \sum_{\substack{n_1 + \cdots + n_p = n \\ n_m > 0}} \frac{1}{p! \, n_1! \cdots n_p!} \overset{\delta}{\rho}_{\alpha_1, n_1}(t, \mathbf{x}) \cdots \qquad \text{(XXIV.28)}$$

$$\cdots \overset{\delta}{\rho}_{\alpha_p, n_p}(t, \mathbf{x}) \mathbb{G}_{\alpha_1 \cdots \alpha_p}^{n_1 \cdots n_p}(\overset{\delta}{\rho}(t, \mathbf{x})) + o(\delta^N),$$

$$\mathbb{G}_{\alpha_1 \cdots \alpha_p}^{n_1 \cdots n_p}(\mathbf{K}) \equiv \partial^p_\psi \mathbb{G}(\mathbf{K})[\mathbf{J}_{\alpha_1}^{n_1}, \ldots, \mathbf{J}_{\alpha_p}^{n_p}].$$

We see from (XXIV.28)$_2$ and (XXIV.17) that this expansion is meaningful only if all polynomials in $\mathbf{r}$ of degree no greater than N lie in the function space $\mathscr{F}$ for gross determiners. For the special choice of $\mathscr{F}$ which we have made in Section (v) of Chapter XXIII this requirement affords us no problem (cf. Footnote 6 to Chapter XXIII).

We now turn to the interpretation of this expansion and its relation to formulae such as (XXIV.2)–(XXIV.4).

(iv) *Approximate gross determiners*

We continue to study the purely analytical problem of approximating functionals, making at first no use of such restrictions upon them as the kinetic theory may provide.

For fixed t and $\mathbf{x}$ the field $\overset{\delta}{\rho}$ enters the right-hand side of (XXIV.28)$_1$ in two different ways. The first is through the vector $\mathbf{K} = \overset{\delta}{\rho}(t, \mathbf{x})$ appearing in the coefficients $\mathbb{G}(\mathbf{K})$ and $\mathbb{G}_{\alpha_1 \cdots \alpha_p}^{n_1 \cdots n_p}(\mathbf{K})$, these being nothing more than $\mathbb{G}$ and its derivatives evaluated at the constant function on $\mathscr{W}$ whose value is $\mathbf{K}$. Second, $\overset{\delta}{\rho}$ enters through its derivatives $\overset{\delta}{\rho}_{,n}(t, \mathbf{x})$, this time explicitly as a polynomial. Thus *the basic expansion of gross determiners expresses a general functional* $\mathbb{G}$, *to within* $o(\delta^N)$, *as a linear combination of certain functions of real variables*. It is just such a result that we wished to obtain since it permits us to study gross determiners with the apparatus of functions of real variables.

To examine the importance of the derivatives of the fields in (XXIV.28)$_1$, we turn to the gross determiners $\mathfrak{P}$ and $\mathbf{q}$ for the pressure deviator and the energy flux, respectively. From the definitions of these as moments of $\mathbb{G}$ we may derive

---

[13] MUNCASTER [1975, Eq. (10.11)].

for them expansions similar to that which we already have for $\mathfrak{G}$:

$$\mathfrak{P}_{km}(\mathfrak{S}\overset{\delta}{\boldsymbol{\rho}}) = \mathfrak{P}_{km}(\overset{\delta}{\boldsymbol{\rho}}) + \sum_{n=1}^{N}\sum_{p=1}^{n}\sum_{\substack{n_1+\cdots+n_p=n \\ n_m>0}} \overset{\delta}{\rho}_{\alpha_1,n_1}\cdots\overset{\delta}{\rho}_{\alpha_p,n_p}\mathfrak{P}_{km\alpha_1\ldots\alpha_p}^{n_1\ldots n_p}(\overset{\delta}{\boldsymbol{\rho}}) + o(\delta^N),$$

$$\mathfrak{q}_k(\mathfrak{S}\overset{\delta}{\boldsymbol{\rho}}) = \mathfrak{q}_k(\overset{\delta}{\boldsymbol{\rho}}) + \sum_{n=1}^{N}\sum_{p=1}^{n}\sum_{\substack{n_1+\cdots+n_p=n \\ n_m>0}} \overset{\delta}{\rho}_{\alpha_1,n_1}\cdots\overset{\delta}{\rho}_{\alpha_p,n_p}\mathfrak{q}_{k\alpha_1\ldots\alpha_p}^{n_1\ldots n_p}(\overset{\delta}{\boldsymbol{\rho}}) + o(\delta^N), \quad \text{(XXIV.29)}$$

in which

$$\mathfrak{P}_{km\alpha_1\ldots\alpha_p}^{n_1\ldots n_p}(\mathbf{K}) \equiv \frac{m}{p!\,n_1!\,\cdots\,n_p!}\int c_{\{k}c_{m\}}\mathfrak{G}_{\alpha_1\ldots\alpha_p}^{n_1\ldots n_p}(\mathbf{K}),$$

$$\mathfrak{q}_{k\alpha_1\ldots\alpha_p}^{n_1\ldots n_p}(\mathbf{K}) \equiv \frac{m}{p!\,n_1!\,\cdots\,n_p!}\int \tfrac{1}{2}c^2 c_k\,\mathfrak{G}_{\alpha_1\ldots\alpha_p}^{n_1\ldots n_p}(\mathbf{K}); \quad \text{(XXIV.30)}$$

$\mathbf{c}$ is the random velocity determined by $\mathbf{K}$, namely, $\mathbf{c} = \mathbf{v} - \mathbf{u}$, and ρ, $\mathbf{u}$, and ε are constants defined as follows in terms of $\mathbf{K}$:

$$\mathbf{K} = (\rho,\,\rho\mathbf{u},\,\rho(2\varepsilon + u^2)). \quad \text{(XXIV.31)}$$

Let us write out explicitly the expansion for $\mathfrak{P}$ for small values of N. If $N=0$, all terms on the right-hand side of $(\text{XXIV.29})_1$ except for the first are lumped into a single term that approaches 0 as $\delta \to 0$:

$$\mathfrak{P}_{km}(\mathfrak{S}\overset{\delta}{\boldsymbol{\rho}}) = \mathfrak{P}_{km}(\overset{\delta}{\boldsymbol{\rho}}) + o(1). \quad \text{(XXIV.32)}$$

When $N=1$, n and p are both 1, the list $n_1,\ldots,n_p$ reduces to n_1 alone, and moreover $n_1 = 1$, so the list that the multi-index $\mathbf{n}_1$ represents contains just a single index, say a. Thus

$$\mathfrak{P}_{km}(\mathfrak{S}\overset{\delta}{\boldsymbol{\rho}}) = \mathfrak{P}_{km}(\overset{\delta}{\boldsymbol{\rho}}) + \mathfrak{P}_{km\alpha}^{a}(\overset{\delta}{\boldsymbol{\rho}})\overset{\delta}{\rho}_{\alpha,a} + o(\delta). \quad \text{(XXIV.33)}$$

In the same way we find that when $N=2$, $\mathfrak{P}$ has the expansion

$$\mathfrak{P}_{km}(\mathfrak{S}\overset{\delta}{\boldsymbol{\rho}}) = \mathfrak{P}_{km}(\overset{\delta}{\boldsymbol{\rho}}) + \mathfrak{P}_{km\alpha}^{a}(\overset{\delta}{\boldsymbol{\rho}})\overset{\delta}{\rho}_{\alpha,a} \quad \text{(XXIV.34)}$$
$$+ \mathfrak{P}_{km\alpha}^{ab}(\overset{\delta}{\boldsymbol{\rho}})\overset{\delta}{\rho}_{\alpha,ab} + \mathfrak{P}_{km\alpha\beta}^{ab}(\overset{\delta}{\boldsymbol{\rho}})\overset{\delta}{\rho}_{\alpha,a}\overset{\delta}{\rho}_{\beta,b} + o(\delta^2).$$

Of course there are similar expansions for $\mathfrak{q}$.

If a molecular density is grossly determined, so is any expectation calculated from it. We have illustrated this fact in connection with the pressure deviator $\mathbf{P}$ and the energy flux vector $\mathbf{q}$; we may do the same for BOLTZMANN's function h and for its flux $\mathbf{s}$:

$$h(t,\mathbf{x}) = \mathfrak{h}(\mathfrak{S}\boldsymbol{\rho}(t,\mathbf{x})), \qquad s_k(t,\mathbf{x}) = \mathfrak{s}_k(\mathfrak{S}\boldsymbol{\rho}(t,\mathbf{x})), \quad \text{(XXIV.35)}$$

(iv) APPROXIMATE GROSS DETERMINERS   443

the mappings $\mathfrak{h}$ and $\mathfrak{s}$ being defined as follows:

$$\mathfrak{h}(\mathfrak{S}\rho) \equiv \frac{\mathfrak{m}}{\rho}\int \mathbb{G}(\mathfrak{S}\rho) \log \mathbb{G}(\mathfrak{S}\rho), \qquad \mathfrak{s}_k(\mathfrak{S}\rho) \equiv \mathfrak{m}\int c_k \, \mathbb{G}(\mathfrak{S}\rho) \log \mathbb{G}(\mathfrak{S}\rho).$$
(XXIV.36)

We can derive expansions for these fields in the same way as we derived (XXIV.29). Here the forms of the general terms are more complicated, so we rest content to write down these expansions to within $o(\delta^2)$.

Just as we arrived at (XXIV.9), we may derive the following expansion:

$$g(\delta, \delta) \log g(\delta, \delta) = g(\delta, 0) \log g(\delta, 0)$$
$$+ \left[(1 + \log g(\delta, 0))\, \partial_\delta g(\delta, 0)\right] \delta$$
$$+ \left[(1 + \log g(\delta, 0))\, \partial_\delta^2 g(\delta, 0) + \frac{(\partial_\delta g(\delta, 0))^2}{g(\delta, 0)}\right] \frac{\delta^2}{2!}$$
$$+ o(\delta^2).$$
(XXIV.37)

The derivatives $\partial_\delta^n g(\delta, 0)$ can be read off from (XXIV.25) and (XXIV.27). In particular,

$$g(\delta, 0) = \mathbb{G}(\overset{\circ}{\rho}),$$
$$\delta \, \partial_\delta g(\delta, 0) = \mathbb{G}_\alpha^a(\overset{\circ}{\rho})\overset{\circ}{\rho}_{\alpha,a},$$
(XXIV.38)
$$\delta^2 \, \partial_\delta^2 g(\delta, 0) = \mathbb{G}_\alpha^{ab}(\overset{\circ}{\rho})\overset{\circ}{\rho}_{\alpha,ab} + \mathbb{G}_{\alpha\beta}^{ab}(\overset{\circ}{\rho})\overset{\circ}{\rho}_{\alpha,a}\overset{\circ}{\rho}_{\beta,b}.$$

If we place these into (XXIV.37) and then substitute the result into (XXIV.36), we obtain the expansions

$$\mathfrak{h}(\mathfrak{S}\overset{\circ}{\rho}) = \mathfrak{h}(\overset{\circ}{\rho}) + \mathfrak{h}_\alpha^a(\overset{\circ}{\rho})\overset{\circ}{\rho}_{\alpha,a} + \mathfrak{h}_\alpha^{ab}(\overset{\circ}{\rho})\overset{\circ}{\rho}_{\alpha,ab} + \mathfrak{h}_{\alpha\beta}^{ab}(\overset{\circ}{\rho})\overset{\circ}{\rho}_{\alpha,a}\overset{\circ}{\rho}_{\beta,b} + o(\delta^2),$$

$$\mathfrak{s}_k(\mathfrak{S}\overset{\circ}{\rho}) = \mathfrak{s}_k(\overset{\circ}{\rho}) + \mathfrak{s}_{k\alpha}^a(\overset{\circ}{\rho})\overset{\circ}{\rho}_{\alpha,a} + \mathfrak{s}_{k\alpha}^{ab}(\overset{\circ}{\rho})\overset{\circ}{\rho}_{\alpha,ab} + \mathfrak{s}_{k\alpha\beta}^{ab}(\overset{\circ}{\rho})\overset{\circ}{\rho}_{\alpha,a}\overset{\circ}{\rho}_{\beta,b} + o(\delta^2),$$
(XXIV.39)

the coefficients in these being given as follows by the coefficients in (XXIV.28)$_1$:

$$\mathfrak{h}_\alpha^k(\mathbf{K}) \equiv \frac{\mathfrak{m}}{\rho}\int (1 + \log \mathbb{G}(\mathbf{K}))\mathbb{G}_\alpha^k(\mathbf{K}),$$

$$\mathfrak{h}_\alpha^{km}(\mathbf{K}) \equiv \frac{\mathfrak{m}}{2\rho}\int (1 + \log \mathbb{G}(\mathbf{K}))\mathbb{G}_\alpha^{km}(\mathbf{K}),$$

$$\mathfrak{h}_{\alpha\beta}^{km}(\mathbf{K}) \equiv \frac{\mathfrak{m}}{2\rho}\int \left\{(1 + \log \mathbb{G}(\mathbf{K}))\mathbb{G}_{\alpha\beta}^{km}(\mathbf{K}) + \frac{\mathbb{G}_\alpha^k(\mathbf{K})\mathbb{G}_\beta^m(\mathbf{K})}{\mathbb{G}(\mathbf{K})}\right\}, \quad \text{(XXIV.40)}$$

$$\mathfrak{s}_{k\alpha}^{m}(\mathbf{K}) \equiv \mathfrak{m} \int c_k(1 + \log \mathbb{G}(\mathbf{K}))\mathbb{G}_{\alpha}^{m}(\mathbf{K}),$$

$$\mathfrak{s}_{k\alpha}^{mr}(\mathbf{K}) \equiv \tfrac{1}{2}\mathfrak{m} \int c_k(1 + \log \mathbb{G}(\mathbf{K}))\mathbb{G}_{\alpha}^{mr}(\mathbf{K}),$$

$$\mathfrak{s}_{k\alpha\beta}^{mr}(\mathbf{K}) \equiv \tfrac{1}{2}\mathfrak{m} \int c_k \left\{ (1 + \log \mathbb{G}(\mathbf{K}))\mathbb{G}_{\alpha\beta}^{mr}(\mathbf{K}) + \frac{\mathbb{G}_{\alpha}^{m}(\mathbf{K})\mathbb{G}_{\beta}^{r}(\mathbf{K})}{\mathbb{G}(\mathbf{K})} \right\}.$$

Implicit in our use of (XXIV.37) is the requirement that $g(\delta, \delta) > 0$, and so the expansions (XXIV.39) are subject to the objections which we have noted already in Footnote 11 to Chapter VIII.

Let us glance back now at (XXIV.32)–(XXIV.34). The terms of (XXIV.33) that do not appear explicitly in (XXIV.32) are linear in grad $\overset{\delta}{\rho}$; the terms of (XXIV.34) that do not appear explicitly in (XXIV.33) are linear in grad^2 $\overset{\delta}{\rho}$ and quadratic in grad $\overset{\delta}{\rho}$; and so on. Apart from the error terms, these formulae resemble (XXIV.2)–(XXIV.4) but are less specific. Indeed, the functional dependences allowed by the expansions we have just obtained are far more general than those which the kinetic theory determines because up to now we have treated only analytical problems. When we bring to bear the conditions imposed by the kinetic theory, we expect that the coefficients such as $\mathfrak{P}_{\alpha}^{k}$, $\mathfrak{q}_{\alpha}^{k}$, $\mathfrak{h}_{\alpha}^{k}$, and $\mathfrak{s}_{\alpha}^{k}$ will reduce to special forms such as to permit comparison with (XXIV.3), (XXIV.4), and to other formulae of that kind, and in Sections (ix) and (x) of this chapter we shall show that they do. Anticipating that fact, we lay down the following definition: *Let the functional* $\mathbb{G}$ *satisfy the equations of gross determinism, namely* (XXIII.25) *and* (XXIII.26). *Then an approximate gross determiner of order N is the function that results by omitting the error term* $o(\delta^N)$ *from the right-hand sides of expansions such as* (XXIV.29)$_1$ *and* (XXIV.29)$_2$. The words "such as" in the definition allude to the fact that gross determiners for expectations other than $\mathbf{P}$ and $\mathbf{q}$ may be expanded similarly. The only other examples we shall need are (XXIV.39)$_1$ and (XXIV.39)$_2$, which are written out only as far as order 2.

Early workers were quick to interpret these functions incorrectly as expressing constitutive relations, for, as MAXWELL remarked[14] "... all the equations are expressed in the forms of the differential calculus, in which the phenomena at a given place are connected with the spatial variations of certain quantities at that place, but in which no quantity appears which explicitly involves the condition of things at a finite distance from that place". That is, *approximate gross determiners obey the principle of local action*, expressed in Chapter I. It does not follow that the same is true of $\mathfrak{P}$ and $\mathbf{q}$ themselves, for the terms $o(\delta^N)$ that have been dropped are still functionals of the field $\overset{\delta}{\rho}$.

---

[14] MAXWELL [1879, Appendix].

(v) EXPANSION COEFFICIENTS   445

A parallel from elementary mathematics may help here. When we come to expand sin x in a Taylor series, we find that the sum of any finite number of terms in that series is unbounded as $x \to \pm\infty$. This fact does not represent any property of sin x; rather, it reflects merely the choice of functions in terms of which we decided at the outset to approximate sin x.

(v) *The expansion coefficients*

Still considering purely analytical problems associated with the expansion of grossly determined quantities, we now observe and record some identities satisfied by the coefficients $\mathbb{G}_{\alpha_1 \cdots \alpha_p}^{n_1 \cdots n_p}$ in the basic expansion (XXIV.28)$_1$. These coefficients are functions of the five real variables K_α, $\alpha = 0, 1, \ldots, 4$, and are defined by the derivatives of $\mathbb{G}$ according to the rule (XXIV.28)$_2$. It is natural to ask whether these functions inherit any regularity from that we have already assumed of $\mathbb{G}$. We prove now that they are smooth as functions of $\mathbf{K}$, and, what is more important for later calculations, we derive formulae for their derivatives in terms of the derivatives of $\mathbb{G}$.

By specializing (XXIV.12) to the case in which ψ and ξ are constant functions on $\mathscr{W}$ with values $\mathbf{K}$ and $\mathbf{H}$, respectively, we find that

$$\mathbb{G}(\mathbf{K} + \mathbf{H}) = \mathbb{G}(\mathbf{K}) + \sum_{n=1}^{N} \frac{1}{n!} \partial_\psi^n \mathbb{G}(\mathbf{K})[\mathbf{1}_{\alpha_1}, \ldots, \mathbf{1}_{\alpha_n}] H_{\alpha_1} \cdots H_{\alpha_n} + o(|\mathbf{H}|^N).$$
(XXIV.41)

We have used the fact that $\partial_\psi^n \mathbb{G}(\mathbf{K})[\cdot, \ldots, \cdot]$ is linear in each of its n arguments, and we have written

$$\mathbf{H} = H_\beta \mathbf{1}_\beta, \qquad \text{(XXIV.42)}$$

$\mathbf{1}_\beta$ being the vector with components $1_{\alpha\beta}$, $\alpha = 0, 1, \ldots, 4$, given by

$$1_{\alpha\beta} \equiv \begin{cases} 1 & \text{if } \beta = \alpha, \\ 0 & \text{if } \beta \neq \alpha. \end{cases} \qquad \text{(XXIV.43)}$$

The expansion (XXIV.41) tells us that the function $\mathbf{K} \mapsto \mathbb{G}(\mathbf{K})$ is N times differentiable, and it gives us the following simple formula for its derivatives:

$$\partial_{K_{\alpha_1}} \cdots \partial_{K_{\alpha_n}} \mathbb{G}(\mathbf{K}) = \partial_\psi^n \mathbb{G}(\mathbf{K})[\mathbf{1}_{\alpha_1}, \ldots, \mathbf{1}_{\alpha_n}]. \qquad \text{(XXIV.44)}$$

There is a similar formula for the general function $\mathbb{G}_{\alpha_1 \cdots \alpha_p}^{n_1 \cdots n_p}$. To obtain it, we note that since $\mathbb{G}$ is N times differentiable, $\partial_\psi^p \mathbb{G}$ is $N - p$ times differentiable, so for given functions $\xi^1, \ldots, \xi^p$ on $\mathscr{W}$ we may replace N by $N - p$ and $\mathbb{G}$ by $\partial_\psi^p \mathbb{G}(\cdot)[\xi^1, \ldots, \xi^p]$ in (XXIV.12). If in addition we choose ψ and ξ to be the constants $\mathbf{K}$ and $\mathbf{H}$ again, we obtain the expansion

$$\partial_\psi^p \mathbb{G}(\mathbf{K} + \mathbf{H})[\xi^1, \ldots, \xi^p] = \partial_\psi^p \mathbb{G}(\mathbf{K})[\xi^1, \ldots, \xi^p]$$
$$+ \sum_{n=1}^{N-p} \frac{1}{n!} \partial_\psi^{n+p} \mathbb{G}(\mathbf{K})[\mathbf{1}_{\alpha_1}, \ldots, \mathbf{1}_{\alpha_n}, \xi^1, \ldots, \xi^p] H_{\alpha_1} \cdots H_{\alpha_n} \quad \text{(XXIV.45)}$$
$$+ o(|\mathbf{H}|^{N-p}).$$

This tells us that the function $\mathbf{K} \mapsto \partial_\psi^p \mathsf{G}(\mathbf{K})[\xi^1, \ldots, \xi^p]$ is $N - p$ times differentiable for each fixed set of functions $\xi^1, \ldots, \xi^p$, and it gives us the following formula for its derivatives:

$$\partial_{K_{\alpha_1}} \cdots \partial_{K_{\alpha_n}} \partial_\psi^p \mathsf{G}(\mathbf{K})[\xi^1, \ldots, \xi^p] = \partial_\psi^{n+p} \mathsf{G}(\mathbf{K})[\mathbf{1}_{\alpha_1}, \ldots, \mathbf{1}_{\alpha_n}, \xi^1, \ldots, \xi^p]. \quad \text{(XXIV.46)}$$

This result is slightly more general than we shall need for later applications. We specialize it now by choosing $\xi^1 = \mathbf{J}_{\beta_1}^{n_1}, \ldots, \xi^p = \mathbf{J}_{\beta_p}^{n_p}$. In view of (XXIV.28)$_2$ we obtain the following generalization of (XXIV.44):

$$\partial_{K_{\alpha_1}} \cdots \partial_{K_{\alpha_n}} \mathsf{G}_{\beta_1 \ldots \beta_p}^{n_1 \ldots n_p}(\mathbf{K}) = \partial_\psi^{n+p} \mathsf{G}(\mathbf{K})[\mathbf{1}_{\alpha_1}, \ldots, \mathbf{1}_{\alpha_n}, \mathbf{J}_{\beta_1}^{n_1}, \ldots, \mathbf{J}_{\beta_p}^{n_p}]. \quad \text{(XXIV.47)}$$

This formula and (XXIV.44) provide us with a simple rule for "eliminating ones" from the linear arguments of derivatives of G. For each such $\mathbf{1}_\alpha$ that appears we simply drop it and the corresponding derivative with respect to ψ from the notation and apply instead the derivative ∂_{K_α}. This rule will be used many times in the remainder of the chapter, in which we derive the basic equations in the iterative procedure for approximating gross determiners.

For a simple illustration of this rule let us look back at the definition (XXIV.28)$_2$ of the coefficients in the basic expansion. According to this definition $n_1 > 0, \ldots, n_p > 0$, since only these possibilities appear in the summands in (XXIV.28)$_1$. However, by (XXIV.17), (XXIV.43), and (XVIB.4), $\mathbf{J}_\alpha^n$ reduces to $\mathbf{1}_\alpha$ when $n = 0$, and so (XXIV.28)$_2$ is meaningful for any non-negative integers $n_1, \ldots, n_p$, not only for those which are positive. However, by use of the rule for eliminating ones we may express the coefficients in this more general set in terms of those in the original set. For example, if $n_1 = 0$, then from (XXIV.28)$_2$ and (XXIV.47) we see that

$$\begin{aligned} \mathsf{G}_{\alpha_1 \alpha_2 \ldots \alpha_p}^{n_1 n_2 \ldots n_p}(\mathbf{K}) &= \partial_\psi^p \mathsf{G}(\mathbf{K})[\mathbf{1}_{\alpha_1}, \mathbf{J}_{\alpha_2}^{n_2}, \ldots, \mathbf{J}_{\alpha_p}^{n_p}], \\ &= \partial_{K_{\alpha_1}} \mathsf{G}_{\alpha_2 \ldots \alpha_p}^{n_2 \ldots n_p}(\mathbf{K}). \end{aligned} \quad \text{(XXIV.48)}$$

Thus, we simply drop the block $\binom{n_1}{\alpha_1}$ from the function in question and effect instead a derivative with respect to K_{α_1}. By applying this rule repeatedly we may eliminate all blocks $\binom{n_k}{\alpha_k}$ in which $n_k = 0$ and so obtain a repeated derivative with respect to $\mathbf{K}$ of one of the coefficients in (XXIV.28)$_1$.

(vi) Derivation of the iterative system for the gross determiner when $\mathbf{b} = 0$

The expansion (XXIV.28) is valid for any functional G and any field $\boldsymbol{\rho}$ provided both be sufficiently smooth. It expresses nothing more than one way to approximate $\mathsf{G}(\overset{\delta}{\mathfrak{S}}\boldsymbol{\rho})$, whether or not G be subject to such restrictions as the kinetic theory provides[15]. However, that the corresponding relations (XXIV.32)–(XXIV.34) should reduce to such simple and special forms as those in (XXIV.2)–(XXIV.4), is a consequence of further conditions on $\mathbb{G}$, namely the requirement that it satisfy the equations of gross determinism. So far we have not used this fact. We do so now.

Throughout most of the rest of this chapter we shall suppose that $\mathbf{b} = \mathbf{0}$. Then the equations of gross determinism are (XXIII.25) and (XXIII.26). In Section (x) we shall restore $\mathbf{b}$, and so there we use the more general system (XXIII.59).

We wish to deduce from (XXIII.25) and (XXIII.26) an iterative system for $\mathbb{G}(\mathbf{K})$ and $\mathbb{G}_{\alpha_1 \ldots \alpha_p}^{n_1 \ldots n_p}(\mathbf{K})$. To begin, we rewrite these governing equations in the somewhat more explicit form

$$\mathbb{D}\mathbb{G} = \mathbb{C}\mathbb{G},$$
$$\psi_\alpha(0) = \mathfrak{m} \int I_\alpha \mathbb{G}(\psi), \qquad \alpha = 0, 1, \ldots, 4; \tag{XXIV.49}$$

here $\mathbb{D}$ is the integro-differential operator defined as follows:

$$\mathbb{D}\,\mathbb{G}(\psi) \equiv \partial_{\psi} \underset{\mathbf{r}}{\mathbb{G}}(\psi(\mathbf{r})) \left[v_d \psi_{,d}(\mathbf{r}) - \mathfrak{m} \int v_d \mathbf{I} \, \partial_\psi \underset{\mathbf{s}}{\mathbb{G}}(\psi(\mathbf{r}+\mathbf{s}))[\psi_{,d}(\mathbf{r}+\mathbf{s})] \right]. \tag{XXIV.50}$$

If $\boldsymbol{\phi}$ is any smooth field, and if we set $\psi(\mathbf{r}) = \boldsymbol{\phi}(\mathbf{x}+\mathbf{r})$, $\mathbf{x}$ being a fixed place, then by (XXIII.22) and (XXIII.11) we see that (XXIV.49) becomes (XXIII.25) and (XXIII.26). In the same way we can show that (XXIII.25) and (XXIII.26) imply (XXIII.49). It is the latter system that we shall use.

Let us consider the necessary condition on $\mathbb{G}$ obtained by evaluating the system (XXIV.49) at a constant field $\psi = \mathbf{K}$. Since grad $\psi = \mathbf{0}$ in this case, the fact that $\mathbb{D}\mathbb{G}$ is linear in grad ψ implies that

$$\mathbb{C}\mathbb{G}(\mathbf{K}) = 0, \qquad K_\alpha = \mathfrak{m} \int I_\alpha \mathbb{G}(\mathbf{K}). \tag{XXIV.51}$$

By $(VI.21)_2$ and (VIII.7) the solution of these equations is

$$\mathbb{G}(\mathbf{K}) = F_M, \tag{XXIV.52}$$

F_M being the unique Maxwellian density whose principal moment is $\mathbf{K}$. Of course this conclusion asserts the universal property of all gross determiners that we deduced in Section (iv) of Chapter XXIII.

We obtained (XXIV.51) by choosing for ψ in (XXIV.49) a constant. We can derive similar necessary conditions on $\mathbb{G}$ by differentiating each side of (XXIV.49) with respect to ψ and then setting $\psi = \mathbf{K}$ once again. A convenient

---

[15] Those already familiar with the kinetic theory will realize that this contrasts sharply with the procedure of ENSKOG. There the Maxwell-Boltzmann equation is exploited from the very outset of the iterative scheme, and equations of the same forms as (XXIV.32)–(XXIV.34) arise only after considerable calculation.

way to do so is to replace ψ by $\psi + \lambda \xi$, ξ being fixed, then differentiate each side of (XXIV.49) with respect to λ and set $\lambda = 0$. This gives us

$$\partial_\psi^2 \underset{r}{G}(\psi(r))\left[\xi(r), v_d \psi_{,d}(r) - \mathfrak{m} \int v_d I \, \partial_\psi \underset{s}{G}(\psi(r+s))[\psi_{,d}(r+s)]\right]$$

$$+ \partial_\psi \underset{r}{G}(\psi(r))\left[v_d \xi_{,d}(r) - \mathfrak{m} \int v_d I \, \partial_\psi^2 \underset{s}{G}(\psi(r+s))[\xi(r+s), \psi_{,d}(r+s)]\right.$$

$$\left. - \mathfrak{m} \int v_d I \, \partial_\psi \underset{s}{G}(\psi(r+s))[\xi_{,d}(r+s)]\right]$$

$$= 2\mathbb{C}(\underset{r}{G}(\psi(r)), \partial_\psi \underset{r}{G}(\psi(r))[\xi(r)]), \qquad \text{(XXIV.53)}$$

$$\xi_\alpha(0) = \mathfrak{m} \int I_\alpha \, \partial_\psi \underset{r}{G}(\psi(r))[\xi(r)].$$

We have two fields here, namely ψ and ξ, which we may choose freely. Let us note that the first derivative $\partial_\psi G(\psi)[\xi]$ appears explicitly in these equations. Let us note also, from (XXIV.28)$_2$, that $G_\alpha^n(K)$ is simply $\partial_\psi G(K)[J_\alpha^n]$. This suggests that we may obtain a system of equations for G_α^n by choosing $\psi = K$ and $\xi = J_\alpha^n$ in (XXIV.53). We illustrate the calculation when $\psi = K$ and $\xi = J_\alpha^k$, this being the case in which $n = 1$, and hence the multi-index n reduces to a single index k.

By (XXIV.17) and (XXIV.43) we see that $J_\alpha^k(r) = 1_\alpha r_k$. Thus

$$\psi_{\beta,k}(r) = 0, \qquad \xi_\beta(0) = 0,$$
$$\xi_\beta(r+s) = 1_{\beta\alpha} r_k + 1_{\beta\alpha} s_k, \qquad \xi_{\beta,m}(r) = \xi_{\beta,m}(r+s) = 1_{\beta\alpha} \delta_{mk}. \qquad \text{(XXIV.54)}$$

We place these into (XXIV.53) and note that the non-vanishing terms on the left-hand side of the first may be simplified as follows:

$$\partial_\psi \underset{r}{G}(\psi(r))\left[v_d \xi_{,d}(r) - \mathfrak{m} \int v_d I \, \partial_\psi \underset{s}{G}(\psi(r+s))[\xi_{,d}(r+s)]\right]$$

$$= \partial_\psi \underset{r}{G}(K)\left[v_k 1_\alpha - \mathfrak{m} \int v_k I_\beta 1_\beta \partial_\psi \underset{s}{G}(K)[1_\alpha]\right], \qquad \text{(XXIV.55)}$$

$$= v_k \, \partial_\psi G(K)[1_\alpha] - \partial_\psi G(K)[1_\beta] \mathfrak{m} \int v_k I_\beta \, \partial_\psi G(K)[1_\alpha].$$

Using (XXIV.52), the definition of G_α^k, and the rule (XXIV.44) for "eliminating ones", we find that (XXIV.53) simplifies to

$$v_k \, \partial_{K_\alpha} F_M - \partial_{K_\beta} F_M \mathfrak{m} \int v_k I_\beta \, \partial_{K_\alpha} F_M = 2\mathbb{C}(F_M, G_\alpha^k(K)),$$
$$\text{(XXIV.56)}$$
$$0 = \mathfrak{m} \int I_\beta \, G_\alpha^k(K).$$

(vi) ITERATIVE SYSTEM WHEN **b** = 0   449

The first of these is simply a linear integral equation for G_α^k. The second equation tells us that the principal moment of G_α^k is null. Before discussing the existence and uniqueness of solutions of this system, we shall indicate how similar equations may be derived for each of the coefficients $G_{\alpha_1 \cdots \alpha_p}^{n_1 \cdots n_p}$.

We may generalize the preceding calculation in two ways. First we may retain (XXIV.53) and simply make different choices of ψ and ξ. For example, if we choose $\psi = \mathbf{K}$ and $\xi(\mathbf{r}) = \mathbf{J}_\alpha^{km}(\mathbf{r}) = 1_\alpha r_k r_m$, we obtain a system of equations for G_α^{km}. This system is

$$v_k G_\alpha^m + v_m G_\alpha^k - G_\beta^k \mathfrak{m} \int v_m I_\beta \, \partial_{K_\alpha} F_M - G_\beta^m \mathfrak{m} \int v_k I_\beta \, \partial_{K_\alpha} F_M$$

$$- \partial_{K_\beta} F_M \mathfrak{m} \int v_k I_\beta G_\alpha^m - \partial_{K_\beta} F_M \mathfrak{m} \int v_m I_\beta G_\alpha^k \qquad \text{(XXIV.57)}$$

$$= 2\mathfrak{C}(F_M, G_\alpha^{km}),$$

$$0 = \mathfrak{m} \int I_\beta G_\alpha^{km}.$$

The first equation is a linear integral equation for G_α^{km} if we presume that G_α^k is known. The second equation tells us that the principal moment of G_α^{km} is also null.

Another way to proceed is to differentiate each side of (XXIV.53) with respect to ψ, ξ being held fixed, and so obtain a system generalizing (XXIV.53) in the same way as (XXIV.53) generalizes (XXIV.49). Specifically, we replace ξ by ξ^1 and ψ by $\psi + \lambda \xi^2$ in (XXIV.53), then differentiate with respect to λ and set $\lambda = 0$. The resulting equations involve the second derivative $\partial_\psi^2 G(\psi)[\xi^1, \xi^2]$ explicitly. Since $G_{\alpha\beta}^{km}(\mathbf{K}) = \partial_\psi^2 G(\mathbf{K})[\mathbf{J}_\alpha^k, \mathbf{J}_\beta^m]$, we may obtain a system for $G_{\alpha\beta}^{km}$ by choosing $\psi = \mathbf{K}$, $\xi^1 = \mathbf{J}_\alpha^k$, $\xi^2 = \mathbf{J}_\beta^m$. This system is

$$v_k \partial_{K_\alpha} G_\beta^m + v_m \partial_{K_\beta} G_\alpha^k - \partial_{K_\gamma} G_\beta^m \mathfrak{m} \int v_k I_\gamma \, \partial_{K_\alpha} F_M - \partial_{K_\gamma} G_\alpha^k \mathfrak{m} \int v_m I_\gamma \, \partial_{K_\beta} F_M$$

$$- G_\gamma^m \mathfrak{m} \int v_k I_\gamma \, \partial_{K_\beta} \partial_{K_\alpha} F_M - G_\gamma^k \mathfrak{m} \int v_m I_\gamma \, \partial_{K_\beta} \partial_{K_\alpha} F_M \qquad \text{(XXIV.58)}$$

$$- \partial_{K_\gamma} F_M \mathfrak{m} \int v_k I_\gamma \, \partial_{K_\alpha} G_\beta^m - \partial_{K_\gamma} F_M \mathfrak{m} \int v_m I_\gamma \, \partial_{K_\beta} G_\alpha^k$$

$$= 2\mathfrak{C}(F_M, G_{\alpha\beta}^{km}) + 2\mathfrak{C}(G_\alpha^k, G_\beta^m),$$

$$0 = \mathfrak{m} \int I_\gamma G_{\alpha\beta}^{km}.$$

If G_α^k is known, the first of these is a linear integral equation for $G_{\alpha\beta}^{km}$. The second equation expresses the fact that the principal moment of $G_{\alpha\beta}^{km}$ is null.

450   XXIV. THE METHOD OF STRETCHED FIELDS

This procedure can be applied repeatedly so as to generate a general iterative system for all of the functions $\mathbb{G}^{n_1\cdots n_p}_{\alpha_1\cdots\alpha_p}$. The calculations, though elementary and straightforward, are lengthy, so we shall only set down the final results, relegating to Appendix A the steps involved in arriving at them:

$$\sum\nolimits^1 v_d\, \mathbb{G}^{a n_2 \cdots n_p}_{\alpha_1 \alpha_2 \cdots \alpha_p} - \sum\nolimits^2 \mathbb{G}^{c n_{q+1} \cdots n_p}_{\beta a_{q+1} \cdots \alpha_p} \mathfrak{m} \int v_d I_\beta\, \mathbb{G}^{b_1 \cdots b_q}_{\alpha_1 \cdots \alpha_q}$$

$$- \sum\nolimits^3 \mathbb{C}(\mathbb{G}^{n_1 \cdots n_q}_{\alpha_1 \cdots \alpha_q},\, \mathbb{G}^{n_{q+1} \cdots n_p}_{\alpha_{q+1} \cdots \alpha_p}) = 2\mathbb{C}(F_{\mathbf{M}},\, \mathbb{G}^{n_1 \cdots n_p}_{\alpha_1 \cdots \alpha_p}),$$

$$0 = \mathfrak{m} \int I_\beta\, \mathbb{G}^{n_1 \cdots n_p}_{\alpha_1 \cdots \alpha_p}. \qquad (\text{XXIV}.59)$$

The second equation here tells us that the principal moment of each of the functions $\mathbb{G}^{n_1\cdots n_p}_{\alpha_1\cdots\alpha_p}$ is null. If the left-hand side of the first equation were known, perhaps by solutions of preceding equations in the system, then we should have a linear integral equation to solve for $\mathbb{G}^{n_1\cdots n_p}_{\alpha_1\cdots\alpha_p}$. In the next section we will prove this fact, and also we shall examine the questions of existence and uniqueness for such integral equations.

The symbols $\sum^1$, $\sum^2$, and $\sum^3$ denote operators of summation. They are somewhat complicated but do not involve any of the coefficient functions. In order to detail them we require some additional notation. If f is a function of m objects $\xi_1, \ldots, \xi_m$, then we set

$$\operatorname*{sym}_{\xi_1,\ldots,\xi_m} f(\xi_1, \ldots, \xi_m) \equiv \frac{1}{m!} \sum\nolimits_P f(\eta_1, \ldots, \eta_m), \qquad (\text{XXIV}.60)$$

$\sum_P$ being a sum over all permutations $(\eta_1, \ldots, \eta_m)$ of $(\xi_1, \ldots, \xi_m)$. We call sym the *symmetrizer*. For any function f, sym f is symmetric in its arguments, and if f is already symmetric, then sym $f = f$. In terms of the symmetrizer, the operators $\sum^1$, $\sum^2$, and $\sum^3$ may be written

$$\sum\nolimits^1 \equiv \operatorname*{sym}_{\binom{n_1}{\alpha_1},\ldots,\binom{n_p}{\alpha_p}} pn_1 \operatorname*{sym}_{n_1},$$

$$\sum\nolimits^2 \equiv \operatorname*{sym}_{\binom{n_1}{\alpha_1},\ldots,\binom{n_p}{\alpha_p}} \sum_{q=1}^{p} \binom{p}{q} q\, \operatorname*{sym}_{\binom{n_1}{\alpha_1},\ldots,\binom{n_q}{\alpha_q}} \operatorname*{sym}_{n_1} \cdots$$

$$\cdots \operatorname*{sym}_{n_q} \sum_{a_1=0}^{n_1} \cdots \sum_{a_q=0}^{n_q} \binom{n_1}{a_1} \cdots \binom{n_q}{a_q} a_1\, \operatorname*{sym}_{a_1}, \qquad (\text{XXIV}.61)$$

$$\sum\nolimits^3 \equiv \operatorname*{sym}_{\binom{n_1}{\alpha_1},\ldots,\binom{n_p}{\alpha_p}} (1 - \delta_{p1}) \sum_{q=1}^{p-1} \binom{p}{q}.$$

(vi) ITERATIVE SYSTEM WHEN $b = 0$

Connecting these with (XXIV.59) we have the following decompositions of multi-indices (*cf.* Appendix B of Chapter XVI):

$$n_1 = da,$$
$$= a_1 b_1,$$
$$n_2 = a_2 b_2,$$
$$\vdots \qquad\qquad\qquad\qquad\qquad\text{(XXIV.62)}$$
$$n_p = a_p b_p,$$
$$a_1 = dc,$$
$$e = ca_2 \ldots a_q.$$

We have derived the equations in the iterative system (XXIV.59) as necessary conditions that must be satisfied by a solution of the equations of gross determinism. It is important to ask whether, or in what sense, they might also be sufficient. A more natural way of exploiting the equations of gross determinism, one which is more closely related to our method of approximating G, is to let ψ be an arbitrary field, then evaluate each side of (XXIV.49) at $\overset{\delta}{\psi}$ as defined by (XXIV.10)$_2$ so as to give

$$\mathbb{D}\, G(\overset{\delta}{\psi}) = \mathbb{C}\, G(\overset{\delta}{\psi}), \qquad \overset{\delta}{\psi}_\alpha(0) = \mathfrak{m}\int I_\alpha\, G(\overset{\delta}{\psi}), \qquad\text{(XXIV.63)}$$

and finally to expand each side of this system, to within $o(\delta^N)$, in a power series in δ. Since $\mathbb{D}\, G$ is a gross determiner, we may derive for it an expansion analogous to (XXIV.28)$_1$, with coefficients $\mathbb{D}^{n_1 \ldots n_p}_{\alpha_1 \ldots \alpha_p}$, say, rather than $G^{n_1 \ldots n_p}_{\alpha_1 \ldots \alpha_p}$. Using this expansion and setting $K = \psi(0)$, we obtain the following expanded form of (XXIV.63):

$$\mathbb{C}\, G(K) + \sum_{n=1}^{N}\sum_{p=1}^{n}\sum_{\substack{n_1+\cdots+n_p=n \\ n_m>0}} \frac{\delta^n}{p!\, n_1! \cdots n_p!}\psi_{\alpha_1,n_1}(0)\cdots$$
$$\cdots \psi_{\alpha_p,n_p}(0)[2\mathbb{C}(G(K), G^{n_1 \ldots n_p}_{\alpha_1 \ldots \alpha_p}(K))$$
$$+ \sum{}^3 \mathbb{C}(G^{n_1 \ldots n_q}_{\alpha_1 \ldots \alpha_q}(K), G^{n_{q+1}\ldots n_p}_{\alpha_{q+1}\ldots \alpha_p}(K)) - \mathbb{D}^{n_1 \ldots n_p}_{\alpha_1 \ldots \alpha_p}(K)] = o(\delta^N), \qquad\text{(XXIV.64)}$$
$$\mathfrak{m}\int I_\beta\, G(K) + \sum_{n=1}^{N}\sum_{p=1}^{n}\sum_{\substack{n_1+\cdots+n_p=n \\ n_m>0}} \frac{\delta^n}{p!\, n_1! \cdots n_p!}\psi_{\alpha_1,n_1}(0)\cdots$$
$$\cdots \psi_{\alpha_p,n_p}(0)\mathfrak{m}\int I_\beta\, G^{n_1 \ldots n_p}_{\alpha_1 \ldots \alpha_p}(K) = K_\beta + o(\delta^N).$$

Since ψ is arbitrary, K and the derivatives $\psi_{\alpha,n}(0)$ are arbitrary real variables, subject of course to the natural symmetry of mixed derivatives. As a result of this, (XXIV.64) is equivalent, to within $o(\delta^N)$, to (XXIV.51) and the system

$$2\mathbb{C}(G, G^{n_1 \ldots n_p}_{\alpha_1 \ldots \alpha_p}) = \mathbb{D}^{n_1 \ldots n_p}_{\alpha_1 \ldots \alpha_p} - \sum{}^3 \mathbb{C}(G^{n_1 \ldots n_q}_{\alpha_1 \ldots \alpha_q}, G^{n_{q+1}\ldots n_p}_{\alpha_{q+1}\ldots \alpha_p}),$$
$$0 = \mathfrak{m}\int I_\beta\, G^{n_1 \ldots n_p}_{\alpha_1 \ldots \alpha_p}. \qquad\text{(XXIV.65)}$$

A detailed expansion of $\mathbb{D}\mathbb{G}$ shows that $\mathbb{D}_{\alpha_1 \ldots \alpha_p}^{n_1 \ldots n_p}$ reduces to the first two terms on the left-hand side of (XXIV.59). Thus, *the general iterative system* (XXIV.51) *and* (XXIV.59) *is equivalent, to within* $o(\delta^N)$, *to the equations of gross determinism.* Indeed, this was the way MUNCASTER first derived the iterative system[16]. We prefer, however, to proceed as we did at the beginning of the section since we thus obtain in a simple way the first few equations in the general system.

(vii)   *Structure of the iterative system. Proof of effectiveness*

Because of the complexity of the operators $\sum^1$, $\sum^2$, and $\sum^3$ and the multi-index decompositions (XXIV.62), the general iterative system is quite complicated. It is not at all apparent how the successive equations determine the functions $\mathbb{G}_{\alpha_1 \ldots \alpha_p}^{n_1 \ldots n_p}$, if indeed they do. We prove now that *the iterative procedure is effective:* Each equation in the system (XXIV.59) has a unique solution.

We define the *order* of $\mathbb{G}_{\alpha_1 \ldots \alpha_p}^{n_1 \ldots n_p}$ as $(p, n_1, \ldots, n_p)$, it being assumed that each n_k is positive. Moreover, we say that the order of $\mathbb{G}_{\beta_1 \ldots \beta_q}^{m_1 \ldots m_q}$ is *less than* that of $\mathbb{G}_{\alpha_1 \ldots \alpha_p}^{n_1 \ldots n_p}$ if (a) $q < p$, or (b) if $p = q$, $m_1 \leq n_1, \ldots, m_p \leq n_p$ and for at least one integer k, $m_k < n_k$. For completeness we call $\mathbb{G}(\mathbf{K}) = F_M$ the iterate of 0th order.

The iterates of order $(p, n_1, \ldots, n_p)$ appear explicitly in (XXIV.59)$_2$ and on the right-hand side of (XXIV.59)$_1$. Let us see if any iterates of this order appear on the left-hand side of (XXIV.59)$_1$. The function $\mathbb{G}_{\alpha_1 \alpha_2 \ldots \alpha_p}^{a n_2 \ldots n_p}$ will be one only if $a = n_1$. However, (XXIV.62)$_1$ shows that $a = n_1 - 1$, so the first term on the left-hand side of (XXIV.59)$_1$ is determined by iterates of lower order. The order of $\mathbb{G}_{\alpha_1 \ldots \alpha_q}^{b_1 \ldots b_q}$ will be $(p, n_1, \ldots, n_p)$ only if $q = p$ and $b_1 = n_1, \ldots, b_p = n_p$. However, (XXIV.62)$_5$ shows that $a_1 \geq 1$, and so by (XXIV.62)$_2$ we see that $b_1 = n_1 - a_1 \leq n_1 - 1$. Hence this function is determined by iterates of lower order. Next we examine $\mathbb{G}_{\beta \alpha_{q+1} \ldots \alpha_p}^{e n_{q+1} \ldots n_p}$. This is an iterate of order $(p, n_1, \ldots, n_p)$ only if $q = 1$ and $e = n_1$. But when $q = 1$, (XXIV.62)$_6$ shows that $e = c$. Moreover, (XXIV.62)$_5$ shows that $c = a_1 - 1$ and (XXIV.62)$_2$ shows that $a_1 \leq n_1$. Hence $e \leq n_1 - 1$, and so this function is also an iterate of lower order. Finally we note from (XXIV.61)$_3$ that $1 \leq q \leq p - 1$ in the third term on the left-hand side of (XXIV.59)$_1$, and so this term is also determined by iterates of order less than $(p, n_1, \ldots, n_p)$. We conclude that we may write the iterative system in the simple form

$$\mathbb{C}(F_M, \mathbb{G}_{\alpha_1 \ldots \alpha_p}^{n_1 \ldots n_p}) = H_{\alpha_1 \ldots \alpha_p}^{n_1 \ldots n_p}, \quad \int I_\beta \mathbb{G}_{\alpha_1 \ldots \alpha_p}^{n_1 \ldots n_p} = 0, \quad \text{(XXIV.66)}$$

with $H_{\alpha_1 \ldots \alpha_p}^{n_1 \ldots n_p}$ determined completely by iterates of order less than $(p, n_1, \ldots, n_p)$. Therefore, when we come to solve (XXIV.66) iteratively, we may take $H_{\alpha_1 \ldots \alpha_p}^{n_1 \ldots n_p}$ as known.

The system (XXIV.66) is precisely the same as one that we considered already in HILBERT's work, namely (XXII.17). Our analysis of that problem

---

[16] MUNCASTER [1975, Chapter VIII].

(vii)   PROOF OF EFFECTIVENESS   453

shows that *for the molecular models specified in the theorem of Chapter XXII,* (XXIV.66) *has a unique solution* $\mathbb{G}_{\alpha_1 \cdots \alpha_p}^{n_1 \cdots n_p}$ *if and only if*

$$\int I_\beta H_{\alpha_1 \cdots \alpha_p}^{n_1 \cdots n_p} = 0. \tag{XXIV.67}$$

In order to complete the proof of effectiveness we must show that this condition is satisfied. In HILBERT's procedure it was not satisfied identically but instead gave rise to the field equations (XXII.16). In the present procedure we shall see that this is not the case. Indeed we prove now the **Lemma**: (XXIV.67) *is satisfied identically.*

Proof. Since each of the iterates $\mathbb{G}_{\beta_1 \cdots \beta_r}^{m_1 \cdots m_r}$ has null principal moment, and since the principal moment of F_M is K, we see from (XXIV.48) that

$$\mathfrak{m} \int I_\alpha \, \mathbb{G}_{\beta a_{q+1} \cdots \alpha_p}^{e n_{q+1} \cdots n_p}(\mathbf{K}) = \begin{cases} \mathfrak{m} \int I_\alpha \, \partial_{K_\beta} F_M & \text{if } q = p, \ e = 0, \\ 0 & \text{otherwise,} \end{cases} \tag{XXIV.68}$$

$$= \begin{cases} \delta_{\alpha\beta} & \text{if } q = p, \ e = 0, \\ 0 & \text{otherwise.} \end{cases}$$

When $q = p$ and $e = 0$, we see from (XXIV.62)$_6$ that $c = a_2 = \ldots = a_p = 0$ and so by (XXIV.62)$_5$, $a_1 = 1$. Therefore the term in $\sum^2$ corresponding to these choices is

$$\binom{n_1}{\alpha_1}, \ldots, \binom{n_p}{\alpha_p} \quad \text{sym} \quad pn_1 \quad \text{sym}, \tag{XXIV.69}$$

and this is precisely $\sum^1$. Thus, since $a_1 = 1, a_2 = \ldots = a_p = 0$ imply that $b_1 = a$, $b_2 = n_2, \ldots, b_p = n_p$, we see that

$$\int I_\alpha \sum{}^2 \, \mathbb{G}_{\beta a_{q+1} \cdots \alpha_p}^{e n_{q+1} \cdots n_p}(\mathbf{K}) \mathfrak{m} \int v_d I_\beta \, \mathbb{G}_{\alpha_1 \cdots \alpha_q}^{b_1 \cdots b_q}(\mathbf{K})$$

$$= \sum{}^1 \delta_{\alpha\beta} \int v_d I_\beta \, \mathbb{G}_{\alpha_1 \hat{\alpha}_2 \cdots \alpha_p}^{a n_2 \cdots n_p}(\mathbf{K}), \tag{XXIV.70}$$

$$= \int I_\alpha \sum{}^1 v_d \, \mathbb{G}_{\alpha_1 \hat{\alpha}_2 \cdots \alpha_p}^{a n_2 \cdots n_p}(\mathbf{K}).$$

In view of (VII.22), the last result is equivalent to (XXIV.67).   △

The crucial step in our proof of effectiveness has been the verification of (XXIV.67). It might seem surprisingly convenient that that condition is satisfied identically, but in fact it should not be. Indeed, we can easily trace the origins of (XXIV.67) back to an important property of $\mathbb{D}$ which we have already discussed in Chapter XXIII. Since $H_{\alpha_1 \cdots \alpha_p}^{n_1 \cdots n_p}$ is one-half the right-hand side of (XXIV.65)$_1$, we see from (VII.22) that (XXIV.67) is equivalent to the statement that the principal moment of $\mathbb{D}_{\alpha_1 \cdots \alpha_p}^{n_1 \cdots n_p}$ vanishes. Moreover, since the functions $\mathbb{D}_{\alpha_1 \cdots \alpha_p}^{n_1 \cdots n_p}$ are the coefficients in the basic expansion of $\mathbb{D}\mathbb{G}$, obtained by replacing $\mathbb{G}$ by $\mathbb{D}\mathbb{G}$ in (XXIV.28), we see that (XXIV.67) is equivalent, to within $o(\delta^N)$, to the statement that the principal moment of $\mathbb{D}\mathbb{G}$ vanishes. But that is just what (XXIV.49)$_2$ and the result stated after (XXIII.64) assert. The result (XXIV.67) is nothing more than an iterative statement of this property of $\mathbb{D}$.

The result just obtained casts powerful light on the difference between HILBERT's method and the one we are presently developing so as to derive all results that ENSKOG's method, if it could be rendered fully explicit, would deliver. Not only does HILBERT's method aim to determine a class of solutions of the Maxwell-Boltzmann equation, but also it requires at each stage that we solve an initial-value problem for a system of partial differential equations. In effect it proceeds through solutions of more and more complicated problems of gas flow. Only in surveying its final results do we see that those problems pertain to grossly determined solutions. The method of stretched fields, on the contrary, does not solve any problem of gas dynamics. It aims to approximate gross determiners directly, setting aside for later consideration such calculation of gas flows as we may desire to carry out. It approximates the solutions of the equations of gross determinism in terms of simpler grossly determined functions, namely, functions of the gradients of the principal moment. *Thus it deals directly with grossly determined functions, and with them alone. The nasty problem of compatibility that arises at each stage of HILBERT's process is outflanked from the start.*

We might seek to use the method of stretched fields to prove an existence theorem for the equations of gross determinism. To do so, we should replace the finite sum in (XXIV.28)$_1$ by an infinite sum in which n ranges from 1 to ∞ and then look for a solution G of this form. The proof of effectiveness ensures that each of the coefficients $\mathsf{G}_{\alpha_1 \cdots \alpha_p}^{n_1 \cdots n_p}$ exists and is unique, and so we reduce the problem to proof that the infinite series converges. Concerning this last step, unfortunately, just as little is presently known[17] as for the procedures of HILBERT, ENSKOG, and GRAD.

Any systematic iterative procedure may be interpreted as giving the formal steps to be used in a proof that the quantities which it is designed to approximate do exist. Unfortunately most of these procedures in the kinetic theory depend critically on this interpretation; for most of them it will not be until such a proof of existence shall have been given that we shall know what is being approximated. The method of stretched fields, however, is different. From the outset we know what we desire to approximate, namely, gross determiners. Moreover, unlike all other iterative procedures, the method of stretched fields

---

[17] Recently the convergence of a restricted version of ENSKOG's process for a model equation has been proved by J. SCHRÖTER in "The complete Chapman-Enskog procedure for the Fokker-Planck equation", *Archive for Rational Mechanics and Analysis* **66**, 183–199 (1977). Although his results do not bear directly on the subject of this book, they are important because the limit of ENSKOG's process which SCHRÖTER has found is grossly determined, the corresponding gross determiner being given explicitly by his Eq. (32). As far as we are aware, his is the first analysis of any kind in which gross determiners have been shown rigorously to exist, and other than the development in this book it is the only published analysis in which the concept of a grossly determined solution is introduced. In regard to the moments the idea is much older; for molecular densities in the special case when $\mathbf{b} = \mathbf{0}$ MUNCASTER introduced it in his unpublished thesis of 1975. *Cf.* Footnotes 14 and 15 to Chapter VIII.

need not entail a proof of convergence. Indeed, we are offered an alternative: We may seek to prove directly that the equations of gross determinism have a solution and that it is smooth in a neighborhood of the constant fields. If so, we may approximate it using nothing more than a finite Taylor expansion. It is in this spirit that we have approached the entire problem of approximation.

At present a proof of existence is wanting. Thus we must fall back upon the iterative procedure for such conclusions as we hope to obtain. One conclusion is suggested by the fact that the coefficients $\mathbb{G}_{\alpha_1 \cdots \alpha_p}^{n_1 \cdots n_p}$ are unique. We state it as the

Conjecture on Uniqueness. *Within the class of functionals $\mathbb{G}$ which are analytic in $\mathbf{\rho}$ in a neighborhood of the constant fields, there is at most one solution of the equations of gross determinism.*

In Section (i) of Chapter XXVI we shall examine some of the implications of this conjecture.

(viii) Properties of some of the expansion coefficients

Recall that (XXIV.56) forms a system of linear integral equations for $\mathbb{G}_\alpha^k$. We shall not be able to solve these explicitly for general molecular models, but we can display some properties of the solution, properties that serve to simplify considerably the approximate gross determiners (XXIV.32)–(XXIV.34).

We begin by rewriting the system in terms of ρ, $\mathbf{u}$, and ε rather than $\mathbf{K}$, these two sets of variables being related through (XXIV.31). The functions $\mathbb{G}_\alpha^k$ are those coefficients in the basic expansion (XXIV.28)$_1$ that determine the part of $\mathbb{G}$ that is linear in the gradient grad $\mathbf{\rho}$. We can express this part of $\mathbb{G}$ as a linear function of grad ρ, grad $\mathbf{u}$, and grad ε by introducing functions $\mathbb{F}_\alpha^k$ that satisfy

$$\mathbb{G}_\alpha^a \rho_{,a} = \mathbb{F}_0^a \rho_{,a} + \mathbb{F}_b^a u_{b,a} + \mathbb{F}_4^a \varepsilon_{,a} \qquad \text{(XXIV.71)}$$

identically for all ρ, $\mathbf{u}$, ε, and $\mathbf{\rho}$ related through (VI.23). We find that

$$\mathbb{F}_0^k = \mathbb{G}_0^k + u_a \mathbb{G}_a^k + (2\varepsilon + u^2)\mathbb{G}_4^k,$$

$$\mathbb{F}_m^k = \rho \mathbb{G}_m^k + 2\rho u_m \mathbb{G}_4^k, \qquad \text{(XXIV.72)}$$

$$\mathbb{F}_4^k = 2\rho \mathbb{G}_4^k.$$

Since each of these functions is a linear combination of the $\mathbb{G}_\alpha^k$, by forming the same linear combinations of the equations in (XXIV.56) we obtain a system for $\mathbb{F}_\alpha^k$. The calculation is routine but long, so we simply present the final result. For those who may wish to verify it, in Appendix B we list explicit formulae for many of the terms in (XXIV.56). The linear integral equations for determining $\mathbb{F}_\alpha^k$ are

$$2\mathbb{C}(F_M, \mathbb{F}_0^k) = 0,$$

$$2\mathbb{C}(F_M, \mathbb{F}_m^k) = \frac{3}{2\varepsilon} c_{\{k} c_{m\}} F_M,$$

$$2\mathbb{C}(F_M, \mathbb{F}_4^k) = \frac{3}{2\varepsilon}\left(\frac{c^2}{2\varepsilon} - \frac{5}{3}\right) c_k F_M, \qquad \text{(XXIV.73)}$$

$$\int I_\beta \mathbb{F}_\alpha^k = 0.$$

We note first that the unique solution for $\mathbb{F}_0^k$ is

$$\mathbb{F}_0^k = 0. \qquad \text{(XXIV.74)}$$

Next we note that since the right-hand side of (XXIV.73)$_2$ is symmetric and traceless, $\mathbb{F}_k^m - \mathbb{F}_m^k$ and $\mathbb{F}_a^a$ satisfy the same integral equation as does $\mathbb{F}_0^k$. Therefore

$$\mathbb{F}_k^m = \mathbb{F}_m^k,$$
$$\mathbb{F}_a^a = 0, \qquad \text{(XXIV.75)}$$

so $\mathbb{F}_m^k$ is also symmetric and traceless. Next we let $\mathbf{Q}$ be any orthogonal tensor. Since a Maxwellian density depends upon $\mathbf{c}$ only through its magnitude, we know that $F_M \circ \mathbf{Q} = F_M$, and so from the orthogonal invariance of $\mathbb{C}(\, , \,)$, as expressed by (VII.29), we see that

$$\mathbb{C}(F_M, \mathbb{F}_k^m \circ \mathbf{Q}) = \mathbb{C}(F_M \circ \mathbf{Q}, \mathbb{F}_k^m \circ \mathbf{Q}),$$
$$= \mathbb{C}(F_M, \mathbb{F}_k^m) \circ \mathbf{Q},$$
$$= \frac{3}{4\varepsilon}((Q_{ka}c_a)(Q_{mb}c_b) - \tfrac{1}{3}c^2 \, \delta_{km})F_M,$$
$$= Q_{ka}Q_{mb}\mathbb{C}(F_M, \mathbb{F}_a^b),$$
$$= \mathbb{C}(F_M, Q_{ka}Q_{mb}\mathbb{F}_a^b); \qquad \text{(XXIV.76)}$$
$$\mathbb{C}(F_M, \mathbb{F}_4^k \circ \mathbf{Q}) = \mathbb{C}(F_M \circ \mathbf{Q}, \mathbb{F}_4^k \circ \mathbf{Q}),$$
$$= \mathbb{C}(F_M, \mathbb{F}_4^k) \circ \mathbf{Q},$$
$$= \frac{3}{4\varepsilon}\left(\frac{c^2}{2\varepsilon} - \frac{5}{3}\right)Q_{ka}c_a F_M,$$
$$= Q_{ka}\mathbb{C}(F_M, \mathbb{F}_4^a),$$
$$= \mathbb{C}(F_M, Q_{ka}\mathbb{F}_4^a).$$

(viii) PROPERTIES OF THE COEFFICIENTS 457

Therefore $\mathbb{F}_k^m \circ \mathbf{Q} - Q_{ka} Q_{mb} \mathbb{F}_a^b$ and $\mathbb{F}_4^k \circ \mathbf{Q} - Q_{ka} \mathbb{F}_4^a$ satisfy the same integral equation as does $\mathbb{F}_0^k$, and so each is null. This shows that

$$\mathbb{F}_k^m(\mathbf{Qc}) = Q_{ka} Q_{mb} \mathbb{F}_a^b(\mathbf{c}),$$
$$\mathbb{F}_4^k(\mathbf{Qc}) = Q_{ka} \mathbb{F}_4^a(\mathbf{c}),$$
(XXIV.77)

for all orthogonal tensors $\mathbf{Q}$, and so[18] $\mathbb{F}_k^m$ *is a symmetric traceless isotropic tensor-valued function of* $\mathbf{c}$, while $\mathbb{F}_4^k$ *is an isotropic vector-valued function of* $\mathbf{c}$. From representation theorems for isotropic functions[19] we conclude that there are real-valued functions A and B, depending upon $\mathbf{c}$ only through its magnitude, such that

$$\mathbb{F}_k^m = c_{\{k} c_{m\}} B,$$
$$\mathbb{F}_4^k = c_k A.$$
(XXIV.78)

These representations reduce the problem of determining the fifteen functions $\mathbb{F}_\alpha^k$ to determining the two functions A and B. We may do so in various ways by using two appropriate combinations of the integral equations (XXIV.73).

There are several other properties of $\mathbb{F}_\alpha^k$ that we can deduce without solving the integral equations explicitly. In general, $\mathbb{F}_\alpha^k$ depends upon ρ, $\mathbf{u}$, and ε as well as $\mathbf{c}$. We have already used the fact that these functions will be independent of $\mathbf{u}$, simply because the right-hand sides of (XXIV.73)$_{1,2,3}$ are not functions of $\mathbf{u}$. This means that A and B are functions of ρ, ε, and c^2. We note next that the

---

[18] These two simple conclusions appear at one stage or another in almost all presentations of CHAPMAN's work or ENSKOG's. There they are not derived by appeal to integral equations such as (XXIV.73) but instead are imposed as additional assumptions, motivated by physical reasoning. CHAPMAN [1916, *1*, §2(E)] phrased such an assumption at the very outset of his analysis: "Clearly the form of F cannot depend upon any special choice of axes of reference ..., so that F is an invariant with respect to any orthogonal transformation of the co-ordinate axes." Reasoning of this type, however, is unsound. Once a governing equation such as (XXIV.73) has been derived, it is a matter for proof, not assumption, that its solution satisfies (XXIV.77). As far as we are aware, ours is the first complete proof of (XXIV.77). CHAPMAN's "clearly" is superseded by our proof above, which rests upon the theorem of WANG presented in Section (v) of Chapter VII.

[19] For (XXIV.78)$_2$ see W. NOLL, "Representation of certain isotropic tensor functions", *Archiv der Mathematik* **21**, 87–90 (1970). For (XXIV.78)$_1$ see Eq. (3.43) of C.-C. WANG, "A new representation theorem for isotropic functions: An answer to Professor G. F. Smith's criticism of my papers on representations for isotropic functions. Part 2. Vector-valued isotropic functions, symmetric tensor-valued isotropic functions, and skew-symmetric tensor-valued isotropic functions", *Archive for Rational Mechanics and Analysis* **36**, 198–223 (1970). Representations derived only for polynomial functions were obtained earlier by others but are obviously insufficient for use here, as a main requirement of a sound derivation is the *proof* that the $\mathbb{F}_\alpha^k$ are indeed scalar multiples of polynomials in $\mathbf{c}$, multiples whose scalar multipliers generally are *not* polynomial functions of their one scalar variable.

Here, too, authors in the kinetic theory have simply guessed the answer, not proved it. For example, CHAPMAN & COWLING [1939, §7.31] obtain (XXIV.78) by repeating "must be", "the only vector which can be formed", *etc.*

density ρ appears in the integral equations (XXIV.73) only because the Maxwellian density F_M is proportional to it. Thus we may cancel ρ from these equations, and so we conclude that

$$A = A(\varepsilon, c^2), \qquad B = B(\varepsilon, c^2). \qquad (\text{XXIV.79})$$

The fact that A and B are independent of ρ propagates such independence throughout all of the formulae for approximate gross determiners. It generalizes MAXWELL's celebrated conclusion that for a gas of Maxwellian molecules μ and M are functions of ε alone, a conclusion we have presented in Chapter XIII in MAXWELL's own fashion. We have already obtained one generalization of it in Chapter XIV, where we showed that the exact viscometric functions of a Maxwellian gas depend upon ρ in a specific way, one effect of which is to make μ and M free of it. In the remainder of this chapter we shall encounter further specific dependences upon the density and further coefficients that are independent of it.

From (XXIV.73) and (XXIV.78) we easily see that the physical dimensions of both A and B are $T^6 L^{-8}$. By applying the Pi-theorem as in Chapter IV we conclude that

$$A = \begin{cases} \dfrac{1}{d^2 \varepsilon^3} f\left(\dfrac{c^2}{\varepsilon}, \dfrac{fd}{m\varepsilon}\right) & \text{for general molecules,} \\[1em] \dfrac{1}{\varepsilon^3}\left(\dfrac{m\varepsilon}{g}\right)^{2/(k-1)} f\left(\dfrac{c^2}{\varepsilon}\right) & \text{for inverse } k^{\text{th}}\text{-power molecules,} \\[1em] \dfrac{1}{d^2 \varepsilon^3} f\left(\dfrac{c^2}{\varepsilon}\right) & \text{for ideal spheres;} \end{cases} \qquad (\text{XXIV.80})$$

the several functions f are dimensionless functions of their denoted arguments and also of such dimensionless constants as may serve to specify the molecular model. The function B has representations of just the same forms as those we have written out for A.

The conclusions (XXIV.74), (XXIV.78), (XXIV.79), and (XXIV.80) are the ones we shall need in analyzing the approximate gross determiners $\mathfrak{P}^{(1)}$ and $\mathfrak{q}^{(1)}$.

To simplify the coefficients of $\mathrm{grad}^2\,\rho$ and $\mathrm{grad}\,\rho \otimes \mathrm{grad}\,\rho$ in the expressions for $\mathfrak{P}$ and $\mathfrak{q}$, namely

$$\mathfrak{P}_{km\,\alpha}^{rs} = \tfrac{1}{2}\mathfrak{m} \int c_{\{k} c_{m\}} \mathfrak{G}_\alpha^{rs}, \qquad \mathfrak{P}_{km\,\alpha\beta}^{rs} = \tfrac{1}{2}\mathfrak{m} \int c_{\{k} c_{m\}} \mathfrak{G}_{\alpha\beta}^{rs},$$

$$\mathfrak{q}_{k\,\alpha}^{mr} = \tfrac{1}{4}\mathfrak{m} \int c^2 c_k \mathfrak{G}_\alpha^{mr}, \qquad \mathfrak{q}_{k\,\alpha\beta}^{mr} = \tfrac{1}{4}\mathfrak{m} \int c^2 c_k \mathfrak{G}_{\alpha\beta}^{mr}, \qquad (\text{XXIV.81})$$

we need $\mathfrak{G}_\alpha^{km}$ and $\mathfrak{G}_{\alpha\beta}^{km}$. The integral equations governing these have been set down previously in (XXIV.57) and (XXIV.58). Just as we did for (XXIV.56) in

(viii) PROPERTIES OF THE COEFFICIENTS 459

reaching (XXIV.73), we may write them in terms of ρ, **u**, and ε rather than **K**. First we introduce new functions $\mathbb{F}^{km}_\alpha$ and $\mathbb{F}^{km}_{\alpha\beta}$ that satisfy the relations

$$\mathbb{G}^{ab}_\alpha \rho_{\alpha,ab} + \mathbb{G}^{ab}_{\alpha\beta} \rho_{\alpha,a} \rho_{\beta,b} = \mathbb{F}^{ab}_0 \rho_{,ab} + \mathbb{F}^{ab}_c u_{c,ab}$$
$$+ \mathbb{F}^{ab}_4 \varepsilon_{,ab} + \mathbb{F}^{ab}_{00} \rho_{,a}\rho_{,b} \qquad\qquad \text{(XXIV.82)}$$
$$+ \mathbb{F}^{ab}_{0c} \rho_{,a} u_{c,b} + \mathbb{F}^{ab}_{04} \rho_{,a} \varepsilon_{,b} + \mathbb{F}^{ab}_{cd} u_{c,a} u_{d,b}$$
$$+ \mathbb{F}^{ab}_{c4} u_{c,a} \varepsilon_{,b} + \mathbb{F}^{ab}_{44} \varepsilon_{,a} \varepsilon_{,b}$$

identically in fields ρ, **u**, ε, and **ρ**, which are related to each other through (VI.23). Each side of this equation represents twice the part of the basic expansion of $\mathbb{G}$ that is linear in second gradients of the fields and quadratic in the first gradients. Substituting **ρ** as given by (VI.23) into the left-hand side of (XXIV.82) and equating coefficients of similar gradients of ρ, **u**, and ε on each side, we find that

$$\mathbb{F}^{km}_0 = \mathbb{G}^{km}_0 + u_a \mathbb{G}^{km}_a + (2\varepsilon + u^2)\mathbb{G}^{km}_4,$$

$$\mathbb{F}^{km}_r = \rho \mathbb{G}^{km}_r + 2\rho u_r \mathbb{G}^{km}_4,$$

$$\mathbb{F}^{km}_4 = 2\rho \mathbb{G}^{km}_4,$$

$$\mathbb{F}^{km}_{00} = \mathbb{G}^{km}_{00} + u_a \mathbb{G}^{km}_{0a} + u_a \mathbb{G}^{km}_{a0} + (2\varepsilon + u^2)\mathbb{G}^{km}_{04} + (2\varepsilon + u^2)\mathbb{G}^{km}_{40}$$
$$+ u_a u_b \mathbb{G}^{km}_{ab} + (2\varepsilon + u^2)u_a \mathbb{G}^{km}_{a4} + (2\varepsilon + u^2)u_a \mathbb{G}^{km}_{4a} + (2\varepsilon + u^2)^2 \mathbb{G}^{km}_{44},$$

$$\mathbb{F}^{km}_{0r} = \mathbb{G}^{km}_r + \mathbb{G}^{mk}_r + 2u_r \mathbb{G}^{km}_4 + 2u_r \mathbb{G}^{mk}_4 + \rho \mathbb{G}^{km}_{0r} + \rho \mathbb{G}^{mk}_{r0} + 2\rho u_r \mathbb{G}^{km}_{04} + 2\rho u_r \mathbb{G}^{mk}_{40}$$
$$+ \rho u_a \mathbb{G}^{km}_{ar} + \rho u_a \mathbb{G}^{mk}_{ra} + 2\rho u_r u_a \mathbb{G}^{km}_{a4} + 2\rho u_r u_a \mathbb{G}^{mk}_{4a} + \rho(2\varepsilon + u^2)\mathbb{G}^{km}_{4r}$$
$$+ \rho(2\varepsilon + u^2)\mathbb{G}^{mk}_{r4} + 2\rho(2\varepsilon + u^2)u_r \mathbb{G}^{km}_{44} + 2\rho(2\varepsilon + u^2)u_r \mathbb{G}^{mk}_{44}, \quad \text{(XXIV.83)}$$

$$\mathbb{F}^{km}_{04} = 2\mathbb{G}^{km}_4 + 2\mathbb{G}^{mk}_4 + 2\rho \mathbb{G}^{km}_{04} + 2\rho \mathbb{G}^{mk}_{40} + 2\rho u_a \mathbb{G}^{km}_{a4} + 2\rho u_a \mathbb{G}^{mk}_{4a}$$
$$+ 2\rho(2\varepsilon + u^2)\mathbb{G}^{km}_{44} + 2\rho(2\varepsilon + u^2)\mathbb{G}^{mk}_{44},$$

$$\mathbb{F}^{km}_{rs} = 2\rho\, \delta_{rs} \mathbb{G}^{km}_4 + \rho^2 \mathbb{G}^{km}_{rs} + 2\rho^2 u_s \mathbb{G}^{km}_{r4} + 2\rho^2 u_r \mathbb{G}^{km}_{4s} + 4\rho^2 u_r u_s \mathbb{G}^{km}_{44},$$

$$\mathbb{F}^{km}_{r4} = 2\rho^2 \mathbb{G}^{km}_{r4} + 2\rho^2 \mathbb{G}^{mk}_{4r} + 4\rho^2 u_r \mathbb{G}^{km}_{44} + 4\rho^2 u_r \mathbb{G}^{mk}_{44},$$

$$\mathbb{F}^{km}_{44} = 4\rho^2 \mathbb{G}^{km}_{44}.$$

To obtain integral equations for $\mathbb{F}^{km}_\alpha$ and $\mathbb{F}^{km}_{\alpha\beta}$, we form linear combinations of the corresponding integral equations for the functions $\mathbb{G}^{km}_\alpha$ and $\mathbb{G}^{km}_{\alpha\beta}$, which are given in (XXIV.57) and (XXIV.58). This process is straightforward though lengthy. For those who wish to verify[20] the result, in Appendix B we list

---

[20] We have preserved the calculations, which fill forty pages of manuscript. We will make them available to anyone who requests them.

formulae for many of the terms in (XXIV.57) and (XXIV.58). Here we simply record the equations that one obtains for $\mathbb{F}_\alpha^{km}$ and $\mathbb{F}_{\alpha\beta}^{km}$:

$$2\mathbb{C}(F_M, \mathbb{F}_0^{km}) = -\frac{4}{3}\frac{\varepsilon}{\rho}c_{\{k}c_{m\}}B,$$

$$2\mathbb{C}(F_M, \mathbb{F}_r^{km}) = (c_k c_{\{m} c_{r\}} + c_m c_{\{k} c_{r\}})B - \tfrac{2}{3}\varepsilon(c_k \delta_{mr} + c_m \delta_{kr})A$$
$$- (c_{\{r} \delta_{k\}m} + c_{\{r} \delta_{m\}k})\frac{F_M{}^m}{5\rho\varepsilon}\int c^4 B,$$

$$2\mathbb{C}(F_M, \mathbb{F}_4^{km}) = -\tfrac{4}{3}c_{\{k}c_{m\}}B + 2c_k c_m A - \delta_{km}\left(\frac{c^2}{2\varepsilon} - 1\right)\frac{F_M{}^m}{2\rho\varepsilon}\int c^4 A,$$

$$2\mathbb{C}(F_M, \mathbb{F}_{00}^{km}) = \frac{4}{3}\frac{\varepsilon}{\rho^2}c_{\{k}c_{m\}}B,$$

$$2\mathbb{C}(F_M, \mathbb{F}_{0r}^{km}) = \frac{8}{3}\frac{\varepsilon}{\rho}c_{\{r}\delta_{m\}k}B + \frac{8}{3}\frac{\varepsilon}{\rho}c_k c_{\{m}c_{r\}}\partial_{c^2}B,$$

$$2\mathbb{C}(F_M, \mathbb{F}_{04}^{km}) = -\frac{4}{3\rho}c_{\{k}c_{m\}}B + \frac{4}{3}\frac{\varepsilon}{\rho}\delta_{km}A + \frac{8}{3}\frac{\varepsilon}{\rho}c_k c_m \partial_{c^2}A,$$

$$2\mathbb{C}(F_M, \mathbb{F}_{rs}^{km}) = (-2c_{\{k}c_{s\}}\delta_{mr} - 2c_{\{m}c_{r\}}\delta_{ks} + \tfrac{2}{3}c_{\{k}c_{r\}}\delta_{ms}$$
$$+ \tfrac{2}{3}c_{\{m}c_{s\}}\delta_{kr} - 2c_{\{k}c_{m\}}\delta_{rs} - \tfrac{4}{3}c^2 \delta_{r\{m}\delta_{s\}k})B$$
$$- 2(c_k c_r c_{\{m}c_{s\}} + c_m c_s c_{\{k}c_{r\}})\partial_{c^2}B \qquad\qquad (XXIV.84)$$
$$- \tfrac{2}{3}\varepsilon(c_{\{k}c_{r\}}\delta_{ms} + c_{\{m}c_{s\}}\delta_{kr})\partial_\varepsilon B$$
$$- 2\mathbb{C}(c_{\{k}c_{r\}}B, c_{\{m}c_{s\}}B)$$
$$- (\delta_{r\{s}\delta_{m\}k} + \delta_{s\{r}\delta_{k\}m})\left(\frac{c^2}{2\varepsilon} - 1\right)\frac{F_M{}^m}{5\rho\varepsilon}\int c^4 B,$$

$$2\mathbb{C}(F_M, \mathbb{F}_{r4}^{km}) = \tfrac{8}{3}c_{\{r}\delta_{k\}m}B + \tfrac{8}{3}c_m c_{\{k}c_{r\}}\partial_{c^2}B + 2c_m c_{\{k}c_{r\}}\partial_\varepsilon B$$
$$- 4(c_k \delta_{mr} + \tfrac{1}{3}c_m \delta_{kr})A - 4c_m c_k c_r \partial_{c^2}A - \tfrac{4}{3}c_m \delta_{kr}\varepsilon \partial_\varepsilon A$$
$$- 4\mathbb{C}(c_m A, c_{\{k}c_{r\}}B) - c_{\{r}\delta_{k\}m}\frac{2F_M{}^m}{5\rho\varepsilon}\int c^4 \partial_\varepsilon B,$$

$$2\mathbb{C}(F_M, \mathbb{F}_{44}^{km}) = \tfrac{4}{3}\delta_{km}A + \tfrac{8}{3}c_k c_m \partial_{c^2}A + 2c_k c_m \partial_\varepsilon A$$
$$- 2\mathbb{C}(c_k A, c_m A) - \delta_{km}\left(\frac{c^2}{2\varepsilon} - 1\right)\frac{F_M{}^m}{2\rho\varepsilon}\int c^4 \partial_\varepsilon A,$$

$$\int I_\gamma \mathbb{F}_\alpha^{km} = 0, \qquad \int I_\gamma \mathbb{F}_{\alpha\beta}^{km} = 0.$$

Notice that the collisions operator enters the last three right-hand sides explicitly. The functions $\mathbb{F}_\alpha^{km}$ and $\mathbb{F}_{\alpha\beta}^{km}$ as delivered by solving these integral equations depend upon ρ in a specific way that is independent of the molecular model.

We could, if we so wished, analyze the 9 sets of integral equations (XXIV.84) in the same way as we did (XXIV.73) and so express the functions $\mathbb{F}_\alpha^{km}$ and $\mathbb{F}_{\alpha\beta}^{km}$ in terms of a much smaller number of unknowns[21], each depending upon only ε and c^2. Fortunately we do not have to do this in order to calculate the coefficients (XXIV.81). As we shall show next, all that is needed is the explicit forms of the right-hand sides of (XXIV.84).

(ix) The formulae of Enskog, Burnett, Chapman & Cowling, and Boltzmann

Now that we have analyzed the functions $\mathbb{F}_\alpha^k$ and have written out the integral equations for $\mathbb{F}_\alpha^{km}$ and $\mathbb{F}_{\alpha\beta}^{km}$, we can calculate reduced forms of some approximate gross determiners. Since the 0th approximation to $\mathbb{G}$ is the Maxwellian density F_M, the 0th approximations to $\mathfrak{P}$ and $\mathfrak{q}$ are the pressure deviator and energy flux, respectively, of such a density:

$$\mathfrak{P}_{km}^{(0)} = 0,$$
$$\mathfrak{q}_k^{(0)} = 0. \qquad\text{(XXIV.85)}$$

For the first approximation we use the fact that each side of (XXIV.71) is that part of the basic expansion of gross determiners that is linear in gradients of the basic fields. If we add F_M to the right-hand side of this equation and use the result as a molecular density for calculating $\mathfrak{P}^{(1)}$ and $\mathfrak{q}^{(1)}$, we obtain

$$\mathfrak{P}_{km}^{(1)} = P_{km0}^{\;\;a}\rho_{,a} + P_{kmb}^{\;\;a}u_{b,a} + P_{km4}^{\;\;a}\varepsilon_{,a},$$
$$\mathfrak{q}_k^{(1)} = q_{k0}^{\;\;a}\rho_{,a} + q_{kb}^{\;\;a}u_{b,a} + q_{k4}^{\;\;a}\varepsilon_{,a}, \qquad\text{(XXIV.86)}$$

in which

$$P_{km\alpha}^{\;\;\;r} \equiv \mathfrak{m}\int c_{\{k}c_{m\}}\,\mathbb{F}_\alpha^r,$$
$$q_{k\alpha}^{\;\;m} \equiv \mathfrak{m}\int \tfrac{1}{2}c^2 c_k\,\mathbb{F}_\alpha^m. \qquad\text{(XXIV.87)}$$

Into these expressions for the coefficients we substitute the reduced forms (XXIV.74) and (XXIV.78) of $\mathbb{F}_\alpha^k$. By (XXIV.74) we see that $P_{km0}^{\;\;r}$ and $q_{k0}^{\;\;m}$ vanish.

---

[21] [LENNARD-]JONES [1923, §8(d)] has given such an analysis for the functions $\mathbb{F}_\alpha^{km}$.

By (XXIV.78) $\mathbb{F}_m^k$ is an even function of $\mathbf{c}$ while $\mathbb{F}_4^k$ is an odd function, and therefore $P_{km4}^{\ r}$ and q_{kr}^m vanish. The remaining coefficients are determined as follows in terms of B and A:

$$P_{kms}^{\ r} = \mathfrak{m} \int c_{\{k} c_{m\}} c_{\{r} c_{s\}} B(\varepsilon, c^2),$$

$$q_{k4}^m = \mathfrak{m} \int \tfrac{1}{2} c^2 c_k c_m A(\varepsilon, c^2). \qquad \text{(XXIV.88)}$$

It is well known that the integral over $\mathscr{V}$ of a function f of c^2 multiplied by $c_{k_1} \cdots c_{k_{2n}}$ is completely determined by the integral $\int_{\mathscr{V}} c^{2n} f(c^2)\, d\mathbf{c}$. In (XXIVB.20) we list some of the simplest formulae illustrating this fact. If we use (XXIVB.20)$_{1,5}$ to simplify the two integrals above, we find that

$$P_{kms}^{\ r} = \tfrac{2}{15} \delta_{r\{k} \delta_{m\}s}\, \mathfrak{m} \int c^4 B,$$

$$q_{k4}^m = \tfrac{1}{6} \delta_{km}\, \mathfrak{m} \int c^4 A. \qquad \text{(XXIV.89)}$$

Placing these into (XXIV.86), we obtain

$$\mathfrak{P}_{km}^{(1)} = -2\mu E_{km},$$

$$\mathfrak{q}_k^{(1)} = -\mathrm{M}\mu \varepsilon_{,k}, \qquad \text{(XXIV.90)}$$

the coefficients μ and M being given in terms of A and B by the following celebrated formulae[22] of ENSKOG:

$$\mu = \mu(\varepsilon) = -\tfrac{1}{15}\mathfrak{m} \int c^4 B,$$

$$\mathrm{M} = \mathrm{M}(\varepsilon) = \frac{5 \int c^4 A}{2 \int c^4 B}. \qquad \text{(XXIV.91)}$$

They provide us with expressions for the viscosity and the Maxwell number of the kinetic gas for a general molecular model.

By appeal to (XXIV.80) we easily confirm the conclusions (IV.6) and (IV.2), the latter of course specialized, as MAXWELL's kinetic theory demands it must be, by requiring that μ not depend upon ρ. In addition, we see that *for ideal spheres the Maxwell number is a constant; for inverse* k^{th}*-power molecules it is a function of* k *alone*. Not only ε but also the dimensional constants $\mathfrak{m}$, $\mathfrak{d}$, and $\mathfrak{g}$ fail to affect M for these two models.

---

[22] ENSKOG [1917, Eqs. (68) and (69)] actually expressed μ and M in terms of the collisions operator $\mathbb{C}$. However, by use of the integral equations (XXIV.73) we may reduce his results immediately to the simpler formulae (XXIV.91). CHAPMAN did not obtain μ and M in this form but rather as certain ratios of infinite determinants. We shall discuss this aspect of the procedure in Section (xii) of this chapter.

(ix)   FORMULAE OF ENSKOG AND OTHERS   463

That M has very nearly the same value, namely $\frac{5}{2}$, for all molecular models, is a characteristic feature of MAXWELL's second kinetic theory. MAXWELL's first theory left great play for the imagination or "intuition" of the theorist who chose to develop it. Indeed, by appropriate arguments it is possible[23] to obtain any desired positive number as the value of M, and some which differed very much from $\frac{5}{2}$ were obtained from one or another plausible suggestion about mean free paths[24].

To calculate μ and M explicitly, it is necessary to solve the integral equations for the functions A and B. We shall see shortly that for a gas of Maxwellian molecules this can be done explicitly. For other molecular models we know of no explicit solutions. The traditional method of obtaining further information is to expand A and B, considered as functions of c^2, in infinite series of orthogonal polynomials and so reduce the integral equations to two infinite systems of algebraic equations. We shall pursue this aspect of the analysis of μ and M in Section (xii) of this chapter.

By returning to the integral equations governing A and B, it is possible to show that μ and M are both positive, thus conforming with $(I.10)_{1,3}$. To do so, we note that

$$c_{\langle a}c_{b\rangle}c_{\langle a}c_{b\rangle} = \tfrac{2}{3}c^4, \qquad (XXIV.92)$$

so

$$\int c^4 B = \tfrac{3}{2}\int c_{\langle a}c_{b\rangle}c_{\langle a}c_{b\rangle}B,$$

$$= 2\varepsilon \int \frac{\mathbb{F}_a^b}{F_M}\frac{3}{4\varepsilon}c_{\langle a}c_{b\rangle}F_M,$$

$$= 2\varepsilon \int \frac{\mathbb{F}_a^b}{F_M}\mathbb{C}(F_M, \mathbb{F}_a^b), \qquad (XXIV.93)$$

$$= -\tfrac{1}{4}\varepsilon \iint\int_{\mathscr{S}} wF_M F_{M*}(g_a^{b'} + g_{a*}^{b'} - g_a^b - g_{a*}^b)(g_a^{b'} + g_{a*}^{b'} - g_a^b - g_{a*}^b),$$

in which

$$g_k^m \equiv \frac{\mathbb{F}_k^m}{F_M} = c_{\langle k}c_{m\rangle}\frac{B}{F_M}. \qquad (XXIV.94)$$

In going from $(XXIV.93)_2$ to $(XXIV.93)_3$ we have used $(XXIV.73)_2$, and to obtain $(XXIV.93)_4$ we have used $(VII.20)_2$ and the fact that $F_M' F_{M*}' = F_M F_{M*}$. Since none of the functions g_k^m is a summational invariant, we see that $\int c^4 B < 0$. In a similar manner, one can show that $\int c^4 A < 0$, and so $\mu > 0$ and $M > 0$. △

---

[23] It suffices to modify the method given by GRAD [1950, Appendix 3]. Instead of assuming that "the last collision occurred exactly one mean free path away," we may assume that it occurred exactly k mean free paths away, k being any positive number.

[24] CHAPMAN [1967], TRUESDELL [1975, §8].

Since each side of (XXIV.82) is twice that part of the basic expansion of $\mathbb{G}$ that is linear in second gradients of the fields and quadratic in first gradients, we can write the second approximations to $\mathfrak{P}$ and $\mathfrak{q}$ in the forms

$$\mathfrak{P}_{km}^{(2)} = -2\mu E_{km} + P_{km0}^{ab}\rho_{,ab} + P_{kmc}^{ab}u_{c,ab} + P_{km4}^{ab}\varepsilon_{,ab}$$

$$+ P_{km00}^{ab}\rho_{,a}\rho_{,b} + P_{km0d}^{ab}\rho_{,a}u_{d,b} + P_{km04}^{ab}\rho_{,a}\varepsilon_{,b}$$

$$+ P_{kmcd}^{ab}u_{c,a}u_{d,b} + P_{kmc4}^{ab}u_{c,a}\varepsilon_{,b} + P_{km44}^{ab}\varepsilon_{,a}\varepsilon_{,b}, \quad \text{(XXIV.95)}$$

$$\mathfrak{q}_k^{(2)} = -M\mu\varepsilon_{,k} + q_{k0}^{ab}\rho_{,ab} + q_{kc}^{ab}u_{c,ab} + q_{k4}^{ab}\varepsilon_{,ab}$$

$$+ q_{k00}^{ab}\rho_{,a}\rho_{,b} + q_{k0d}^{ab}\rho_{,a}u_{d,b} + q_{k04}^{ab}\rho_{,a}\varepsilon_{,b}$$

$$+ q_{kcd}^{ab}u_{c,a}u_{d,b} + q_{kc4}^{ab}u_{c,a}\varepsilon_{,b} + q_{k44}^{ab}\varepsilon_{,a}\varepsilon_{,b},$$

in which

$$P_{km\alpha}^{rs} \equiv \tfrac{1}{2}\mathfrak{m}\int c_{\{k}c_{m\}}\mathbb{F}_\alpha^{rs}, \qquad P_{km\alpha\beta}^{rs} \equiv \tfrac{1}{2}\mathfrak{m}\int c_{\{k}c_{m\}}\mathbb{F}_{\alpha\beta}^{rs},$$

$$q_{k\alpha}^{mr} \equiv \tfrac{1}{4}\mathfrak{m}\int c^2 c_k \mathbb{F}_\alpha^{mr}, \qquad q_{k\alpha\beta}^{mr} \equiv \tfrac{1}{4}\mathfrak{m}\int c^2 c_k \mathbb{F}_{\alpha\beta}^{mr}. \quad \text{(XXIV.96)}$$

We have not calculated reduced forms of $\mathbb{F}_\alpha^{km}$ and $\mathbb{F}_{\alpha\beta}^{km}$, but for the purposes of calculating the coefficients in (XXIV.95) we need only the integral equations (XXIV.84). To see why, we make use of (XXIV.73)$_2$ and (VII.20)$_2$ so as to write

$$P_{km\alpha}^{rs} = \frac{2\mathfrak{m}\varepsilon}{3}\int \frac{\mathbb{F}_\alpha^{rs}}{F_M}\frac{3}{4\varepsilon}c_{\{k}c_{m\}}F_M,$$

$$= \frac{2\mathfrak{m}\varepsilon}{3}\int \frac{\mathbb{F}_\alpha^{rs}}{F_M}\mathbb{C}(F_M, \mathbb{F}_m^k),$$

$$= -\frac{2\mathfrak{m}\varepsilon}{3}\frac{1}{8}\iiint_{\mathscr{G}} wF_M F_{M*}\left(\frac{\mathbb{F}_\alpha^{rs\prime}}{F_M'} + \frac{\mathbb{F}_{\alpha*}^{rs\prime}}{F_{M*}'} - \frac{\mathbb{F}_\alpha^{rs}}{F_M} - \frac{\mathbb{F}_{\alpha*}^{rs}}{F_{M*}}\right) \quad \text{(XXIV.97)}$$

$$\times \left(\frac{\mathbb{F}_m^{k\prime}}{F_M'} + \frac{\mathbb{F}_{m*}^{k\prime}}{F_{M*}'} - \frac{\mathbb{F}_m^k}{F_M} - \frac{\mathbb{F}_{m*}^k}{F_{M*}}\right),$$

$$= \frac{\mathfrak{m}\varepsilon}{3}\int \frac{\mathbb{F}_m^k}{F_M}2\mathbb{C}(F_M, \mathbb{F}_\alpha^{rs}).$$

Similarly we find that

$$P_{km\alpha\beta}^{rs} = \frac{\mathfrak{m}\varepsilon}{3}\int \frac{\mathbb{F}_m^k}{F_M}2\mathbb{C}(F_M, \mathbb{F}_{\alpha\beta}^{rs}),$$

$$q_{k\alpha}^{mr} = \frac{\mathfrak{m}\varepsilon^2}{3}\int \frac{\mathbb{F}_4^k}{F_M}2\mathbb{C}(F_M, \mathbb{F}_\alpha^{mr}), \quad \text{(XXIV.98)}$$

$$q_{k\alpha\beta}^{mr} = \frac{\mathfrak{m}\varepsilon^2}{3}\int \frac{\mathbb{F}_4^k}{F_M}2\mathbb{C}(F_M, \mathbb{F}_{\alpha\beta}^{mr}).$$

(ix) SECOND APPROXIMATION

In these formulae $\mathbb{F}^{km}_{\alpha}$ and $\mathbb{F}^{km}_{\alpha\beta}$ appear only through the values $2\mathbb{C}(F_M, \mathbb{F}^{km}_{\alpha})$ and $2\mathbb{C}(F_M, \mathbb{F}^{km}_{\alpha\beta})$ of the bilinear form, and the integral equations (XXIV.84) give us these values explicitly in terms of the functions A and B. This technique can easily be generalized so as to show that *to calculate the nth approximations to $\mathfrak{P}$ and $\mathbf{q}$, we need only the iterates to $\mathbb{G}$ of orders $0, 1, \ldots, n-1$ as well as the integral equations for those of order n.*

Each of the coefficients $\mathbb{F}^k_m$ and $\mathbb{F}^k_4$ and each of the right-hand sides of (XXIV.84), considered as a function of $\mathbf{c}$, is either even or odd. When one of the former and one of the latter are of different parities, the corresponding integrand in (XXIV.97)$_4$ or (XXIV.98) is odd, and therefore the corresponding coefficient $P^{\cdots}_{\cdots}$ or $q^{\cdots}_{\cdots}$ vanishes. Thus

$$P_{km\,t}^{rs} = 0, \qquad q_{k\,0}^{mr} = 0, \qquad q_{k\,4}^{mr} = 0,$$
$$P_{km0t}^{rs} = 0, \qquad P_{kmt4}^{rs} = 0, \qquad q_{k00}^{mr} = 0, \qquad \text{(XXIV.99)}$$
$$q_{k04}^{mr} = 0, \qquad q_{k\,st}^{mr} = 0, \qquad q_{k44}^{mr} = 0.$$

All the other coefficients are linear combinations of integrals of a product of $c_{k_1} \cdots c_{k_{2n}}$ and a function of c^2, or in some cases also a collisions integral. Each of these may be simplified by use of (XXIVB.20) and (XXIVB.22), and so we obtain the following expressions for the remaining coefficients:

$$P_{km\,0}^{rs} = -\frac{8}{135}\frac{\mathfrak{m}\varepsilon^2}{\rho}\delta_{r\{k}\delta_{m\}s}\int c^4\frac{B^2}{F_M},$$

$$P_{km\,4}^{rs} = \tfrac{2}{45}\mathfrak{m}\varepsilon\,\delta_{r\{k}\delta_{m\}s}\left(-\tfrac{4}{3}\int c^4\frac{B^2}{F_M} + 2\int c^4\frac{AB}{F_M}\right),$$

$$q_{k\,s}^{mr} = \tfrac{2}{45}\mathfrak{m}\varepsilon^2(\delta_{k\{s}\delta_{m\}r} + \delta_{k\{s}\delta_{r\}m})\int c^4\frac{AB}{F_M}$$
$$- \tfrac{2}{27}\mathfrak{m}\varepsilon^3(\delta_{km}\delta_{rs} + \delta_{kr}\delta_{ms})\int c^2\frac{A^2}{F_M},$$

$$P_{km00}^{rs} = \frac{8}{135}\frac{\mathfrak{m}\varepsilon^2}{\rho^2}\delta_{r\{k}\delta_{m\}s}\int c^4\frac{B^2}{F_M},$$

$$P_{km04}^{rs} = -\frac{8}{135}\frac{\mathfrak{m}\varepsilon}{\rho}\delta_{r\{k}\delta_{m\}s}\left(\int c^4\frac{B^2}{F_M} - 2\varepsilon\int c^4\frac{B\,\partial_{c^2}A}{F_M}\right),$$

$$P_{kmtu}^{rs} = -\tfrac{4}{45}\mathfrak{m}\varepsilon(\delta_{st}\delta_{r\{k}\delta_{m\}u} + \delta_{ru}\delta_{s\{k}\delta_{m\}t})\int c^4\frac{B^2}{F_M}$$
$$- \tfrac{4}{45}\mathfrak{m}\varepsilon\,\delta_{tu}\delta_{r\{k}\delta_{m\}s}\int c^4\frac{B^2}{F_M} \qquad \text{(XXIV.100)}$$
$$+ \tfrac{4}{135}\mathfrak{m}\varepsilon(\delta_{su}\delta_{r\{k}\delta_{m\}t} + \delta_{rt}\delta_{s\{k}\delta_{m\}u})\left(\int c^4\frac{B^2}{F_M} - \varepsilon\int c^4\frac{B\,\partial_\varepsilon B}{F_M}\right)$$

$$\begin{aligned}&-\tfrac{4}{315}\mathfrak{m}\varepsilon\big(\delta_{r\{k}\,\delta_{m\}s}\,\delta_{tu}+\delta_{r\{k}\,\delta_{m\}u}\,\delta_{ts}+\delta_{t\{k}\,\delta_{m\}s}\,\delta_{ru}\\&+\delta_{t\{k}\,\delta_{m\}u}\,\delta_{rs}+\delta_{s\{k}\,\delta_{m\}u}\,\delta_{rt}-\tfrac{4}{3}\delta_{su}\,\delta_{r\{k}\,\delta_{m\}t}\\&+\delta_{s\{k}\,\delta_{m\}r}\,\delta_{tu}+\delta_{s\{k}\,\delta_{m\}t}\,\delta_{ur}+\delta_{u\{k}\,\delta_{m\}r}\,\delta_{st}\\&+\delta_{u\{k}\,\delta_{m\}t}\,\delta_{sr}+\delta_{r\{k}\,\delta_{m\}t}\,\delta_{su}-\tfrac{4}{3}\delta_{rt}\,\delta_{s\{k}\,\delta_{m\}u}\big)\int c^{6}\frac{B\,\partial_{c^{2}}B}{F_{\mathrm{M}}}\\&-\tfrac{2}{35}\mathfrak{m}\varepsilon\big(\delta_{s\{k}\,\delta_{m\}r}\,\delta_{tu}+\delta_{s\{k}\,\delta_{m\}t}\,\delta_{ru}+\delta_{u\{k}\,\delta_{m\}r}\,\delta_{st}\\&+\delta_{u\{k}\,\delta_{m\}t}\,\delta_{rs}-\tfrac{4}{3}\delta_{su}\,\delta_{r\{k}\,\delta_{m\}t}-\tfrac{4}{3}\delta_{rt}\,\delta_{s\{k}\,\delta_{m\}u}\big)\\&\times\int\frac{B}{F_{\mathrm{M}}}c_{\{a}c_{b\}}\mathbb{C}(c_{\{b}c_{d\}}B,c_{\{d}c_{a\}}B),\end{aligned}$$

$$\begin{aligned}P_{km44}^{\ r\ s}=&\tfrac{4}{45}\mathfrak{m}\varepsilon\,\delta_{r\{k}\,\delta_{m\}s}\bigg(\int c^{4}\frac{B\,\partial_{\varepsilon}A}{F_{\mathrm{M}}}+\tfrac{4}{3}\int c^{4}\frac{B\,\partial_{c^{2}}A}{F_{\mathrm{M}}}\bigg)\\&-\tfrac{2}{15}\mathfrak{m}\varepsilon\,\delta_{r\{k}\,\delta_{m\}s}\int\frac{B}{F_{\mathrm{M}}}c_{\{a}c_{b\}}\mathbb{C}(c_{a}A,c_{b}A),\end{aligned}$$

$$q_{k0m}^{\ r\ s}=\frac{8}{27}\frac{\mathfrak{m}\varepsilon^{3}}{\rho}\,\delta_{r\{s}\,\delta_{m\}k}\bigg(\int c^{2}\frac{AB}{F_{\mathrm{M}}}+\tfrac{2}{5}\int c^{4}\frac{A\,\partial_{c^{2}}B}{F_{\mathrm{M}}}\bigg),$$

$$\begin{aligned}q_{km4}^{\ r\ s}=&\tfrac{8}{27}\mathfrak{m}\varepsilon^{2}\,\delta_{k\{r}\,\delta_{m\}s}\bigg(\int c^{2}\frac{AB}{F_{\mathrm{M}}}+\tfrac{2}{5}\int c^{4}\frac{A\,\partial_{c^{2}}B}{F_{\mathrm{M}}}+\tfrac{3}{10}\int c^{4}\frac{A\,\partial_{\varepsilon}B}{F_{\mathrm{M}}}\bigg)\\&-\tfrac{4}{9}\mathfrak{m}\varepsilon^{2}(\delta_{kr}\,\delta_{sm}+\tfrac{1}{3}\delta_{ks}\,\delta_{rm})\int c^{2}\frac{A^{2}}{F_{\mathrm{M}}}-\tfrac{4}{27}\mathfrak{m}\varepsilon^{3}\,\delta_{rm}\,\delta_{ks}\int c^{2}\frac{A\,\partial_{\varepsilon}A}{F_{\mathrm{M}}}\\&-\tfrac{4}{45}\mathfrak{m}\varepsilon^{2}(\delta_{kr}\,\delta_{sm}+\delta_{ks}\,\delta_{rm}+\delta_{km}\,\delta_{rs})\int c^{4}\frac{A\,\partial_{c^{2}}A}{F_{\mathrm{M}}}\\&-\tfrac{4}{15}\mathfrak{m}\varepsilon^{2}\,\delta_{k\{r}\,\delta_{m\}s}\int\frac{A}{F_{\mathrm{M}}}c_{a}\mathbb{C}(c_{b}A,c_{\{a}c_{b\}}B).\end{aligned}$$

Finally we place (XXIV.99) and (XXIV.100) into the general second approximations (XXIV.95) and simplify the formulae that result. We obtain

$$\begin{aligned}\mathfrak{P}_{km}^{(2)}=&-2\mu E_{km}+\varpi_{1}\frac{\mu^{2}}{p}EE_{km}-\varpi_{2}\frac{\mu^{2}}{p}\bigg[\bigg(\frac{1}{\rho}p_{,\{k}\bigg)_{,m\}}+u_{\{a,k}u_{m,a\}}+2u_{\{a,k}E_{ma\}}\bigg]\\&+\varpi_{3}\frac{\mu^{2}}{\rho\varepsilon}\varepsilon_{,\{km\}}+\varpi_{4}\frac{\mu^{2}}{\rho\rho\varepsilon}p_{,\{k}\varepsilon_{,m\}}+\varpi_{5}\frac{\mu^{2}}{\rho\varepsilon^{2}}\varepsilon_{,\{k}\varepsilon_{,m\}}+\varpi_{6}\frac{\mu^{2}}{p}E_{\{ak}E_{ma\}},\end{aligned}$$
(XXIV.101)
$$\begin{aligned}q_{k}^{(2)}=&-M\mu\varepsilon_{,k}+\theta_{1}\frac{\mu^{2}}{\rho\varepsilon}EE_{,k}-\theta_{2}\frac{\mu^{2}}{\rho\varepsilon}[\tfrac{2}{3}(\varepsilon E)_{,k}+2\varepsilon_{,a}u_{a,k}]\\&+\theta_{3}\frac{\mu^{2}}{\rho p}E_{ka}p_{,a}+\theta_{4}\frac{\mu^{2}}{\rho}E_{ka,a}+\theta_{5}\frac{\mu^{2}}{\rho\varepsilon}E_{ka}\varepsilon_{,a};\end{aligned}$$

the coefficients $\varpi_1, \varpi_2, \ldots, \theta_5$ are functions of ε only and are determined by the functions A and B as follows:

$$\varpi_1 = -\tfrac{80}{9}\varepsilon^2 \int^* c^4 B(\varepsilon\, \partial_\varepsilon B + c^2\, \partial_{c^2} B),$$

$$\varpi_2 = \tfrac{40}{3}\varepsilon^2 \int^* c^4 B^2,$$

$$\varpi_3 = 20\varepsilon^2 \int^* c^4 AB,$$

$$\varpi_4 = \tfrac{80}{3}\varepsilon^3 \int^* c^4 B\, \partial_{c^2} A,$$

$$\varpi_5 = 30\varepsilon^3 \left(\tfrac{2}{3}\int^* c^4 B\, \partial_\varepsilon A - \int^* c_{\{a} c_{b\}}\, B\mathbb{C}(c_a A, c_b A)\right),$$

$$\varpi_6 = -\tfrac{80}{21}\varepsilon^2 \left(4\int^* c^6 B\, \partial_{c^2} B + 9\int^* c_{\{a} c_{b\}}\, B\mathbb{C}(c_{\{b} c_{d\}} B, c_{\{d} c_{a\}} B)\right),$$

$$\theta_1 = -\tfrac{100}{3}\varepsilon^3 \int^* c^2 A(\varepsilon\, \partial_\varepsilon A + c^2\, \partial_{c^2} A), \qquad\qquad (\text{XXIV.102})$$

$$\theta_2 = 50\varepsilon^3 \int^* c^2 A^2,$$

$$\theta_3 = \tfrac{40}{3}\varepsilon^3 \int^* c^2 A(5B + 2c^2\, \partial_{c^2} B),$$

$$\theta_4 = 20\varepsilon^2 \int^* c^4 AB,$$

$$\theta_5 = 20\varepsilon^3 \left(\int^* c^4 A(\partial_\varepsilon B - 2\, \partial_{c^2} A) - 3\int^* c_a A\mathbb{C}(c_b A, c_{\{a} c_{b\}} B)\right),$$

in which the notation $\int^*$ is defined as follows:

$$\int^* \cdots \equiv \frac{1}{\left(\int c^4 B\right)^2} \int \frac{1}{\dfrac{m}{\rho} F_M} \cdots. \qquad\qquad (\text{XXIV.103})$$

The formula $(\text{XXIV.101})_1$ for $\mathfrak{P}^{(2)}$ is the special case of $(\text{XXIV.4})_1$ in which the body force is null, while $(\text{XXIV.101})_2$ for $\mathbf{q}^{(2)}$ agrees exactly with $(\text{XXIV.4})_2$. It is a remarkable consequence of these formulae that $\mathfrak{P}^{(2)}$ need not vanish when $\mathbf{E} = \mathbf{0}$ and $E = 0$, and that $\mathbf{q}^{(2)}$ need not vanish when $\operatorname{grad} \varepsilon = \mathbf{0}$.

The formal agreement between (XXIV.101) and (XXIV.4) obscures the difference in their meanings, a difference denoted by our use of $\mathbf{P}^{\langle 2 \rangle}$ and $\mathbf{q}^{\langle 2 \rangle}$ for the latter, $\mathfrak{P}^{(2)}$ and $\mathfrak{q}^{(2)}$ for the former. On the right-hand sides of (XXIV.4) such quantities as E_{km} and $\varepsilon_{,k}$ are *functions of time* derived from a molecular density that satisfies the Maxwell–Boltzmann equation. In (XXIV.101) those same symbols E_{km} and $\varepsilon_{,k}$ denote fields calculated from an arbitrary *independent variable* ρ.

To our knowledge the full list of exact evaluations (XXIV.102) has not been displayed explicitly before[25]. Using them, we may derive the following three formulae, which express some of the functions in terms of others. They are *universal relations* in the sense that *they hold for all molecular models*[26]:

$$\varpi_3 = \theta_4,$$

$$\theta_1 = \frac{2}{3}\left(\frac{7}{2} - \frac{\mu'\varepsilon}{\mu}\right)\theta_2 - \tfrac{1}{3}\varepsilon\, \partial_\varepsilon \theta_2, \qquad \text{(XXIV.104)}$$

$$\varpi_1 = \frac{2}{3}\left(\frac{7}{2} - \frac{\mu'\varepsilon}{\mu}\right)\varpi_2 - \tfrac{1}{3}\varepsilon\, \partial_\varepsilon \varpi_2.$$

These relations show that the 8 *coefficients* $\varpi_2, \varpi_3, \varpi_4, \varpi_5, \varpi_6, \theta_2, \theta_3, \theta_5$ determine *all* 11 *second-order coefficients* in (XXIV.101), μ and M being assumed known.

By appeal to (XXIV.80) in the formulae set forth in (XXIV.102) we easily conclude that *for a gas of ideal spheres the coefficients* $\varpi_1, \varpi_2, \ldots, \theta_5$ *are independent of* $\mathfrak{d}$ *and* ε; *for a gas of inverse* $\mathfrak{k}^{th}$-*power molecules, they are independent of* ε. Of course, for inverse $\mathfrak{k}^{th}$-power molecules the coefficients are functions of $\mathfrak{k}$ which may or may not reduce to constants. The same conclusions may be reached more easily by a direct dimensional analysis parallel to those given in Chapter IV. Referring back to (IV.6), we may reduce the universal relations (XXIV.104)$_{2,3}$ to the forms

$$\frac{\theta_1}{\theta_2} = \frac{\varpi_1}{\varpi_2} = \begin{cases} \dfrac{2}{3}\dfrac{3\mathfrak{k}-5}{\mathfrak{k}-1} & \text{for inverse } \mathfrak{k}^{th}\text{-power molecules,} \\ 2 & \text{for ideal spheres.} \end{cases} \qquad \text{(XXIV.105)}$$

---

[25] The formulae for $\theta_2, \theta_4, \varpi_2$, and ϖ_3 were published by WANG CHANG [1948, *1*, Eq. (2)]. What she denotes by θ_4 is our $\theta_4 \mu^2/\rho$, and what she denotes by $\varpi_3 T$ is our $\varpi_3 \mu^2/\rho$. Her remarks indicate she may have obtained the full list (XXIV.102).

For references to previous determinations of these "Burnett coefficients"—exact for a gas of Maxwellian molecules, alleged to be approximate for other kinds of molecules—see Footnotes 33 and 34 to this chapter and also Footnote 22 to Chapter XXV.

[26] CHAPMAN & COWLING [1939, Eq. 15.41.5] obtained (XXIV.104)$_{2,3}$, though only for inverse $\mathfrak{k}^{th}$-power molecules and with $\varpi_1, \varpi_2, \theta_1$, and θ_2 replaced by their first approximations as described in Section (xii) of this chapter. The universal relations (XXIV.104), which to the best of our knowledge are new, are exact.

But this is not all. We can calculate also some approximate gross determiners for BOLTZMANN's function h and its flux s. The 0th approximation is the leading term in each of (XXIV.39)$_{1,2}$, these being h and s calculated using the Maxwellian density whose principal moment is ρ:

$$\mathfrak{h}^{(0)} = \log \frac{n}{(\frac{4}{3}\pi\varepsilon)^{\frac{3}{2}}} - \tfrac{3}{2} = h_{\text{M}},$$

$$\mathfrak{s}_k^{(0)} = 0.$$
(XXIV.106)

For the next approximation we must take account of the terms in (XXIV.39) that are linear in the gradients of the fields. Since $\mathbb{G}(\mathbf{K}) = F_{\text{M}}$, it follows from (XXIV.40)$_{1,4}$, (XXIV.30)$_2$, and (XXIV.56)$_2$ that

$$\mathfrak{h}_\alpha^k = \frac{\mathfrak{m}}{\rho}\int\left(1 - \frac{3c^2}{4\varepsilon} + \log\frac{n}{(\frac{4}{3}\pi\varepsilon)^{\frac{3}{2}}}\right)\mathbb{G}_\alpha^k = 0,$$

$$\mathfrak{s}_{k\alpha}^m = \mathfrak{m}\int c_k\left(1 - \frac{3c^2}{4\varepsilon} + \log\frac{n}{(\frac{4}{3}\pi\varepsilon)^{\frac{3}{2}}}\right)\mathbb{G}_\alpha^m,$$

$$= -\frac{3}{2\varepsilon}\mathfrak{m}\int \tfrac{1}{2}c_k c^2 \mathbb{G}_\alpha^m,$$

$$= -\frac{3}{2\varepsilon}\mathfrak{q}_{k\alpha}^m,$$
(XXIV.107)

whence we find that

$$\mathfrak{h}^{(1)} = \log\frac{n}{(\frac{4}{3}\pi\varepsilon)^{\frac{3}{2}}} - \tfrac{3}{2} = h_{\text{M}},$$

$$\mathfrak{s}_k^{(1)} = -\frac{1}{\frac{2}{3}\varepsilon}\mathfrak{q}_k^{(1)}.$$
(XXIV.108)

To first order $-\mathfrak{r}h$ is the caloric of an ideal perfect gas for which $\gamma = \tfrac{5}{3}$, and s is such as to make the formal broad H-theorem and the Clausius-Duhem inequality become one and the same statement.

For the second approximation we must take account of all the coefficients listed in (XXIV.40). By reasoning as in the proof of (XXIV.107) we find

$$\mathfrak{h}_\alpha^{km} = 0, \qquad \mathfrak{h}_{\alpha\beta}^{km} = \frac{\mathfrak{m}}{2\rho}\int\frac{\mathbb{G}_\alpha^k \mathbb{G}_\beta^m}{F_{\text{M}}},$$

$$\mathfrak{s}_{k\ \alpha}^{mr} = -\frac{3}{2\varepsilon}\mathfrak{q}_{k\ \alpha}^{mr}, \qquad \mathfrak{s}_{k\alpha\beta}^{mr} = -\frac{3}{2\varepsilon}\mathfrak{q}_{k\alpha\beta}^{mr} + \tfrac{1}{2}\mathfrak{m}\int c_k\frac{\mathbb{G}_\alpha^m \mathbb{G}_\beta^r}{F_{\text{M}}}.$$
(XXIV.109)

Using (XXIV.71), (XXIV.74), and (XXIV.78), we see that

$$\frac{\mathfrak{G}_\alpha^a \mathfrak{G}_\beta^b}{F_M} \rho_{\alpha,a} \rho_{\beta,b} = \frac{1}{F_M} (\mathbb{F}_d^a u_{d,a} + \mathbb{F}_4^a \varepsilon_{,a})(\mathbb{F}_e^b u_{e,b} + \mathbb{F}_4^b \varepsilon_{,b}),$$

$$= \frac{B^2}{F_M} c_{\{a} c_{d\}} c_{\{b} c_{e\}} u_{d,a} u_{e,b} + 2\frac{AB}{F_M} c_{\{a} c_{d\}} c_b u_{d,a} \varepsilon_{,b} \quad \text{(XXIV.110)}$$

$$+ \frac{A^2}{F_M} c_a c_b \varepsilon_{,a} \varepsilon_{,b}.$$

From this result and the reduction formulae (XXIVB.20) we obtain

$$\frac{\mathfrak{m}}{2\rho} \int \frac{\mathfrak{G}_\alpha^a \mathfrak{G}_\beta^b}{F_M} \rho_{\alpha,a} \rho_{\beta,b} = \frac{1}{15} \frac{\mathfrak{m}}{\rho} \delta_{b\{a} \delta_{d\}e} u_{d,a} u_{e,b} \int c^4 \frac{B^2}{F_M}$$

$$+ \frac{1}{6} \frac{\mathfrak{m}}{\rho} \delta_{ab} \varepsilon_{,a} \varepsilon_{,b} \int c^2 \frac{A^2}{F_M}, \quad \text{(XXIV.111)}$$

$$\tfrac{1}{2}\mathfrak{m} \int c_k \frac{\mathfrak{G}_\alpha^a \mathfrak{G}_\beta^b}{F_M} \rho_{\alpha,a} \rho_{\beta,b} = \tfrac{2}{15}\mathfrak{m} \, \delta_{k\{a} \delta_{d\}b} u_{d,a} \varepsilon_{,b} \int c^4 \frac{AB}{F_M},$$

and if we place these into (XXIV.109) and the resulting coefficients into (XXIV.39), by use of (XXIV.102) we may show that the second approximations to the gross determiners of h and $\mathbf{s}$ are[27]

$$\mathfrak{h}^{(2)} = \log \frac{n}{(\tfrac{4}{3}\pi\varepsilon)^{\tfrac{3}{2}}} - \tfrac{3}{2} + \frac{\mu^2}{2p^2} \varpi_2 E_{ab} E_{ab} + \frac{\mu^2}{3p^2\varepsilon} \theta_2 \varepsilon_{,a} \varepsilon_{,a},$$

$$\mathfrak{s}_k^{(2)} = -\frac{1}{\tfrac{2}{3}\varepsilon} \mathfrak{q}_k^{(2)} + \frac{\mu^2}{p\varepsilon} \theta_4 E_{ka} \varepsilon_{,a}.$$
(XXIV.112)

---

[27] BOLTZMANN [1896, Eq. (250)] obtained (XXIV.112)$_1$ for Maxwellian molecules, but his coefficients of $E_{ab} E_{ab}$ and $\varepsilon_{,a} \varepsilon_{,a}$ are not correct. To the best of our knowledge its general form has not been published before.
  PRIGOGINE [1949] presented arguments alleged to indicate that at this stage of iteration it would be found that $\mathbf{s} \neq -\mathbf{q}/(\tfrac{2}{3}\varepsilon)$. MÜLLER [1966, Eq. 5.85] by appeal to the linearized Boltzmann equation concluded that for a gas of Maxwellian molecules

$$s_k + \frac{q_k}{\tfrac{2}{3}\varepsilon} = \tfrac{2}{5}\rho_0 \frac{P_{ka} q_a}{p_0^2},$$

in which p_0 and ρ_0 are the constant pressure and mass-density of the uniform Maxwellian molecular density about which perturbations are effected. To compare this result with (XXIV.112)$_2$, we replace $\mathbf{P}$ by $\mathfrak{P}^{(1)}$ and $\mathbf{q}$ by $\mathbf{q}^{(1)}$, and we replace p_0 and ρ_0 by p and ρ. Then MÜLLER's result becomes

$$s_k + \frac{q_k}{\tfrac{2}{3}\varepsilon} \approx \tfrac{4}{5}\rho M \frac{\mu^2}{p^2} E_{ka} \varepsilon_{,a}.$$

(continued)

Therefore, at this stage of approximation h differs from the functional of ρ to which it reduces when the molecular density is Maxwellian.

In terms of the truncation number Tr, defined by (IV.16), and the heat-transfer number H, defined in Footnote 7 to Chapter VIII, (XXIV.112)$_1$ reads

$$\mathfrak{h}^{(2)} - h_M = \tfrac{1}{4}\varpi_2 \operatorname{Tr}^2 + \tfrac{1}{3}\theta_2 H^2/M^2. \tag{XXIV.113}$$

The older authors were prone to describe results of this kind as evidence of some failing in thermodynamics, or even to conclude from them that a gas could not justly be regarded as a continuum. Nowadays there is nothing surprising at all about a caloric that depends constitutively upon the temperature gradient as well as the velocity gradient; far more general calorics are commonly considered in the theory of materials with memory.

The result (XXIV.113) conforms with the theorem on h expressed by (VIII.51): h_M is the minimum value of h for given ρ; appealing to (XXIV.102)$_{2,8}$, we see that $\varpi_2 > 0$ and $\theta_2 > 0$, so the minimum is achieved if and only if Tr $= 0$ and H $= 0$. At the end of Chapter XI we have seen that the formal broad H-theorem implies no specific trend in the difference $h - h_M$ for a general solution. The result (XXIV.113) allows us to draw a simple and definite conclusion at this stage of the method of stretched fields. If we approximate M and ϖ_2 and θ_2 by constants (cf. (XVII.62) and (XXIV.172)$_{2,8}$, below), we see that *a sufficient condition for the difference* $\mathfrak{h}^{(2)} - h_M$ *to decrease is that both* Tr *and* H *shall decrease*, while *a sufficient condition for the difference to increase is that both* Tr *and* H *shall increase*. It is also possible that $\mathfrak{h}^{(2)} - h_M$ *shall remain constant, a sufficient condition being that* Tr *and* H *remain constant*. Such is the case in all the exact solutions given in Chapters XIV and XV. There H $= 0$ in all cases, while T $=$ const. The reader must recall that the results of the method of stretched fields can pertain only to the principal solutions in those chapters. Thus the properties of the corresponding $\mathfrak{h}^{(2)}$ can cast little if any light upon the meaning of the H-theorem, which is thought to reflect some tendency of general solutions to approach something like a corresponding principal solution. For the exact solutions in Chapters XIV and XV we have analysed such trends in explicit detail, but without any knowledge regarding h.

More serious is the fact that $\mathbf{s}$ no longer equals $-\mathbf{q}/(\tfrac{2}{3}\varepsilon)$. If we retain the identifications introduced in Chapter VIII, we conclude that the kinetic theory ceases at this stage of the iteration to support the Clausius–Duhem inequality of continuum thermodynamics. The fact that $\mathbf{s}^{(2)}$ is not only proportional to the gradient of ε but contains also a term which is an isotropic vector function of E_{km} and $\varepsilon_{,k}$ was only to be expected from the known representation of an isotropic vector-valued function of one symmetric tensor and one vector. We conclude from it that so far as the second stage, at least, in the method of

---

This formula agrees with (XXIV.112)$_2$ if $\tfrac{2}{3}\theta_4 = \tfrac{4}{5}$M. In Section (xi) of this chapter we shall see that for a gas of Maxwellian molecules M $= \tfrac{5}{2}$ and $\theta_4 = 3$. Thus, for such a gas, our result (XXIV.112)$_2$ confirms and generalizes MÜLLER's. To the best of our knowledge (XXIV.112)$_2$ with the general coefficient θ_4 and as an outcome of a systematic procedure applied to the true Maxwell–Boltzmann equation has not been published before.

In Footnotes 11 and 12 to Chapter VIII we have presented approximate formulae published by GRAD in 1949 for $h - h_M$ and $\mathbf{s} + \mathbf{q}/(\tfrac{2}{3}\varepsilon)$. Those approximations, when suitably corrected according to the remarks in Footnote 11 to that chapter and then specialized by replacing $\mathbf{P}$ and $\mathbf{q}$ by $\mathfrak{P}^{(1)}$ and $\mathfrak{q}^{(1)}$, respectively, also agree with (XXIV.112), when ϖ_2 and θ_2 are evaluated in first approximation by (XXIV.172)$_{2,8}$.

stretched fields the Clausius–Duhem inequality is supported if and only if $\mathbf{E}\,\mathrm{grad}\,\varepsilon = \mathbf{0}$; such is the case, in particular, when the temperature field is uniform or the gross flow is a dilatation. We remark that in the circumstances just specified neither $\mathbf{P}$ nor $\mathbf{q}$ as given by (XXIV.101) need vanish. In general $|\mathbf{s}|$ may be greater or less than $|\mathbf{q}/\tfrac{2}{3}\varepsilon|$, while s_k and $-q_k/\tfrac{2}{3}\varepsilon$ may have the same sign or opposite signs. These remarks refer only to (XXIV.112) and (XXIV.101) as they stand, disregarding such restrictions as the equations of balance may impose upon the fields which appear here as independent variables.

While the Clausius–Duhem inequality has proved again and again its efficiency as a mathematical basis for proofs of plausible and useful results, it is not an essential building block for continuum thermodynamics. I. MÜLLER, guided by the aspect of the kinetic theory that we have just presented, has adopted as a statement of the Second Law an inequality in which the flux of entropy is not necessarily $\mathbf{q}/\theta$ but is set up as a constitutive field along with $\mathbf{P}$, $\mathbf{q}$, and η. Cf. his paper, "On the entropy inequality", *Archive for Rational Mechanics and Analysis* **26**, 118–141 (1967). For those who regard the kinetic theory as a touchstone for assessing continuum theories, MÜLLER's is not the only way to reconcile the two. Instead, they may reconsider the traditional identification of $-\tau h$ with η, as we have remarked in Chapter VIII.

(x) *Extension to take account of the body force*

In Section (vii) of Chapter XXIII we extended the general theory of gross determiners to take account of the body force. The new gross determiner $\mathbb{Z}$ was required there to satisfy the generalized equations of gross determinism (XXIII.59). To extend the method of stretched fields from $\mathbb{G}$ to $\mathbb{Z}$, let us note that at the heart of the method is the universal property of $\mathbb{G}$ expressed by (XXIV.52). This property is precisely what suggested an expansion of $\mathbb{G}$ about a set of uniform fields. The corresponding result (XXIII.62) for $\mathbb{Z}$ suggests that we try once again an expansion about a uniform field, namely $\chi = (\mathbf{K}, \mathbf{0})$, but in particular a uniform field whose last three components vanish. In order to obtain such an expansion we replace the general argument $(\rho, \mathbf{b})$ of $\mathbb{Z}$ by a new argument $(\overset{\delta}{\rho}, \overset{[\delta]}{\mathbf{b}})$, $\overset{\delta}{\rho}$ being the stretched field corresponding to ρ as defined by (XXIV.5), and $\overset{[\delta]}{\mathbf{b}}$ being the following field determined by $\mathbf{b}$:

$$\overset{[\delta]}{b_k}(\mathbf{x}) \equiv \delta \overset{\delta}{b_k}(\mathbf{x}) = \delta b_k(\mathbf{x}_0 + \delta(\mathbf{x} - \mathbf{x}_0)). \qquad \text{(XXIV.114)}$$

As this field is proportional to a stretched field, certainly it becomes uniform as $\delta \to 0$, but more important than this is the fact that it approaches $\mathbf{0}$ as $\delta \to 0$. Hence we shall look to extending the method of stretched fields to $\mathbb{Z}$ by expanding $\mathbb{Z}(\mathfrak{S}\overset{\delta}{\rho}, \mathfrak{S}\overset{[\delta]}{\mathbf{b}})$ in a power series near $\delta = 0$. We may expect then to recover (XXIII.62) at the first stage in this new iterative procedure.

(x) EXTENSION TO TAKE ACCOUNT OF b

Extending the method used in Section (iii) for deriving the basic expansion of $\mathbb{G}$, we set

$$g(\lambda, \delta) \equiv \underset{\mathbf{r}}{\mathbb{Z}}(\rho(t, \mathbf{x}_0 + \lambda(\mathbf{x} - \mathbf{x}_0) + \delta\mathbf{r}), \delta\mathbf{b}(\mathbf{x}_0 + \lambda(\mathbf{x} - \mathbf{x}_0) + \delta\mathbf{r})),$$
$$= \mathbb{Z}(\overset{\delta}{\psi}, \delta\overset{\delta}{\mathbf{f}}), \qquad (XXIV.115)$$

in which

$$\overset{\delta}{\psi}_\alpha(\mathbf{r}) \equiv \rho_\alpha(t, \mathbf{x}_0 + \lambda(\mathbf{x} - \mathbf{x}_0) + \delta\mathbf{r}), \qquad \overset{\delta}{f}_k(\mathbf{r}) \equiv b_k(\mathbf{x}_0 + \lambda(\mathbf{x} - \mathbf{x}_0) + \delta\mathbf{r}), \qquad (XXIV.116)$$

t, $\mathbf{x}$, and λ being held fixed. We see from (XXIV.115)$_1$ and the definitions of $\overset{[\delta]}{\rho}$ and $\overset{[\delta]}{\mathbf{b}}$ that

$$g(\delta, \delta) = \mathbb{Z}(\mathfrak{S}\overset{\delta}{\rho}, \mathfrak{S}\overset{[\delta]}{\mathbf{b}}), \qquad (XXIV.117)$$

and so it is $g(\delta, \delta)$ that we wish to approximate. We use again the simple expansion formula (XXIV.9). All that remains to be done is to calculate the coefficients $\partial_\delta^n g(\delta, 0)$ using (XXIV.115) and then place these into (XXIV.9). In order to illustrate what emerges we carry out the calculation explicitly for the values $n = 0, 1, 2$. Using the chain rule repeatedly, we find that

$$g(\lambda, 0) = \mathbb{Z}(\overset{0}{\psi}, \mathbf{0}),$$

$$\partial_\delta g(\lambda, 0) = \partial_\psi \mathbb{Z}(\overset{0}{\psi}, \mathbf{0}) \left[\frac{d}{d\delta} \overset{\delta}{\psi} \bigg|_{\delta=0} \right] + \partial_\mathbf{f} \mathbb{Z}(\overset{0}{\psi}, \mathbf{0})[\overset{0}{\mathbf{f}}],$$

$$\partial_\delta^2 g(\lambda, 0) = \partial_\psi \mathbb{Z}(\overset{0}{\psi}, \mathbf{0}) \left[\frac{d^2}{d\delta^2} \overset{\delta}{\psi} \bigg|_{\delta=0} \right] + \partial_\mathbf{f} \mathbb{Z}(\overset{0}{\psi}, \mathbf{0}) \left[2 \frac{d}{d\delta} \overset{\delta}{\mathbf{f}} \bigg|_{\delta=0} \right] \qquad (XXIV.118)$$

$$+ \partial_\psi^2 \mathbb{Z}(\overset{0}{\psi}, \mathbf{0}) \left[\frac{d}{d\delta} \overset{\delta}{\psi} \bigg|_{\delta=0}, \frac{d}{d\delta} \overset{\delta}{\psi} \bigg|_{\delta=0} \right] + \partial_\mathbf{f}^2 \mathbb{Z}(\overset{0}{\psi}, \mathbf{0})[\overset{0}{\mathbf{f}}, \overset{0}{\mathbf{f}}]$$

$$+ 2 \partial_\psi \partial_\mathbf{f} \mathbb{Z}(\overset{0}{\psi}, \mathbf{0}) \left[\frac{d}{d\delta} \overset{\delta}{\psi} \bigg|_{\delta=0}, \overset{0}{\mathbf{f}} \right].$$

Moreover, using (XXIV.116) and the definitions of $\overset{\delta}{\rho}$ and $\overset{[\delta]}{\mathbf{b}}$, we obtain

$$\overset{0}{\psi}_\alpha = \rho_\alpha(t, \mathbf{x}_0 + \lambda(\mathbf{x} - \mathbf{x}_0)) = \overset{\lambda}{\rho}_\alpha(t, \mathbf{x}),$$

$$\frac{d}{d\delta} \overset{\delta}{\psi}_\alpha \bigg|_{\delta=0} = \rho_{\alpha,a}(t, \mathbf{x}_0 + \lambda(\mathbf{x} - \mathbf{x}_0)) r_a = \frac{1}{\lambda} \overset{\lambda}{\rho}_{\alpha,a}(t, \mathbf{x}) r_a,$$

$$\frac{d^2}{d\delta^2} \overset{\delta}{\psi}_\alpha \bigg|_{\delta=0} = \rho_{\alpha,ab}(t, \mathbf{x}_0 + \lambda(\mathbf{x} - \mathbf{x}_0)) r_a r_b = \frac{1}{\lambda^2} \overset{\lambda}{\rho}_{\alpha,ab}(t, \mathbf{x}) r_a r_b, \qquad (XXIV.119)$$

$$\overset{0}{f}_k = b_k(\mathbf{x}_0 + \lambda(\mathbf{x} - \mathbf{x}_0)) = \frac{1}{\lambda} \overset{[\lambda]}{b}_k(\mathbf{x}),$$

$$\frac{d}{d\delta} \overset{\delta}{f}_k \bigg|_{\delta=0} = b_{k,a}(\mathbf{x}_0 + \lambda(\mathbf{x} - \mathbf{x}_0))r_a = \frac{1}{\lambda^2} \overset{[\lambda]}{b}_{k,a}(\mathbf{x})r_a.$$

Note that the fields $\overset{\lambda}{\rho}$ and $\overset{[\lambda]}{b}$ and their derivatives here are all evaluated at t and $\mathbf{x}$, and so they do not depend upon $\mathbf{r}$. Thus we may take them outside of the linear arguments of the derivatives of $\mathbb{Z}$ appearing in (XXIV.118). Hence, placing (XXIV.119) into (XXIV.118) and setting $\lambda = \delta$, we obtain

$$g(\delta, 0) = \mathbb{Z}(\overset{\delta}{\rho}, 0),$$

$$\delta \, \partial_\delta g(\delta, 0) = \partial_\psi \mathbb{Z}(\overset{\delta}{\rho}, 0)[1_\alpha r_a]\overset{\delta}{\rho}_{\alpha,a} + \partial_\mathbf{f} \mathbb{Z}(\overset{\delta}{\rho}, 0)[1_a]\overset{[\delta]}{b}_a,$$

$$\delta^2 \, \partial_\delta^2 g(\delta, 0) = \partial_\psi \mathbb{Z}(\overset{\delta}{\rho}, 0)[1_\alpha r_a r_b]\overset{\delta}{\rho}_{\alpha,ab} + 2\partial_\mathbf{f} \mathbb{Z}(\overset{\delta}{\rho}, 0)[1_c r_a]\overset{[\delta]}{b}_{c,a}$$

$$+ \partial_\psi^2 \mathbb{Z}(\overset{\delta}{\rho}, 0)[1_\alpha r_a, 1_\beta r_b]\overset{\delta}{\rho}_{\alpha,a}\overset{\delta}{\rho}_{\beta,b} \quad \text{(XXIV.120)}$$

$$+ \partial_\mathbf{f}^2 \mathbb{Z}(\overset{\delta}{\rho}, 0)[1_a, 1_c]\overset{[\delta]}{b}_a \overset{[\delta]}{b}_c$$

$$+ 2 \, \partial_\psi \partial_\mathbf{f} \mathbb{Z}(\overset{\delta}{\rho}, 0)[1_\alpha r_a, 1_c]\overset{\delta}{\rho}_{\alpha,a} \overset{[\delta]}{b}_c,$$

in which t and $\mathbf{x}$ have been suppressed. Finally we place these expressions into (XXIV.9). Using (XXIV.120), (XXIII.61), and (XXIV.28)$_2$, we arrive at the following three successively more refined expansions of $\mathbb{Z}$:

$$\mathbb{Z}(\mathfrak{S}\overset{\delta}{\rho}, \mathfrak{S}\overset{[\delta]}{b}) = \mathbb{G}(\overset{\delta}{\rho}) + o(1),$$

$$= \mathbb{G}(\overset{\delta}{\rho}) + \mathbb{G}_\alpha^a(\overset{\delta}{\rho})\overset{\delta}{\rho}_{\alpha,a} + \mathbb{Z}_a(\overset{\delta}{\rho})\overset{[\delta]}{b}_a + o(\delta),$$

$$= \mathbb{G}(\overset{\delta}{\rho}) + \mathbb{G}_\alpha^a(\overset{\delta}{\rho})\overset{\delta}{\rho}_{\alpha,a} + \mathbb{Z}_a(\overset{\delta}{\rho})\overset{[\delta]}{b}_a$$

$$+ \mathbb{G}_\alpha^{ab}(\overset{\delta}{\rho})\overset{\delta}{\rho}_{\alpha,ab} + \mathbb{Z}_c^a(\overset{\delta}{\rho})\overset{[\delta]}{b}_{c,a} \quad \text{(XXIV.121)}$$

$$+ \mathbb{G}_{\alpha\beta}^{ab}(\overset{\delta}{\rho})\overset{\delta}{\rho}_{\alpha,a}\overset{\delta}{\rho}_{\beta,b} + \mathbb{Z}_{ac}(\overset{\delta}{\rho})\overset{[\delta]}{b}_a \overset{[\delta]}{b}_c$$

$$+ \mathbb{Z}_{\alpha c}^a(\overset{\delta}{\rho})\overset{\delta}{\rho}_{\alpha,a}\overset{[\delta]}{b}_c + o(\delta^2).$$

The new coefficients appearing here are special values of derivatives of $\mathbb{Z}$ with respect to $\mathbf{f}$ and $\mathbf{\psi}$ and are defined as follows:

$$\mathbb{Z}_k(\mathbf{K}) \equiv \partial_\mathbf{f} \mathbb{Z}(\mathbf{K}, 0)[1_k], \qquad \mathbb{Z}_m^k(\mathbf{K}) \equiv \partial_\mathbf{f} \mathbb{Z}(\mathbf{K}, 0)[1_m r_k],$$

$$\text{(XXIV.122)}$$

$$\mathbb{Z}_{km}(\mathbf{K}) \equiv \tfrac{1}{2} \partial_\mathbf{f}^2 \mathbb{Z}(\mathbf{K}, 0)[1_k, 1_m], \qquad \mathbb{Z}_{\alpha m}^k(\mathbf{K}) \equiv \partial_\psi \partial_\mathbf{f} \mathbb{Z}(\mathbf{K}, 0)[1_\alpha r_k, 1_m].$$

By no more than elementary calculations we can write a similar expansion for $\mathbb{Z}$ to within $o(\delta^N)$ for any given choice of N. For general N we call the resulting formula *the basic expansion of the gross determiner* $\mathbb{Z}$, generalizing our previous result (XXIV.28) for the special case described by $\mathbb{G}$.

The leading term in each of (XXIV.121)$_{1,2,3}$ is the function $\mathbf{K} \mapsto \mathbb{G}(\mathbf{K})$, which by (XXIV.51)$_2$ and (XXIV.52) is the unique Maxwellian density with principal moment $\mathbf{K}$. Moreover we have already analyzed the functions $\mathbb{G}_\alpha^k$, $\mathbb{G}_\alpha^{km}$, and $\mathbb{G}_{\alpha\beta}^{km}$ that appear in (XXIV.121). Therefore our problem is reduced to examination of the new coefficients (XXIV.122), using of course the fact that $\mathbb{Z}$ is a solution of the generalized equations of gross determinism (XXIII.59). Just as in Section (vi), we write these in the explicit form

$$\mathbb{D}\mathbb{Z} = \mathbb{C}\mathbb{Z},$$

$$\psi_\alpha(0) = \mathfrak{m} \int I_\alpha \mathbb{Z}(\chi),$$

(XXIV.123)

in which $\chi = (\psi, \mathbf{f})$ and the functional-differential operator $\mathbb{D}$, which is defined by (XXIII.60), may be written by use of (XXIII.11) in the explicit form

$$\mathbb{D}\,\mathbb{Z}(\chi) = \partial_\psi \underset{\mathbf{r}}{\mathbb{Z}}(\chi(\mathbf{r})) \bigg[v_a \psi_{,a}(\mathbf{r}) - \mathfrak{m} \int_{\mathbf{s}} v_a I \,\partial_\psi \mathbb{Z}(\chi(\mathbf{r}+\mathbf{s}))[\psi_{,a}(\mathbf{r}+\mathbf{s})]$$

$$- \mathfrak{m} \int_{\mathbf{s}} v_a I \,\partial_{\mathbf{f}} \mathbb{Z}(\chi(\mathbf{r}+\mathbf{s}))[\mathbf{f}_{,a}(\mathbf{r}+\mathbf{s})]$$

$$- \mathfrak{m} f_a(\mathbf{r}) \int_{\mathbf{s}} I \,\partial_{v_a} \mathbb{Z}(\chi(\mathbf{r}+\mathbf{s})) \bigg]$$

(XXIV.124)

$$+ \partial_{\mathbf{f}} \underset{\mathbf{r}}{\mathbb{Z}}(\chi(\mathbf{r}))[v_a \mathbf{f}_{,a}(\mathbf{r})] + f_a(0) \,\partial_{v_a} \underset{\mathbf{r}}{\mathbb{Z}}(\chi(\mathbf{r})).$$

Let us recall how we constructed the iterative systems governing $\mathbb{G}_\alpha^k$, $\mathbb{G}_\alpha^{km}$, and $\mathbb{G}_{\alpha\beta}^{km}$. The first two of these functions are special values of $\partial_\psi \mathbb{G}$, or equivalently special values of $\partial_\psi \mathbb{Z}$. The integral equations they satisfy were obtained by differentiating the equations of gross determinism once with respect to ψ and then specializing appropriately the arguments of $\partial_\psi \mathbb{G}$. Similarly $\mathbb{G}_{\alpha\beta}^{km}$ is a special value of $\partial_\psi^2 \mathbb{G}$, and we obtained the integral equation for it by differentiating the equations of gross determinism twice with respect to ψ and then choosing special arguments. Carrying this idea a step further, we differentiate (XXIV.123) once with respect to $\mathbf{f}$ and then specialize arguments. Since $\mathbb{Z}_k$ and $\mathbb{Z}_m^k$ are special values of $\partial_{\mathbf{f}} \mathbb{Z}$, we should expect in this way to obtain integral equations governing them. Specifically we consider the equations

$$2\mathbb{C}(\mathbb{Z}(\chi), \partial_{\mathbf{f}} \mathbb{Z}(\chi)[\mathbf{h}]) = \frac{d}{d\lambda} \mathbb{D}\mathbb{Z}(\psi, \mathbf{f} + \lambda \mathbf{h}) \bigg|_{\lambda=0},$$

$$0 = \mathfrak{m} \int I_\alpha \,\partial_{\mathbf{f}} \mathbb{Z}(\chi)[\mathbf{h}],$$

(XXIV.125)

obtained from (XXIV.123) by replacing $\mathbf{f}$ by $\mathbf{f} + \lambda\mathbf{h}$, differentiating with respect to λ and then setting $\lambda = 0$. Now we choose $\chi = (\psi, \mathbf{f}) = (\mathbf{K}, 0)$ and $\mathbf{h} = \mathbf{1}_k$ as suggested by (XXIV.122)$_1$ and then simplify the right-hand side of (XXIV.125)$_1$. The steps involved in this simplification are long though only straightforward, and we have already presented a number of calculations of this type in Section (vi); more may be found in Appendix A. The results, however, are particularly simple:

$$2\mathbb{C}(F_\mathbf{M}, \mathbb{Z}_k) = 0, \qquad 0 = \mathfrak{m} \int I_\alpha \mathbb{Z}_k. \qquad \text{(XXIV.126)}$$

We know by the proof of effectiveness in Section (vii) that these equations have a unique solution. Clearly this solution is

$$\mathbb{Z}_k = 0. \qquad \text{(XXIV.127)}$$

Next we set $\chi = (\psi, \mathbf{f}) = (\mathbf{K}, 0)$ and $\mathbf{h} = \mathbf{1}_m r_k$ in (XXIV.125) so as to obtain a system of integral equations for $\mathbb{Z}_m^k$. Again the calculations are long, but they yield a simple result:

$$2\mathbb{C}(F_\mathbf{M}, \mathbb{Z}_m^k) = \mathbb{F}_m^k, \qquad 0 = \mathfrak{m} \int I_\alpha \mathbb{Z}_m^k. \qquad \text{(XXIV.128)}$$

In these equations $\mathbb{F}_m^k$ is given by (XXIV.78)$_1$ and may be considered known.

We see from (XXIV.122)$_3$ that $\mathbb{Z}_{km}$ is a special value of $\partial_\mathbf{f}^2 \mathbb{Z}$. This suggests that we look at the equations obtained by differentiating (XXIV.125) with respect to $\mathbf{f}$, or equivalently by differentiating the generalized equations of gross determinism twice with respect to $\mathbf{f}$. To do this, we replace $\mathbf{h}$ by $\mathbf{h}^1$ and $\mathbf{f}$ by $\mathbf{f} + \nu\mathbf{h}^2$ in (XXIV.125), then differentiate with respect to ν, and finally set $\nu = 0$. We obtain the system

$$2\mathbb{C}(\mathbb{Z}(\chi), \partial_\mathbf{f}^2 \mathbb{Z}(\chi)[\mathbf{h}^1, \mathbf{h}^2]) = \frac{d}{d\nu}\frac{d}{d\lambda} \mathbb{D}\mathbb{Z}(\psi, \mathbf{f} + \lambda\mathbf{h}^1 + \nu\mathbf{h}^2)|_{\lambda=0}|_{\nu=0}$$
$$- 2\mathbb{C}(\partial_\mathbf{f} \mathbb{Z}(\chi)[\mathbf{h}^1], \partial_\mathbf{f} \mathbb{Z}(\chi)[\mathbf{h}^2]), \qquad \text{(XXIV.129)}$$

$$0 = \mathfrak{m} \int I_\alpha \partial_\mathbf{f}^2 \mathbb{Z}(\chi)[\mathbf{h}^1, \mathbf{h}^2].$$

Next we set $\chi = (\psi, \mathbf{f}) = (\mathbf{K}, 0)$, $\mathbf{h}^1 = \mathbf{1}_k$, and $\mathbf{h}^2 = \mathbf{1}_m$ as suggested by (XXIV.122)$_3$ and then simplify. We get the following system of integral equations for $\mathbb{Z}_{km}$:

$$2\mathbb{C}(F_\mathbf{M}, \mathbb{Z}_{km}) = 0, \qquad 0 = \mathfrak{m} \int I_\alpha \mathbb{Z}_{km}. \qquad \text{(XXIV.130)}$$

As with (XXIV.126) the unique solution of this system is

$$\mathbb{Z}_{km} = 0. \qquad \text{(XXIV.131)}$$

Since $\mathbb{Z}^k_{\alpha m}$ is a special value of $\partial_\psi \partial_f \mathbb{Z}$, we look at the equations obtained from (XXIV.125) by differentiating with respect to ψ. These equations are

$$2\mathbb{C}(\mathbb{Z}(\chi), \partial_\psi \partial_f \mathbb{Z}(\chi)[\xi, \mathbf{h}]) = \frac{d}{dv}\frac{d}{d\lambda}\mathbb{D}\mathbb{Z}(\psi + v\xi, \mathbf{f} + \lambda\mathbf{h})|_{\lambda=0}|_{v=0}$$
$$- 2\mathbb{C}(\partial_\psi \mathbb{Z}(\chi)[\xi], \partial_f \mathbb{Z}(\chi)[\mathbf{h}]), \qquad \text{(XXIV.132)}$$

$$0 = \mathfrak{m}\int I_\alpha \partial_\psi \partial_f \mathbb{Z}(\chi)[\xi, \mathbf{h}].$$

Next we set $\chi = (\psi, \mathbf{f}) = (\mathbf{K}, 0)$, $\xi = 1_\alpha r_k$, and $\mathbf{h} = 1_m$, and then simplify the result. We obtain

$$2\mathbb{C}(F_{\mathbf{M}}, \mathbb{Z}^k_{\alpha m}) = 0, \qquad 0 = \mathfrak{m}\int I_\gamma \mathbb{Z}^k_{\alpha m}. \qquad \text{(XXIV.133)}$$

As before, we see that the unique solution of this system is

$$\mathbb{Z}^k_{\alpha m} = 0. \qquad \text{(XXIV.134)}$$

Let us look back now at the expansions of $\mathbb{Z}$ in the light of the solutions (XXIV.127), (XXIV.131), and (XXIV.134). From (XXIV.121)$_2$ we see that to within terms of order $o(\delta)$ the basic expansions of $\mathbb{G}$ and $\mathbb{Z}$ agree. If we carry this fact over to the gross determiners $\mathfrak{P}$ and $\mathfrak{q}$, we find that $\mathbb{G}$ and $\mathbb{Z}$ both deliver the 0th and first approximations (XXIV.85) and (XXIV.90), respectively: *To order $o(\delta)$ the functionals $\mathfrak{P}$ and $\mathfrak{q}$ are independent of the body force.*

Of the four new terms entering (XXIV.121)$_3$, three are null. The fourth, that multiplying grad $\mathbf{b}$, does not vanish, and so we see that to order $o(\delta^2)$ we may not neglect $\mathbf{b}$. Let us see what changes occur as a result in our formulae for $\mathfrak{P}^{(2)}$ and $\mathfrak{q}^{(2)}$. Because (XXIV.121)$_3$ differs from the corresponding expansion for $\mathbb{G}$ only by the term $\mathbb{Z}^{[\delta]a}_c b_{c,a}$, we may obtain the approximate gross determiners $\mathfrak{P}^{(2)}$ and $\mathfrak{q}^{(2)}$ corresponding to $\mathbb{Z}$ by adding to the right-hand sides of (XXIV.101)$_1$ and (XXIV.101)$_2$ the terms

$$\mathfrak{m}\int c_{\{k}c_{m\}}\mathbb{Z}^a_c b_{c,a} \qquad \text{(XXIV.135)}$$

and

$$\mathfrak{m}\int \tfrac{1}{2}c^2 c_k \mathbb{Z}^a_c b_{c,a}, \qquad \text{(XXIV.136)}$$

respectively. To evaluate these, we use the integral equations (XXIV.128).

Recall that $\mathbb{F}^k_m$ is a symmetric, traceless, isotropic tensor-valued function of $\mathbf{c}$. Using this fact, we may analyze (XXIV.128) just as we did (XXIV.73)$_2$ and show that $\mathbb{Z}^k_m$ is also symmetric, traceless, and isotropic. Hence $\mathbb{Z}^k_m$ has the form $c_{\{k}c_{m\}}E(\varepsilon, c^2)$ for some function E. In particular, we note that $\mathbb{Z}^k_m$ is an even function of $\mathbf{c}$, whence we conclude that the term (XXIV.136) is null. In order to evaluate (XXIV.135), we apply the technique illustrated in (XXIV.97), and we find that

$$\mathfrak{m} \int c_{\{k} c_{m\}} \mathbb{Z}^r_s = \frac{2\varepsilon\mathfrak{m}}{3} \int \frac{\mathbb{Z}^r_s}{F_M} 2\mathbb{C}(F_M, F^m_k) = \frac{2\varepsilon\mathfrak{m}}{3} \int \frac{F^m_k}{F_M} 2\mathbb{C}(F_M, \mathbb{Z}^r_s). \quad \text{(XXIV.137)}$$

It follows now from (XXIV.128), (XXIV.78)$_1$, (XXIVB.20)$_5$, (XXIV.91)$_1$, and (XXIV.102)$_2$ that

$$\begin{aligned}
\mathfrak{m} \int c_{\{k} c_{m\}} \mathbb{Z}^r_s &= \frac{2\varepsilon\mathfrak{m}}{3} \int c_{\{k} c_{m\}} c_{\{r} c_{s\}} \frac{B^2}{F_M}, \\
&= \frac{2\varepsilon\mathfrak{m}}{3} \frac{2}{15} \delta_{r\{k} \delta_{m\}s} \int c^4 \frac{B^2}{F_M}, \quad \text{(XXIV.138)} \\
&= \frac{\mu^2}{p} \varpi_2 \delta_{r\{k} \delta_{m\}s}.
\end{aligned}$$

Placing this result into (XXIV.135), we obtain the expression

$$\frac{\mu^2}{p} \varpi_2 b_{\{k,m\}}. \quad \text{(XXIV.139)}$$

Recalling that (XXIV.136) is null, we conclude that *the second approximations to $\mathfrak{P}$ and $\mathfrak{q}$ calculated using the gross determiner $\mathbb{Z}$ are given by the formula* (XXIV.4), which has been obtained heretofore by ENSKOG's procedure.

We may calculate also the new approximate gross determiners for $\mathfrak{h}$ and $\mathfrak{s}$. By placing (XXIV.120) into (XXIV.37) and using (XXIV.122), (XXIV.127), (XXIV.131), and (XXIV.134) we find that to order $o(\delta)$ the expansion for $\mathbb{Z} \log \mathbb{Z}$ agrees with that of $\mathbb{G} \log \mathbb{G}$, but to order $o(\delta^2)$ we must add to the expansion of $\mathbb{G} \log \mathbb{G}$ the term $(1 + \log F_M) \mathbb{Z}^a_c b^{[\delta]}_{c,a}$. This means that $\mathbb{G}$ and $\mathbb{Z}$ both give the same 0th and first approximations (XXIV.106) and (XXIV.108), respectively. To obtain $\mathfrak{h}^{(2)}$ and $\mathfrak{s}^{(2)}$ for $\mathbb{Z}$, we must add to the right-hand sides of (XXIV.112)$_1$ and (XXIV.112)$_2$ the terms

$$\frac{\mathfrak{m}}{\rho} \int (1 + \log F_M) \mathbb{Z}^a_c b_{c,a} \quad \text{(XXIV.140)}$$

and

$$\mathfrak{m} \int c_k (1 + \log F_M) \mathbb{Z}^a_c b_{c,a}, \quad \text{(XXIV.141)}$$

respectively. Since $1 + \log F_M$ is a summational invariant, we see from (XXIV.128)$_2$ that the first of these two terms is null. Since $\mathbb{Z}^r_s$ and $1 + \log F_M$ are both even functions of $\mathbf{c}$, the second term is also null. *To order $o(\delta^2)$ the* *gross determiners* $\mathfrak{h}$ *and* $\mathfrak{s}$ *are independent of the body force.* Thus $\mathfrak{h}^{(2)}$ and $\mathfrak{s}^{(2)}$ are given by our formulae (XXIV.112) even if $\mathbf{b} \neq \mathbf{0}$.

The reader will have noticed that the choice of limit process for $\mathbf{b}$, a process defined in terms of (XXIV.114), is not dictated by that used for ρ. ENSKOG's method was devised expressly so as to make the Stokes–Kirchhoff–Fourier constitutive relations, neither more nor less, emerge at the first stage. We have selected a particular method of stretching $\mathbf{b}$ in such a way as to make the second stage according to our process deliver the results obtained at the second stage by ENSKOG's process, neither more nor less. ENSKOG might have chosen a different iterative scheme; we might have chosen different ways of stretching both ρ and $\mathbf{b}$. We could even stretch ρ and $\mathbf{b}$ through two different parameters and so derive a whole family of systems of approximation. GRAD's criticism of ENSKOG's process, quoted above in Footnote 9, should remind the reader that in it there is more than a little arbitrariness.

(xi) *Explicit results for Maxwellian molecules*

For a gas of Maxwellian molecules we can exhibit explicitly the solutions of the integral equations for A and B, and using these compute values for each of the functions μ, M, ϖ_1, …, θ_5 defined previously for general molecular models. To obtain these solutions we calculate first some values of the bilinear form $\mathbb{C}(\ ,\ )$. By using (VII.13) and the fact that $F'_M F'_{M*} = F_M F_{M*}$ we obtain

$$\begin{aligned}\mathbb{C}(F_M, F_M g) &= \tfrac{1}{2} \int \int_{\mathscr{S}} w F_M F_{M*} (g' + g'_* - g - g_*), \\ &= \tfrac{1}{2} F_M \int F_{M*}\, \mathbb{B}g,\end{aligned} \quad (\text{XXIV}.142)$$

$\mathbb{B}$ being the operator defined by $(\text{XII}.13)_2$. Since 1, c_k, and c^2 are summational invariants, we see that

$$\begin{aligned}\mathbb{C}(F_M, F_M) &= 0, \\ \mathbb{C}(F_M, F_M c_k) &= 0, \\ \mathbb{C}(F_M, F_M c^2) &= 0.\end{aligned} \quad (\text{XXIV}.143)$$

When g is not a summational invariant, $\mathbb{C}(F_M, F_M g) \neq 0$ in general, but we can still calculate some of its values. Using (XII.21) specialized to the case in which $\mathbb{k} = 5$, we may calculate $\mathbb{C}(F_M, F_M c_k c_m)$. In view of the integration formula (XXIVB.21) we obtain

$$\mathbb{C}(F_M, F_M c_k c_m) = -\tfrac{3}{4}\rho\mathfrak{a}\sqrt{\frac{2\mathfrak{g}}{\mathfrak{m}^3}}\, c_{\{k} c_{m\}} F_M. \quad (\text{XXIV}.144)$$

In the same way we may use (XII.22)$_2$, also specialized to the case in which $\mathbb{k} = 5$, to show that[28]

$$\mathbb{C}(F_M, F_M c^2 c_k) = -\rho \varepsilon \mathfrak{a} \sqrt{\frac{2\mathfrak{g}}{\mathfrak{m}^3}} c_k \left(\frac{c^2}{2\varepsilon} - \frac{5}{3}\right) F_M. \qquad \text{(XXIV.145)}$$

In view of (XXIV.143), (XXIV.144), and (XXIV.145), one solution of the integral equations (XXIV.73)$_{2,3}$ is

$$\mathbb{F}_k^m = -\frac{1}{\rho \varepsilon \mathfrak{a}} \sqrt{\frac{\mathfrak{m}^3}{2\mathfrak{g}}} c_{\{m} c_{k\}} F_M,$$

$$\mathbb{F}_4^k = -\frac{3}{2\rho \varepsilon \mathfrak{a}} \sqrt{\frac{\mathfrak{m}^3}{2\mathfrak{g}}} c_k \left(\frac{c^2}{2\varepsilon} - \frac{5}{3}\right) F_M. \qquad \text{(XXIV.146)}$$

(The reader should reconsider (VIII.33) and the remarks following it, which show that the terms of (XXIV.71), rendered explicit by (XXIV.146), approximate $\mathbb{G}$ by a function which is not a molecular density.) The general solution is found by adding to each of these a summational invariant. From our general analysis of the iterative system we know that there is a unique solution satisfying also the conditions (XXIV.73)$_4$. Using (XXIVB.21), we can show that the unique solution is in fact (XXIV.146), and so

$$B = -\frac{1}{\varepsilon \mathfrak{a}} \sqrt{\frac{\mathfrak{m}^3}{2\mathfrak{g}}} \frac{F_M}{\rho} = -\frac{\mathfrak{m}}{2\varepsilon \mathbb{A}_{2|2}} \frac{F_M}{\rho},$$

$$A = -\frac{3}{2\varepsilon \mathfrak{a}} \sqrt{\frac{\mathfrak{m}^3}{2\mathfrak{g}}} \left(\frac{c^2}{2\varepsilon} - \frac{5}{3}\right) \frac{F_M}{\rho} = -\frac{3\mathfrak{m}}{4\varepsilon \mathbb{A}_{2|2}} \left(\frac{c^2}{2\varepsilon} - \frac{5}{3}\right) \frac{F_M}{\rho}. \qquad \text{(XXIV.147)}$$

If we place these functions into (XXIV.91) and (XXIV.102) and carry out the integrations indicated using (XXIVB.21) and (XXIVB.27), we obtain finally the expressions

$$\mu = \frac{1}{3\mathfrak{a}} \sqrt{\frac{2\mathfrak{m}^3}{\mathfrak{g}}} \frac{2}{3} \varepsilon; \qquad \text{(XIII.14)}_{3r}$$

$$M = \tfrac{5}{2}; \qquad \text{(XIII.18)}_r$$

$$\varpi_1 = \tfrac{10}{3}, \quad \varpi_2 = 2, \quad \varpi_3 = 3, \quad \varpi_4 = 0, \quad \varpi_5 = 3, \quad \varpi_6 = 8,$$
$$\theta_1 = \tfrac{75}{8}, \quad \theta_2 = \tfrac{45}{8}, \quad \theta_3 = -3, \quad \theta_4 = 3, \quad \theta_5 = \tfrac{117}{4}. \qquad \text{(XXIV.148)}$$

As we should expect, the values of μ and M are the same as those which we obtained by MAXWELL's method in Chapter XIII; here they have been derived as special values of formulae valid for any spherically symmetric molecular model.

---

[28] Formulae such as these are common in studies of the linearized collisions operator $g \mapsto \mathbb{C}(F_M, F_M g)$. WANG CHANG & UHLENBECK [1952, §3] have generalized them so as to exhibit explicitly the complete set of proper functions for this operator in a gas of Maxwellian molecules.

(xii) *Explicit first approximations for general molecular models*

We know of no exact solutions for A and B for any molecular model other than the Maxwellian one. We can, however, approximate these functions by using a procedure akin to GRAD's method of truncation, which we have described in Section (iii) of Chapter XVII. To begin with we introduce expansions similar to (XVII.9) for A and B. We assume given two sets of functions $\psi_0, \psi_1, \ldots$ and $\phi_0, \phi_1, \ldots$ such that A and B may be written in the forms

$$B(\varepsilon, c^2) = F_M \sum_{n=0}^{\infty} B_n(\varepsilon)\psi_n(\kappa^2),$$

$$A(\varepsilon, c^2) = F_M \sum_{n=0}^{\infty} A_n(\varepsilon)\phi_n(\kappa^2),$$

(XXIV.149)

κ being the dimensionless random velocity

$$\kappa_k \equiv \sqrt{\frac{3}{2\varepsilon}} c_k.$$

(XVII.6)$_r$

As suggested by GRAD's analysis, it is particularly useful to work with orthogonal sets of functions[29], and to this end we introduce the Sonine polynomials

$$S_r^{(n)}(y) \equiv \sum_{p=0}^{n} (-y)^p \frac{(r+n)_{n-p}}{p!(n-p)!}, \qquad n = 0, 1, \ldots, \qquad \text{(XXIV.150)}$$

p_q denoting the product $p(p-1) \cdots (p-q+1)$. For each positive number r these form a complete orthogonal set, the orthogonality condition being

$$\int_0^{\infty} y^r e^{-y} S_r^{(m)}(y) S_r^{(n)}(y) \, dy = \frac{\Gamma(r+n+1)}{n!} \delta_{mn}.$$

(XXIV.151)

---

[29] CHAPMAN [1916, *1*] and ENSKOG [1917] did not perceive this fact; the former chose

$$\psi_n(\kappa^2) = \phi_n(\kappa^2) = \kappa^{2n}, \qquad n = 0, 1, \ldots,$$

and the latter

$$\psi_n(\kappa^2) = \kappa^{2n}, \qquad n = 0, 1, \ldots,$$

$$\phi_n(\kappa^2) = \kappa^{2n} - (n + \tfrac{3}{2})(n + \tfrac{1}{2}) \cdots \tfrac{5}{2}, \qquad n = 0, 1, \ldots.$$

This difference of choice led, of course, to different approximations of μ and M (see Footnote 5). Here we follow CHAPMAN & COWLING [1939, §§7.5, 7.51, 7.52] in using Sonine polynomials.

In the expansion for B we choose

$$\psi_n(\kappa^2) = S_{\frac{3}{2}}^{(n)}(\tfrac{1}{2}\kappa^2), \qquad n = 0, 1, \ldots, \qquad \text{(XXIV.152)}$$

while for A we choose

$$\phi_n(\kappa^2) = S_{\frac{3}{2}}^{(n)}(\tfrac{1}{2}\kappa^2), \qquad n = 0, 1, \ldots. \qquad \text{(XXIV.153)}$$

These special choices are suggested, in retrospect, by our subsequent analysis of A and B.

The statements $(XXIV.73)_2$ and $(XXIV.73)_3$ are each equivalent to a single scalar integral equation. To obtain these, we place (XXIV.78) into $(XXIV.73)_{2,3}$ and then multiply the former by $c_{\{k}c_{m\}}$ and the latter by c_m. Using (XXIV.92) and (XVII.6), we obtain[30]

$$2\kappa_{\{a}\kappa_{b\}}\mathbb{C}(F_\mathrm{M}, \kappa_{\{a}\kappa_{b\}}B) = \frac{1}{\varepsilon}\kappa^4 F_\mathrm{M}, \qquad 2\kappa_a \mathbb{C}(F_\mathrm{M}, \kappa_a A) = \frac{1}{2\varepsilon}(\kappa^2 - 5)\kappa^2 F_\mathrm{M}.$$

(XXIV.154)

Moreover, by (XXIVB.12), (XXIV.72), and (XVII.6), the integrability conditions $(XXIV.73)_4$ reduce to

$$\int \kappa^2 A = 0. \qquad \text{(XXIV.155)}$$

Using the general expansion of A, we may write this last condition in the equivalent form

$$\sum_{n=0}^{\infty} A_n \alpha^n = 0, \qquad \text{(XXIV.156)}$$

α^n being defined as follows:

$$\alpha^n \equiv \int \kappa^2 \phi_n F_\mathrm{M}. \qquad \text{(XXIV.157)}$$

---

[30] PEKERIS & ALTERMAN [1957] have shown, for a gas of ideal spheres, how one may reduce each of these scalar integral equations to a differential equation. They observed that BOLTZMANN [1881, Part III, Eq. (38)] had arrived at an erroneous equation of this kind for B, "the derivation covering 160 pages, of which over 40 are pure mathematics, in the sense that they are entirely word-pure or are adorned with just one sentence." A preceding paper by PEKERIS [1955] on self-diffusion contains further interesting historical remarks.

By integrating these differential equations numerically PEKERIS & ALTERMAN provided approximate values of μ and M without recourse to the method of truncation which we are presenting here. *Cf.* Footnote 34 to this chapter.

(xii) APPROXIMATIONS FOR GENERAL GASES   483

Next we multiply (XXIV.154)$_{1,2}$ by ψ_m and ϕ_m, respectively, integrate with respect to **c**, and finally substitute the expansions (XXIV.149). We obtain, then, the equivalent conditions

$$\sum_{n=0}^{\infty} \beta_{mn} B_n = \beta_m, \quad m = 0, 1, \ldots,$$

$$\sum_{n=0}^{\infty} \alpha_{mn} A_n = \alpha_m, \quad m = 0, 1, \ldots,$$
(XXIV.158)

in which

$$\beta_{mn} \equiv 2 \int \psi_m \kappa_{\{a} \kappa_{b\}} \mathbb{C}(F_M, F_M \psi_n \kappa_{\{a} \kappa_{b\}}), \quad \alpha_{mn} \equiv 2 \int \phi_m \kappa_a \mathbb{C}(F_M, F_M \phi_n \kappa_a),$$

$$\beta_m \equiv \frac{1}{\varepsilon} \int \psi_m \kappa^4 F_M, \quad \alpha_m \equiv \frac{1}{2\varepsilon} \int \phi_m (\kappa^2 - 5) \kappa^2 F_M.$$
(XXIV.159)

The basic equations for A and B are now (XXIV.156) and (XXIV.158).

Once functions ψ_m and ϕ_m have been chosen, the quantities defined by (XXIV.157) and (XXIV.159) may be calculated explicitly. For example, since $S_{\frac{3}{2}}^{(0)}(y) = 1$, using the specific choice (XXIV.153) and the orthogonality condition (XXIV.151) we find that

$$\alpha^n = \int \kappa^2 S_{\frac{3}{2}}^{(0)}(\tfrac{1}{2}\kappa^2) S_{\frac{3}{2}}^{(n)}(\tfrac{1}{2}\kappa^2) F_M,$$

$$= \frac{4\rho}{\sqrt{\pi}\, \mathfrak{m}} \int_0^\infty (\tfrac{1}{2}\kappa^2)^{\frac{3}{2}} e^{-\frac{1}{2}\kappa^2} S_{\frac{3}{2}}^{(0)}(\tfrac{1}{2}\kappa^2) S_{\frac{3}{2}}^{(n)}(\tfrac{1}{2}\kappa^2) d(\tfrac{1}{2}\kappa^2), \quad \text{(XXIV.160)}$$

$$= \begin{cases} 3\rho/\mathfrak{m} & \text{if } n = 0, \\ 0 & \text{if } n \neq 0. \end{cases}$$

Therefore (XXIV.156) becomes nothing more than

$$A_0 = 0. \qquad \text{(XXIV.161)}$$

Since $S_{\frac{3}{2}}^{(1)}(\tfrac{1}{2}\kappa^2) = (5 - \kappa^2)/2$, from (XXIV.153) and (XXIV.151) we get

$$\alpha_m = -\frac{1}{\varepsilon} \int S_{\frac{3}{2}}^{(1)}(\tfrac{1}{2}\kappa^2) S_{\frac{3}{2}}^{(m)}(\tfrac{1}{2}\kappa^2) \kappa^2 F_M,$$

$$= -\frac{4\rho}{\sqrt{\pi}\, \mathfrak{m}\varepsilon} \int_0^\infty (\tfrac{1}{2}\kappa^2)^{\frac{3}{2}} e^{-\frac{1}{2}\kappa^2} S_{\frac{3}{2}}^{(1)}(\tfrac{1}{2}\kappa^2) S_{\frac{3}{2}}^{(m)}(\tfrac{1}{2}\kappa^2) d(\tfrac{1}{2}\kappa^2),$$
(XXIV.162)

$$= \begin{cases} -\dfrac{15\rho}{2\mathfrak{m}\varepsilon} & \text{if } m = 1, \\ 0 & \text{if } m \neq 1. \end{cases}$$

Similarly we find

$$\beta_m = \frac{1}{\varepsilon} \int \kappa^4 S^{(0)}_{\frac{3}{2}}(\tfrac{1}{2}\kappa^2) S^{(m)}_{\frac{3}{2}}(\tfrac{1}{2}\kappa^2) F_M,$$

$$= \frac{8\rho}{\sqrt{\pi}\, \mathfrak{m}\varepsilon} \int_0^\infty (\tfrac{1}{2}\kappa^2)^{\frac{3}{2}} e^{-\frac{1}{2}\kappa^2} S^{(0)}_{\frac{3}{2}}(\tfrac{1}{2}\kappa^2) S^{(m)}_{\frac{3}{2}}(\tfrac{1}{2}\kappa^2)\, d(\tfrac{1}{2}\kappa^2), \qquad \text{(XXIV.163)}$$

$$= \begin{cases} \dfrac{15\rho}{\mathfrak{m}\varepsilon} & \text{if } m = 0, \\ 0 & \text{if } m \ne 0. \end{cases}$$

The coefficients β_{mn} and α_{mn} are analogues here of GRAD's coefficients $\mathbb{A}_{rmn}$. Each is a function of ε and the kinetic constitutive quantities, and we may express them, just as we have done in GRAD's analysis, in terms of the functions $\mathbb{A}_{p|m|n}$ defined by (XVII.31). For example, from (XXIVB.27)$_1$ and (XVII.6) we may write

$$\beta_{00} = 2 \int \kappa_{\{a}\kappa_{b\}} \mathbb{C}(F_M, F_M \kappa_{\{a}\kappa_{b\}}) = -30n^2 \mathbb{A}_{2|2|2}. \qquad \text{(XXIV.164)}$$

A number of the coefficients α_{mn} vanish identically. Since $\phi_0 = S^{(0)}_{\frac{3}{2}} = 1$, we see that $\phi_0 \mathbf{c}$ is a summational invariant, and so from (VII.22) and (XXIV.159)$_2$ we find that

$$\alpha_{0n} = 0, \qquad n = 0, 1, \ldots. \qquad \text{(XXIV.165)}$$

The remaining coefficients may be expressed in terms of the functions $\mathbb{A}_{p|m|n}$ in exactly the way we followed to obtain (XXIV.164). For example, from (XXIVB.27)$_2$ and (XVII.6) we find that

$$\alpha_{11} = \tfrac{1}{2} \int \kappa_a(5 - \kappa^2) \mathbb{C}(F_M, F_M \kappa_a(5 - \kappa^2)) = -15n^2 \mathbb{A}_{2|2|2}. \qquad \text{(XXIV.166)}$$

In (XXIV.158) we have two infinite systems of linear algebraic equations, one system for each of the two sets of coefficients A_n and B_n. Using somewhat formally the theory of finite linear systems, we may express each A_n and B_n as a ratio of two infinite-dimensional determinants, and from these and (XXIV.149) we may construct A and B. This path leads, of course, to similar expressions for μ and M in terms of infinite determinants[31]. To avoid expressions of this type, we introduce a method of truncation. First we truncate the expansions for A

---

[31] CHAPMAN [1916, 2, Eqs. (140), (237), (248)] obtained expressions for μ and $M\mu$ of this type without first deriving the integral equations (XXIV.73). As a result he did not prove existence and uniqueness of the solutions of the formally equivalent systems (XXIV.158) but simply presumed that a solution existed and proceeded to approximate it. ENSKOG [1917, Eqs. (61), (68), (69)], in contrast, derived (XXIV.73) first, and only after giving a formal proof of existence for it did he proceed to approximate its solutions.

(xii)   APPROXIMATIONS FOR GENERAL GASES

and B after M terms and view the resulting quantities $A^{(M)}$ and $B^{(M)}$ as approximate solutions[32]. In view of (XXIV.161), then, we set

$$B^{(M)} \equiv F_M \sum_{n=0}^{M-1} \hat{B}_n \psi_n, \qquad A^{(M)} \equiv F_M \sum_{n=1}^{M} \hat{A}_n \phi_n. \qquad \text{(XXIV.167)}$$

To determine the M coefficients $\hat{B}_0, \ldots, \hat{B}_{M-1}$, we set $B_M = B_{M+1} = \ldots = 0$ in (XXIV.158)$_1$ and then retain only the first M equations from the system. To determine the M coefficients $\hat{A}_1, \ldots, \hat{A}_M$, we set $A_{M+1} = A_{M+2} = \ldots = 0$ in (XXIV.158)$_2$ and then retain only the equations corresponding to $m = 1, \ldots, M$. (We note from (XXIV.165) and (XXIV.162)$_4$ that the equation in (XXIV.158)$_2$ corresponding to $m = 0$ is satisfied identically.) Thus $\hat{B}_0, \ldots, \hat{A}_M$ are determined by the finite systems

$$\sum_{n=0}^{M-1} \beta_{mn} \hat{B}_n = \beta_m, \qquad m = 0, \ldots, M-1,$$

$$\sum_{n=1}^{M} \alpha_{mn} \hat{A}_n = \alpha_m, \qquad m = 1, \ldots, M. \qquad \text{(XXIV.168)}$$

Rather than analyse (XXIV.168) in detail for general M, we shall rest content to compute explicitly $A^{(1)}$ and $B^{(1)}$. According to (XXIV.167)

$$B^{(1)} = F_M \hat{B}_0 \psi_0, \qquad A^{(1)} = F_M \hat{A}_1 \phi_1, \qquad \text{(XXIV.169)}$$

and $\hat{B}_0$ and $\hat{A}_1$ are determined by the following special case of (XXIV.168):

$$\beta_{00} \hat{B}_0 = \beta_0, \qquad \alpha_{11} \hat{A}_1 = \alpha_1. \qquad \text{(XXIV.170)}$$

In view of (XXIV.162), (XXIV.163), (XXIV.164), (XXIV.166), and (XXIV.169) we obtain

$$B^{(1)} = \frac{\beta_0}{\beta_{00}} F_M = -\frac{\mathfrak{m}}{2\varepsilon \mathbb{A}_{2|2|2}} \frac{F_M}{\rho},$$

$$A^{(1)} = \frac{\alpha_1}{\alpha_{11}} \frac{(5 - \kappa^2)}{2} F_M = -\frac{3\mathfrak{m}}{4\varepsilon \mathbb{A}_{2|2|2}} \left(\frac{c^2}{2\varepsilon} - \frac{5}{3} \right) \frac{F_M}{\rho}. \qquad \text{(XXIV.171)}$$

If we recall that for a gas of Maxwellian molecules $\mathbb{A}_{2|2|2} = \mathbb{A}_{2|2}$, we see that these reduce for such a gas to the exact solutions (XXIV.147). Finally, we may use these approximations to compute approximate values of μ, M, $\theta_1, \ldots, \varpi_6$.

---

[32] For various molecular models BURNETT [1935] considered the question of whether the expansions (XXIV.149) were valid and whether $A^{(M)}$ and $B^{(M)}$ converged to A and B, respectively, as $M \to \infty$.

This may be done in precisely the way we obtained (XXIV.148) from (XXIV.147), and it leads to the following values[33]:

$$\mu^{(1)} = \frac{2\mathfrak{m}\varepsilon}{9\mathbb{A}_{2|2|2}};\qquad\text{(XVII.61)}_r$$

$$\mathbf{M}^{(1)} = \tfrac{5}{2};\qquad\text{(XVII.62)}_r$$

$$\varpi_1^{(1)} = \frac{4}{3}\left(\frac{7}{3} - \frac{\varepsilon\mu^{(1)\prime}}{\mu^{(1)}}\right),\quad \varpi_2^{(1)} = 2,\quad \varpi_3^{(1)} = 3,\quad \varpi_4^{(1)} = 0,$$

$$\varpi_5^{(1)} = 3\frac{\varepsilon\mu^{(1)\prime}}{\mu^{(1)}} + \frac{567\,(\mathbb{A}_{4|2|2} - 2\mathbb{A}_{3|2|2} + \mathbb{A}_{2|2|2})}{64\,\mathbb{A}_{2|2|2}},$$

$$\varpi_6^{(1)} = 8 - 2\frac{(\mathbb{A}_{3|2|2} - \mathbb{A}_{2|2|2})}{\mathbb{A}_{2|2|2}},\qquad\text{(XXIV.172)}$$

$$\theta_1^{(1)} = \frac{15}{4}\left(\frac{7}{2} - \frac{\varepsilon\mu^{(1)\prime}}{\mu^{(1)}}\right),\quad \theta_2^{(1)} = \tfrac{45}{8},\quad \theta_3^{(1)} = -3,\quad \theta_4^{(1)} = 3,$$

$$\theta_5^{(1)} = 3\left(\frac{\varepsilon\mu^{(1)\prime}}{\mu^{(1)}} + \frac{35}{4}\right) - \frac{297\,(\mathbb{A}_{3|2|2} - \mathbb{A}_{2|2|2})}{32\,\mathbb{A}_{2|2|2}};$$

a prime here denotes a derivative with respect to ε. These formulae are valid for general spherically symmetric molecular models. The approximations $\mu^{(1)}$ and $M^{(1)}$ are precisely those which we obtained previously in Chapter XVII in connection with GRAD's method of truncation, though our derivation of them here is quite different.

Using (XVII.32) and (XVII.33), we may evaluate these coefficients for ideal spheres and inverse k^{th}-power molecules. Expressions for $\mu^{(1)}$ have been given already in (XVII.63) and (XVII.64). For ideal spheres the coefficients in

---

[33] CHAPMAN & COWLING [1952, Eqs. 15.4.4, 15.4.5, 15.4.6 (and the line just before it), 15.41.1, 15.41.3 (and the two lines just before it)]. WANG CHANG & UHLENBECK [1948, Tables 1 and 2] recalculated the first approximations to $\varpi_1, \ldots, \theta_5$ and obtained values of θ_2 and θ_5 different from those first given by CHAPMAN & COWLING [1939]. In the second edition of their book, cited at the beginning of this footnote, CHAPMAN & COWLING adopt WANG CHANG & UHLENBECK's value of θ_5 but retain their original value for θ_2. We have calculated all the coefficients by the method of stretched fields, and in Chapter XXV we shall calculate them again through another procedure. In both cases we obtain the values listed finally by CHAPMAN & COWLING.

(XXIV.172) which are not already numerical have the values[34]

$$\varpi_1^{(1)} = \tfrac{10}{3} \times \tfrac{6}{5}, \qquad \varpi_5^{(1)} = 3 \times \tfrac{29}{64}, \qquad \varpi_6^{(1)} = 8 \times \tfrac{27}{28},$$
$$\theta_1^{(1)} = \tfrac{75}{8} \times \tfrac{6}{5}, \qquad \theta_5^{(1)} = \tfrac{117}{4} \times \tfrac{1973}{2184}.$$
(XXIV.173)

With the exception of $\varpi_5^{(1)}$ these values are close to those given in (XXIV.148) for Maxwellian molecules. For inverse k^{th}-power molecules we obtain

$$\varpi_1^{(1)} = \frac{10}{3}\left(1 + \frac{1}{5}\frac{k-5}{k-1}\right), \qquad \varpi_5^{(1)} = 3\left(1 - \frac{1}{2}\frac{k-5}{k-1} - \frac{3}{64}\frac{(k-5)(k+3)}{(k-1)^2}\right),$$

$$\varpi_6^{(1)} = 8\left(1 - \frac{1}{28}\frac{k-5}{k-1}\right), \qquad\qquad\qquad\qquad\qquad\qquad\text{(XXIV.174)}$$

$$\theta_1^{(1)} = \frac{75}{8}\left(1 + \frac{1}{5}\frac{k-5}{k-1}\right), \qquad \theta_5^{(1)} = \frac{117}{4}\left(1 - \frac{211}{2184}\frac{k-5}{k-1}\right).$$

---

[34] For ideal spheres PEKERIS & ALTERMAN [1957] calculated numerically the deviations of μ and $M\mu$ from their first approximations by first transforming the integral equations for A and B into ordinary differential equations (cf. Footnote 30 above). They obtained

$$\frac{\mu}{\mu^{(1)}} = 1.016034, \qquad \frac{M\mu}{M^{(1)}\mu^{(1)}} = 1.025218.$$

REVERCOMB & BRUCH [1972, §IV], using variational arguments closely connected with the method of truncation employed here, obtained the following bounds for these ratios:

$$1.01603 \leq \frac{\mu}{\mu^{(1)}} \leq 1.01606, \qquad 1.02521 \leq \frac{M\mu}{M^{(1)}\mu^{(1)}} \leq 1.02525.$$

Earlier WANG CHANG & UHLENBECK [1948, Tables 1 and 2] recorded third approximations to $\varpi_1, \ldots, \theta_5$ for a gas of ideal spheres. In terms of their first approximations their results after correction by WANG CHANG [1948, 1, §II] are

$$\varpi_1^{(3)} = 1.014\varpi_1^{(1)}, \qquad \varpi_2^{(3)} = 1.014\varpi_2^{(1)}, \qquad\qquad \varpi_3^{(3)} = 0.806\varpi_3^{(1)},$$

$$\varpi_4^{(3)} = 0.681, \qquad \varpi_5^{(3)} = 0.806 \times \frac{3\varepsilon\mu^{(1)\prime}}{\mu^{(1)}} = 0.990, \qquad \varpi_6^{(3)} = 0.928\varpi_6^{(1)},$$

$$\theta_1^{(3)} = 1.035\theta_1^{(1)}, \qquad \theta_2^{(3)} = 1.035\theta_2^{(1)}, \qquad\qquad \theta_3^{(3)} = 1.030\theta_3^{(1)},$$

$$\theta_4^{(3)} = 0.806\theta_4^{(1)}, \qquad \theta_5^{(3)} = \tfrac{105}{4} \times 0.918 + 3 \times 0.806\frac{\varepsilon\mu^{(1)\prime}}{\mu^{(1)}} - 0.150.$$

For $\varpi_1, \ldots, \varpi_6$ these approximations agree exactly with values which had been obtained by BURNETT [1936, §9].

The foregoing list gives us some idea of the degree to which the first approximations (XXIV.172) are accurate for general molecular models. At the first approximation ϖ_4 is 0 for all gases. For Maxwellian molecules this value is exact, while for ideal spheres the third approximation $\varpi_4^{(3)} = 0.681$ indicates a significant error in the first approximation. In addition, if we set $\varepsilon\mu^{(1)\prime}/\mu^{(1)} = \frac{1}{2}$, as follows from (XVII.63) for ideal spheres, the third approximation to ϖ_5 becomes $\varpi_5^{(3)} = 0.161\varpi_5^{(1)}$, again indicating a significant error in the first approximation.

In conformity with the results of the dimensional analysis preceding (XXIV.105), each of these values is independent of ε, while those in (XXIV.173) are independent of both ε and $đ$.

(xiii) Retrospect

The method of stretched fields provides a *systematic, explicit, clear procedure* that delivers all of ENSKOG's and CHAPMAN & COWLING's results and as many more of the same kind as the student has leisure and patience to calculate, without recourse to any "ad hoc recipe for juggling". We attempt to leave no doubt about *what the method of stretched fields seeks to approximate* and *how to calculate the general stage according to that method*. Thus we have cleared a path through the swamp of divination whence ENSKOG's method sprang and where it yet resides, a path that debouches upon two *definite mathematical problems*:

(a) To prove the existence of grossly determined solutions, possibly also the uniqueness of the several gross determiners.

(b) To prove the convergence of the method of stretched fields to the gross determiners or to some subset of them.

These problems are for the future.

In the following chapter we present an entirely different method of calculation: differential iteration. This method rests upon repeated differentiation; it avoids the use of integral equations entirely.

Appendix A

Derivation of the Iterative System

We show here how equation (XXIV.59) may be derived. To do so, we shall apply repeatedly to (XXIV.49) the operations of replacing ψ by $\psi + \lambda\xi$, then differentiating with respect to λ and finally setting $\lambda = 0$. These operations may be expressed concisely in terms of the Gâteaux derivative of a function, so we begin with a brief summary of the theory of such derivatives.

Let G be a real-valued function defined on an open set $\mathcal{U}$ in a Banach space $\mathcal{M}$. We say that G is *Gâteaux differentiable* on $\mathcal{U}$ if for each $\psi \in \mathcal{U}$ and each $\xi \in \mathcal{M}$ the limit on the right-hand side of the following expression exists:

$$\delta_\psi G(\psi)[\xi] \equiv \lim_{\lambda \to 0} \frac{1}{\lambda}(G(\psi + \lambda\xi) - G(\psi)). \qquad \text{(XXIVA.1)}$$

APPENDIX A. THE ITERATIVE SYSTEM

The mapping $\delta_\psi G$ that delivers this limit is called the Gâteaux derivative of G. For example, suppose that G is differentiable on $\mathcal{U}$. Then by the chain rule

$$\delta_\psi G(\psi)[\xi] = \frac{d}{d\lambda} G(\psi + \lambda\xi)\big|_{\lambda=0} = \partial_\psi G(\psi)[\xi], \qquad \text{(XXIVA.2)}$$

whence G is Gâteaux differentiable and its Gâteaux derivative agrees with its ordinary derivative. As another example, let L be a linear function on $\mathcal{M}$. Then

$$\lim_{\lambda \to 0} \frac{1}{\lambda}(L(\psi + \lambda\xi) - L(\psi)) = \lim_{\lambda \to 0} \frac{1}{\lambda}(L(\psi) + \lambda L(\xi) - L(\psi)) = L(\xi), \qquad \text{(XXIVA.3)}$$

so L is Gâteaux differentiable and its Gâteaux derivative is simply

$$\delta_\psi L(\psi)[\xi] = L(\xi). \qquad \text{(XXIVA.4)}$$

We define the pth Gâteaux derivative of a function inductively by the formula

$$\delta_\psi^p G(\psi)[\xi^1, \ldots, \xi^p] \equiv \lim_{\lambda \to 0} \frac{1}{\lambda}(\delta_\psi^{p-1} G(\psi + \lambda\xi^p)[\xi^1, \ldots, \xi^{p-1}]$$
$$- \delta_\psi^{p-1} G(\psi)[\xi^1, \ldots, \xi^{p-1}]), \qquad \text{(XXIVA.5)}$$

presuming that the limit exists. If G is p times differentiable on $\mathcal{U}$, one may easily show that G is also p times Gâteaux differentiable on $\mathcal{U}$, and (XXIVA.2)$_2$ generalizes to

$$\delta_\psi^p G(\psi)[\xi^1, \ldots, \xi^p] = \partial_\psi^p G(\psi)[\xi^1, \ldots, \xi^p]. \qquad \text{(XXIVA.6)}$$

Also, since (XXIVA.4) shows that the Gâteaux derivative of a linear function of ψ is independent of ψ, linear functions are Gâteaux differentiable any number of times, and

$$\delta_\psi^p L(\psi)[\xi^1, \ldots, \xi^p] = 0 \quad \text{if} \quad p \geq 2. \qquad \text{(XXIVA.7)}$$

The Gâteaux derivative obeys most of the formal rules usually associated with a derivative. We illustrate this in one important case, namely, the rule for the derivative of a product of functions. Let G and H be two real-valued continuous functions defined on $\mathcal{U}$ that are both p times Gâteaux differentiable. Denoting their product by GH, we have

$$\delta_\psi GH(\psi)[\xi] = \lim_{\lambda \to 0} \frac{1}{\lambda}(G(\psi + \lambda\xi)H(\psi + \lambda\xi) - G(\psi)H(\psi)),$$
$$= \lim_{\lambda \to 0} \frac{1}{\lambda}(G(\psi + \lambda\xi) - G(\psi)) \lim_{\nu \to 0} H(\psi + \nu\xi) \qquad \text{(XXIVA.8)}$$
$$+ G(\psi) \lim_{\lambda \to 0} \frac{1}{\lambda}(H(\psi + \lambda\xi) - H(\psi)),$$
$$= \delta_\psi G(\psi)[\xi]H(\psi) + G(\psi)\,\delta_\psi H(\psi)[\xi].$$

By applying this formula first to the product $\delta_\psi G(\cdot)[\xi]H(\cdot)$ and then to the product $G(\cdot)\,\delta_\psi H(\cdot)[\xi]$, ξ being held fixed, we obtain

$$\delta_\psi^2 GH(\psi)[\xi^1, \xi^2] = \delta_\psi^2 G(\psi)[\xi^1, \xi^2]H(\psi) + \delta_\psi G(\psi)[\xi^1]\,\delta_\psi H(\psi)[\xi^2]$$
$$+ \delta_\psi G(\psi)[\xi^2]\,\delta_\psi H(\psi)[\xi^1] + G(\psi)\,\delta_\psi^2 H(\psi)[\xi^1, \xi^2].$$
(XXIVA.9)

Note that the right-hand side here is a symmetric function of ξ^1 and ξ^2. In general the pth Gâteaux derivative of a product is given by

$$\delta_\psi^p GH(\psi)[\xi^1, \ldots, \xi^p] = \underset{\xi^1, \ldots, \xi^p}{\text{sym}} \sum_{q=0}^{p} \binom{p}{q} \delta_\psi^q G(\psi)[\xi^1, \ldots, \xi^q]$$
$$\times \delta_\psi^{p-q} H(\psi)[\xi^{q+1}, \ldots, \xi^p], \quad \text{(XXIVA.10)}$$

sym being the symmetrizer defined by (XXIV.60). When $p = 1$ and $p = 2$, this formula reduces to (XXIVA.8)$_3$ and (XXIVA.9), respectively, and by applying the Gâteaux derivative once again one may show by induction that it holds in general. We leave the details to the reader.

The procedure we have used in deriving (XXIV.53) from (XXIV.49) is nothing more than that of applying the Gâteaux derivative with respect to ψ to each side of (XXIV.49). We generalize this procedure now by applying the pth Gâteaux derivative. First we write the equations of gross determinism in the form

$$\partial_\psi \mathbb{G}(\psi)[\mathbf{L}_1(\psi) - \mathbf{T}(\psi)] = \mathbb{C}(\mathbb{G}(\psi), \mathbb{G}(\psi)),$$
(XXIVA.11)

$$\mathbf{L}_3(\psi) = \mathfrak{m} \int \mathbb{IG}(\psi),$$

in which

$$\mathbf{T}(\psi) \equiv \mathfrak{m} \int v_d \mathbf{I}\,\partial_\psi\, \mathbb{G}(\psi(\mathbf{r} + \mathbf{s}))[\mathbf{L}_2(\psi)],$$
(XXIVA.12)

$$\mathbf{L}_1(\psi) \equiv v_d \psi_{,d}(\mathbf{r}), \qquad \mathbf{L}_2(\psi) \equiv \psi_{,d}(\mathbf{r} + \mathbf{s}), \qquad \mathbf{L}_3(\psi) \equiv \psi(0).$$

For the linear functions $\mathbf{L}_1$, $\mathbf{L}_2$, and $\mathbf{L}_3$ we have the simple formulae (XXIVA.4) and (XXIVA.7). Apart from them the only function appearing in (XXIVA.11) is $\mathbb{G}$, and since it is smooth, we may use (XXIVA.2)$_2$ and (XXIVA.6) in evaluating its Gâteaux derivatives.

APPENDIX A. THE ITERATIVE SYSTEM

By applying the pth and q^{th} Gâteaux derivatives to each side of (XXIVA.11) and (XXIVA.12)$_1$, respectively, we obtain

$$\operatorname*{sym}_{\xi^1,\ldots,\,\xi^p} \sum_{q=0}^{p} \binom{p}{q} \partial_\psi^{p-q+1} \mathbb{G}(\psi)[\delta_\psi^q \mathbf{L}_1(\psi)[\xi^1,\ldots,\xi^q]$$

$$- \delta_\psi^q \mathbf{T}(\psi)[\xi^1,\ldots,\xi^q], \xi^{q+1},\ldots,\xi^p]$$

$$= \operatorname*{sym}_{\xi^1,\ldots,\,\xi^p} \sum_{q=0}^{p} \binom{p}{q} \mathbb{C}(\partial_\psi^q \mathbb{G}(\psi)[\xi^1,\ldots,\xi^q], \partial_\psi^{p-q} \mathbb{G}(\psi)[\xi^{q+1},\ldots,\xi^p]),$$

(XXIVA.13)

$$\delta_\psi^q \mathbf{T}(\psi)[\xi^1,\ldots,\xi^q] = \operatorname*{sym}_{\xi^1,\ldots,\,\xi^q} \sum_{a=0}^{q} \binom{q}{a} \mathfrak{m} \int v_d \mathbf{I}$$

$$\times \partial_\psi^{q-a+1} \mathbb{G}(\psi(\mathbf{r}+\mathbf{s}))[\delta_\psi^a \mathbf{L}_2(\psi)[\xi^1,\ldots,\xi^a], \xi^{a+1}(\mathbf{r}+\mathbf{s}),\ldots,\xi^q(\mathbf{r}+\mathbf{s})],$$

$$\delta_\psi^p \mathbf{L}_3(\psi)[\xi^1,\ldots,\xi^p] = \mathfrak{m} \int \mathbf{I}\, \partial_\psi^p \mathbb{G}(\psi)[\xi^1,\ldots,\xi^p].$$

By placing (XXIVA.13)$_2$ into (XXIVA.13)$_1$ and using (XXIVA.12) and (XXIVA.7) we obtain the following generalization of (XXIV.53):

$$\partial_\psi^{p+1} \mathbb{G}(\psi(\mathbf{r}))[v_d \psi_{,d}(\mathbf{r}), \xi^1(\mathbf{r}),\ldots,\xi^p(\mathbf{r})]$$

$$+ \operatorname*{sym}_{\xi^1,\ldots,\,\xi^p} p\, \partial_\psi^p \mathbb{G}(\psi(\mathbf{r}))[v_d \xi^1_{,d}(\mathbf{r}), \xi^2(\mathbf{r}),\ldots,\xi^p(\mathbf{r})]$$

$$- \partial_\psi^{p+1} \mathbb{G}(\psi(\mathbf{r}))\left[\mathfrak{m}\int v_d \mathbf{I}\, \partial_\psi \mathbb{G}(\psi(\mathbf{r}+\mathbf{s}))[\psi_{,d}(\mathbf{r}+\mathbf{s})], \xi^1(\mathbf{r}),\ldots,\xi^p(\mathbf{r})\right]$$

$$- \operatorname*{sym}_{\xi^1,\ldots,\,\xi^p} \sum_{q=1}^{p} \binom{p}{q} \partial_\psi^{p-q+1} \mathbb{G}(\psi(\mathbf{r}))\left[\mathfrak{m}\int v_d \mathbf{I}\right.$$

$$\times \partial_\psi^{q+1} \mathbb{G}(\psi(\mathbf{r}+\mathbf{s}))[\psi_{,d}(\mathbf{r}+\mathbf{s}), \xi^1(\mathbf{r}+\mathbf{s}),\ldots,\xi^q(\mathbf{r}+\mathbf{s})] \quad \text{(XXIVA.14)}$$

$$+ q \operatorname*{sym}_{\xi^1,\ldots,\,\xi^q} \mathfrak{m}\int v_d \mathbf{I}\, \partial_\psi^q \mathbb{G}(\psi(\mathbf{r}+\mathbf{s}))$$

$$\left.[\xi^1_{,d}(\mathbf{r}+\mathbf{s}), \xi^2(\mathbf{r}+\mathbf{s}),\ldots,\xi^q(\mathbf{r}+\mathbf{s})], \xi^{q+1}(\mathbf{r}),\ldots,\xi^p(\mathbf{r})\right]$$

$$= \operatorname*{sym}_{\xi^1,\ldots,\,\xi^p} \sum_{q=0}^{p} \binom{p}{q} \mathbb{C}(\partial_\psi^q \mathbb{G}(\psi)[\xi^1,\ldots,\xi^q], \partial_\psi^{p-q} \mathbb{G}(\psi)[\xi^{q+1},\ldots,\xi^p]),$$

$$\mathfrak{m}\int I_\gamma\, \partial_\psi^p \mathbb{G}(\psi)[\xi^1,\ldots,\xi^p] = \begin{cases} \xi^1_\gamma(0) & \text{if } p=1, \\ 0 & \text{if } p\geq 2. \end{cases}$$

Recall now that we obtained (XXIV.51) from (XXIV.49) and (XXIV.56) from (XXIV.53) by making special choices of ψ and ξ. The corresponding choice here, as suggested by (XXIV.28)$_2$, is

$$\psi_\gamma = K_\gamma, \qquad \xi^1_\gamma = J^{n_1}_{\gamma\alpha_1}, \ldots, \qquad \xi^p_\gamma = J^{n_p}_{\gamma\alpha_p}. \qquad \text{(XXIVA.15)}$$

Since ψ is now a constant, the terms on the left-hand side of (XXIVA.14)$_1$ that are linear in grad ψ vanish. Since J^n_α is a homogeneous function, $J^n_\alpha(0) = 0$, and so the right-hand side of (XXIVA.14)$_2$ is null. Hence, using (XXIV.28)$_2$ we find that (XXIVA.14) simplifies with this choice of arguments to

$$\underset{\binom{n_1}{\alpha_1}\ldots\binom{n_p}{\alpha_p}}{\text{sym}} p\, \partial^p_\psi\, \mathsf{G}(\mathbf{K})[v_d\, \mathbf{J}^{n_1}_{\alpha_1,d}(\mathbf{r}), \mathbf{J}^{n_2}_{\alpha_2}(\mathbf{r}), \ldots, \mathbf{J}^{n_p}_{\alpha_p}(\mathbf{r})]$$

$$- \underset{\binom{n_1}{\alpha_1}\ldots\binom{n_p}{\alpha_p}}{\text{sym}} \sum_{q=1}^{p} \binom{p}{q} \partial^{p-q+1}_\psi \mathsf{G}(\mathbf{K}) \bigg[q \underset{\binom{n_1}{\alpha_1}\ldots\binom{n_q}{\alpha_q}}{\text{sym}} \mathfrak{m} \int v_d \mathbf{I}\, \partial^q_\psi\, \mathsf{G}(\mathbf{K})$$

$$[\mathbf{J}^{n_1}_{\alpha_1,d}(\mathbf{r}+\mathbf{s}), \mathbf{J}^{n_2}_{\alpha_2}(\mathbf{r}+\mathbf{s}), \ldots, \mathbf{J}^{n_q}_{\alpha_q}(\mathbf{r}+\mathbf{s})], \mathbf{J}^{n_{q+1}}_{\alpha_{q+1}}(\mathbf{r}), \ldots, \mathbf{J}^{n_p}_{\alpha_p}(\mathbf{r}) \bigg]$$

$$= \underset{\binom{n_1}{\alpha_1}\ldots\binom{n_p}{\alpha_p}}{\text{sym}} \sum_{q=0}^{p} \binom{p}{q} \mathsf{C}(\mathsf{G}^{n_1\ldots n_q}_{\alpha_1\ldots\alpha_q}(\mathbf{K}), \mathsf{G}^{n_{q+1}\ldots n_p}_{\alpha_{q+1}\ldots\alpha_p}(\mathbf{K})), \qquad \text{(XXIVA.16)}$$

$$\mathfrak{m} \int I_\gamma\, \mathsf{G}^{n_1\ldots n_p}_{\alpha_1\ldots\alpha_p}(\mathbf{K}) = 0.$$

Before we can simplify the left-hand side of the first of these equations, we need to express $v_d\, \mathbf{J}^n_{\alpha,d}(\mathbf{r})$, $\mathbf{J}^n_\alpha(\mathbf{r}+\mathbf{s})$, and $v_d\, \mathbf{J}^n_{\alpha,d}(\mathbf{r}+\mathbf{s})$ in terms of $\mathbf{J}^m_\beta(\mathbf{r})$ and $\mathbf{J}^p_\gamma(\mathbf{s})$. We indicate now how this may be done. According to (XXIV.17) and (XXIV.43), $\mathbf{J}^n_\alpha(\mathbf{r}) = \mathbf{1}_\alpha r_n = \mathbf{1}_\alpha r_{k_1} \cdots r_{k_n}$ is a symmetric function of the indices $k_1, \ldots, k_n$. Therefore, using the symmetrizer, we may write

$$\mathbf{J}^n_{\gamma\alpha}(\mathbf{r}) = \underset{n}{\text{sym}}\, \mathbf{J}^n_{\gamma\alpha}(\mathbf{r}) = \mathbf{1}_{\gamma\alpha} \underset{k_1,\ldots,k_n}{\text{sym}}\, r_{k_1} \cdots r_{k_n}. \qquad \text{(XXIVA.17)}$$

The right-hand side here is more convenient to use than $\mathbf{J}^n_\alpha$ itself since in it we may rearrange at will the indices represented by n. For example, it is not difficult to see that

$$\mathbf{J}^n_{\gamma\alpha}(\mathbf{r}+\mathbf{s}) = \mathbf{1}_{\gamma\alpha} \underset{k_1,\ldots,k_n}{\text{sym}}\, (r_{k_1}+s_{k_1}) \cdots (r_{k_n}+s_{k_n}),$$

$$= \mathbf{1}_{\gamma\alpha} \underset{k_1,\ldots,k_n}{\text{sym}}\, \{r_{k_1} \cdots r_{k_n} + n r_{k_1} \cdots r_{k_{n-1}} s_{k_n}$$

$$+ \binom{n}{2} r_{k_1} \cdots r_{k_{n-2}} s_{k_{n-1}} s_{k_n} + \cdots + s_{k_1} \cdots s_{k_n}\}, \qquad \text{(XXIVA.18)}$$

$$= \underset{n}{\text{sym}} \sum_{a=0}^{n} \binom{n}{a} \mathbf{J}^a_{\alpha\beta}(\mathbf{r}) \mathbf{J}^b_{\gamma\beta}(\mathbf{s}),$$

$$n = ab.$$

APPENDIX A. THE ITERATIVE SYSTEM 493

We may also use (XXIVA.17)$_2$ to show that

$$v_d J^n_{\gamma\alpha,d}(\mathbf{r}) = 1_{\gamma\alpha} \operatorname*{sym}_{k_1,\ldots,k_n} v_d(r_{k_1} \cdots r_{k_n})_{,d},$$

$$= 1_{\gamma\alpha} \operatorname*{sym}_{k_1,\ldots,k_n} \sum_{a=1}^{n} r_{k_1} \cdots r_{k_{a-1}} v_{k_a} r_{k_{a+1}} \cdots r_{k_n},$$

$$= n 1_{\gamma\alpha} \operatorname*{sym}_{k_1,\ldots,k_n} v_{k_1} r_{k_2} \cdots r_{k_n}, \qquad \text{(XXIVA.19)}$$

$$= n \operatorname*{sym}_{n} v_d J^a_{\gamma\alpha}(\mathbf{r}),$$

$$\mathsf{n} = d\mathsf{a}.$$

By combining this last result with (XXIVA.18) we obtain a formula for $v_d J^n_{\alpha,d}(\mathbf{r}+\mathbf{s})$:

$$v_d J^n_{\gamma\alpha,d}(\mathbf{r}+\mathbf{s}) = \operatorname*{sym}_{n} \sum_{a=0}^{n} \binom{n}{a} v_d J^a_{\alpha\beta,d}(\mathbf{r}) J^b_{\gamma\beta}(\mathbf{s}),$$

$$= \operatorname*{sym}_{n} \sum_{a=0}^{n} \binom{n}{a} a \operatorname*{sym}_{a} v_d J^c_{\alpha\beta}(\mathbf{r}) J^b_{\gamma\beta}(\mathbf{s}), \qquad \text{(XXIVA.20)}$$

$$\mathsf{n} = \mathsf{a}\mathsf{b}, \qquad \mathsf{a} = d\mathsf{c}.$$

When $a = 0$ the multi-index c is undefined, but for this value of a the corresponding term in the sum in (XXIVA.20)$_2$ vanishes anyway.

The formulae (XXIVA.18)–(XXIVA.20) are all we require in order to complete the derivation of (XXIV.59). If we place these into (XXIVA.16) and use the fact that $\partial^m_{\psi} G(\psi)[\cdot, \ldots, \cdot]$ is linear in each of its m arguments, we obtain

$$\operatorname*{sym}_{\binom{n_1}{\alpha_1}\ldots\binom{n_p}{\alpha_p}} p n_1 \operatorname*{sym}_{n_1} v_d \, \mathbb{G}^{a n_2 \ldots n_p}_{\alpha_1 \alpha_2 \ldots \alpha_p}(\mathbf{K})$$

$$- \operatorname*{sym}_{\binom{n_1}{\alpha_1}\ldots\binom{n_p}{\alpha_p}} \sum_{q=1}^{p} \binom{p}{q} q \operatorname*{sym}_{\binom{n_1}{\alpha_1}\ldots\binom{n_q}{\alpha_q}} \operatorname*{sym}_{n_1} \cdots \operatorname*{sym}_{n_q} \sum_{a_1=0}^{n_1} \cdots \sum_{a_q=0}^{n_q} \binom{n_1}{a_1} \cdots \binom{n_q}{a_q} a_1$$

$$\times \operatorname*{sym}_{a_1} \mathbb{G}^{e n_{q+1} \ldots n_p}_{\beta \alpha_{q+1} \ldots \alpha_p}(\mathbf{K}) \mathfrak{m} \int v_d I_\beta \, \mathbb{G}^{b_1 \ldots b_q}_{\alpha_1 \ldots \alpha_q}(\mathbf{K}) \qquad \text{(XXIVA.21)}$$

$$= \operatorname*{sym}_{\binom{n_1}{\alpha_1}\ldots\binom{n_p}{\alpha_p}} \sum_{q=0}^{p} \binom{p}{q} \mathbb{C}(\mathbb{G}^{n_1 \ldots n_q}_{\alpha_1 \ldots \alpha_q}(\mathbf{K}), \mathbb{G}^{n_{q+1} \ldots n_p}_{\alpha_{q+1} \ldots \alpha_p}(\mathbf{K}))$$

$$\mathfrak{m} \int I_\gamma \, \mathbb{G}^{n_1 \ldots n_p}_{\alpha_1 \ldots \alpha_p}(\mathbf{K}) = 0,$$

in which

$$n_1 = da,$$
$$ = a_1 b_1,$$
$$n_2 = a_2 b_2,$$
$$\vdots \qquad\qquad\qquad\qquad\qquad \text{(XXIV.62)}_r$$
$$n_p = a_p b_p,$$
$$a_1 = dc,$$
$$e = ca_2 \ldots a_q.$$

This is precisely the iterative system (XXIV.59).

Appendix B

Computational Formulae

Many of the calculations in this chapter make use of standard formulae for the integrals and derivatives of principal moments, Maxwellian densities, and other expressions. So as to simplify these calculations, we list here the standard formulae that are required. In some cases the results are less obvious than others, and for these we describe some of the intermediate steps.

(i) Principal moments

$$\rho_\alpha = \begin{cases} \rho, & \alpha = 0, \\ \rho u_k, & \alpha = k, \\ \rho(2\varepsilon + u^2), & \alpha = 4. \end{cases} \qquad \text{(XXIVB.1)}$$

$$\rho_{\alpha,r} = \begin{cases} \rho_{,r}, & \alpha = 0, \\ u_k \rho_{,r} + \rho u_{k,r}, & \alpha = k, \\ (2\varepsilon + u^2)\rho_{,r} + 2\rho \varepsilon_{,r} + 2\rho u_a u_{a,r}, & \alpha = 4. \end{cases} \qquad \text{(XXIVB.2)}$$

$$\rho_{\alpha,rs} = \begin{cases} \rho_{,rs}, & \alpha = 0, \\ u_k \rho_{,rs} + u_{k,s}\rho_{,r} + \rho_{,s} u_{k,r} + \rho u_{k,rs}, & \alpha = k, \\ (2\varepsilon + u^2)\rho_{,rs} + 2\rho_{,r}\varepsilon_{,s} + 2\rho_{,r} u_a u_{a,s} + 2\rho\varepsilon_{,rs} & \\ \quad + 2\rho_{,s}\varepsilon_{,r} + 2\rho_{,s} u_a u_{a,r} + 2\rho u_{a,s} u_{a,r} + 2\rho u_a u_{a,rs}, & \alpha = 4. \end{cases} \qquad \text{(XXIVB.3)}$$

APPENDIX B. COMPUTATIONAL FORMULAE

$$\partial_{\rho_\alpha} = \begin{cases} \dfrac{1}{\rho}(\rho\,\partial_\rho - u_a\,\partial_{u_a} + \tfrac{1}{2}(u^2 - 2\varepsilon)\,\partial_\varepsilon), & \alpha = 0, \\ \dfrac{1}{\rho}(\partial_{u_k} - u_k\,\partial_\varepsilon), & \alpha = k, \\ \dfrac{1}{2\rho}\partial_\varepsilon, & \alpha = 4. \end{cases} \quad \text{(XXIVB.4)}$$

(ii) Maxwellian densities

$$F_M = \frac{\rho}{\mathfrak{m}(\tfrac{4}{3}\pi\varepsilon)^{\tfrac{3}{2}}} e^{-\tfrac{3}{4}c^2/\varepsilon}. \quad \text{(XXIVB.5)}$$

$$\frac{\rho}{F_M}\partial_{\rho_\alpha}F_M = \begin{cases} 1 - \dfrac{3}{2\varepsilon}c_a u_a + \dfrac{1}{2}(u^2 - 2\varepsilon)\left(-\dfrac{3}{2\varepsilon} + \dfrac{3c^2}{4\varepsilon^2}\right), & \alpha = 0, \\ \dfrac{3}{2\varepsilon}c_k - u_k\left(-\dfrac{3}{2\varepsilon} + \dfrac{3c^2}{4\varepsilon^2}\right), & \alpha = k, \\ \dfrac{1}{2}\left(-\dfrac{3}{2\varepsilon} + \dfrac{3c^2}{4\varepsilon^2}\right), & \alpha = 4. \end{cases} \quad \text{(XXIVB.6)}$$

$$\mathfrak{m}\int v_m I_\alpha F_M = \begin{cases} \rho u_m, & \alpha = 0, \\ \rho u_m u_k + p\,\delta_{mk}, & \alpha = k, \\ 5pu_m + \rho u_m u^2, & \alpha = 4. \end{cases} \quad \text{(XXIVB.7)}$$

$$\mathfrak{m}\int v_m I_\alpha \partial_{\rho_\beta} F_M = \begin{cases} 0, & \alpha = 0,\ \beta = 0, \\ \delta_{mp}, & \alpha = 0,\ \beta = p, \\ 0, & \alpha = 0,\ \beta = 4, \\ -u_{\{m}u_{k\}}, & \alpha = k,\ \beta = 0, \\ 2u_{\{m}\delta_{k\}p}, & \alpha = k,\ \beta = p, \\ \tfrac{1}{3}\delta_{mk}, & \alpha = k,\ \beta = 4, \\ -\tfrac{10}{3}\varepsilon u_m - \tfrac{1}{3}u_m u^2, & \alpha = 4,\ \beta = 0, \\ (\tfrac{10}{3}\varepsilon + u^2)\delta_{mp} - \tfrac{4}{3}u_m u_p, & \alpha = 4,\ \beta = p, \\ \tfrac{5}{3}u_m, & \alpha = 4,\ \beta = 4. \end{cases} \quad \text{(XXIVB.8)}$$

$$\mathfrak{m}\int v_m I_\alpha \, \partial_{\rho\beta} \, \partial_{\rho\gamma} F_{\mathbf{M}} = \begin{cases} 0, & \alpha = 0, \text{ all } \beta, \text{ all } \gamma, \\[4pt] \dfrac{2}{\rho} u_{\{m} u_{k\}}, & \alpha = k, \ \beta = 0, \ \gamma = 0, \\[4pt] -\dfrac{2}{\rho} u_{\{k} \delta_{m\}q}, & \alpha = k, \ \beta = 0, \ \gamma = q, \\[4pt] \dfrac{2}{\rho} \delta_{m\{q} \delta_{p\}k}, & \alpha = k, \ \beta = p, \ \gamma = q, \\[4pt] 0, & \alpha = k, \ \beta = 4, \text{ all } \gamma, \\[4pt] \dfrac{20}{3}\dfrac{\varepsilon}{\rho} u_m - \dfrac{2}{3\rho} u_m u^2, & \alpha = 4, \ \beta = 0, \ \gamma = 0, \\[4pt] -\dfrac{1}{3\rho}(10\varepsilon + u^2)\delta_{mq} + \dfrac{8}{3\rho} u_m u_q, & \alpha = 4, \ \beta = 0, \ \gamma = q, \\[4pt] -\dfrac{5}{3\rho} u_m, & \alpha = 4, \ \beta = 0, \ \gamma = 4, \\[4pt] -\dfrac{4}{3\rho}(u_p \delta_{mq} + u_m \delta_{pq} + u_q \delta_{mp}), & \alpha = 4, \ \beta = p, \ \gamma = q, \\[4pt] \dfrac{5}{3\rho} \delta_{mp}, & \alpha = 4, \ \beta = p, \ \gamma = 4, \\[4pt] 0, & \alpha = 4, \ \beta = 4, \ \gamma = 4. \end{cases} \tag{XXIVB.9}$$

(iii) First iterates

$$\mathbb{F}_k^m = c_{\{m} c_{k\}} B, \qquad \mathbb{F}_4^k = c_k A. \tag{XXIVB.10}$$

$$\mathbb{G}_\alpha^m = \begin{cases} -\dfrac{1}{\rho} u_a c_{\{m} c_{a\}} B + \dfrac{u^2 - 2\varepsilon}{2\rho} c_m A, & \alpha = 0, \\[4pt] \dfrac{1}{\rho} c_{\{m} c_{k\}} B - \dfrac{1}{\rho} u_k c_m A, & \alpha = k, \\[4pt] \dfrac{1}{2\rho} c_m A, & \alpha = 4. \end{cases} \tag{XXIVB.11}$$

$$\int I_\gamma \, \mathbb{G}_\alpha^k = 0 \quad \Leftrightarrow \quad \int c^2 A = 0. \tag{XXIVB.12}$$

APPENDIX B. COMPUTATIONAL FORMULAE

$$\int v_m I_\beta \, G^r_\alpha = \begin{cases} 0, & \beta = 0, \text{ all } \alpha, \\ -\dfrac{2}{15\rho} u_{\{m} \delta_{p\}r} \int c^4 B, & \beta = p, \quad \alpha = 0, \\ \dfrac{2}{15\rho} \delta_{m\{r} \delta_{k\}p} \int c^4 B, & \beta = p, \quad \alpha = k, \\ 0, & \beta = p, \quad \alpha = 4, \\ -\dfrac{2}{15\rho} (\tfrac{1}{3} u_m u_r + u^2 \delta_{mr}) \int c^4 B + \dfrac{u^2 - 2\varepsilon}{6\rho} \delta_{mr} \int c^4 A, & \beta = 4, \quad \alpha = 0, \\ \dfrac{4}{15\rho} u_{\{k} \delta_{r\}m} \int c^4 B - \dfrac{1}{3\rho} u_k \delta_{mr} \int c^4 A, & \beta = 4, \quad \alpha = k, \\ \dfrac{1}{6\rho} \delta_{mr} \int c^4 A, & \beta = 4, \quad \alpha = 4. \end{cases}$$

(XXIVB.13)

In evaluating these integrals we have used (XXIVB.12), (XXIVB.20), and the facts that A and B are functions of c^2. By applying (XXIVB.4) to (XXIVB.11), we obtain

$$\partial_{\rho_\beta} G^r_k = \begin{cases} -\dfrac{u^2 - 2\varepsilon}{2\rho^2} u_a c_{\{r} c_{a\}} \partial_\varepsilon B + \dfrac{(u^2 - 2\varepsilon)^2}{4\rho^2} c_r \partial_\varepsilon A + \dfrac{2}{\rho^2} u_a c_{\{r} c_{a\}} B \\ \quad - \dfrac{2}{\rho^2} u_a c_{\{r} c_{a\}} u_b c_b \partial_{c^2} B - \dfrac{2}{\rho^2} u_a u_{\{r} c_{a\}} B - \dfrac{2(u^2 - \varepsilon)}{\rho^2} c_r A \\ \quad + \dfrac{u^2 - 2\varepsilon}{2\rho^2} u_r A + \dfrac{u^2 - 2\varepsilon}{\rho^2} c_r c_a u_a \partial_{c^2} A, & \beta = 0, \\[4pt] \dfrac{1}{\rho^2} u_k u_a c_{\{r} c_{a\}} \partial_\varepsilon B - \dfrac{u^2 - 2\varepsilon}{2\rho^2} u_k c_r \partial_\varepsilon A - \dfrac{1}{\rho^2} c_{\{r} c_{k\}} B \\ \quad + \dfrac{2}{\rho^2} u_a \delta_{k\{r} c_{a\}} B + \dfrac{2}{\rho^2} c_k c_{\{r} c_{a\}} u_a \partial_{c^2} B + \dfrac{2}{\rho^2} u_k c_r A \\ \quad - \dfrac{u^2 - 2\varepsilon}{2\rho^2} \delta_{rk} A - \dfrac{u^2 - 2\varepsilon}{\rho^2} c_r c_k \partial_{c^2} A, & \beta = k, \\[4pt] -\dfrac{1}{2\rho^2} u_a c_{\{r} c_{a\}} \partial_\varepsilon B + \dfrac{u^2 - 2\varepsilon}{4\rho^2} c_r \partial_\varepsilon A - \dfrac{1}{2\rho^2} c_r A, & \beta = 4. \end{cases}$$

(XXIVB.14)

$$\partial_{\rho_\beta} \mathbb{G}_0^r = \begin{cases} \dfrac{u^2 - 2\varepsilon}{2\rho^2} c_{\{r} c_{k\}} \partial_\varepsilon B - \dfrac{u^2 - 2\varepsilon}{2\rho^2} u_k c_r \partial_\varepsilon A - \dfrac{1}{\rho^2} c_{\{r} c_{k\}} B \\[6pt] + \dfrac{2}{\rho^2} u_{\{r} c_{k\}} B + \dfrac{2}{\rho^2} c_{\{r} c_{k\}} c_a u_a \partial_{c^2} B + \dfrac{2}{\rho^2} u_k c_r A \\[6pt] - \dfrac{1}{\rho^2} u_k u_r A - \dfrac{2}{\rho^2} u_k c_r u_a c_a \partial_{c^2} A, & \beta = 0, \\[10pt] - \dfrac{1}{\rho^2} u_p c_{\{r} c_{k\}} \partial_\varepsilon B + \dfrac{1}{\rho^2} u_p u_k c_r \partial_\varepsilon A - \dfrac{2}{\rho^2} c_{\{k} \delta_{r\}p} B \\[6pt] - \dfrac{2}{\rho^2} c_p c_{\{k} c_{r\}} \partial_{c^2} B + \dfrac{1}{\rho^2} (u_k \delta_{rp} - c_r \delta_{kp}) A \\[6pt] + \dfrac{2}{\rho^2} u_k c_r c_p \partial_{c^2} A, & \beta = p, \\[10pt] \dfrac{1}{2\rho^2} c_{\{r} c_{k\}} \partial_\varepsilon B - \dfrac{1}{2\rho^2} u_k c_r \partial_\varepsilon A, & \beta = 4. \end{cases} \quad \text{(XXIVB.15)}$$

$$\partial_{\rho_\beta} \mathbb{G}_4^r = \begin{cases} -\dfrac{1}{2\rho^2} c_r A + \dfrac{1}{2\rho^2} u_r A + \dfrac{1}{\rho^2} u_a c_a c_r \partial_{c^2} A \\[6pt] + \dfrac{u^2 - 2\varepsilon}{4\rho^2} c_r \partial_\varepsilon A, & \beta = 0, \\[10pt] -\dfrac{1}{2\rho^2} \delta_{rp} A - \dfrac{1}{\rho^2} c_r c_p \partial_{c^2} A - \dfrac{1}{2\rho^2} u_p c_r \partial_\varepsilon A, & \beta = p, \\[10pt] \dfrac{1}{4\rho^2} c_r \partial_\varepsilon A, & \beta = 4. \end{cases} \quad \text{(XXIVB.16)}$$

APPENDIX B. COMPUTATIONAL FORMULAE

If now we either apply the integral $\int v_m \mathbf{I}$ to each of the last three equations or apply (XXIVB.4) to (XXIVB.13), we obtain

$$\int v_m I_0 \, \partial_{\rho_\gamma} \mathbb{G}^r_\beta = 0, \quad \text{all } \beta, \text{ all } \gamma. \tag{XXIVB.17}$$

$$\int v_m I_q \, \partial_{\rho_\gamma} \mathbb{G}^r_\beta = \begin{cases} \dfrac{4}{15\rho^2} u_{\{m} \delta_{q\}r} \int c^4 B - \dfrac{2}{15\rho} \dfrac{u^2 - 2\varepsilon}{2\rho} u_{\{m} \delta_{q\}r} \int c^4 \, \partial_\varepsilon B, \\ \hspace{6cm} \beta = 0, \quad \gamma = 0, \\[4pt] -\dfrac{2}{15\rho^2} \delta_{p\{m} \delta_{q\}r} \int c^4 B + \dfrac{2}{15\rho^2} u_p u_{\{m} \delta_{q\}r} \int c^4 \, \partial_\varepsilon B, \\ \hspace{6cm} \beta = 0, \quad \gamma = p, \\[4pt] -\dfrac{1}{15\rho^2} u_{\{m} \delta_{q\}r} \int c^4 \, \partial_\varepsilon B, \hspace{1.5cm} \beta = 0, \quad \gamma = 4, \\[4pt] \hspace{6cm} \text{(XXIVB.18)} \\[4pt] -\dfrac{2}{15\rho^2} \delta_{m\{r} \delta_{k\}q} \int c^4 B + \dfrac{2}{15\rho} \dfrac{u^2 - 2\varepsilon}{2\rho} \delta_{m\{r} \delta_{k\}q} \int c^4 \, \partial_\varepsilon B, \\ \hspace{6cm} \beta = k, \quad \gamma = 0, \\[4pt] -\dfrac{2}{15\rho^2} u_p \, \delta_{m\{r} \delta_{k\}q} \int c^4 \, \partial_\varepsilon B, \hspace{1cm} \beta = k, \quad \gamma = p, \\[4pt] \dfrac{1}{15\rho^2} \delta_{m\{r} \delta_{k\}q} \int c^4 \, \partial_\varepsilon B, \hspace{1.5cm} \beta = k, \quad \gamma = 4, \\[4pt] 0, \hspace{6cm} \beta = 4, \quad \text{all } \gamma. \end{cases}$$

$$\int v_m I_4 \, \partial_{\rho_\gamma} \mathbb{G}^r_\beta$$

$$= \begin{cases} \dfrac{2}{5\rho^2}(\tfrac{1}{3}u_r u_m + u^2 \delta_{mr}) \int c^4 B - \dfrac{2}{15\rho} \dfrac{u^2 - 2\varepsilon}{2\rho} (\tfrac{1}{3}u_m u_r + u^2 \delta_{mr}) \int c^4 \partial_\varepsilon B \\[2pt]
\quad - \dfrac{2(u^2 - \varepsilon)}{3\rho^2} \delta_{mr} \int c^4 A + \dfrac{(u^2 - 2\varepsilon)^2}{12\rho^2} \delta_{mr} \int c^4 \partial_\varepsilon A, \hfill \beta = 0, \quad \gamma = 0, \\[8pt]
\dfrac{2}{15\rho^2} u_p(\tfrac{1}{3}u_m u_r + u^2 \delta_{mr}) \int c^4 \partial_\varepsilon B - \dfrac{u^2 - 2\varepsilon}{6\rho^2} u_p \delta_{mr} \int c^4 \partial_\varepsilon A \\[2pt]
\quad - \dfrac{2}{15\rho^2}(\tfrac{1}{3}u_r \delta_{mp} + \tfrac{1}{3}u_m \delta_{pr} + 2u_p \delta_{mr}) \int c^4 B + \dfrac{2}{3\rho^2} u_p \delta_{mr} \int c^4 A, \\
\hfill \beta = 0, \quad \gamma = p, \\[8pt]
- \dfrac{1}{15\rho^2}(\tfrac{1}{3}u_m u_r + u^2 \delta_{mr}) \int c^4 \partial_\varepsilon B - \dfrac{1}{6\rho^2} \delta_{mr} \int c^4 A \\[2pt]
\quad + \dfrac{u^2 - 2\varepsilon}{12\rho^2} \delta_{mr} \int c^4 \partial_\varepsilon A, \hfill \beta = 0, \quad \gamma = 4, \\[8pt]
- \dfrac{8}{15\rho^2} u_{\{k} \delta_{r\}m} \int c^4 B + \dfrac{2}{3\rho^2} u_k \delta_{mr} \int c^4 A \hfill \text{(XXIVB.19)} \\[2pt]
\quad + \dfrac{2(u^2 - 2\varepsilon)}{15\rho^2} u_{\{k} \delta_{r\}m} \int c^4 \partial_\varepsilon B - \dfrac{u^2 - 2\varepsilon}{6\rho^2} u_k \delta_{rm} \int c^4 \partial_\varepsilon A, \\
\hfill \beta = k, \quad \gamma = 0, \\[8pt]
- \dfrac{4}{15\rho^2} u_p u_{\{k} \delta_{r\}m} \int c^4 \partial_\varepsilon B + \dfrac{1}{3\rho^2} u_p u_k \delta_{rm} \int c^4 \partial_\varepsilon A \\[2pt]
\quad + \dfrac{4}{15\rho^2} \delta_{p\{k} \delta_{r\}m} \int c^4 B - \dfrac{1}{3\rho^2} \delta_{pk} \delta_{mr} \int c^4 A, \hfill \beta = k, \quad \gamma = p, \\[8pt]
\dfrac{2}{15\rho^2} u_{\{k} \delta_{r\}m} \int c^4 \partial_\varepsilon B - \dfrac{1}{6\rho^2} u_k \delta_{rm} \int c^4 \partial_\varepsilon A, \hfill \beta = k, \quad \gamma = 4, \\[8pt]
- \dfrac{1}{6\rho^2} \delta_{mr} \int c^4 A + \dfrac{u^2 - 2\varepsilon}{12\rho^2} \delta_{mr} \int c^4 \partial_\varepsilon A, \hfill \beta = 4, \quad \gamma = 0, \\[8pt]
- \dfrac{1}{6\rho^2} u_p \delta_{mr} \int c^4 \partial_\varepsilon A, \hfill \beta = 4, \quad \gamma = p, \\[8pt]
\dfrac{1}{12\rho^2} \delta_{mr} \int c^4 \partial_\varepsilon A, \hfill \beta = 4, \quad \gamma = 4. \end{cases}$$

The formulae (XXIVB.11), (XXIVB.13), (XXIVB.6), and (XXIVB.8) contain all the expressions in the integral equations (XXIV.57) for $\mathbb{G}^{km}_\alpha$. For the integral equations (XXIV.58) for $\mathbb{G}^{km}_{\alpha\beta}$ we require also the formulae (XXIVB.9), (XXIVB.14)–(XXIVB.16), and (XXIVB.17)–(XXIVB.19).

(iv) *Definite integrals*

The following list refers to vectors **c** in a 3-dimensional vector-space $\mathscr{V}$.

$$\int_\mathscr{V} c_k c_m f(c^2)\, d\mathbf{c} = \tfrac{1}{3} \delta_{km} \int_\mathscr{V} c^2 f(c^2)\, d\mathbf{c},$$

$$\int_\mathscr{V} c_{\{k} c_{m\}} f(c^2)\, d\mathbf{c} = 0,$$

$$\int_\mathscr{V} c_k c_m c_r c_s f(c^2)\, d\mathbf{c} = \tfrac{1}{15}(\delta_{km}\delta_{rs} + \delta_{kr}\delta_{ms} + \delta_{ks}\delta_{mr}) \int_\mathscr{V} c^4 f(c^2)\, d\mathbf{c},$$

$$\int_\mathscr{V} c_{\{k} c_{m\}} c_r c_s f(c^2)\, d\mathbf{c} = \tfrac{2}{15} \delta_{r\{k} \delta_{m\}s} \int_\mathscr{V} c^4 f(c^2)\, d\mathbf{c},$$

$$\int_\mathscr{V} c_{\{k} c_{m\}} c_{\{r} c_{s\}} f(c^2)\, d\mathbf{c} = \tfrac{2}{15} \delta_{r\{k} \delta_{m\}s} \int_\mathscr{V} c^4 f(c^2)\, d\mathbf{c},$$

$$\int_\mathscr{V} c_k c_m c_r c_s c_u c_v f(c^2)\, d\mathbf{c} = \tfrac{1}{105}(\delta_{km}\delta_{rs}\delta_{uv} + \delta_{km}\delta_{ru}\delta_{sv} + \delta_{km}\delta_{rv}\delta_{su}$$
$$+ \delta_{kr}\delta_{ms}\delta_{uv} + \delta_{kr}\delta_{mu}\delta_{sv} + \delta_{kr}\delta_{mv}\delta_{su}$$
$$+ \delta_{ks}\delta_{mr}\delta_{uv} + \delta_{ks}\delta_{mu}\delta_{rv} + \delta_{ks}\delta_{mv}\delta_{ru}$$
$$+ \delta_{ku}\delta_{mr}\delta_{sv} + \delta_{ku}\delta_{ms}\delta_{rv} + \delta_{ku}\delta_{mv}\delta_{rs}$$
$$+ \delta_{kv}\delta_{mr}\delta_{su} + \delta_{kv}\delta_{ms}\delta_{ru} + \delta_{kv}\delta_{mu}\delta_{rs})$$
$$\times \int_\mathscr{V} c^6 f(c^2)\, d\mathbf{c}, \hspace{2cm} \text{(XXIVB.20)}$$

$$\int_\mathscr{V} c_{\{k} c_{m\}} c_r c_s c_u c_v f(c^2)\, d\mathbf{c} = \tfrac{2}{105}(\delta_{r\{k}\delta_{m\}s}\delta_{uv} + \delta_{r\{k}\delta_{m\}u}\delta_{sv} + \delta_{r\{k}\delta_{m\}v}\delta_{su}$$
$$+ \delta_{s\{k}\delta_{m\}u}\delta_{rv} + \delta_{s\{k}\delta_{m\}v}\delta_{ru} + \delta_{u\{k}\delta_{m\}v}\delta_{rs})$$
$$\times \int_\mathscr{V} c^6 f(c^2)\, d\mathbf{c},$$

$$\int_{\mathscr{V}} c_{\{k} c_{m\}} c_r c_s c_{\{u} c_{v\}} f(c^2) \, d\mathbf{c} = \tfrac{2}{105}(\delta_{r\{k}\delta_{m\}u}\delta_{sv} + \delta_{r\{k}\delta_{m\}v}\delta_{su} + \delta_{s\{k}\delta_{m\}u}\delta_{rv}$$

$$+ \delta_{s\{k}\delta_{m\}v}\delta_{ru} + \delta_{u\{k}\delta_{m\}v}\delta_{rs} - \tfrac{4}{3}\delta_{uv}\delta_{r\{k}\delta_{m\}s})\int_{\mathscr{V}} c^6 f(c^2) \, d\mathbf{c},$$

$$\int_0^\infty c^{2n} e^{-\alpha c^2} \, dc = \frac{\sqrt{\pi}}{2}\frac{(2n-1)!!}{2^n}\alpha^{-(2n+1)/2}. \qquad \text{(XXIVB.21)}$$

(v) Collisions integrals

In analogy to the definite integrals (XXIVB.20) we have the following formulae for collisions integrals:

$$\int c_{\{k} c_{m\}}\frac{B}{F_M}\mathbb{C}(F_M, c_{\{r} c_{s\}} B) = \frac{1}{5}\delta_{r\{k}\delta_{m\}s}\int c_{\{a} c_{b\}}\frac{B}{F_M}\mathbb{C}(F_M, c_{\{a} c_{b\}} B),$$

$$\int c_k \frac{A}{F_M}\mathbb{C}(F_M, c_m A) = \frac{1}{3}\delta_{km}\int c_a \frac{A}{F_M}\mathbb{C}(F_M, c_a A),$$

$$\int c_{\{k} c_{m\}}\frac{B}{F_M}\mathbb{C}(c_r A, c_s A) = \frac{1}{5}\delta_{r\{k}\delta_{m\}s}\int c_{\{a} c_{b\}}\frac{B}{F_M}\mathbb{C}(c_a A, c_b A),$$

$$\int c_k \frac{A}{F_M}\mathbb{C}(c_s A, c_{\{r} c_{m\}} B) = \frac{1}{5}\delta_{k\{r}\delta_{m\}s}\int c_a \frac{A}{F_M}\mathbb{C}(c_b A, c_{\{a} c_{b\}} B), \qquad \text{(XXIVB.22)}$$

$$\int c_{\{k} c_{m\}}\frac{B}{F_M}\mathbb{C}(c_{\{r} c_{s\}} B, c_{\{u} c_{v\}} B) = \frac{3}{35}(\delta_{u\{k}\delta_{m\}r}\delta_{sv} + \delta_{u\{k}\delta_{m\}s}\delta_{rv} + \delta_{v\{k}\delta_{m\}r}\delta_{us}$$

$$+ \delta_{v\{k}\delta_{m\}s}\delta_{ru} - \tfrac{4}{3}\delta_{uv}\delta_{r\{k}\delta_{m\}s} - \tfrac{4}{3}\delta_{rs}\delta_{u\{k}\delta_{m\}v})$$

$$\times \int c_{\{a} c_{b\}}\frac{B}{F_M}\mathbb{C}(c_{\{b} c_{d\}} B, c_{\{d} c_{a\}} B).$$

These are valid for arbitrary functions A and B of c^2 and ε alone. They are derived in the same way as is (XXIVB.20), but for completeness we describe the procedure with (XXIVB.22)$_3$. Let us denote by K_{kmrs} the left-hand side of (XXIVB.22)$_3$. We see that this quantity satisfies the symmetry relations

$$K_{kmrs} = K_{mkrs} = K_{mksr}, \qquad K_{aars} = 0. \qquad \text{(XXIVB.23)}$$

Moreover, from the orthogonal invariance of the collisions operator and the fact that A, B, and F_M are functions of c^2 we see that

$$K_{kmrs} = Q_{ka} Q_{mb} Q_{rc} Q_{sd} K_{abcd} \qquad \text{(XXIVB.24)}$$

for all orthogonal tensors **Q**. Hence K_{kmrs} are the components of a fourth-order isotropic tensor, and so from a standard representation theorem for such tensors we know that K_{kmrs} is a linear combination of $\delta_{km}\delta_{rs}$, $\delta_{kr}\delta_{ms}$, and $\delta_{ks}\delta_{mr}$. The only combination of these that also satisfies (XXIVB.23) is

$$K_{kmrs} = \beta\,\delta_{r\{k}\delta_{m\}s}, \qquad \text{(XXIVB.25)}$$

in which β is a constant. By contracting now on k and r and on m and s we obtain a formula for β in terms of K_{kmrs}, and placing this into (XXIVB.25) we arrive at

$$K_{kmrs} = \tfrac{1}{5}K_{abab}\,\delta_{r\{k}\delta_{m\}s}. \qquad \text{(XXIVB.26)}$$

This formula is precisely (XXIVB.22)$_3$. If instead we denote the left-hand side of (XXIVB.20)$_4$ by K_{kmrs}, the same reasoning leads us again to (XXIVB.26). But now we see that

$$K_{abab} = \int_{\mathscr{V}} c_{\{a}c_{b\}}c_a c_b\, f(c^2)\, d\mathbf{c} = \tfrac{2}{3}\int_{\mathscr{V}} c^4 f(c^2)\, d\mathbf{c},$$

and so we obtain (XXIVB.20)$_4$.

Finally we record the following special evaluations of some collisions integrals:

$$\int c_{\{a}c_{b\}}\mathbb{C}(F_M, F_M c_{\{a}c_{b\}}) = -\frac{20}{3}n^2\varepsilon^2 \mathbb{A}_{2|2|2},$$

$$\int c_a c^2 \mathbb{C}\left(F_M, F_M c_a\left(\frac{c^2}{2\varepsilon}-\frac{5}{3}\right)\right) = -\frac{40}{9}n^2\varepsilon^2 \mathbb{A}_{2|2|2},$$

$$\int c_{\{a}c_{b\}}\mathbb{C}\left(F_M c_a\left(\frac{c^2}{2\varepsilon}-\frac{5}{3}\right), F_M c_b\left(\frac{c^2}{2\varepsilon}-\frac{5}{3}\right)\right) \qquad \text{(XXIVB.27)}$$

$$= \frac{35}{3}n^2\varepsilon^2(\mathbb{A}_{4|2|2} - 2\mathbb{A}_{3|2|2} + \mathbb{A}_{2|2|2}),$$

$$\int c_a c^2 \mathbb{C}\left(F_M c_b\left(\frac{c^2}{2\varepsilon}-\frac{5}{3}\right), F_M c_{\{a}c_{b\}}\right) = -\frac{110}{9}n^2\varepsilon^3(\mathbb{A}_{3|2|2} - \mathbb{A}_{2|2|2}),$$

$$\int c_{\{a}c_{b\}}\mathbb{C}(F_M c_{\{b}c_{d\}}, F_M c_{\{d}c_{a\}}) = -\frac{140}{27}n^2\varepsilon^3(\mathbb{A}_{3|2|2} - \mathbb{A}_{2|2|2}),$$

the functions $\mathbb{A}_{p|m|n}$ being defined by (XVII.31). These may be obtained from the general list of collisions integrals given in (XVII.34). To illustrate how to do so, we present the steps in obtaining (XXIVB.27)$_5$. If we set $A^{(3)}_{0|kmr} = 0$ and

$A_n = 0$, $n = 4, 5, \ldots$ in GRAD's expansion (XVII.9), we obtain from (XVII.6), (XVII.44)$_1$, and (XVII.46) the function

$$F = F_M(1 + \tfrac{1}{2}H_{ab}(\kappa)A_{ab} + \tfrac{1}{10}H_a^{(3)}(\kappa)A_a^{(3)}), \quad \text{(XXIVB.28)}$$

$$= F_M\left(1 + \frac{3}{4\varepsilon}c_{\{a}c_{b\}}A_{ab} + \frac{3}{10}\sqrt{\frac{3}{2\varepsilon}}c_a\left(\frac{c^2}{2\varepsilon} - \frac{5}{3}\right)A_a^{(3)}\right).$$

The collisions integral $m\bar{C}H_{km}$ for this particular F is found by specializing (XVII.34)$_3$ in the same way, and it is given by

$$m\bar{C}H_{km} = -3n\rho\mathbb{A}_{2|2|2}A_{km} - \tfrac{3}{2}n\rho(\mathbb{A}_{3|2|2} - \mathbb{A}_{2|2|2})A_{\{ek}A_{me\}}$$
$$+ \tfrac{189}{400}n\rho(\mathbb{A}_{4|2|2} - 2\mathbb{A}_{3|2|2} + \mathbb{A}_{2|2|2})A_{\{k}^{(3)}A_{m\}}^{(3)}. \quad \text{(XXIVB.29)}$$

Another expression for this same collisions integral is obtained by placing F, as given by (XXIVB.28)$_2$, into the definition (VII.15)$_1$ and then expanding and using the bilinearity of $\mathbb{C}(\,,\,)$. The result is

$$m\bar{C}H_{km} = m\left(\frac{3}{2\varepsilon}\right)^2 \int c_{\{k}c_{m\}}\mathbb{C}(F_M, F_M c_{\{a}c_{b\}})A_{ab}$$

$$+ m\,\frac{9}{10\varepsilon}\sqrt{\frac{3}{2\varepsilon}}\int c_{\{k}c_{m\}}\mathbb{C}\left(F_M, F_M c_a\left(\frac{c^2}{2\varepsilon} - \frac{5}{3}\right)\right)A_a^{(3)}$$

$$+ 2m\left(\frac{3}{4\varepsilon}\right)^3 \int c_{\{k}c_{m\}}\mathbb{C}(F_M c_{\{a}c_{b\}}, F_M c_{\{d}c_{e\}})A_{ab}A_{de}$$

$$+ m\,\frac{9}{100}\left(\frac{3}{2\varepsilon}\right)^2 \int c_{\{k}c_{m\}}\mathbb{C}\left(F_M c_a\left(\frac{c^2}{2\varepsilon} - \frac{5}{3}\right),\right. \quad \text{(XXIVB.30)}$$

$$\left.F_M c_b\left(\frac{c^2}{2\varepsilon} - \frac{5}{3}\right)\right)A_a^{(3)}A_b^{(3)}$$

$$+ m\,\frac{3}{10}\left(\frac{3}{2\varepsilon}\right)^2\sqrt{\frac{3}{2\varepsilon}}\int c_{\{k}c_{m\}}\mathbb{C}\left(F_M c_a\left(\frac{c^2}{2\varepsilon} - \frac{5}{3}\right),\right.$$

$$\left.F_M c_{\{b}c_{d\}}\right)A_a^{(3)}A_{bd}.$$

These two expressions agree for all choices of A_{km} and $A_k^{(3)}$. Let us equate, then, the coefficients of $A_{ab}A_{de}$. To do this we note from (XXIVB.22)$_5$, specialized by choosing $B = F_M$, that

$$\int c_{\{k}c_{m\}}\mathbb{C}(F_M c_{\{a}c_{b\}}, F_M c_{\{d}c_{e\}})A_{ab}A_{de}$$
$$= \tfrac{12}{35}A_{e\{k}A_{m\}e}\int c_{\{a}c_{b\}}\mathbb{C}(F_M c_{\{b}c_{d\}}, F_M c_{\{d}c_{a\}}). \quad \text{(XXIVB.31)}$$

Placing this formula into (XXIVB.30) and then equating the coefficients of $A_{e\{k} A_{m\}e}$ in (XXIVB.29) and (XXIVB.30), we arrive at (XXIVB.27)$_5$. These same formulae deliver (XXIVB.27)$_{1,3}$ in the same way, while for (XXIVB.27)$_{2,4}$ we must repeat the analysis using the collisions integral $\mathfrak{m}\bar{\mathbb{C}}H_{kaa}$.

We recall that for a gas of Maxwellian molecules $\mathbb{A}_{p|2|2}$ is independent of p, and so for such a gas the three collisions integrals (XXIVB.27)$_{3,4,5}$ are all null. This is to be expected, since, as the above calculation shows, the values of these three collisions integrals arise from terms in GRAD's evaluations (XVII.34)$_{3,4}$ for which $r \neq m + n$, and for a gas of Maxwellian molecules, IKENBERRY's theorem shows that these terms are null.

Chapter XXV

The Maxwellian Iteration of Ikenberry & Truesdell

In Chapters XXII and XXIV we have studied two formal iterative procedures designed to select grossly determined quantities: the method of HILBERT, which aims for grossly determined *solutions*, and the method of stretched fields (designed to replace the method of ENSKOG), which aims directly for *gross determiners*. Both of these methods rely upon solution of integral equations. When applied to a gas of Maxwellian molecules neither method simplifies at all until the very end, at which, for such a gas, the solutions of the integral equations for determining the scalar coefficients of approximate gross determiners may be exhibited.

In Chapters XVI and XVII we have seen that for Maxwellian molecules the collisions integrals of polynomials can be evaluated explicitly as finite sums of moments of F, the coefficients of which are functions of m and g alone and are the same functions for all molecular densities F. In this sense, F itself need not be determined if our aim is to study gross relations. IKENBERRY & TRUESDELL constructed a method[1] designed especially to exploit from the outset the simplicity of the Maxwellian molecule. It does not involve integral equations. Instead, it employs nothing more than successive differentiations. It is a family of methods rather than a single one. One member of this family delivers approximate gross determiners for P and q that are the same, as far as can be told from the results presently worked out, as those of ENSKOG's method and the method of stretched fields. IKENBERRY & TRUESDELL called their basic method *Maxwellian iteration* because it rendered general some of the ideas MAXWELL initiated in his memoir on rarefied gases. In that work he was stopped by difficulties concerning the fourth moments, but the method of IKENBERRY & TRUESDELL overcomes these and delivers through routine algebra as many iterative steps as be desired, provided only that enough collisions integrals shall have been calculated first. IKENBERRY & TRUESDELL laid

---

[1] IKENBERRY & TRUESDELL [1956, Chapter I].

emphasis on determining not so much the coefficients appearing in the first few iterates as the character of the iterates of high orders. We have stated already that one variant of the method, the variant called *atemporal Maxwellian iteration*, delivers results which, at least for the first few stages, are the same as those obtained by ENSKOG's method and by the method of stretched fields, specialized to Maxwellian molecules.

More important than the agreement and the simplicity just described is the basic method, Maxwellian iteration itself.

The iterates it delivers are meant to approximate **P** and **q** not by functions of the gross condition (ρ, **b**) and its spatial derivatives but rather by functions of the derivatives of the principal moment ρ alone—*derivatives not only with respect to place but also with respect to time*. The body force **b**, to take account of which offers some complication to the method of stretched fields, plays no part at all in Maxwellian iteration. It is eliminated from the start by use of the equation of balance of linear momentum. That explains a fact we shall soon observe: In place of the space derivatives of **b** that appear in the results of ENSKOG's process, the results of Maxwellian iteration are functions whose arguments include higher and higher time derivatives of ρ and its space derivatives. Functions of this kind we agreed in Section (vi) of Chapter VIII to call *momentally determined*. In general, a field B is momentally determined if

$$B(t, \mathbf{x}) = \mathfrak{B}(\rho(t + \cdot, \mathbf{x} + \cdot)), \qquad (\text{VIII.73})_{2r}$$

and $\mathfrak{B}$ is its *momental determiner*. We may say, then, that Maxwellian iteration aims to approximate momental determiners of **P** and **q**, on the presumption that they exist.

Although IKENBERRY & TRUESDELL limited their work to a gas of Maxwellian molecules, TRUESDELL immediately[2] suggested several ways to extend it. One of these we now work out and incorporate in our presentation. Namely, we adopt GRAD's *approximation* to the collisions integrals but otherwise use exact formulae. As we explained in Section (v) of Chapter XVII, GRAD's approximation is exact for Maxwellian molecules, and reasons have been given in support of it when other kinetic constitutive relations are adopted.

(i) Exact results to which Maxwellian iteration is applied

Maxwellian iteration rests upon the exact equation of transfer for a function g of **c** alone which we obtained in Chapter IX:

$$\mathfrak{L}g = \mathfrak{m}\bar{\mathbb{C}}g, \qquad (\text{IX.13})_r$$

---

[2] TRUESDELL [1956, §56].

(i) REVIEW OF EXACT RESULTS

the operator $\mathfrak{L}$ being defined as follows:

$$\mathfrak{L}g \equiv (\rho\bar{g})\dot{} + \rho E\bar{g} + \overline{\rho \mathbf{c} \otimes \operatorname{grad} g} \cdot (\operatorname{grad} \mathbf{u})^{\mathrm{T}} - \overline{\operatorname{grad} g} \cdot \operatorname{div} \mathbf{M} + \operatorname{div}(\rho\overline{g\mathbf{c}}).$$

(IX.14)$_r$

For g we take IKENBERRY's spherical harmonics $\mathbf{Y}_{2r|s}$, defined by (XVI.26), and we recall the spherical moments $\mathbf{P}_{2r|s}$, defined by (XVI.28). Because the set of spherical moments $\mathbf{P}_{2r|s}$ of order $2r + s$ and the set of general moments $^{2r+s}\mathbf{M}$ of order $2r + s$ determine one another uniquely, the finite system of equations of transfer for the general moments of this order is equivalent to the system

$$\mathfrak{L} Y_{2r|s} = \mathfrak{m}\overline{\mathbb{C}}\, Y_{2r|s},$$

(XVI.40)$_r$

with the sole exception of the pair $r = 0$, $s = 1$.

The general system of equations for moments is formally equivalent to the Maxwell–Boltzmann equation; more correctly, its solutions $^n\mathbf{M}$, $n = 0, 1, \ldots$, are generalized solutions, quantities some presumably very large subset of which will be the moments of a solution F of the Maxwell–Boltzmann equation. As was shown in Chapter IX, the field equation expressing balance of linear momentum corresponds to the choice $n = 1$. In contrast, IKENBERRY & TRUESDELL's system (XVI.40) does not imply that equation. The member that corresponds to the choice $r = 0$, $s = 1$, is $\mathbf{0} = \mathbf{0}$. IKENBERRY & TRUESDELL advisedly chose to begin from (IX.13) because it is free of the body force $\mathbf{b}$.

Results based upon (XVI.40) alone may be expected to play a part parallel to that of constitutive relations in continuum mechanics. Just as the constitutive relations assumed in continuum mechanics never take $\mathbf{b}$ as an independent variable, *results that follow from* (XVI.40) *are valid for an arbitrary body force* $\mathbf{b}$. By taking (XVI.40) as their starting point, IKENBERRY & TRUESDELL sought to find relations in the kinetic theory *that might be compared with the constitutive relations that continuum mechanics introduces as generic axioms*.

The parallel, however, is not strict, for while the arguments of constitutive relations (Chapter I) are independent variables, all results that follow from (XVI.40) *necessarily satisfy the equation of balance of mass* (IX.3) *and the equation of balance of energy* (IX.5). Moreover, while (XVI.40) itself does not imply the principle of linear momentum, the kinetic theory delivers (XVI.40) only for fields that do satisfy that principle. Thus from (XVI.40) we cannot demonstrate constitutive relations. For example, although (XVI.40) itself makes sense for an arbitrary body force $\mathbf{b}$ (because $\mathbf{b}$ does not enter it at all), it itself is an *extension* to arbitrary ρ and $\mathbf{b}$ of results the kinetic theory justifies *only* for fields ρ and $\mathbf{b}$ such that $\rho\dot{\mathbf{u}} = -\operatorname{grad} p - \operatorname{div} \mathbf{P} + \rho\mathbf{b}$.

For convenience we repeat here the list of explicit $\mathfrak{L} Y_{2r|s}$ we obtained in Chapter XVI:

$$\mathfrak{L} Y_{0|0} = \dot{\rho} + \rho E,$$

$$\mathfrak{L} Y_{0|k} = 0,$$

$$\mathfrak{L} Y_{2|0} = 3\dot{p} + 5pE + 2P_{ab}E_{ab} + 2q_{a,a},$$

$$\mathfrak{L} Y_{0|km} = \dot{P}_{km} + P_{km}E + 2P_{\{ak}u_{m,a\}} + 2pE_{km} + P_{0|kma,a} + \tfrac{4}{5}q_{\{k,m\}},$$

$$\mathfrak{L} Y_{2|k} = 2\dot{q}_k + \tfrac{10}{3}q_k E + 2P_{0|kab}E_{ab} + 2q_a u_{k,a} + \tfrac{8}{5}q_a E_{ka}$$
$$- \frac{5p}{\rho}M_{ka,a} - \frac{2}{\rho}P_{ka}M_{ab,b} + P_{2|ka,a} + \tfrac{1}{3}P_{4|0,k},$$

$$\mathfrak{L} Y_{0|kmr} = \dot{P}_{0|kmr} + P_{0|kmr}E + 3P_{0|\{akm}u_{r,a\}} + \tfrac{12}{5}q_{\{k}E_{mr\}}$$
$$- \frac{3}{\rho}P_{\{km}M_{ra,a\}} + P_{0|kmra,a} + \tfrac{3}{7}P_{2|\{km,r\}},$$

$$\mathfrak{L} Y_{4|0} = \dot{P}_{4|0} + \tfrac{7}{3}P_{4|0}E + 4P_{2|ab}E_{ab} - \frac{8}{\rho}q_a M_{ab,b} + P_{4|a,a},$$

$$\mathfrak{L} Y_{2|km} = \dot{P}_{2|km} + \tfrac{5}{3}P_{2|km}E + 2P_{0|kmab}E_{ab} + 2P_{2|\{ak}u_{m,a\}} + \tfrac{8}{7}P_{2|\{ak}E_{ma\}}$$
$$+ \tfrac{14}{15}P_{4|0}E_{km} - \frac{2}{\rho}P_{0|kma}M_{ab,b} \qquad (XVI.41)_r$$
$$- \frac{28}{5\rho}q_{\{k}M_{ma,a\}} + P_{2|kma,a} + \tfrac{2}{5}P_{4|\{k,m\}},$$

$$\mathfrak{L} Y_{0|kmrs} = \dot{P}_{0|kmrs} + P_{0|kmrs}E + 4P_{0|\{akmr}u_{s,a\}} + \tfrac{12}{7}P_{2|\{km}E_{rs\}}$$
$$- \frac{4}{\rho}P_{0|\{kmr}M_{sa,a\}} + P_{0|kmrsa,a} + \tfrac{4}{9}P_{2|\{kmr,s\}},$$

$$\mathfrak{L} Y_{4|k} = \dot{P}_{4|k} + \tfrac{7}{3}P_{4|k}E + P_{4|a}u_{k,a} + 4P_{2|kab}E_{ab} + \tfrac{8}{5}P_{4|a}E_{ak}$$
$$- \frac{7}{3\rho}P_{4|0}M_{ka,a} - \frac{4}{\rho}P_{2|ka}M_{ab,b} + P_{4|ka,a} + \tfrac{1}{3}P_{6|0,k},$$

$$\mathfrak{L} Y_{2|kmr} = \dot{P}_{2|kmr} + \tfrac{5}{3}P_{2|kmr}E + 2P_{0|kmrab}E_{ab} + \tfrac{4}{3}P_{2|\{akm}E_{ra\}} + 3P_{2|\{akm}u_{r,a\}}$$
$$+ \tfrac{54}{35}P_{4|\{k}E_{mr\}} - \frac{2}{\rho}P_{0|kmra}M_{ab,b}$$
$$- \frac{27}{7\rho}P_{2|\{km}M_{ra,a\}} + P_{2|kmra,a} + \tfrac{3}{7}P_{4|\{km,r\}}.$$

Of course these results hold for all molecular models. As always in this book, braces around a set of indices indicate the totally symmetric traceless part, formed after all repeated indices have been summed. The simplest examples have been listed as (XVI.38).

(i) REVIEW OF EXACT RESULTS 511

For a gas of Maxwellian molecules the collisions integrals corresponding to (XVI.41) are delivered by IKENBERRY's theorem (XVI.31); we have displayed them in (XVI.37). For the developments in this chapter we prefer to use the results of GRAD's formal calculation after converting them to expressions in terms of IKENBERRY's spherical harmonics and spherical moments:

$$m\bar{C}Y_{0|0} = 0,$$

$$m\bar{C}Y_{0|k} = 0,$$

$$m\bar{C}Y_{2|0} = 0,$$

$$m\bar{C}Y_{0|km} = -3n\mathbb{A}_{2|2|2}P_{km} + R_{0|km},$$

$$m\bar{C}Y_{2|k} = -4n\mathbb{A}_{2|2|2}q_k + R_{2|k},$$

$$m\bar{C}Y_{0|kmr} = -\frac{9n}{2}\mathbb{A}_{2|2|2}P_{0|kmr} + R_{0|kmr},$$

$$m\bar{C}Y_{4|0} = -\frac{2n}{\rho}\mathbb{A}_{2|2|2}(\rho P_{4|0} - 15p^2 + P_{ab}P_{ab}) + R_{4|0}, \qquad \text{(XVIIA.5)}_r$$

$$m\bar{C}Y_{2|km} = -\frac{n}{8}(27\mathbb{A}_{4|2|2} - 42\mathbb{A}_{3|2|2} + 43\mathbb{A}_{2|2|2})P_{2|km}$$

$$-\frac{n}{4\rho}(27\mathbb{A}_{4|2|2} - 42\mathbb{A}_{3|2|2} + 23\mathbb{A}_{2|2|2})P_{\{ak}P_{ma\}}$$

$$+\frac{7n}{8\rho}(27\mathbb{A}_{4|2|2} - 42\mathbb{A}_{3|2|2} + 19\mathbb{A}_{2|2|2})pP_{km} + R_{2|km},$$

$$m\bar{C}Y_{0|kmrs} = \frac{n}{4}(35\mathbb{A}_{4|4|4} - 10\mathbb{A}_{4|2|2} - 18\mathbb{A}_{2|2|2})P_{0|kmrs} + \frac{3n}{4\rho}(35\mathbb{A}_{4|4|4}$$

$$- 10\mathbb{A}_{4|2|2} + 6\mathbb{A}_{2|2|2})P_{\{km}P_{rs\}} + R_{0|kmrs}.$$

In Appendix A to Chapter XVII we list some conversion formulae by which one may derive this list from (XVII.34). The coefficients $\mathbb{A}_{p|n|m}$ are defined by (XVII.31), following which may be found the special cases appropriate to ideal spheres and inverse k^{th}-power molecules. IKENBERRY's theorem states that for a gas of Maxwellian molecules *the remainders* $\mathbf{R}_{2r|s}$ *are null* and that the resulting formulae are valid *with no proviso* except that the left-hand members shall exist. Conditions sufficient for the existence of these integrals can be established on the basis of Theorem 4 in Chapter XIX. For other molecular models the results are only formal, as was explained in Section (ii) of Chapter XVII, and the remainders, which include the terms Q_{km}, Q_{kmr}, and Q_{kmrs} in (XVII.34), are infinite series whose terms are proportional to products $P_{2r_1|s_1}P_{2r_2|s_2}$ such that $2r_1 + s_1 + 2r_2 + s_2 \neq 2r + s$. In what follows, we shall simply drop all those remainders, just as GRAD did in his method of approximation, which we have presented in Chapter XVII.

The great advantage of GRAD's approximation for collisions integrals is that when converted to expressions in terms of spherical moments *it can be stated in the same form as* IKENBERRY's *theorem*:

$$\mathfrak{m}\bar{C}Y_{2r|s} = -c_{2r|s}P_{2r|s} + Q_{2r|s}, \qquad 2r + s \geqq 1, \qquad \text{(XVI.31)}_r$$

$Q_{2r|s}$ being a bilinear function of the spherical moments whose orders are less than $2r + s$. For a gas of Maxwellian molecules the scalar $c_{2r|s}$ is given by (XVI.33). By use of (XVII.25) it would be possible to give an explicit formula for $c_{2r|s}$ for general molecules, but the result would be complicated. The particular $-c_{2r|s}$ we shall need in this chapter are the leading coefficients of the right-hand sides of (XVIIA.5).

Once we have IKENBERRY's theorem (XVI.31) for Maxwellian molecules or GRAD's corresponding approximations for general molecules, we can write the basic equation of transfer (XVI.40) in the form

$$-c_{2r|s}P_{2r|s} = -Q_{2r|s} + \mathfrak{L}Y_{2r|s}. \qquad \text{(XXV.1)}$$

Maxwellian iteration applies directly to this equation of transfer. The only difference here between a gas of Maxwellian molecules and a general one lies in the interpretation: For Maxwellian molecules, Maxwellian iteration aims to approximate certain solutions of *exact and demonstrated equations* for moments; for other molecular models, certain solutions of *merely formal approximations to equations* for moments. This property, which results directly from GRAD's truncation of the infinite series (XVII.24), Maxwellian iteration shares with GRAD's method of approximating the system of equations for the moments. In every other regard Maxwellian iteration is altogether different from GRAD's method.

(ii) The scheme of Maxwellian iteration

The method proposed by IKENBERRY & TRUESDELL presumes that $\mathfrak{L}Y_{2r|s}$ and $\mathfrak{m}\bar{C}Y_{2r|s}$ shall have been evaluated already for all orders $2r + s$ up to a certain number, a number which depends on how many iterative steps we desire to carry out. The results (XVI.41) and (XVIIA.5) are specimens of that evaluation, and they will suffice for the purposes of this chapter. We notice once and for all that $\mathfrak{L}Y_{2r|s}$ is a polynomial in the moments of orders up to and including $2r + s$ and in the time derivatives and space derivatives of the moments of orders up to and including $2r + s + 1$. Thus the forward coupling of the infinite system for the moments results from the nature of $\mathfrak{L}Y_{2r|s}$, not from $\mathfrak{m}\bar{C}Y_{2r|s}$.

We seek approximate momental determiners of **P** and **q**. Thus we agree to regard the principal moment field **ρ** as an independent variable. The equations of balance (IX.3), (IX.4), and (IX.5) are set aside for consideration later; not entering the iterative scheme, they will be used only afterward, just as if the approximate momental determiners were constitutive relations of continuum mechanics. Thus Maxwellian iteration shares with ENSKOG's method and the method of stretched fields the aim of eliminating the higher moments entirely so as to obtain, finally, approximate partial differential equations like those of a continuum theory of gases: just five scalar equations, not more.

Accordingly, in the iterative scheme to be based on (XXV.1) we leave aside entirely the choices $(r, s) = (0, 0)$ and $(r, s) = (1, 0)$, for these choices reduce (XXV.1) to (IX.3) and (IX.5), respectively. Of course the choice $(r, s) = (0, 1)$ reduces (XXV.1) to $\mathbf{0} = \mathbf{0}$, as we have already remarked.

This much settled, we now turn back to (XXV.1) and base the following iterative scheme upon it:

$$-\mathfrak{c}_{2r|s} P^{(n+1)}_{2r|s} \equiv -Q^{(n+1)}_{2r|s} + \mathfrak{L}^{(n)} Y_{2r|s}, \quad 2r + s \geq 2, \quad (r, s) \neq (1, 0). \quad (XXV.2)$$

The symbol $\mathfrak{L}^{(n)}Y_{2r|s}$ denotes the result of replacing in $\mathfrak{L}Y_{2r|s}$ each spherical moment $\mathbf{P}_{2q|u}$ by its nth iterate $\mathbf{P}^{(n)}_{2q|u}$. At the $n + 1^{\text{st}}$ step we thus presume that the nth iterates of all spherical moments of orders up to and including $2r + s$ have been determined already. The symbol $\mathbf{Q}^{(n+1)}_{2r|s}$ denotes the result of replacing in $\mathbf{Q}_{2r|s}$ each spherical moment $\mathbf{P}_{2q|u}$ by its $n + 1^{\text{st}}$ iterate $\mathbf{P}^{(n+1)}_{2q|u}$. Since $\mathbf{Q}_{2r|s}$ is a bilinear function of the spherical moments of orders less than $2r + s$, we can regard $\mathbf{Q}^{(n+1)}_{2r|s}$ as a known quantity if the iteration has been carried already $n + 1$ steps for all spherical moments of orders less than $2r + s$. The iteration is in a sense a double one: To proceed one step ahead for any given moment, we must already have completed all steps up to and including that one for all moments of lower order. As the iteration proceeds, it makes use of more and more members of the infinite system of equations for moments.

As the notation indicates, the value of the right-hand side of (XXV.2)$_1$ defines the left-hand side. If $\mathfrak{c}_{2r|s} \neq 0$, the $n + 1^{\text{st}}$ iterate $\mathbf{P}^{(n+1)}_{2r|s}$ is thus determined uniquely. For Maxwell molecules we have the estimate (XVIII.18), which is more than enough to show that $\mathfrak{c}_{2r|s} > 0$. It is not obvious that the general $\mathfrak{c}_{2r|s}$ appearing in (XVIIA.5) are positive, but we shall not enter into the question.

We begin with the simplest case: $(r, s) = (0, 2)$. In its simplicity it is not typical: The bilinear function $\mathbf{Q}_{0|2}$ vanishes, as we see from (XVIIA.5)$_4$, use of which and of (XVI.41)$_4$ yields at once

$$-3 \mathbb{A}_{2|2|2} n P^{(n+1)}_{km} \equiv 2pE_{km} + \dot{P}^{(n)}_{km} + P^{(n)}_{km} E + 2P^{(n)}_{\{ak} u_{m,a\}} + P^{(n)}_{0|kma,a} + \tfrac{4}{5} q^{(n)}_{\{k,m\}}.$$

(XXV.3)

As we remarked in connection with (XVII.31), $\mathbb{A}_{2|2|2} > 0$, so (XXV.3) delivers $\mathbf{P}^{(n+1)}$ if $\mathbf{P}^{(n)}$, $\mathbf{P}^{(n)}_{0|3}$, and $\mathbf{q}^{(n)}$ have been determined previously. Likewise, calling upon (XVIIA.5)$_5$ and (XVI.41)$_5$, we obtain

$$-4\mathbb{A}_{2|2|2}nq_k^{(n+1)} \equiv 2\dot{q}_k^{(n)} + \tfrac{10}{3}q_k^{(n)}E + 2P^{(n)}_{0|kab}E_{ab} + 2q_a^{(n)}u_{k,a} + \tfrac{8}{5}q_a^{(n)}E_{ka}$$

$$-\frac{5p}{\rho}M^{(n)}_{ka,a} - \frac{2}{\rho}P^{(n)}_{ka}M_{ab,b} + P^{(n)}_{2|ka,a} + \tfrac{1}{3}P^{(n)}_{4|0,k}, \qquad \text{(XXV.4)}$$

which delivers $\mathbf{q}^{(n+1)}$ if $\mathbf{q}^{(n)}$, $\mathbf{P}^{(n)}_{0|3}$, $\mathbf{P}^{(n)}$, $\mathbf{P}^{(n)}_{2|2}$, and $\mathbf{P}^{(n)}_{4|0}$ have been determined previously.

To begin the iteration, we need the iterates of order 0. We shall take these fields as being the *spherical moments of a Maxwellian density, expressed as functions of the principal moment* $\boldsymbol{\rho}$. In view of (XVI.30)$_2$, that means

$$P^{(0)}_{2r|s} \equiv \begin{cases} 0 & \text{if } s \neq 0, \\ & \quad 2r+s \geq 2, \quad (r,s) \neq (1,0). \\ (2r+1)!!\,\rho(p/\rho)^r = (2r+1)!!\,\rho(\tfrac{2}{3}\varepsilon)^r & \text{if } s = 0, \end{cases} \qquad \text{(XXV.5)}$$

The scheme defined by (XXV.2) with this choice of initial iterates is called **Maxwellian Iteration**. Other choices of initial iterates and variants of the iterative scheme or supplements to it are possible; we shall mention some of these below.

(iii) *Illustration of the idea of Maxwellian iteration, applied to an ordinary differential equation*

Maxwellian iteration calculates the successive iterates by repeated differentiation. Since repeated differentiation is altogether untypical of methods for solving differential equations, we now turn aside for the moment from the kinetic theory and illustrate the method by applying it to a very simple example, a single ordinary differential equation:

$$\tau\dot{P} = -P, \qquad \text{(XXV.6)}$$

in which τ is a positive constant. Of course the general solution of the initial-value problem for this equation is

$$P = P(0)e^{-t/\tau}. \qquad \text{(XXV.7)}$$

We recognize the result as being a one-dimensional counterpart of Maxwell's relaxation theorem, presented in Chapter XIII. The solution is defined uniquely by the initial value $P(0)$. Moreover, no matter what be the initial value $P(0)$, the solution $P(t) \to 0$ as $t/\tau \to \infty$. The particular solution determined by the initial condition $P(0) = 0$, namely, 0, is the principal solution; all solutions tend to it for each fixed τ as $t \to \infty$, or for each fixed t as $\tau \to 0$.

(iv) THE FIRST TWO ITERATES 515

The counterpart of Maxwellian iteration for (XXV.6) is the scheme

$$-P^{(n+1)} \equiv \tau \dot{P}^{(n)}, \qquad P^{(0)} = 0. \qquad \text{(XXV.8)}$$

Thus

$$P^{(n)} = (-\tau)^n \frac{d^n}{dt^n} P^{(0)} = 0. \qquad \text{(XXV.9)}$$

Every Maxwellian iterate is thus the principal solution itself. If instead of the particular initial iterate $(XXV.8)_2$ we consider initial iterates belonging to certain classes, we can obtain various sequences $P^{(n)}$ that converge to the principal solution. For example, if $P^{(0)}$ is a polynomial in t, the early iterates $P^{(n)}$ are not 0, but the iterative process converges to 0 in a finite number of steps. If $P^{(0)}$ is $e^{-t/k}$ multiplied by a polynomial, the process converges to the principal solution if $k > \tau$ but diverges otherwise. Processes of this class may be called *differential iterations*. The example shows that *a differential iteration may select out of all possible solutions one that is specially smooth* and suggests that *a differential iteration may converge to a special solution, the principal one*.

TRUESDELL[3] explored the properties of differential iterations by means of a model, an infinite, forward-coupled system of ordinary differential equations chosen so as to reflect some properties of the equations for moments in the kinetic theory. He showed by example that such an iteration might provide an asymptotic representation of a principal solution even when the process itself diverged. He constructed also an example in which if the 0th iterate is chosen as the principal solution, the iteration may first depart from it but ultimately converge back to it.

We have promised that in this book we shall not go into models at all. TRUESDELL's model is no exception. The only reason we mention it is that it may cast some light upon a central problem of the kinetic theory.

(iv) *The first two stages of Maxwellian iteration: The Maxwell second approximation to* **P** *and its companion for* **q**

Returning now to Maxwellian iteration in the kinetic theory, we substitute the appropriate special cases of (XXV.5) into (XVI.41) and so obtain[4]

---

[3] TRUESDELL [1956, Chapter III].

[4] IKENBERRY & TRUESDELL [1956, §9] simplified their results by using the exact equation $\dot{\rho} + \rho E = 0$. Here we have chosen to set that equation aside for later consideration, and as a result our $(XXV.10)_4$ contains one more term than theirs. The effect of this extra term propagates, of course, through the iterative scheme, giving us different expressions for some of the iterates. This difference simply illustrates the remarks in the fine print just before (XVI.41) in Section (i) of this chapter: Both sets of approximate momental determiners, one which we present here and the other obtained by IKENBERRY & TRUESDELL, are extensions to arbitrary fields **ρ** of results which the kinetic theory can deliver only for **ρ** satisfying the equations of balance. As such, one has the same status as the other, the two being equal for such **ρ** as satisfy (IX.3)–(IX.5).

$$\mathfrak{L}^{(0)}Y_{0|km} = 2pE_{km},$$

$$\mathfrak{L}^{(0)}Y_{2|k} = \tfrac{10}{3}p\varepsilon_{,k},$$

$$\mathfrak{L}^{(0)}Y_{0|kmr} = 0,$$

$$\mathfrak{L}^{(0)}Y_{4|0} = 20\frac{p^2}{\rho}\dot{\beta} + 35\frac{p^2}{\rho^2}(\dot{\rho} + \rho E),$$

$$\mathfrak{L}^{(0)}Y_{2|km} = 14\frac{p^2}{\rho}E_{km}, \tag{XXV.10}$$

$$\mathfrak{L}^{(0)}Y_{0|kmrs} = 0,$$

$$\mathfrak{L}^{(0)}Y_{4|k} = \frac{140}{3}\frac{p^2}{\rho}\varepsilon_{,k},$$

$$\mathfrak{L}^{(0)}Y_{2|kmr} = 0,$$

in which

$$\beta \equiv \tfrac{3}{2}\log p\rho^{-5/3}. \tag{XXV.11}$$

By use of (XXV.3), (XXV.4), (XXV.10), and (III.15) we obtain the first two of the following first iterates:

$$-3n\mathbb{A}_{2|2|2}P^{(1)}_{km} = 2pE_{km},$$

$$-3n\mathbb{A}_{2|2|2}q^{(1)}_k = \tfrac{5}{2}p\varepsilon_{,k}, \tag{XXV.12}$$

$$P^{(1)}_{0|kmr} = 0.$$

The third one is an immediate consequence of (XXV.10)$_3$, (XVIIA.5)$_6$, and (XXV.5). For Maxwellian molecules the first two reduce to MAXWELL's (XIII.13) and (XIII.17) with the difference that now his formulae for **P** and **q** appear, not as the result of casting away terms believed to be small, *but as the first iterates in a systematic method of formal calculation*. It is convenient to write these same formulae as

$$P^{(1)}_{km} = -2\mu E_{km}, \qquad \mu = \mu^{(1)} \equiv \frac{p}{3n\mathbb{A}_{2|2|2}},$$

$$q^{(1)}_k = -M\mu\varepsilon_{,k}, \qquad M = M^{(1)} \equiv \tfrac{5}{2}. \tag{XXV.13}$$

For Maxwellian molecules $\mu^{(1)}$ and $M^{(1)}$ of course agree with μ and M as given by (XIII.14) and (XIII.18). For general molecules $\mu^{(1)}$ and $M^{(1)}$ agree with GRAD's determinations (XVII.61) and (XVII.62).

The first iterates, given by (XXV.13), obviously are both grossly and momentally determined. For Maxwellian molecules they are exactly the same as those obtained by ENSKOG's method and by the method of stretched fields,

(iv) THE MAXWELL SECOND APPROXIMATION

namely (XXIV.90). For general molecules they agree with ENSKOG's first approximation (*cf.* Footnote 17 to Chapter XVII). They have been obtained by routine algebraic substitutions of derivatives of 0th iterates; there have been no integral equations to solve. As we shall see now, this simplicity carries on to every stage of the iteration. However, the succeeding iterates will not be grossly determined.

For $\mathbf{P}^{(2)}$ there is no difficulty. We need only substitute $(XXV.13)_{1,4}$ into the right-hand side of (XXV.3) when $n = 1$. The result is the *Maxwell Second Approximation* to $\mathbf{P}$:

$$P_{km}^{(2)} = -2\mu^{(1)}E_{km} + \frac{2\mu^{(1)}\mu}{p}\left[\frac{1}{\mu}(\mu E_{km})^{\cdot} + EE_{km} + 2E_{\{ak}u_{m,a\}}\right] + \frac{4\mu^{(1)}}{5p}(M\mu\varepsilon_{,\{k\}})_{,m\}}.$$
(XXV.14)

MAXWELL provided apparatus sufficient to obtain this formula, but he was interested only in the term written last here[5], and only a part of that, so he did not work out the rest. In addition he made some numerical errors. Here we see that the whole formula follows straight off by Maxwellian iteration.

The iterative process has delivered the Stokes–Kirchhoff formulae $(XXV.13)_{1,4}$ only subject to the restrictions $(XXV.13)_{2,5}$. However, in calculating (XXV.14) by the iterative process we made no use of those restrictions. We should get exactly the same result if we let μ and M be any scalar functions of ε, even also of ρ. The iterative scheme is then understood *to begin from* $\mathbf{P}^{(1)}$ *and* $\mathbf{q}^{(1)}$ *as given by the Stokes–Kirchhoff theory*. The initial iterates to the remaining spherical moments remain given by (XXV.5), of course. For the rest of this section we shall continue the iteration in this spirit.

We have not yet encountered the characteristic difficulties of the kinetic theory, which begin at the equations for fourth moments. We wish now to calculate the components $q_k^{(2)}$. In order to get them from (XXV.4), we require the second iterates of the moments of all orders less than 3, namely, $P_{0|km}^{(2)}$, and the first iterates of most of the moments of orders up to 4, namely, not only $P_{0|km}^{(1)}$, $q_k^{(1)}$, $P_{0|kmr}^{(1)}$ but also $P_{4|0}^{(1)}$ and $P_{2|km}^{(1)}$. We have the former; to obtain the latter we apply the prescription (XXV.2) with $n = 0$ and $2r + s = 4$, calling upon (XVI.31) and $(XVIIA.5)_{7,8}$ for the appropriate values of $c_{2r|s}$ and $\mathbf{Q}_{2r|s}$, and referring to $(XXV.10)_{4,5}$ for $\mathfrak{L}^{(0)}\mathbf{Y}_{2r|s}$. The results are

$$P_{4|0}^{(1)} = \frac{1}{\rho}\left[15p^2 - 4\mu^2 E_{ab}E_{ab} - 30\mu^{(1)}p\dot{\beta} - \frac{105p\mu^{(1)}}{2\rho}(\dot{\rho} + \rho E)\right],$$
(XXV.15)
$$P_{2|km}^{(1)} = -\frac{2\mu}{\rho}[7\lambda_1 pE_{km} + 4\lambda_2 \mu E_{\{ak}E_{ma\}}],$$

---
[5] MAXWELL [1879, Eq. (53)].

in which the dimensionless coefficients λ_1 and λ_2 have the following values:

$$\lambda_1 = \frac{27 A_{4|2|2} - 42 A_{3|2|2} + 19 A_{2|2|2} + 8p/(n\mu)}{27 A_{4|2|2} - 42 A_{3|2|2} + 43 A_{2|2|2}},$$

$$\lambda_2 = \frac{27 A_{4|2|2} - 42 A_{3|2|2} + 23 A_{2|2|2}}{27 A_{4|2|2} - 42 A_{3|2|2} + 43 A_{2|2|2}}.$$

(XXV.16)

We have assumed that the denominators do not vanish; for Maxwellian molecules we know they do not. If we use (XXV.13)$_2$ to determine μ, we find that

$$\lambda_1 = 1. \qquad (\text{XXV}.17)$$

For Maxwellian molecules $\lambda_2 = \frac{2}{7}$. From the prescription (XXV.4) with $n = 1$, by use of (XXV.13) and (XXV.15) we obtain the **Companion of the Maxwell Second Approximation** (XXV.14):

$$
\begin{aligned}
q_k^{(2)} = & -\mu^{(1)} M^{(1)} \bigg\{ \varepsilon_{,k} - \frac{3}{2p M^{(1)}} (M \mu \varepsilon_{,k})^\cdot \\
& - \frac{M\mu}{p} \bigg[\varepsilon_{,k} E + \frac{3}{5} \varepsilon_{,a} u_{k,a} + \frac{6}{5}\bigg(\frac{2}{5} + \frac{7\lambda_1}{3M}\bigg) \varepsilon_{,a} E_{ka} \bigg] \bigg\} \\
& + \frac{3\mu^{(1)}}{2\rho} (7\lambda_1 - 5)(\mu E_{ka})_{,a} - \frac{3\mu\mu^{(1)}}{p\rho} E_{ka} p_{,a} \\
& + \frac{15\mu^{(1)}}{2\rho\varepsilon} (\mu^{(1)} \varepsilon \dot{\beta})_{,k} + \frac{21}{2} \lambda_{1,a} \frac{\mu\mu^{(1)}}{\rho} E_{ka} \\
& + \frac{6\mu^{(1)}}{p} \bigg(\lambda_2 \frac{\mu^2}{\rho} E_{ka} E_{ab} \bigg)_{,b} + \frac{\mu^{(1)}}{p} \bigg[(1 - 2\lambda_2) \frac{\mu^2}{\rho} E_{ab} E_{ab} \bigg]_{,k} \\
& + \frac{6\mu\mu^{(1)}}{p\rho} E_{ka}(\mu E_{ab})_{,b} + \frac{105\mu^{(1)}}{8p} \bigg[\frac{p\mu^{(1)}}{\rho^2} (\dot{\rho} + \rho E) \bigg]_{,k}.
\end{aligned}
$$

(XXV.18)

(v) *Comments on the results, origin, and nature of Maxwellian iteration*

In Chapter XIII we have presented MAXWELL's roughshod approach to calculation of μ and then the more subtle method he invented so as to arrive at a value of M. In his later work on rarefied gases he began from a polynomial approximation[6] to F/F_M but used it only so as to infer relations among the

---

[6] MAXWELL [1879, Eqs. (12), (21), and (22)], "with Boltzmann".

moments. He certainly had (XXV.14) in his hands, but his expressed ideas are not sufficient to obtain its companion (XXV.18). IKENBERRY & TRUESDELL abstracted and generalized one common aspect of MAXWELL's three attempts, each of which is subtler than its predecessor, to obtain approximations to **P** and **q** in terms of the derivatives of **ρ**. For this reason[7] they named their method "Maxwellian iteration".

The steps already carried out enable us to form a good idea of the nature of Maxwellian iteration. It proceeds by successive differentiation. Using (I.6) to express as a combination of ∂_t and grad the material derivative, denoted by the superimposed dot, we see that both the Maxwell second approximation (XXV.14) and its companion (XXV.18) are functions of **ρ**, ∂_t**ρ**, grad **ρ**, ∂_t grad **ρ**, and grad grad **ρ**. In this way we choose to adopt the completely spatial standpoint customary in studies of gas dynamics. Looking at the general prescription, we see that the 0th iterates of all moments are taken as functions of **ρ** alone, functions given explicitly by (XXV.5). The coefficients of the collisions integrals (XVIIA.5) are functions of **ρ**. The differential operator $\mathfrak{L}$, defined by (IX.14), is a linear combination of the derivatives ∂_t and grad, their coefficients and arguments alike being moments. Thus Maxwellian iteration delivers functions $\mathfrak{P}^{(n)}$ and $\mathfrak{q}^{(n)}$ such that

$$P_{km}^{(n)} = \mathfrak{P}_{km}^{(n)}(\rho, \partial_t \rho, \text{grad } \rho, \ldots, \partial_t^n \rho, \partial_t^{n-1} \text{grad } \rho, \ldots, \text{grad}^n \rho),$$
$$q_k^{(n)} = \mathfrak{q}_k^{(n)}(\rho, \partial_t \rho, \text{grad } \rho, \ldots, \partial_t^n \rho, \partial_t^{n-1} \text{grad } \rho, \ldots, \text{grad}^n \rho).$$
(XXV.19)

It is understood here that both sides are functions of t and **x**. That is, $\mathbf{P}^{(n)}$ and $\mathbf{q}^{(n)}$ depend upon t and **x** only through the dependence of **ρ** and its derivatives up to order n upon t and **x**. (The reader will not confuse the mappings denoted here by $\mathfrak{P}^{(n)}$ and $\mathfrak{q}^{(n)}$ with those denoted by the same symbols in the preceding chapter, since the argument functions are different.)

If Maxwellian iteration converges, *the limits $\mathfrak{P}^{\infty}$ and $\mathfrak{q}^{\infty}$ are mappings of the principal moment field **ρ** restricted to an arbitrarily small space-time neighborhood*. The corresponding limits $\mathbf{P}^{\infty}$ and $\mathbf{q}^{\infty}$ at $(t, \mathbf{x})$ are thus momentally determined. We remind the reader of the difference between gross determinism and momental determinism:

$$P_{km}(t, \mathbf{x}) = \begin{cases} \mathfrak{P}_{km}^G(\rho(t, \mathbf{x}+\cdot), \mathbf{b}(\mathbf{x}+\cdot)) & \text{if grossly determined,} \\ \mathfrak{P}_{km}^M(\rho(t+\cdot, \mathbf{x}+\cdot)) & \text{if momentally determined;} \end{cases}$$
$$q_k(t, \mathbf{x}) = \begin{cases} \mathfrak{q}_k^G(\rho(t, \mathbf{x}+\cdot), \mathbf{b}(\mathbf{x}+\cdot)) & \text{if grossly determined,} \\ \mathfrak{q}_k^M(\rho(t+\cdot, \mathbf{x}+\cdot)) & \text{if momentally determined.} \end{cases}$$
(XXV.20)

---

[7] *Cf.* IKENBERRY & TRUESDELL [1956, §2]: "... we carry to its natural conclusion what seems to us MAXWELL's idea, an idea which he himself exploited only in part and disguised by unnecessary and in fact ultimately complicating approximations."

The mappings $\mathfrak{P}^G$ and $\mathbf{q}^G$ are *gross determiners*, while the mappings $\mathfrak{P}^M$ and $\mathbf{q}^M$ are *momental determiners*. In the preceding chapters we denoted the former by $\mathfrak{P}$ and $\mathbf{q}$ without superscript.

Because **P** and **q** for a Maxwellian molecular density are momentally determined, the class of solutions yielding momentally determined **P** and **q** is not empty. It is equally plain that the momentally determined class, if it exists, is *a special class, exempt from relaxation*, for if ρ has a constant value in a space-time neighborhood of $(t, \mathbf{x})$, so does the value of any mapping applied to $\rho(t + \cdot, \mathbf{x} + \cdot)$ in that neighborhood; in particular, **P** as given by $(XXV.20)_2$ is constant not only in space but also in time.

It would be worthwhile to study momentally determined solutions at the level of generality on which we have studied grossly determined solutions in Chapter XXIII. The two classes might even coincide with each other and with HILBERT's gas-dynamic solutions. Certainly $\mathfrak{P}^G$ and $\mathfrak{P}^M$ in (XXV.20) are different mappings, but the requirement that the argument ρ for the latter and the arguments ρ and **b** for the former shall correspond to a solution of the Maxwell-Boltzmann equation might cause the classes of solutions to which they lead to be identical or essentially so.

The role of the body force **b** is central in distinguishing momentally determined solutions from grossly determined ones. As we have remarked in Section (i) of this chapter, IKENBERRY & TRUESDELL advisedly used as the basis of their method an equation of transfer *from which* **b** *had been eliminated*, namely (IX.13). Noting the strange appearance of **b** in the result (XXIV.4) obtained by ENSKOG's process—a dependence foreign to the general ideas of continuum mechanics, in which contact forces and body forces are conceived as entities of different kinds[8]—they found a method of approximation that leads to results formally similar to the basic assumptions made in some continuum theories of materials.

Failing a general theory of solutions with momentally determined **P** and **q**, we must fall back upon the iterative scheme for such conjectures as we may hazard regarding such solutions. Of course we wish to regard as approximations to $\mathfrak{P}^M$ and $\mathbf{q}^M$ the iterates that Maxwellian iteration delivers. Thus we call $\mathfrak{P}^{(n)}$ and $\mathbf{q}^{(n)}$ *Approximate Momental Determiners.*

As we have remarked already, Maxwellian iteration uses at any given stage the differential equations for the moments of order greater by 1 than those of highest order it used in the preceding stage. For example, to calculate $\mathbf{q}^{(2)}$ we had recourse to the equations of 4th moments so as to obtain $^4\mathbf{M}^{(1)}$. Before calculating $\mathbf{q}^{(3)}$ we shall need those same equations so as to obtain $^4\mathbf{M}^{(2)}$, but in order to do that we shall have first to use the equations of 5th moments so as to

---

[8] See, for example, Chapter III of the book by TRUESDELL, cited in Footnote 1 to Chapter I.

obtain $^5M^{(1)}$. *At any one iterative stage Maxwellian iteration uses only a finite number of the equations for moments, but the successive stages use more and more of those equations.* No matter how large is n, Maxwellian iteration will at some stage in approximating **P** and **q** call upon the equation of transfer for ^{n}**M**, and it will continue to use that equation in all later stages.

In this respect Maxwellian iteration contrasts with GRAD's method, which we have presented in Chapter XVII. GRAD's method, too, begins with the infinite system of equations for moments, but it truncates that system and simply disregards the infinitely many remaining members. GRAD's method produces finite systems of partial differential equations designed to approximate the differential equations for the moments of lowest order. Maxwellian iteration uses the equations for moments to obtain approximations to momental determiners $\mathfrak{P}^M$ and $\mathfrak{q}^M$, presumed to exist. Unlike the methods of HILBERT and ENSKOG, it does not have recourse to the intermediary of an approximate solution of the Maxwell-Boltzmann equation. Rather, developing MAXWELL's approach[9], it *outflanks that equation and goes straight after approximations to* $\mathfrak{P}^M$ *and* $\mathfrak{q}^M$.

The iterates $\mathfrak{P}^{(n)}$ and $\mathfrak{q}^{(n)}$ are intended to replace $\mathfrak{P}^M$ and $\mathfrak{q}^M$, *just as if they were constitutive functions in continuum mechanics*. If we calculate $\mathbf{P}^{(n)}$ and $\mathbf{q}^{(n)}$ by (XXV.19) and substitute the results for **P** and **q** in the field equations (IX.3)–(IX.5), we obtain *a system of differential equations* for ρ, in fact 5 scalar equations. Each iterative step, as we see from (XXV.2) and (IX.14), introduces one more differentiation with respect to t. The result when $n = 2$, which we have written out as (XXV.14) and (XXV.18), when substituted into the field equations yields a system of order 1 in t, suggesting that at that order ρ is determined by $\rho(0, \cdot)$ and boundary conditions. For general n the order is $n - 1$. Thus we are led to expect that *use of Maxwellian iteration replaces the infinite system of equations of moments by a system of 5 scalar equations for the 5 components* ρ_α, *of order* $n - 1$ *in* t. Thus, formally, we expect that *the Maxwellian iterates of order n will determine* ρ *when boundary conditions and the initial values* $\rho(0, \cdot)$, $\dot{\rho}(0, \cdot)$, ..., $^{(n-1)}\rho(0, \cdot)$ *are prescribed*. Again the method differs sharply from GRAD's, which results in a system of first order for a certain finite set of moments ^{n}**M** and hence suggests that initial values of those same moments must be prescribed if we are to determine a unique solution.

In Table XXV.1 we survey the aims and results of the iterative procedures so far discussed.

---

[9] TRUESDELL [1956, §55] wrote in this regard, "Indeed, in spirit this memoir passes over all developments in the kinetic theory since 1879 and goes back for its source and inspiration to what MAXWELL left us." The same may be said of the work of CHAPMAN [1916, *1*, §2], which in every other regard is entirely different.

XXV. MAXWELLIAN ITERATION

TABLE XXV.1

Method	Unknowns at n^{th} stage	Initial values to be prescribed if result at n^{th} stage is adopted as final	Type of gas flows ρ sought to approximate when n is large	Relaxation
GRAD	Finite set of moments, increasing with n	All moments that appear	General	Included
Maxwellian iteration	Principal moment ρ	$\rho, \dot\rho, \ldots, \overset{(n-1)}{\rho}$	Momentally determined	Excluded
HILBERT	n^{th} summand $\rho^{(n)}$ to ρ	$\rho^{(n)}$	Gas-dynamic, shown to be grossly determined	Excluded
ENSKOG, stretched fields	Principal moment ρ	ρ	Grossly determined, thought to be gas-dynamic	Excluded

In Chapter I we have presented (I.25) as defining "materials of the differential type" in continuum mechanics. The results of Maxwellian iteration in the kinetic theory seem to be special cases of (I.25) but are different for two reasons[10]. First, they contain special numerical coefficients, as is seen by a glance at (XXV.14) and (XXV.18). Secondly, those relations if interpreted as applying to arbitrary fields ρ do not satisfy the principle of material frame-indifference.

This fact has given rise to misunderstandings. While the function on the right-hand side of (XXV.14) makes sense for arbitrary ρ, *the kinetic theory gives no status to that sense*. The reader who doubts this statement need only notice that that function is well defined when $\rho < 0$ or $\varepsilon < 0$, but no solution in the kinetic theory ever yields negative fields ρ and ε. More important is the requirement imposed by MAXWELL's consistency theorem: Insofar as (XXV.14) can *follow* from the kinetic theory, *its arguments must be restricted* to those satisfying (IX.3), (IX.4), and (IX.5). Thus (XXV.14) interpreted as applying to an arbitrary ρ is an *extension* of a statement in the kinetic theory. There are infinitely many other extensions. For example, we see that μ appears in (XXV.14), and hence $\dot\varepsilon$ appears as an argument. If the kinetic theory supports (XXV.14), it supports equally the result of replacing $\dot\varepsilon$ in all its appearances therein by the expression that (IX.5) gives for $\dot\varepsilon$ as a functional not only of ρ but also of **P** and **q**, the very quantities that we are seeking to approximate. *Cf.* also Footnote 4, above.

We cannot tell whether (XXV.14) does or does not approximate the restriction of some frame-indifferent mapping to arguments compatible with (IX.3), (IX.4), and (IX.5). It is senseless

---

[10] M. BRILLOUIN [1900] attempted to survey the general form of these relations by what was essentially an approach through continuum mechanics, but that subject was not then sufficiently developed to provide much more than a heap of complicated formulae. Study of BRILLOUIN's work as well as MAXWELL's provided one of the impetus for TRUESDELL's first attempts at a theory of constitutive relations, published in the late 1940s and early 1950s.

(vi) THIRD ITERATES   523

to ask whether the kinetic theory does or does not support the principle of material frame-indifference[11]. The kinetic theory gives no status to *any* relation in which the field ρ is free of the differential restrictions (IX.3), (IX.4), and (IX.5).

Having made these comments on the nature of the first few Maxwellian iterates, we turn back now to the process itself and carry it one stage further.

(vi) *The third stage of Maxwellian iteration*

With the use of the explicit formulae (XVI.41), (XVI.37), and (XXV.5) it is a routine if long and tedious matter to go one step further in Maxwellian iteration. We give the results only for Maxwellian molecules because only for these have all the necessary collisions integrals been evaluated explicitly. In addition we shall simplify the results by imposing immediately the equation of balance of mass, $\dot\rho + \rho E = 0$. Thus, for example, the final terms in (XXV.10)$_4$, (XXV.15)$_1$, and (XXV.18) are null. The definition of $P^{(3)}_{km}$ in terms of $P^{(2)}_{km}$, $q^{(2)}_k$, and $P^{(2)}_{0|kmr}$ is of course immediate from (XXV.3):

$$P^{(3)}_{km} \equiv -\frac{\mu}{p}(2pE_{km} + \dot{P}^{(2)}_{km} + P^{(2)}_{km}E + 2P^{(2)}_{\{ak}u_{m,a\}} + P^{(2)}_{0|kma,a} + \tfrac{4}{5}q^{(2)}_{\{k,m\}}), \quad \text{(XXV.21)}$$

in which $P^{(2)}_{km}$ and $q^{(2)}_k$ are given by (XXV.14) and (XXV.18), respectively. The expression for $P^{(2)}_{0|kmr}$ is quite complicated. We shall record here only the terms proportional to μ^2, indicating those proportional to μ^3 by dots[12]:

$$P^{(2)}_{0|kmr} = \frac{6\mu^2}{\rho\varepsilon}\varepsilon_{,\{k}E_{mr\}} + 4\mu\left(\frac{\mu}{\rho}E_{\{km}\right)_{,r\}} + \cdots. \quad \text{(XXV.22)}$$

Similarly (XXV.4) yields

$$q^{(3)}_k = -\frac{3\mu}{4p}\bigg(2\dot{q}^{(2)}_k + \frac{10}{3}q^{(2)}_k E + 2P^{(2)}_{0|kab}E_{ab} + 2q^{(2)}_a u_{k,a}$$

$$+ \frac{8}{5}q^{(2)}_a E_{ka} - \frac{5p}{\rho}P^{(2)}_{ka,a} \qquad\qquad\qquad \text{(XXV.23)}$$

$$-\frac{2}{\rho}P^{(2)}_{ka}P^{(2)}_{ab,b} - \frac{5p}{\rho}p_{,k} - \frac{2}{\rho}P^{(2)}_{ka}p_{,a} + P^{(2)}_{2|ka,a} + \frac{1}{3}P^{(2)}_{4|0,k}\bigg),$$

---

[11] TRUESDELL [1977].
[12] The full formula may be found in the paper of IKENBERRY & TRUESDELL [1956, Eq. (11.3)].

in which we replace $P^{(2)}_{km}$, $q^{(2)}_k$, and $P^{(2)}_{0|kmr}$, respectively, by the right-hand sides of (XXV.14), (XXV.18), and (XXV.22); the expressions for $P^{(2)}_{2|km}$ and $P^{(2)}_{4|0}$, with terms proportional to μ^3 omitted[13], are

$$P^{(2)}_{2|km} = \frac{\mu}{\rho}\bigg[-14pE_{km} + 14(\mu E_{km})^{\cdot} + 32\mu\dot{\beta}E_{km} + 14\mu EE_{km}$$

$$+ 28\mu E_{\{ak}u_{m,a\}} + \frac{80\mu}{7}E_{\{ak}E_{ma\}}$$

$$-\frac{18\mu}{\rho\varepsilon}\varepsilon_{,\{k}P_{,m\}} + \frac{36\mu p}{\rho\varepsilon^2}\varepsilon_{,\{k}\varepsilon_{,m\}} + \frac{39p}{\rho\varepsilon}(\mu\varepsilon_{,\{k)},m\}}\bigg] + \cdots, \quad \text{(XXV.24)}$$

$$P^{(2)}_{4|0} = \frac{15p^2}{\rho} - \frac{30\mu p}{\rho}\dot{\beta} + \frac{45\mu}{\rho\varepsilon}(\mu\varepsilon\dot{\beta})^{\cdot} + \frac{105\mu^2}{\rho}\dot{\beta}E + \frac{80\mu^2}{\rho}E_{ab}E_{ab}$$

$$-\frac{45\mu^2}{\rho^2\varepsilon}\varepsilon_{,a}P_{,a} + \frac{315\mu p}{2\rho^2\varepsilon^2}(\mu\varepsilon\varepsilon_{,a})_{,a} + \cdots.$$

It might be thought that formulae as long and elaborate as these would be of no use or interest to anybody. On the contrary, some physicists[14] have put out enormous drudgery to calculate expressions of this kind and to apply them to problems concerning rarefied gases. In doing so they have labored with the complications and obscurities of ENSKOG's method, which makes the calculations incomparably more uncertain as well as more difficult than those we have just outlined.

(vii) Proof of effectiveness

We have illustrated the method of Maxwellian iteration by working out its first three stages. We now prove that the method is effective in the sense that a finite number of differentiations and algebraic operations suffice to calculate

---

[13] The full formulae were obtained by IKENBERRY & TRUESDELL [1956, Eqs. (11.4) and (11.5)] but with some numerical errors due to the incorrect coefficients they had obtained in their expression for $m\bar{C}Y_{4|k}$, noted in Footnote 5 to Chapter XVI. Those errors affect mainly the terms indicated here by dots. We will provide the entire formulae to anyone who requests them.

[14] WANG CHANG & UHLENBECK [1948, end of §III], WANG CHANG [1948, 2, §IIIB IV], and later authors. For a gas of Maxwellian molecules WANG CHANG published the numerical coefficients of the terms linear in the third spatial derivatives in $\mathbf{P}^{\langle 3 \rangle}$ and $\mathbf{q}^{\langle 3 \rangle}$. To compare her results with ours we should have to calculate the third atemporal iterates $\mathbf{P}^{[3]}$ and $\mathbf{q}^{[3]}$ as defined and explained below in Section (ix), but we have not thought the matter of sufficient importance to carry it through.

MAXWELL [1879, added note at end of text, before Appendix] stated that he had calculated some of the terms in what we here call $P^{(3)}_{km}$ and $P^{(4)}_{km}$, but his remarks about the rates of decay suggest that he may have made errors; cf. Footnote 5 to our Chapter XVIII.

$^m\mathbf{M}^{(n)}$ for any positive integers m and n. Here "algebraic operation" means addition, subtraction, and multiplication only. The proof applies exactly to a gas of Maxwellian molecules; to other gases it applies approximately in the sense that the remainders $\mathbf{R}_{2r|s}$ in (XVIIA.5) and its extensions to moments of higher orders are neglected. For Maxwellian molecules $\mathfrak{c}_{2r|s} > 0$, and we assume here that the same is true for other molecular models; then division by $\mathfrak{c}_{2r|s}$ may be effected at the outset in (XXV.2) and so not mentioned. Of course no extraction of roots or solution of linear systems is required. Since the $\mathbf{Y}_{2r|s}$ form a complete set of polynomials, it suffices to prove that $\mathbf{P}_{2r|s}^{(n)}$ may be calculated when $2r + s \geqq 2$.

Let q be of the order of $\mathbf{Y}_{2r|s}$. That is, $q = 2r + s$. Then the polynomials $\mathbf{Y}_{2r|s,k}$, $c_k \mathbf{Y}_{2r|s,m}$, and $c_k \mathbf{Y}_{2r|s}$ are linear combinations of the polynomials $\mathbf{Y}_{2u|v}$ of orders $q - 1$, q, and $q + 1$, respectively. The definitions (XVI.28) and (IX.14) show that $\mathfrak{L}\mathbf{Y}_{2r|s}$ is a function only of ρ, grad $\mathbf{u}$, ε, of the $\mathbf{P}_{2u|v}$ of orders 2, $q - 1$, q, and $q + 1$, and of the first derivatives of the $\mathbf{P}_{2u|v}$ of those orders. The definition of $\mathfrak{L}^{(n)}$ shows then that $\mathfrak{L}^{(n)}\mathbf{Y}_{2r|s}$ is a function only of ρ, grad $\mathbf{u}$, ε, of the $\mathbf{P}_{2u|v}^{(n)}$ of orders 2, $q - 1$, q, and $q + 1$, and of the first derivatives of the $\mathbf{P}_{2u|v}^{(n)}$ of those orders. Because $\mathbf{Q}_{2r|s}^{(n+1)}$ is a function of p, ρ, μ, and the $\mathbf{P}_{2u|v}^{(n+1)}$ of orders less than q, (XXV.2) delivers $\mathbf{P}_{2r|s}^{(n+1)}$ uniquely. We have established the following

Lemma. *Let* $\bar{\mathbb{C}}\mathbf{Y}_{2r|s}$ *have been calculated in terms of the* $\mathbf{P}_{2u|v}$. *Then a finite number of algebraic and differential operations yield* $\mathbf{P}_{2r|s}^{(n+1)}$ *as a function of* $\boldsymbol{\rho}$, *the* $\mathbf{P}_{2u|v}^{(n+1)}$ *of lesser orders, the* $\mathbf{P}_{2u|v}^{(n)}$ *of orders not more than one greater, and the derivatives of these quantities with respect to* t *and* $\mathbf{x}$.

This lemma enables us to determine how many iterates are needed to calculate the nth iterate to a $\mathbf{P}_{2u|v}$ of order q. First we need the nth iterates to certain of the $\mathbf{P}_{2u|v}$ of lesser orders. Next we need the $n - 1^{\text{st}}$ iterates to certain of the $\mathbf{P}_{2u|v}$ of orders $q + 1$; the $n - 2^{\text{nd}}$ iterates to the $\mathbf{P}_{2u|v}$ of orders $q + 2$; ...; the 0th iterates to the $\mathbf{P}_{2u|v}$ of order $q + n$. Thus to calculate $\mathbf{P}_{2r|s}^{(n)}$ we carry out the following steps.

-1. Calculate $\mathfrak{L}\mathbf{Y}_{2u|v}$ and $\bar{\mathbb{C}}\mathbf{Y}_{2u|v}$ for all $\mathbf{Y}_{2u|v}$ such that $2u + v \leqq n + 2r + s - 1$.

0. Write out the explicit form of (XXV.2) for the $\mathbf{Y}_{2u|v}$ specified in Step -1.

1. Using the formulae obtained in Step 0, calculate $\mathbf{P}_{2u|v}^{(1)}$ for all u and v such that $2u + v \leqq n + 2r + s - 1$.

2. Using the results of Step 1, calculate $\mathbf{P}_{2u|v}^{(2)}$ for all u and v such that $2u + v \leqq n + 2r + s - 2$.

$\vdots$

n. Using the results of Step $n - 1$, calculate $\mathbf{P}_{2r|s}^{(n)}$.

By induction we easily see that this prescription holds for all n. Hence we have established the ***Theorem of Effectiveness***[15]: *If the collisions integrals for all* $Y_{2u|v}$ *of orders not greater than* $2r + s + n$ *have been evaluated, then a finite number of differentiations and algebraic operations determine* $P^{(n)}_{2r|s}$. The prescription itself makes it clear that $P^{(n)}_{2r|s}$ *at* (t, x) *is a uniquely determined function of* ρ *and a finite number of its derivatives with respect to t and* x, *each evaluated at* (t, x).

Maxwellian iteration as presented here rests on use of exact collisions integrals for a gas of Maxwellian molecules, truncated collisions integrals for other gases. Within that limitation, the theorem of effectiveness makes its status at least as secure as that of any other method of successive approximation proposed so far. As we have seen, up to the present time there is no proof that HILBERT's n^{th} step can actually be found explicitly according to the prescription laid down. A proof of effectiveness for the method of stretched fields has been provided in Chapter XXIV. We may charitably presume that ENSKOG's method is carried along thereby.

Whether Maxwellian iteration converges, and, if it does, whether the limit functions correspond to a solution, and, if they do, what is the status of that solution, are unknown. When it comes to convergence, all methods but one are on a par at null. The one exception is the method of stretched fields. This is the only method so far proposed that takes as its basis *a precise characterization of what it desires to approximate*. The reflections of this fact upon questions of convergence have been discussed at the end of Section (vii) of Chapter XXIV.

(viii) Example: Homo-energetic simple shearing of a gas of Maxwellian molecules

We can grasp the kinds of results to be expected from Maxwellian iteration by applying its first few stages to homo-energetic simple shearing of a gas of Maxwellian molecules[16]:

$$u_1 = Kx_2, \quad u_2 = 0, \quad u_3 = 0, \quad K = \text{const.} \qquad (\text{II}.57)_r$$

To do so, we use again the notations

$$T \equiv \tau K; \quad \tau \equiv \frac{1}{3n\mathbb{A}_{2|2}}, \qquad (\text{II}.60)_r;\ (\text{XIII}.3)_{1r}$$

---

[15] IKENBERRY & TRUESDELL [1956, §13].
[16] The analysis here includes and extends that of TRUESDELL [1956, §43].

(viii) EXAMPLE: SIMPLE SHEARING

set
$$s \equiv t/\tau, \qquad \text{(XXV.25)}$$

and recall that for a given flow of this kind both T and τ are constants. From (XXV.3) we obtain the iterative systems

$$-P_{12}^{(n+1)} = \dot{P}_{12}^{(n)} + \mathsf{T}(p + P_{22}^{(n)}), \qquad -P_{22}^{(n+1)} = \dot{P}_{22}^{(n)} - \tfrac{2}{3}\mathsf{T}P_{12}^{(n)}, \quad \text{(XXV.26)}$$

in which the superimposed dot denotes d/ds. It is instructive to compare the members of this system with their counterparts in the exact system (XIV.1). As $\mathbf{P}^{(0)} = \mathbf{0}$, we can easily calculate $P_{12}^{(n)}$ for n as large as we wish. Introducing a new variable as follows:

$$\alpha \equiv \dot{\varepsilon}/\varepsilon, \qquad \text{(XXV.27)}$$

we quickly calculate the Maxwellian iterates of orders 1, 2, and 3:

$$\begin{aligned}
-P_{12}^{(1)}/p &= \mathsf{T}, & P_{22}^{(1)} &= 0, \\
-P_{12}^{(2)}/p &= \mathsf{T}(1-\alpha), & -P_{22}^{(2)}/p &= \tfrac{2}{3}\mathsf{T}^2, \\
-P_{12}^{(3)}/p &= \mathsf{T}[1 - \tfrac{2}{3}\mathsf{T}^2 - (1-\alpha)\alpha + \dot{\alpha}], & -P_{22}^{(3)}/p &= \tfrac{2}{3}\mathsf{T}^2(1 - 2\alpha).
\end{aligned} \qquad \text{(XXV.28)}$$

It is clear that $P_{12}^{(n)}/p$ is a polynomial in $\alpha, \dot{\alpha}, \ldots, \overset{(n-2)}{\alpha}$, linear in the last. To obtain the solution corresponding to the nth iterate, we must substitute this expression for $P_{12}^{(n)}/p$ in place of P_{12}/p in the equation of balance of energy (XIV.1)$_1$:

$$\alpha = -\tfrac{2}{3}\mathsf{T}P_{12}^{(n)}/p. \qquad \text{(XXV.29)}$$

The result is a differential equation of order $n-2$ for α. Once this equation has been solved, we shall know $P_{12}^{(n)}$ and $P_{22}^{(n)}$ as functions of t/τ, and we may calculate ε by integrating (XXV.27):

$$\varepsilon = \varepsilon(0) \exp \int_0^{t/\tau} \alpha(s)\, ds. \qquad \text{(XXV.30)}$$

If $n=1$, we find from (XXV.29) and (XXV.28)$_1$ that

$$\alpha = \tfrac{2}{3}\mathsf{T}^2, \qquad \varepsilon = \varepsilon(0) \exp(\tfrac{2}{3}\mathsf{T}^2 t/\tau). \qquad \text{(XXV.31)}$$

If $n=2$, then

$$\alpha = \frac{\tfrac{2}{3}\mathsf{T}^2}{1 + \tfrac{2}{3}\mathsf{T}^2}, \qquad \varepsilon = \varepsilon(0) \exp\left(\frac{\tfrac{2}{3}\mathsf{T}^2}{1 + \tfrac{2}{3}\mathsf{T}^2} \frac{t}{\tau}\right). \qquad \text{(XXV.32)}$$

At the stage $n=3$ we obtain the first typical example:

$$-\dot{\alpha} - \alpha^2 + \left(1 + \frac{1}{\tfrac{2}{3}\mathsf{T}^2}\right)\alpha = 1 - \tfrac{2}{3}\mathsf{T}^2. \qquad \text{(XXV.33)}$$

Therefore

$$\alpha(s) = \frac{1}{Ce^{Ss} - 1/S} + \tfrac{1}{2}\left(S + 1 + \frac{1}{\tfrac{2}{3}T^2}\right),$$
(XXV.34)

$$C = \frac{1}{S} + \frac{1}{\alpha(0) - \tfrac{1}{2}\left(S + 1 + \frac{1}{\tfrac{2}{3}T^2}\right)}, \quad S = \pm\sqrt{\left(1 + \frac{1}{\tfrac{2}{3}T^2}\right)^2 - 4(1 - \tfrac{2}{3}T^2)},$$

whence ε may be obtained by the aid of (XXV.30).

If we compare these results with their exact counterparts in Chapter XIV, we see that Maxwellian iteration delivers something intermediate between the general dependence on time provided by (XIV.6)$_1$ and the purely exponential dependence of the dominant part (XIV.8)$_1$. The dominant part is determined by T and by the two ratios of initial values $P_{12}(0)/p(0)$ and $P_{22}(0)/p(0)$. In the solution at the stage $n = 1$ of Maxwellian iteration we obtain the ε that corresponds to the principal solution with R approximated by the first term in the expansion (XIV.5) in powers of T. At the stage $n = 2$ we obtain in (XXV.32) the same thing but with a much better approximation[17] to R, not a power series but a rational function that is fairly close to R for all values of $\tfrac{2}{3}T^2$ in the interval [0, 1]. At the stage $n = 3$ the solution $\varepsilon/\varepsilon(0)$ is determined by a single parameter $\dot\varepsilon(0)/\varepsilon(0)$, and higher stages will introduce $\ddot\varepsilon(0)/\varepsilon(0)$ and the initial values of still higher derivatives of ε. Assignment of these derivatives at 0 may compensate somewhat for the fact that the dominant part of an exact solution is not generally determined by its own initial values but rather by the initial values of the particular solution of which it is an asymptotic form. *Cf.* the third main theorem in Section (x) of Chapter XIV.

(ix) *Atemporal Maxwellian iteration*

Once the method of Maxwellian iteration has been grasped, various alternatives suggest themselves to the student.

One of these is to modify the prescriptions as follows: Writing the equations of moments in a certain order, in the iterative scheme replace each $\mathbf{P}_{2r|s}$ by the iterate $\mathbf{P}_{2r|s}^{(q)}$ with the largest q for which it is available. That is, each iterative result is used as soon as we have it. This iteration causes some terms to appear at earlier stages than they do in Maxwellian iteration.

Another scheme, called *iteration by powers*, is designed to do just the opposite: produce at the nth stage a summand proportional to μ^n. It begins by writing the basic equation of transfer (IX.13) in the form $\mathfrak{L}g = \lambda\mathfrak{m}\bar{\mathbb{C}}g$, λ being a parameter which afterward will be set equal to 1. (This scheme somewhat recalls HILBERT's.) $\mathbf{P}_{2r|s}^{(n)}$ is then defined as the coefficient of λ^n in an expansion of $\mathbf{P}_{2r|s}$ in powers of λ. Then $\sum_{k=0}^{n}\mathbf{P}_{2r|s}^{(k)}$ is the same as what we should obtain by taking the corresponding Maxwellian iterate $\mathbf{P}_{2r|s}^{(n)}$ and simply discarding all terms proportional to powers of μ greater than

---

[17] *Cf.* TRUESDELL [1956, Eqs. (33.7) and (35.3) and Figure 35.1].

the nth. If we look at (XXV.14), we see that it contains no terms that would fail to appear at the second stage in iteration by powers, but the last three summands before the last in (XXV.18) are not of that kind, being proportional to μ^3. The terms we have indicated by dots in (XXV.22) and (XXV.24) are the terms Maxwellian iteration delivers but iteration by powers does not deliver at the same stage. Iteration by powers conforms to widespread, vague prejudices about "order". It is a neater process than Maxwellian iteration, and shorter in the reckoning, but definition of the iterates through the insertion of a parameter afterward set equal to 1, though neither more nor less rigorous than the equally formal prescriptions that define HILBERT's process and Maxwellian iteration, is suggested by nothing but hindsight, beginning from the desired form of the result and working toward it.

HILBERT's method, ENSKOG's method, and the method of stretched fields all start from a Maxwellian density. This is not a matter of choice. The schemes themselves force this start: Their very nature makes them produce some kind of perturbation from a locally Maxwellian flow. For Maxwellian iteration, on the contrary, the choice of $\mathbf{P}^{(0)}_{2r|s}$ given by (XXV.5) is arbitrary. The formal method of differential iteration would go through equally well *for any other choice of* 0th *iterates*. It can thus be visualized as a method of perturbation about *any* gas flow. We have already illustrated the idea in the mathematical example (XXV.8). We have illustrated it also in the calculations leading to (XXV.14) and (XXV.18), in which we forgot about some of the 0th iterates and simply set $P^{(1)}_{km} \equiv -2\mu E_{km}$ and $q^{(1)}_k \equiv -M\mu\varepsilon_{,k}$; that is, we used an arbitrary Stokes-Kirchhoff flow to provide $\mathbf{P}^{(1)}$ and $\mathbf{q}^{(1)}$ but otherwise adhered to the original scheme. A much more interesting possibility[18] is to take for $\mathbf{P}^{(0)}_{2r|s}$ the corresponding spherical moments of some "free molecular flow" (*cf.* Section (vii) of Chapter XIV).

These and other possibilities have been discussed and illustrated by IKENBERRY & TRUESDELL[19]. In this book we shall present the results for only one of their variants, the one they called *atemporal Maxwellian iteration*.

ENSKOG's process eliminates all time derivatives, expressing them approximately in terms of space derivatives. The method of stretched fields achieves the same end more clearly and more cleanly by seeking at the outset to approximate gross determiners. The reasons for doing so are, first, to avoid having to face HILBERT's successive systems of partial differential equations and the initial-value problems for them, and, second, to get formulae similar to the constitutive equations of continuum mechanics. The former reason has no bearing on Maxwellian iteration, and Maxwellian iteration is so conceived as to obtain directly approximate momental determiners. However, we certainly can eliminate time derivatives formally if we wish to. In order to obtain expressions that could be compared with those calculated by ENSKOG's process, IKENBERRY & TRUESDELL formulated the scheme of atemporal Maxwellian iteration. That scheme of theirs applies to the results of Maxwellian iteration the same formally approximate elimination of time derivatives as is used in most presentations[20] of ENSKOG's procedure. That elimination starts from the

---

[18] IKENBERRY & TRUESDELL [1956, §22] worked out some of the details when the gross flow is a plane free expansion.

[19] IKENBERRY & TRUESDELL [1956, Chapter II].

[20] *E.g.* CHAPMAN & COWLING [1939, §§7.14, 15.2, 15.3]. IKENBERRY & TRUESDELL formulated the method in some generality, and in the next-to-last draught of this book we provided a fully general treatment. As we now consider the confusion introduced by iterative approximation of time derivatives altogether superfluous, we do not employ it at all in this book.

exact expressions for time derivatives in terms of space derivatives provided by the equations of balance of mass, momentum, and energy. In this way IKENBERRY & TRUESDELL obtained at the first and second stages precisely the formulae already familiar as products of the procedure of ENSKOG, specialized to Maxwellian molecules. More important than this is the fact that atemporal Maxwellian iteration, because of its close connection with Maxwellian iteration itself, is a differential iteration. It allows us to calculate as many iterates as we may wish without having to encounter any integral equations. In the next chapter we shall see an example of this advantage when we calculate the nth atemporal Maxwellian iterates for **P** and **q** for affine flows. As a result, of all the iterative procedures we present in this book, atemporal Maxwellian iteration is by far the most useful in what it tells us about the status of approximate gross determiners. That status will form the subject of Chapter XXVI. Here we will formulate one way to get the various iterates; we shall calculate the first two explicitly. The final products, of course, are approximate gross determiners.

There is no need to follow the complications of IKENBERRY & TRUESDELL's original treatment. It is better now to appeal directly to the general theory of gross determiners, which we have presented in Chapter XXIII, and to apply Maxwellian iteration within that theory. Since the time derivatives have been eliminated at the start, once and for all, we expect this scheme to deliver at once what IKENBERRY & TRUESDELL obtained by eliminating time derivatives through a second iteration. Now we shall use the very same *mathematical method* as in the foregoing sections of this chapter: Maxwellian iteration. The scheme differs only in that we apply the method to the *special* equation of transfer (XXIII.75) for gross determiners rather than to the general equation of transfer (IX.13). Because we regard this new scheme as superseding entirely the one IKENBERRY & TRUESDELL formulated toward the same end, we give it the name they gave theirs: ***Atemporal Maxwellian Iteration.***

If we denote the left-hand side of (XXIII.75) by $\mathfrak{A}g$, then the operator $\mathfrak{A}$ will play the same role in atemporal iteration as the operator $\mathfrak{L}$ played in Maxwellian iteration. In analogy with (XXV.2) we lay down the iterative scheme

$$-c_{2r|s} \mathfrak{P}_{2r|s}^{[n+1]} \equiv -Q_{2r|s}^{[n+1]} + \mathfrak{A}^{[n]} Y_{2r|s}, \qquad 2r + s \geq 2, \quad (r, s) \neq (1, 0), \qquad \text{(XXV.35)}$$

and in analogy with (XXV.5) we choose as the initial iterates $\mathfrak{P}_{2r|s}^{[0]}$ the gross determiners of the spherical moments of a general Maxwellian density:

$$\mathfrak{P}_{2r|s}^{[0]}(\mathfrak{S}\rho, \mathfrak{S}\mathbf{u}, \mathfrak{S}p, \mathfrak{S}\mathbf{b}) \equiv \begin{cases} 0 & \text{if} \quad s \neq 0, \\ & 2r + s \geq 2, \quad (r, s) \neq (1, 0). \\ (2r + 1)!! \rho (p/\rho)^r & \text{if} \quad s = 0, \end{cases} \qquad \text{(XXV.36)}$$

Of course $\mathfrak{A}^{[n]} Y_{2r|s}$ denotes the result of replacing each gross determiner in $\mathfrak{A} Y_{2r|s}$ by its nth iterate, and $Q_{2r|s}^{[n+1]}$ denotes the result of replacing each spherical moment $\mathbf{P}_{2u|v}$ in $\mathbf{Q}_{2r|s}$ by the $n + 1^{\text{st}}$ iterate $\mathfrak{P}_{2u|v}^{[n+1]}$ of its gross determiner.

(ix) ATEMPORAL MAXWELLIAN ITERATION

The relations between $\mathfrak{A}$ and $\mathfrak{L}$ are quite simple: $\mathfrak{L}g$ is a field, while $\mathfrak{A}g$ is a gross determiner. If all expectations in $\mathfrak{L}g$ are calculated using a grossly determined molecular density, then $\mathfrak{L}g$ itself will be grossly determined, and by comparing (IX.14) with the left-hand side of (XXIII.75) we see that its gross determiner is just $\mathfrak{A}g$. Because of this fact, $\mathfrak{L}$ and $\mathfrak{A}$ have the same structure insofar as that structure enters the proof of effectiveness given in Section (vii). Therefore, we obtain immediately a *Proof of Effectiveness* for the iterative scheme based upon (XXV.35) and (XXV.36). This much having been dispensed with, we may proceed now to calculate some of the iterates.

When $2r + s \leq 4$, we may read off from the list (XVIIA.5) expressions for the coefficients $\mathfrak{c}_{2r|s}$ and the functions $\mathbf{Q}_{2r|s}$. We also require the following list of expressions for $\mathfrak{A}Y_{2r|s}$; it plays the same role here as the list (XVI.41) played in Maxwellian iteration:

$$\mathfrak{A}Y_{0|0} = 0,$$

$$\mathfrak{A}Y_{0|k} = 0,$$

$$\mathfrak{A}Y_{2|0} = 0,$$

$$\mathfrak{A}Y_{0|km} = \partial_\delta \mathfrak{P}_{km}[\mathfrak{S}A] + \partial_{v_a} \mathfrak{P}_{km}[\mathfrak{S}g_a] + \partial_\varpi \mathfrak{P}_{km}[\mathfrak{S}B] + u_a \mathfrak{P}_{km,a}$$
$$+ \mathfrak{P}_{km} E + 2\mathfrak{P}_{\{ak} u_{m,a\}} + 2pE_{km} + \mathfrak{P}_{0|kma,a} + \tfrac{4}{5} \mathfrak{q}_{\{k,m\}},$$

$$\mathfrak{A}Y_{2|k} = 2\,\partial_\delta \mathfrak{q}_k[\mathfrak{S}A] + 2\,\partial_{v_a} \mathfrak{q}_k[\mathfrak{S}g_a] + 2\,\partial_\varpi \mathfrak{q}_k[\mathfrak{S}A] + 2u_a \mathfrak{q}_{k,a}$$
$$+ \tfrac{10}{3}\mathfrak{q}_k E + 2\mathfrak{P}_{0|kab} E_{ab} + 2\mathfrak{q}_a u_{k,a} + \tfrac{8}{5}\mathfrak{q}_a E_{ka}$$
$$- \frac{5p}{\rho} \mathfrak{M}_{ka,a} - \frac{2}{\rho} \mathfrak{P}_{ka} \mathfrak{M}_{ab,b} + \mathfrak{P}_{2|ka,a} + \tfrac{1}{3}\mathfrak{P}_{4|0,k},$$

$$\mathfrak{A}Y_{0|kmr} = \partial_\delta \mathfrak{P}_{0|kmr}[\mathfrak{S}A] + \partial_{v_a} \mathfrak{P}_{0|kmr}[\mathfrak{S}g_a] + \partial_\varpi \mathfrak{P}_{0|kmr}[\mathfrak{S}B]$$
$$+ u_a \mathfrak{P}_{0|kmr,a} + \mathfrak{P}_{0|kmr} E + 3\mathfrak{P}_{0|\{akm} u_{r,a\}} \qquad \text{(XXV.37)}$$
$$+ \tfrac{12}{5}\mathfrak{q}_{\{k} E_{mr\}} - \frac{3}{\rho} \mathfrak{P}_{\{km} \mathfrak{M}_{ra,a\}} + \mathfrak{P}_{0|kmra,a} + \tfrac{3}{7}\mathfrak{P}_{2|\{km,r\}},$$

$$\mathfrak{A}Y_{4|0} = \partial_\delta \mathfrak{P}_{4|0}[\mathfrak{S}A] + \partial_{v_a} \mathfrak{P}_{4|0}[\mathfrak{S}g_a] + \partial_\varpi \mathfrak{P}_{4|0}[\mathfrak{S}B] + u_a \mathfrak{P}_{4|0,a}$$
$$+ \tfrac{7}{3}\mathfrak{P}_{4|0} E + 4\mathfrak{P}_{2|ab} E_{ab} - \frac{8}{\rho} \mathfrak{q}_a \mathfrak{M}_{ab,b} + \mathfrak{P}_{4|a,a},$$

$$\mathfrak{A}Y_{2|km} = \partial_\delta \mathfrak{P}_{2|km}[\mathfrak{S}A] + \partial_{v_a} \mathfrak{P}_{2|km}[\mathfrak{S}g_a] + \partial_\varpi \mathfrak{P}_{2|km}[\mathfrak{S}B] + u_a \mathfrak{P}_{2|km,a}$$
$$+ \tfrac{5}{3}\mathfrak{P}_{2|km} E + 2\mathfrak{P}_{0|kmab} E_{ab} + 2\mathfrak{P}_{2|\{ak} u_{m,a\}} + \tfrac{8}{7}\mathfrak{P}_{2|\{ak} E_{ma\}}$$
$$+ \tfrac{14}{15}\mathfrak{P}_{4|0} E_{km} - \frac{2}{\rho} \mathfrak{P}_{0|kma} \mathfrak{M}_{ab,b} - \frac{28}{5\rho} \mathfrak{q}_{\{k} \mathfrak{M}_{ma,a\}} + \mathfrak{P}_{2|kma,a}$$
$$+ \tfrac{2}{5}\mathfrak{P}_{4|\{k,m\}};$$

the quantities A, $\mathbf{g}$, and B are the expressions for the time derivatives of ρ, $\mathbf{u}$, and p delivered by the equations of balance, namely

$$A = -u_a \rho_{,a} - \rho E,$$

$$g_k = -u_a u_{k,a} - \frac{1}{\rho} p_{,k} - \frac{1}{\rho} \mathfrak{P}_{ka,a} + b_k, \qquad \text{(XXV.38)}$$

$$B = -u_a p_{,a} - \tfrac{5}{3} p E - \tfrac{2}{3} \mathfrak{P}_{ab} E_{ab} - \tfrac{2}{3} q_{a,a},$$

and $\mathfrak{M}$ is the gross determiner of the pressure tensor $\mathbf{M}$:

$$\mathfrak{M}_{km} \equiv \mathfrak{P}_{km} + p\, \delta_{km}. \qquad \text{(XXV.39)}$$

We have indicated in Section (vii) of Chapter XXIII two different methods of calculating values of $\mathfrak{A}g$, and we have illustrated the calculation already in the three cases recorded in (XXIII.76). The full list (XXV.37) is obtained in the same way.

To apply the prescription (XXV.35), we must calculate $\mathfrak{A}^{[n]} Y_{2r|s}$, and we see from (XXV.37) that this involves calculating the derivatives of $\mathfrak{P}^{[n]}_{2u|v}$ with respect to δ, $\mathbf{v}$, ϖ, and $\mathbf{x}$. It is useful first to indicate generally how this may be done. Let K be a grossly determined field, with gross determiner $\mathfrak{K}$:

$$K(t, \mathbf{x}) = \mathfrak{K}(\mathfrak{S}\rho(t, \mathbf{x}), \mathfrak{S}\mathbf{u}(t, \mathbf{x}), \mathfrak{S}p(t, \mathbf{x}), \mathfrak{S}\mathbf{b}(\mathbf{x})). \qquad \text{(XXV.40)}$$

We shall be concerned here only with the case in which $\mathfrak{K}$ reduces to a function of the fields and a finite number of their spatial derivatives, each evaluated at $(t, \mathbf{x})$:

$$\mathfrak{K}(\mathfrak{S}\rho, \mathfrak{S}\mathbf{u}, \mathfrak{S}p, \mathfrak{S}\mathbf{b}) = \mathfrak{k}(\rho, \mathbf{u}, p, \mathbf{b}, \operatorname{grad} \rho, \operatorname{grad} \mathbf{u}, \operatorname{grad} p,$$

$$\operatorname{grad} \mathbf{b}, \operatorname{grad}^2 \rho, \ldots). \qquad \text{(XXV.41)}$$

For such a gross determiner we calculate the derivatives with respect to δ, $\mathbf{v}$, and ϖ by means of the chain rule. In terms of a general set of fields ξ, ζ, and η, that rule takes the form

$$\partial_\delta \mathfrak{K}[\mathfrak{S}\xi] = \xi\, \partial_\rho \mathfrak{k} + \xi_{,a}\, \partial_{\rho_{,a}} \mathfrak{k} + \xi_{,ab}\, \partial_{\rho_{,ab}} \mathfrak{k} + \cdots,$$

$$\partial_{v_a} \mathfrak{K}[\mathfrak{S}\zeta_a] = \zeta_a\, \partial_{u_a} \mathfrak{k} + \zeta_{a,c}\, \partial_{u_{a,c}} \mathfrak{k} + \zeta_{a,cd}\, \partial_{u_{a,cd}} \mathfrak{k} + \cdots, \qquad \text{(XXV.42)}$$

$$\partial_\varpi \mathfrak{K}[\mathfrak{S}\eta] = \eta\, \partial_p \mathfrak{k} + \eta_{,a}\, \partial_{p_{,a}} \mathfrak{k} + \eta_{,ab}\, \partial_{p_{,ab}} \mathfrak{k} + \cdots.$$

Moreover, derivatives with respect to $\mathbf{x}$ are also calculated using the chain rule, as expressed by

$$\mathfrak{K}_{,k} = \rho_{,k}\, \partial_\rho \mathfrak{k} + u_{a,k}\, \partial_{u_a} \mathfrak{k} + p_{,k}\, \partial_p \mathfrak{k} + b_{a,k}\, \partial_{b_a} \mathfrak{k} + \rho_{,ak}\, \partial_{\rho_{,a}} \mathfrak{k}$$

$$+ u_{a,ck}\, \partial_{u_{a,c}} \mathfrak{k} + p_{,ak}\, \partial_{p_{,a}} \mathfrak{k} + b_{a,ck}\, \partial_{b_{a,c}} \mathfrak{k} + \cdots. \qquad \text{(XXV.43)}$$

(ix)   FIRST ATEMPORAL ITERATES   533

By use of (XXV.37), (XXV.38), (XXV.42), (XXV.43), and the initial iterates (XXV.36) we obtain the following list of values of $\mathfrak{U}^{[0]}Y_{2r|s}$:

$$\mathfrak{U}^{[0]}Y_{0|km} = 2pE_{km}, \qquad \mathfrak{U}^{[0]}Y_{0|kmr} = 0,$$
$$\mathfrak{U}^{[0]}Y_{2|k} = \tfrac{10}{3}p\varepsilon_{,k}, \qquad \mathfrak{U}^{[0]}Y_{4|0} = 0, \qquad\qquad \text{(XXV.44)}$$
$$\mathfrak{U}^{[0]}Y_{2|km} = 14\frac{p^2}{\rho}E_{km}.$$

We illustrate the calculation involved by working out $\mathfrak{U}^{[0]}Y_{4|0}$ explicitly. From (XXV.36) we see that

$$\mathfrak{P}^{[0]}_{km} = 0, \qquad \mathfrak{q}^{[0]}_k = 0,$$
$$\mathfrak{P}^{[0]}_{4|0} = 15\frac{p^2}{\rho}, \qquad \mathfrak{P}^{[0]}_{2|km} = 0, \qquad \mathfrak{P}^{[0]}_{4|k} = 0, \qquad \text{(XXV.45)}$$

and so by (XXV.38) the initial iterates of A, $\mathbf{g}$, and B are

$$A^{[0]} = -u_a\rho_{,a} - \rho E, \quad g^{[0]}_k = -u_a u_{k,a} - \frac{1}{\rho}p_{,k} + b_k, \quad B^{[0]} = -u_a p_{,a} - \tfrac{5}{3}pE.$$
$$\text{(XXV.46)}$$

Thus by use of (XXV.37)$_7$, (XXV.42), (XXV.43), and (XXV.45) we obtain

$$\mathfrak{U}^{[0]}Y_{4|0} = \partial_\delta \mathfrak{P}^{[0]}_{4|0}[\mathfrak{S}A^{[0]}] + \partial_{v_a}\mathfrak{P}^{[0]}_{4|0}[\mathfrak{S}g^{[0]}_a] + \partial_\varpi \mathfrak{P}^{[0]}_{4|0}[\mathfrak{S}B^{[0]}]$$
$$+ u_a \mathfrak{P}^{[0]}_{4|0,a} + \tfrac{7}{3}\mathfrak{P}^{[0]}_{4|0}E,$$
$$= -\frac{15p^2}{\rho^2}A^{[0]} + \frac{30p}{\rho}B^{[0]} + u_a\left(-\frac{15p^2}{\rho^2}\rho_{,a} + \frac{30p}{\rho}p_{,a}\right)$$
$$+ \frac{7}{3}\frac{15p^2}{\rho}E, \qquad\qquad \text{(XXV.47)}$$
$$= 0,$$

namely (XXV.44)$_4$.

From (XXV.44), (XXV.36), (XVIIA.5), and the general prescription (XXV.35) we obtain immediately the first atemporal iterates:

$$\mathfrak{P}^{[1]}_{km} = -2\mu E_{km},$$
$$\mathfrak{q}^{[1]}_k = -M\mu\varepsilon_{,k},$$
$$\mathfrak{P}^{[1]}_{0|kmr} = 0, \qquad\qquad \text{(XXV.48)}$$
$$\mathfrak{P}^{[1]}_{4|0} = \frac{1}{\rho}(15p^2 - 4\mu^2 E_{ab}E_{ab}),$$
$$\mathfrak{P}^{[1]}_{2|km} = -\frac{2\mu}{\rho}[7\lambda_1 pE_{km} + 4\lambda_2 \mu E_{\{ak}E_{ma\}}],$$

μ, M, λ_1, and λ_2 being given again by (XXV.13)$_{2,5}$ and (XXV.16). For comparisons with the results of ENSKOG's procedure we shall always consider the possibility that $\mu = \mu^{(1)}$ and $M = M^{(1)}$, and in this case we see from (XXV.17) that $\lambda_1 = 1$. Then from the first iterates just obtained we recover the first approximations of $\mathfrak{P}$ and $\mathfrak{q}$ delivered by ENSKOG's procedure, namely (XXIV.90), though with μ and M given not by the general formulae (XXIV.91) but by their first approximations $\mu^{(1)}$ and $M^{(1)}$, derived in Section (xii) of Chapter XXIV.

To obtain second iterates of $\mathfrak{P}$ and $\mathfrak{q}$, we note from (XXV.48)$_1$ and (XXV.42), recalling that μ is a function of $\tfrac{3}{2}p/\rho$ alone, that

$$\partial_\delta \mathfrak{P}^{[1]}_{km}[\mathfrak{S}\xi] = \frac{3p\mu'}{\rho^2}\xi E_{km}, \qquad \partial_{v_a}\mathfrak{P}^{[1]}_{km}[\mathfrak{S}\zeta_a] = -2\mu\zeta_{\{k,m\}},$$

$$\partial_\varpi \mathfrak{P}^{[1]}_{km}[\mathfrak{S}\eta] = -\frac{3\mu'}{\rho}\eta E_{km}.$$

(XXV.49)

Moreover, since M is a constant and since

$$\operatorname{grad} \varepsilon = \frac{3}{2\rho}\operatorname{grad} p - \frac{3p}{2\rho^2}\operatorname{grad}\rho,$$

we obtain from (XXV.48)$_2$ and (XXV.42)

$$\partial_\delta \mathfrak{q}^{[1]}_k[\mathfrak{S}\xi] = \frac{M}{\rho}(\mu'\varepsilon + 2\mu)\xi\varepsilon_{,k} + \frac{M\mu\varepsilon}{\rho}\xi_{,k} - \frac{M\mu\varepsilon}{p\rho}\xi p_{,k},$$

$$\partial_{v_a}\mathfrak{q}^{[1]}_k[\mathfrak{S}\zeta_a] = 0, \qquad\qquad\qquad\qquad\qquad\qquad\text{(XXV.50)}$$

$$\partial_\varpi \mathfrak{q}^{[1]}_k[\mathfrak{S}\eta] = \frac{M\mu\varepsilon}{p^2}\eta p_{,k} - \frac{M}{p}(\mu'\varepsilon + \mu)\eta\varepsilon_{,k} - \frac{M\mu\varepsilon}{p}\eta_{,k}.$$

In these we replace ξ, ζ, and η by the first approximations $A^{[1]}$, $\mathbf{g}^{[1]}$, and $B^{[1]}$, which from (XXV.38) and (XXV.48)$_{1,2}$ are

$$A^{[1]} = -u_a \rho_{,a} - \rho E,$$

$$g^{[1]}_k = -u_a u_{k,a} - \frac{1}{\rho}p_{,k} + \frac{2}{\rho}(\mu'\varepsilon_{,a}E_{ka} + \mu E_{ka,a}) + b_k, \qquad \text{(XXV.51)}$$

$$B^{[1]} = -u_a p_{,a} - \tfrac{5}{3}pE + \frac{4\mu}{3}E_{ab}E_{ab} + \tfrac{2}{3}M(\mu'\varepsilon_{,a}\varepsilon_{,a} + \mu\varepsilon_{,aa}).$$

After some simplification we arrive at

$$\mathfrak{A}^{[1]}Y_{0|km} = 2pE_{km} - 2\mu\left[\left(1 - \frac{2}{3}\frac{\varepsilon\mu'}{\mu}\right)EE_{km} - u_{\{k,a}u_{a,m\}} - \left(\frac{1}{\rho}p_{,\{k}\right)_{,m\}} + b_{\{k,m\}}\right]$$

$$- 4\mu E_{\{ak}u_{m,a\}} - \tfrac{4}{5}M(\mu\varepsilon_{,\{k})_{,m\}} + \cdots,$$

(ix) SECOND ATEMPORAL ITERATES 535

$$\mathfrak{A}^{[1]}Y_{2|k} = \tfrac{10}{3}p\varepsilon_{,k} + \tfrac{4}{3}\mathbf{M}(\mu\varepsilon E)_{,k} + 2\mathbf{M}\mu u_{a,k}\varepsilon_{,a} - \tfrac{2}{3}\mathbf{M}\mu(5\varepsilon_{,k}E \quad \text{(XXV.52)}$$

$$+ 3\varepsilon_{,a}u_{k,a} + 8\varepsilon_{,a}E_{ka}) - \frac{4p}{\rho}(\mu E_{ka})_{,a} + \frac{4\mu}{\rho}E_{ka}p_{,a} + \cdots,$$

where the dots denote terms proportional to powers of μ greater than the first. Henceforth we shall simply omit these terms. IKENBERRY & TRUESDELL also omitted these terms in their original scheme of atemporal Maxwellian iteration; more generally, at the nth stage they cast away all terms multiplied by a power of μ greater than n. This omission does not affect the limit functions, if they exist, because all terms introduced in atemporal Maxwellian iteration at the nth stage are proportional to μ^n or to higher powers of μ, never to lower ones[21]. The omission is not at all necessary, but, as we shall see, it produces results such as to be comparable at each stage with those of ENSKOG's process and the method of stretched fields.

If we now apply the prescription (XXV.35) and use (XXV.52)$_1$, we obtain the following analogue of the Maxwell second approximation (XXV.14):

$$\mathfrak{P}^{[2]}_{km} = -2\mu E_{km} + \frac{2\mu^2}{p}\left[\left(1 - \frac{2}{3}\frac{\varepsilon\mu'}{\mu}\right)EE_{km} - u_{\{k,a}u_{a,m\}} + 2E_{\{ak}u_{m,a\}}\right.$$

$$\left. - \left(\frac{1}{\rho}p_{,\{k}\right)_{,m\}} + b_{\{k,m\}}\right] + \frac{4}{5}\mathbf{M}\frac{\mu}{p}(\mu\varepsilon_{,\{k})_{,m\}}. \quad \text{(XXV.53)}$$

This formula is precisely the second atemporal Maxwellian iterate for **P** that IKENBERRY & TRUESDELL obtained. It is complicated as it stands, but a little effort can make it more so. By using the identity

$$E_{\{ak}u_{m,a\}} = 2E_{\{ka}E_{am\}} - E_{\{ka}u_{a,m\}} + \tfrac{2}{3}EE_{km} \quad \text{(XXV.54)}$$

we can elongate it into the Burnett formula (XXIV.4)$_1$ with coefficients as follows:

$$\varpi_1 = \frac{4}{3}\left(\frac{7}{2} - \frac{\varepsilon\mu^{(1)\prime}}{\mu^{(1)}}\right), \qquad \varpi_2 = 2,$$

$$\varpi_3 = \frac{6}{5}\mathbf{M}^{(1)} = 3, \qquad \varpi_4 = 0, \quad \text{(XXV.55)}$$

$$\varpi_5 = \frac{6}{5}\mathbf{M}^{(1)}\frac{\varepsilon\mu^{(1)\prime}}{\mu^{(1)}} = 3\frac{\varepsilon\mu^{(1)\prime}}{\mu^{(1)}}, \qquad \varpi_6 = 8.$$

---

[21] This fact was proved by IKENBERRY & TRUESDELL [1956, §13]. In Section (i) of Chapter XXVI the reader will see an example in which it becomes obvious.

To obtain the counterpart of (XXV.18) for the energy flux, namely the second atemporal Maxwellian iterate $\mathbf{q}^{[2]}$, we apply (XXV.35) again, this time using (XXV.52)$_2$:

$$q^{[2]}_k = -M\mu\left[\varepsilon_{,k} + \frac{3}{2\rho\varepsilon}(\mu\varepsilon E)_{,k} + \frac{9\mu}{4\rho\varepsilon}u_{a,k}\varepsilon_{,a}\right.$$

$$\left. - \frac{3\mu}{4\rho\varepsilon}(5\varepsilon_{,k}E + 3\varepsilon_{,a}u_{k,a} + 8\varepsilon_{,a}E_{ka})\right]$$

$$+ \frac{3\mu}{\rho}(\mu E_{ka})_{,a} - \frac{3\mu^2}{\rho\rho}E_{ka}p_{,a}. \qquad \text{(XXV.56)}$$

If we use the identity $u_{k,m} = -u_{m,k} + 2E_{km} + \frac{2}{3}E\delta_{km}$, after some rearrangement we may write (XXV.56) in CHAPMAN & COWLING's form (XXIV.4)$_2$ with coefficients as follows:

$$\theta_1 = \frac{3}{2}M^{(1)}\left(\frac{7}{2} - \frac{\varepsilon\mu^{(1)'}}{\mu^{(1)}}\right) = \frac{15}{4}\left(\frac{7}{2} - \frac{\varepsilon\mu^{(1)'}}{\mu^{(1)}}\right),$$

$$\theta_2 = \tfrac{45}{8}, \qquad \theta_3 = -3, \qquad \theta_4 = 3, \qquad \text{(XXV.57)}$$

$$\theta_5 = \frac{21}{2}M^{(1)} + \frac{3\varepsilon\mu^{(1)'}}{\mu^{(1)}} = 3\left(\frac{35}{4} + \frac{\varepsilon\mu^{(1)'}}{\mu^{(1)}}\right).$$

For a gas of Maxwellian molecules $\varepsilon\mu^{(1)'}/\mu^{(1)} = 1$; this substitution reduces (XXV.55) and (XXV.57) to the result we have already established for such a gas by using the method of stretched fields, namely (XXIV.148). For a general molecular model (XXV.55) and (XXV.57) are exactly the results obtained as approximations by CHAPMAN & COWLING[22]. Using the method of stretched fields, in Chapter XXIV we obtained the exact values of these coefficients,

---

[22] CHAPMAN & COWLING [1952, §§15.4 and 15.41] obtained the more general expressions listed in (XXIV.172) (cf. Footnote 33 to Chapter XXIV), but in their final results they record only (XXV.55) and (XXV.57) (they write μ rather than $\mu^{(1)}$, but as they do not fully specify how they got their admittedly approximate results, it is not certain that μ would be different from $\mu^{(1)}$ for them). They note that for a gas of Maxwellian molecules the additional terms in (XXIV.172)$_{5,6,11}$ are null, and so they conjecture that those terms could be neglected for more general molecular models. According to the last comment in Appendix B of Chapter XXIV, these additional contributions to ϖ_5, ϖ_6, and θ_5 arise from terms in GRAD's evaluation of collisions integrals (XVII.24)$_2$ for which $r \neq m + n$, and since we have agreed from the outset in this chapter to discard such terms, it is not surprising that here we have obtained (XXV.55) and (XXV.57) rather than the more general list (XXIV.172).

namely (XXIV.102). There the cost in labor was enormously greater than that paid here. The immediately preceding analysis shows that to get the approximate coefficients which nearly all previous authors seem to have regarded as sufficient for iterates past the first, there was no point at all in using ENSKOG's method.

In the next chapter we shall apply atemporal Maxwellian iteration as reformulated here to the three special cases in which the gross condition of the gas corresponds to the special solutions known explicitly and presented in Chapters XIV and XV. In those cases we shall see that for small enough values of the tension number T atemporal Maxwellian iteration for **P**, **q**, and $\mathbf{P}_{0|3}$ converges to the corresponding exact gross determiners.

(*x*) *Use of differential iteration to generate and improve Grad's method of truncation*

Maxwellian iteration divides the equations for moments into two classes: the equations of balance of mass, momentum, and energy, which are let stand as exact, and the remainder, to which an iteration designed to select a particular class of solutions is applied. Calling these two classes (S) and (I), we write them out as follows:

$$(S) \begin{cases} \dot{\rho} = -\rho E; & (IX.3)_r \\ \rho \dot{\mathbf{u}} = -\operatorname{grad} p - \operatorname{div} \mathbf{P} + \rho \mathbf{b}; & (IX.4)_r \\ \rho \dot{\varepsilon} + pE = -\mathbf{P} \cdot \mathbf{E} - \operatorname{div} \mathbf{q}; & (IX.5)_{2r} \end{cases}$$

(I) $\mathfrak{L} Y_{2r|s} = \mathfrak{m}\overline{\mathbb{C}} Y_{2r|s},\qquad 2r+s \geqq 2,\quad (r,s) \neq (1,0).\qquad (XXV.58)$

The differential iteration is applied only to the iterative class (I), while the standing class (S) is set aside for use after the iteration has been carried as far as is desired. To start the iteration, we choose as 0th iterates the functions of ρ to which the iterated moments reduce when $F = F_M$. They are given by (XXV.5).

We may modify the scheme by enlarging the class (S), say by moving some equations from (I) into (S), while leaving the method otherwise unchanged. The number of dependent variables in the equations obtained by substituting into the new (S) the results of the iteration based on the new (I) in thereby increased.

For example, recalling the program of GRAD, we may design to get a final system of field equations in which just ρ, **P**, and **q** are the dependent variables. The example of a gas of Maxwellian molecules suffices to illustrate the method.

For the class (S) we take the exact equations of balance and the exact equations for **P** and **q**:

$$(S) \begin{cases} \dot\rho = -\rho E; & \text{(IX.3)}_r \\ \rho\dot{\mathbf u} = -\operatorname{grad} p - \operatorname{div}\mathbf P + \rho\mathbf b; & \text{(IX.4)}_r \\ \rho\dot\varepsilon + pE = -\mathbf P\cdot\mathbf E - \operatorname{div}\mathbf q; & \text{(IX.5)}_r \\ \dot P_{km} = -P_{km}E - 2P_{\{ak}u_{m,a\}} - 2pE_{km} \\ \qquad - P_{0|kma,a} - \tfrac{4}{5}q_{\{k,m\}} - 3n\mathbb{A}_{2|2}P_{km}, \\ \dot q_k = -\tfrac{5}{3}q_k E - P_{0|kab}E_{ab} - q_a u_{k,a} - \tfrac{4}{5}q_a E_{ka} \\ \qquad + \dfrac{5p}{2\rho}M_{ka,a} + \dfrac{1}{\rho}P_{ka}M_{ab,b} \\ \qquad - \tfrac{1}{2}P_{2|ka,a} - \tfrac{1}{6}P_{4|0,k} - 2n\mathbb{A}_{2|2}q_k. \end{cases} \quad \text{(XXV.59)}$$

The remaining equations for moments form the iterative system (I) and serve as the basis for the iterative scheme:

$$-\mathbb{C}_{2r|s}P^{(n+1)}_{2r|s} \equiv -\mathcal{Q}^{(n+1)}_{2r|s} + \mathfrak{L}^{(n)}Y_{2r|s}, \quad 2r+s \ge 3,\ (r,s) \ne (1,1), \quad \text{(XXV.60)}$$

with the 0th iterates chosen as follows:

$$P^{(0)}_{2r|s} \equiv \begin{cases} 0 & \text{if } s \ne 0, \quad 2r+s \ge 3,\ (r,s) \ne (1,1). \\ (2r+1)!!\,\rho(p/\rho)^r & \text{if } s = 0, \end{cases} \quad \text{(XXV.61)}$$

We are to iterate for the moments $P_{0|kmr}$, $P_{2|km}$, and $P_{4|0}$. After carrying the iteration as far as we consider sufficient, we substitute the result into the standing system (S). The result is a system of 13 scalar partial differential equations in which the unknowns are the 13 components of ρ, **P**, and **q**. That is, no matter how far the iteration is carried, the final outcome is similar to the equations defining a theory of the rate type in continuum mechanics.

From (XXV.61) we see at once that

$$P^{(0)}_{0|kmr} = 0, \qquad P^{(0)}_{2|km} = 0, \qquad P^{(0)}_{4|0} = 15\frac{p^2}{\rho}. \quad \text{(XXV.62)}$$

If we put these evaluations in place of the quantities they are desired to approximate in (XXV.59), by comparison with (XVII.52) we see that *the result is* GRAD's *13-moment system*.

Now putting $n = 1$ in (XXV.60), after some elementary calculations we obtain

$$P^{(1)}_{0|kmr} = -\frac{\mu}{p}\left[\frac{8}{5}q_{\{k}E_{mr\}} - \frac{2}{\rho}P_{\{km}M_{ra,a\}}\right],$$

$$P^{(1)}_{2|km} = \frac{p}{\rho}P_{km} - \frac{4}{7\rho}P_{\{ka}P_{am\}} - 12\frac{\mu}{p}\left[\frac{p^2}{\rho}E_{km} - \frac{2}{5\rho}q_{\{k}M_{ma,a\}}\right], \quad \text{(XXV.63)}$$

$$P^{(1)}_{4|0} = 15\frac{p^2}{\rho} - \frac{1}{\rho}P_{ab}P_{ab} - \frac{\mu}{p}\left[\frac{45}{2}\left(\frac{p^2}{\rho}\right)^{\cdot} + \frac{105}{2}\frac{p^2}{\rho}E - \frac{12}{\rho}q_a M_{ab,b}\right].$$

When these evaluations are substituted for the quantities they are designed to approximate in (XXV.59), higher spatial derivatives are introduced, but the result is still a system of first order in the time.

While this new method and GRAD's lead to the same result at the first stage, no such agreement occurs thereafter. GRAD's method *simply discards* the equations for moments above a certain order. To pass to a higher stage, *it introduces more moments as unknowns* and thus requires the user to prescribe *a greater number of initial values*. To improve upon the 13-moment system, GRAD passes to the 20-moment system. The present method, on the contrary, *leaves fixed* the number of unknowns. The result of putting (XXV.63) into (XXV.59) *is still a system for* 13 *unknowns*, but the original 13-moment system has been corrected by use of some information gotten from the equations of 4th moments. As with Maxwellian iteration, *each stage calls upon use of equations for moments of a higher order* than any used before, and no equation, no matter how high its order, is ever discarded.

(xi) *Retrospect upon formal methods of approximation*

The power of the method of differential iteration has been illustrated by four examples in this chapter. The two called "Maxwellian" have been shown to yield the kind of formal approximations that most studies in the kinetic theory have taken heretofore as sufficient. To this extent we may justly disagree with HILBERT's opinion, which we have quoted at the beginning of Chapter XXII: To calculate the usual first approximations to transport coefficients and the functional relations in which they appear, *the theory of integral equations provides only an unnecessary detour*.

For ultimate precision we have replaced ENSKOG's method by the method of stretched fields. There, indeed, integral equations play an essential part. However, except for a gas of Maxwellian molecules, where they are not ever needed at all, those integral equations themselves are solved by some process of approximation. TRUESDELL[23] suggested that instead of approximating the formally exact integral equations it might be easier to approximate GRAD's explicit and formally exact infinite series (XVII.24) by a method of successively less and less severe truncations, so adjusted as to include at a given iterative stage only moments for which iterates from a preceding stage were available. To such truncated series Maxwellian iteration can be applied with only a little complication. A method of this kind should deliver for the transport coefficients approximations as accurate as desired. Again there would be no integral equations to solve. The only real difficulty lies in evaluating GRAD's Hermite coefficients. However, this avenue now would be less rewarding that it seemed twenty years ago, for the method of stretched fields has removed all the obscurities that dogged ENSKOG's method, and machine calculation should be able to obviate recourse to the sort of merely numerical approximation that fills most of the older books on the kinetic theory.

In this book we have chosen to present and elaborate two methods of formal calculation peculiar to the kinetic theory, namely HILBERT's and GRAD's, and two techniques of approximation in analysis, namely IKENBERRY & TRUESDELL's differential iteration and MUNCASTER's method of stretched fields, which were conceived for use in the kinetic theory and can be applied to various equations that arise in it. Each of these methods has one or another formal appeal. More than that, today, none can have—except in two special cases, for which both existence of the desired mapping and convergence to it have been demonstrated and delimited. That is the subject of the last chapter.

---

[23] TRUESDELL [1956, §56].

Chapter XXVI

Convergence and Divergence of Atemporal Maxwellian Iteration in Flows for Which an Exact Solution Is Known. Failure of the Higher Iterates to Improve the Asymptotic Approximation

In Chapters XIV and XV we determined the explicit solutions for the pressure tensor **P** in homo-energetic simple shearing, dilatation, and extension of a gas of Maxwellian molecules subject to null body force. For these particular gross conditions the infinite system of equations for moments breaks up into a sequence of systems of finite order, systems which can be solved one after the other. We shall see now that for these same special flows atemporal Maxwellian iteration likewise degenerates into groups of iterative systems which can be solved easily. We shall be able, consequently, to calculate $\mathfrak{P}^{[n]}$, $\mathfrak{q}^{[n]}$, and $\mathfrak{P}_{0|3}^{[n]}$ explicitly, to determine their character, and to relate them to the corresponding exact solutions in each case.

The reader will recall always that the atemporal Maxwellian iterates at the first and second stages are the same as corresponding iterates delivered by ENSKOG's process and by the method of stretched fields, so the results in this chapter certainly apply to those procedures as far as they have been carried and very likely apply to them at all stages.

We shall suppose that $\mathbf{b} = \mathbf{0}$, as we did in treating these flows in Chapters XIV and XV, and we shall restrict attention to a gas of Maxwellian molecules. We recall the definition

$$\tau \equiv \frac{1}{3n\mathbb{A}_{2|2}} = \frac{1}{3\mathbb{a}}\sqrt{\frac{2\mathbb{m}^3}{\mathbb{g}}}\frac{1}{\rho}, \qquad \text{(XIII.3)}_r$$

in terms of which

$$\mu = p\tau. \qquad \text{(XIII.14)}_{1r}$$

(i) Homo-energetic affine flows in general

Each of the flows we shall examine is homo-energetic and affine:

$$u_k = G_{ka}(t)r_a + g_k(t), \qquad (\text{II.41})_{2r}$$

and ρ, p, ε, **P**, ^{3}**M**, **g**, and **G** are functions of t alone. In Section (viii) of Chapter XXIII we reduced the general theory of gross determiners to the special form it takes for these flows. Of course, there ρ, p, **G**, and **g** were constants, namely, the initial values of the corresponding functions in affine flow. We saw that the gross determiner $\mathbb{Z}$ of the molecular density, while in general a function of **v** and a functional of the fields ρ, **u**, p, and **b**, could be taken to be simply a function $\mathbb{Z}^G$ of a finite number of real variables, namely, ρ, p, the components c_k of the random velocity, and the components G_{km} of the velocity gradient. As a result, any expectation calculated using $\mathbb{Z}$ reduced to a function of ρ, p, and **G** alone. In particular, the gross determiners $\mathfrak{P}$, $\mathfrak{q}$, and $\mathfrak{P}_{0|3}$ reduced to the functions **P**, **q**, and $\mathbf{P}_{0|3}$ defined as follows:

$$P_{km}(\rho, \mathbf{G}, p) \equiv \mathfrak{m} \int c_{\{k}c_{m\}} Z^G(\mathbf{c}; \rho, \mathbf{G}, p),$$

$$q_k(\rho, \mathbf{G}, p) \equiv \mathfrak{m} \int \tfrac{1}{2}c^2 c_k\, Z^G(\mathbf{c}; \rho, \mathbf{G}, p), \qquad (\text{XXVI.1})$$

$$P_{0|kmr}(\rho, \mathbf{G}, p) \equiv \mathfrak{m} \int Y_{0|kmr}(\mathbf{c})\, Z^G(\mathbf{c}; \rho, \mathbf{G}, p).$$

Corresponding simplifications occurred in the general equation of transfer (XXIII.75). The equations of transfer for $\mathfrak{P}$, $\mathfrak{q}$, and $\mathfrak{P}_{0|3}$, specialized to a gas of Maxwellian molecules, are

$$-\rho E\, \partial_\rho P_{km} - G_{ab}G_{bc}\, \partial_{G_{ac}}P_{km} - (\tfrac{5}{3}pE + \tfrac{2}{3}P_{ab}E_{ab})\, \partial_p P_{km}$$
$$+ P_{km}E + 2P_{\{ak}G_{ma\}} + 2pE_{km} = -P_{km}/\tau; \qquad (\text{XXIII.92})_r$$

$$-\rho E\, \partial_\rho q_k - G_{ab}G_{bc}\, \partial_{G_{ac}}q_k - (\tfrac{5}{3}pE + \tfrac{2}{3}P_{ab}E_{ab})\, \partial_p q_k + \tfrac{5}{3}q_k E + P_{0|kab}E_{ab}$$
$$+ q_a G_{ka} + \tfrac{4}{5}q_a E_{ka} = -2q_k/3\tau, \qquad (\text{XXIII.93})_r$$

$$-\rho E\, \partial_\rho P_{0|kmr} - G_{ab}G_{bc}\, \partial_{G_{ac}}P_{0|kmr} - (\tfrac{5}{3}pE + \tfrac{2}{3}P_{ab}E_{ab})\, \partial_p P_{0|kmr} + P_{0|kmr}E$$
$$+ P_{0|a(km}G_{r)a} - \tfrac{6}{5}P_{0|ab(k}\,\delta_{mr)}E_{ab} + \tfrac{12}{5}q_{(k}E_{mr)}$$
$$- \tfrac{24}{25}q_a E_{a(k}\,\delta_{mr)} = -3P_{0|kmr}/2\tau.$$

As was explained in Chapter XXIII when these equations were derived, in them $E_{km} \equiv G_{\{km\}}$ and $E \equiv G_{aa}$.

In Section (ix) of Chapter XXV we defined and developed atemporal Maxwellian iteration. That procedure takes as its basis the exact equation of transfer (XXIII.75) for gross determiners and applies to it the prescription of

Maxwellian iteration. For affine flows the atemporal iterates may be obtained by applying that same prescription directly to the reduced equations of transfer appropriate to such flows. For example, from (XXIII.92) we see that one member of the iterative system is

$$-P_{km}^{[n+1]}/\tau \equiv 2pE_{km} - \rho E \, \partial_\rho P_{km}^{[n]} - \left(\tfrac{5}{3}pE + \tfrac{2}{3}P_{ab}^{[n]}E_{ab}\right)\partial_p P_{km}^{[n]}$$
$$- G_{ab}G_{bc}\,\partial_{G_{ac}}P_{km}^{[n]} + EP_{km}^{[n]} + 2P_{\{ak}^{[n]}G_{ma\}}. \qquad \text{(XXVI.2)}$$

This is a system for determining the iterates $\mathbf{P}^{[n]}$ alone, with no reference to the gross determiners of higher moments. The 0th iterate, of course, is $\mathbf{P}^{[0]} = \mathbf{0}$. There is no need to write out the iterative system which follows from (XXIII.93), because if we replace both $\mathbf{q}$ and $\mathbf{P}_{0|3}$ on the left-hand sides of (XXIII.93) by their n^{th} iterates and on the right-hand side by their $n+1^{\text{st}}$ iterates and then recall that $\mathbf{q}^{[0]} = \mathbf{0}$ and $\mathbf{P}_{0|3}^{[0]} = \mathbf{0}$, we see immediately that for all n

$$q_k^{[n]} = 0, \qquad P_{0|kmr}^{[n]} = 0. \qquad \text{(XXVI.3)}$$

Our objective here is to prove, as far as $\mathbf{P}$, $\mathbf{q}$, and $\mathbf{P}_{0|3}$ are concerned, that for certain affine flows atemporal Maxwellian iteration converges, and moreover that *it converges to the exact gross determiners* for the flows in question. We must first solve the exact equations of transfer (XXIII.92) and (XXIII.93). To do so, we invoke the *conjecture on uniqueness* presented in Section (vii) of Chapter XXIV, rephrased now so as to apply to the present case: *Within the class of all functions* $\mathbf{P}$, $\mathbf{q}$, *and* $\mathbf{P}_{0|3}$ *which are analytic in* ρ, p, *and* $\mathbf{G}$ *in a neighborhood of* $\mathbf{G} = \mathbf{0}$, *there is at most one solution of the equations of transfer* (XXIII.92) *and* (XXIII.93). We denote these unique solutions by $\mathbf{P}^G$, $\mathbf{q}^G$, and $\mathbf{P}_{0|3}^G$. Be this conjecture true, we see immediately that the unique solution of (XXIII.93) is

$$q_k^G = 0, \qquad P_{0|kmr}^G = 0. \qquad \text{(XXVI.4)}$$

Comparing this with (XXVI.3) we see, trivially, that *for all affine flows the gross determiners* $\mathbf{q}^G$ *and* $\mathbf{P}_{0|3}^G$ *are null, and atemporal Maxwellian iteration converges to these exact values.* Corresponding conclusions for $\mathbf{P}$ are not so easily obtained, and in fact we shall be able to draw such conclusions here only for certain affine flows, namely, those for which we have presented exact solutions in Chapters XIV and XV.

For those particular flows we may use the exact solutions to prove that the conjecture on uniqueness is true. For other affine flows, so far as we can say, it might not hold. Then arguments of the kind we are about to present would show only that atemporal Maxwellian iteration converged to *some* solution of (XXIII.92) and (XXIII.93).

We begin by analysing in detail the iterates defined by (XXVI.2). For a gas of Maxwellian molecules τ is a function of ρ alone; since $P_{km}^{[1]} = -2\mu E_{km}$, it follows from (XIII.14)$_1$ that $P_{km}^{[1]}/p$ is independent of p, so $\partial_p P_{km}^{[1]} = P_{km}^{[1]}/p$; from

(XXVI.2) we see that $P^{[2]}_{km}/p$ is independent of p, and so by induction we may define as follows a set of functions $\varpi^{[n]}_{km}$ of ρ and $\mathbf{G}$ alone:

$$\varpi^{[n]}_{km} \equiv P^{[n]}_{km}/p. \tag{XXVI.5}$$

In terms of $\boldsymbol{\varpi}^{[n]}$ the iterative scheme (XXVI.2) becomes

$$-\varpi^{[n+1]}_{km}/\tau \equiv 2E_{km} - \rho E \, \partial_\rho \varpi^{[n]}_{km} - \tfrac{2}{3}(E + \varpi^{[n]}_{ab} E_{ab})\varpi^{[n]}_{km}$$
$$- G_{ab} G_{bc} \, \partial_{G_{ac}} \varpi^{[n]}_{km} + 2\varpi^{[n]}_{\{ak} G_{ma\}}, \tag{XXVI.6}$$

and of course in atemporal Maxwellian iteration $\varpi^{[0]}_{km} \equiv 0$. Hence $\varpi^{[1]}_{km} = -2\rho\tau E_{km}/\rho$. From (XIII.3) we see that $\rho\tau = $ const. Hence $\boldsymbol{\varpi}^{[1]}$ equals a function of the quantities G_{rs}/ρ alone. Let us set

$$\alpha_{rs} \equiv G_{rs}/\rho. \tag{XXVI.7}$$

If we assume as hypothesis of induction that $\boldsymbol{\varpi}^{[n]}$ for some fixed n equals a function of $\boldsymbol{\alpha}$ alone, then

$$\rho \, \partial_\rho \varpi^{[n]}_{km} = (\rho \, \partial_{\alpha_{ab}} \varpi^{[n]}_{km})\left(-\frac{G_{ab}}{\rho^2}\right),$$
$$= -\alpha_{ab} \, \partial_{\alpha_{ab}} \varpi^{[n]}_{km}, \tag{XXVI.8}$$
$$\partial_{G_{rs}} \varpi^{[n]}_{km} = \partial_{\alpha_{rs}} \varpi^{[n]}_{km}/\rho.$$

Therefore (XXVI.6) can be written in the form

$$-\varpi^{[n+1]}_{km}/(\rho\tau) = 2\alpha_{\{km\}} + (\alpha_{cc}\alpha_{ab} - \alpha_{ac}\alpha_{cb}) \, \partial_{\alpha_{ab}} \varpi^{[n]}_{km}$$
$$- \tfrac{2}{3}(\alpha_{cc} + \varpi^{[n]}_{ab}\alpha_{\{ab\}})\varpi^{[n]}_{km} + 2\varpi^{[n]}_{\{ak}\alpha_{ma\}}. \tag{XXVI.9}$$

The hypothesis of induction makes the right-hand side obviously a function of $\boldsymbol{\alpha}$ alone. As $\rho\tau$ is a constant, $\boldsymbol{\varpi}^{[n+1]}$ is also a function of $\boldsymbol{\alpha}$ alone. Thus we have shown that $\boldsymbol{\varpi}^{[n]}$ is a function of $\boldsymbol{\alpha}$ alone for every n. At the same time, (XXVI.9) is established as a definitive form of the iterative scheme for all homo-energetic affine flows.

From (XXVI.9) it is easy to prove by induction that

$$\varpi^{[n+1]}_{km} - \varpi^{[n]}_{km} = \tau^{n+1}[\ldots], \tag{XXVI.10}$$

the bracketed dots indicating a polynomial in τ. Thus the coefficients of the terms proportional to $\tau^0, \tau^1, \ldots, \tau^n$ are obtained once and for all at the nth stage. This fact illustrates a more general theorem, proved by IKENBERRY & TRUESDELL, which we cited in Footnote 21 to Chapter XXV. In Section (ix) of that chapter, in our presentation of atemporal Maxwellian iteration in general we used that theorem to justify dropping in $\mathfrak{P}^{[n]}$ all summands proportional to a power of μ higher than the nth. We did so partly to avoid long formulae, partly to facilitate comparisons with the results of ENSKOG's method and the method of stretched fields. If we do the same here, from (XXVI.9) we obtain $\varpi^{[n]}_{km}$ as a *uniquely determined polynomial of degree n in* $\rho\tau$; *the coefficients are functions of the nine quantities* α_{rs} *alone*. The calculation involves algebraic and differential operations only. In the present context the formulae are simpler, and the treatment is neater if we retain all terms. Thus in proving theorems we shall use (XXVI.9) alone, just as it stands. When, later, we exhibit some explicit iterates, we shall revert to shortened formulae as in Chapter XXV.

A glance at (XXVI.9) shows that a sufficient condition for the differential operations to drop out altogether is

$$G_{aa}G_{km} - G_{ka}G_{am} = 0. \tag{XXVI.11}$$

That is, $EG = G^2$. This condition is satisfied by all steady affine flows since for them $\mathbf{G}^2 = \mathbf{0}$; some unsteady ones satisfy it, and some do not. If it does hold, the iterative system (XXVI.9) reduces to

$$-\varpi_{km}^{[n+1]}/\tau = 2E_{km} - \tfrac{2}{3}(E + \varpi_{ab}^{[n]}E_{ab})\varpi_{km}^{[n]} + 2\varpi_{\{ak}^{[n]}G_{ma\}}. \tag{XXVI.12}$$

The dependence on ρ has dropped out altogether. From (XXVI.12) it is clear that $\varpi_{km}^{[n]}$ is a dimensionless polynomial in the dimensionless quantities τG_{11}, $\tau G_{12}, \ldots, \tau G_{33}$.

Henceforth we shall consider only the special class of homo-energetic affine flows that results when (XXVI.11) holds. Thus we shall be justified in taking (XXVI.12) as the iterative scheme for $\varpi^{[n]}$.

Because $\varpi^{[n]}$ is a polynomial in τ satisfying (XXVI.10), *the sequence* $\varpi^{[n]}$ *can converge as* $n \to \infty$ *only to an analytic function of* τ. If it does converge, say

$$\varpi_{km}^{[n]} \to \varpi_{km}^{\infty} \quad \text{as} \quad n \to \infty, \tag{XXVI.13}$$

then the limit ϖ^{∞} is a function ϖ of τG such as to satisfy the equation

$$-\varpi_{km}/\tau = 2E_{km} - \tfrac{2}{3}(E + \varpi_{ab}E_{ab})\varpi_{km} + 2\varpi_{\{ak}G_{ma\}}. \tag{XXVI.14}$$

We can look at this equation directly. If we can prove that it has one and only one real solution ϖ that has a convergent expansion in powers of τ, then necessarily $\varpi^{\infty} = \varpi$ for values of τ such that the expansion for ϖ converges. We may summarize this conclusion as the **Basic Convergence Lemma**. *If* (XXVI.14) *has one and only one real solution* ϖ *which has an expansion in powers of* τ *with radius of convergence a, then the iterates* $\varpi^{[n]}$ *converge to* ϖ *if* $|\tau| < a$, *diverge if* $|\tau| > a$. We can actually say more. If we look for a solution of the equation of transfer (XXIII.92) for the gross determiner $\mathbf{P}$ of the form $\mathbf{P} = p\varpi(\tau\mathbf{G})$, we find that ϖ must satisfy (XXVI.14), at least for those affine flows that satisfy (XXVI.11). Therefore, if we can prove that (XXVI.14) has a unique solution which is analytic in $\mathbf{G}$ near $\mathbf{G} = \mathbf{0}$, we shall conclude from the uniqueness conjecture that the corresponding solution $\mathbf{P}$ of (XXIII.92) is the unique solution $\mathbf{P}^G$. Thus not only will the iterates $\mathbf{P}^{[n]}$ converge, but their limit will be the exact gross determiner of the flow, as delivered by the general theory of gross determiners in Chapter XXIII. In this way we can determine the convergence or divergence of the atemporal Maxwellian iterates by determining the only possible analytic limit function they may have, and we can identify that function.

While we cannot solve this problem in general, we can do so for all three classes of flows for which we have obtained the exact solutions for the pressures. For each of these (XXVI.11) holds. Hence we may take (XXVI.12) as the iterative scheme for calculating $\varpi^{[n]}$; moreover, for each we obtain a rigorous theorem of convergence and divergence.

In the homo-energetic flows for which (XXVI.11) does not hold, generally the terms involving differentiation do not drop out of (XXVI.9). Proof along the above lines that for them, too, the iteration converges would face two added difficulties. First, we should have to show that the limits as $n \to \infty$ commute with the differentiations. Second, the formal limit equation generalizing (XXVI.14) is a differential equation, which will have, in general, infinitely many solutions. The iterates $\varpi^{[n]}$ are unique and can converge to at most one of those solutions. Which that one should be, is not obvious.

(ii) Homo-energetic dilatation

$\mathbf{E} = \mathbf{0}$ for any dilatation, so the unique solution of the iterative system (XXVI.2) is $\mathbf{P}^{[n]} = \mathbf{0}$ for every n. Gross rest, which we discussed in Chapter XIII, is included as one special case, while GALKIN's solutions for certain dilatations, which we presented in Chapter XV, furnish another. The gross determiner $\mathbf{P}^G = \mathbf{0}$. For all dilatations *atemporal Maxwellian iteration delivers their gross determiners, exactly, at every stage.*

(iii) Homo-energetic simple shearing[1]

The velocity field of a homo-energetic simple shearing is

$$u_1 = Kx_2, \quad u_2 = 0, \quad u_3 = 0, \quad K = \text{const.} \tag{II.57}_r$$

To describe the solution, we introduce the tension number T:

$$\mathsf{T} \equiv \tau K. \tag{II.60}_r$$

Since this is a steady flow, the condition (XXVI.11) is satisfied. The system (XXVI.12) reduces to

$$-\varpi_{km}^{[n+1]}/\mathsf{T} = \delta_{k1}\delta_{m2} + \delta_{m1}\delta_{k2} - \tfrac{2}{3}\varpi_{12}^{[n]}\varpi_{km}^{[n]} - \tfrac{2}{3}\varpi_{12}^{[n]}\delta_{km} + \varpi_{2k}^{[n]}\delta_{m1} + \varpi_{2m}^{[n]}\delta_{k1},$$

$$\tag{XXVI.15}$$

---

[1] The analysis for simple shearing was given by TRUESDELL [1956, §36]. That for extension, which we present in the following section, is of the same kind.

and the system (XXVI.14) for determining the components of the limit function ϖ is

$$-\varpi_{km}/\mathsf{T} = \delta_{k1}\delta_{m2} + \delta_{m1}\delta_{k2} - \tfrac{2}{3}\varpi_{12}\varpi_{km} - \tfrac{2}{3}\varpi_{12}\delta_{km} + \varpi_{2k}\delta_{m1} + \varpi_{2m}\delta_{k1};$$
(XXVI.16)

component by component,

$$\begin{aligned}
-\varpi_{11}/\mathsf{T} &= -\tfrac{2}{3}\varpi_{12}\varpi_{11} + \tfrac{4}{3}\varpi_{12}, \\
-\varpi_{22}/\mathsf{T} &= -\tfrac{2}{3}\varpi_{12}\varpi_{22} - \tfrac{2}{3}\varpi_{12}, \\
-\varpi_{12}/\mathsf{T} &= 1 - \tfrac{2}{3}\varpi_{12}^2 + \varpi_{22}, \\
-\varpi_{13}/\mathsf{T} &= -\tfrac{2}{3}\varpi_{12}\varpi_{13} + \varpi_{23}, \\
-\varpi_{23}/\mathsf{T} &= -\tfrac{2}{3}\varpi_{12}\varpi_{23}.
\end{aligned}$$
(XXVI.17)

It is easy to show that $\tfrac{2}{3}\mathsf{T}\varpi_{12} \neq 1$ and hence necessarily

$$\varpi_{13} = \varpi_{23} = 0, \qquad \varpi_{11} + 2\varpi_{22} = 0. \tag{XXVI.18}$$

Solving (XXVI.17)$_2$ for ϖ_{22} and substituting the result into (XXVI.17)$_3$, we obtain an equation for ϖ_{12} alone:

$$(1 - \tfrac{2}{3}\mathsf{T}\varpi_{12})(\varpi_{12} + \mathsf{T} - \tfrac{2}{3}\mathsf{T}\varpi_{12}^2) = -\tfrac{2}{3}\mathsf{T}^2\varpi_{12}. \tag{XXVI.19}$$

If $\chi = -\tfrac{2}{3}\mathsf{T}\varpi_{12}$, this equation assumes the form

$$\chi(\chi + 1)^2 = \tfrac{2}{3}\mathsf{T}^2, \tag{XIV.4}_r$$

an equation we have shown in Chapter XIV to have one and only one real root R. Thus

$$\varpi_{12} = -\frac{R}{\tfrac{2}{3}\mathsf{T}}, \qquad \varpi_{22} = -\frac{R}{1+R}. \tag{XXVI.20}$$

Our analysis has shown that if there be any real solutions ϖ_{km} of (XXVI.16), they can be nothing else than (XXVI.18) and (XXVI.20); conversely, we easily verify that these formulae do indeed solve (XXVI.16). Thus (XXVI.16) has a unique real solution. Furthermore, in Chapter XIV we have shown that R has a power series expansion in T, the interval of convergence of which is $|\mathsf{T}| < \sqrt{2/3}$, and the leading term of which is $\tfrac{2}{3}\mathsf{T}^2$. Therefore, the right-hand side of (XXVI.20)$_1$ has an expansion in powers of T with interval of convergence $|\mathsf{T}| < \sqrt{2/3}$. Because the condition $\mathsf{T} < \sqrt{2/3}$ is sufficient that $R < 1$, the interval of convergence of the power series expansion of (XXVI.20)$_2$ is at least $|\mathsf{T}| < \sqrt{2/3}$. From (XXVI.20)$_2$ we see that that interval cannot be greater than the one for the expansion of R, namely $|\mathsf{T}| < \sqrt{2/3}$. Comparison of (XXVI.18) and (XXVI.20) with (XIV.39) shows that the limit function of the sequence $\mathbf{P}^{[n]}$ is the gross determiner $\mathbf{P}^G$. Comparing the statements just proved and recalling the basic convergence lemma, we obtain the following **Convergence Theorem**: *If the flow is a homo-energetic simple shearing, atemporal Maxwellian iteration for* **P** *converges to the gross determiner* **P**^{G} *if* $\mathsf{T} < \sqrt{2/3}$; *if* $\mathsf{T} > \sqrt{2/3}$, *the process diverges.*

(iv) *Homo-energetic extension*

The velocity field of homo-energetic extension is

$$u_1 = \xi(t)(x_1 - x_0), \qquad u_2 = 0, \qquad u_3 = 0; \tag{II.64}_r$$

$$\xi(t) = \frac{1}{t+T}. \tag{II.66}_{2r}$$

To describe the solution, we introduce the number T:

$$\mathsf{T} \equiv \frac{4}{3}\tau\xi = \frac{4}{3}\frac{\tau(0)}{T}. \tag{II.72}_r, \ (II.76)_r$$

As we stated shortly after (IV.13), T is the tension number if it is positive, while if $\mathsf{T} < 0$, the tension number is $-\tfrac{1}{2}\mathsf{T}$. For this flow it is obvious that (XXVI.11) is satisfied, so the iterative scheme (XXVI.12) applies and reduces to

$$-\tfrac{4}{3}\varpi_{km}^{[n+1]}/\mathsf{T} = 2(\delta_{k1}\delta_{m1} - \tfrac{1}{3}\delta_{km}) - \tfrac{2}{3}(1 + \varpi_{11}^{[n]})\varpi_{km}^{[n]} + \varpi_{1k}^{[n]}\delta_{m1}$$
$$+ \varpi_{1m}^{[n]}\delta_{k1} - \tfrac{2}{3}\varpi_{11}^{[n]}\delta_{km}. \tag{XXVI.21}$$

The limit functions ϖ_{km}, if they exist, must satisfy

$$-\tfrac{4}{3}\varpi_{km}/\mathsf{T} = 2(\delta_{k1}\delta_{m1} - \tfrac{1}{3}\delta_{km}) - \tfrac{2}{3}(1 + \varpi_{11})\varpi_{km}$$
$$+ \varpi_{1k}\delta_{m1} + \varpi_{1m}\delta_{k1} - \tfrac{2}{3}\varpi_{11}\delta_{km}. \tag{XXVI.22}$$

The 1–1 component of this system is

$$-\tfrac{1}{2}\varpi_{11}^2 + \left(\frac{1}{2} + \frac{1}{\mathsf{T}}\right)\varpi_{11} + 1 = 0. \tag{XXVI.23}$$

If $\varpi_{11} \equiv -\tfrac{1}{2}(3\chi + 5)$, we see that (XXVI.23) assumes the form

$$\chi^2 + 4\chi\left(1 + \frac{1}{3\mathsf{T}}\right) + 3 + \frac{20}{9\mathsf{T}} = 0, \tag{XV.19}_r$$

and therefore χ must be either R or S as given by (XV.20). Only R gives rise to a solution analytic near $\mathsf{T} = 0$, and this solution is

$$\varpi_{11} = -\frac{3R + 5}{2} \tag{XXVI.24}$$

It is easy to show from (XXVI.22) that

$$\varpi_{33} = \varpi_{22} = -\tfrac{1}{2}\varpi_{11}, \qquad \varpi_{12} = \varpi_{23} = \varpi_{13} = 0. \qquad \text{(XXVI.25)}$$

In Chapter XV we have shown that the power series expansion for R as a function of T converges if $|\mathsf{T}| < \tfrac{2}{3}$ and diverges if $|\mathsf{T}| > \tfrac{2}{3}$. The basic convergence lemma enables us to conclude that the iterative scheme (XXVI.21) converges or diverges as does the series for R. Collecting the statements just made and comparing the results with (XV.37), we obtain the following **Convergence Theorem**[2]: *If the flow is a homo-energetic extension, atemporal Maxwellian iteration for* $\mathbf{P}$ *converges if* $|\mathsf{T}| < \tfrac{2}{3}$, *diverges if* $|\mathsf{T}| > \tfrac{2}{3}$. *The limit function is the gross determiner* $\mathbf{P}^G$.

(v) Failure of the classical approach to approximate solution[3]

The *classical approach* to approximate solution in the kinetic theory may be described as follows:

(a) Find approximations to the gross determiners $\mathfrak{P}$ and $\mathbf{q}$.

(b) Substitute those approximations for $\mathbf{P}$ and $\mathbf{q}$ in the equations of balance of mass, momentum, and energy. For the system of differential equations so obtained, set up and solve an initial-value problem to determine, presumably approximately, the fields ρ, $\mathbf{u}$, and p, using their *exact* initial values.

In a word, the classical approach first seeks approximate formulae similar to constitutive equations of continuum mechanics and then proceeds to use them as if they were indeed constitutive equations.

The approximative process commonly applied is that of ENSKOG as presented by CHAPMAN & COWLING and others. Those who use that process seem never to question the justice of their steps. Because the process itself is extremely complicated, any consideration of its general term has always seemed beyond reach. In Chapter XXV we have seen that for a gas of Maxwellian molecules the results of atemporal Maxwellian iteration agree with those of the method of ENSKOG as far as the latter has been carried. Atemporal Maxwellian iteration exploits the simplifications afforded by Maxwellian molecules and is easy to carry through to any stage desired. In Chapters XIV and XV we have obtained general solutions for the pressures in three particular classes of homo-energetic flow of a Maxwellian gas: simple shearing, dilatation, and extension. We have thus provided everything necessary *to estimate precisely the worth of the classical approach*, though only for these three flows of a Maxwellian gas.

---

[2] GALKIN [1964] asserted that for these flows atemporal Maxwellian iteration converged "at least" when $|\mathsf{T}| < 2(\sqrt{10} - 1)/9 \approx 0.48$, but he supplied no proof.

[3] The analysis for simple shearing was given by TRUESDELL [1956, §§40–42]. The results we give for extension are of the same kind.

In homo-energetic dilatation we have shown that atemporal Maxwellian iteration provides the gross determiner exactly, from the very first step on. Since this determiner is $\mathbf{P}^G = \mathbf{0}$, $\mathbf{q}^G = \mathbf{0}$, it corresponds exactly to the Stokes–Kirchhoff theory and hence provides an example in which the classical approach leads to correct results, but trivial ones.

In Chapters XIV and XV we have shown that in homo-energetic simple shearing and homo-energetic extension each *exact solution* for p and $\mathbf{P}$ has a *dominant part* p^D and $\mathbf{P}^D$. These dominant parts are determined by the initial values $p(0)$ and $\mathbf{P}(0)$ through a constant A whose representation in terms of these initial values, they being held constant, is given as $\mathsf{T} \to 0$ by

$$A = \begin{cases} 1 - \dfrac{2}{3}\dfrac{P_{12}(0)}{p(0)}\mathsf{T} + O(\mathsf{T}^2) & \text{for shearing;} & \text{(XIV.48)}_r \\ 1 - \dfrac{1}{2}\dfrac{P_{11}(0)}{p(0)}\mathsf{T} + O(\mathsf{T}^2) & \text{for extension.} & \text{(XV.46)}_r \end{cases}$$

We showed also that for each of these flows there was a class of *principal solutions* p^P and $\mathbf{P}^P$, solutions to which all exact solutions were asymptotic as $t \to \infty$. We may summarize the relation between an arbitrary set of dominant parts and an arbitrary set of principal solutions by writing (XIV.8), (XIV.43), (XV.34), and (XV.39) in the forms

$$p^D(t) = A \frac{p(0)}{p^P(0)} p^P(t),$$

$$P^D_{km}(t) = A \frac{p(0)}{p^P(0)} P^P_{km}(t) = A \frac{p(0)}{p^P(0)} p^P(t)\varpi^G_{km}(\mathsf{T});$$

(XXVI.26)

here $\varpi^G \equiv \mathbf{P}^G/p$. We see that to obtain the principal solution which agrees with a given set of dominant parts, we should have to take the initial value $p^P(0)$ to be given as follows *by the full set of initial values* $p(0)$ *and* $\mathbf{P}(0)$:

$$p^P(0) = Ap(0). \qquad \text{(XV.41)}_r$$

Were we to attempt to use the principal solutions in carrying out the classical approach to approximate solution, we should not choose the initial value of p^P according to (XV.41), but rather in agreement with the exact data of the problem:

$$p^P(0) = p(0). \qquad \text{(XXVI.27)}$$

Indeed, we could scarcely do otherwise, since A is determined through initial values that the traditional approach of gas dynamics would regard as unnecessary to determine or perhaps even undeterminable in practice. Because $A \neq 1$ except for special choices of those initial values, namely, those that determine the principal solution itself, we see from (XXVI.26) that *in the classical*

(v) FAILURE OF THE CLASSICAL APPROACH

approach the principal solution is almost always in error by a multiplicative factor. However, if we make an asymptotic comparison as $\mathsf{T} \to 0$ at a fixed value of t, from (XIV.48) and (XV.46) we see that $A \to 1$ with an error generally $O(\mathsf{T})$. Since in this same limit ϖ^G is asymptotic to its value according to the Stokes-Kirchhoff constitutive relation, namely, $-2(\mu/p)\mathbf{E}$, *the Stokes-Kirchhoff solution is valid asymptotically; the error is not 0 except for specially selected initial conditions; generally it is $O(\mathsf{T})$ and not $O(\mathsf{T}^2)$*.

All this we obtained in Chapters XIV and XV. When interpreted in terms of the iterative procedures for approximating $\mathbf{P}^G$ and $\mathbf{q}^G$, it suggests that *the classical approach yields false results except insofar as it confirms the Stokes-Kirchhoff theory*—correct results if $n = 1$, false results if $n > 1$. We now address ourselves to proof of this fact.

For homo-energetic simple shearing the first few iterates delivered by atemporal Maxwellian iteration[4] (XXVI.15) are as follows[5]:

$$\varpi_{12}^{[1]} = -\mathsf{T}, \qquad \varpi_{22}^{[1]} = 0,$$

$$\varpi_{12}^{[2]} = -\mathsf{T}, \qquad \varpi_{22}^{[2]} = -\tfrac{2}{3}\mathsf{T}^2, \qquad \text{(XXVI.28)}$$

$$\varpi_{12}^{[3]} = -\mathsf{T}(1 - \tfrac{4}{3}\mathsf{T}^2), \qquad \varpi_{22}^{[3]} = -\tfrac{2}{3}\mathsf{T}^2.$$

The corrections added at each stage alternate between the shearing pressure and the deviatoric normal pressure. The first stage is the Stokes-Kirchhoff approximation; the second stage is the Burnett approximation. As we remarked in Section (i), $\mathbf{q}^{[n]} = \mathbf{0}$ for all n.

The classical approach to approximate solution substitutes $\varpi^{[n]}$ and $\mathbf{q}^{[n]}$ into the equations of balance of mass, momentum, and energy and solves the resulting initial-value problem. For homo-energetic simple shearing the equations of balance reduce to (II.61) alone. Writing $^{[n]}p$ for the solution of that equation when P_{12}/p is replaced by $\varpi_{12}^{[n]}(\mathsf{T})$, we have

$$^{[n]}p = p(0)\,\exp[-\tfrac{2}{3}\mathsf{T}\varpi_{12}^{[n]}(\mathsf{T})t/\tau]. \qquad \text{(XXVI.29)}$$

Thus the corresponding solution $^{[n]}\mathbf{P}$ for $\mathbf{P}$ is given by

$$^{[n]}P_{km} = {}^{[n]}p(t)\varpi_{km}^{[n]}(\mathsf{T}),$$
$$= p(0)\varpi_{km}^{[n]}(\mathsf{T})\,\exp[-\tfrac{2}{3}\mathsf{T}\varpi_{12}^{[n]}(\mathsf{T})t/\tau]. \qquad \text{(XXVI.30)}$$

---

[4] Now, for brevity, we drop from the expression for $\varpi^{[n]}$ delivered by (XXVI.15) all terms of higher order in τ than τ^n.

[5] We omit here the analysis of other variants of Maxwellian iteration carried out by TRUESDELL [1956, §§35, 43]. He pointed out that some of them provide much better approximations at the first two or three steps. In Section (viii) of Chapter XXV we have presented and extended his remarks on the Maxwellian iterates with time derivatives left just as they come. Other variant iterations yield second iterates which give the correct qualitative behavior of all components of ϖ in the entire range $0 \leq \mathsf{T} < \infty$. *Cf.* in particular TRUESDELL's Figures 35.1 and 35.2.

We compare these results with the corresponding dominant parts of the exact solution, given by (XIV.8), and so obtain

$$\frac{^{[n]}p}{p^D} = \frac{1}{A} \exp\{[-\tfrac{2}{3}T\varpi_{12}^{[n]}(T) - R]t/\tau\},$$

$$\frac{^{[n]}P_{12}}{P_{12}^D} = -\frac{\tfrac{2}{3}T\varpi_{12}^{[n]}(T)}{AR} \exp\{[-\tfrac{2}{3}T\varpi_{12}^{[n]}(T) - R]t/\tau\}, \qquad \text{(XXVI.31)}$$

$$\frac{^{[n]}P_{22}}{P_{22}^D} = -\frac{(1+R)\varpi_{22}^{[n]}(T)}{AR} \exp\{[-\tfrac{2}{3}T\varpi_{12}^{[n]}(T) - R]t/\tau\}.$$

For the case in which $n = 1$ we have obtained these results already, recording the second of them as (XIV.16)$_2$. In doing so we remarked that $R < \tfrac{2}{3}T^2$ and hence the exponential on the right-hand side tends to ∞ as $t \to \infty$. The same is true of (XXVI.31)$_1$ and (XXVI.31)$_3$ as well. It is not difficult to see that $-\tfrac{2}{3}T\varpi_{12}^{[n]}(T) - R$ is positive when n, for some positive integer m, has the value $4m - 3$ or $4m - 2$; negative when $n = 4m - 1$ or $4m$. Thus we can summarize the behavior of the approximation $^{[n]}\mathbf{M}$ as $t \to \infty$ in a single formula: For those components M_{km} that are not identically zero, as $t \to \infty$

$$\left| \frac{^{[n]}M_{km}(t)}{M_{km}(t)} \right| \to \text{either } \infty \text{ or } 0. \qquad \text{(XXVI.32)}$$

We have written $\mathbf{M}$ in the denominator instead of $\mathbf{M}^D$ because the two differ by a quantity that tends to 0 as $t \to \infty$. To avoid trivial exceptions, we assume also that $T \neq 0$. According to (XXVI.32), *for each fixed n the proportional error of the "approximate" solution $^{[n]}\mathbf{M}(t)$ becomes infinite as $t \to \infty$.* If $T < \sqrt{2/3}$, we know that $-\tfrac{2}{3}T\varpi_{12}^{[n]}(T) - R \to 0$ as $T \to 0$, so the interval of t in which the dependence of $^{[n]}\mathbf{M}(t)$ upon time is of the right kind becomes longer as n grows larger.

Even so, there is the disturbing factor A, which depends upon the initial conditions and so differs from one true solution to another. To try to eliminate it, we hold t/τ fixed and consider the limit as $T \to 0$, of course making the choice $^{[n]}p(0) = p(0)$. The result for the first stage we have obtained already and recorded as (XIV.19); we may rewrite it as follows: As $T \to 0$,

$$\frac{^{[1]}p(t)}{p^D(t)} \to 1, \qquad \frac{^{[1]}P_{12}(t)}{P_{12}^D(t)} \to 1. \qquad \text{(XXVI.33)}$$

The same cannot be said of all components of $\mathbf{P}$, since $^{[1]}P_{22}(t) = 0$ for all t, but generally $P_{22}(t) \neq 0$. We expect to find adjustment for this fact at the next stage, and we do:

$$\frac{^{[2]}P_{22}}{P_{22}^D} = -\frac{(1+R)(-\tfrac{2}{3}T^2)}{AR} \exp[O(T^4)t/\tau]. \qquad \text{(XXVI.34)}$$

(v) FAILURE OF THE CLASSICAL APPROACH

Use of (XIV.5) and (XIV.48) shows that

$$\frac{{}^{[2]}P_{22}}{P_{22}^{D}} \to 1 \quad \text{as} \quad \mathsf{T} \to 0. \tag{XXVI.35}$$

The same may be shown for the other normal pressures. Of course we may replace ${}^{[1]}P_{12}$ by ${}^{[2]}P_{12}$ in (XXVI.33)$_2$, because $\varpi_{12}^{[2]} = \varpi_{12}^{[1]}$, as (XXVI.28)$_{1,3}$ show.

We cannot get results any better than these by appeal to the higher iterates. Indeed, if we substitute (XIV.48) into (XXVI.31)$_1$, we obtain as $\mathsf{T} \to 0$

$$\frac{{}^{[n]}p}{p^D} = 1 + \frac{2}{3}\frac{P_{12}(0)}{p(0)}\mathsf{T} + O(\mathsf{T}^2). \tag{XXVI.36}$$

To increase n does not improve the asymptotic approximation. A glance at (XXVI.31)$_{2,3}$ shows that the same conclusion holds for the components P_{km} as soon as we reach an n such that $\varpi_{km}^{[n]} \neq 0$. To within $O(\mathsf{T}^2)$, *the initial conditions for the pressure deviator do not drop out of the problem.* For the component P_{12} the n in question is 1; for the components P_{22}, P_{33}, and P_{11} it is 2, and for that reason the Burnett approximation does give an improvement over the Stokes-Kirchhoff approximation when applied to simple shearing. No higher iterate will, as far as the dependence upon the initial conditions through the ratio $P_{12}(0)/p(0)$ is concerned. The foregoing conclusions are summarized in the following **Theorem**: *If* ${}^{[2]}p(0) = p(0)$ *and if the ratios* $M_{22}(0)/p(0)$, $P_{12}(0)/p(0)$, *and* t/τ *are held fixed, then the " Burnett" solution* ${}^{[2]}\mathbf{M}(t)$ *of the initial-value problem for homo-energetic simple shearing is asymptotically correct as* $\mathsf{T} \to 0$ *with fractional error* $O(\mathsf{T})$; *the solution* ${}^{[n]}\mathbf{M}(t)$ *if* $n > 2$ *has exactly the same asymptotic status, no better.*

Turning now to homo-energetic extension, from (XXVI.21) we find the first two iterates:

$$\varpi_{11}^{[1]} = -\mathsf{T}, \qquad \varpi_{12}^{[1]} = 0,$$
$$\varpi_{11}^{[2]} = -\mathsf{T} + \tfrac{1}{2}\mathsf{T}^2, \qquad \varpi_{12}^{[2]} = 0. \tag{XXVI.37}$$

Now the corrections provided at the succeeding stages do not alternate between shear pressures and normal pressures as they did for shearing, so we cannot expect any improvement in the solution as a function of time by passing from the Stokes-Kirchhoff approximation to the Burnett approximation. To obtain that solution, we substitute $\varpi_{11}^{[n]}$ for P_{11}/p in (II.70)$_2$ after putting $\gamma = \tfrac{5}{3}$. The result is

$$^{[n]}\varepsilon = \varepsilon(0)\left(1 + \frac{t}{T}\right)^{-\tfrac{2}{3}[1 + \varpi_{11}^{[n]}(\mathsf{T})]} \tag{XXVI.38}$$

With the aid of (II.67) we obtain

$$^{[n]}p = p(0)\left(1 + \frac{t}{T}\right)^{-[(5/3)+(2/3)\varpi_{11}{}^{[n]}(T)]} \tag{XXVI.39}$$

Dividing this equation by (XV.34)$_1$ yields

$$\frac{^{[n]}p}{p^D} = \frac{1}{A}\left(1 + \frac{t}{T}\right)^{-[(5/3)+(2/3)\varpi_{11}{}^{[n]}(T)+R]} \tag{XXVI.40}$$

From this formula we may draw conclusions similar to those based on (XXVI.31)$_1$. The dependence upon time becomes more and more nearly correct for small values of t/T as $n \to \infty$, but the error ultimately becomes infinite; here "ultimately" means as $t/T \to \infty$ if $T > 0$, as $t/T \to -1$ if $T < 0$. The asymptotic formula as $T \to 0$ when t/T is fixed follows from (XV.46):

$$\frac{^{[n]}p}{p^D} = 1 + \frac{1}{2}\frac{P_{11}(0)}{p(0)}T + O(T^2), \tag{XXVI.41}$$

similar to (XXVI.36), and again *to increase n does not improve the asymptotic approximation*. From (XXVI.21) it is clear that for no choice of k and m such that $\varpi_{km}^{[1]} = 0$ is $\varpi_{km}^{[n]} \neq 0$ for any n, so the improved approximation at the Burnett stage which we found for simple shearing has no counterpart here. The foregoing conclusions are summarized in the following **Theorem**: *If* $^{[1]}p(0) = p(0)$ *and if the ratios* $P_{11}(0)/p(0)$ *and* t/T *are held fixed, then the Stokes–Kirchhoff solution* $^{[1]}\mathbf{M}(t)$ *of the initial-value problem for homo-energetic extension is asymptotically correct as* $T \to 0$ *with fractional error* $O(T)$; *the solution* $^{[n]}\mathbf{M}(t)$ *if* $n > 1$ *has exactly the same asymptotic status, no better.* The first part of this theorem merely reaffirms a result obtained in Chapter XV.

The convergence theorems proved in Sections (iii) and (iv) of this chapter show that as $n \to \infty$

$$\begin{array}{llll} \text{for shearing} & -\tfrac{2}{3}T\varpi_{12}^{[n]}(T) - R \to 0 & \text{if} & T < \sqrt{2/3}, \\ \text{for extension} & -\tfrac{5}{3} - \tfrac{2}{3}\varpi_{11}^{[n]}(T) - R \to 0 & \text{if} & |T| < \tfrac{2}{3}. \end{array} \tag{XXVI.42}$$

Looking back at (XXVI.31) and (XXVI.40), we see that since the functions $x \mapsto e^x$ and $x \mapsto (1 + t/T)^x$ are continuous, for both shearing and extension it follows from (XXVI.42) that for fixed t and fixed initial conditions, provided T be small enough,

$$\lim_{n \to \infty} {}^{[n]}M_{km}(t) = \frac{1}{A} M_{km}^D(t) = M_{km}^P(t), \tag{XXVI.43}$$

the solution $\mathbf{M}^P$ being the principal solution as given by (XIV.43), (XV.39), and (XXVI.27), namely, the solution that corresponds to initial values so chosen that $A = 1$ and, for shearing, $B = C = 0$, with the special initial conditions (XXVI.27). If T is too large, on the other hand, (XXVI.42) fails, and the quantities written before the arrows approach no limits. Thus (XXVI.43) also fails.

(v) FAILURE OF THE CLASSICAL APPROACH   555

Summarizing these facts, we have the following **Convergence Theorem**: *For homo-energetic simple shearing and for homo-energetic extension there are constants* T_0 *such that if* $|T| < T_0$ *the respective functions* $^{[n]}P(t)$ *that correspond to the atemporal iterate* $\varpi^{[n]}(T)$ *converge to the principal solutions* $P^P(t)$ *as* $n \to \infty$; *if* $|T| > T_0$, *the functions* $^{[n]}P(t)$ *do not approach limits as* $n \to \infty$. *For the two flows* $T_0 = \sqrt{2/3}$ *and* $\frac{2}{3}$, *respectively.*

There is nothing surprising about this theorem. It merely asserts legitimate the interchange of two limits, one being the limit as the iteration proceeds, the other being the integral that expresses the solution of the differential equation of energy when $\varpi^{[n]}$ is substituted in it for the exact ratio ϖ. The result itself rounds out nicely the program stated at the beginning of this section, for it shows that the failure of the classical approach lies not in the iterative process but in the fact that what it delivers is a special solution. What the classical approach seeks to calculate, namely, a solution of gas-dynamic type such as to approximate *all* the solutions that correspond to a given flow, *does not exist* except in the limit of an asymptotic process as $\mathfrak{m} \to 0$. What this section has shown is that in general *the best possible results in that limit are obtained already at the Stokes-Kirchhoff stage*—that use of higher iterates does not improve them[6].

---

[6] What may be a contrary opinion has been proclaimed by GRAD [1963, *1*, p. 149]:
The apparent ineffectiveness of the Burnett equations in describing departures from ordinary continuum mechanics is evidently a consequence of the fact that the Chapman-Enskog expansion is asymptotic rather than convergent (see end of Sec. XI). The series must always be truncated. Given the value of ε [the parameter we have denoted by δ in the Hilbert expansion (XXII.1)], there is an optimum number of terms which number increases as ε is taken smaller. Therefore the Burnett and higher approximations do *not* serve to extend the range of validity farther from the fluid regime. On the contrary, the higher approximations can only be used when the Navier-Stokes equations are already very good (ε is small) in order to give an ultra-refined fluid description.

We confess ourselves unable to understand what this statement means or to find a mathematical basis from which we could conjecture an assertion of this kind. We claim to have proved that both in homo-energetic simple shearing and in homo-energetic extension atemporal Maxwellian iteration for **P** *converges* for small enough **T**; that it converges to the gross determiner of **P**; and that use of any finite iterate $P^{[n]}$ as if it were a constitutive relation of continuum mechanics does *not* yield for the corresponding time-dependent *solution* a result any better asymptotically as $\mathfrak{m} \to 0$ than does $P^{[1]}$. For dilatation the iteration converges trivially.

GRAD's opinion has propagated. According to FERZIGER & KAPER [1972, §5.8], GRAD [1958] "has shown that the Chapman-Enskog process generates a series which is asymptotic in ... the ratio of the mean free-path and a characteristic dimension of the system ..." On the contrary, GRAD himself [1958, §22] describes his own considerations in this regard [1958, §26] as "a mathematically heuristic asymptotic evaluation". FERZIGER & KAPER continue: "It is likely, though unproven, that this series does not converge for any ε."

Of course we do not claim to have proved anything on this subject except for the two special flows. Our results are theorems. Like all theorems, they present precisely delimited conclusions from precisely stated assumptions. We feign no generality. We print no heuristics.

On the other hand, unless some error shall be found in our analysis, we cannot see justice in blanket assertions of what seems to be just the opposite of the behavior we have demonstrated in two special cases.

(vi) Retrospect

Circles in which "physical intuition" is a generic substitute for mathematical proof mumble a Pavlovian comfit to the effect that the iterative processes of the kinetic theory "cannot converge" but are "only asymptotic". To what they cannot converge and to what they are only asymptotic is said to be a "solution" but is otherwise left unspecified.

In this book we have distinguished all along between a *solution* of the Maxwell–Boltzmann equation, the *gross determiner* of a solution or a class of solutions, and the *momental determiner*. The first is a function of time, presumably rendered unique by specifying certain initial values. The second is a *time-independent mapping of the fields of principal moment and body force*; in the special cases treated in this chapter, that mapping reduces to a function of ρ, ε, and grad **u**.

The theorems on the existence of limits proved in the preceding sections of this chapter are *the only presently published convergence theorems for any iterative process in the kinetic theory*. For the extremely special flows of special gases they concern, they provide a certain status for atemporal Maxwellian iteration, which, as we have seen in Chapter XXV, seems to be an alternative formulation of the method of stretched fields, with which we have replaced ENSKOG's process. *That status refers to the gross determiner alone*. Specifically, the atemporal iterates $\mathbf{P}^{[n]}(\rho, \text{grad } \rho)$ *converge*[7] to the gross determiner $\mathbf{P}^G(\rho, \text{grad } \rho)$ if ɱ is sufficiently small.

The classical approach to the kinetic theory regards the iterates $\mathbf{P}^{[n]}(\rho, \text{grad } \rho)$ as approximate constitutive relations of continuum mechanics. Treating them as such, it places them into the equations of balance of mass, momentum, and energy so as to obtain "approximate" solutions $^{[n]}\mathbf{M}(t)$ to the initial-value problem of the kinetic theory. We have shown that such results are grievously in error in two different senses:

(a) As t increases, the ratio of every non-zero component $^{[n]}M_{km}(t)$ to any corresponding non-null exact solution $M_{km}(t)$ becomes either null or in absolute value infinite. Thus the iterative *solutions* $^{[n]}\mathbf{M}(t)$ do *not* represent well the behavior of exact solutions over *long* periods of time. On the contrary, as n increases the ratio $^{[n]}M_{km}(t)/M_{km}(t)$ becomes more nearly constant over any given *short* period of time, provided ɱ be small enough. These conclusions are strictly contrary to the remarks often offered in support of ENSKOG's process.

---

[7] In physical circles the HILBERT expansion also is often regarded as "only asymptotic", yet most of the manipulations GRAD employed so as to draw conclusions from it, manipulations we have followed and extended in Chapter XXII, presume that expansion convergent for all values of δ. Such is indeed the case in the one non-trivial example of a complete HILBERT expansion ever produced, namely (XXII.45), and for some choices of the constants $\rho^{(n)}(0)$ in the essentially trivial example (XXII.44).

(b) At a *fixed* time, in the limit as $\mathfrak{m} \to 0$, the initial conditions being held fixed, we find that the ratio of any non-vanishing component $^{[1]}M_{km}$ to the dominant part of the corresponding exact solution, if that is not 0, tends to 1. That is, *the Stokes-Kirchhoff constitutive relation* $\mathbf{P}^{[1]}(\mathbf{\rho}, \text{grad } \mathbf{\rho})$ *provides solutions* $^{[1]}\mathbf{M}(t)$ *that are universally asymptotic to all corresponding exact solutions as* $\mathfrak{m} \to 0$ *while t is fixed*. The same is true of $^{[n]}M_{km}$, but *the degree of approximation is not improved by increasing n*. Indeed, the principal solution $\mathbf{M}^P(t)$, which corresponds to initial conditions compatible with the gross determiner $\mathbf{P}^G$, does not approximate the exact solutions of the problem except to within an error of order 1 in $\mathfrak{m}$ as $\mathfrak{m} \to 0$. The "corrections" to the solution according to ordinary gas dynamics which the kinetic theory is often claimed to provide are simply impossible, for *ordinary gas-dynamic data are insufficient to determine the dominant parts* $\mathbf{P}^D(t)$ *of the exact solutions*. Thus *the objective of the classical approach to the kinetic theory is unattainable*, not from any defect in the methods of approximation proposed, but from inherent properties of the kinetic theory itself.

Conclusions of this kind, based on analysis of simple shearing, were published by TRUESDELL in 1956. The literature of the kinetic theory has ignored them[8]. Thinking that the reason might be no more than kind reluctance to point out some error, we have checked the details carefully; we have found not only that they are correct but also that the same conclusions follow from analysis of GALKIN's solution for extension.

---

[8] *Cf.* p. 54 of S. G. BRUSH, *Kinetic Theory* 3, Oxford *etc.*, Pergamon, 1972: "Truesdell's campaign against the classical ... approach ... has not won very many converts. It has been largely ignored by physicists ...."
  Indeed, GRAD has simply upstaged the analysis we have reviewed and extended in this chapter:
  1. GRAD [1964, *1*, p. 8]:
     ... we recall that almost all the extrapolations that were drawn from a very ingenious explicit solution of an exact nonlinear Boltzmann equation have been found to be nonrepresentative of the behavior of the typical solution.
  The reference adduced to support this "recall" is to an entire article in the *Handbuch der Physik*: GRAD [1958]. There the only notices of TRUESDELL's work are on pp. 264 and 291; both of these present aspects of that work correctly, and neither of them claims or gives any evidence that the "extrapolations" were "found to be nonrepresentative". The brief references to TRUESDELL's work by GRAD [1960, pp. 116 and 121] do not mention anything "nonrepresentative"
  2. GRAD [1964, *2*, p. 6]:
     But, after development of a comprehensive theory of the linear equation, it is found that the conclusions that were drawn on the basis of this explicit solution are completely misrepresentative of the general case.
  The reference adduced this time is GRAD [1963, *1*]. The only notice given to TRUESDELL's work there is to support the following sentence in §1, which we have already quoted in Footnote 15 to Chapter XIV:
     As observed by Hilbert, the quantities which arise in his theory are formally causal; their initial values uniquely determine their future. But it is a misinterpretation (found in the usual presentations of the Chapman-Enskog theory) to deduce from this that the fluid state is causal, even asymptotically.

(continued)

One counterexample suffices to disprove a general claim. Here we have two. Of course, counterexamples may refer to degenerate or exceptional cases only, as does the fact that the even number 2 is a prime, while all other primes are odd. We cannot deny the possibility that for flows more typical than homo-energetic shearing and homo-energetic extension the objections we have found to the results of the iterative processes might fail. On the other hand, usually a general property is trivially true, or almost so, for the simplest examples; the simplest examples often point the way to the general case; and the failure of a general claim when applied to simple cases often indicates that the claim is false.

Be all this as it may, the literature of this aspect of the subject is rich in affirmation and intuition, but beyond the results presented in this chapter and the single example provided in Chapter XXII we have been able to find nothing else rigorously proved in regard to the convergence, divergence, and significance of processes of approximate calculation of gross or momental determiners in the kinetic theory.

With one interpretation of the vague word "causal", this conclusion from TRUESDELL's work is correct except for the final phrase "even asymptotically".

In a later paper GRAD [1969, p. 305, cf. also p. 279] speaks of "an endless variety of boundary layers and nonuniform limits" but does not mention TRUESDELL's work.

As in his condemnations GRAD does not even cite correctly the paper he criticizes, some doubt as to the care with which he has studied it might be raised. Nevertheless, secondary authors have echoed his conclusions.

Epilogue

Among the few theories that have proved their worth again and again by furnishing decent models for flows of real gases, MAXWELL's kinetic theory stands as one of the few whose basic structures spring directly from a molecular picture: a throng of tiny, rare mass-points whose motions, randomly initiated and very swift, are governed by Newtonian mechanics. Its singular beauty grows from the connection it provides between the simplicity of the molecule and the complex variety of gas flow.

We have treated MAXWELL's theory as a branch of rational mechanics. We have made searching analysis of the basic structures prerequisite to probing the theory for such conclusions as it may provide. Only upon that foundation can the student expect to trace those conclusions to their roots, to extend them, and to refine them.

Continuum thermomechanics and the kinetic theory spring from their respective generic axioms. For each, these embrace all bodies which it is designed to model; they incorporate both its generality and its inherent limitations; and so they give birth to the discipline based upon it. In continuum thermomechanics they assert the balance of mass, linear momentum, rotational momentum, and energy, channeled by some statement of irreversibility such as a dissipation inequality. In the kinetic theory they are the equation of evolution (III.38) and MAXWELL's choice (VII.2) for the collisions operator $\mathbb{C}$. Each theory adjoins to these axioms certain further conditions which define ideal materials, of which bodies are composed. In continuum mechanics these take the form of constitutive relations which connect the scalar pressure p, the pressure deviator $\mathbf{P}$, the energy-flux vector $\mathbf{q}$, the energetic ε, and the caloric η with the fields of density ρ, velocity $\mathbf{u}$, and temperature θ. In the kinetic theory the constitutive quantities are the molecular mass $\mathbb{m}$, the encounter operator $\mathbb{E}$, and the cross-section $\mathscr{S}$. Both theories, however, have the same ultimate objective, namely, to determine the fields ρ, $\mathbf{u}$, and θ appropriate to particular circumstances: typically, particular initial conditions and particular confining walls.

In this book we have treated only a simple gas: a gas consisting of molecules which are single mass-points, all alike. As this case is the easiest to handle mathematically as well as the simplest in concept, we have reason to hope that many results concerning it will have analogues for mixed gases of more complicated molecules. It might be expected that a comprehensive treatise on the simplest case would provide solutions of all the major problems of the theory. The opposite, however, is true. Though some problems have surrendered to analysis in the 113 years since MAXWELL shaped and promulged this theory, the major ones remain open today. Besides assembling, organizing, and extending what is now known, we have striven also and above all to discern and indicate them. Therefore it is fitting that we recall them here, state them here in general terms, and for some of them conjecture here, on the basis of the partial knowledge gathered in the preceding pages and there exposed, what may be their solutions. We call them "main", for any competent reader will have noticed many lacunae and many problems of the kinds more or less common in analysis today, and we expect that publication of this book will draw attention to them with no further remark on our part.

First Main Open Problem: *Existence theory.* The first main open problem is to prove that positive, classical solutions of the Maxwell–Boltzmann equation corresponding to specified initial conditions and boundary conditions exist and are unique. Of the problems we state in this epilogue, this one has drawn most attention from mathematicians and has yielded most to analysis. For spatially homogeneous solutions ARKERYD's results, which we have presented and developed in Chapter XXI, go a long way toward solving the problem when it is restricted to a gas of molecules with a cut-off. TRUESDELL's existence theorem in Chapter XVIII concerns a gas of Maxwellian molecules; it refers only to the equations for the moments, assumed spatially homogeneous. The solutions of major interest, or course, are place-dependent. For gas flows in infinite space, weak solutions have been proved to exist for a finite interval of time and to be unique (Chapter XX), but again only for molecules with a cut-off. Whether or not they are classical solutions remains unknown in general. GRAD's solutions, described in Chapter XX but not presented in detail, are regular, but they pertain only to gas flows which are nearly uniform. Some results have been published very recently for flows within confining vessels; these, too, we have cited in Footnote 1 to Chapter XVIII and in Section (ii) of Chapter XX.

Much effort has been spent toward proof that place-dependent solutions exist for all time. Some results of this kind we have cited in Section (ii) of Chapter XX in connection with GRAD's work. However, that in general the interval of existence be infinite, is too much to expect. We know exact solutions which fail to exist after a finite time (*cf.* (XVIIB.8), (II.39), and (XVIIB.13) when $T < 0$). There is nothing surprising here, for the same is true in ordinary gas dynamics. The main problem really is to discover and specify the circumstances that give rise to solutions which persist forever. Only after having done that can we expect to construct proofs that such solutions exist, are unique, and are regular. This problem touches upon the theories of stability, bifurcation, singularity, and blow-up. As far as we know, analysis of this kind for the kinetic theory wants altogether. Despite much progress for incompressible Navier–Stokes fluids, such analysis wants equally for the Navier–Stokes theory of viscous gases, a theory presumed to be less obstinate to mathematics than is the kinetic theory.

For molecules with intermolecular forces of infinite range, apart from the very special results given in Chapter XVIII and PAO's analysis of the operator $\mathbb{C}(F_M, \cdot)$ (cf. Footnote 6 to Chapter XXII), nearly all analytical questions remain open. Only too often these are swept away by a claim that such molecular models lie outside the basic physical motivation commonly put forward in setting up the theory (cf. Footnote 4 to Chapter VII). It is to them that most calculations by physicists and chemists refer, including all of those by MAXWELL himself, and to exclude them would therewith exclude also nearly all explicit, exact results so far achieved, namely, those for a gas of Maxwellian molecules (Chapters XII–XV).

Despite the body of results we now have, existence theory for the Maxwell–Boltzmann equation may still be in its infancy. Much of the recent analysis is very technical; often we struggle in vain to see clearly, if at all, which properties be responsible for such existence, uniqueness, and regularity as it does establish. The very fact that the mathematics is so difficult, so intricate, and so beset by niggling conditions suggests that perhaps analysis has not yet found the best language in which to describe these problems. Perhaps the main stumbling block, namely the complexity of the operator $\mathbb{C}$, may in time be overcome by reduction to simple statements about the regularity, integrability, *etc.*, of some general function defined on a suitable function space. The convergence theorems of Chapter XIX provide examples of work in this direction, but perhaps the best is GRAD's determination of the structure of the linearized collisions operator (Footnote 6 to Chapter XXII), which a number of people have used as a stepping stone toward analysis of $\mathbb{C}$ itself. Certainly work of this kind makes it easier for analysts to get a handle on the kinetic theory and perhaps to sense its peculiar charm.

Second Main Open Problem: *Asymptotic trend to a grossly determined state.* It is widely believed that the kinetic theory includes and delivers a continuum theory of fluids. Indeed, in forming the kinetic theory MAXWELL had no other purpose than to extract gross information from it. Older authors reconciled their view with contrary indications by appeal to MAXWELL's relaxation theorem (Section (ii) of Chapter XIII), upon which, it seems, they based one or another vague claim that the kinetic gas if "left to itself" for a while would approach a state determined by its own fields of density, gross velocity, and temperature. Hoping to prepare the way toward framing such a conjecture precisely, we have introduced the concept of a *grossly determined solution*: a solution which is determined at any given instant by the gross condition of the gas at that instant. Chapter XXIII develops MUNCASTER's general theory of gross determiners, the mappings whose values are grossly determined solutions. This theory is one of the two main departures that this book provides from all previous work on the kinetic theory.

HILBERT's formal iterative procedure supports the existence of a broad class of grossly determined solutions (Section (vi) of Chapter XXII). In Chapter XXIV we have introduced a method of functional approximation, the method of stretched fields. This is our second main departure. When applied to the theory of gross determiners developed in Chapter XXIII, it supports the existence of those determiners, at least when ρ, **u**, and ε are nearly uniform and **b** is near to **0** (Sections (vii) and (x) of Chapter XXIV). The exact solutions for moments which we have presented in Chapters XIII–XV provide examples of such solutions when **b** = **0** and moreover suggest that for small enough values of the tension number T general solutions are asymptotic in time to some member of a grossly determined class. Since the tension number is small whenever the field **u** is nearly uniform (cf. (IV.10), and in the examples ρ and ε are precisely uniform), we may be led to extrapolate these results into the following

Principal Conjectures. *If the fields of density, gross velocity, and energetic corresponding to a solution F of the Maxwell–Boltzmann equation lie in sufficiently small neighborhoods of constant fields, and if the body force is sufficiently close to* **0**, *then*

1. *For large times the fields* **P**, **q**, *h, and* **s** *for F have asymptotic forms which are grossly determined.*

2. *To at least one of the sets of four asymptotic forms which Conjecture 1 asserts to exist correspond infinitely many solutions F which have asymptotically a common gross condition* ρ, **u**, ε, *and* **b**.

3. *Among the infinitely many solutions to which Conjecture 2 refers there is exactly one, a grossly determined one, for which the asymptotic approximations which Conjecture 1 asserts to exist are also the exact values of* **P**, **q**, *h, and* **s**, *respectively, for every t in the interval for which the particular solution exists*[1].

---

[1] The cautious wording in these conjectures reflects what we have learned from examples, some of which the reader should recall here. The "principal solutions" in Chapters XIV and XV are examples of the particular solutions to which Conjecture 3 refers. It might be that F itself would have the properties asserted by the three conjectures, but such need not be the case. The free locally Maxwellian solutions (Chapter X) are grossly determined and approach grossly determined limits, but if ε is not constant, those limits are not molecular densities. MUNCASTER's exact solution exhibited in Appendix B to Chapter XVII is not grossly determined and with

The principal conjectures are both less pontifical and more specific than the claims of earlier authors, but nevertheless they are still vague. We prefer to conjecture that an asymptotic form exist for "large times" rather than in the limit as $t \to \infty$ because we have exhibited solutions which exist only during a finite interval of time yet still have grossly determined dominant parts (cf. the solution for homo-energetic extension when $T < 0$). If we let the known explicit solutions teach us what to expect, "large times" here should mean times large in comparison with some time characteristic of the flow. The crude estimate of τ given just after (XIII.14) suggests that such times may be quite short on the scale of human perception, even for solutions that exist in an interval of time large on the same scale, though not infinite.

Although the principal conjectures are supported by all evidence we have been able to assemble—and by evidence we mean rigorously proved statements—general proofs for them are nowhere to be found. They remain open questions. Perhaps the vagueness of a "large" but not necessarily infinite time calls most loudly here for a precise replacement. In presenting the exact solutions for moments in Chapters XIV and XV we have avoided this difficulty by speaking of the *dominant part* of a solution. While we have made that term precise and useful in the analysis of particular affine flows, our attempts to formulate it in general have remained frustrate.

The importance of the principal conjectures is clear. Could we substantiate them, we could thereafter replace the kinetic theory, at least in some asymptotic sense and for a broad class of flows, by a certain continuum theory, a continuum theory which the kinetic theory would deliver: the continuum theory that would result from use of the gross determiners.

Third Main Open Problem: *Interpretation of* BOLTZMANN's *H-theorem, and its bearing on the trend to equilibrium.* "If a body of gas be confined in a rigid, perfectly conducting, stationary vessel, to which it adheres at the walls, then, regardless of the initial disturbance, the gas will approach in time a state of gross rest." This statement typifies, if somewhat crudely, the famous open problem of the *trend to equilibrium*. It also points out, if again only crudely, the critical role played by the properties of the vessel in ensuring that a state of equilibrium be reached. These properties enter the theory through boundary conditions (Section (iii) of Chapter XI). If the vessel is continually

---

increasing time does not approach a grossly determined limit, yet **M** and **q** associated with it are always those of a Maxwellian density and hence are grossly determined. In Conjecture 2 we have allowed the possibility that gross determiners may fail to be unique. However, for each of the exact solutions exhibited in Chapters XIII–XV the grossly determined asymptotic forms are unique in the class of solutions considered. Moreover, the proof of effectiveness for MUNCASTER's method of stretched fields (Section (vii) of Chapter XXIV) shows that such gross determiners as have a finite Taylor expansion in terms of the gross condition of the gas have one and the same expansion in common.

shaken or if from its walls it constantly supplies energy to the gas, we ought not expect a trend in time one way or the other. Indeed, implicit in this famous open problem is the task of *delimiting circumstances in which a trend to equilibrium may be expected*. BOLTZMANN's H-theorem may provide some clues to finding such circumstances. It might be conjectured that *the validity of the narrow H-theorem is a necessary condition for an approach to equilibrium*. In Chapter XI we have described some classes of boundary conditions for which such a necessary condition would be satisfied at least formally. However, specific examples show that this condition itself is not enough. In Section (v) and (vi) of Chapter XI we have discussed at length this and other aspects of the approach to equilibrium.

The trend to equilibrium has been established in some cases. TRUESDELL's theorem in Chapter XVIII not only delivers it for the solutions of the equations for moments of a Maxwellian gas when they are assumed to be spatially homogeneous but also estimates their individual rates of decrease. For spatially homogeneous solutions in a gas of molecules with a cut-off, ARKERYD's theorems in Chapter XXI ensure it for quite general initial values, though only in the sense of weak convergence. CARLEMAN demonstrated it in the sense of uniform convergence, but only for ideal spheres and even then only for solutions of a very special type. For place-dependent solutions some statements of this kind have been proved (Section (ii) of Chapter XX), but only for particular boundary conditions and once again only for a class of molecules with a cut-off. Today we stand only at the doorway of precise statement and general proof of the trend to equilibrium according to the kinetic theory.

For circumstances more general than those in which a trend to equilibrium could reasonably be expected, the factors that produce that trend when they can do so must still manifest themselves, must effect some irreversibility of a different kind. To find it, we look at the formal broad H-theorem, but what does it mean? What does it tell us about gas flows? In Section (iv) of Chapter XI we have proved the kinetic theory consistent with continuum thermomechanics in regard to the total calory of a body confined by a vessel with linear walls, but that is far from sufficient to show even that the gas within such a vessel will tend to a state of gross rest. For gas flows not confined by walls we have scarcely any idea what kind of irreversibility the formal broad H-theorem implies (Section (vii) of Chapter XI). An example in Section (ix) of Chapter XIV shows that it does not always imply non-negative dissipation of gross energy.

We must note the relation between the principal conjectures and the trend to equilibrium. The former concern the asymptotic forms of solutions for times which are large compared with some time characteristic of the flow; the latter concerns the limit of solutions as $t \to \infty$, presuming of course that they exist for all time. Solutions might settle into a grossly determined state, at least approximately, after a time which is short on the scale of human perception and thereafter approach a state of equilibrium. Thus the principal conjectures and the trend to equilibrium concern two distinct properties of solutions, both of which may pertain in some circumstances. The behavior we have just described suggests one way in which we might prove the trend to equilibrium: First prove the existence of a grossly determined asymptotic form for the solution, thereafter use the corresponding gross determiner like a constitutive relation and thereby construct a proof of approach to equilibrium within the framework of continuum thermomechanics. The traditional objective of the kinetic theory, however, is just the opposite, namely, to establish the trend by using the kinetic theory directly, with no recourse to ideas deriving from continuum mechanics. Such few and very special theorems as we have so far are of this kind.

Fourth Main Open Problem: *Asymptotic status of the Stokes–Kirchhoff theory, and replacements for it when it is no longer valid.* A second traditional objective of the kinetic theory has been to confirm, in some asymptotic sense, classical gas dynamics. Conventionally this meant deriving, by some scheme of approximation, the Navier–Stokes–Fourier constitutive relations (I.7) and (I.8), but with the shear viscosity μ and the thermal conductivity κ delivered by the kinetic theory as certain explicit functions of temperature, functions determined by the kinetic constitutive quantities $\mathbb{m}$, $\mathbb{E}$, and $\mathscr{S}$. It was precisely for this reason that many of the iterative procedures of the kinetic theory (Part G) were first devised, and in a formal sense they were successful in achieving this aim.

However, with the continual development of more powerful methods of analysis, our expectations and hopes for the kinetic theory have risen. We cannot deny the essential service which the Stokes–Kirchhoff theory has provided in the past. By virtue of its simplicity and explicitness it has permitted solution of many problems which in the kinetic theory even today seem beyond reach. We may, however, question the Stokes–Kirchhoff theory and try to determine the range of validity not only of its constitutive relations but also of the solutions that correspond to them. Rigorous results in this direction are few. Once again we must revert to the exact solutions to find a basis for such conjectures as we might make. All of these support the expectation that *for fixed values of t and* **x**, *not only the Stokes–Kirchhoff constitutive relations but also the solutions according to the Stokes–Kirchhoff theory emerge from the kinetic theory asymptotically as the tension number* **T** *approaches* 0. This conjecture is meaningful for affine flows but requires some adjustment for more general flows (*cf.* the remark concerning **T** in Footnote 11 to Chapter XXIV). Since by (IV.10) $\mathsf{T} \to 0$ as ρ, **u**, and ε become uniform, we might expect to obtain the Stokes–Kirchhoff constitutive relations as we approach a uniform state. This conjecture is borne out formally by the method of stretched fields (*cf.* (XXIV.90)). Alternatively, $\mathsf{T} \to 0$ as $\mu \to 0$, so we might also expect agreement as μ becomes small. This is confirmed formally by atemporal Maxwellian iteration since for it the iterates are polynomial functions of μ (Footnote 21 to Chapter XXV). A third sense of asymptotic agreement is also possible. If we adopt ENSKOG's first approximation (XVII.61) for μ, then $\mu \to 0$ as we consider successively lighter gases at a fixed time: $\mathfrak{m} \to 0$. Asymptotic agreement in this sense is borne out well by all the explicit solutions which we have presented (cf. Section (iii) of Chapter XIII, Section (vi) of Chapter XIV, Section (v) of Chapter XV, and Appendix B of Chapter XVII).

We may also ask for improvements or replacements of the Stokes–Kirchhoff theory in circumstances beyond its range of validity. We might naively adopt one of the series of continuum theories which result when the successively higher iterates in some procedure for approximating gross determiners are used as if they were constitutive relations. The convergence theorems of Section (v) of Chapter XXVI, which are the only rigorous results we now have on this subject, do not bear out such an expectation. Whether the fault lie in the ground theory of gross determiners or in the special gas flows to which the theorems apply or in the way in which gross determiners are conventionally approximated, is not presently known. If the third of these possibilities does hold, we may try to improve the Stokes–Kirchhoff theory by using as constitutive relations the exact gross determiners corresponding to general grossly determined solutions. This would certainly conform with our hopes for the principal conjectures, but it may be difficult to implement in practice since it would require us to have already found exact solutions of the equations of gross determinism or of the associated equations of transfer. Only in certain special cases are we presently able to do that (Sections (ii), (iii), and (iv) of Chapter XXVI).

Beyond these possibilities we may look to the momentally determined approximations of IKENBERRY & TRUESDELL or to the 13-moment and 20-moment systems of GRAD. Their precise status in the kinetic theory are still obscure (cf. the instability of the energy flux in homo-energetic simple shearing, Section (viii) of Chapter XIV).

In other domains of mathematics both the suggestive value of descending to special cases and the danger inherent in doing so are well known. In contrast, specialists and expositors in the kinetic theory have hitherto bestowed upon the body of special explicit solutions and partial solutions published from 1956 onward scant and superficial attention if any. Esteeming the proved worth of exemplary cases in other physical theories, we have developed, completed, and extended the explicit solutions in detail (Chapters X, XIII–XV, Appendix B to Chapter XVII). More than that, we have examined their counterparts for the Stokes–Kirchhoff theory (Section (v) of Chapter II), themselves little if at all known heretofore; we have compared these counterparts with the exact solutions and other approximate solutions for the kinetic theory (Section (vi) of Chapter XIV, Sections (ii) and (v) of Chapter XV,

Section (iv) of Chapter XVII); we have developed the special case of the theory of gross determiners which applies to them (Section (viii) of Chapter XXIII); and we have examined the convergence of some of the iterative procedures of the kinetic theory when specialized to them (Chapter XXVI). In each case the calculations are explicit, the results simple and suggestive; they have provided us essential clues toward our formulation of the main problems and principal conjectures.

Each of the four main problems is not so much a single problem as a domain of inquiry, each surely rich enough to afford a lifetime of research.

List of Works Cited

The following list, not a bibliography of the kinetic theory, is limited to works which seem to us to have made a contribution, if in some cases only peripheral, to the specific subject of this book. The year under which a work is *listed* is the year of *actual* first publication. Other years mentioned in the entry are nominal years or years of publication of translations or reprints.

Works on analysis and continuum mechanics are not included here. They are cited in footnotes, as occasion arises to use them.

Roman numerals in parentheses following an entry indicate the chapters and sections of the text in which the entry is cited. Subscript 0 indicates the introduction to a chapter, before Section (i).

1867

J. C. MAXWELL, "On the dynamical theory of gases", *Philosophical Transactions of the Royal Society* (London) **157** (1866), 49–88 = (with corrections) *Philosophical Magazine* (4) **35**, 129–145, 185–217 (1868) = *Scientific Papers* **2**, 26–78 = pp. 23–87 of *Kinetic Theory* **2**, ed. S. G. BRUSH, Pergamon Press, 1966. ($I_{i,iii,iv}$, III_0, V_{iv}, $VII_{i,iv}$, VIII_{ii}, IX, X, XII, $XIII_{i,ii,iv}$)

1868

L. BOLTZMANN, "Studien über das Gleichgewicht der lebendigen Kraft zwischen bewegten materiellen Punkten", *Sitzungsberichte der Akademie der Wissenschaften, Wien* **58**$_2$, 517–560 = *Wissenschaftliche Abhandlungen* **1**, 49–96. (X)

1871

L. BOLTZMANN, "Über das Wärmegleichgewicht zwischen mehratomigen Gasmolekülen", *Sitzungsberichte der Akademie der Wissenschaften, Wien* **63**$_2$, 397–418 = *Wissenschaftliche Abhandlungen* **1**, 237–258. (X)

1872

L. BOLTZMANN, "Weitere Studien über das Wärmegleichgewicht unter Gasmolekülen", *Sitzungsberichte der Akademie der Wissenschaften, Wien* **66**$_2$, 275–370 = *Wissenschaftliche Abhandlungen* **1**, 316–402. Transl., "Further studies on the thermal equilibrium of gas molecules", pp. 88–174 of *Kinetic Theory* **2**, ed. S. G. BRUSH, Pergamon Press, 1966. ($VII_{i,iv}$, $VIII_{i,iii}$, IX, XI_i, XII, XIII_{iv})

1873

J. C. MAXWELL, "Clerk-Maxwell's kinetic theory of gases", *Nature* **8**, 85 [not reprinted in his *Scientific Papers*]. (X)

1875
L. BOLTZMANN, "Über das Wärmegleichgewicht von Gasen, auf welche äussere Kräfte wirken", *Sitzungsberichte der Akademie der Wissenschaften, Wien* **72**$_2$, 427-457 = *Wissenschaftliche Abhandlungen* **2**, 1-30. (VI$_{ii}$, VII$_i$, VIII$_i$, XI$_{i,iv,v}$)

1876
L. BOLTZMANN, "Über die Aufstellung und Integration der Gleichungen, welche die Molekularbewegungen in Gasen bestimmen", *Sitzungsberichte der Akademie der Wissenschaften, Wien* **74**$_2$, 503-552 = *Wissenschaftliche Abhandlungen* **2**, 55-102. (II$_v$, VI$_{ii}$, VIII$_i$, X)

1879
J. C. MAXWELL, "On stresses in rarified gases arising from inequalities of temperature", *Philosophical Transactions of the Royal Society* (London) **170**, 231-256 = *Scientific Papers* **2**, 680-712. (IV$_{ii}$, VII$_i$, VIII$_{iii}$, XI$_{iii}$, XIII$_{i,iii}$, XVI$_{i,ii}$, XVIII$_{iii}$, XXIV$_{ii,iv,v}$, XXV$_{iv,vi}$)

1880
L. BOLTZMANN, "Zur Theorie der Gasreibung I", *Sitzungsberichte der Akademie der Wissenschaften, Wien* **81**$_2$, 117-158 = *Wissenschaftliche Abhandlungen* **2**, 389-430. (VIII$_{iii}$, XXIV$_i$)

1881
L. BOLTZMANN, "Zur Theorie der Gasreibung II, III", *Sitzungsberichte der Akademie der Wissenschaften, Wien* **84**$_2$, 40-135, 1230-1263 = *Wissenschaftliche Abhandlungen* **2**, 431-556. (XXIV$_{i,xii}$)

1893
H. POINCARÉ, "Sur une objection à la théorie cinétique des gas", *Comptes Rendus Hebdomadaires des Séances, Académie des Sciences* (Paris) **116**, 1017-1021 = *Oeuvres* **10**, 240-243. (XII, XIII$_{iv}$)

1895
L. BOLTZMANN, "On Maxwell's method of deriving the equations of hydrodynamics from the kinetic theory of gases," *Report of the British Association for the Advancement of Science 1894*, 579 = *Wissenschaftliche Abhandlungen* **3**, 526-527. (XIII, XVI)

1896
L. BOLTZMANN, *Vorlesungen über Gastheorie*, I. Theil, Leipzig, Barth. (VI$_{iii}$, VII$_{i,iv}$, XI$_{i,iv}$, XXIV$_{ix}$)

1900
M. BRILLOUIN, "Théorie moléculaire des gas. Diffusion du mouvement et de l'énergie", *Annales de Chimie* (7) **20**, 440-485. (XXV$_v$)

J. W. STRUTT (Lord RAYLEIGH), "On the viscosity of Argon as affected by temperature", *Proceedings of the Royal Society* (London) **66**, 68-74 = *Scientific Papers* **4**, 452-458. (IV$_{ii}$)

1911
D. ENSKOG
1. "Über eine Verallgemeinerung der zweiten Maxwellschen Theorie der Gase", *Physikalische Zeitschrift* **12**, 56-60. (XXIV$_i$)
2. "Bemerkungen zu einer Fundamentalgleichung in der kinetischen Gastheorie", *Physikalische Zeitschrift* **12**, 533-539. (IX)

1912
D. HILBERT, "Begründung der kinetischen Gastheorie", *Mathematische Annalen* **72**, 562-577 = Kap. XXII of *Gründzüge einer allgemeinen Theorie der linearen Integralgleichungen*, Leipzig, 1912. Transl. [inaccurate] J. KOPP, "Foundations of the kinetic theory of gases", pp. 89-101 of *Kinetic Theory* **3**, ed. S. G. BRUSH, Oxford, *etc.*, Pergamon Press, 1972. (V$_{iv}$, VII$_{iv}$, XXII$_{i,ii}$, XXIII$_0$)

1913
A. C. LUNN, [Abstract of a paper on HILBERT's method of integrating the Maxwell-Boltzmann equation], *Bulletin of the American Mathematical Society* **19**, 455. (XXII_i, XXIII_0)

1915
S. BOGUSLAWSKI, "Zum Problem der inneren Reibung in der kinetischen Theorie", *Mathematische Annalen* **76**, 431–437. (XXIII_0)

T. H. GRONWALL, "A functional equation in the kinetic theory of gases", *Annals of Mathematics* (2) **17**, 1–4. (VI_{ii})

1916
S. CHAPMAN
1. "The kinetic theory of simple and composite gases: Viscosity, thermal conduction, and diffusion", *Proceedings of the Royal Society* (London) **A93** (1916/17), 1–20 = pp. 102–119 of *Kinetic Theory* **3**, ed. S. G. BRUSH, Oxford etc., Pergamon Press, 1972. [In 1912 CHAPMAN had published some steps toward the results of this paper.] (XVII_v, $XXIV_{i,viii,xii}$)
2. "On the law of distribution of molecular velocities, and on the theory of viscosity and thermal conduction, in a non-uniform simple monatomic gas", *Philosophical Transactions of the Royal Society* (London) **A216**, 279–348. (XVII_v, XXIV_i)

T. H. GRONWALL, "Sur une équation fonctionnelle dans la théorie cinétique des gaz", *Comptes Rendus Hebdomadaires des Séances de l'Académie des Sciences* (Paris) **162**, 415–418. (VI_{ii})

1917
D. ENSKOG, *Kinetische Theorie der Vorgänge in mässig verdünnten Gasen, I. Allgemeiner Teil*, Uppsala, Almqvist & Wiksell. Transl. J. KOPP, "Kinetic theory of processes in dilute gases", pp. 125–225 of *Kinetic Theory* **3**, ed. S. G. BRUSH, Oxford etc., Pergamon Press, 1972. (VIII_{iii}, XVII_v, $XXIV_{i,viii,xii}$)

1921
D. ENSKOG, "Die numerische Berechnung der Vorgänge in mässig verdünnten Gasen", *Arkiv för Matematik, Astronomi och Fysik* **16**, N:o 16, 60 pp. (XXIV_i)

1922
E. HECKE, "Über die Integralgleichungen der kinetischen Gastheorie", *Mathematische Zeitschrift* **12**, 274–286. (XXII_{ii}, XXIII_0)

1923
J. E. [LENNARD-] JONES, "On the velocity distribution function and on the stress in a non-uniform rarefied monatomic gas", *Philosophical Transactions of the Royal Society* (London) **A223**, 1–33. ($XXIV_{i,viii}$)

1928
D. ENSKOG, "Über die Grundgleichungen in der kinetischen Theorie der Flüssigkeiten und der Gase", *Arkiv för Matematik Astronomi og Fysik* **21A**, 1–28. (III_{iv})

1933
T. CARLEMAN, "Sur la théorie de l'équation intégrodifférentielle de Boltzmann", *Acta Mathematica* **60**, 91–146. (XXI_i)

1935
D. BURNETT, "The distribution of velocities in a slightly non-uniform gas", *Proceedings of the London Mathematical Society* (2) **39**, 385–430. ($XXIV_{i,xii}$)

1936
D. BURNETT, "The distribution of molecular velocities and the mean motion in a non-uniform gas", *Proceedings of the London Mathematical Society* (2) **40**, 382–435. ($XXIV_{i,xii}$)

1938

R. C. TOLMAN, *The Principles of Statistical Mechanics*, Oxford, Clarendon Press. (VIII_v)

1939

S. CHAPMAN & T. G. COWLING, *The Mathematical Theory of Non-Uniform Gases*, Cambridge University Press. (VIII_i, IX, $X_{i,iv,v}$, XII, XIV_x, XVII_v, XIX_{ii}, XXII_{ii}, XXIII_0, $XXIV_{i,ix,xii}$, XXV_{ix}) This book is quoted in Chapter XII and Chapter XIV, Section (x) by permission. Copyright 1939 by Cambridge University Press.

1948

C. S. WANG CHANG & G. E. UHLENBECK, *On the transport phenomena in rarified gases*, Applied Physics Laboratory Report No. APL/JHU CM-443, February 20 = *Studies in Statistical Mechanics* **5**, 1–16 (1970). (VIII_{ii}, XXIV_{xii}, XXV_{vi})

C. S. WANG CHANG

1. *The dispersion of sound in helium*, Applied Physics Laboratory Report No. APL/JHU CM-467, UMH-3F, May 1 = *Studies in Statistical Mechanics* **5**, 17–26 (1970). (XXIV_{xii})
2. *On the theory of thickness of weak shock waves*, Applied Physics Laboratory Report No. APL/JHU CM-503, August 19 = *Studies in Statistical Mechanics* **5**, 27–42 (1970). (XXV_{vi})

1949

H. GRAD, "On the kinetic theory of rarified gases", *Communications on Pure and Applied Mathematics* (New York University) **2**, 331–407. (VI_{iii}, $VIII_{iii,v}$, $XVII_{0,i,ii,iii,v}$, XXII_{iv})

I. PRIGOGINE, "Le domaine de la validité de la thermodynamique des phénomènes irréversibles", *Physica* **15**, 272–284. (XXIV_{ix})

1950

H. GRAD, *Kinetic Theory and Statistical Mechanics*, Mimeographed notes, Institute of Mathematical Sciences, New York University. (XXIV_{ix})

1951

E. WILD, "On Boltzmann's equation in the kinetic theory of gases", *Proceedings of the Cambridge Philosophical Society* **47**, 602–609. (III_{iv}, XII, XXI_i)

1952

S. CHAPMAN & T. G. COWLING, Corrected re-issue of CHAPMAN & COWLING [1939]. (XIV_x, XXIV_{ix})

C. TRUESDELL, "On the viscosity of fluids according to the kinetic theory", *Zeitschrift für Physik* **131**, 273–289. ($IV_{ii,iii}$)

C. S. WANG CHANG & G. E. UHLENBECK, *On the propagation of sound in monatomic gases*, University of Michigan Engineering Research Institute Report, Project M999, October = *Studies in Statistical Mechanics* **5**, 43–75 (1970). (XVIII_{iii}, XXIV_{xi})

1953

C. S. WANG CHANG & G. E. UHLENBECK, *The heat transport between two parallel plates as functions of the Knudsen number*, University of Michigan Engineering Research Instute Report, Project M999, September. (XVI_{ii})

1954

D. MORGENSTERN, "General existence and uniqueness proof for spatially homogeneous solutions of the Maxwell–Boltzmann equation in the case of Maxwellian molecules", *Proceedings of the National Academy of Sciences* (U.S.A.) **40**, 719–721. (XXI_i)

1955

C. L. PEKERIS, "Solution of the Boltzmann–Hilbert integral equation", *Proceedings of the National Academy of Sciences* (U.S.A.) **41**, 661–669. (XXIV_i)

1956

В. С. ГАЛКИН (V. S. GALKIN), «Об одном решений кинетического уравнения», *Прикладная Математика и Механика* **20**, 445–446. (XIV_i, XV)

E. IKENBERRY & C. TRUESDELL, "On the pressures and the flux of energy in a gas according to Maxwell's kinetic theory, I", *Journal of Rational Mechanics and Analysis* **5**, 1–54. (VIII_{vi}, IX, XII, XIII_i, XVI_{ii}, XVIA, XVIII_{iii}, $XXV_{0,iv,v,vi,vii,ix}$)

C. TRUESDELL, "On the pressures and the flux of energy in a gas according to Maxwell's kinetic theory, II", *Journal of Rational Mechanics and Analysis* **5**, 55–128. ($XIV_{i,ii,vii,viii,ix,x}$, XVII_{iv}, XVIII_{ii}, $XXV_{0,iii,v,viii,xi}$, $XXVI_{iii,v}$)

1957

T. CARLEMAN, *Problèmes Mathématiques dans la Théorie Cinétique des Gaz*, Uppsala, Almqvist & Wiksell. [Posthumous work, with important sections and appendices by L. CARLESON and O. FROSTMAN.] (VI_{ii}, XI_{iv}, XXII_{ii})

C. L. PEKERIS & Z. ALTERMAN, "Solution of the Boltzmann-Hilbert integral equation II. The coefficients of viscosity and heat conduction", *Proceedings of the National Academy of Sciences* (U.S.A.) **43**, 998–1007. (XXIV_{xii})

1958

В. С. ГАЛКИН (V. S. GALKIN), «Об одном классе решений уравнений кинетических моментов Града», *Прикладная Математика и Механика* **22**, 386–389. Transl., "On a class of solutions of Grad's moment equations", *PMM* **22**, 532–536. (II_v, XV_i)

H. GRAD, "Principles of the kinetic theory of gases", FLÜGGE's *Handbuch der Physik* **12**, 205–294. Berlin etc., Springer. (III_{iv}, VII_i, $XI_{i,iii}$, XIV_x, XIX_{ii}, XX_{ii}, $XXII_{i,iii,iv}$, XXIV_i, XXVI_v) This article is quoted in Chapter XIV, Section (x) by permission. Copyright 1958 by Springer-Verlag.

1960

H. GRAD, "Theory of rarefied gases" (1958), pp. 100–138 of *Rarefied Gas Dynamics*, London etc., Pergamon Press. (XXVI_{vi})

R. E. STREET, *Shock-Wave Structure based on Ikenberry-Truesdell approach to Kinetic Theory of Gases*, National Aeronautics and Space Administration Technical Note D-365, Washington, February. (XVI_{ii})

1962

Z. ALTERMAN, K. FRANKOWSKI, & C. L. PEKERIS, "Eigenvalues and eigenfunctions of the linearized Boltzmann collision operator for a Maxwell gas and for a gas of rigid spheres", *The Astrophysical Journal*, Supplement Series **7** (1962/3), 291–331. (XVI_{ii}, XVIII_{iii})

M. ŁUNC, "Les équations de transport des quantités moléculaires pareilles à l'entropie", *Archiwum Mechaniki Stosowanej* **14**, 561–564. (VIII_{iii})

А. Я. ПОВЗНЕР (A. YA. POVZNER), «Об уравнении Больцмана кинетическом теории газов», *Математический Сборник* (n.s.) **58**(100), 65–86. Transl. A. A. BROWN, "The Boltzmann equation in the kinetic theory of gases", *American Mathematical Society Translations* (2) **47**, 193–216. (III_{iv}, XIX_{ii}, XXI_i)

1963

H. GRAD

1. "Asymptotic theory of the Boltzmann equation", *Physics of Fluids* **6**, 147–181. (XIV_x, XXVI_v)

2. "Asymptotic theory of the Boltzmann equation, II", pp. 26–59 of *Rarefied Gas Dynamics* I, ed. J. A. LAURMANN, New York, Academic Press. (XXII_{ii})

M. ŁUNC, "A criterion for the degree of departure from equilibrium and its application to rarefied gases", pp. 94–101 of *Rarefied Gas Dynamics* I, ed. J. A. LAURMANN, New York, Academic Press. (IV_{iv}, VIII_{iii})

А. А. НИКОЛЬСКИЙ (A. A. Nikol'skii)
1. «Простейшие точные решения уравнения Больцмана для движений разреженного газа», *Доклады Академии Наук СССР* **151**, 299–301. Transl., "The simplest exact solutions of the Boltzmann equation for the motion of a rarefied gas", *Soviet Physics-Doklady* **8**, 633–635 (1964). (XVIIB)
2. «Трехмерное однородное расширение-сжатие разреженного газа со степенными функциями взаимодействия», *Доклады Академии Наук СССР* **151**, 522–524. Transl., "The three-dimensional expansion-contraction of a rarefied gas with power interaction functions", *Soviet Physics-Doklady* **8**, 639–641 (1964). (XVIIB)

1964

В. С. ГАЛКИН (V. S. Galkin), «Одномерное нестационарное решений уравнений кинетических моментов одноатомного газа», *Прикладная Математика и Механика* **28**, 186–188. Transl., "One-dimensional unsteady solution of the equation for the kinetic moments of a monatomic gas", *PMM* **28**, 226–229. (XV_i, XXVI_{iv})

H. Grad
1. *Theory of the Boltzmann Equation*, AFOSR report, Courant Institute of Mathematical Sciences, New York University, August = pp. 741–748 of *Applied Mechanics, Proceedings of the Eleventh International Congress of* (1964), Berlin–Heidelberg–New York, Springer-Verlag, 1966. (XXVI_{vi})
2. *Accuracy and Limits of Applicability of Solutions of the Equations of Transport; Dilute Monatomic Gases*, AEC Report, Courant Institute of Mathematical Sciences, New York University, September 30 = pp. 39–55 of *Proceedings of the International Symposium on the Transport Properties of Gases*, Providence, Brown University. (XXVI_{vi})

1965

L. Finkelstein, "Structure of the Boltzmann collision operator", *Physics of Fluids* **8**, 431–436. (XIX_{iii})

H. Grad
1. "Asymptotic equivalence of the Navier–Stokes and non-linear Boltzmann equation", *Proceedings of the American Mathematical Society Symposia on Applied Mathematics* **17**, 154–183. (XX_{ii})
2. "Solution of the Boltzmann equation in an unbounded domain", *Communications on Pure and Applied Mathematics* (New York University) **18**, 345–354. (XX_{ii})
3. "On Boltzmann's H-theorem", *Journal of the Society for Industrial and Applied Mathematics* **13**, 259–277. (VIII_{iv}, XI_{vi})

J.-P. Guiraud, "Sur l'adhérence d'un gaz à la paroi", *Comptes Rendus Hebdomadaires des Séances, Académie des Sciences* (Paris) **261**, 3739–3740. (XI_{iv})

А. А. НИКОЛЬСКИЙ (A. A. Nikol'skii)
1. «Однородные движения сдвига одноатомного разреженного газа», *Инженерный Журнал* **5**, 752–755. Transl., "Uniform shear motions of a rarefied monatomic gas", *Soviet Engineering Journal* **5**, 584–585. (XIV_{vii})
2. «Об общем классе однородных движений сплошных сред и разреженных газов», *Инженерный Журнал* **5**, 1044–1050. Transl., "On a general class of uniform motions of continuous media and rarefied gases", *Soviet Engineering Journal* **5**, 757–760. (II_v)

1966

J. Darrozes & J.-P. Guiraud, "Généralisation formelle du théorème H en présence de parois. Applications", *Comptes Rendus Hebdomadaires des Séances, Académie des Sciences* (Paris) **A262**, 1368–1371. ($XI_{iii,iv}$)

В. С. ГАЛКИН (V. S. Galkin), «О точных решениях уравнений кинетических моментов смеси одноатомных газов», *Известия Академий СССР. Механики Жидкости и Газов* **1**, 41–50. Transl., "Exact solutions of the kinetic-moment equations of a mixture of monatomic gases", *Fluid Dynamics* **1**, 29–34. (XV_i)

I. Müller, *Zur Ausbreitungsgeschwindigkeit von Störungen in kontinuierlichen Medien*, Dissertation, Rhein-Westfälische Technische Hochschule. (XXIV$_{ix}$)

1967

S. Chapman, "The kinetic theory of gases fifty years ago", pp. 1-13 of *Kinetic Theory (Lectures in Theoretical Physics 9C)*, ed. W. Brittin, New York etc., Gordon & Breach. (XXII$_{ii}$, XXIII$_0$, XXIV$_{i,ix}$)

1969

C. Cercignani, *Mathematical Methods in Kinetic Theory*, New York, Plenum Press. (XI$_{iv}$)

H. Grad, "Singular and nonuniform limits of solutions of the Boltzmann equation" (1967), *Transport Theory* (Volume 1, SIAM-AMS Proceedings), 269-308. (XXVI$_{vi}$)

C. Truesdell

1. *Rational Thermodynamics, A Course of Lectures on Selected Topics*, New York, etc., McGraw-Hill. (X, XIV$_{iii}$)
2. "A precise upper limit for the correctness of the Navier-Stokes theory with respect to the kinetic theory", *Journal of Statistical Physics* **1**, 313-318. (II$_v$, IV$_{iv}$, VIII)

1970

S. Chapman & T. G. Cowling, with the assistance of D. Burnett, Third edition of Chapman & Cowling [1939] [1952]. (XI$_v$, XIV$_x$, XVII$_v$, XXV)

This book is quoted in Chapter XIV, Section (x) by permission. Copyright 1970 by Cambridge University Press.

J.-P. Guiraud, "Problème aux limites intérieur pour l'équation de Boltzmann linéaire", *Journal de Mécanique* **9**, 443-490. (XI$_{iv}$)

1971

C. Cercignani & M. Lampis, "Kinetic models for gas-surface interactions", *Transport Theory and Statistical Physics* **1**, 101-114. (XI$_{iv}$)

1972

L. Arkeryd

1. "On the Boltzmann equation. Part I: Existence", *Archive for Rational Mechanics and Analysis* **45**, 1-16. (VI$_{iv}$, XXI$_i$)
2. "On the Boltzmann equation. Part II: The full initial value problem, *Archive for Rational Mechanics and Analysis* **45**, 17-34. (XXI$_{i,iv,v,vi}$)

C. Cercignani, "Scattering kernels for gas-surface interactions", *Transport Theory and Statistical Physics* **2**, 27-53. (XI$_{iv}$)

G. Di Blasio, "Strong solution for Boltzmann equation in the spatially homogeneous case", *Bollettino della Unione Matematica Italiana* (4) **8**, 127-136. (XXI$_i$)

J. H. Ferziger & H. G. Kaper, *Mathematical Theory of Transport Processes in Gases*, Amsterdam etc., North-Holland Co.-American Elsevier. (XXVI$_v$)

A. Glikson

1. "On the existence of general solutions of the initial value problem for the non-linear Boltzmann equation with a cut-off", *Archive for Rational Mechanics and Analysis* **45**, 35-46. (III$_{iv}$, XX$_{ii,v}$)
2. "On solution of the non-linear Boltzmann equation with a cut-off in an unbounded domain", *Archive for Rational Mechanics and Analysis* **47**, 389-394; corrections, *ibid.* **51**, 387 (1973). (XX$_{ii,v}$)

H. Grad, *Singular Limits of Solutions of Boltzmann's Equation*, AFOSR Report, Courant Institute of Mathematical Sciences, New York University, October = pp. 37-53 of *Eighth International Symposium on Rarefied Gas Dynamics* (1972), N.Y. & London, Academic Press, 1974.

J.-P. Guiraud, "Problème aux limites intérieur pour l'équation de Boltzmann en régime stationnaire, faiblement non linéaire", *Journal de Mécanique* **11**, 183-231. (XI$_{iv}$)

H. E. REVERCOMB & L. W. BRUCH, "Variational solution of the linearized Boltzmann equation with applications to helium", *Journal of Chemical Physics* **57**, 5530–5541. (XXIV$_{xii}$)

1973

C. CERCIGNANI, "Comment to Professor Uhlenbeck's paper", *Acta Physica Austriaca* Supplement **10** (*The Boltzmann equation, theory and applications*), 121–123. (XIX$_{iv}$)

J.-P. GUIRAUD, "The Boltzmann equation in kinetic theory a survey of mathematical results", 50 pp., in [*Proc.*] *XIth Symposium on Advanced Problems and Methods in Fluid Mechanics*, Instytut Podstawowych Problemow Techniki PAN, Warsaw. (XI$_{iv}$, XVIII$_i$)

C. TRUESDELL, *Mathematical Aspects of the Kinetic Theory of Gases*, Notas de Matemática Fisica **3**, Instituto de Matemática da Universidade Federal do Rio de Janeiro. (VI$_{iv}$, X, XI$_i$)

1974

G. DI BLASIO, "Differentiability of spatially homogeneous solutions of the Boltzmann equation in the non-Maxwellian case", *Communications in Mathematical Physics* **38**, 331–340. (XXI$_i$)

R. G. MUNCASTER, "Instability of the heat flux in a gas of Maxwellian molecules", *Istituto Lombardo Rendiconti, Classe di Scienze Matematiche* (A) **108**, 433–440. (XIV$_{viii}$)

Y.-P. PAO

1. "Boltzmann collision operator with inverse-power intermolecular potentials, I", *Communications on Pure and Applied Mathematics* **27**, 407–428. (XXII$_{ii}$)
2. "Boltzmann collision operator with inverse-power intermolecular potentials, II", *Communications on Pure and Applied Mathematics* **27**, 559–581. (XXII$_{ii}$)

S. UKAI, "On the existence of global solutions of mixed problem for non-linear Boltzmann equation", *Proceedings of the Japan Academy* **50**, 179–184. (XVIII$_i$, XX$_{ii}$)

1975

C. CERCIGNANI, *Theory and Application of the Boltzmann Equation*, New York, Elsevier. (XI$_{iii,iv}$)

J.-P. GUIRAUD, "An *H*-theorem for a gas of rigid spheres in a bounded domain", pp. 29–58 of *Colloques Internationales du Centre National de Recherches Scientifiques* No. 236 (1974). (XI$_{iv}$)

R. G. MUNCASTER, *Constitutive Relations in the Kinetic Theory of Gases*, Dissertation, The Johns Hopkins University, vi + 112 pp. (VIII$_{vi}$, XXIII$_{iv}$, XXIV$_{iii,vi}$)

J. SCHNUTE, "Entropy and kinetic energy for a confined gas", *Canadian Journal of Mathematics* **27**, 1271–1315. (VI$_{iii}$)

C. TRUESDELL, "Early kinetic theories of gases", *Archive for History of Exact Sciences* **15**, 1–66. (Prologue, III$_{ii,iv}$, VII$_{iii}$, XII, XVII$_v$, XXIV$_{ix}$)

1976

M. KROOK & T. T. WU, "Formation of Maxwellian tails", *Physical Review Letters* **36**, 1107–1109. (XVIIB)

S. UKAI, "Les solutions globales de l'équation non-linéaire de Boltzmann dans l'espace tout entier et dans le demiespace", *Comptes Rendus Hebdomadaires des Séances, Académie des Sciences* (Paris) **A282**, 317–320. (XVIII$_i$, XX$_{ii}$)

C.-C. WANG, "A transformation property of the Boltzmann equation", *Quarterly of Applied Mathematics* **33** (1975/6), 369–375. (VII$_v$)

1977

S. KANIEL, "On the derivation of thermohydrodynamic equations from the Boltzmann equation", *Journal of Statistical Physics* **16**, 415–450. (XII, XVIA, XVIII$_{iii}$)

M. KROOK & T. T. WU, "Exact solutions of the Boltzmann equation", *The Physics of Fluids* **20**, 1589–1595. (XVIIB)

T. NISHIDA & K. IMAI, "Global solutions to the initial value problem for the nonlinear Boltzmann equation", *Publications of the Research Institute for Mathematical Sciences, Kyoto University* **12**, 229–239. (XVIII$_i$, XX$_{ii}$)

Y. Shizuta & K. Asano, "Global solutions of the Boltzmann equation in a bounded convex domain", *Proceedings of the Japan Academy Ser. A. Math. Sci.* **53**, 3–5. (XVIII$_i$)

C. Truesdell, "Correction of two errors in the kinetic theory that have been used to cast unfounded doubt upon the principle of material frame-indifference", *Meccanica* **11** (1976), 196–199. (IX)

1978

S. Kaniel & M. Shinbrot, "The Boltzmann equation, I: Uniqueness and local existence", *Communications in Mathematical Physics* **58**, 65–84. (XVIII$_i$, XX$_{ii}$)

1979

R. G. Muncaster, "On generating exact solutions of the Maxwell–Boltzmann equation", *Archive for Rational Mechanics and Analysis* **70**. (XVIIB)

D. Ray, "On a class of solutions of non-linear Boltzmann equations", *Journal of Statistical Physics* **20**, 115–119. (XVI$_{ii}$)

Y. Shizuta
1. "On the classical solutions of the Boltzmann equation", *Communications on Pure and Applied Mathematics* (New York University). (XVIII$_i$, XX$_i$)
2. "The existence and approach to equilibrium of classical solutions of the Boltzmann equation", *Communications in Mathematical Physics.* (XVIII$_i$, XX$_i$)

Index of Authors Cited

A

ALTERMAN, ZIPORA STEPHANIA (1925–), 247, 249, 300, 482, 487, 573
ARKERYD, LEIF, 84, 85, 337, 338, 349, 356, 359, 363, 364, 365, 366, 560, 564, 575
ASANO, KIYOSHI, 296, 577

B

BATRA, ROMESH C. (1947–), xxi
BOGUSLAWSKI, SERGEI ANATOLEVICH (1883–1923), 397, 571
BOLTZMANN, LUDWIG (1844–1906), 12, 25, 27, 44, 72, 75, 91, 96, 99, 106, 107, 113, 118, 119, 131, 139, 140, 141, 142, 145, 147, 148, 161, 166, 179, 185, 191, 193, 194, 195, 244, 245, 295, 307, 375, 431, 432, 442, 469, 470, 482, 563, 564, 569, 570
BRILLOUIN, MARCEL LOUIS (1854–1948), 522, 570
BRUCH, LUDWIG WALTER (1935–), 487, 576
BRUSH, STEPHEN GEORGE (1935–), 153, 176, 245, 433, 557
BURNETT, DAVID (1900–), 431, 433, 485, 487, 571

C

CARLEMAN, TAGE GILLIS TORSTEN (1892–1949), 72, 163, 336, 337, 364, 365, 382, 564, 571, 573
CARLESON, LENNART AXEL EDWARD (1928–), 72, 161, 163, 167, 573
CARLSON, DONALD EARLE (1938–), 53
CAUCHY, AUGUSTIN-LOUIS (1789–1857), 24

CERCIGNANI, CARLO, 157, 163, 164, 165, 575, 576
CHAPMAN, SYDNEY (1888–1970), 106, 109, 132, 147, 161, 166, 171, 176, 217, 218, 280, 309, 381, 383, 384, 397, 430, 431, 432, 433, 457, 462, 463, 468, 481, 484, 486, 488, 521, 529, 536, 549, 571, 572, 575
CLAUSIUS, RUDOLF JULIUS EMMANUEL (1822–1888), 17, 95, 119, 148, 161, 179
COIMBRA, ALBERTO, xxi
COLEMAN, BERNARD DAVID (1930–), 408, 437
COURANT, RICHARD (1888–1972), 381
COWLING, THOMAS GEORGE (1906–), 109, 132, 147, 161, 166, 171, 176, 217, 218, 280, 309, 381, 383, 384, 397, 431, 432, 433, 457, 468, 481, 486, 488, 529, 536, 549, 572, 575

D

DAFERMOS, CONSTANTINE (1941–), xxii
DARROZES, JEAN, 159, 160, 163, 167, 574
DAVINI, CESARE (1943–), 411
DAY, WILLIAM ALAN (1942–), 9, 13
DI BLASIO, GABRIELLA, 338, 575, 576
DIEUDONNÉ, JEAN ALEXANDRE (1906–), 409, 424

E

EDWARDS, ROBERT EDMUND, 365
ENSKOG, DAVID (1884–1947), 46, 48, 56, 115, 134, 146, 280, 295, 398, 430, 431, 432, 433, 434, 447, 454, 457, 462, 478, 479, 481, 484, 488, 507, 508, 513, 516, 517, 520, 521, 522, 524, 526. 529, 530, 534, 535, 537, 539, 540, 541, 544, 549, 556, 566, 570, 571

ERICKSEN, JERALD LAVERNE (1924–), 58
EULER, LEONHARD (1707– 1783), xvii
EVERITT, C. W. FRANCIS (1934–), 153

F

FERZIGER, JOEL H., 555, 575
FINKELSTEIN, LEIB, 311, 574
FLÜGGE, SIEGFRIED, 9, 15, 21
FOURIER, JOSEPH (1768– 1830), 3, 112, 195
FRANKOWSKI, K., 247, 249, 300, 573
FROSTMAN, OTTO (1907–), 72, 573

G

GALKIN, V. S., 29, 198, 221, 225, 227, 291, 546, 549, 557, 573, 574
GIBBS, JOSIAH WILLARD (1839– 1903), 119
GLIKSON, ALEKSANDER, 48, 322, 327, 328, 332, 575
GRAD, HAROLD (1923–), 48, 75, 91, 92, 110, 115, 119, 120, 121, 147, 159, 169, 189, 218, 261, 262, 265, 266, 268, 269, 270, 271, 272, 273, 274, 275, 276, 277, 278, 279, 280, 281, 284, 302, 309, 320, 321, 322, 375, 381, 386, 388, 433, 434, 454, 463, 471, 479, 481, 484, 486, 504, 505, 508, 511, 512, 516, 521, 522, 536, 537, 538, 539, 540, 555, 556, 557, 558, 560, 561, 566, 572, 573, 574, 575
GRONWALL, THOMAS HAKON (1877– 1932), 72, 571
GUIRAUD, JEAN-PIERRE, 159, 160, 163, 167, 296, 574, 575, 576
GURTIN, MORTON EDWARD (1934–), 127

H

HARTMAN, PHILIP (1915–), 220, 296, 323, 329
HECKE, ERICH (1887– 1947), 100, 382, 398, 571
HERAPATH, JOHN (1790– 1869), 95
HILBERT, DAVID (1862– 1943), 67, 98, 99, 100, 218, 375, 376, 377, 379, 380, 381, 382, 383, 384, 385, 386, 387, 388, 389, 390, 391, 392, 393, 394, 395, 397, 398, 399, 400, 402, 429, 432, 434, 452, 453, 454, 507, 520, 521, 522, 526, 528, 529, 539, 540, 556, 562, 570
HUGONIOT, PIERRE HENRI (1851– 1887), 21
HUYGENS, CHRISTIAAN (1629– 1695), 157

I

IKENBERRY, ERNEST (1913–), 126, 127, 135, 136, 161, 184, 189, 244, 245, 247, 248, 249, 250, 252, 253, 254, 257, 261, 270, 271, 275, 281, 282, 283, 297, 300, 301, 302, 430, 505, 507, 508, 509, 511, 512, 515, 519, 520, 523, 524, 526, 529, 530, 535, 540, 544, 566, 573
IMAI, K., 296, 321, 322, 576

K

KANIEL, SHMUEL, 186, 252, 253, 256, 257, 296, 302, 576, 577
KAPER, HANS G., 555, 575
KELVIN, WILLIAM THOMSON, Lord (1824– 1907), 152, 245
KESTELMAN, HYMAN, 72
KIRCHHOFF, GUSTAV ROBERT (1824– 1887), 21
KORTEWEG, DIEDERIK JOHANNES (1848– 1941), 9, 10, 11
KROOK, MAX (1913–), 290, 291, 299, 301, 303, 393, 576

L

LAMB, HORACE (1849– 1934), 42
LAMPIS, M., 163, 575
LANG, SERGE (1927–), 307, 328, 329, 347, 424
LA PENHA, GUILHERME DE, xxi
LEIBNIZ, GOTTFRIED WILHELM (1646– 1716), xvii
[LENNARD]-JONES, J. E. (1894– 1954), 431, 433, 461, 571
ŁUNC, MICHAŁ (1908–1974), 59, 112, 115, 573
LUNN, ARTHUR CONSTANT (1877– 1942), 376, 398, 571

M

MAAK, WILHELM (1912–), 72
MAN, CHI-SING (1947–), 123
MAXWELL, JAMES CLERK (1831– 1879), xviii, xix, xxii, 3, 7, 11, 12, 14, 15, 16, 39, 41, 44, 45, 46, 50, 52, 54, 56, 57, 58, 64, 67, 69, 91, 92, 93, 96, 98, 99, 100, 102, 103, 106, 107, 113, 127, 131, 132, 133, 134, 136, 138, 139, 141, 142, 152, 153, 154, 155, 156, 157, 160, 161, 165, 175, 176, 181, 184, 185, 189, 190, 191, 192, 193, 194, 195, 198, 199, 201, 202, 204, 208, 212, 213, 216, 223, 238, 244, 245, 249, 252, 257, 275, 276, 279, 280, 291, 295, 298, 299, 300, 302, 309, 375, 395, 402, 415, 430, 431, 432, 433, 435, 444, 458, 462, 463, 480, 507, 514, 516, 517, 518, 519, 521, 522, 524, 559, 560, 561, 569, 570
MENIKOFF, ARTHUR (1947–), 296

INDEX OF AUTHORS CITED

Meyer, Oskar-Emil (1834–1909), 179
Morgenstern, Dietrich (1924–), 336, 337, 338, 572
Müller, Ingo (1940–), 470, 471, 472, 575
Muncaster, Robert Gary (1948–), xxi, xxii, 126, 209, 210, 284, 291, 296, 303, 393, 406, 441, 452, 454, 540, 562, 563, 576, 577

N

Newton, Isaac (1642–1727), xvi, 112, 125, 157
Nikol'skii, A. A., 29, 208, 291, 574
Nishida, Takaaki, 296, 321, 322, 576
Noll, Walter (1925–), 9, 11, 12, 127, 202, 214, 234, 408, 410, 437, 457

P

Pao, Y.-P., 381, 561, 576
Pekeris, Chaim Leib (1908–), 247, 249, 300, 434, 482, 487, 572, 573
Pitteri, Mario (1948–), xxii, 166, 290
Poincaré, Jules Henri (1854–1912), 185, 195, 570
Poisson, Siméon-Denis (1781–1840), 20, 24, 139, 140, 223
Povzner, A. Ya., 48, 311, 337, 348, 353, 369, 370, 573
Prigogine, Ilya (1917–), 470, 572

R

Ray, Dipankar, 297, 577
Rayleigh, John William Strutt, Lord (1842–1919), 55, 570
Revercomb, H. E., 487, 576
Reynolds, Osborne (1842–1912), 152, 153
Royden, Halsey Lawrence (1928–), 306, 340, 343, 347
Rubens, Peter Paul (1577–1640), 152
Rudin, Walter (1921–), 164, 340, 358

S

Schnute, Jon, 79, 576
Schröter, Joachim, 454
Serrin, James Burton, Jr. (1926–), 21, 56
Shinbrot, Marvin (1928–), 296, 577
Shizuta, Yasushi, 296, 321, 322, 577

Stokes, George Gabriel (1819–1903), 193, 245
Street, Robert Elliot (1912–), 248, 573
Strutt, see Rayleigh

T

Tait, Peter Guthrie (1831–1901), 179
Thomson, see Kelvin
Tolman, Richard Chace (1881–1948), 119, 572
Toupin, Richard A. (1926–), 15, 58
Truesdell, Clifford Ambrose, III (1919–), xix, 3, 9, 11, 12, 13, 15, 21, 29, 31, 39, 43, 46, 52, 57, 58, 84, 95, 109, 126, 127, 134, 135, 136, 142, 147, 157, 179, 184, 189, 198, 199, 202, 203, 204, 208, 210, 213, 218, 244, 245, 248, 249, 252, 257, 278, 280, 291, 297, 300, 301, 302, 336, 337, 398, 410, 430, 435, 463, 507, 508, 509, 512, 515, 519, 520, 521, 522, 523, 524, 526, 528, 529, 530, 535, 540, 544, 546, 549, 551, 557, 558, 560, 564, 566, 572, 573, 575, 576, 577

U

Uhlenbeck, George E. (1900–), 109, 247, 302, 480, 486, 487, 524, 572
Ukai, Seiji, 296, 321, 322, 576

W

Wang, Chao-Cheng (1938–), 101, 141, 142, 457, 576
Wang Chang, C. S., 109, 247, 302, 457, 468, 480, 486, 487, 524, 572
Waterston, John James (1811–1883), 43, 95, 179
Widom, Harold (1932–), 351
Wild, Ernest (1936–), 48, 186, 336, 572
Williams, William Orville (1940–), 127
Wu, Tai Tsun (1933–), 290, 291, 299, 301, 303, 393, 576

Y

Yin, Wan-Lee (1941–), 26
Yosida, Kôsaku, 363, 382

Index of Matters Treated

Basic quantities which occur throughout the book are indexed only for their first occurrences or major restatements, *e.g.*, molecular density, number density, Maxwell–Boltzmann equation.

A

Acceleration, 4
Accommodation coefficient, 156–157, 160
Adiabatic boundary (wall), 17, 161
Adiabatic change, 20, 139–140, 223
Aerostatics, 23, 137–138, 141
Affine flow, homo-energetic, *see also* Atemporal Maxwellian iteration
  continuum thermomechanics
    dilatation, 24–28, 36
    extension, 33–36
    general, 28–32
    simple shearing, 32–33, 36
  kinetic theory
    dilatation, 221–223, 234, 291
    extension, 221, 224–234
    fundamental theorem, 220
    general, 219–221
    gross determinism, 420–424
    retrospect, 233–234
    simple shearing, 197–218, 221, 234
Amount of shearing, 32, 197
Analytical dynamics, inconsistency of kinetic theory with, 102–103, 106
Approximate gross determiners, *see* Gross determiners
Approximate momental determiners, *see* Momental determiners
Approximate solution, summary of methods, 521–522
Approximation
  by Maxwellian density, 112–125

  methods of, retrospect upon, 539–540, 556–558
Arkeryd's theorems on existence, H-theorem, and trend to equilibrium, 337–371
Asymptotic, *see also* Trend to
  "only", 556
  status of Stokes–Kirchhoff theory, *see* Stokes–Kirchhoff theory of viscous gases
  velocities of molecules, 70–71, 91–92, 94, 101
Atemporal Maxwellian iteration
  for affine flows
    convergence lemma, 545
    dilatation, 546, 550
    extension, 548–550, 553–555
    general, 542–546
    simple shearing, 546–548, 550–553, 555
  definition and scheme, 528–537
  iterates
    first, 533–534
    second, 535–536
Average number of molecules, 40
Axiomatics, xx

B

Balance, equations of
  continuum thermomechanics, 4–5, 68
  kinetic theory, 132
Banach spaces, calculus in, 424–427
Bilinear form, *see* Hilbert's bilinear form
Binary encounter, 69–71
Body, body-point, 8–9

INDEX OF MATTERS TREATED 583

Body force
  compatible with equilibrium, 23
  compatible with homo-energetic dilatation, 26–27
  in continuum thermomechanics, 3
  effect on gross determiners, 399, 413–419
  effect on moments of the derivative, 66–67
  in existence theory, 323
  influence on Maxwellian iteration, 509, 520
  in kinetic theory, 39, 95
  in method of stretched fields, 472–479
Boltzmann, Boltzmann's
  factor, see Maxwell-Boltzmann-Wang theorem on kinetic equilibrium
  field h, see also H-theorem
    definition, 117–118, 145, 212
    differential equation, 146, 171
    flux of
      definition, 121, 145
      gross determiner
        approximate, 469–471, 478–479
        exact, 442–444
      jump at a wall, 123–125
      relation to $q/\tfrac{1}{3}\varepsilon$
        grossly determined F, 469–472
        ideally rough wall, 161–163
        linear wall, 161–166
        nearly Maxwellian F, 121–122
    gross determiner
      approximate, 469–471, 478–479
      exact, 442–443
    jump at a wall, 122
    relation to caloric
      grossly determined F, 469–471
      Maxwellian F, 117–118
      nearly Maxwellian F, 119–120, 122
  H-theorem, see H-theorem
  monotonicity theorem, 105–107, 147, 308, 358–359, 365
Boltzmann-Gronwall theorem, 71–74
Boundary, wall
  adiabatic, 17, 161
  conditions
    dynamic, 48, 153
    general for kinetic theory, 157
    kinetic, various, 152–166, 296
  energetic at, 124, 149–152
  ideally rough, 155–156, 160–163, 166
  ideally smooth, 154–155, 160–161
  insulated, 17, 161
  jumps of fields at, 111–112, 122–125
  linear, 159–160, 163–166
  stationary, 153

Bounds for h and for its flux s, 119–122
Bulk viscosity, 6, 20, 25, 57, 140, 193
"Burnett coefficients", see Second-order coefficients
Burnett's formula, 432–433, 466, 535, 555

C

Caloric, specific entropy
  continuum thermomechanics, 7–8, 13–14, 20, 22, 117
  equation of state, 5, 12, 19, 118
  ideal gas, 22
  kinetic analogue, see Boltzmann's field h
Calory, entropy
  continuum thermomechanics, 13–18
  kinetic analogue, see H-theorem
Carleman's theorems on existence, H-theorem, and trend to equilibrium, 336–337, 364
Causal, 217–218, 386
Chaos, molecular, 92
"Chapman-Enskog process", 430–432, see also Enskog's procedure; Stretched fields, method of
Classical approach, failure of, 549–555
Classical fluid mechanics, see Euler-Hadamard theory of gases; Navier-Stokes-Fourier theory of viscous gases; Stokes-Kirchhoff theory of viscous gases
Clausius inequality
  continuum thermomechanics, 15–16
  kinetic analogue, see H-theorem, narrow
Clausius-Duhem inequality
  continuum thermomechanics, 13–17, 121, 471–472
  kinetic analogue, 149–151, 171–172, 212, 471–472, see also H-theorem, broad
Coefficients in expansions of gross determiners, 445–488
Coleman's method of retardation, 437
Collision frequency, 96, 178–179, see also frequency operator
Collisional invariant, 71–74, 344
Collisions, see also Cross-section; Encounter; Scattering factor
  integrals
    convergence of
      general molecules, 305–309, 311–315
      inverse k^{th}-power molecules, 316–317
      molecules with a cut-off, 309–311

general molecules, 175, 181–184, 266–272, 282–283, 502–505, 511–512
Grad's approximation, 281, 508, 512
ideal spheres, 183
inverse k^{th}-power molecules, 184
Maxwellian molecules, 184–185, 237–249, 251–257, see also Ikenberry's theorem
operators
alternative forms, 95–97
definition, 50, 91–95
domain of, 305–317
frame-indifference, 102
frequency, 95–97, 178, 310
for ideal spheres, 177–178
for inverse th-power molecules, 179–181
modified, 347–348
for molecules with a cut-off, 95–97
orthogonal invariance, 101–102
partial, 96–97, 310
for pseudomolecules, 186
scattering factor, 97, 177–181
total
definition, 91, 98
transformations of, 98–100, 305–306, 310, 315
Companion of Maxwell's second approximation for pressure tensor, 518
Conductivity, thermal (heat), 5–6, 397, see also Maxwell number
Conjectures, principal, 562–563
Consistency theorem, Maxwell's, 132–133, 275, 402, 415–416, 522
Constitutive constant of an ideal gas, 19, 44
Constitutive equations or relations
continuum thermomechanics, 4–5, 9–12, 24, 133–134, 192, 202, 529, see also Euler-Hadamard theory of gases; Navier-Stokes-Fourier theory of viscous gases; Stokes-Kirchhoff theory of viscous gases
kinetic theory (alleged), 18, 133–134, 444, 509, 521–523
Constitutive quantities for kinetic theory, 93–95, 97, 103, 136, 376
Continuum theories, xvii
Continuum thermomechanics of fluids, 3–36, 39, 51–52, 56–57, 68, 112, 132–133, 275, 277, 410, 521, see also Euler-Hadamard theory of gases; Material; Navier-Stokes-Fourier theory of viscous gases; Stokes-Kirchhoff theory of viscous gases

Convergence and divergence
atemporal Maxwellian iteration, 541–558
integrals, 99, 183, 239, 305–317
Cooling, Newton's law of, 112, 125
Coupling, forward, 136, 250
Cross-section, 81–82, 92–95, 296
Cut-off
angle, 88, 97, 186
hard potential, 321, 381
molecules with a, 69–70, 88, 95–97, 296, 309–311, 319–371, 381
Cylindrical dilatations, 221, 303

D

Damped part, 201, 215
Decomposition, multi-index, 258–259
Density, see Mass density; Maxwellian density, molecular; Molecular density; Number density
Derivative
material, 4, 42
mild, 62
strong
definition, 64
moments of, 66–67
Derivatives, time, elimination of, 405–406, 414, 433–434, 529–530
Determinations, gross, see Grossly determined
Determiners
gross, see Gross determiners
momental, see Momental determiners
Determinism, 8–10, 52, 408, see also Gross determinism; Grossly determined; Momentally determined
Diameter of molecule, 53, see also Ideal spheres
Differential iteration, 515, 537, 540, see also Maxwellian iteration
Differentiation along extrinsic trajectories, 62–63
Dilatation
continuum thermomechanics
general, 5
homo-energetic, 24–28, 36
kinetic theory, 139–140, 221–223, 234, 291, 303, 546, 550
Dimensions, physical, 3, 30, 40, 52–56, 458
Dissipation
general, 8
inequalities, classical, 13, 172, 212–213, see also Clausius-Duhem inequality

INDEX OF MATTERS TREATED

Dissipationless, 8, 20–21, 24–25, 36, *see also* Adiabatic change
Distortion, 4
Distortionless, 5–6
Divergence, *see* Convergence and divergence
Domain, essential, 40
Dominant part, 201–203, 211–214, 230, 563

E

Effectiveness, proof of
  atemporal Maxwellian iteration, 531
  Hilbert's procedure, 380–384
  Maxwellian iteration, 524–526
  method of stretched fields, 452–455
"Elastic" encounter, 54, 69
Elimination of time derivatives, 405–406, 414, 433–434, 529–530
Emitted molecule, 153, 158
Encounter, *see also* collisions
  binary, 69–71
  operator, 84–87, 94–95, 177, 297, 311–312
  parameter, 74–76, 78, 81–82, 87–88
  problem, 71, 74–84, 87–88
Energetic
  continuum thermomechanics, 3–5, 10, 19
  free, 4–5, 10, 22
  kinetic theory
    definition, 43
    wall, 149–151, 155–156, 160, 162–163
Energy
  equation of balance
    continuum thermomechanics, 4, 20
    kinetic theory, 67–68, 132
  flux of, 23
  flux vector
    continuum thermomechanics, 3–5, 10, 20
    kinetic theory
      definition, 42
      formulae for, 108, 195, 210, 214, 461–466, 516–518, 523, 533, 536
      relaxation of, 190, 223, 279
  internal, *see* energetic
  total, 16, 17, 148
Enskog's
  formulae for the viscosity and the Maxwell number
    approximate, 279–280, 486–487, 516, 533–534
    exact, 462–463, 480
  procedure, 430–434, 447, 454, 478, 507–508, 529, 556

Entropy
  specific, *see* Caloric, specific entropy
  total, *see* Calory, entropy
Equilibrium
  continuum thermomechanics
    definition, 22–23
    free, 23
  kinetic theory
    gross, 137–138
    gross free, 137, 167–168, 189
    kinetic, 141, 167–168, 190
    trend to, *see* Trend to equilibrium
Essential domain, 40
Euler–Hadamard theory of gases
  continuum thermomechanics
    definition, 21
    particular flows
      dilatation, 24–25, 36
      simple shearing, 36
  kinetic theory, comparisons with
    general, for nearly Maxwellian densities, 120
    Hilbert's procedure, 379
    particular flows
      dilatation, 58, 223
      gross equilibrium, 137–138
      Krook & Wu's solution, 290
      locally Maxwellian solutions, 139
Evolution, equation of, 49–50, 64–65, 125, 133, *see also* Maxwell–Boltzmann equation
Exact (explicit) solutions
  equations for moments, 190–234
  Maxwell–Boltzmann equation, 138–141, 284–291
  Stokes–Kirchhoff theory, 23–36
Existence, *see also* Initial-value problem
  hypothesis regarding gas-dynamic solutions, 398
  place-dependent solutions, initial-value problem
    discussion, 320–322
    Glikson's theorem, 327–333
    Grad's theorems, 320–322
    prolegomena, 295–297
  spatially homogeneous solutions, initial-value problem
    Maxwellian molecules, Truesdell's theorem, 297–302
    molecules with a cut-off
      Arkeryd's theorems, 337–355
      Carleman's theorem, 336–337

Morgenstern's theorem, 336-337
Povzner's theorem, 337
statements of rigorous propositions, 167
theory, open problems, 560-561
Expansion, 4
Expectation, 41
Expected number of molecules, 40
Extension, homo-energetic
  general, 33-35
  kinetic theory, Maxwellian molecules
    atemporal Maxwellian iterates, 548-550, 553-555
    behavior for large time, 227-231
    comparison with Stokes-Kirchhoff solution, 231-233
    dominant part, 230
    general solution for pressures, 224-227
    gross determiners, 230
    main theorems, 230-231
    principal solution, 231
  Stokes-Kirchhoff solution, 35-36
Extrinsic force, 46, see also Body force
Extrinsic trajectory
  definition, 47-49, 61
  differentiation along, 62-63
  integration along, 63

F

Field equations
  continuum thermomechanics, 4-5, 68
  kinetic theory
    general, 132
    Hilbert's, 379
Fields, basic
  continuum thermomechanics, 3
  kinetic theory, 41-43
  ordinary, 126-127
First Law of Thermodynamics, 16
Flow
  "continuum" regime, 59
  definition
    continuum thermomechanics, 5
    kinetic theory, 42, 401-402
  free molecular, 59, 207-208
  homo-energetic, 24
  homothermal, 24
Fluid, linearly viscous, 4-8, 14, 51, see also Navier-Stokes-Fourier theory of viscous gases; Stokes-Kirchhoff theory of viscous gases
Fluid mechanics, classical, see Euler-Hadamard theory of gases; Navier-Stokes-Fourier theory of viscous gases; Stokes-Kirchhoff theory of viscous gases
Flux, see Boltzmann's field h; Clausius-Duhem inequality; Energy; H-theorem; Heat-bath inequality
Force
  body, see Body force
  extrinsic, 46
  intermolecular, 53, 71, 88
  mutual, 46
Forward coupling, 136, 250
Frame
  change of, 11
  indifference
    collisions operators, 102
    material, 10-11, 134, 275, 523
    time fluxes, 12
Free energetic, 4-5, 10, 22
Free locally Maxwellian, see Maxwellian solutions
Frequency
  collision, 96, 178-179
  operator, 95-96, 178-179
Friction, internal, 20, 36, see also Viscosity

G

Galkin's solutions, see also Affine flow, homo-energetic
  cylindrical dilatation, 221, 303
  dilatation, 221-223, 303
  extension, 221, 224-227, 303, 549
  simple shearing, 198, 221, 303
Gas
  constant, 19, 44
  dynamics, see Euler-Hadamard theory of gases; Navier-Stokes-Fourier theory of viscous gases; Stokes-Kirchhoff theory of viscous gases
  flows, 5, 42, 401-402
  ideal, 19-20, 22, 44, 132
  monatomic, 43
  perfect, 6, 20-21
  simple, 40
  viscous, see Navier-Stokes-Fourier theory of viscous gases; Stokes-Kirchhoff theory of viscous gases
Glikson's existence theorem, 327-333
Grad's
  approximation to collisions integrals, 276-281, 508, 512
  equation of transfer, 264-270
  evaluation of collisions integrals, 266-272

existence theorems, 320-322
expansion, 261-270
method of truncation, 272-277, 512, 537-539
13-moment system
  definition, 275
  improvement of, 538-539
  properties, 277
  relaxation theorem, 278-279
  simple shearing, 277-278
20-moment system, 273-274, 278
Gross causality, Hilbert's assertion, 384-386
Gross condition, 74, 126, 394, 413
Gross determination, *see* Grossly determined
Gross determiners, *see also* Viscosity; Maxwell number
  approximation of, *see* Stretched fields, method of
  definition, 126
  dilatation, 223
  domain of, 407-411
  equation of transfer, 416-419
  extension, 230-231
  general flows
    approximate
      definition, 444
      Enskog's, 432-433
      general molecules, 279-280, 461-472, 477-479, 481-486
      ideal spheres, 468, 486-487
      inverse k^{th}-power molecules, 462, 468, 487
      Maxwellian molecules, 479-480, 485
    exact, 401-411, 413-424, 429, 433, 507, 520, 530, 556, 562
    expansion for, 434-461, 472-479
    simple shearing, 203, 214-216
    uniqueness, conjecture on, 455
Gross determinism
  definition, 126
  equations of, 405-407, 413-415, 444, 447, 451-452, 475
Gross fields, 41, 137
Gross homogeneity, 189-190
Gross statement, 137
Gross velocity, 41
Grossly determined, *see also* Gross determiners
  definition, 126-127, 519
  similarity (delusive) to constitutive relations, 192, *see also* Constitutive equations or relations, kinetic theory (alleged)

solutions
  analysis of, 397-424
  approximation of, *see* Stretched fields, method of
  conjecture regarding, 562
  examples, *see* Principal solutions; Maxwellian solutions, locally
  proof that
    gas-dynamic solutions are, 400-401
    general solutions are not, 52, 192-193
    Hilbert's solutions are, 394-395
    principal solutions are, 214-215, 223, 227-231

H

H-theorem
  broad, 146-152, 171-172, 471
  comments on, 191, 307-308, 564
  narrow, 147-152, 161-163, 166, 168-171, 289
  open problems, 563-564
  ultra-narrow
    statement, 166-168
  theorems
    Arkeryd's, 356-363
    Carleman's, 336
Heat-bath inequality
  continuum thermomechanics, 15-17
  kinetic theory, 149-151, 162-163
Heat-transfer
  condition, 15, 17, 150, 165-166
  number, 115, 435, 471
Hermite coefficients, 263-264, 266, 269-271
Hermite functions, polynomials, 261-262, 281-282
Higher moments, 45
Hilbert, Hilbert's
  assertion of gross causality, 384-386
  bilinear form
    definition, 98
    domain of, 305-317
    transformations of, 98-100, 305-306, 310, 315
  expansion, formal, 376-377
  field equations, 379, 383-384, 453-454
  iterative procedure
    formulation, 377-380
    illustrated by locally Maxwellian solutions, 390-393
    proof of effectiveness, 380-384
    remarks on, 375, 395-398, 402, 429, 432, 434, 454, 522, 529, 540

mapping
  existence of, 386–389
  generating formula, 389–390
  solutions proved to be
    gas dynamic, 384–385, *see also* Hilbert's assertion of gross causality
    grossly determined, 394–395
Homo-energetic, 24, 139, *see also* Affine flow, homo-energetic
Homothermal, 24
Hygiene, xix

I

Ideal gas, 19–20, 22, 44, 132, *see also* Euler–Hadamard theory of gases
Ideal spheres, 54, 56, 69, 177–178, 183, 268, 458, 468, 482, 486–487
Ideally rough wall, 155–156, 161–163, 166
Ideally smooth wall, 154–155, 161
Ikenberry's
  polynomials, 245, 281–282
  theorem, 244–247, 249, 251–257, 270–271
Impact, *see* Collisions; Encounter
Incident molecule, 153, 158
Influence function, 410
Initial-value problem, 50, 64, 133, 167, 296, *see also* Existence
Instability of energy flux in simple shearing, 208–212, 278, 566
Insulated boundary (wall), 17, 161
Integral along extrinsic trajectory, 63
Intermolecular force
  cut-off, 69, 88
  general, 53, 71
  infinite range, 88
  inverse k^{th}-power, 55–56
  spherically symmetric, 83
Internal energy, specific, *see* Energetic
Internal friction, 20, 36
Invariance, *see also* Collisions operators; Frame-indifference; Orthogonal invariance
  hypothesis regarding gas-dynamic solutions, 399–400
Invariant
  collisional, 71
  principal, 73
  subspace, 72
  summational, 71–72, 100
Inverse k^{th}-power molecules, 55–56, 179–181, 183–184, 268, 291, 316–317, 381, 458, 462, 468, 487
Irreversibility 103, 106–107, 171, 191, *see also* Boltzmann's monotonicity theorem; Clausius inequality; Clausius–Duhem inequality; Dissipation; Dissipation inequalities; H-theorem; Heat-bath inequality
Isentropic, isocaloric, 8
Isolated body, 16–17, 161
Isotropic, 12
Iterates, *see* Atemporal Maxwellian iteration; Differential iteration; Hilbert's iterative procedure; Maxwellian iteration

K

Kinetic statement, 137, *see also* Constitutive quantities for kinetic theory; Equilibrium
Knudsen number, 109–110
Korteweg's fluid, 9–11
Krook & Wu's solution, 290, 299, 301, 303

L

Laws of thermodynamics, 14–16
Linear momentum, equation of balance of, 4, 132
Linear wall, 159–160, 163–166
Linearizations, xix
Linearly viscous, *see* Navier–Stokes–Fourier theory of viscous gases; Stokes–Kirchhoff theory of viscous gases
Local action, 10, 408–411
Locally Maxwellian, *see* Maxwellian solutions

M

Mach number, 21–22, 109–110
Mass
  density, 3, 41
  equation of balance, 4, 132
  molecular, 40, 53, 95
Material
  derivative, 4, 42
  differential type, 10–11, 522
  frame-indifference, 10–12, 275, 522–523
  not simple, 9–10
  rate type, 11–12, 275
  simple, 9
  surface, 157
Mathematics, rigorous, in physics, xviii–xix, 103
Matter, discrete and continuous, xv–xviii
Maxwell, Maxwell's
  assertion regarding $CF = 0$, 107
  collisions operator, 50, 91–97, 101–103, 176–181, 305–317
  consistency theorem, 132–133, 139, 275, 402, 415–416, 522

INDEX OF MATTERS TREATED 589

equation of transfer, 134
evaluation of viscosities, 193–194
number
  definition, 7
  Enskog's evaluations
    approximate, 279–280, 486, 516, 533–534
    exact, 462–463, 480
  Maxwell's evaluation, 194–195, 430
  in scaling, 21
  relaxation theorem, 189–193, 298–299
  second approximation for pressure tensor, 517, *see also* Companion of Maxwell's second approximation for pressure tensor
Maxwell–Boltzmann equation, *see also* Evolution, equation of
  general, 46, 131
  grossly determined solutions, 403–405, 414
  integral forms, 319–320, 339
Maxwell–Boltzmann–Wang theorem on kinetic equilibrium, 142
Maxwellian density, molecular
  definition, 106
  jump at wall, 110–112
  properties of, 107–110, 495–496
  role in
    approximation of general expectations, 112–116, 120–122
    existence theory, 320–321, 327
    Grad's expansion, 263
    Hilbert's procedure, 377
    Maxwellian iteration, 514, 529
    method of stretched fields, 447
    theory of gross determiners, 407, 415
  uniform, 142
  wall, 160
Maxwellian iteration
  applied to simple shearing, 526–528
  atemporal
    applied to homo-energetic affine flows, 542–555
    iterates in general, 531–537
    proof of effectiveness, 531
    scheme of, 528–531
  first two stages, 515–518
  illustration of idea, 514–515
  proof of effectiveness, 524–526
  results, origin, and nature, 518–523
  scheme of, 512–514
  third stage, 523–524
Maxwellian molecules, 175, 181, 184–187, 237–243, 261, 277, 284–291, 479–480, 508

Maxwellian solutions
  equilibrium, 137–138
  locally, 139–142, 169–170
  obtained by Hilbert's procedure, 390–393
  uniform, 166–168
Mean free path, 96, 109, 179, 376, 435
Mean normal pressure, 3, 43–44
Mean random speed, 108
Mean value, *see* Expectation
Method of stretched fields, *see* Stretched fields, method of
Mild derivative, 62
Mild solution, 65, 320, 332
Models
  for kinetic theory, xix, 515
  for materials, xvii–xviii
  mathematical, xv–xvi, 103
  molecular, 103
Molecular chaos, 92
Molecular density, 40, *see also* Maxwellian density, molecular
Molecular diameter, 53
Molecular mass, 40, 53, 95
Molecular model, 103, 136
Molecules, *see also* Constitutive quantities for kinetic theory; Cut-off; Ideal spheres; Inverse k^{th}-power molecules; Maxwellian molecules
  emitted, 153, 158
  incident, 153, 158
  oncoming, 76
  spherically symmetric, 83
Moment
  conversion formulae, 246, 264, 282
  definition, 45
  equations for, 187–189, 237, 249–251, 266–270, 298–299
  higher, 45
  of Hermite polynomials, 263–264
  of strong derivative, 66–67
  principal, 74, 113, 411–413, 494, 519, *see also* Gross condition
  relative, 45
  spherical, 246
Momental determiners
  approximate, 516–524
  exact, 127, 508, 519, 556
Momentally determined, 127, 508, 516, 519
Monatomic, 43
Monotonicity theorem, Boltzmann, 105–107, 147, 308, 358–359
Morgenstern's existence theorem, 336–337

Motion, equations of, 36, 402, *see also* Field equations
Multi-indices, 238, 258-259, 263, 440, 451
Muncaster's
    exact (explicit) solutions, 284-291
    method of stretched fields, 429-488,
        see also Stretched fields, method of
    theory of gross determiners, 397-424,
        see also Gross determiners;
        Gross determinism; Grossly determined
Mutual force, 46

N

Navier-Stokes constitutive relations, 4, 11, 103
Navier-Stokes-Fourier theory of viscous gases, 4-8, 14, 51, 57, *see also* Stokes-Kirchhoff theory of viscous gases
"Nearly Maxwellian" functions, 113, 122
Newton's law of cooling, 112, 125
Nikol'skii's theorem on dilatations, 291
Non-conductor, 6
"Normal" solution, 217-218
Normal-stress difference in shearing, 202-205
Number density, 40

O

Oncoming molecule, 76
Open problems, 560-566
Orthogonal invariance, 86-87, 101-102, 456-457

P

Partial collisions operator, 96-97
Perfect fluid, 6, 20-21
Place, 3, 39-40
Poisson's law of adiabatic change, 20, 139-140, 223
Pressure
    deviator, 3-4, 41
    mean normal, 3-4, 43-44
    tensor, 3, 41-42
    thermodynamic, 4-6, 43
Pressures, equation for, 187-188
Principal conjectures, 562-563
Principal invariant, 73
Principal moment, 74, 113, 411-413, 494, 519, *see also* Grossly determined
Principal shearings, 58
Principal solutions
    dilatation, 223
    extension, 227-231, 550-551, 554-555
    general, 218, 562

    shearing, 214-218, 550-551, 554-555
Principal stretchings, 58
Pseudo-Maxwellian, *see* Pseudomolecules
Pseudomolecules, 186, 320, 336, 338

R

Random kinetic energy, 41
Random momentum, 41
Random speed, mean, 108
Random velocity
    definition, 41
    dimensionless, 262
Rarefied gas, 43, 69, 91, 109-110, 132
Rate
    dissipation of energy, 8
    type, 11-12, 275
Rates of approach
    to equilibrium, 190, 299-302
    to grossly determined state, 201, 223, 230, 289-290, *see also* Dominant part; Principal solutions
Ratio of specific heats, 7, 19, 43-44
Reduced viscometric functions, 204-205
Reflection, specular, 153
Relative moment, 45
Relaxation theorem
    Grad's, 278-279
    Maxwell's, 189-193
    Truesdell's, 291, 298-299
Relaxation, times of, 188, 190, 194, 279, 299-302
Retardation of a function, 437
Retrogression, 48, 61-62
Retrogressor, 47-48, 61-62
Reynolds number, 21, 109
Rigor, mathematical, in physics, xviii-xix, 103
Rotational momentum, equation of balance, 4
Rough boundary (wall), 155-156, 161-163, 166

S

Scaling, 21
Scattering factor
    definition, 97, 177
    for ideal spheres, 177-178
    for inverse k^{th}-power molecules, 179-181
Second Law of Thermodynamics, 14-15
Second-order coefficients, 467-468, 478, 480, 486-488, 535-536
Shear stress
    function, 202
    on wall, 154

INDEX OF MATTERS TREATED

Shear viscosity, *see* Viscosity
Shearing
  amount of, 32, 197
  principal, 58
  simple
    general, 32–33, 197–198
    kinetic theory
      asymptotic forms, 207–208
      atemporal Maxwellian iterates,
        546–548, 551–555
      comments on, 234
      comparison with
        Stokes–Kirchhoff solution, 204–207,
          551–555
      13-moment solution, 277–278
      damped part, 201
      dissipation, 212–214
      dominant part, 201–203, 211–212,
        214–215, 550
      general solution for
        energy flux, 208–212
        pressures, 199–201
      gross determiners, 203, 211–212,
        214–215
      instability of energy flux, 211
      main theorems, 214–217
      Maxwellian iterates, 526–528
      principal solution, 215–217
      shear viscosity as limit, 203–204
      tension number, 197
      viscometric functions, 202–205
    Stokes–Kirchhoff solution, 33, 36
Shift of a field, 403
Shifter, 403
Similarity, 21
Simple gas, 40
Simple material, 9
Simple shearing, *see* Shearing, simple
Slip along a wall, 154
Smooth boundary (wall), 154–155, 161
Sonine polynomials, 481–484
Sound, speed of, 21
Spatially homogeneous solutions, survey,
  335–338, *see also* Existence
Specific heats, 7, 19, 43–44
Specular reflection, 153
Speed of sound, 21
Spherical harmonics, 251
Spherical moments, 246
Spherically symmetric molecules, 83
State, equation of
  caloric, 5, 12, 19, 118
  thermal, 5, 12, 19, 23, 116

Statistics, xvii
Stokes–Duhem inequalities, 6, 14
Stokes–Kirchhoff theory of viscous gases
  continuum thermomechanics
    caloric, 22
    constitutive equations, 19–20
    equations of state
      caloric, 19, 22
      thermal, 19
    particular flows
      dilatation, 24–28, 36
      extension, 33–36
      homo-energetic affine flows, 28–32, 36
      simple shearing, 32–33, 36, 198–199
      viscosity proportional to temperature,
        31–33, 35–36
  kinetic theory, comparisons with
    agreement, possible, and disagreement,
      certain, 57–58, 133
    caloric, 117–118
    constitutive equations
      general, 115, 133, 195–196, 565
      particular flows
        dilatation, 223
        extension, 232
        gross homogeneity, 190–193
        Muncaster's solutions, 290
        simple shearing, 204–207
    dissipation, 212–213
    equations of state
      caloric, 118
      thermal, 43, 116
    problem of asymptotic status, 565
    solutions
      general, 565–566
      grossly determined, 193–195, 279, 380,
        397, 462, 557, 566
      momentally determined, 193–195, 279,
        516
      particular flows
        dilatation, 223, 234
        extension, 231–234
        Muncaster's solutions, 290
        simple shearing, 205–207, 217, 234
      using truncated expansions, 551–555,
        565–566
Stokes relation for bulk viscosity, 6, 20, 57,
  193
Stretched field, 436
Stretched fields, method of
  as replacement for Enskog's procedure,
    433–434, 444, 479, 507, 540, 557
  body force, effect on, 472–479

comments on, 429, 444, 454–455, 479, 488, 507–508, 529, 540, 557, 562
expansion
  basic, 437–441, 472–475
  coefficients
    explicit
      approximate, 481–487
      exact, 479–480
    general, 457–458, 461–462, 465–470, 478
    properties, 455–461, 476–477
    regularity, 445–446
  of fields, 441–444
  uniqueness, conjecture on, 455
  iterative system
    derivation, 488–494
    general, 450–452
    particular members, 446–449, 456, 460, 475–477
    structure, 452
  proof of effectiveness, 452–453
  scheme of, 434–436
Stretchings, principal, 58
Strict trend, *see* Trend to
Strong derivative
  definition, 64
  moments of, 66–67
Strong solution, 65
Summational invariants
  Boltzmann–Gronwall theorem on, 72–73
  definition, 71–72
  value of total collisions operator on, 100
Symmetrizer, 450

T

t-continuous, 343
t-differentiable, 343
Temperature
  continuum thermomechanics
    general, 3
    Stokes–Kirchhoff theory, 19, 22
    wall, 14–17
  kinetic theory, *see also* Energetic
    definition, 44
    wall, 111–112, 149–151, 162–165
Tension
  impossible in kinetic theory, 57
  number, *see also* Knudsen number
    definition and properties, 58–60
    in extension, 34–35, 59, 224
    in simple shearing, 33, 59, 197
Theories of fluids, *see* Euler–Hadamard theory of gases; Material; Navier–Stokes–Fourier theory of viscous gases; Stokes–Kirchhoff theory of viscous gases
Theory, role of, xv–xvi
Theory of gross determiners, 397–424, 530, 562, *see also* Gross determiners; Gross determinism; Grossly determined
Thermal conductivity, 5–6, 397, *see also* Maxwell number
Thermal equation of state, 5, 12, 23, 116
Thermodynamic pressure, 4–6
Thermodynamics, thermostatics
  continuum thermomechanics, 7–8, 12–17, 19, 22, 125, 127
  kinetic theory of, *see* Boltzmann's field h; Caloric, specific entropy; Dissipation; State, equation of; H-theorem; Temperature
Time
  concept, 3, 39–40
  derivatives, elimination of, 405–406, 414, 433–434, 529–530
  flux, frame-indifferent, 12
  relaxation, *see* Relaxation, times of
Total collisions operators, 98, *see also* Collisions operators
Transfer, equation of
  basic, 132
  Enskog's, 134
  Grad's, 266–270
  for gross determiners
    affine flows, 423–424
    general, 416–419
  Maxwell's, 134
  variant, 135
Trend to
  equilibrium, *see also* H-theorem
  strict, traditional
    alleged, 166–170
    counter examples, 140–141, 169, 289
    open problem of, 563–565
  ultra-narrow (spatially homogeneous solutions)
    Krook & Wu's example, 290
    statement 167
    theorems
      Arkeryd's, 363–368
      Carleman's, 336
      Truesdell's, 297–302

grossly determined state
  examples and counter examples, 214, 223, 230, 289–290
  open problem of, 561–563
Truesdell's, *see also* Differential iteration; Ikenberry's theorem; Maxwellian iteration
  solution for shearing, 197–218
  theorem of existence and trend to equilibrium, 297–302, 336–337
Truncation
  Grad's method of, 272–277, 537–539
  number, 31, 59–60
  of system for viscosity and Maxwell number, 484–485

U

Uniform Maxwellian density, 142, 166–168
Uniqueness of gross determiners, conjecture on, 455
Universal relations, 71, 195, 468
Universe, 17, 148

V

Velocity
  continuum thermomechanics, 3
  molecules, dynamics of
    of approach, 76
    asymptotic, 70–71
  kinetic theory
    gross, 41
  molecular, 40
  random
    definition, 41
    dimensionless, 262
Viscometric functions, 202–205
Viscosity
  continuum thermomechanics
    bulk, 6, 20, 25
    shear
      definition, 6
      effects if proportional to temperature, 31–32, 35–36
      viscometric functions, 202–205
  kinetic theory, *see also* Gross determiners
    bulk, 57, 140, 193
    shear
      dimensional analysis, 53–56
      Enskog's evaluation
        approximate, 279–280, 486–487, 516, 533–534
        exact, 462–463, 480
      evaluation as limit in shearing, 203–205
      Maxwell's evaluation, 193–194, 198
Viscous, 6, 19, *see also* Navier–Stokes–Fourier theory of viscous gases; Stokes–Kirchhoff theory of viscous gases

W

Wall, *see* Boundary, wall
Weak convergence, 363–364
Weak solution, 65, 336

Pure and Applied Mathematics

A Series of Monographs and Textbooks

Editors **Samuel Eilenberg and Hyman Bass**
Columbia University, New York

RECENT TITLES

I. MARTIN ISAACS. Character Theory of Finite Groups
JAMES R. BROWN. Ergodic Theory and Topological Dynamics
C. TRUESDELL. A First Course in Rational Continuum Mechanics: Volume 1, General Concepts
GEORGE GRATZER. General Lattice Theory
K. D. STROYAN AND W. A. J. LUXEMBURG. Introduction to the Theory of Infinitesimals
B. M. PUTTASWAMAIAH AND JOHN D. DIXON. Modular Representations of Finite Groups
MELVYN BERGER. Nonlinearity and Functional Analysis: Lectures on Nonlinear Problems in Mathematical Analysis
CHARALAMBOS D. ALIPRANTIS AND OWEN BURKINSHAW. Locally Solid Riesz Spaces
JAN MIKUSINSKI. The Bochner Integral
THOMAS JECH. Set Theory
CARL L. DEVITO. Functional Analysis
MICHIEL HAZEWINKEL. Formal Groups and Applications
SIGURDUR HELGASON. Differential Geometry, Lie Groups, and Symmetric Spaces
ROBERT B. BURCKEL. An Introduction to Classical Complex Analysis: Volume 1
JOSEPH J. ROTMAN. An Introduction to Homological Algebra
C. TRUESDELL AND R. G. MUNCASTER. Fundamentals of Maxwell's Kinetic Theory of a Simple Monatomic Gas: Treated as a Branch of Rational Mechanics

IN PREPARATION

LOUIS HALLE ROWEN. Polynominal Identities in Ring Theory
ROBERT B. BURCKEL. An Introduction To Classical Complex Analysis: Volume 2
BARRY SIMON. Functional Integration and Quantum Physics
DRAGOS M. CVETLOVIC, MICHAEL DOOB, AND HORST SACHS. Spectra of Graphs
DAVID KINDERLEHRER and GUIDO STAMPACCHIA. Introduction to Variational Inequalities and Their Applications.
HERBERT SEIFERT AND W. THRELFALL. Seifert and Threlfall's Textbook on Topology
GRZEGORZ ROZENBERG AND ARTO SALOMAA. The Mathematical Theory of L Systems.
DONALD W. KAHN. Introduction to Global Analysis

RAYMOND H. FOGLER LIBRARY